D1228601

ELECTRONIC MATERIALS AND PROCESSES HANDBOOK

Electronic Packaging and Interconnection Series
Charles Harper, Series Advisor

Published Books

Classon • SURFACE MOUNT TECHNOLOGY FOR CONCURRENT ENGINEERING AND MANUFACTURING

Harper • ELECTRONIC PACKAGING AND INTERCONNECTION HANDBOOK

Ginsberg and Schnorr • MULTICHIP MODULES AND RELATED TECHNOLOGIES

Related Books of Interest

Boswell • SUBCONTRACTING ELECTRONICS

Boswell and Wickam • SURFACE MOUNT GUIDELINES FOR PROCESS CONTROL, QUALITY, AND RELIABILITY

Byers • PRINTED CIRCUIT BOARD DESIGN WITH MICROCOMPUTERS

Capillo • SURFACE MOUNT TECHNOLOGY

Chen • COMPUTER ENGINEERING HANDBOOK

Coombs • PRINTED CIRCUITS HANDBOOK

Di Giacomo • DIGITAL BUS HANDBOOK

Di Giacomo • VLSI HANDBOOK

Fink and Christiansen • ELECTRONICS ENGINEERS' HANDBOOK

Ginsberg • PRINTED CIRCUIT BOARD DESIGN

Harper • HANDBOOK OF PLASTICS, ELASTOMERS, AND COMPOSITES

Johnson • ANTENNA ENGINEERING HANDBOOK

Juran and Gryna • JURAN'S QUALITY CONTROL HANDBOOK

Kaufman and Seidman • HANDBOOK OF ELECTRONICS CALCULATIONS

Lenk • MCGRAW-HILL ELECTRONIC TESTING HANDBOOK

Lenk • MCGRAW-HILL CIRCUIT ENCYCLOPEDIA AND TROUBLESHOOTING GUIDE

Lenk • LENK'S DIGITAL HANDBOOK

Manko • SOLDERS AND SOLDERING

Perry • VHDL

Rao • MULTILEVEL INTERCONNECTION TECHNOLOGY

Sze • VLSI TECHNOLOGY

Tuma • ENGINEERING MATHEMATICS HANDBOOK

Van Zant • MICROCHIP FABRICATION

Waynant • ELECTRO-OPTICS HANDBOOK

ELECTRONIC MATERIALS AND PROCESSES HANDBOOK

Charles A. Harper Editor
Technology Seminars, Inc. Lutherville, Maryland

Ronald N. Sampson Editor
Technology Seminars, Inc. Murrysville, Pennsylvania

Second Edition

McGraw-Hill, Inc.
New York San Francisco Washington, D.C. Auckland Bogotá
Caracas Lisbon Madrid Mexico City Milan
Montreal New Delhi San Juan Singapore
Sydney Tokyo Toronto

Library of Congress Cataloging-in-Publication Data

Electronic materials and processes handbook / Charles A. Harper, editor, Ronald N. Sampson, editor — 2nd ed.
p. cm. — (Electronic packaging and interconnection series)
Includes index
ISBN 0-07-054299-6
1. Electronics—Materials—Handbooks, manuals, etc. 2. Manufacturing processes—Handbooks, manuals, etc. 3. Electronic apparatus and appliances—Materials—Handbooks, manuals, etc. I. Harper, Charles A. II. Sampson, Ronald N. III. Series.
TK7836.E4653 1993
621.381—dc20 93-41644
CIP

The first edition of this book was published in 1970 under the title *Handbook of Materials and Processes for Electronics.*

1 2 3 4 5 6 7 8 9 0 DOC/DOC 9 9 8 7 6 5 4 3

ISBN 0-07-054299-6

The sponsoring editor for this book was Stephen S. Chapman, the editing supervisor was Paul R. Sobel, and the production supervisor was Pamela A. Pelton. It was set in Times Roman by McGraw-Hill's Professional Book Group composition unit.

Printed and bound by R. R. Donnelley & Sons Company.

This book was printed on acid-free paper.

CONTENTS

CONTRIBUTORS

William M. Alvino *Westinghouse Electric, Science and Technology Center, Pittsburgh, Pennsylvania* (CHAP. 1)

Donald Baudrand *WITCO Allied-Kelite, Melrose Park, Illinois* (CHAP. 9)

Carl T. Brooks *Westinghouse Electric Corp., Electronic Systems, Baltimore, Maryland* (CHAP. 2)

Victor J. Brzozowski *Westinghouse Electric Corp., Electronic Systems, Baltimore, Maryland* (CHAP. 2)

Albert A. Burk, Jr. *Westinghouse Electric, Science and Technology Center, Pittsburgh, Pennsylvania* (CHAP. 4)

Edward J. Croop *Pittsburgh, Pennsylvania* (CHAP. 6)

Jurgen Diekmann *Hockessin, Deleware* (CHAP. 10)

Don E. Harrison *Murrysville, Pennsylvania* (CHAP. 5)

John D. Harrison *Watsonville, California* (CHAP. 5)

Stephen G. Konsowski *Westinghouse Electric Corp., Electronic Systems, Baltimore, Maryland* (CHAP. 11)

James J. Licari *Whittier, California* (CHAP. 8)

Douglas M. Mattox *University of Missouri at Rolla, Rolla, Missouri* (CHAP. 3)

James F. Maguire *Boeing Seattle, Seattle, Washington* (CHAP. 7)

R. Noel Thomas *Westinghouse Electric, Science and Technology Center, Pittsburgh, Pennsylvania* (CHAP. 4)

PREFACE

In the years since the publication of the widely acclaimed first edition of this Handbook, the growth in the electronics industry has increased beyond all expectations. This industry has become the most important and biggest area in industrial history. Based on a constantly increasing number of materials discoveries, packaging concepts, manufacturing techniques and ingenious ideas, the impact of this technology on new and improved products is phenomenal. This second edition of the Handbook brings together the changes which have occurred in electronic materials and process technology in the ensuing years. It presents a modern up-to-date source book of practical material data, guidelines, information and manufacturing methods including cost and fabrication trade-off, design and application criteria, and performance limits.

To update this Handbook we were able to obtain a group of outstanding chapter authors—each an acknowledged expert in their technologies. Combined, they cover these material and process technologies clearly and thoroughly in this broad industry known as electronics. We take pleasure in recognizing these scientists, engineers, and technologists and their contributions to the *Electronic Materials and Processes Handbook.*

To maximize reader convenience the Handbook is designed in three parts. The first five chapters describe the materials used in electronic systems, starting with polymers then printed wiring boards, ceramics, semicondutors, and metals. The next five chapters describe processes used to manufacture and interconnect electronic circuits. These include wiring, metal joining, thick films, thin films, electrodeposition, lithography and photofabrication. The final chapter discusses the most recent advanced packaging concepts, both in materials and processes. Other useful features include updated standards, a revised index, definitions, and new end of chapter reference sources.

We are proud to be associated with these experts and are confident that this Handbook will be even more useful than its parent. *Electronic Materials and Processes Handbook* will join other Handbooks in the Electronic Packaging and Interconnection Series currently being published by McGraw-Hill, Inc.

Ronald N. Sampson
Charles A. Harper

CHAPTER 1

PLASTICS, ELASTOMERS, AND PROCESSING FOR ELECTRONICS

William M. Alvino

1.1 INTRODUCTION

There is no doubt that plastics have become one of the dominant materials of the twentieth century. Prior to 1930 most household goods and industrial components were made of metals, wood, glass, paper, leather, or vulcanized rubber. Since then plastics have made significant advances in the markets of all these materials as well as creating new markets of their own. The widespread use of plastics has been brought about because of their unique combination of properties, such as strength, light weight, low cost, and ease of processing and fabrication. Plastics are not the panacea of industry's material problems, but they offer such a unique combination of properties that they have become one of the important classes of materials and have found widespread use in the electrical and electronics industries. Plastics play a key role in this industry and function in a variety of ways. The most important function of devices in the electronics industry is the conduction of signals through the circuit contributed by metals in the form of wires, contacts, foils, solders, and so on. Of equal importance is the electrical insulation function, which prevents the loss of the signal currents and confines them to the desired paths. Insulation systems exist in a variety of forms—liquids, solids, or gases—and the type of material used determines the life of the device. Plastic materials perform structural roles and support the circuit physically. Plastics are also used to provide environmental protection from such elements as moisture, heat, and radiation to sensitive electronic devices. Continuing improvements in the properties of plastics over the years have made them even more important to the electrical industry by extending their useful range.

It is the purpose of this chapter to present to the reader an overview of the nature of plastic materials. This overview will include those topics related to plastic fundamentals, thermoplastics, thermosets, elastomers, coatings, and polymer processing. The overview pertains only to those plastics that are of significant importance in the electronics industry.

1.2 FUNDAMENTALS

1.2.1 Polymer Definition

Polymers are macromolecules, that is, large molecules, formed by the linking together of large numbers of small molecules called monomers. The process involved in the joining of these monomers is called polymerization. Plastics are a group of synthetic polymers made up of chains of atoms or molecules. The long molecular chains contain various combinations of oxygen, hydrogen, nitrogen, carbon, silicon, chlorine, fluorine, and sulfur. As more repeating units are added, molecular weight of the plastic increases and can reach into the millions, but typically most polymers used for practical applications fall into the molecular weight range of 5000 to 200,000. There are exceptions.

1.2.2 Types of Polymers

Polymers can be differentiated by the way in which their monomers are joined together, that is, addition or condensation polymerization. Furthermore, the molecular chains are linked by the successive addition of one monomer to another. Typical polymers are polyolefins, polystyrenes, acrylics, vinyls, and fluoroplastics.

Condensation polymers are prepared by the reaction of two different molecules, each having two reactive end groups. Molecular weight is built up by the linking together of these end groups and the elimination of a small molecule, which must be removed from the reaction medium to attain a high molecular weight. Examples include nylons, polyesters, polyurethanes, and polyimides.

All polymers can be classified in this manner, but they can also be further subclassified to define their structural and compositional characteristics more accurately. They can be linear, branched, crystalline, amorphous, or liquid crystalline copolymers, elastomers, and alloys. All of these, except the elastomers, can be divided into two major groups—thermoplastics and thermosets. Both types of plastics are fluid enough to be formed and molded at some stage in their conversion to the finished product. Thermoplastics solidify by cooling and can be remelted. Thermoset resins undergo cross-linking to form a three-dimensional network, and unlike thermoplastics, they cannot be remelted and reshaped. With few exceptions polymers are not used in their natural state and are usually mixed with other materials to yield a compounded polymer, which may be in the form of pellets, granules, powder, or liquid. The properties of a polymer can be varied, and the process that is used to polymerize the monomer with one or more different monomers is called copolymerization. These polymers are called copolymers or terpolymers, depending on whether two or three comonomers are used during the copolymerization. Another technique used to vary the properties of polymers is to blend one polymer with another mechanically to form an alloy. The properties of these alloys generally fall between those of the starting polymers. Elastomers differ significantly from plastics. While they are also polymers, elastomers easily undergo very large reversible elongations at relatively low stresses. In order for this to happen, the polymer must be completely amorphous with a low glass transition temperature and low secondary forces so as to obtain high mobility of the polymer chains. Some degree of cross-linking is needed so that the deformation is rapidly reversible. Figure 1.1 illustrates the differences between plastics, fibers, and elastomers by way of a stress-strain plot.

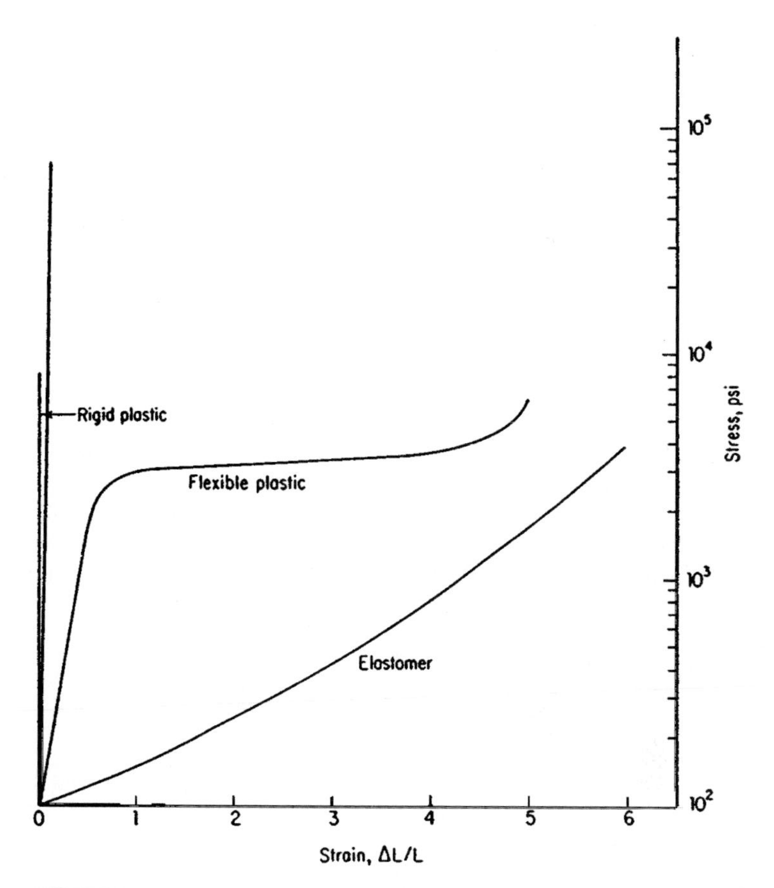

FIGURE 1.1 Stress-strain plots for typical elastomer, plastic, and fiber. (*From Odian*[1]; *reprinted with permission.*)

1.2.3 Structure and Properties

In addition to the broad categories of thermoplastics and thermosets, polymeric materials can be classified in terms of their structure: linear, branched, cross-linked, amorphous, crystalline, and liquid crystalline. If we recall, a polymer molecule consists of monomer molecules that have been linked together in one continuous length. Such a polymer is termed a linear polymer. Branched polymers are those in which there are side branches of linked monomer molecules protruding from various points along the main polymer chain. Cross-linked polymers are those in which adjacent molecules are linked together, resulting in a complex interconnected network. Figure 1.2 is a schematic illustration of these structures.

In some thermoplastics the chemical structure is such that the polymer chains will fold on themselves and pack together in an organized manner (Fig. 1.3). The resulting organized regions show the behavior characteristics of crystals. Plastics that have

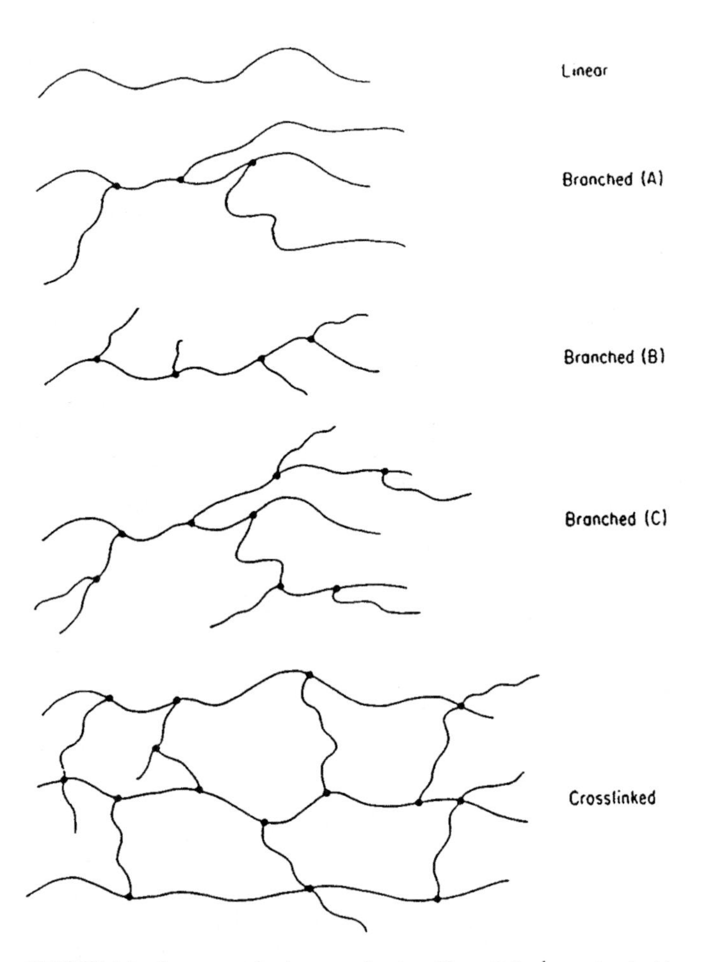

FIGURE 1.2 Structures of polymer molecules. (*From Odian[1]; reprinted with permission.*)

these regions are called crystalline. Plastics without these regions are called amorphous. All of the crystalline plastics have amorphous regions between and connecting the crystalline regions. For this reason, the crystalline plastics are often referred to as semicrystalline in the literature.

Liquid crystalline polymers are best thought of as being a separate and unique class of plastics. The molecules are stiff, rodlike structures which are organized in large parallel arrays or domains in both the melted and the solid states. These large, ordered domains provide liquid crystalline polymers with unique characteristics compared to those of the crystalline and amorphous polymers.

Many of the mechanical and physical property differences between plastics can be attributed to their structure. As a generalization, the ordering of crystalline and liquid crystalline thermoplastics makes them stiffer, stronger, and less impact-resistant than their amorphous counterparts. Also, crystalline and liquid crystalline materials have a

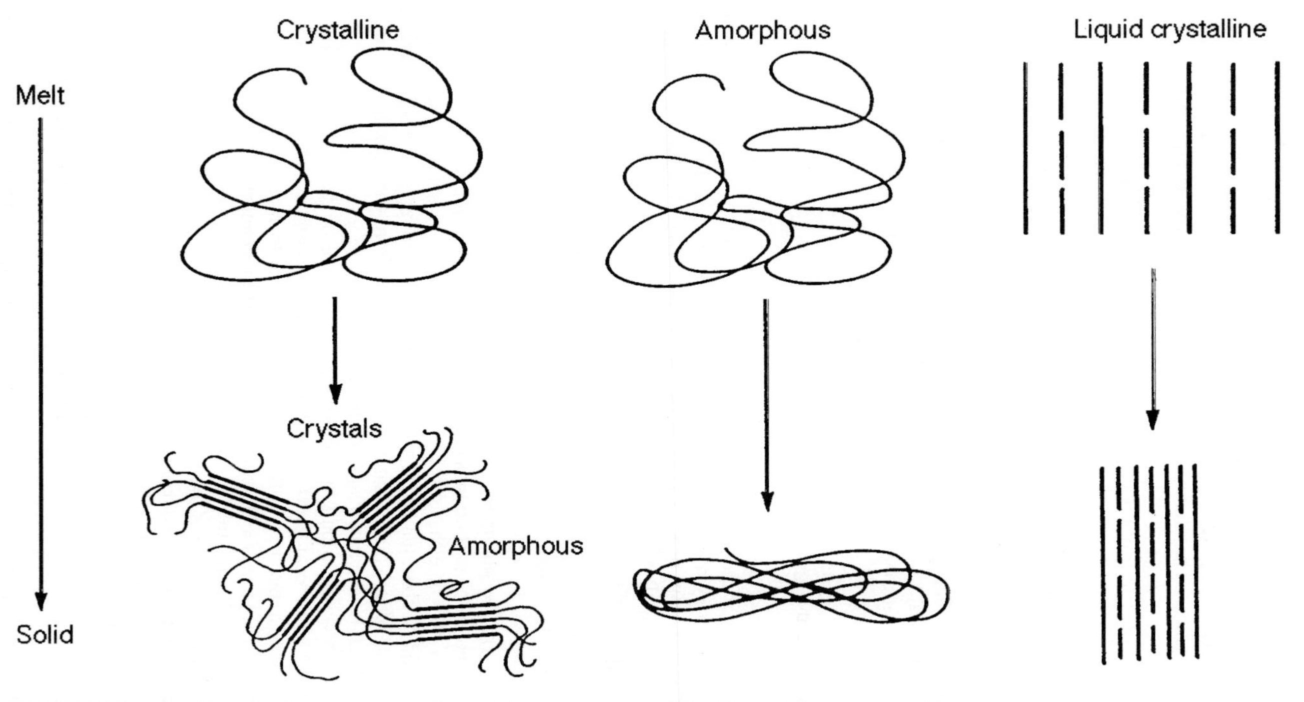

FIGURE 1.3 Two-dimensional representation of crystalline, amorphous, and liquid crystalline structures. (*From Hoechst Celanese; reprinted with permission.*)

higher resistance to creep, heat, and chemicals. However, crystalline materials are more difficult to process because they have higher melt temperatures and tend to shrink and warp more than amorphous polymers. Amorphous polymers soften gradually and continuously as heat is applied, and in the molding process do not flow as easily as do melted crystalline polymers. Liquid crystalline polymers have the high melt temperature of crystalline plastic, but soften gradually and continuously like amorphous polymers. They have the lowest viscosity, warpage, and shrinkage of the thermoplastics.

One of the most important characteristics of a polymer is its molecular weight because the properties of polymers are a consequence of their high molecular weight. Strength does not usually develop in polymers until a minimum molecular weight (5000 to 10,000) is attained. Above this value there is a rapid increase in mechanical properties and a leveling off as the molecular weight increases further. In most instances there is some molecular weight range for which a given polymer property will be optimum for a particular application. Polymers are not all homogeneous and are composed of molecules of different sizes. In order to fully characterize the size of a polymer one should know its molecular weight and its molecular weight distribution. Both of these properties affect processing and strength significantly.

1.2.4 Synthesis

There are four basic methods of producing a polymer. Many factors influence the choice of a particular method of producing a polymer. In many instances the nature of the reaction chemistry dictates the specific method to be used. In other instances the resultant polymer (low or semiviscous liquid, friable or rigid solid) may limit one's choice. The interested reader is referred to any basic organic polymer chemistry text for more detailed descriptions.

Bulk Polymerization. From the point of view of equipment, complexity, and economics, the simplest method is mass or bulk polymerization. This procedure merely allows the monomer to react at a predetermined reaction temperature, with or without catalysts, to form the polymer. Theoretically the monomer can be a gas, liquid, or solid, but in practice almost all mass polymerizations take place in a liquid phase.

The polymer may be either soluble or insoluble in the monomer. If the former, then the mass viscosity continually increases until the final degree of polymerization is obtained. In the latter, the polymer will precipitate from the remaining unreacted monomer and can be separated subsequently.

A serious drawback to bulk polymerization is control of the heat of reaction. The generated exothermic heat tends to stay within the mass and is not easily withdrawn. Stirring the mass helps, but as the viscosity continues to increase, stirring becomes more difficult, with a less efficient heat-dissipation mechanism. This lack of control causes resultant difficulty in the control of the molecular weight and the molecular weight distribution (MWD) of the final polymer.

The method does, however, lend itself for use in small casting or batch production. Gaseous-phase bulk polymerization takes place under pressure, often requiring specific catalysts for conversion.

In summary, mass or bulk polymerization uses simple equipment, is highly exothermic with difficult heat control, and yields a polymer with a broad MWD.

Solution Polymerization. If we carry out the polymerization in a suitable solvent, we can simplify the reaction heat removal, since the solution of solvent, monomer, and polymer is less viscous than molten polymer. This technique is called *solution* polymerization.

If a solvent can be found in which the monomer is soluble but the polymer is insoluble, the resultant polymer precipitation facilitates the separation steps.

In summary, one can control heat more readily in solution polymerization, although higher-molecular-weight polymers are difficult to produce. A solution of the polymer itself may be marketable, but the purification of solid polymer may involve complex procedures.

Emulsion Polymerization. If the monomer can be polymerized in a water emulsion, then we can retain the low viscosity needed for good heat control without the hazards associated with the handling of solvents. Such a procedure is called *emulsion* polymerization.

Reaction rates and molecular weights are usually higher with this method than with mass or solution polymerization. The MWD is often quite narrow, water is cheaper and less hazardous than solvent, and recovery steps are not as complex. However, ingredients must be added to aid emulsification (emulsifying and stabilizing agents). This added contamination and the requirement of a drying step for the polymer constitute significant disadvantages of the process.

Suspension Polymerization. Finally there is suspension polymerization, in which the monomer and the forming polymer globules are maintained in suspension by agitation without the use of an emulsifying agent. The polymer beads are formed by coalescence, and their size is regulated by suspension stabilizers and the amount and intensity of agitation. The final beads must be screened out of the liquid phase, washed, and dried before they can be used, although suspensions can be, and are, marketable. Control of exothermic heat is good, and high-molecular-weight polymers with relatively narrow MWDs are possible.

1.2.5 Terminology

To acquaint those unfamiliar with the language of polymers, Tables 1.1 and 1.2 present terms associated with polymers and their use in the electronics industry.

TABLE 1.1 Definition of Terms for Plastic Materials

Accelerator. A chemical used to speed up a reaction or cure. For example, cobalt naphthenate is used to accelerate the reaction of certain polyester resins. The term *accelerator* is sometimes used interchangeably with the term *promoter.* An accelerator is often used along with a catalyst, hardener, or curing agent.
Adhesive. Broadly, any substance used in promoting and maintaining a bond between two materials.
Aging. The change in properties of a material with time under specific conditions.
Arc resistance. The time required for an arc to establish a conductive path in a material.

B stage. An intermediate stage in the curing of a thermosetting resin. In this state, a resin can be heated and caused to flow, thereby allowing final curing in the desired shape. The term *A stage* is used to describe an earlier stage in the curing resin. Most molding materials are in the B stage when supplied for compression or transfer molding.
Blowing agent. Chemicals that can be added to plastics and that generate inert gases upon heating. This blowing or expansion causes the plastic to expand, thus forming a foam. Also known as *foaming agent.*
Bond strength. The amount of adhesion between bonded surfaces.

Capacitance (capacity). That property of a system of conductors and dielectrics which permits the storage of electricity when potential difference exists between the conductors. Its value is expressed as the ratio of the quantity of electricity to a potential difference. A capacitance value is always positive.
Cast. To embed a component or assembly in a liquid resin, using molds that separate from the part for reuse after the resin is cured. See *Embed, Pot.*
Catalyst. A chemical that causes or speeds up the cure of a resin, but that does not become a chemical part of the final product. Catalysts are normally added in small quantities. The peroxides used with polyester resins are typical catalysts.
Coat. To cover with a finishing, protecting, or enclosing layer of any compound (such as varnish).
Coefficient of expansion. The fractional change in dimension of a material for a unit change in temperature.
Cold flow (creep). The continuing dimensional change that follows initial instantaneous deformation in a nonrigid material under static load.
Compound. Some combination of elements in a stable molecular arrangement.
Contact bonding. A type of adhesive (particularly nonvulcanizing natural rubber adhesives) that bonds to itself on contact although solvent evaporation has left it dry to the touch.
Cross-linking. The forming of chemical links between reactive atoms in the molecular chain of a plastic. It is this cross-linking in thermosetting resins that makes them infusible.
Crystalline melting point. The temperature at which the crystalline structure in a material is broken down.
Cure. To change the physical properties of a material (usually from a liquid to a solid) by chemical reaction, by the action of heat and catalysts, alone or in combination, with or without pressure.
Curing agent. See *Hardener.*
Curing temperature. The temperature at which a material is subjected to curing.
Curing time. In the molding of thermosetting plastics, the time it takes for the material to be properly cured.

Dielectric constant (permittivity or specific inductive capacity). That property of a dielectric which determines the electrostatic energy stored per unit volume for unit potential gradient.
Dielectric loss. The time rate at which electric energy is transformed into heat in a dielectric when it is subjected to a changing electric field.

TABLE 1.1 Definition of Terms for Plastic Materials (*Continued*)

Dielectric loss angle (dielectric phase difference). The difference between 90° and the dielectric phase angle.
Dielectric loss factor (dielectric loss index). The product of the dielectric constant and the tangent of the dielectric loss angle for a material.
Dielectric phase angle. The angular difference in phase between the sinusoidal alternating potential difference applied to a dielectric and the component of the resulting alternating current having the same period as the potential difference.
Dielectric power factor. The cosine of the dielectric phase angle (or sine of the dielectric loss angle).
Dielectric strength. The voltage that an insulating material can withstand before breakdown occurs, usually expressed as a voltage gradient (such as volts per mil).
Dissipation factor (loss tangent, tan δ, approximate power factor). The tangent of the loss angle of the insulating material.

Elastomer. A material which at room temperature stretches under low stress to at least twice its length and snaps back to its original length upon the release of stress. See *Rubber.*
Electric strength (dielectric strength or disruptive gradient). The maximum potential gradient that a material can withstand without rupture. The value obtained for the electric strength will depend on the thickness of the material and the method and conditions of test.
Embed. To encase completely a component or assembly in some material—a plastic for current purposes. See *Cast, Pot.*
Encapsulate. To coat a component or assembly in a conformal or thixotropic coating by dipping, brushing, or spraying.
Exotherm. The characteristic curve of a resin during its cure, which shows heat of reaction (temperature) versus time. Peak exotherm is the maximum temperature on this curve.
Exothermic. A chemical reaction in which heat is given off.

Filler. A material, usually inert, that is added to plastics to reduce cost or modify physical properties.
Film adhesive. A thin layer of dried adhesive. Also describes a class of adhesives provided in dry-film form with or without reinforcing fabric, which are cured by heat and pressure.
Flexibilizer. A material that is added to rigid plastics to make them resilient or flexible. Flexibilizers can be either inert or a reactive part of the chemical reaction. Also called a *plasticizer* in some cases.
Flexural modulus. The ratio, within the elastic limit, of stress to corresponding strain.
Flexural strength. The strength of a material in bending, expressed as the tensile stress of the outermost fibers of a bent test sample at the instant of failure.
Fluorocarbon. An organic compound having fluorine atoms in its chemical structure. This property usually lends stability to plastics. Teflon* is a fluorocarbon.

Gel. The soft, rubbery mass that is formed as a thermosetting resin goes from a fluid to an infusible solid. This is an intermediate state in a curing reaction, and a stage in which the resin is mechanically very weak. *Gel point* is defined as the point at which gelation begins.
Glass transition point. The temperature at which a material loses its glasslike properties and becomes a semiliquid.

Hardener. A chemical added to a thermosetting resin for the purpose of causing curing or hardening. Amines and acid anhydrides are hardeners for epoxy resins. Such hardeners are a part of the chemical reaction and a part of the chemical composition of the cured resin. The terms *hardener* and *curing agent* are used interchangeably. Note that these can differ from catalysts, promoters, and accelerators.
Heat-distortion point. The temperature at which a standard test bar (ASTM D-648) deflects 0.010 in under a stated load of either 66 or 264 lb/in^2.

TABLE 1.1 Definition of Terms for Plastic Materials (*Continued*)

Heat sealing. A method of joining plastic films by simultaneous application of heat and pressure to areas in contact. Heat may be supplied conductively or dielectrically.
Hot-melt adhesive. A thermoplastic adhesive compound, usually solid at room temperature, which is heated to a fluid state for application.
Hydrocarbon. An organic compound having hydrogen atoms in its chemical structure. Most organic compounds are hydrocarbons. Aliphatic hydrocarbons are straight-chained hydrocarbons, and aromatic hydrocarbons are ringed structures based on the benzene ring. Methyl alcohol, trichloroethylene are aliphatic; benzene, xylene, toluene are aromatic.
Hydrolysis. The chemical decomposition of a substance involving the addition of water.
Hygroscopic. Tending to absorb moisture.

Impregnate. To force resin into every interstice of a part. Cloths are impregnated for laminating, and tightly wound coils are impregnated in liquid resin using air pressure or vacuum as the impregnating force.
Inhibitor. A chemical added to resins to slow down the curing reaction. Inhibitors are normally added to prolong the storage life of thermosetting resins.
Insulation resistance. The ratio of applied voltage to total current between two electrodes in contact with a specific insulator.

Modulus of elasticity. The ratio of stress to strain in a material that is elastically deformed.
Moisture resistance. The ability of a material to resist absorbing moisture, either from the air or when immersed in water.
Mold. To form a plastic part by compression, transfer, injection molding, or some other pressure process.

NEMA standards. Property values adopted as standard by the National Electrical Manufacturers Association.

Organic. Composed of matter originating in plant or animal life, or composed of chemicals of hydrocarbon origin, either natural or synthetic. Used in referring to chemical structures based on the carbon atom.

Permittivity. Preferred term of dielectric constant.
pH. A measure of the acid or alkaline condition of a solution. A pH of 7 is neutral (distilled water), pH values below 7 are increasingly acid as pH values go toward 0, and pH values above 7 are increasingly alkaline as pH values go toward the maximum value of 14.
Plastic. An organic resin or polymer.
Plasticizer. A material added to resins to make them softer and more flexible when cured.
Polymer. A high-molecular-weight compound (usually organic) made up of repeated small chemical units. Polymers can be thermosetting or thermoplastic.
Polymerize. To unite chemically two or more monomers or polymers of the same kind to form a molecule with higher molecular weight.
Pot. To embed a component or assembly in a liquid resin, using a shell, can, or case, which remains as an integral part of the product after the resin is cured. See *Embed, Cast.*
Pot life. The time during which a liquid resin remains workable as a liquid after catalysts, curing agents, promoters have been added; roughly equivalent to gel time. Sometimes also called *working life.*
Power factor. The cosine of the angle between the voltage applied and the resulting current.
Promoter. A chemical, itself a feeble catalyst, that greatly increases the activity of a given catalyst.

TABLE 1.1 Definition of Terms for Plastic Materials (*Continued*)

Resin. A high-molecular-weight organic material with no sharp melting point. For current purposes, the terms *resin, polymer,* and *plastic* can be used interchangeably.
Resistivity. The ability of a material to resist passage of electric current either through its bulk or on a surface. The unit of volume resistivity is the ohm-centimeter, and the unit of surface resistivity is the ohm.
Rockwell hardness number. A number derived from the net increase in depth of impression as the load on a penetrator is increased from a fixed minimum load to a higher load and then returned to minimum load. Penetrators include steel balls of several specified diameters and a diamond cone.
Rubber. An elastomer capable of rapid elastic recovery.

Shore hardness. A procedure for determining the indentation hardness of a material by means of a durometer. Shore designation is given to tests made with a specified durometer.
Solvent. A liquid substance that dissolves other substances.
Storage life. The period of time during which a liquid resin or adhesive can be stored and remain suitable for use. Also called *shelf life.*
Strain. The deformation resulting from a stress, measured by the ratio of the change to the total value of the dimension in which the change occurred.
Stress. The force producing or tending to produce deformation in a body, measured by the force applied per unit area.
Surface resistivity. The resistance of a material between two opposite sides of a unit square of its surface. Surface resistivity may vary widely with the conditions of measurement.

Thermal conductivity. The ability of material to conduct heat; the physical constant for the quantity of heat that passes through a unit cube of a material in a unit of time when the difference in temperature of two faces is 1°C.
Thermoplastic. A classification of resin that can be readily softened and resoftened by repeated heating. Hardening is achieved by cooling.
Thermosetting. A classification of resin that cures by chemical reaction when heated and, when cured, cannot be resoftened by heating.
Thixotropic. Describing materials that are gel-like at rest but fluid when agitated.

Vicat softening temperature. A temperature at which a specified needle point will penetrate a material under specified test conditions.
Viscosity. A measure of the resistance of a fluid to flow (usually through a specific orifice).
Volume resistivity (specific insulation resistance). The electrical resistance between opposite faces of a 1-cm cube of insulating material, commonly expressed in ohm-centimeters. The recommended test is ASTM D-257-54T.
Vulcanization. A chemical reaction in which the physical properties of an elastomer are changed by causing it to react with sulfur or other cross-linking agents.

Water absorption. The ratio of the weight of water absorbed by a material to the weight of the dry material.
Wetting. The ability to adhere to a surface immediately upon contact.
Working life. The period of time during which a liquid resin or adhesive, after mixing with a catalyst, solvent, or other compounding ingredients, remains usable. See *Pot life.*

*Trademark of E.I. duPont de Nemours & Co., Inc., Wilmington, Dela.

Source: From Harper.[3] Reprinted with permission.

TABLE 1.2 Significance of Important Electrical Insulation Properties

Property and definition	Significance of values
Dielectric strength	
All insulating materials fail at some level of applied voltage for a given set of operating conditions. The dielectric strength is the voltage that an insulating material can withstand before dielectric breakdown occurs. Dielectric strength is normally expressed in voltage gradient terms, such as volts per mil. In testing for dielectric strength, two methods of applying the voltage (gradual or by steps) are used. Type of voltage, temperature, and any preconditioning of the test part must be noted. Also, the thickness of the piece tested must be recorded because the voltage per mil at which breakdown occurs varies with the thickness of the test piece. Normally, breakdown occurs at a much higher volt-per-mil value in very thin test pieces (a few mils thick) than in thicker sections ($^1/_8$ in thick, for example).	The higher the value, the better the insulator. The dielectric strength of a material (per mil of thickness) usually increases considerably with a decrease in insulation thickness. Materials suppliers can provide curves of dielectric strength versus thickness for their insulating materials.
Resistance and resistivity	
Resistance of insulating material, like that of a conductor, is the resistance offered by the conducting path to passage of electric current. Resistance is expressed in ohms. Insulating materials are very poor conductors, offering high resistance. For insulating materials, the term *volume resistivity* is more commonly applied. Volume resistivity is the electrical resistance between opposite faces of a unit cube for a given material and at a given temperature. The relationship between resistance and resistivity is expressed by the equation $\rho = RA/l$, where ρ = volume resistivity in ohm-centimeters, A = area of the faces, and l = distance between faces of the piece on which measurement is made. This is not resistance per unit volume, which would be ohms per cubic centimeter, although this term is sometimes used erroneously. Other terms are sometimes used to describe a specific application or condition. One such term is surface resistivity, which is the resistance between two opposite edges of a surface film 1 cm square. Since the length and width of the path are the same, the centimeter terms cancel. Thus units of surface resistivity are actually ohms. However, to avoid confusion with usual resistance values, surface resistivity is normally given in ohms per square. Another broadly used term is insulation resistance, which again is a measurement of ohmic resistance for a given condition, rather than a standardized resistivity test. For both surface resistivity and insulation resistance, standardized comparative tests are normally used. Such tests can provide data such as effects of humidity on a given insulating material configuration.	The higher the value, the better; that is, a good insulating material. The resistance value for a given material depends on a number of factors. It varies inversely with temperature, and is affected by humidity, moisture content of the test part, level of the applied voltage, and time during which the voltage is applied. When tests are made on a piece that has been subjected to moist or humid conditions, it is important that measurements be made at controlled time intervals during or after the test condition has been applied, since dry-out and resistance increase occur rapidly. Comparing or interpreting data is difficult unless the test period is controlled and defined.

TABLE 1.2 Significance of Important Electrical Insulation Properties *(Continued)*

Property and definition	Significance of values
Dielectric constant	
The dielectric constant of an insulating material is the ratio of the capacitance of a capacitor containing that particular material to the capacitance of the same electrode system with air replacing the insulation as the dielectric medium. The dielectric constant is also sometimes defined as the property of an insulation which determines the electrostatic energy stored within the solid material. The dielectric constant of most commercial insulating materials varies from about 2 to 10, air having the value 1.	Low values are best for high-frequency or power applications, to minimize electric power losses. Higher values are best for capacitance applications. For most insulating materials, the dielectric constant increases with temperature, especially above a critical temperature region, which is unique for each material. Dielectric constant values are also affected (usually to a lesser degree) by frequency. This variation is also unique for each material.
Power factor and dissipation factor	
Power factor is the ratio of the power (watts) dissipated in an insulating material to the product of the effective voltage and current (volt-ampere) input and is a measure of the relative dielectric loss in the insulation when the system acts as a capacitor. The power factor is nondimensional and is a commonly used measure of insulation quality. It is of particular interest at high levels of frequency and power in such applications as microwave equipment, transformers, and other inductive devices. Low values are favorable, indicating a more efficient system, with lower losses. Dissipation factor is the tangent of the dielectric loss angle. Hence the term *tan* δ (tangent of the angle) is also sometimes used. For the low values ordinarily encountered in insulation, dissipation factor is practically the equivalent of power factor, and the terms are used interchangeably.	Low values are favorable, indicating a more efficient system, with lower losses.
Arc resistance	
Arc resistance is a measure of an electrical breakdown condition along an insulating surface, caused by the formation of a conductive path on the surface. It is a common ASTM measurement, especially used with plastic materials because of the variations among plastics in the extent to which a surface breakdown occurs. Arc resistance is measured as the time, in seconds, required for breakdown along the surface of the material being measured. Surface breakdown (arcing or electrical tracking along the surface) is also affected by surface cleanliness and dryness.	The higher the value, the better. Higher values indicate greater resistance to breakdown along the surface due to arcing or tracking conditions.

TABLE 1.2 Significance of Important Electrical Insulation Properties (*Continued*)

Property and definition	Significance of values
Comparative tracking index	
This is an Underwriters Laboratories test which is run similar to arc resistance except that an electrolyte solution (ammonium chloride) is put on the surface. The CTI is the value of the voltage required to cause a conductive path to form between electrodes.	The test is useful because it measures the arc resistance on a contaminated surface, which is often the case with actual electrical and electronics equipment.

Source: From Harper.[3] Reprinted with permission.

1.3 THERMOPLASTICS

Thermoplastic materials are polymers that can be repeatedly softened when heated and hardened when cooled. Because the high temperatures required for melting can cause degradation of thermoplastics, there is a limit to the number of reheat cycles. Thermoplastics are fabricated by blow molding, extrusion, foaming, injection and rotational molding, stamping, and vacuum forming. Detailed descriptions of all thermoplastics can be found elsewhere.[4–6] Tables 1.3 to 1.6 contain basic property and application information on those thermoplastics used in electrical and electronic applications. Supplier information is given in Green,[4] and a brief description of these materials follows.

1.3.1 Acrylics

Acrylic resins comprise a range of polymers and copolymers that are derived from acrylic acid–esters and their derivatives. The structural unit that characterizes these materials is as follows:

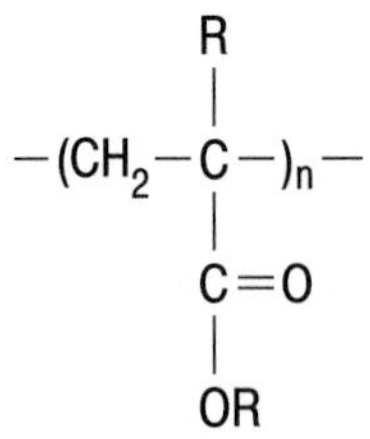

$$-(CH_2-\underset{\underset{\underset{OR}{|}}{\underset{C=O}{|}}}{\overset{\overset{R}{|}}{C}}-)_n-$$

These polymers are made by free-radical polymerization, and modifications are affected by the incorporation of other monomers during polymerization or by blending other resins such as vinyls, butadiene, polyester, or other acrylics in order to alter specific properties of the resin. Acrylics are characterized by their exceptional transparency (92 percent light transmission), with haze measurements ranging from 1 to 3 percent. They exhibit good weatherability and are resistant to solutions of inorganic acids, alkalies, and aliphatic hydrocarbons. Acrylics are attacked by chlorinated, aromatic,

TABLE 1.3 General Characteristics of Thermoplastics

Material	Characteristic properties	Processing*	Electrical/electronic applications
Acrylics	Crystal clarity, good surface hardness, weatherability, chemical and environmental resistance, mechanical stability	1,2,3,4,5,6	Colored electronic display filters, conformal coatings
Fluoroplastics	Heat resistance, superior chemical resistance, low dielectric losses, zero water absorption, low friction coefficient	9,10,11,12,13 Some fluoroplastics can be molded by more conventional methods (2,7)	Wire and cable insulation, electrical components
Ketone plastics	Heat resistance, chemical resistance, high strength, resistance to burning, thermal and oxidative stability, excellent electrical properties, low smoke emission	1,2,8,13	Wire insulation, cable connectors
Liquid crystal polymers	High temperature resistance, chemical resistance, high mechanical strength, low thermal expansion	1	Chip carriers, sockets, connectors, relay cases
Nylon	Mechanical strength, tough, abrasion and wear resistance, low friction coefficient	1,2,3,4,6,8	Connectors, wire jackets, wire ties, coil bobbins
Polyamide-imide	High temperature resistance, superior mechanical properties at elevated temperature, dimensional stability, creep and chemical resistance, radiation resistance	1,2,7,9	Connectors, circuit boards, radomes, films, wire coating
Polyarylate	Ultraviolet stability, dimensional stability, heat resistance, stable electrical properties, flame-retardant, high arc resistance	1,2,3,4	Connectors, coil bobbins, switch and fuse covers, relay housings
Polycarbonate	Clarity, toughness, heat resistance, flame-retardant	1,2,3,4	Connectors, terminal boards, bobbins
Polyesters (PBT, PCT, PET)	Good electrical properties, chemical resistance, high temperature resistance, low moisture absorption	1,2	Connectors, sockets, chip carriers, switches, coil bobbins, relays

TABLE 1.3 General Characteristics of Thermoplastics (*Continued*)

Material	Characteristic properties	Processing*	Electrical/electronic applications
Polyetherimide	Good high temperature strength, dimensional stability, chemical resistance, long-term heat resistance, low smoke generation	1,2,3,7	Connectors, low-loss radomes, printed circuit boards, chip carriers, sockets, bobbins, switches
Polyolefins	Range of strength and toughness, chemical resistance, low friction coefficient, processability, excellent electrical properties	1,2,3,4,7,8	Wire and cable insulation
Polyimides	Superior high temperature properties, radiation resistance, flame resistance, good electrical properties	1,6,7	Insulation for electric motors, magnet wire, flat cable, integrated-circuit applications
Polyphenylene oxide	Low moisture absorption, good electrical properties, chemical resistance	1,2,3,4	Connectors, fuse blocks
Polyphenylene sulfide	Flame resistance, high temperature resistance, dimensional stability, chemical resistance, good electrical properties	1	Connectors
Polyphthalamide	Good combination of mechanical, chemical, and electrical properties	1	Connectors, switches
Styrenes	Range of mechanical, chemical, electrical properties depending on type of styrene polymer, low dielectric losses	1,2,3,4,8	Housings
Polysulfones	High temperature resistance, excellent electrical properties, radiation resistance	1,2	Circuit boards, connectors, TV components
Vinyls	Range of properties depending on type	1,2,3,4	Wire insulation, tubing, sleeving

*1 Injection molding
2 Extrusion
3 Thermoforming
4 Blow molding
5 Machining
6 Casting
7 Compression molding
8 Rotational molding
9 Powder metallurgy
10 Sintering
11 Dispersion coating
12 Fluidized bed coating
13 Electrostatic coating

Source: From Harper[3] and Greene.[4] Reprinted with permission.

TABLE 1.4 Typical Physical Properties of Thermoplastics

Resin material	Coefficient of thermal expansion, 10^{-5} in/in/°C	Thermal conductivity 10^{-4} cal-cm/sec-cm^{2}°C	Water absorption 24 h, %	Flammability,* in/min	Specific gravity
ABS	6–13	4–9	0.2–0.5	1.0–2	1.01–1.07
Acrylic	−9	1.4	0.3	9–1.2	1.18–1.19
Chlorotrifluoroethylene	3.6	4–6	Nil	Nil	2.8–2.2
Fluorinated ethylene propylene	8.3–10.5	5.9	<0.05	Nonflammable	2.16
Polytetrafluoroethylene	2–12	6	<0.01	Nonflammable	2.14–2.28
Nylon 6	1.1–8.0	5.9	1.5	Self-extinguishing	1.07–1.15
Polycarbonate	6.7–7	4.6	0.15	Self-extinguishing	1.2
Polyethylene, low density	10–20	8	<0.01	Slow burning	0.910–0.925
Polyethylene, medium density	10–20	8	<0.01	Slow burning	0.926–0.940
Polyphthalamide	0.8–3.3	1.7–2.6	0.1–0.8	Self-extinguishing	1.13–1.70
Polyamide-imide	3.0	6.2	0.33	Self-extinguishing	1.42
Polyarylate	2.7–4.0	—	0.1–0.2	Self-extinguishing	1.19–1.22
Polybutylene terephthalate	6.0–9.5	4.2–6.9	0.08	—	1.30–1.38
Polyethylene terephthalate	6.5	3.4	0.1–0.2	—	1.29–1.40
Polyethylene, high density	5–11	11–12	<0.01	Slow burning	0.941–0.965
Polyethylene, high molecular weight	7–11	8	<0.01	Slow burning	0.93–0.94
Polyethylene, UHMW	13–20	—	<0.01	Slow burning	0.94
Polyimide	4.5–5.6	2.3–2.6	0.24	—	
Polypropylene	3.8–9	2.8–4	<0.01	Slow burning to nonburning	0.90–1.4
Polystyrene	5–8	3	0.01–0.03	0.5–2.5	1.04–1.05
Polyurethane	10–20	7.4	0.60–0.80	Slow to self-extinguishing	1.11–1.26
Polyvinyl chloride (flexible)	7–25	3–5	0.15–0.75	Self-extinguishing	1.15–1.80
Polyvinyl chloride, rigid	5–10	3–5	0.07–0.40	Self-extinguishing	1.33–1.58
Polyvinyl dichloride, rigid	7–8	3–4	0.07–0.11	Self-extinguishing	1.50–1.54
Styrene acrylonitrile (SAN)	7	3	0.15–0.25	0.4–0.7	1.07–1.08
Polyphenylene oxide	3.8–7.0	3.8	0.06–0.1	Self-extinguishing	1.04
Polysulfone	5.6	6.2	0.3	Self-extinguishing	1.24–1.25
Polyarylsulfone	3.1–4.9	—	0.1	Self-extinguishing	1.37
Polyethersulfone	5.5	3.2–4.4	0.12–1.7	Self-extinguishing	1.37
Polyetheretherketone	4.0–4.7	—	0.1	Self-extinguishing	1.31
Polyetherketone	1.8	10.5	0.05	Self-extinguishing	1.30
Polycyclohexylene dimethyl terephthalate (PCT), 30% glass reinforced	2.0	6.9	0.05		
Polyetherimide	4.7–5.6	1.6	0.25	Self-extinguishing	1.27
Polyphenylene sulfide	2.7–4.9	2.0–6.9	0.05	Self-extinguishing	1.35

*Samples 0.125 in thick.

Source: From Harper[3] and Green.[4] Reprinted with permission.

TABLE 1.5 Typical Physical and Mechanical Properties of Thermoplastics

Resin material	Impact strength, notched (Izod), ft • lb/in, 1/8-in bar	Tensile strength 10^2 lb/in^2	Tensile modulus 10^2 lb/in^2	Elongation, %	Flexural strength 10^2 lb/in^2	Compressive strength 10^2 lb/in^2	Compressive modulus 10^2 lb/in^2	Heat distortion temperature, at 264 lb/in^2 °F	Heat resistance, continuous, °F
ABS	1.5–12	2.5–5.8	120–420	20–100	5–13.5	5–11	120–200	180–245	160–235
Acrylic	0.3–0.4	8.7–11.0	350–450	3–6	2–7	11–19	350–430	167–198	130–195
Chlorotrifluoroethylene	2.5–5.0	6	150–300	80–250	7.4–11	4.6–7.4	180	160–170	390
Fluorinated ethylene propylene	No break	2–3.2	50	250–350	—	2.2	70	124	400
Polytetrafluoroethylene	No break	2–5	50–80	200–400	—	1.7	70–90	132	500
Nylon 6		0.9–4	9.5–12.4	100–380	25–300	5.8–15.7	13–16	347	150–175
Polycarbonate	12–16	8–9.5	345	110–120	13.5	10–12.5	350	265–290	250
Polyethylene, low density	No break	1–2.4	14–38	100–965	—	—	—	—	140–175
Polyethylene, medium density	No break	1.7–2.8	50–80	100–965	—	—	—	—	150–180
Polyethylene, high density	0.4–4.0	2.8–5	75–200	10–1200	1–4	2.7–3.6	50–110	110–125	180–225
Polyethylene, high molecular weight	4.5	2.3–5.4	136	170–800	3.5	2.4	110	120	180–225
Polyimide	1.5	5–14.0	300	8–10	19–28.8	30–40	—	680	500–600
Polyethylene	0.5–1.5	4.5–6.0	150–650	100–600	6.0	5–8	—	140–205	250
Polystyrene	0.40	5.2–7.5	400–500	1.5–2.5	10–15	11.5–16	300–560	160–215	150–190
Polyurethane	No break	4.5–8	1–3.7	60–120	0.–1	>20	85	—	190
Polyvinyl chloride, flexible	Varied	1–4	—	200–450	—	—	—	—	150–175
Polyvinyl chloride, rigid	0.4–22	7.5	200–600	40–80	10–15	10–11	300–400	140–175	160–165
Polyvinyl dichloride, rigid	1.0–6.0	7.5–9.0	360–450	160–240	14.2–17	13–22	—	212–235	195–210
Styrene acrylonitrile (SAN)	0.4–0.6	10	475–560	2–3	11–19	15–17.5	650	200–218	170–210
Polyphenylene oxide	3–6	6.8–7.8	380	50	8.3–12.8	15	380	375	250
Polysulfone	0.6–1.0	10.2	360	50–100	15.4	15.4	370	345	300

Polyphthalamide	1.0	>1.5	—	—	23.3	—	—	248	300
Polyamide-imide	2.7	22	650	7	27.4–34.9	32	—	532	430
Polybutylene terephthalate	0.7–1.0	8.2	280–430	50–300	12–16.7	8.6–14.5	—	122–185	270
Polyethylene terephthalate	0.25–0.7	7–10	400–600	30–300	14–18	11–15	—	70–100	—
Polyethylene, UHMW	No break	5.6–7.0	—	420–525	—	—	—	110–120	—
Polyarylsulfone	1.2	9.0	310–380	40–60	12.4–16.1	—	374	400	—
Polyethersulfone	1.4	9.8–13.8	350	6–80	17–18.7	11.8–15.6	—	395	350
Polyetherketone	1.6	13.5	520	50	24.5	20	—	323	—
Polyetheretherketone	2.0	10–15	—	30–150	16	18	—	—	—
Polyetherimide	1.1	14	430	60	22	20	420	390	330–350
Polyphenylene sulfide	<0.5	7–12	480	1–3	14–20	16	—	250	350–400
Polycyclohexylene di-methyl terephthalate (PCT), 30% glass reinforced	1.7	18–19	—	1.9	24–28	—	—	500	300

Source: From Harper[3] and Green.[4] Reprinted with permission.

TABLE 1.6 Typical Electrical Properties of Thermoplastics

Resin material	Volume resistivity, Ω • cm	Dielectric constant at 60 Hz	Dielectric strength, ST, $^1/_8$-in thickness, V/mil	Dissipation or power factor at 60 Hz	Arc resistance, s
ABS	10^{15}–10^{17}	2.6–3.5	300–450	0.003–0.007	45–90
Acrylic	>10^{14}	3.3–3.9	400	0.04–0.05	No tracking
Chlorotrifluoroethylene	10^{18}	2.65	450	0.015	>360
Fluorinated ethylene propylene	>10^{18}	2.1	500	0.0002	>165
Polytetrafluoroethylene	>10^{18}	2.1	400	<0.0001	No tracking
Nylon 6	10^{14}–10^{15}	6.1	300–400	0.4–0.6	140
Polycarbonate	6.1×10^{15}	2.97	410	0.0001–0.0005	10–120
Polyethylene, low density	10^{15}–10^{18}	2.98	450–1,000	0.006	Melts
Polyethylene, medium density	10^{15}–10^{18}	2.3	450–1,000	0.0001–0.0005	Melts
Polyethylene, high density	6×10^{15}–10^{18}	2.3	450–1,000	0.002–0.0003	Melts
Polyethylene, high molecular weight	>10^{16}	2.3–2.6	500–710	0.0003	Melts
Polyimide	10^{16}–10^{17}	3.5	400	0.002–0.003	230
Polypropylene	10^{15}–10^{17}	2.1–2.7	450–650	0.005–0.0007	36–136
Polystyrene	10^{17}–10^{21}	2.5–2.65	500–700	0.0001–0.0005	60–100
Polyurethane	2×10^{11}	6–8	850–1,100	0.276	—
Polyvinyl chloride, flexible	10^{11}–10^{15}	5–9	300–1,000	0.08–0.15	60–80
Polyvinyl chloride, rigid	10^{12}–10^{16}	3.4	425–1,040	0.01–0.02	—
Polyvinyl dichloride, rigid	10^{15}	3.08	1,200–1,550	0.018–0.0208	—
Styrene acrylonitrile (SAN)	10^{15}	2.8–3	400–500	0.006–0.008	100–500
Polyphenylene oxide	10^{17}	2.58	400–500	0.00035	75
Polysulfone	5×10^{16}	2.82	425	0.008–0.0056	122
Polyarylate	2×10^{14}	3.08	610	0.002	125
Polybutylene terephthalate (PBT)	1.4×10^{15}	3.3	420	0.002	190
Polycyclohexylene dimethyl terephthalate (PCT)	2×10^{15}	3.2	470–530	0.0018	68–136
Polyethylene terephthalate (PET)	1×10^{15}	3.8	650	0.0059	123
Polyphenylene sulfide, 40% glass	10^{16}	3.5–3.8	340–450	0.0012	34
Polyetherimide	6.7–10^{17}	3.15	750–831	0.0013	126
Polyetherketone	10^{17}	3.5	—	0.002	—

Source: From Harper[3] and Green.[4] Reprinted with permission.

ester, and ketone solvents. The maximum service temperature is about 200°F, but it can go as high as 313°F with the new acrylic-imide copolymers. The excellent arc and track resistance of the acrylics has made them a good choice in some high-voltage applications such as circuit breakers. Acrylics are one of the few plastics that exhibit an essentially linear decrease in dielectric constant and dissipation factor with increasing frequency. The acrylics are produced in many forms, including film, rod, sheet, tube, powder, solutions, and reactive syrups. Acrylics can also be formed into thermosetting resins, which are discussed in the appropriate section of thermosetting resins.

1.3.2 Fluoropolymers

These materials are a class of hydrocarbon polymers that has some or all of its hydrogens replaced by fluorine or chlorine. They are polytetrafluoroethylene (PTFE), fluorinated ethylene-propylene (FEP), perfluoroalkoxy (PFA), ethylene-tetrafluoroethylene (ETFE), polyvinylidene fluoride (PVDF), polychlorotrifluoroethylene (PCTFE), ethylene-chlorotrifluoroethylene (ECTFE), and polyvinyl fluoride (PVF). The structural units of these polymers are as follows:

Polymer	Structure
PTFE	$-(-CF_2-CF_2-)_n-$
FEP	$-(-CF_2-CF_2-CF_2-\underset{\displaystyle CF_3}{\underset{\vert}{CF}}-$
PFA	$-(-CF_2-CF_2-\underset{\displaystyle OR}{\underset{\vert}{CF}}-CF_2-CF_2-$
PCTFE	$-(-CF_2-CFCL-)_n-$
ECTFE	$-(-CF_2-CF_2-CF_2-CFCL-)_n-$
PVDF	$-(-CF_2-CF_2-)_n-$

The fluoropolymers and copolymers are synthesized by free-radical polymerization techniques. The polymers are characterized by their unique combination of chemical, electrical, mechanical, and thermal properties. They do not support flame propagation, are unaffected by most chemicals, and have excellent arc resistance, low dielectric losses, and essentially zero water absorption. Their limitation is that these polymers are relatively soft, difficult to process, expensive, and subject to creep. Service temperatures for these materials range from 110 to 260°C, depending on the type of polymer. Some specific properties of these materials are given in Table 1.7. Fluoropolymers are used in electronic circuits as coatings, films, tubing, fibers, and tapes.

TABLE 1.7 Properties of Fluoropolymers

	PTFE	FEP	PFA	ETFE	PVDF	PCTFE	ECTFE	PVF
% flourine	76.0	76.0	76.0	59.4	59.4	48.9	39.4	41.3
Melting point, °C	327	265	310	270	160	218	245	200
Upper use temperature, °C	260	200	260	180	150	204	170	110
Density, g/cm^3	2.13	2.15	2.12	1.7	1.78	2.13	1.68	1.38
Oxygen index, %	>95	95	95	28–32	44	—	48–64	—
Arc resistance, seconds	>240	>300	>300	72	60	2.4	18	—
Dielectric constant	2.1	2.1	2.1	2.6	9–10	2.5	2.5	9
Dissipation factor	0.0002	0.0002	0.0002	0.0008	0.02–0.02	0.02	0.003	0.002
Tensile strength, lb/in^2	5000	3100	4300	7000	6200	6000	—	—
Specific gravity	2.2	2.17	2.17	1.7	1.78	2.2	1.68	—
Water absorption, % 24 h	0	0	0.03	0.03	0.06	0	—	—
Electrical strength, V/mil	480	600	500	400	280	600	—	—

Source: From Harper.[7] Reprinted with permission.

1.3.3 Ketone Resins

These materials are partially crystalline polymers and are characterized by the presence of the phenyl ring and both ether (—O—) and ketone (R_2—C=O) groups in the polymer chain. There are several types of ketone polymers which are synthesized by condensation polymerization. The major ketone polymers are polyetherketone (PEK), polyetheretherketone (PEEK), and polyetherketonetherketoneketone (PEKEKK). Their structures are as follows:

Polymer | Structure

PEK

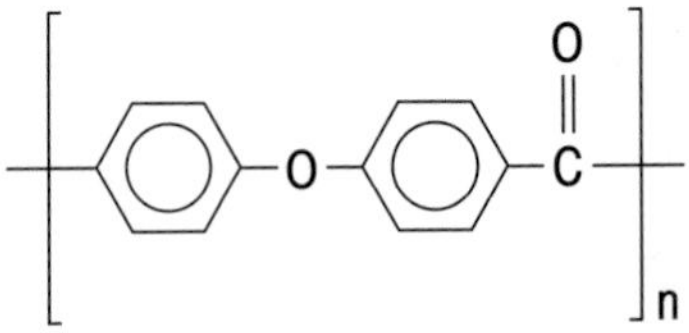

PEEK

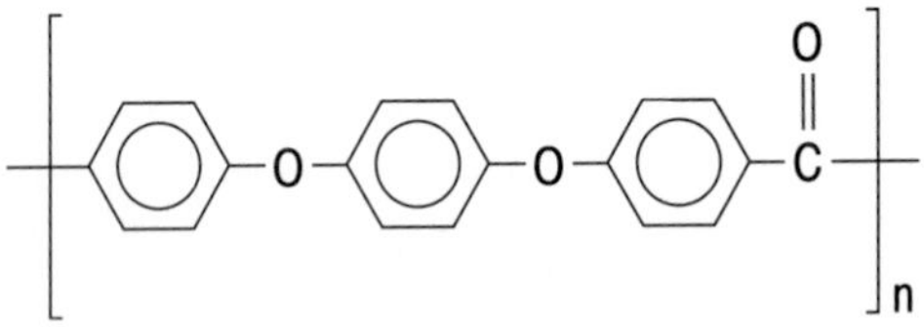

PEKEK

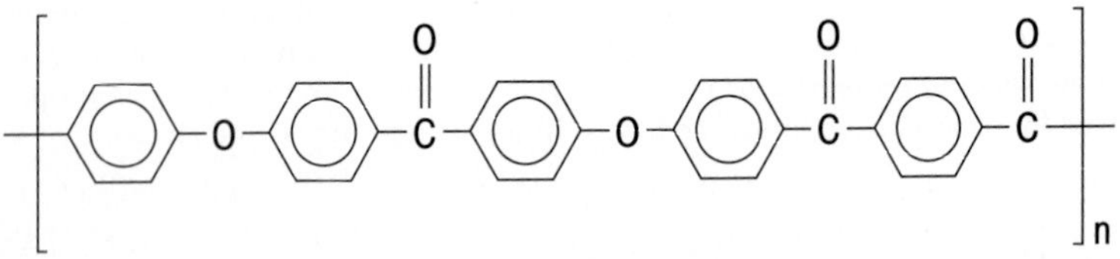

These polymers are characterized by their excellent elevated-temperature properties, with continuous service temperatures approaching 260°C. The polymers are tough and have high impact strength as well as good dielectric strength, volume, and surface resistivity. Only concentrated, anhydrous, or strong oxidizing acids have an effect on these polymers. Common organic solvents do not attack these polymers, and they are highly resistant to hot-water hydrolysis. Their main use is for wire and cable insulation and connectors.

1.3.4 Liquid Crystal Polymers

These polymers belong to a class of materials that exhibits a highly ordered structure in both melt and solid states. Because of the high degree of molecular ordering, liquid crystal polymers (LCPs) exhibit a high degree of anisotropy. If the liquid crystalline phase forms on melting the polymer, it is known as a thermotropic liquid crystal and if it forms in solution, as the result of solvent addition, it is known as lyotropic. Condensation polymerization has been used to prepare these polymers. A number of polymers exhibit liquid crystalline behavior, but the three commercially important polymers are Xydar (Amoco Performance Products, Inc.), Vectra (Hoechst Celanese Corp.), and the HX series (E.I. duPont de Nemours & Co). There is no one chemical structure that characterizes LCPs; however, all LCPs have these common characteristics: the molecular shape has a large aspect ratio (length or diameter to width or thickness), the molecule has a large polarizability along the rigid chain axis compared to the transverse direction, and the molecule must have good molecular parallelism of the rigid units comprising its structure. In order to meet these requirements an LCP should possess a rigid molecular structure.[8] Both Xydar and Vectra meet these criteria, and their structures are as follows:

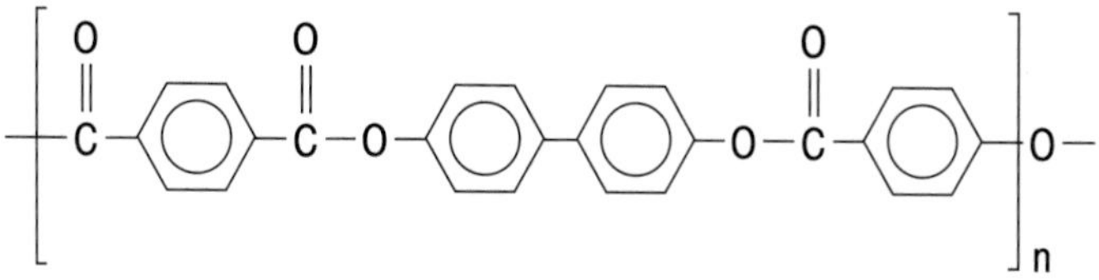

(Xydar)

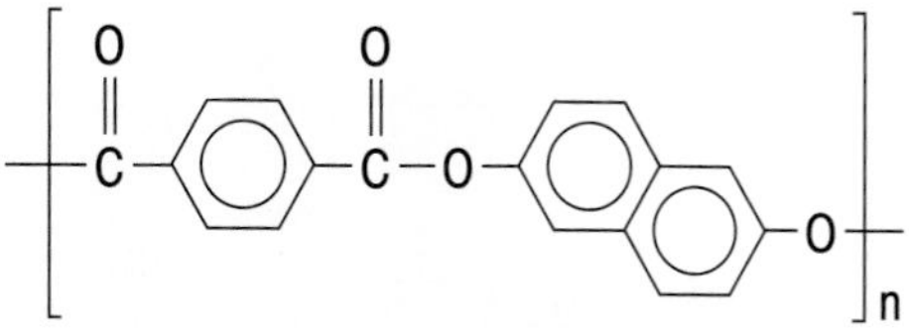

(Vectra)

The major properties that characterize LCPs are low melt viscosity, exceptional tensile, compressive, and modulus values, and outstanding chemical, radiation, and thermal stability. A general comparison of flexural moduli and mold shrinkage for LCPs and other polymers is given in Figs. 1.4 and 1.5, while Table 1.8 presents a comparison of specific properties of filled LCPs. The main uses for these polymers in the electronics industry are for the molding of high-precision complex parts, chip carriers, sockets, connectors, pin-grid arrays, bobbins, and relay cases.

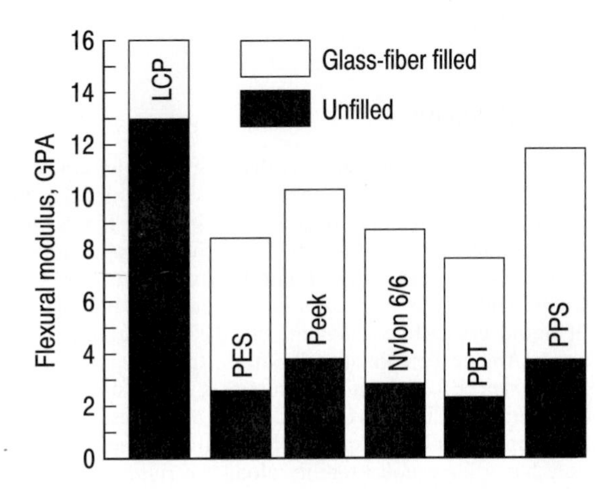

FIGURE 1.4 Comparison of flexural moduli of selected thermoplastics. (*From Klein[9]; reprinted with permission.*)

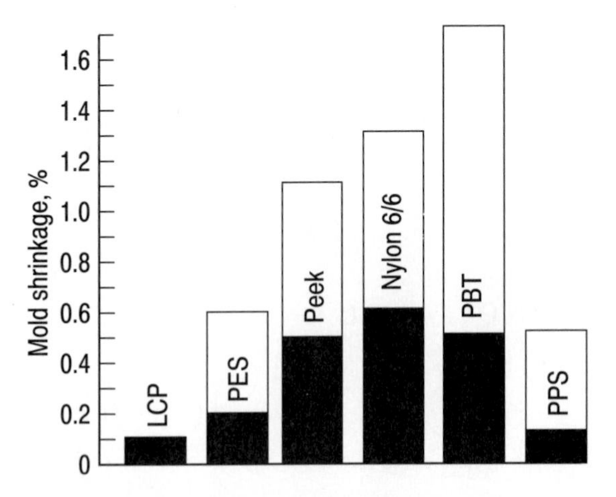

FIGURE 1.5 Comparison of mold shrinkage of selected thermoplastics. (*From Klein[9]; reprinted with permission.*)

TABLE 1.8 Typical Properties of LCP Molding Compounds

Grade	Vectra* Al30	Vectra B230	Vectra A420	Xydar† G-930	Xydar NG-350	DuPont HX4100	DuPont HX6130
Fillers	30% glass fiber	30% carbon fiber	50% glass/ mineral, 5% graphite flake	30% glass fiber	50% mineral/ glass	30% glass fiber	30% glass fiber
Tensile strength 10^3 lb/in^2	30	35	21.5	19.6	14.2	23.0	21
@ temperature	1.9 @ 400°F	—	1.5 @ 400°F	2.8 @ 500°F	2.3 @ 500°F	5.8 @ 392°F	—
Elongation, %	2.2	1.0	1.4	1.6	2.3	1.0	2.5
Flex modulus, 10^6 lb/in^2	2.1	4.6	2.9	2.26	1.73	2.7	1.9
@ temperature	0.85 @ 428°F	0.41 @ 428°F	—	0.46 @ 500°F	0.29 @ 500°F	0.90 @ 392°F	—
Impact strength, notched, ft • lb/in	2.8	1.4	1.9	1.8	1.9	1.1	2.6
Coefficient of thermal expansion, ppm/°F							
Flow direction	2.8	−1.7	3.9	2.7	5.7	3.0	4.0
Transverse	36	25	28	43	21	54	12
Dielectric strength, V/mil	1100‡	—	—	900	900	1250¶	—
Heat deflection temperature @ 264 lb/in^2, °F	446	440	437	520	493	520	482
Dielectric constant at 1 MHz			3.4–3.9				
Dissipation factor at 1 MHz			0.009–0.34		0.39		
Volume, Ω • cm			10^{14}–10^{15}				

*Hoechst Celanese.

†Amoco Performance Products.

‡At 1.50-mm thickness.

¶At 1.59-mm thickness.

Source: From Hunt.[10] Reprinted with permission.

1.3.5 Nylon

These materials, also known as polyamides, are characterized by having the amide group (—CONH—) as an integral part of the polymer structure. While this chemical unit is present in all nylons, the multiplicity of monomers that can be used to prepare nylons has led to a wide variety of materials with different properties. Presently there are 11 types of nylons available,[10] nine are aliphatic and two are aromatic. Nylons are synthesized by both condensation polymerization (types 6/6, 6/9, 6/10, 6/12) and addition polymerization (types 6, 11, 12). Most nylons are partially crystalline polymers. The nylons can be modified by the addition of additives or copolymerized with other monomers to produce a wide range of materials with different properties. In addition some blending of nylon polymers can be done with acrylonitrile-butadiene-styrene and polyphenylene ether polymers. Transparent nylon is available, and unlike the other grades of nylon polymers, this is amorphous. The nylons are strong, tough thermoplastics having good tensile, flexural, friction, and impact properties. Nylons can operate satisfactorily over the temperature range of 0 to 149°C. All nylons are hygroscopic, although the degree of water absorption decreases with increasing hydrocarbon chain length. This moisture absorption affects some properties; for example, it has a plasticizing effect on the polymer and increases flexural and impact strength while decreasing tensile strength. The electrical properties of nylons are quite sensitive to moisture and deteriorate with increasing water content. Nevertheless, these properties are quite adequate to allow the use of nylons in most 60-Hz power applications. Nylons have good chemical resistance to hydrocarbons and aromatic and aliphatic solvents, and are attacked by strong acids, bases, and phenols. Elevated temperature and ultraviolet radiation exposure will degrade nylon depending on the duration and level of the exposure. The nylons can be processed by almost all of the common thermoplastic fabrication techniques. The reader is directed to the references for additional information and specific properties.[4,6,10] Applications for nylons include card guides, connectors, terminal blocks, antenna mounts, coil bobbins, and receptacle plugs.

High-Temperature Nylon. In addition to the aforementioned nylons there is another class of polyamides that is based on the presence of an aromatic ring in the chemical structure. Two materials comprise this class: Nomex [(poly)1,3-phenylene isophthalamide] and Kevlar [(poly)1,4-(poly)1,4-phenylene terephthalamide] (both E.I. duPont de Nemours & Co.).

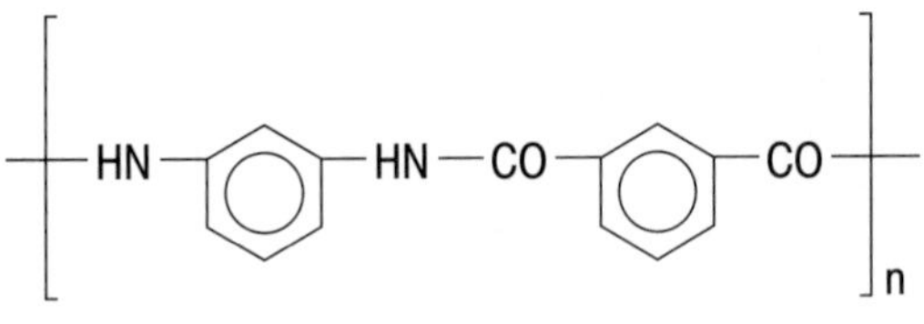

$$\left[-\mathrm{HN}-\bigcirc-\mathrm{HN}-\mathrm{CO}-\bigcirc-\mathrm{CO}- \right]_n$$

(I)

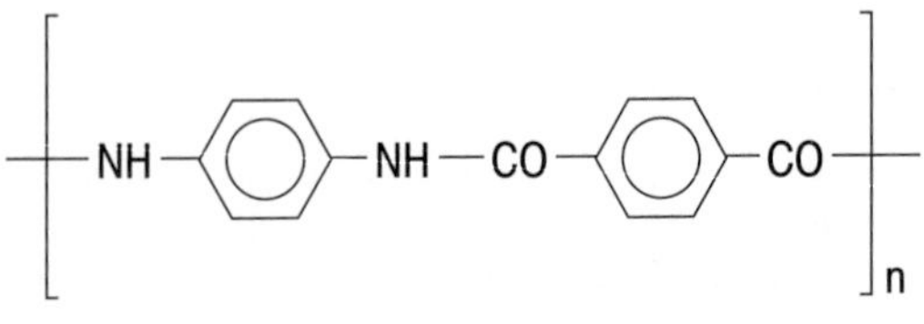

$$\left[-\mathrm{NH}-\bigcirc-\mathrm{NH}-\mathrm{CO}-\bigcirc-\mathrm{CO}- \right]_n$$

(II)

TABLE 1.9 Electrical Properties of Nomex*

Nomex type	Thickness, mil	Dielectric strength, V/mil, ASTM D-149	Dielectric constant, at 60 Hz, ASTM D-150	Dissipation factor at 60 Hz, ASTM D-150	Volume resistivity, Ω • cm, ASTM D-257
410	3	540	1.6	0.005	10^{16}
411	5	230	1.2	0.003	—
414	3.4	530	1.7	0.005	10^{16}
418	3	730	2.9	0.006	10^{16}
419	7	325	2.0	—	—
992	125	380	1.7	0.020	10^{17}
993	120	540	2.6	0.015	10^{17}
994	250		3.5	0.010	10^{16}

*Unless otherwise noted, the Nomex properties are typical values measured by air under "standard" conditions (in equilibrium at 23°C, 50% relative humidity) and should not be used as specification limits. The dissipation factors of types 418, 419, 992, and 993 and all of the volume resistivities are measured under dry conditions. Nomex is a registered trademark of E.I. duPont de Nemours & Co. for its aramid products.

Source: From duPont Co.[11] Reprinted with permission.

Kevlar is spun into fiber and is mostly used in composite applications, while Nomex is processed into fiber, paper, sheet, and press-board and is used extensively in the electrical industry as insulation for transformer coils and motor stators. Nomex is recognized by Underwriters Laboratories as a 220°C material. Table 1.9 gives some electrical properties of Nomex.

1.3.6 Polyamide-Imides

These polymers are amorphous materials produced by the condensation polymerization of trimellitic anhydride and aromatic diamines. The characteristic chemical groups in the polymer chain are the amide linkage (—CONH—) and the imide linkage (—CONCO—). These polymers can be solution cast into film or converted into powders for further processing into various forms. The polymers possess outstanding high-temperature (260°C) and radiation stability (10^9 rads) as well as excellent mechanical properties, low dielectric losses, and good wear resistance. The polymers are useful over a wide temperature range (−195 to +260°C) and are inherently fire-resistant (oxygen index of 43 and UL-94 rating of V–0). Their chemical resistance is excellent, but they are attacked by hot caustic and acid as well as steam. Applications for molded parts include electronic connectors and jet engine component generator parts. The solution form of the polymer is used for wire enamels and a variety of electronics applications.

1.3.7 Polyimides

These materials are derived from the solution condensation polymerization of aromatic dianhydrides and diamines and are characterized by the presence of only the imide linkage (—CONCO—). The general structure that depicts these polymers is as follows:

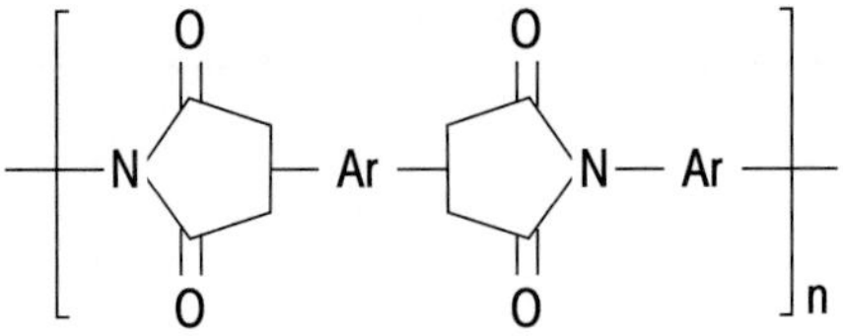

The polyimides are characterized by high glass transition temperatures, excellent radiation resistance, toughness, good electrical properties, and good flame resistance. The properties of polyimides can be modified by adjusting both the type and the ratio of the monomers. Fillers have also been added to polyimides to alter their properties. These modifications have produced a variety of polyimide materials. The polyimides can be processed in solution or powder form and can be converted into film, molding powders, tape, and varnishes. Polyimides can be compression and injection molded, but considerable expertise is required because of the high glass transition temperatures and melt viscosities of these polymers. Kapton is perhaps the most widely known of the polyimide family, but other types include Vespel, Pyralin, Pyre ML, Envex, Skybond 700, FM-34, Untratherm, Eymyd, Duramid, Avimid, Cypac, and Matrimid.* Although there are differences in the properties of the various polyimides, the properties of the polyimide family are illustrated with those of Kapton film. Tables 1.10 and 1.11 show these properties.

1.3.8 Polyetherimide

Although this material belongs to the polyimide family of resins and has similar properties as the all-aromatic polyimides, it has lower thermal stability. (It is UL-rated for 170°C continuous use, compared to 220°C for Kapton polyimide.) It is an amorphous polyimide having aromatic imide and ether repeating units in its molecular chain as follows:

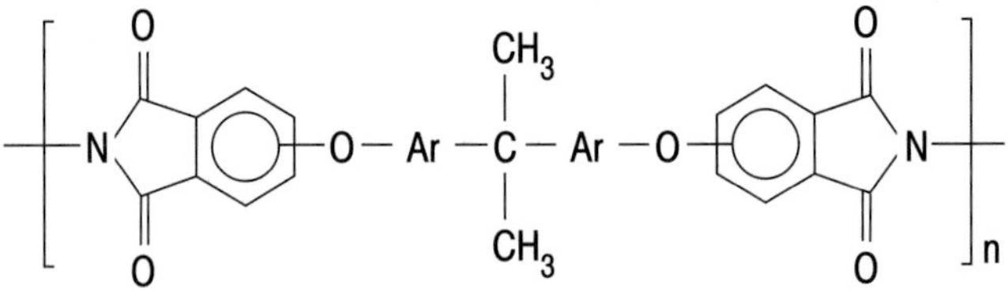

*These polyimide products are registered trademarks of the following corporations: Vespel, Pyralin, Pyre ML, and Avimid—E.I. duPont de Nemours & Co.; Envex and Duramid—Rogers Corp.; Skybond 700—Monsanto Corp.; Untratherm—P.D. George Co.; Eymyd—Ethyl Corp.; Cypac—American Cyanamid Co.; and Matrimid—Ciba-Geigy Corp. FM-34 is a product of the American Cyanamid Co.

TABLE 1.10 Physical and Electrical Properties of Kapton 100 HN Film

	Typical values		
Physical properties	23°C (73°F)	200°C (392°F)	Test method
Ultimate tensile (MD)* strength, MPa (lb/in^2)	231 (33,500)	139 (20,000)	ASTM D-882-83, method A†
Yield point (MD) at 3%, MPa (lb/in^2)	69 (10,000)	41 (6,000)	ASTM D-882-81
Stress to produce (MD) 5% elongation, MPa (lb/in^2)	90 (13,000)	61 (9,000)	ASTM D-882-81
Ultimate elongation (MD), %	72	83	ASTM D-882-81
Tensile modulus (MD), GPa (lb/in^2)	2.5 (370,000)	2.0 (290,000)	ASTM D-882-81
Impact strength, kg • cm (ft • lb)	8 (0.58)		DuPont pneumatic impact test
Folding endurance‡, cycles	285,000		ASTM D-2176-69 (1982)
Tear strength (MD), propagating (Elmendorf), g	7		ASTM D-1922-67 (1978)
Density, g/cm^3	1.42		ASTM D-1505-68 (1979)
Coefficient of friction, kinetic (film-to-film)	0.48		ASTM D-1894-78
Coefficient of friction, static (film-to-film)	0.63		ASTM D-1894-78
Refractive index (Becke line)	1.66		ASTM D-542-50 (1977)
Poisson's ratio	0.34		Avg 3 samples elongated at 5%, 7%, 10%
Low-temperature flex life	Pass		IPC-TM-650, method 2.6.18
Dielectric strength at 60 Hz, V/mil	7,700		ASTM D-149-81
Dielectric constant at 1 kHz	3.4		ASTM D-150-81
Dissipation factor	0.0018		ASTM D-150-81
Volume resistivity, Ω • cm	10^{17}		ASTM D-257-78

*MD—Machine direction

†Specimen size: 25 by 170 mm (1 by 6 in); jaw separation: 100 mm (4 in); jaw speed: 50 mm/min (2 in/min); ultimate refers to tensile strength and elongation measured at break.

‡Massachusetts Institute of Technology Test.

Source: From duPont.[12] Reprinted with permission.

It processes much better on conventional thermoplastic equipment compared to the completely aromatic polyimides and is easily molded into complex shapes. It has a UL-94 V–O flame resistance rating and an oxygen index of 47. Chemical resistance is good against hydrocarbons, and it resists mild acid and base for short exposure times. It has excellent ultraviolet and gamma radiation resistance (94 percent retention of tensile strength after exposure to 400 Merads of cobalt irradiation). The polymer

TABLE 1.11 Thermal Properties of Kapton 100 HN Film

Thermal properties	Typical values	Test condition	Test method
Melting point	None	None	ASTM E-794-81
Thermal coefficient of expansion	20 ppm/°C (11 ppm/°F)	−14 to 38°C (7 to 100°F)	ASTM D-696-79
Coefficient of thermal conductivity, W/m • K	0.12	296 K	University of Delaware Physics Department method
$\frac{\text{cal}}{\text{cm} \cdot \text{s} \cdot {}^\circ\text{C}}$	2.87×10^{-4}	23°C	
Specific heat	$\frac{1.09 \text{ J/g} \cdot \text{K}}{(0.261) \text{ cal/g} \cdot {}^\circ\text{C}}$		Differential calorimetry
Flammability	94 V–0		UL-94 (2-8-85)
Shrinkage, %	0.17 1.25	30 min @ 150°C 120 min @ 400°C	IPC-TM-650, method 2.2.4A
Heat sealability	Not heat-sealable		
Limiting oxygen index, %	37		ASTM D-2863-77
Solder float	Pass	IPC-TM-650,	Method 2.4.13A
Smoke generation	DM <1	NBS smoke chamber	NFPA-258
Glass transition temperature T_g	A second-order transition occurs in Kapton between 360°C (680°F) and 410°C (770°F) and is assumed to be the glass transition temperature. Different measurement techniques produce different results within the above temperature range.		

Source: From duPont.[12] Reprinted with permission.

retains 85 percent of its tensile strength after 10^4 h in boiling water. Electrical properties show very good stability under various conditions of temperature, humidity, and frequency. Its low dissipation factor makes it transparent to microwaves. Applications include low-loss radomes, printed-wiring boards, IC chip carriers, bobbins, and infrared switches.

1.3.9 Polyarylate and Polyesters

The thermoplastic polyesters include polyarylate (PA), polybutylene terephthalate (PBT), polyethylene terephthalate (PT), and polycyclohexylene dimethylene terephthalate (PCT). These linear polyesters range from amorphous to crystalline materials and have the characteristic ester functional group (—COOR—) present along the polymer chain. Except for the polyarylates, the other polyesters are made by a transesterification of the appropriate alcohol and ester monomers. Polyarylate is prepared from the reaction of bisphenol A and a mixture of iso and terephthalic acids. Because of the variety of alcohols, acids, or esters that are available for reaction, polyesters with a broad range of properties can be synthesized.

Polyarylate (PA) resins are aromatic, linear, amorphous polyesters with excellent toughness, ultraviolet resistance, flex strength, dimensional stability, flame retardance, and electrical properties. Electrical properties are fairly constant over a broad temperature range. These polymers are susceptible to stress cracking when exposed to ketone, aromatic hydrocarbon, ester, and chlorinated solvents. If polyarylates are alloyed with other polymers, the stress crack resistance is improved. Polyarylates are processed by most conventional melt processes such as injection extrusion, flow molding, and thermoforming. Electrical applications include connectors, relay housings, coil bobbins, and switch and fuse covers.

Polybutylene terephthalate (PBT) is a linear, semicrystalline, aliphatic polyester, although it does have some aromatic character. It has excellent chemical and temperature resistance and good electrical properties which are unaffected by humidity. PBT is unaffected by water, weak acids and bases, and common organic solvents at room temperature. It cannot be solvent bonded because of its solvent resistance. PBT resins are processed mostly by injection molding.

Polyethylene terephthalate (PET) is also a linear, semicrystalline, aliphatic polyester that is produced in a standard- and an engineering-grade material. The latter grade has superior properties (strength, stiffness, dimensional stability, chemical and heat resistance, and electrical properties) and is therefore preferred for electrical applications. The material is processed by injection molding and is used for lamp sockets, coil forms (audio/video transformers), connectors, and terminal blocks.

Polycyclohexylene dimethylene terephthalate (PCT) is a linear high-temperature semicrystalline material. Its high heat resistance distinguishes it from PET and PBT. PCT has a melting point of 290°C, compared to 224°C for PBT and 250°C for PET. This high temperature resistance makes PCT useful for surface-mount electronic components. The material has an excellent balance of physical, chemical, electrical, mechanical, and thermal properties. Injection molding is the preferred processing method. Applications include sockets, chip carriers, pin grid arrays, coil bobbins, and surface-mount components.

1.3.10 Polycarbonates

Polycarbonate is a linear, amorphous material having the following repeat unit in the polymer chain:

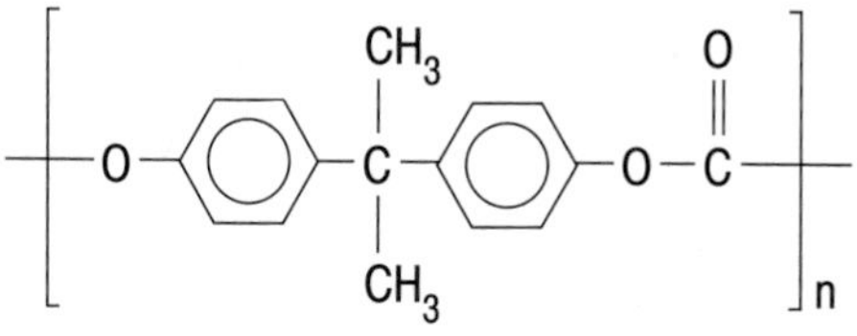

It is synthesized by interfacial polymerization. The characteristic properties of this polymer are excellent transparency, toughness, and high-temperature properties (up to 140°C). The polymer has very good electrical properties and is essentially self-extinguishing. The electrical properties are good with a stable dielectric constant over a wide temperature and frequency range. The ultraviolet and chemical resistance of polycarbonates is limited. They are attacked by alkali, amines, ketones, esters, and aromatic hydrocarbons. Stressed polycarbonate parts are sensitive to many solvents and will crack upon exposure. Polycarbonate is processed by most conventional thermoplastic processing methods. Applications include connectors, circuit breaker boxes, and bobbins.

1.3.11 Polyolefins

This class of materials includes the polymers and copolymers of polyethylene and polypropylenes. The grades include a range of densities (0.83 g/cm^3 for polymethylpentene) to 0.96 g/cm^3 for the high-density polymers) and a range of molecular weights, including ultrahigh molecular weight. These polymers are characterized by the —(—CH_2— —CHR)— repeat unit, where R represents hydrogen or an alkyl radical. There are a multitude of materials available, but the common bond that joins all these materials is their low dielectric constant, low dissipation factor, low water absorption, low coefficient of friction, and excellent chemical resistance (resistant to acids, alkalis, and most solvents). Special grades of polyolefins, such as the higher-molecular-weight material, exhibit increased toughness, abrasion resistance, and freedom from environmental stress cracking. Wire and cable insulation are the main uses for these materials in the electrical industry.

1.3.12 Polyphenylene Oxide

Polyphenylene oxide (PPO) is a linear amorphous polymer made by a procedure called oxidative coupling. The structural unit characteristic of the polymer is as follows:

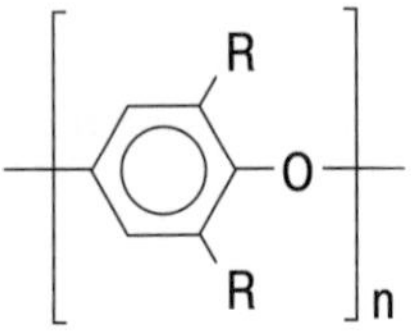

The glass transition temperature of pure PPO is 210°C. This polymer is not used in its neat form, but rather is blended with polystyrene to produce the commercial materials Noryl and Prevex (both General Electric Co.). This modification reduces the glass transition temperature of the blend. The PPO alloys have good resistance to acids and alkalis, but are attached by some aromatic and chlorinated solvents. A number of grades are available, all of which can be easily processed on conventional thermoplastic molding equipment. Applications include computers, connectors, fuse blocks, relays, and bus bar insulation.

1.3.13 Polyphenylene Sulfide

Polyphenylene sulfide (PPS) is a semicrystalline material that is inherently flame-retardant. The structure that characterizes this polymer is as follows:

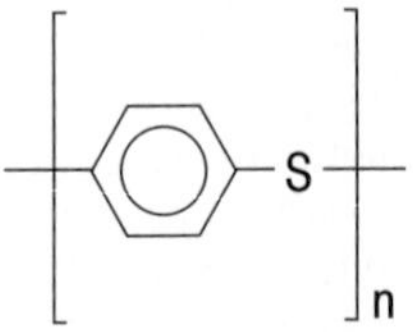

Two forms of PPS are available, a linear and a branched polymer. The former has better strength and melt viscosity properties. PPS is inert to all solvents except hot nitric acid. It is processed like most thermoplastics and can be injection and compression molded. Cross-linkable grades are available with exceptional heat resistance. The PPS polymers are rated UL-94 V–0. The polymers have stable electrical properties over broad temperature, frequency, and humidity ranges. Both volume and insulation resistance are excellent in wet and dry environments and the arc resistance is good. The primary uses of PPS resins are in electric connectors, coil forms, bobbins, yokes, and terminal boards.

1.3.14 Styrenics

The styrene polymers include acrylonitrile-butadiene-styrene (ABS), acrylic-styrene-acrylonitrile (ASA), polystyrene (PS), styrene-acrylonitrile (SAN), styrene-butadiene (SB), and styrene–maleic anhydride (SMA). The properties of these polymers are dependent on the ratio of the monomeric components, which leads to a broad range of material properties. The structural unit common to all these polymers is as follows:

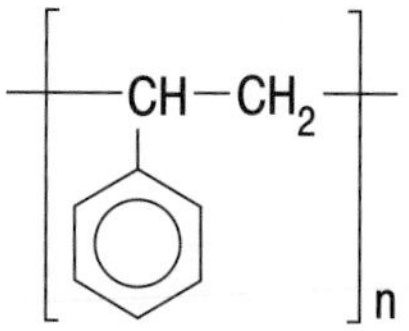

$$\left[-CH(C_6H_5)-CH_2- \right]_n$$

These polymers are generally not used in electrical applications except as electronic housings. Polystyrene has very low dielectric losses ($e' = 2.45$, $\tan \delta = 0.001$), and its chief use in electronics is in striplines. The styrenics are amenable to all forms of thermoplastic processing.

1.3.15 Polysulfones

Polysulfones are amorphous polymers of high temperature resistance. These materials contain the arylsulfone group as part of their chemical structure:

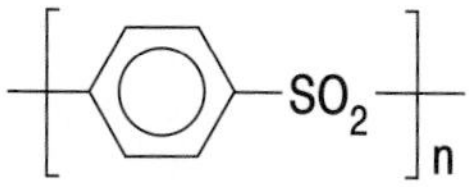

$$\left[-C_6H_4-SO_2- \right]_n$$

There are three important sulfone polymers, all of which have excellent electrical properties: polysulfone, polyarylsulfone, and polyethersulfone. These materials are resistant to acid and alkaline hydrolysis, but can exhibit stress crazing when exposed to certain chemicals (esters, ketones, and some hydrocarbons). All the polymers have excellent creep and radiation resistance. Processing is carried out on standard thermoplastic equipment. Applications include printed-wiring-board substrates, television components, coil bobbins, connectors, and switch housings.

1.3.16 Vinyls

The vinyl polymers include polyvinylchloride (PVC), polyvinylidene chloride (PVDC), and chlorinated polyvinylchloride (CPVC). The vinyl resins contain the structural unit —(—CH_2—CHCL—)— in the polymer chain, although the PVDC and CPVC contain extra chlorine atoms. All the resins show good chemical and flame resistance, with PVC being the weakest member of the family with regard to these properties. PVC is the most versatile of the plastics because of its wide blending capability. CPVC has excellent chemical resistance, rigidity, strength, and weatherability. PVDC has low permeability to gases and liquids, good barrier properties, and also good chemical resistance. All the vinyl resins are processed on conventional thermoplastic equipment. Electrical applications include wire insulation, cable jackets, sleeving, and tubing.

1.4 THERMOPLASTIC ALLOYS AND BLENDS

The ability of polymers to be mixed together to form a new polymer mixture with greatly expanded properties has given rise to this new group of materials. Alloys are synergistic polymer combinations with real property advantages derived from a high level of thermodynamic compatibility between the components. Alloys exhibit strong intermolecular forces and form single-phase systems with one glass transition temperature.[4] Noryl, an alloy of polystyrene and PPO, is a typical example. Blends do not have the exacting thermodynamic capabilities that alloys have and exhibit discrete phases and multiple glass transition temperatures. In general, the properties of a blend reflect the weighted average of the properties of the components of the blend. Alloys and blends are processed on all conventional thermoplastic equipment. A list of common alloys and blends is given in Table 1.12 together with supplier names and important properties. More detailed information on alloys and blends can be obtained in Albee.[13]

1.5 THERMOSETS

Unlike thermoplastics, thermoset materials are polymers that form three-dimensional cross-linked networks of polymer chains which cannot be softened or reheated for additional use. In general, these materials can provide higher temperature capability than the thermoplastic materials. Thermoset materials, before being cured, are fabricated by casting, compression molding, filament winding, laminating, pultrusion, or injection and transfer molding. Most thermoset materials, before they are cross-linked, are considerably more fluid than thermoplastics during processing. A thermoset material must contain a functionality greater than 2 to facilitate cross-linking, that is, the polymer chain must have enough reactive sites to form a three-dimensional network. Difunctional materials form linear or branched uncross-linked polymers but can be cross-linked by forming more reactive sites by the addition of either a catalyst or a curing agent. These ingredients promote the formation of active sites for further reaction. The curing or cross-linking reaction is an exothermic reaction, and consideration must be given to control the temperature rise to prevent a runaway reaction. Thermoset materials usually shrink when they are cross-linked, but the shrinkage can be con-

TABLE 1.12 Common Alloys and Blends

Alloy or blend	Trade name	Supplier	Key properties
PPO/PS	Noryl	GE Plastics	Heat resistance, toughness, improved processing, low cost, low moisture absorption
Nylon/ABS	Triax 1000	Monsanto	Heat and chemical resistance, good flow, low-temperature impact
Nylon/elastomers	Zytel ST Nylon 7000 Capron Grilon	DuPont Hoechst Celanese Allied EMS-American Grilon	Improved toughness and fatigue resistance, chemical and heat resistance
PPO/nylon	Noryl GTX	GE Plastics	High temperature resistance, chemical resistance, low moisture absorption, dimensional stability
ABS/PC	Triax 200 Pulse Bayblend	Monsanto Dow Mobay	Heat resistance, processability, low-temperature impact
PPO/PBT	Gemax	GE Plastics	High temperature resistance, dimensional stability, processability, class A surface
Polyester/elastomer	Rynite SST Valox VCT Duraloy Pocan	DuPont GE Plastics Hoechst Celanese Mobay	Stiffness, toughness, high temperature and solvent resistance
PC/PBT	Xenoy	GE Plastics	High impact and modulus over wide temperature range, chemical resistance
ASA/PC	GX 200 Terblend S	GE Plastics BASF	Impact, thermal stability, weatherability
TPE alloy	Bexloy V	DuPont	Can tailor stiffness, impact strength, control shrinkage
PET/PBT	Valox E1 Series Celanex	GE Plastics Thermofil Hoechst Celanese	Heat resistance, fast molding, low cost, improved gloss
Acetate/elastomer	Delrin Duraloy	DuPont Hoechst Celanese	Stiffness, toughness, fatigue, wear
PC/PET	Makroblend	Mobay	Low-temperature impact, chemical and UV resistance
SMA/ABS	Cadon	Monsanto	Heat resistance, impact, low cost

TABLE 1.12 Common Alloys and Blends (*Continued*)

Alloy or blend	Trade name	Supplier	Key properties
Polysulfone/ABS	Mindel	Amoco	Processability, low cost, heat resistance, can be plated
PEEK/PES	Victrex	ICI	High HDT, easy processing, low cost, impact
PC/TPU	Texin	Mobay	Stiffness, wear, low-temperature impact
Nylon/PE	Selar RB	DuPont	Heat/chemical resistance, barrier properties, wear

Source: From Green.[4] Reprinted with permission.

trolled with additives such as fillers and reinforcing fibers or fabrics. The conversion of these materials to the thermoset state can be accomplished at room or elevated temperature, the latter giving a faster and more complete cure of the resin. The reader is referred to other texts for more detailed descriptions of all thermosetting plastics.[4–6,14] Properties of thermosets for electronics applications are given in Table 1.13.

1.5.1 Allyl Resins

The allyl resins are thermosetting polyester materials which retain their desirable physical and electrical properties on prolonged exposure to severe environmental conditions such as high temperature and humidity. These resins have good chemical resistance and can withstand between 10^4 and 10^{12} rads of gamma radiation.[14] These polymers have the allyl radical CH_2—$CH{=}CH_2$ as part of their chemical structure. The principal allyl resins are based on diallyl phthalate (DAP) and isophthalate (DAIP) monomers and prepolymers. There are other resins that are used alone or in combination with DAP and DAIP. They are diethylene glycol bis(allyl carbonate), allyl methacrylate, diallyl fumarate and maleate, as well as triallyl cyanurate. The allyl resins are converted to thermoset materials by heat and by the addition of free-radical sources such as benzoyl peroxide and *t*-butyl perbenzoate to the resin formulation. Curing of these resins is slow below 150°C. These resins are used as cross-linking agents in other polyester systems and as molding compounds, preimpregnated glass cloth, sealants, insulating coatings, and decorative laminates. Most critical electronics applications requiring high reliability under adverse conditions use allyl resins, such as connectors in communications, computers, and aerospace systems, insulator switches, chip carriers, and circuit boards. The allyl resins have a low loss factor, high volume and surface resistivity, and high arc resistance, and these properties are retained under high-humidity conditions. The allyl resins can be compression, transfer, and injection molded and they can also be used in prepregging operations. In general DAP compounds are designed for continuous operation at about 176°C, while DAIP can operate at about 232°C.

TABLE 1.13 Properties of Thermosetting Plastics

	DAP	Epoxy		Phenolics			Alkyds		Poly-	Polyi-	Poly-	Mineral-	Bismale	Cyanate	Benzo-
	(GDI-30)	Glass-filled	Mineral-filled	General-purpose	Glass-filled	Mineral-filled	MAG	MAI-60	ester GPO-3	imide	urethane	filled silicone	imides	esters	cyclobutene
Dielectric constant, D-150															
60 Hz	4.2	5.0	4.0	12.0	50.0	6.0	6.3	5.6	4.5	3.5	6	3.6	—	—	—
10^6 Hz	3.5	4.6	5.0	6.0	6.0	10.0	4.7	4.6	—	3.4	3	3.7	3.5	2.66–3.10	2.65–2.70
Dissipation factor, D-150															
60 Hz	0.004	0.01	0.01	0.3	0.3	0.07	0.04	0.10	0.05	0.0025	0.1	0.005	—	—	—
10^6 Hz	0.01	0.01	0.01	0.7	0.8	0.10	0.02	0.02	—	0.01	0.04	0.003	0.007	0.01–0.005	10.0008–0.002
Dielectric strength, D-149; V/mil	400	360	400	400	350	400	400	375	300	6500	500	425	480–508		
Volume resistivity, D-257; Ω-cm	10^{13}	3.8×10^{15}	9×10^{15}	10^{13}	10^{13}	10^{14}	10^{14}	10^{13}	—	10^{18}	10^{14}	10^{15}	—	10^{16}	9×10^{19}
Arc resistance, D-495; seconds	140	140	180	50	70	180	>180	180	>180	230	120	240	—	Excellent	—
Specific gravity, D-792	1.7	1.8	2.1	1.45	1.95	1.83	2.24	2.07	1.95	1.4	1.1	2.05	1.30	1.10–1.43	
Water absorption, D-570; % 24 h	<0.2	0.2	0.04	0.7	0.5	0.5	0.08	0.07	0.5	2.9	0.2	0.15	4.0–4.4*	0.6–2.5*	<0.1
Heat deflection temperature, D-648; at 264 lb/in^2, °F	500	400	250	340	400	500	350	>400	—	680	190	>500	520	480	—
Tensile strength, D-638; lb/in^2	10,000	30,000	15,000	10,000	7000	11,000	3000	6000	9000	17,000	1000	6500	12,000	13,000	—
Impact strength (Izod), D-256; ft • lb/in	5.0	10	0.4	0.3	3.5	15.0	0.3	9.5	8.0	1.5	25	0.5	0.3–0.5	0.7–0.9	—
Coefficient of thermal expansion, D-696; 10^{-5}/°F	2.6	1.7	2.2	2.5	—	0.88	3	2	2	2.8	25	2.8	40 ppm/°C	50 ppm/°C	42–70 ppm/°C
Thermal conductivity, C-177; Btu • in/(h • in • ft^2 • °F)	—	6	—	0.3	0.34	0.2	7.2	3.6	4	6.8	0.1	3.1	—	—	—

*500-h water boil.

Source: From Harper.[7] Reprinted with permission.

1.5.2 Bismaleimides

Within the polyimide family of resins there is a class of thermosetting polymers that have a preimidized structure and form a three-dimensional network via addition polymerization without the evolution of volatile material. These materials are classified as bismaleimides (BMIs), and the monomers and prepolymers are prepared by the reaction of maleic anhydride and diamines. The chemical unit present in these polymers is as follows:

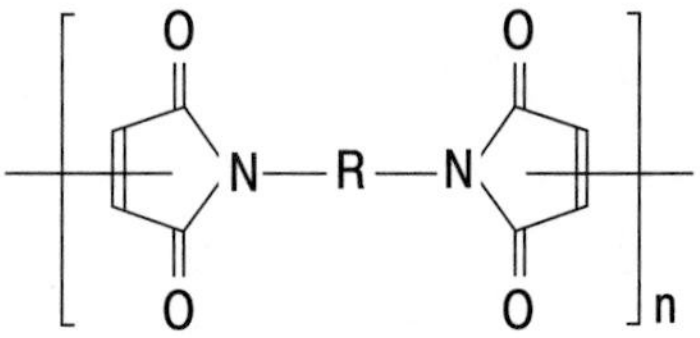

$$\left[\text{(maleimide)}\text{N—R—N}\text{(maleimide)} \right]_n$$

The material is very reactive and can be homopolymerized or copolymerized to produce a wide variety of thermosetting resins. The polymers are characterized as having the processing ease of epoxy resins but superior elevated-temperature performance properties. Epoxies operate in the 150°C temperature range, while the BMIs operate in the 200 to 232°C range. Compression, transfer, and injection molding, filament winding, and prepregging are the normal processing methods for bismaleimides. Bismaleimides are sold as powders or as solutions in polar solvents. These materials are primarily used in printed-wiring-board substrates.

1.5.3 Epoxy Resins

Epoxy resins are characterized by the presence of the epoxy (oxirane) ring:

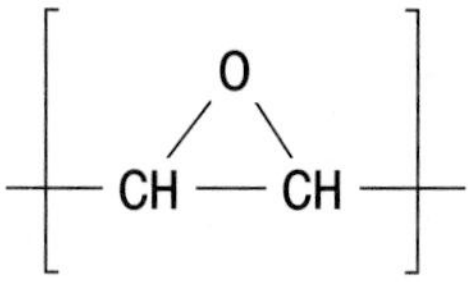

$$\left[\overset{\text{O}}{\text{CH}—\text{CH}} \right]$$

Most commercial epoxy resins are derived from bisphenol A and epichlorohydrin, but there are many other types based on the epoxidation of multifunctional molecules that give rise to epoxy resins with a broad range of properties. Epoxy resins can be liquids or solids. Curing of these resins is accomplished by reaction through the epoxide and hydroxyl functional groups. Curing agent type and amount and temperature determine the condition of cure and the final properties of the resin. Typical curing agents include the aliphatic amines and amides for ambient-temperature cure, and the anhydrides, organic acids, aromatic amines, and various phenolic condensation products for elevated-temperature cure. Most common epoxy resins are solventless (100 percent solids). However, higher-molecular-weight and multifunctional epoxies are solid and are usually processed in solution form. The curing reaction is exothermic, which may be necessary to control in large-batch operations. The cured resins have an excellent combination of physical, chemical, mechanical, and electrical properties and are used extensively in many electrical and nonelectrical applications. Epoxies can be compression and transfer molded and filament wound. They are used in casting, prepregging, and laminating operations. Epoxies can be formulated to produce conformal coatings, adhesives, and varnishes and are used in the electrical industry as bobbins, connectors, and chip carriers, and as the matrix resin in printed-wiring-board substrates.

1.5.4 Phenolic Resins

Phenolic resins are the reaction product of phenol and formaldehyde. Two kinds of phenolics are produced: the resols (alkaline condensation products) and the novolacs (acid condensation products). The chemical structure of a resol is composed of methylene and ether bridges, as follows:

$$-HOH_2C-C_6H_4-CH_2-C_6H_4-CH_2-O-CH_2-C_6H_4-CH_2OH$$

For a novolac the structure is as follows:

$$C_6H_4(OH)-CH_2-C_6H_3(OH)-CH_2-C_6H_4(OH)$$

The basic difference between a resol and a novolac is the presence of one or more free methylol groups on the resol. The resins are heat-cured to form a dense cross-linked network which gives the phenolic resins their high heat resistance and dimensional stability. The phenolic resins have poor arc resistance. Phenolic resins are available in solution form or as powders and can be converted to molding compounds, varnishes, and laminates. They are processed by injection, compression, and transfer molding. Phenolics are used as chip carriers, connectors, and bobbins, and as matrix resins for printed-wiring-board substrates.

1.5.5 Polyesters

Polyester resins are versatile materials and are available as low-viscosity liquids to thick pastes. Included within the polyester family are alkyd resins, unsaturated resins, vinyl esters, and the allyl resins discussed in Sec. 1.5.1. The characteristic functional group present in these resins is the ester group (—COOR—), but the composition of these polyester resins can be varied tremendously to produce resins having widely dif-

TABLE 1.14 Unsaturated Polyester Components

Components	Ingredients	Characteristics
Unsaturated anhydrides and dibasic acids	Maleic anhydride	Lowest cost, moderately high heat deflection temperature (HDT)
	Fumaric acid	Highest reactivity (cross-linking), higher HDT, more rigidity
Saturated anhydrides and dibasic acids	Phthalic (ortho-phthalic) anhydride	Lowest cost, moderately high HDT, provides stiffness, high flexible and tensile strength
	Isophthalic acid	Higher tensile and flexible strength, better chemical and water resistance
	Adipic acid, azelaic acid, sebacic acid	Flexibility (toughness, resilience, impact strength); adipic acid is lowest in cost of flexibilizing acids
	Chlorendic anhydride	Flame retardance
	Nadic methyl anhydride	Very high HDT
	Tetrachlorophthalic	Flame retardance
Glycols	Propylene glycol	Lowest cost, good water resistance and flexibility, compatibility with styrene
	Dipropylene glycol	Flexibility and toughness
	Ethylene glycol	High heat resistance, tensile strength, low cost
	Diethylene glycol	Greater toughness, impact strength, and flexibility
	Bisphenol-A adduct	Corrosion resistance, high HDT, high flexible and tensile strength
	Hydrogenated bisphenol-A adduct	Corrosion resistance, high HDT, high flexible and tensile strength
Monomers	Styrene	Lowest cost, high reactivity, fairly good HDT, high flexible strength
	Diallyl phthalate	High heat resistance, long shelf life, low volatility
	Methyl methacrylate	Light stability, good weatherability, fairly high HDT
	Vinyl toluene	Low volatility, more flexibility, high reactivity
	Triallyl cyanurate	Very high HDT, high reactivity, high flexible and tensile strength
	Methyl acrylate	Light stability, good weatherability, moderate strength

Source: From Schwartz and Goodman.[6] Reprinted with permission.

ferent properties. Table 1.14 lists the various components available for preparing polyester resins. Cross-linking of these materials is accomplished by the addition of polyfunctional acids or alcohols, unsaturated monomers, and a peroxide catalyst. Curing is done from room temperature to about 160°C. Fillers, pigments, and fibers can be mixed with the resins. Characteristic properties include ease of processing, low cost, good electrical properties, and high arc resistance. Applications include bobbins, terminal boards, connectors, and housings. Polyester resins can be compression or transfer molded, laminated, pultruded, and filament wound.

1.5.6 Polyurethanes

These polymers are derived from the reaction of polyfunctional isocyanates and polyhydroxy (polyether and polyester polyols) compounds which yield linear or branched polymers. The basic chemical unit of polyurethanes is the urethane (—RNHCOOR—). These resins are produced as castable liquids (prepolymers) and are cross-linked by adjusting the stoichiometry and functionality of the isocyanate or polyol. Catalysts are added to enhance the rate of reaction. A variety of other ingredients (active hydrogen compounds) can be added to produce polyurethanes with different properties, ranging from elastomeric to rigid polymers. The polyether urethanes are more hydrolytically stable than the polyester urethanes, but the latter give better strength and abrasion resistance. Polyurethanes have poor solvent resistance. They are sensitive to chlorinated and aromatic solvents as well as to acids and bases. Urethanes are used as conformal coatings to encapsulate sensitive electronic components. They are processed by reaction injection molding (RIM), compression molding, and casting.

1.5.7 Silicones

Silicones are polymers that consist of alternating silicon and oxygen atoms along the backbone of the polymer chain. The backbone is modified by attaching organic side groups to the silicon atom and in so doing imparts the unique properties found in these polymers. The structure of the silicones is as follows:

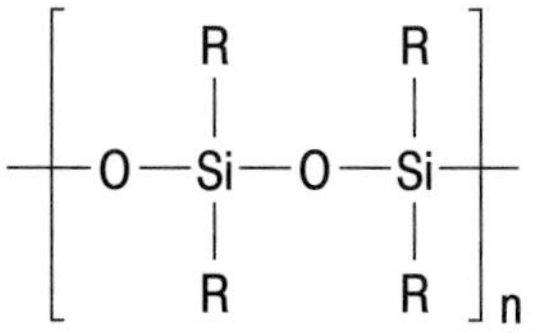

$$\left[-O-\underset{\displaystyle R}{\overset{\displaystyle R}{\underset{|}{\overset{|}{Si}}}}-O-\underset{\displaystyle R}{\overset{\displaystyle R}{\underset{|}{\overset{|}{Si}}}}- \right]_n$$

The silicones can be produced in the form of liquids, greases, elastomers, and hard resins. The organic group attached to the silicon atom can be aliphatic, aromatic, or vinyl, which affects the properties of the final silicone polymer. The silicone fluids are low-molecular-weight polymers where the organic group on the silicone is methyl or phenyl, or a mixture of both. The silicone resins are branched polymers that cure to a solid while the elastomers are linear oils or higher-molecular-weight silicones that are reinforced with a filler and then vulcanized (cross-linked). The elastomers come in three forms: heat-cured rubber, two-component liquid injection molding compounds, and room-temperature vulcanizing (RTV) products. The conversion of silicones into cross-linked elastomers can be accomplished by free-radical condensation, addition, and ultraviolet radiation curing techniques. The silicones are characterized by their useful properties over a broad temperature range (−65 to 248°C). They exhibit excellent weatherability, arc and track resistance, and impact, abrasion, and chemical resistance. Silicones can also be copolymerized with other polymers to produce materials with a variety of interesting properties, such as silicone-polyimide, silicone-EPDM, and silicone polycarbonate. Electronics applications include wire enamels, laminates, sleeving and heat-shrinkable tubing, potting of electronic components, conformal coatings, and varnishes. Properties of various silicone polymers are listed in Tables 1.15 to 1.18.

1.5.8 Cross-Linked Thermoplastics

Overall property enhancement is the underlying principle for the commercial development of cross-linked thermoplastics. This enhancement manifests itself in improved resistance to thermal degradation of physical properties, stress cracking, creep, and other environmental effects. Thermoplastics are cross-linked by radiation and chemical techniques. The techniques, which include X rays, gamma rays, high-energy electrons, and organic peroxides, under controlled conditions, can be used to produce beneficial changes in the properties of irradiated polymers. Typical polymers capable of being cross-linked include the polyolefins, fluoroplastics, vinyls, neoprene, and silicone. Electrical applications include shrink-fit tubing, underground cable insulation, and microwave insulation. Table 1.19 lists cross-linked thermoplastic products and their applications.

1.5.9 Cyanate Ester Resins

Cyanate ester resins are bisphenol derivatives containing the cyanate —O—C≡N— functional group. These monomers and polymers cyclotrimerize on heating to form a cross-linked network of oxygen-linked triazine rings via addition polymerization. The cyanate ester resins range from liquids to solids and are characterized by superior dielectric properties, adhesion, low moisture absorption, flame resistance, high-temperature capability, and excellent dimensional stability. Glass transition temperatures range from 250 to 290°C. Several grades are available and can be formulated to produce laminating varnishes for impregnating inorganic and organic reinforcements. The formulations can be homopolymers, blends with other cyanate esters or with bismaleimides, and epoxy resins. Some properties of the neat resins are shown in Figs. 1.6 to 1.9[15] and of E-glass laminates in Tables 1.20 and 1.21.[15] Cyanate ester resins can be processed by melt polymerization, prepregging, and lamination operations. Applications in the electrical industry include printed-wiring-board substrates and radome structures. Cyanate ester resins can be toughened with thermoplastics such as polyethersulfone, polyetherimide, polyarylates, polyimides, and methylethylketone-soluble copolyesters and elastomers.

TABLE 1.15 Approximate Physical Properties at 25°C of Methylpolysiloxane Fluids (Rhodorsil Oil 47V)

Viscosity, cst	VTC*	Specific gravity	Flash point, °C	Freezing point, °C	Surface tension, dynes/cm	Vapor† pressure, mm Hg	VCE,‡ $cm^3/cm^3 \cdot °C$	Dielectric constant¶	Dielectric strength, kV/mm
5	0.55	0.910	136	−65	19.7	—	1.05×10^{-3}	2.59	—
10	0.57	0.930	162	−65	20.1	—	1.08×10^{-3}	2.63	13
20	0.59	0.950	230	−60	20.6	1×10^{-2}	1.07×10^{-3}	2.68	—
50	0.59	0.959	280	−55	20.7	1×10^{-2}	1.05×10^{-3}	2.8	15
100	0.60	0.965	>300	−55	20.9	1×10^{-2}	0.05×10^{-3}	2.8	16
300	0.62	0.970	>300	−50	21.1	1×10^{-2}	0.95×10^{-3}	2.8	16
500	0.62	0.970	>300	−50	21.1	1×10^{-2}	0.95×10^{-3}	2.8	16
1000	0.62	0.970	>300	−50	21.1	1×10^{-2}	0.95×10^{-3}	2.8	16
5000 to 2,500,000	0.62	0.973	>300	−45	21.1	1×10^{-2}	0.95×10^{-3}	2.8	18

*Viscosity/temperature coefficient = 1 − (viscosity at 99°C/viscosity at 38°C).

†At 200°C.

‡Volume coefficient of expansion between 25 and 100°C.

¶ Between 0.5 and 100 kHz.

Source: From Goodman.[14] Reprinted with permission.

TABLE 1.16 Typical Properties of Condensation Cure Methylphenyl RTV Silicone Rubber Products Cured at Room Temperature

Viscosity, cst	Specific gravity	Hardness, Shore A	Tensile strength, lb/in^2	Useful temperature range, °F	Dielectric strength, V/mil	Dielectric constant at 1 kHz	Dissipation factor at 1 kHz
16,000	1.21	42	380	−175–400	520	3.6	0.005
30,000	1.42	55	690	−175–500	540	3.9	0.02
700,000	1.35	48	440	−175–500	470	3.9	0.02

Source: From Goodman.[14] Reprinted with permission from *General Electric Silicones.*

TABLE 1.17 Typical Properties of Addition Cure Clear RTV Silicone Rubber Products (Heat-Accelerated Cure)

Viscosity, cst	Specific gravity	Hardness, Shore A	Tensile strength, lb/in^2	Useful temperature range, °F	Dielectric strength, V/mil	Dielectric constant at 1 kHz	Dissipation factor at 1 kHz
4,000	1.02	44	920	−75 to 400	520	2.7	0.0006
5,200	1.04	45	920	−175 to 400	530	2.69	0.0004

Source: From Goodman.[14] Reprinted with permission from *General Electric Silicones.*

TABLE 1.18 Estimated Useful Life of Silicone Rubber at Elevated Temperatures

Service temperature, °F	Useful life*
250	10–20 years
300	5–10 years
400	2–5 years
500	3 months
600	2 weeks

*Retention of 50 percent elongation.

Source: From Goodman.[14] Reprinted with permission from *General Electric Silicones.*

TABLE 1.19 Heat-Shrinkable Insulation and Encapsulation Tubings

	Product	Description	Typical applications
Flexible polyolefins	RNF-100 type 1	General-purpose, flame-retarded, flexible polyolefin	Insulation of wire bundles; cable and wire identification; terminal and component insulation, protection, and identification
	RNF-100 type 2	General-purpose, flexible, transparent polyolefin	Transparent coverings for components such as resistors, capacitors, and cables where markings must be protected and remain legible
	RT-876	Highly flame-retarded, very flexible polyolefin with low shrink temperature	Coverings for cables and components where excellent flexibility and outstanding flame retardance are needed
	RT-102	Highly flexible, flame-retarded, polyolefin with very low shrink temperature	Flexible material for general-purpose protection and insulation; especially effective for low-temperature use
	RVW-1	Highly flame-retarded, flexible polyolefin	Lightweight harness insulation, terminal insulation, wire strain relief and general-purpose component packaging and insulation where a UL recognized product with a VW-1 (FR-1) rating is needed
Semirigid polyolefins	CRN type 1	General-purpose, flame-retarded, semirigid polyolefin	Insulation and strain relief of soldered or crimped terminations; protection of delicate components; cable and component identification
	RT-3	Semirigid, flame-retarded, opaque polyolefin	Particularly suited for automated application systems to insulate and strain relieve crimped or soldered terminals; furnished in cut pieces
Dual-wall polyolefins	SCL	Meltable inner walls, selectively cross-linked semirigid polyolefin	Encapsulation of components, splices, terminations, requiring moisture resistance, mechanical protection and shrink ratios as high as 6:1.
	TAT	Flexible, dual-wall adhesive tubing	Insulates and seals electrical splices, bimetallic joints, and components from moisture and corrosion
	ATUM	Semiflexible, high expansion, heavy dual-wall, adhesive tubing	Environmental protection for a wide variety of electrical components, including wire splices and harness breakouts

TABLE 1.19 Heat-Shrinkable Insulation and Encapsulation Tubings (*Continued*)

	Product	Description	Typical applications
Fluoroplastics	Kynar*	High-temperature, flame-resistant, clear, semirigid fluoroplastic	Transparent insulation, mechanical protection of wires, solder joints, terminals, connections, and component covering
	Convolex	Convoluted, flexible, irradiated polyvinylidene fluoride	Mechanical protection of cable harnesses; excellent flexibility and chemical resistance; good high-temperature performance
	RT-218	Semirigid, white, high-temperature, low-outgassing fluoroplastic tubing	Insulation of splices and terminations in aircraft and mass transit markets; cable and wire identification
Vinyl	PVC	Flexible, flame-resistance, polyvinyl chloride	Insulation and covering of cables, components, terminals, handles
Elastomers	NT (neoprene)	Heavy-duty, flexible, abrasion-resistant, flame-retarded elastomer	Insulation and abrasion protection of wire bundles and cable harnesses
	SFR (silicone)	Highly flexible, flame-retarded, heat or cold shock-resistant	Cable and harness protection requiring maximum flexibility and resistance to extreme temperatures; ablative protection for cables in rocket blast
	Viton† (fluoroelastomer)	Flexible, flame-retarded, heat and chemical resistant fluoroelastomer	Insulation and protection of cables exposed to high temperature and/or solvents such as jet fuel
Caps	PD	Semirigid polyolefin, meltable inner wall	Encapsulation of stub splices, especially fractional-horsepower motor windings

*Registered trademark of Pennwalt Corp.

†Registered trademark of E.I. duPont de Nemours & Co.

Source: From Goodman.[14] Reprinted with permission from RayChem Corp.

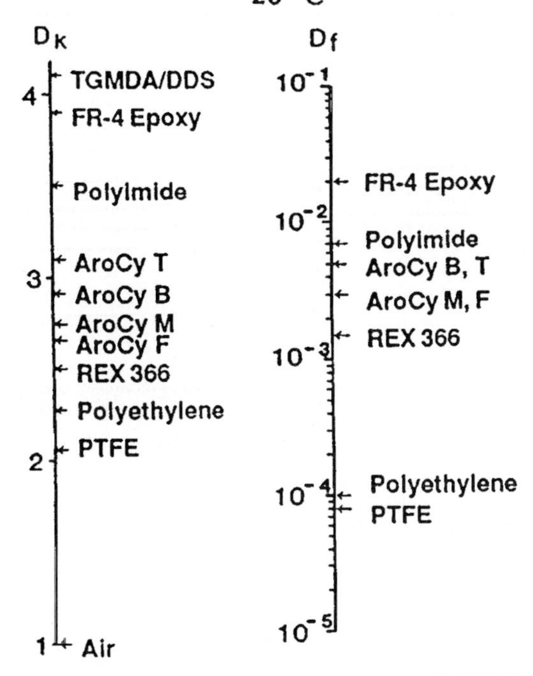

FIGURE 1.6 Comparison of dielectric constant and dissipation factor values measured at 25°C and 1 MHz for representative thermoset and thermoplastic polymers. (*From Shimp[15]; reprinted with permission.*)

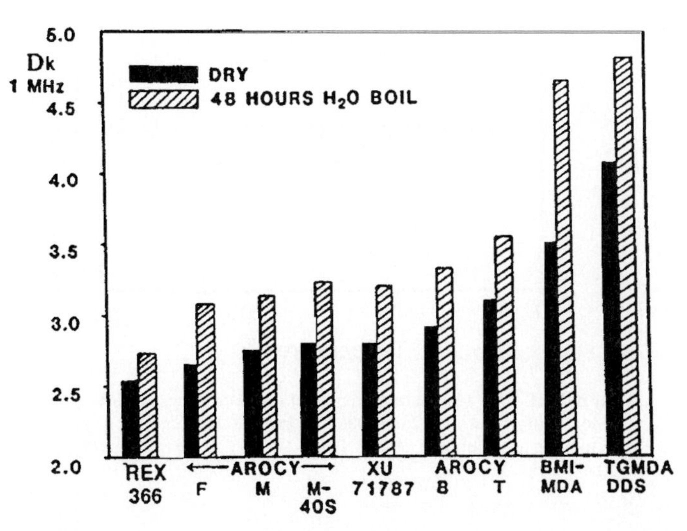

FIGURE 1.7 Effect of moisture conditioning on dielectric constant of several thermoset resins. (*From Shimp[15]; reprinted with permission.*)

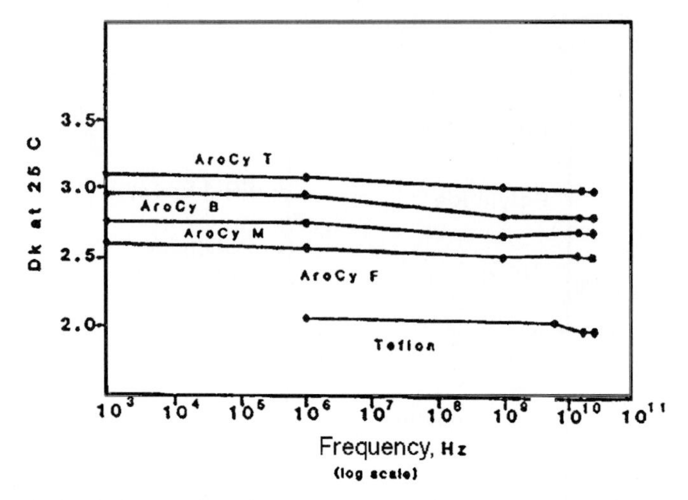

FIGURE 1.8 Flat dielectric constant response of cyanate ester homopolymers to increasing test frequency. (*From Shimp*[15]*; reprinted with permission.*)

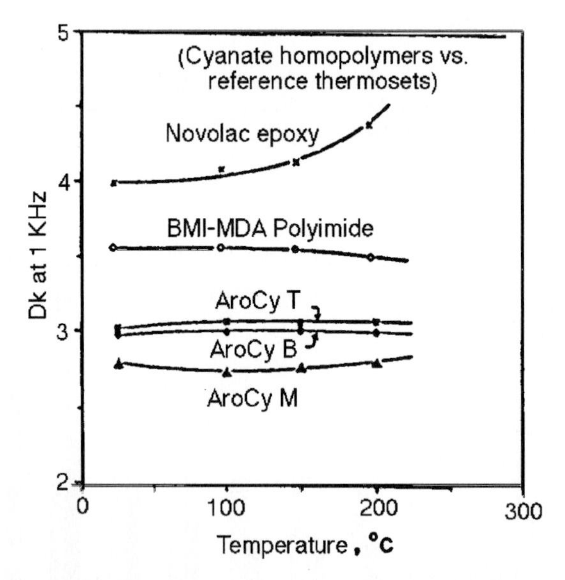

FIGURE 1.9 Flat dielectric constant response of cyanate ester homopolymers over 25 to 200°C temperature range. Epoxy novolac is reference resin. (*From Shimp*[15]*; reprinted with permission.*)

TABLE 1.20 Comparison of AroCy B-40S Laminate Properties with 60 Percent Epoxy Modification and FR-4 Reference*

Laminate property	100% cyanate‡	60% epoxy† 40% cyanate‡	100% epoxy†
Press cure, h/°C	1/177	1/177	1/177
Postcure, h/°C	3/225	—	—
T_g (T_{MAS}), °C	225	183	130
CTE (Z), ppm/°C	44	55	60
Steam/solder, min	120	120	45
Flammability, UL-94	Burns	V–0	V–0
Peel strength, lb/in			
25°C	12.3	11.4	12.0
200°C	9.4	8.5	4.2
D_k, 1 MHz	4.05	4.2	4.8
D_k, 1 MHz	0.003	0.008	0.020

*Laminates are 8-ply, style 7628 E-glass reinforced; 55 ± 2% resin by volume.

† Brominated hard epoxy, WPE 500, 27% Br.

‡ AroCy B-40S.

Source: From Shimp.[15] Reprinted with permission.

TABLE 1.21 Comparison of E-glass Laminate Properties for Several Resin Systems at Equal Resin Volume Content*

Resin	D_k, 1 MHz, vol % 70	D_k, 1 MHz, vol % 55	D_f (10^{-3})	DMA T_g, °C	TGA onset, °C	Flammability rating UL-94	Peel strength, lb/in 25°C	Peel strength, lb/in 200°C	Pressure cooker, min
Cyanate ester									
ArOCy F-40S	3.5	3.9	2	290	400	V–0	11	9	120
AroCy M-40S	3.6	4.0	2	290	415	VI	12	10	120
XU 71787	3.6	4.0	3	255	426	†	8	6	120
ArOCy B-40S	3.7	4.1	3	290	405	†	12	10	120
Polyimide									
BMI-MDA	4.1	4.5	9	312	400	VI	9	6	120
Epoxy FR-4	4.5	4.9	20	145	300	V–0	12	4	45

*Except for D_k measurements on 70 vol % resin laminates, tests were performed on 55 vol %, 0.060-in, 8-ply laminates prepared with 7628 E-glass and postcured 4 h at 225–235°C.

† Burn times exceed self-extinguishing classifications.

Source: From Shimp.[15] Reprinted with permission.

1.5.10 Benzocyclobutenes

Benzocyclobutenes (BCB) are derived from monomers of the following generic form:

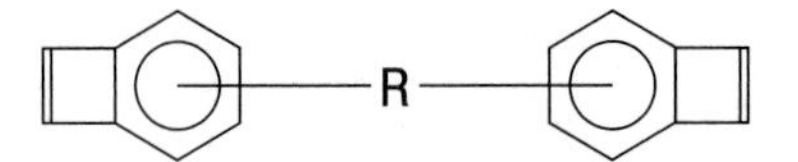

The BCB is thermally polymerized (addition polymerization) to produce a cross-linked resin without the evolution of volatiles. The BCB is supplied as a prepolymer partially polymerized in hydrocarbon solvents such as toluene or mesitylene (20 to 70 percent solids). The BCB resins have low dielectric constant, low water absorption, good thermal stability, high adhesion, and good planarization and chemical resistance. Properties of several BCB resins are listed in Table 1.22.[16]

TABLE 1.22 Properties of BCB-Based Resins

Property	BCB XU13005	BCB XU130026	BCB XU130028
Glass transition temperature (TMA, DMA), °C	>350	>350	>350
Flexural modulus, 10^3 lb/in^2	480	540	747
Linear coefficient of thermal expansion 25 to 300°C, TMA, ppm/°C	65–70	42	27
Water absorption, 24-h water boil, %	0.25	0.52	0.87
Weight loss at 350°C, nitrogen, 2 h, 20-μm film on silicon wafer, %	2	1	0
Dielectric constant at 1 MHz, +1 to 0.1	2.7	2.7	2.7
Dissipation factor at 1 MHz	0.0008	0.0006	0.0004

Source: From Burdeaux et al.[16] Reprinted with permission.

1.6 ELASTOMERS[7]

Elastomers are considered apart from other polymeric materials because of their special properties. The distinguishing characteristics of elastomer materials are their ability of sustain large deformations (5 to 10 times the unstretched dimensions) and their capacity to spontaneously recover nearly all of that deformation without rupturing. The unique structural feature of all rubberlike substances is the presence of long polymer chains interwoven and joined together through cross-linkages. Generally elastomers are not as widely used as plastics for electronics applications, and only a brief review of elastomer types, properties, and their applications will be presented. A compilation of electrical property information on thermoplastics, thermosets, and elastomers is given in Ku.[17] For a detailed listing of the properties the reader is referred to Greene[4] and Babbit.[18]

1.6.1 Properties

Elastomers are almost always used in the compounded state. The neat material is blended with a variety of additives to cure and enhance the properties of the elastomer.

Aging. Elastomers are affected by the environment more than other polymers. Thermal aging of the elastomer increases stiffness and hardness and decreases elongation. Radiation has a similar effect. Elastomers are sensitive to oxidation and in particular to the effects of ozone. Ultraviolet radiation acts similarly to ionizing radiation, so some elastomers do not weather well. Environmental effects are especially noted on highly stressed parts, and some elastomers are particularly affected by hydrolysis.

Creep. Creep of elastomers refers to a change in strain when stress is held constant. Special terms are used for elastomers. Compression set (ASTM D-395[19]) is creep that occurs when the elastomer has been held at either constant strain or constant stress in compression. Constant strain is most common and is recorded as a percent of permanent creep divided by original strain. Strain of 25 percent is common. Permanent set is deformation remaining after a stress is released.

Hardness. The hardness of elastomers is a measure of the resistance to deformation measured by pressing an instrument into the elastomer surface. Special instruments have been developed, the most common being the Shore durometer. Figure 1.10 shows the hardness of elastomers and plastics.

Hysteresis. Hysteresis is energy loss per loading cycle. This mechanical loss of energy is converted into heat in elastomers and is caused by internal friction of the molecular chains moving against each other. The effect causes a heat buildup in the elastomer, increasing its temperature, changing its properties, and aging it. A similar electrical effect can occur at high frequencies when the dissipation factor of the elastomer is high.

Low-Temperature Properties. As temperatures are decreased, elastomers tend to become stiffer and harder. Each material exhibits a stiffening range and a brittle point at the glass transition temperature. These effects are usually time-dependent.

Tear Resistance. This is a measure of the stress needed to continue rupturing a sheet elastomer after an initiating cut or notch. Elastomers vary widely in their ability to withstand tearing.

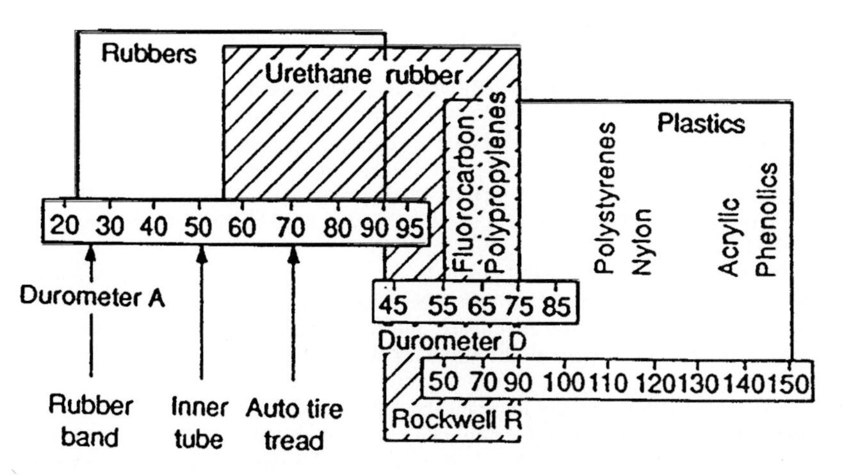

FIGURE 1.10 Hardness of rubber and plastics. (*From Harper.*[7])

Tensile Strength, Elongation, and Modulus. In tension, metals behave in accordance with Hooke's law, and in strain, they react linearly to the yield point. Polymers and plastics (unreinforced) deviate somewhat from linearity (logarithmically). Special tensile tests are used for elastomers per ASTM D-412.[20] Elastomers are not generally designed for tensile service, but many other physical properties of the elastomers correlate with tensile strength.

1.6.2 Types of Elastomers

A large number of chemically different elastomers exist. ASTM D-1418[21] describes many of these. Tables 1.23 and 1.24 list elastomers and their properties.

Natural rubber (NR) is still used in many applications. It is not one uniform product, but varies with the nature of the plant producing the sap, the weather, the locale, the care in producing the elastomer, and many other factors. A variant of NR, gutta-percha, was used in most of the early electrical products, especially cable. It has excellent electrical properties, as shown in Table 1.24, low creep, and high tear strength. On aging it reverts to the gum.

Isoprene rubber (IR) is similar in chemistry to NR, but it is produced synthetically. Polyisoprene constitutes 97 percent of its composition. It is more consistent and much easier to process than NR.

Acrylic elastomer (ABR) has a heat resistance that is almost as good as that of fluorinated compounds and silicones. It also ages well but is sensitive to water. Its chief use is in contact with oils.

Butadiene elastomer (BR) is used to copolymerize with SBR and NR in tire stocks.

Epichlorohydrin elastomer (CO, ECO) are flame-retardant because of the presence of chlorine. Their electrical properties are modest, but they age well and resist most chemicals. Dissipation factors are high.

Carboxylic elastomer (COX) has good low-temperature performance, excellent weather resistance, and extremely good wear resistance. Electrical properties are average.

Neoprene (CR) (chloroprene) was the first synthetic elastomer and is widely used in industry. It is nonflammable and resists ozone, weather, chemicals, and radiation. However, it is highly polar and has a high dissipation factor and dielectric constant.

Chlorosulfonated polyethylene (CSM) is similar to CR, with some improvement in electrical properties and better heat resistance. It is available in colors and often used in high-voltage applications.

Ethylene-propylene terpolymer (EPDM) is synthesized from ethylene, propylene, and a third monomer, a diene. The diene permits conventional sulfur vulcanization. The elastomer is exceptionally resistant to radiation and heat. The glass transition temperature is −60°C and electrical properties are good.

Ethylene-propylene copolymer (EPM) which was often used as a wire insulation, is being replaced by EPDM because processing qualities are somewhat inferior to those of EPDM.

Fluorinated elastomers (FPM) include several types—fluorocarbons, fluorosilicones, and fluoroalkoxy phosphazenes. The elastomers can be used to 600°F, do not burn, are unaffected by most chemicals, and have excellent electrical properties. In thermal stability and aging only the silicones are better. Physical properties are high, but so is the cost.

Butyl rubber (IIR) is highly impermeable to water vapor. Its nonpolar nature gives it good electrical properties. Compounded with aluminum oxide trihydrate it has exceptional arc and track resistance. Butyl has good aging characteristics and good flexibility at low temperatures.

TABLE 1.23 Chemical Description of Elastomers

ASTM D-1418	Chemical type	Properties
NR	Natural rubber polyisoprene	Excellent physical properties
IR	Polyisoprene synthetic	Same as NR, but more consistency and better water resistance
ABR	Arylate butadiene	Mechanical elastomer; excellent heat and ozone resistance
BR	Polybutadiene	Copolymerizes with NR and SBR; abrasion resistance
CO	Epichlorohydrin	Chemical resistance
COX	Butadiene-acrylonitrite	Used with NBR to improve low-temperature performance
CR	Chloroprene, neoprene	Withstands weathering, flame-retardant, chemical resistance
CSM	Chlorosulfonated polyethylene	Colors available, weathering and chemical resistance, poor electrical properties
EPDM	Ethylenepropylene terpolymer	Similar to EPM, good electrical properties, resists water and steam
EPM	Ethylene-propylene copolymer	Similar to EPDM, good heat resistance, wire insulation
FPM	Fluorinated copolymers	Outstanding heat and chemical resistance
IIR	Isobutyleneisoprene, butyl	Outstanding weather resistance, low physical properties, track resistance
NBR	Butadiene-acrylonitrite, nitrile, Buna N	General-purpose elastomer, poor electrical properties
PVC/NBR	Polyvinyl chloride and NBR	Colors available, weather, chemical, and ozone resistance
SBR	Styrene: butadiene, GRS, Buna S	General-purpose elastomer, good physical properties, poor oil and weather resistance
SI (FS1, PS1, VS1, PVS1)	Silicone copolymers	Outstanding at high and low temperatures, arc- and track-resistance, resist weather and ozone, excellent electrical properties, poor physical properties
T	Polysulfide	Excellent weather resistance and solvent resistance
U	Polyurethane	High physical and electrical properties

Source: From Harper.[7] Reprinted with permission.

TABLE 1.24 Electrical Properties of Elastomers

ASTM elastomer	Dielectric strength, V/mil	Dissipation factor tan δ	Dielectric constant	Volume resistivity, $\Omega \cdot$ cm
COX	500	0.05	10	10^{15}
CR	700	0.03	8	10^{11}
CSM	700	0.07	8	10^{14}
EPDM	800	0.007	3.5	10^{16}
FPM	700	0.04	18	10^{13}
IIR	600	0.003	2.4	10^{17}
MR	800	0.0025	3	10^{16}
SBR	800	0.003	3.5	10^{15}
SI	700	0.001	3.6	10^{15}
T	700	0.005	9.5	10^{12}
U	500	0.03	5	10^{12}

Source: From Harper.[7] Reprinted with permission.

Nitrile rubber (NBR) is resistant to most chemicals, but its polarity gives it poor electrical properties, so its major use is in mechanical applications.

Polyvinyl chloride copolymers (PVC/NBR) are similar to NBR. They can be colored and are used in wire and cable jackets.

GRS (SBR) stands for government rubber, styrene, a nomenclature derived during World War II when natural rubber was not available in the West. It is used in mechanical applications.

Silicone elastomers (SI), which are composed of silicon and oxygen atom backbones, have the highest temperature ability (600°F), a large temperature range (−100 to 600°F), and excellent electrical properties. They do not burn and are arc resistant. Physical properties are modest.

Polysulfides (T) weather best of all and are highly chemical-resistant. Dissipation factors are excellent (as low as 0.001); physical properties are modest.

Polyurethanes (U) are either ester- or ether-based. Ester-based elastomers are poor in water resistance. They are excellent in electrical applications, with outstanding physical properties. Abrasion resistance is particularly high. They become stiff at low temperatures. They can be compounded like regular elastomers, used as cast elastomers, or injection molded like thermoplastics.

1.6.3 Thermoplastic Elastomers (TPE)

These materials have the functional requirements of elastomers (extensibility and rapid retraction) and the processability of thermoplastics. The principal advantages of the TPEs compared to vulcanized rubber are (1) reduction in compounding requirements, (2) easier and more efficient processing cycles, (3) scrap recycling, and (4) availability of thermoplastic processing methods. There are six generic classes of TPEs: styrenic block copolymers, polyolefin blends (TPO), elastomeric alloys, thermoplastic polyurethanes (TPU), thermoplastic copolyesters, and thermoplastic polyamides. TPEs are processed almost exclusively by extrusion and injection molding but can be blow molded, thermoformed, and heat welded. None of these methods are available to thermoset-type elastomers. The TPE materials are identified in Table 1.25. Additional information can be obtained from Greene.[4]

TABLE 1.25 Thermoplastic Elastomers

Generic class	Product name	Supplier	Electrical application
Elastomeric alloys	Alcryn	DuPont	Wire and cable insulation
	Santoprene	Monsanto	
	Geolast	Monsanto	
Engineering	Hytrel, Berloy	DuPont	None
TPEs, copolyesters	Ecdel	Eastman Chemicals	
	Riteflex	Hoechst Celanese	
	Lomad	GE Plastics	
Olefinic (TPO)		D&S Plastics	None
		Himont	
		Monsanto	
		A. Schulman	
		CCT	
Styrenic	Kraton	Shell Chemical Co.	Wire and cable insulation
	C-Flex	Concept Polymer Tech.	
	Finaprene	Fira Oil & Chemical	
	Europrene	Enimont	
	Solprene	Housmex Inc.	

1.7 ORGANIC COATINGS

Coatings are applied to a variety of substrates primarily to protect that substrate from deterioration due to the action of outside agents. It gives the substrate an extra level of protection against chemical, radiation, thermal, and oxidative attack. A detailed list of all types of organic coatings can be found in Stevens.[10] While most organic polymers can be used as coatings in one form or another, only those polymers that can be converted into formulations for conformal applications are discussed in this section. Conformal coatings are generally liquid resin formulations used in the protection of assembled printed-wiring boards from a variety of environmental effects. These resins conform to the topography of the board and the components thereon and are cured to form a relatively thin (1 to 10 mil) protective coating. While the main function of the coating is to provide a moisture barrier for circuit traces and components, secondary benefits are provided against dust, other contaminants, chemicals, abrasion, and some degree of shock and vibration. There is no coating that will totally resist the effects of environmental stresses, and so these coatings do have a finite time of protection and are designed to operate under the requirements of the system in which they are used. Humidity and process contaminants can lead to serious degradation of electrical components, causing lower insulation resistance between conductors, premature high-voltage breakdown, corrosion of conductors, and even short circuits. As a result the coating material chosen must have an excellent combination of physical, chemical, and electrical properties in addition to ease of application.

1.7.1 Coating Types

A variety of conformal coating materials are available to meet specific application needs. MIL-I-46058 defines five classes of polymers for conformal coatings. They are

TABLE 1.26 Typical Characteristics of Various Coating Materials

Properties	Acrylic	Urethane	Epoxy	Silicone	Polyimide	DAP
Volume resistivity (50% RH, 23°C), Ω • cm	10^{15}	11×10^{14}	10^{12}–10^{17}	2×10^{16}	10^{16}	1.8×10^{16}
Dielectric constant						
60 Hz	3–4	5.4–7.6	3.5–5.0	2.7–3.1	3.4	3.6
1 kHz	2.5–3.5	5.5–7.6	3.5–4.5		3.4	3.6
1 MHz	2.2–3.2	4.2–5.1	3.3–4.0	2.6–2.7	3.4	3.4
Dissipation (power) factor						
60 Hz	0.02–0.04	0.015–0.048	0.002–0.010	0.007–0.001	—	0.010
1 kHz	0.02–0.04	0.04–0.060	0.002–0.02	—	0.002	0.009
1 MHz	2.5–3.5	0.05–0.07	0.030–0.050	0.001–0.002	0.005	0.011
Thermal conductivity, 10^{-4} cal/(s • cm^3 • °C)	3–6	1.7–7.4	4–5	3.5–7.5	—	4–5
Thermal expansion, 10^{-3}/°C	5–9	10–20	4.5–6.5	6–9	4.0–5.0	—
Resistance to heat, continuous, °F	250	250	250	400	500	350
Effect of weak acids	None	Slight to dissolve	None	Little or none	Resistant	None
Effect of weak alkalis	None	Slight to dissolve	None	Little or none	Slow attack	None
Effect of organic solvents	Attacked by ketones, aromatics, and chlorinated hydrocarbons	Resists most	Generally resistant	Attacked by some	Very resistant	Resistant

Source: Form Coombs.[22] Reprinted with permission.

TABLE 1.27 Coating Selection Chart*

	Acrylic	Urethane	Epoxy	Silicone	Polyimide	DAP
Application	A	B	C	C	C	C
Removal (chemically)	A	B	—	C	—	—
Removal (burn through)	A	B	C	—	—	—
Abrasion resistance	C	B	A	B	A	B
Mechanical strength	C	B	A	B	B	B
Temperature resistance	D	D	D	B	A	C
Humidity resistance	A	A	B	A	A	A
Humidity resistance (extended period)	B	A	C	B	A	A
Potlife	A	B	D	D	C	C
Optimum cure	A	B	B	C	C	C
Room-temperature curing	A	B	B	C	—	—
Elevated-temperature curing	A	B	B	C	C	C

*Property ratings (A–D) are in descending order, A being optimum.

Source: From Coombs.[22] Reprinted with permission.

acrylics, epoxies, polyurethanes, silicones, and paraxylylene polymers. While not defined in the mil specifications, other polymer types that could be considered include the polyimides, diallyl phthalate resins, and the benzocyclobutenes. The properties of these materials are shown in Table 1.26. These coatings can be solvent-based, water-based, or solventless systems. They can be applied as liquids, solids (powder), or film (vapor deposition), and the coatings can be cured either thermally or with radiation. They can be applied by brushing, spraying, dipping, or flow coating. A relative coating selection chart is given in Table 1.27. For more detailed information on conformal coatings refer to Chap. 2.

1.8 POLYMER PROCESSING

The fundamental differences among the classes of polymers, namely, thermoplastics, thermosets, and elastomers, dictate the processing method to be used. Furthermore, within each class, the differences in the thermal and melt properties of the polymers also dictate what processing methods are best suited for a given material. This section is designed to acquaint the reader with some basic information about polymer processing. The reader is directed to the references for a more detailed description of each of the processing methods.[4–6] It is recommended that the plastics suppliers be used as a resource for guidance in both design and processing of polymers. The process sequence for all polymers involves heating the polymer to soften it, forcing the softened polymer into a mold or through a die to shape it, and then cooling or curing the molten polymer into its final shape. While polymers are not necessarily all solids (some are liquids), heating facilitates their processing. Tables 1.28 and 1.29 are summaries of the processing methods used for all polymers.

TABLE 1.28 Descriptions and Guidelines for Plastic Processing Methods

Process	Descriptions	Key advantage	Notable limitations
Blow molding	An extruded tube (parison) of heated thermoplastic is placed between two halves of an open split mold and expanded against the sides of the closed mold by air pressure. The mold is opened, and the part is ejected.	Low tool and tie costs; rapid production rates; ability to mold relatively complex hollow shapes in one piece.	Limited to hollow or tubular parts.
Calendering	Doughlike thermplastic mass is worked into a sheet of uniform thickness by passing it through and over a series of heated or cooled rolls. Calenders also are used to apply plastic covering to the back of other materials.	Low cost; sheet materials are virtually free of molded-in stresses.	Limited to sheet materials; very thin films not possible.
Casting	Liquid plastic (usually thermoset) is poured into a mold (without pressure), cured, and removed from the mold. Cast thermoplastic films are made by depositing the material, ether in solution or in hot-melt form, against a highly polished supporting surface.	Low mold cost; ability to produce large parts with thick cross sections; good surface finish; suitable to low-volume production.	Limited to relatively simple shapes; except for cast films, becomes uneconomical at high-volume production levels; most thermoplastics not suitable.
Compression molding	A thermoplastic or partially polymerized thermosetting resin compound is placed in a heated mold cavity; the mold is closed, heat and pressure are applied, and the material flows and fills the mold cavity. Heat completes polymerization, and the mold is opened to remove the part.	Little waste of material and low finishing costs; large, bulky parts are possible.	Extremely intricate parts involving undercuts, undercuts, side draws, small holes, delicate inserts, etc., not practical; very close tolerances difficult to produce.
Cold forming	Similar to compression molding in that material is charged into split mold; it differs in that it uses no heat, only pressure. Part is cured in an oven in a separate operation. Some thermoplastic sheet material and billets are cold-formed in process similar to drop hammer-die forming of metals.	Ability to form heavy or tough-to-mold materials; simple; inexpensive; often has rapid production rate.	Limited to relatively simple shapes; few materials can be processed in this manner

Extrusion	Thermoplastic or thermoset molding compound is fed from a hopper to a screw and barrel where it is heated to plasticity and then forwarded, usually by a rotating screw, through a nozzle having the desired cross-section configuration.	Low tool cost; great many complex profile shapes possible; very rapid production rates; can apply coatings or jacketing to core materials, such as wire.	Limited to sections of uniform cross section.
Filament winding	Continuous filaments, usually glass, in form of rovings are saturated with resin and machine-wound onto mandrels having shape of desired finished part. Once winding is completed, part and mandrel are placed in oven for curing. Mandrel is then removed through porthole at end of wound part.	High-strength reinforcements are oriented precisely in direction where strength is needed; exceptional strength-to-weight ratio; good uniformity of resin distribution of finished part.	Limited to shapes of positive curvature; openings and holes reduce strength.
Injection molding	Thermoplastic or thermoset molding compound is heated to plasticity in cylinder at controlled temperature; then forced under pressure through a nozzle into sprues, runners, gates, and cavities of mold. The resin solidifies rapidly, the mold is opened, and the part(s) ejected. In modified version of process—runnerless molding—runners are part of mold cavity.	Extremely rapid production rates, hence low cost per part; little finishing required; good dimensional accuracy; ability to produce relatively large, complex shapes; very good surface finish.	High initial tool and die costs; not practical for small runs.
Laminating	Material, usually in form of reinforcing cloth, paper, foil, etc., preimpregnated or coated with thermoset resin (sometimes a thermoplastic), is molded under pressure into sheet, rod, tube, or other simple shape.	Excellent dimensional stability of finished product; very economical in large production of parts.	High tool and die costs; limited to simple shapes and cross-section profiles.

TABLE 1.28 Descriptions and Guidelines for Plastic Processing Methods (*Continued*)

Process	Descriptions	Key advantage	Notable limitations
Matched-die molding	A variation of conventional compression molding, this process uses two metal molds having a close-fitting, telescoping area to seal in the plastic compound being molded and to trim the reinforcement. The reinforcement, usually mat or preform, is positioned in the mold, and the mold is closed and heated (pressures generally vary between 150 and 400 lb/in^2). Mold is then opened and part lifted out.	Rapid production rates; good quality and reproducibility of parts.	High mold and equipment costs; parts often require extensive surface finishing such as sanding.
Rotational molding	A predetermined amount of powdered or liquid thermoplastic or thermoset material is poured into mold. Mold is closed, heated, and rotated in the axis of two planes until contents have fused to inner walls of mold. The mold is opened and part removed.	Low mold cost; large hollow parts in one piece can be produced; molded parts are essentially isotropic.	Limited to hollow parts; in general, production rates are slow.
Slush molding	Powdered or liquid thermoplastic material is poured into a mold to capacity. Mold is closed and heated for a predetermined time to achieve a specified buildup of partially cured material on mold walls. Mold is opened, and unpolymerized material is poured out. Semifused part is removed from mold and fully polymerized in oven.	Very low mold costs; very economical for small production runs.	Limited to hollow parts; production rates are very slow; limited choice of materials that can be processed.
Thermoforming	Heat-softened thermoplastic sheet is placed over male or female mold. Air is evacuated from between sheet and mold, causing sheet to conform to contour of mold.	Tooling costs generally are low; produces large parts with thin sections; often economical for limited production of parts.	In general, limited to parts of simple configuration; limited number of materials to choose from; high scrap.
Transfer molding	Thermoset molding compound is fed from hopper into a transfer chamber where it is heated to plasticity. It is then fed by means of a plunger through sprues, runners, and gates of closed mold into mold cavity. Mold is opened and the part ejected.	Good dimensional accuracy; rapid production rate; very intricate parts can be produced.	Molds are expensive; high material loss in sprues and runners; size of parts is somewhat limited.

Wet lay-up or contact molding	Number of layers, consisting of a mixture of reinforcement (usually glass cloth) and resin (thermosetting), are placed in mold and contoured by roller to mold's shape. Assembly is allowed to cure (usually in an oven) without application of pressure. In modification of process, called spray molding, resin systems and chopped fibers are sprayed simultaneously from spray gun against mold surface; roller assist also is used. Wet lay-up parts sometimes are cured under pressure, using vacuum bag, pressure bag, or autoclave.	Very low cost; large parts can be produced; suitable for low-volume production of parts.	Not economical for large-volume production; uniformity of resin distribution very difficult to control; mainly limited to simple shapes.
Pultrusion	Fiber bundles are pulled through a liquid resin and then through a forming die to preshape the material and finally into a heated die.	Fully automated process; low tooling costs; low labor costs; produces continuous cross-sectional composite profiles.	Shapes are limited.
Reaction injection molding	Reactive liquid components are pumped separately into a mixing head. The mixed resin is forced into a heated mold where it cures.	Low cost; large parts can be obtained.	Limited to a few resins; difficult to control process; low physical properties.

Source: From Harper.[3] Reprinted with permission.

TABLE 1.29 Basic Considerations of the Various Embedding Processes

Method	Advantages	Limitations	Material requirements	Applications
Casting consists of pouring a catalyzed or hardenable liquid into a mold. The hardened cast part takes the shape of the mold, and the mold is removed for reuse.	Requires a minimum of equipment and facilities; is ideal for short runs.	For large-volume runs; molds, mold handling, and maintenance can be expensive; assemblies must be positioned so they do not touch the mold during casting; patching or surface defects can be difficult.	Viscosity must be controlled so that the embedding material completely flows around all parts in the assembly at the processing temperature and pressure.	Most mechanical or electromechanical assemblies within certain size limitations can be cast.
Polling is similar to casting except that the catalyzed or hardenable liquid is poured into a shell or housing which remains as an integral part of the unit.	Excellent for large-volume runs; tooling is minimal; presence of a shell or housing assures no exposed components, as can occur in casting.	Some materials do not adhere to shell or housing; electrical short-circuiting to the housing can occur if the housing is metal.	Same material requirements as for casting, except that materials that will bond to shells or housings are required.	Most mechanical or electromechanical assemblies; subject to certain size limitations and housing complexity limitations.
Impregnating consists of completely immersing a part in a liquid so that the interstices are thoroughly soaked and wetted; usually accomplished by vacuum and/or pressure.	The most positive method for obtaining total embedding in deep or dense assembly sections such as transformer coils.	Requires vacuum or pressure equipment which can be costly; in curing, the impregnating material tends to run out of the assembly, creating internal voids unless on encapsulating coating has first been applied to the outside of the assembly.	Low-viscosity materials are required for the most efficient and most thorough impregnation.	Dense assemblies which must be thoroughly soaked; electric coils are primary examples.

Encapsulating consists of coating (usually by dipping) a part with a curable or hardenable coating; coatings are relatively thick compared with varnish coatings.	Requires a minimum of equipment and facilities.	Obtaining a uniform, drip-free coating is difficult; specialized equipment for applying encapsulating coatings by spray techniques overcomes this problem, however.	Must be both high viscosity and thixotropic, that is, material must not run off the part during the cure.	Parts requiring a thick outer coating, such as transformers.
Transfer molding is the process of transferring a catalyzed or hardenable material, under pressure, from a pot or container into the mold which contains the part to be embedded.	Economical for large-volume operations.	Initial facility and mold costs are high; requires care so that parts of assemblies are not exposed; some pressure is required, and processing temperatures are often higher than for other embedding operations.	Should be moldable at the lowest possible pressure and temperature, and should cure in the shortest possible time for lowest processing cost.	For embedding small electronic assemblies in large-volume operations.

REFERENCES

1. G. Odian, *Principles of Polymerization,* McGraw-Hill, New York, 1970.
2. "Designing with Plastic (The Fundamentals)," Design Manual TDM-1, Hoechst Celanese Corp., 1989.
3. C. A. Harper (ed.), *Handbook of Materials and Processes for Electronics,* McGraw-Hill, New York, 1970.
4. R. Greene (ed.), *Modern Plastics Encyclopedia,* McGraw-Hill, New York, 1992.
5. H. F. Mark et al. (eds.), *Encyclopedia of Polymer Science and Engineering,* 2d ed., vols. 1–17, Wiley, New York, 1985–1990.
6. S. S. Schwartz and S. H. Goodman, *Plastics Materials and Processes,* Van Nostrand Reinhold, New York, 1982.
7. C. A. Harper (ed.), *Electronic Packing and Interconnection Handbook,* McGraw-Hill, New York, 1991.
8. T. S. Chung, *Polymer Eng. and Sci.,* vol. 26, no. 13, p. 901, July 1986.
9. A. J. Klein, "Liquid Crystal Polymers Gain Momentum," *Plastics Design Forum* (Edgell Communications), Jan./Feb., 1989.
10. T. Stevens, *Mater. Eng.,* vol. 102, Jan. 1991.
11. "Properties of Nomex," DuPont Co. Bull. H-22368 to H-22375, Mar. 1990.
12. "Kapton Summary of Properties," DuPont Co. Bull. E-93189, Mar. 1988.
13. N. Albee, "Polymer Blending for Property Tailoring," *Plastics Compounding* (Edgell Communications), July/Aug. 1992.
14. S. H. Goodman (ed.), *Handbook of Thermoset Plastics,* Noyes Publ., Parkridge, N.J., 1986.
15. D. A. Shimp, "Cyanate Ester Resins—Chemistry, Properties and Applications," Hi-Tek Polymers, Inc., Jan. 1990.
16. D. Burdeaux et al., "Benzocyclobutenes—Dielectrics for the Fabrication of High Density, Thin Film Multichip Modules," *J. Electro. Mater.,* vol. 19, no. 12, 1990.
17. Chen C. Ku et al., *Electrical Properties of Polymers,* Hanser Publ., New York, 1987.
18. R. O. Babbit (ed.), *The Vanderbilt Rubber Handbook,* R. T. Vanderbilt Co., Norwalk, Conn., 1978.
19. ASTM D-395, "Test Method for Rubber Property—Compression Set," Am. Soc. for Testing and Materials, Philadelphia, Pa., 1989.
20. ASTM D-412, "Test Methods for Rubber Properties in Tension," Am. Soc. for Testing and Materials, Philadelphia, Pa., 1987.
21. ASTM D-1418, "Practices for Rubber and Rubber Lattices—Nomenclature," Am. Soc. for Testing and Materials, Philadelphia, Pa., 1990.
22. C. F. Coombs, *Printed Circuits Handbook,* 3d ed., McGraw-Hill, New York, 1988.

CHAPTER 2

PRINTED-WIRING BOARDS

Victor J. Brzozowski and Carl T. Brooks

2.1 INTRODUCTION

Printed-wiring boards, commonly called PWBs, are sometimes referred to as the baseline in electronic packaging. Electronic packaging is fundamentally an interconnection technology and the PWB is the baseline building block of this technology. It serves a wide variety of functions. Foremost, it contains the wiring required to interconnect the components electrically, and it acts as the primary structure to support those components. In some instances it is also used to conduct away heat generated by the components. In summary, the PWB is the interconnection medium upon which electronic components are formed into electronic systems.

One frequent cliché for describing PWBs is that PWB technology is a technology that is constantly in turmoil. This is perhaps also correct, since continued and constant progress in component parts (especially semiconductor components), coupled with requirements for continuing higher performance in electronic systems, results in continued and constant pressure for improvements in all aspects of PWB technology. Some of these trends are discussed in Sec. 2.1.1. Based on the critical importance of PWBs as the interconnecting link between individual component parts and the formulation of electronic systems, this chapter is devoted to presenting an understanding of the many important facets of PWB materials and processes which must be fully understood to achieve success in electronic systems.

2.1.1 Critical Factors in PWB Technology

The constant pressures for improvements in PWB technology arise in all aspects of this technology. Electrically, the increase in high-speed and high-frequency electronic systems creates demand for PWBs having lower electrical losses. In addition, higher operating voltages increasingly require PWBs with greater resistance to voltage breakdown, high-voltage tracking, and corona.

Aside from the requirements for higher electrical performance of PWBs, higher electronic-system functional densities and the resultant higher thermal densities create demand for lower thermal resistance of PWB materials. In an increasing number of applications, the old industry standard epoxy-glass PWB is thermally inadequate.

New developments in component technology in the 1960s and 1970s, with the movement away from through-hole technology to the higher-density surface-mount technology (SMT) using leadless chip carrier (LCC) components, have forced innovations in new PWB materials and processes to control the end-item printed-wiring-assembly (PWA) coefficient of thermal expansion (CTE). SMT has caused the solder joint to act not only as the electrical interconnect between the component and the PWB but also as the main structural element attaching the component to the PWB. This PWA CTE control is required to reduce stresses imparted into the solder joint due to thermal operational excursions. Without it, solder joint failure occurs.

The constant trend toward higher-functionality integrated-circuit (IC) components with higher input-output (I/O) pin counts of the IC packages has resulted in increasing demand for finer-line PWBs. High I/O count package technology and fine-line PWB technology are often called fine-pitch technology (FPT).

Basically, the problem is that advances in IC technology have caused significant increases in the complexity of component packaging technology, which are solved and many times implemented without determining the impacts being imparted to PWB technology. PWB technology advances are many times implemented after the fact instead of in conjunction with IC technology advances. It has been estimated that the IC/PWB complexity factor was 30:1 in 1965, 50:1 in 1975, and 100:1 in 1985. With this trend, standard subtractive PWB technology often becomes inadequate, leading to new fine-line circuit interconnection forms, such as additive PWB circuits, microwire interconnections, multichip modules, and chip-on-board. A more in-depth discussion of many of these topics can be found in Chap. 11 and in Harper.[1]

2.1.2 Terms and Definitions

Terms and definitions associated with PWB technology and the complete subject of electronic packaging and interconnection have been published by the Institute for Interconnecting and Packaging Electronic Circuits (IPC) in IPC-T-50 and in Miller.[2] Applicable military and industry standards are listed at the end of this chapter.

2.2 PRINTED-WIRING-BOARD SYSTEM TYPES

PWBs can be classified into several categories based on their dielectric material or their fabrication technique. This section describes PWBs fabricated using organic dielectrics. Ceramic dielectric boards are discussed in Chaps. 3 and 8. Other PWB types such as discrete wire, wire wrap, and newer interconnection techniques are covered in Chap. 7 and in other publications, such as Harper.[1]

Organic PWBs are fabricated using an organic dielectric material, with copper usually forming the conductive paths. The organic-based boards can be subdivided into the following classifications: rigid, flexible, rigid-flexible, combining the attributes of both rigid and flexible boards in one unit, and molded. Molded PWBs will not be covered in this chapter, but are discussed in Harper.[1] Each of these classifications, except for molded, can be further subdivided into single-sided, double-sided, and multilayer PWBs.

The circuit interconnection pattern is created by imaging the conductor pattern on copper sheets using a photoresist material and one of two image-transfer techniques: screen printing or photo imaging. The resist acts as a protective cover, defining the conductor patterns, while unwanted copper is etched away. In 1985 58 percent of the PWBs produced used the screen-printing process and 42 percent the photo-imaged process.

These techniques can be used with a variety of dielectric materials to achieve various mechanical and electrical characteristics in the final product. Among the most common dielectric materials are epoxy/E-glass (electronic-grade glass) laminates used in the fabrication of rigid PWBs and polyimide film used in the fabrication of flexible printed wiring. The rigid-flexible boards use a combination of these two materials. Section 2.3 addresses the wide range of flexible materials and their characteristics that may be used for a given application. The following sections describe in detail the construction of each organic board type.

2.2.1 Rigid PWBs

The rigid PWB is fabricated from copper-clad dielectric materials. The dielectric consists of an organic resin reinforced with fibers. The most commonly used fiber materials are paper and E-glass. The fibers are either chopped (usually paper) or woven into a fabric. The organic media can be of a wide formulation and include flame-retardant phenolic, epoxy, polyfunctional epoxy, and polyimide resins. Built within the laminate structure may be low-CTE clad metals such as copper-Invar-copper or copper-molybdenum-copper for decreasing the CTE of the overall PWB structure in SMT applications. Detailed information on base materials is included in Sec. 2.3.

As the name implies, rigid PWBs consist of layers of the organic laminates that are laminated through heat and pressure into a rigid interconnection structure. This structure is usually sufficiently rigid in nature to be able to support the components which are mounted to it. Specialized applications may require the PWB to be mounted to a support structure. The support structure may be used to remove heat generated by the components, decrease the movement of the PWB under extreme vibration, or decrease the CTE of the PWB in SMT applications. Further details on PWB-supporting structural materials are given in Sec. 2.3 and in Chap. 11.

The rigid PWB interconnection structure may be further subdivided by the number of wiring layers contained within the structure into the following three categories: single-sided, double-sided, or multi-layered. Figure 2.1 shows a cross-sectional view of each type. The following sections briefly explain the differences in these three PWB structures and highlights the steps used to produce each. Section 2.4 discusses the process procedures in further detail.

Single-Sided PWBs. A single-sided PWB consists of a single layer of copper interconnection (usually on the component side of the PWB). The rigid dielectric material is fabricated from multiple layers of unclad laminate material pressed to the final end-use thickness. A single layer of copper cladding is applied to one of the outside layers during this process. In some instances double-sided copper cladding may be used, with the copper on one face being completely etched away during the processing.

The base laminate of single-sided boards can be of woven or paper (unwoven) materials with copper foil, usually of 1- or 2-oz weight, clad to one side. It should be noted here that copper cladding is most often referred to by its weight rather than by its thickness (1 oz/ft^2 equals 0.00137 in thick). The raw clad laminate is first cut into working panels suited to the equipment that will handle the subsequent operations. The panel is then drilled or punched to provide a registration system. Laminate flatness is important in achieving a good registration baseline. This is critical in an automated print and etch system because the panel tends to warp after the copper is removed during etching. This warping allows stresses built into the material during its fabrication to be relieved. Excessively warped panels may not register properly for subsequent operations.

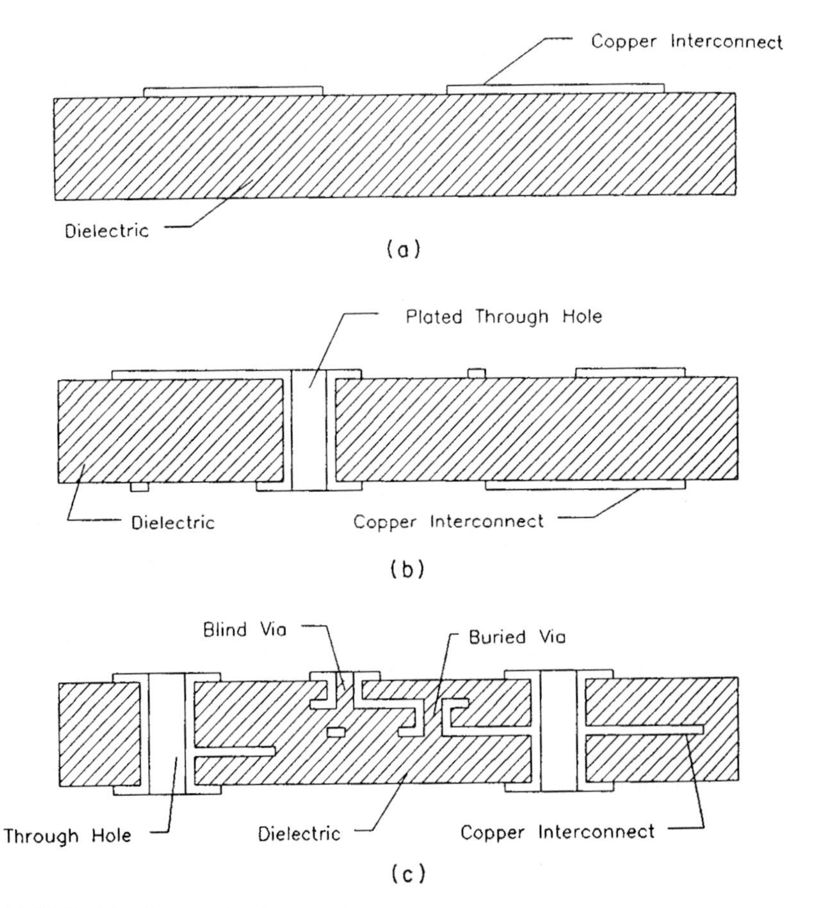

FIGURE 2.1 Cross-sectional views of organic PWBs. (*a*) Single-sided. (*b*) Double-sided. (*c*) Multilayer.

The individual artworks which define the conductor patterns are then arranged or panelized so that one or more PWBs will be produced from a single panel. This is accomplished by stepping and repeating the patterns into a panel phototool. Once the panel layout is established, the panel can be drilled or punched to produce the final hole pattern. Holes required are either drilled in glass-reinforced products or punched in paper-reinforced products. Registration of the conductor pattern to holes is accomplished through either the right-angle edge of the panel or on pilot holes contained in opposite corners of the panel. Drilling of holes is usually done after the panels are first cut, punching of holes as the last operation.

Following the drilling operation, the etch resist is applied and the circuit pattern formed. This pattern can be made by printing a liquid resist or photo imaging of a film or liquid. The next step is to etch away the unwanted copper from the laminate, leaving only the desired circuit pattern. Finally the resist is stripped and the single-sided board is complete in panel form. At this point additional processes such as platings or solder masks may be performed, or the individual boards may be sheared or routed from the panel.

While single-sided boards with their simplicity might seem doomed due to the emergence of modern complex electronics, they continue to have a substantial market, especially where cost is a strong driver. Their use with fairly complex IC devices can be seen in many places, such as watches, cameras, and automobiles. This will probably continue to be the case for the foreseeable future.

Double-Sided PWBs. From a historical perspective the double-sided board is probably the most often designed type of all PWBs, although the increased complexities of today's electronics are causing a change to multilayer technology as the dominant form of PWB technology. It retains much of the production simplicity of the single-sided board, but allows circuit complexities far in excess of 2:1 over its simpler cousin. This is the case because it allows basic x and y routing of the circuit on its two outer faces, thus improving the routing efficiency and the circuit density. Interconnection of the two conductor patterns is accomplished through drilling and subsequent plating or filling with a silver conductive ink the interconnection holes called vias. The conductive ink is normally used only in low-cost consumer products. The most widely used method is to plate the vias with copper.

Double-sided boards are fabricated from laminates with copper (usually 1 oz) cladded to both outside layers. The copper may be clad to a variety of materials, as discussed in Sec. 2.3. Here, as with the single-sided board, the material is purchased from a laminator who specializes in providing laminates to the electronic industry.

Once the raw laminate is cut into panels, the fabrication operation begins with the interconnection hole drilling. The via holes may also serve as mounting holes for the components to be soldered into. After the via hole pattern has been drilled, the holes may be filled with the conductive ink, or the panel is electroless copper plated in preparation for subsequent plating by either of two methods: pattern plating or panel plating (see Sec. 2.4.8).

The conductor image is formed similar to single-sided boards, except that the photoresist application and imaging take place on both sides of the panel. Obviously, the registration of the photo images from one side of the panel to the other is critical. The circuit pattern on one side must be properly registered to the pattern on the other side, or the plated-through holes will not properly connect between the two sides.

The next step is to etch away the copper laminate, leaving only the desired circuit pattern. At this point additional processes such as resist stripping, platings, or solder masks may be performed and the individual PWBs then sheared or routed from the panel.

Multilayer PWBs. Multilayer boards are those PWBs having three or more conductive layers, including any pads-only layers. Since 1985 demand for multilayer PWB products has been a driving force in the worldwide PWB industry. In 1991 North American multilayer market sales totaled over $4.3 billion, capturing over 65 percent of the total rigid PWB market. Sales totals in other world markets for 1991 include $2.36 billion in Japan, $1.88 billion in Western Europe, and $372 million in Southeast Asia.

The typical modern multilayer board will have from 4 to 10 layers of circuitry, with some high-density applications requiring upward of 50 layers. Most multilayer boards are fabricated by laminating single- or double-clad, preetched, patterned sheets of thin (<0.005 in) laminate together using partially cured resin in a carrier fabric. The materials commonly used for this purpose are discussed at length in Sec. 2.3.

The single- or double-clad laminate material is processed similarly to the single- or double-sided PWB, except that the via or component holes are usually not drilled until after lamination. It should be noted here that the importance of registration is amplified as the layer count increases. Increased pad sizes may be required to minimize via hole breakout due to misregistration. The same requirement may limit the size of panels due to run out of the circuit features.

Following the fabrication of the individual layers or layer pair, a "book" of layers and their interposed B-stage bonding layers are stacked together in a particular sequence to achieve the required lay-up. This book is then laminated under heat and pressure to the appropriate thickness for the final board. The outer layers are not preetched so that the laminated book appears the same as a thick double-sided copper-clad laminate of comparable thickness. After lamination, the book is processed the same as a thick double-sided board. The book is drilled to add the via holes and then processed as if it were a double-sided board using plated-through holes. A cross section of a typical multilayer board is shown in Fig. 2.1.

In some cases standardized layers such as power or ground distribution can be "mass"-laminated into the raw laminate. This is a very cost-effective means of achieving multilayer density at near double-sided board cost since the outer layer processing and via drilling are identical to double-sided PWB processing. As a result, four-layer boards are the most prevalent multilayer boards.

Where circuit density requirements override cost considerations, techniques such as blind or buried vias are used to increase the interconnection wiring density on a given layer. Where these techniques are used, the inner layer pairs are fabricated as double-sided boards, complete with plated vias, and then are assembled into books for processing into multilayer boards. Thus the inner layers may be interconnected without a via hole through the entire board. Similarly, blind vias may connect to the first or subsequent buried layer on each side of the board without penetrating the entire board (see Fig. 2.1).

The multilayer board has now achieved a cost and reliability level which allows its use in any level of electronics. It is no longer the exclusive tool of the mainframe computer, telecommunications, or military electronics. It is often seen even in toys.

2.2.2 Flexible PWBs

As defined by the IPC, flexible printed wiring is a random arrangement of printed wiring, utilizing flexible base material with or without cover layers. Interconnection systems consisting of flat cables, collated cable, ribbon cable, and sometimes wiring harnesses are sometimes confused with flexible printed wiring. Flexible printed wiring is used in applications requiring continuous or periodic movement of the circuit as part of the end-product function and in those applications where the wiring cannot be planar and is moved only for servicing. Flexible wiring can be used as interconnect cabling harnesses between various system circuit-card assemblies or connectors as well as to interconnect individual electrical components. Figure 2.2 shows a typical flexible printed-wiring interconnect.

Visually, flexible printed wiring looks similar to rigid printed wiring. The main difference in the products is the base or dielectric material. Flexible printed wiring is manufactured using ductile copper foil bonded to thin, flexible dielectrics. The dielectric materials include polyimide (Kapton), polyester terephthalate (Mylar), random-fiber aramid (Nomex), polyamide-imide Teflon TFE and FEP, and polyvinyl chloride (PVC). Dielectric materials are discussed in further detail in Sec. 2.3.2.

As with their rigid printed-wiring brethren, the flexible printed wiring may be manufactured in single-sided, double-sided, or multilayer configurations (Fig. 2.3). The conductor patterns are formed in a manner similar to rigid PWBs, using either screen printing or photo imaging of a resist to form the conductor pattern and then etching

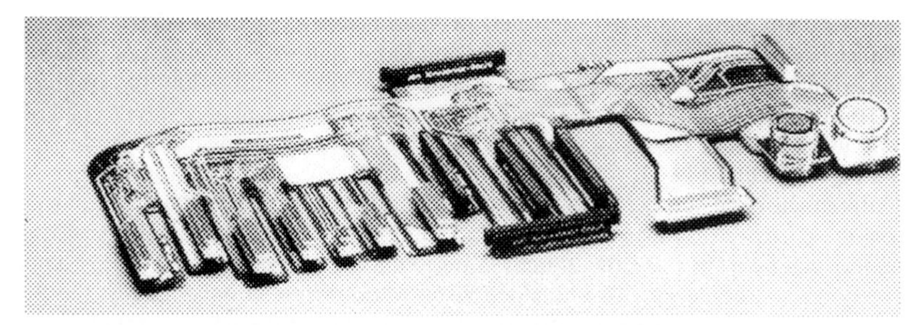

FIGURE 2.2 Typical flexible PWB interconnect in connector assembly.

the unwanted copper. A variety of adhesive materials are used in their manufacture to bond the various layers together, including acrylics, epoxies, phenolic butyrals, polyesters, and polytetrafluoroethylene (PTFE). In addition, newer processes have been developed to laminate the conductor directly to the dielectric film without the use of an adhesive layer. Adhesive materials are discussed in Sec. 2.3.2.

On single- and double-sided flexible wiring, cover layers are often used to protect the etched copper circuitry. When a film cover layer is employed, the access holes to

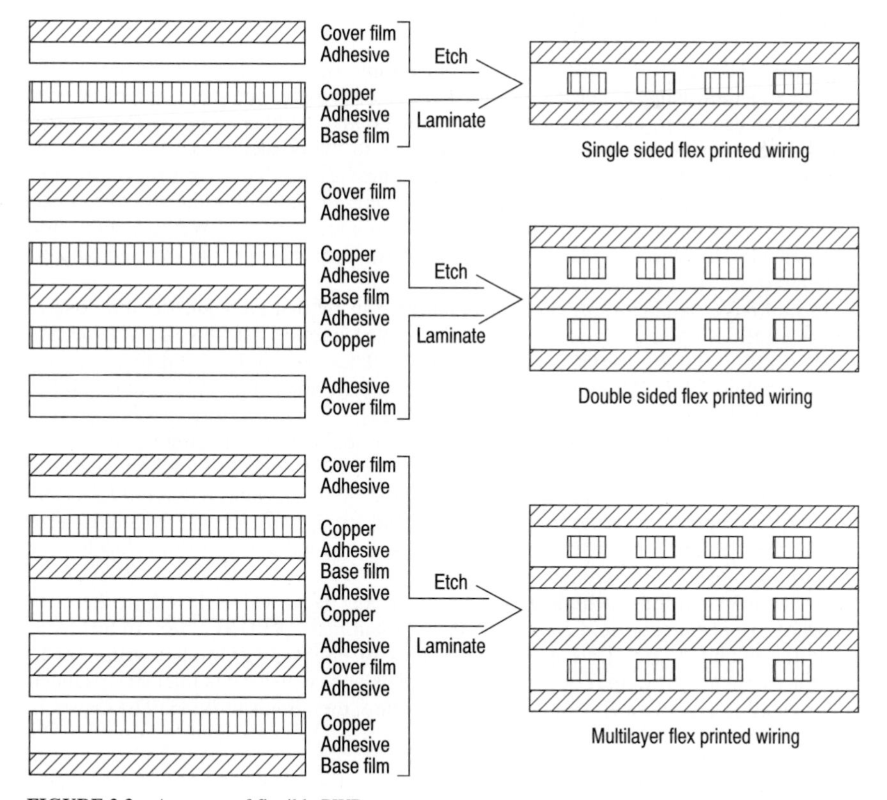

FIGURE 2.3 Anatomy of flexible PWB.

the circuitry are either punched or drilled in the adhesive-coated film. The cover layer is then mechanically aligned to the wiring and laminated in a platen press under heat and pressure. The adhesive systems must be of a no-flow or low-flow formulation to keep from contaminating the interconnection pads. Screen-printable cover coats may also be used. These are usually ultraviolet curable and give a moderately pinhole-free insulating surface.

Unlike their rigid counterparts, the outline features of flexible wiring are not routed but cut using either soft or hard tooling. Soft tooling involves a rule die. In its simplest form, this can be a steel die and X-acto knife to cut the outline pattern. More complex dies may consist of steel strips imbedded in plywood. When pressed against the flexible wiring, they act as cookie cutters to form the outline. Rule dies are dimensionally less accurate and less expensive than hard tools. Hard tools are punches, die plates, and strippers manufactured from hardened steel. They can hold tolerances within the die of ±0.001 in.

Due to the extreme flimsiness of flexible wiring, when components are to be mounted, adequate reinforcement must be added to the flexible wiring to eliminate stress points at the component circuit interfaces. Reinforcements typically used are simple pieces of unclad rigid laminates or complex formed, cast, or machined metals or plastics to which the flexible wiring is laminated.

2.2.3 Rigid-Flexible PWBs

Rigid-flexible circuitry consists of single or multiple flexible wiring layers selectively bonded together using either a modified acrylic adhesive or an epoxy prepreg bond film. Cap layers of rigid-core copper-clad laminate may be bonded to the top and bottom surfaces of the circuit to add further stability to the bonded areas (Fig. 2.4). This technique is in wide use in the industry today to provide rigid areas at connector interfaces (Fig. 2.5). In other instances, PWBs have incorporated within the structure flexible layers to interconnect one PWB to another or to interconnect a PWB to a connector. The rigid-flexible wiring exhibits the lowest profile form factor of all the interconnect system types.

In multilayer applications, problems can arise in fabricating rigid-flexible assemblies with the Kapton*/acrylic film layers integrated into the hard-board PWB section which would contain epoxy/E-glass materials. The high moisture absorption rate (4 percent) and CTE (400 to 600 ppm/°F) of the acrylic adhesive can lead to reliability problems in the finished product under thermal uses.

To allow for increased reliability in multilayer applications, the amount of acrylic adhesive in the rigidized area must be limited. There are a couple of methods in use in the industry today to accomplish this. One method involves using a polyimide/acrylic base with a prepreg covercoat in the rigid area and a polyimide/acrylic selective cover coat in the flexible ares. This technique forces the manufacturer to develop tooling techniques to overlap the separate dielectrics.

A second technique, which was developed and patented by Teledyne Electro-Mechanisms, is called a rigid epoxy glass acrylic laminate (REGAL) flex. In this technique the manufacturer starts with a base stock of epoxy prepreg clad with copper. The traces are etched in the copper. The base stock epoxy prepreg is then encapsulated in the flexible area with flexible dielectric to allow the circuit to bend. The traces in

*Kapton is a trade name of E.I. duPont de Nemours & Co.

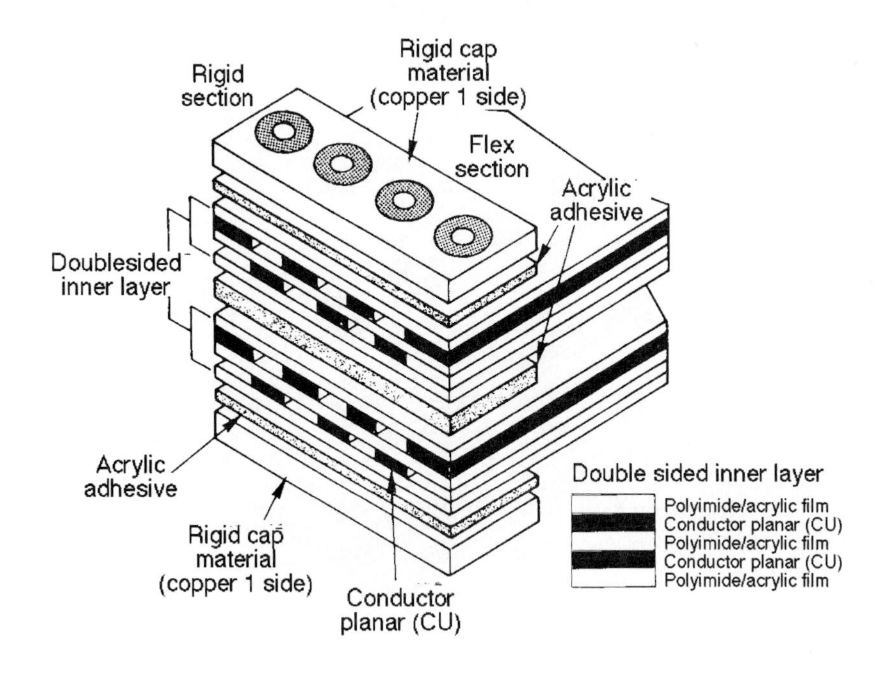

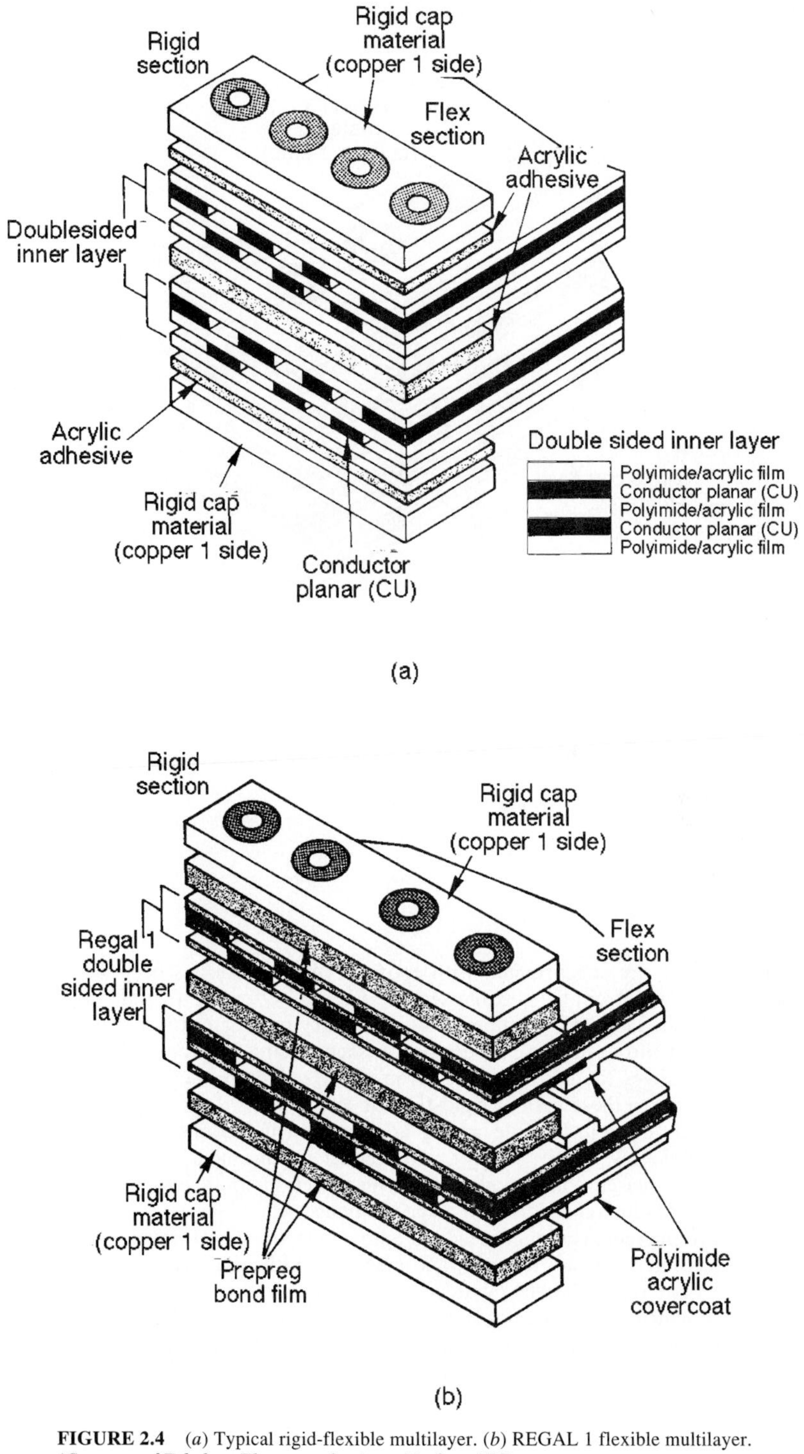

(a)

Rigid section
Rigid cap material (copper 1 side)
Flex section
Regal 1 double sided inner layer
Rigid cap material (copper 1 side)
Prepreg bond film
Polyimide acrylic covercoat

(b)

FIGURE 2.4 (*a*) Typical rigid-flexible multilayer. (*b*) REGAL 1 flexible multilayer. (*Courtesy of Teledyne Electromechanisms, Nashua, NH.*)

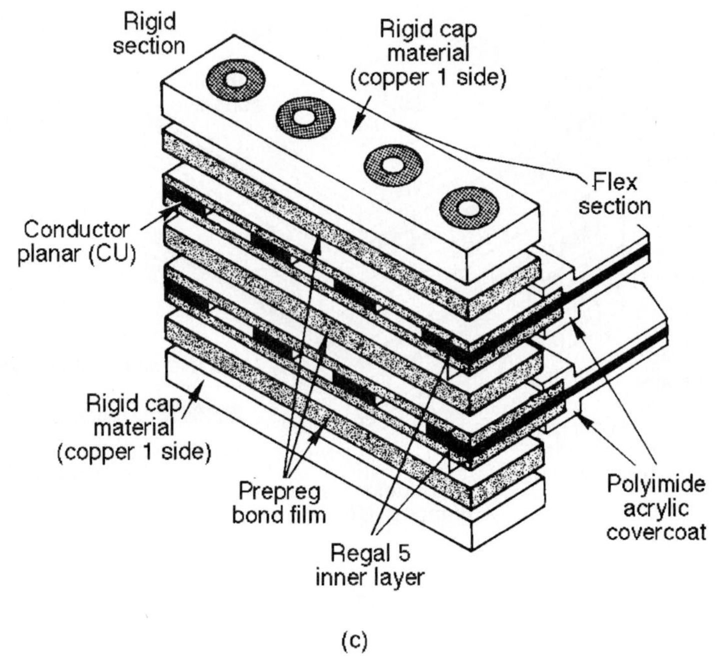

FIGURE 2.4 (*Continued*) (*c*) REGAL 5 flexible multilayer. (*Courtesy of Teledyne Electromechanisms, Nashua, NH.*)

the rigid section are encapsulated with epoxy prepreg which has been windowed to remove the prepreg from the flexible section. Each cover-coated flexible layer can be selectively bonded together with prewindowed epoxy prepreg to form a rigidized area for subsequent through-hole processing. A second construction technique, called REGAL 5, removes the epoxy prepreg from the flexible area. Figure 2.4 shows the differences between these various rigid-flexible techniques.

The benefits of rigid-flexible wiring are apparent in the design, manufacturing, installation, assembly, quality control, and product enhancement of the end item. The designer has increased conceptual freedom in the end-item design. Conformability, three-dimensional interconnects, and a space-saving form factor are benefits. In many cases reduced interconnect length leads to optimal electrical performance. Mechanical and electrical interfaces are reduced, and mechanical, thermal, and electrical characteristics are more repeatable than with conventionally wired systems.

In manufacturing their use leads to reduced assembly costs within a totally unitized interconnect system. There are increased opportunities for automation. In addition, reduced system interconnect errors and improved system interconnect yields occur.

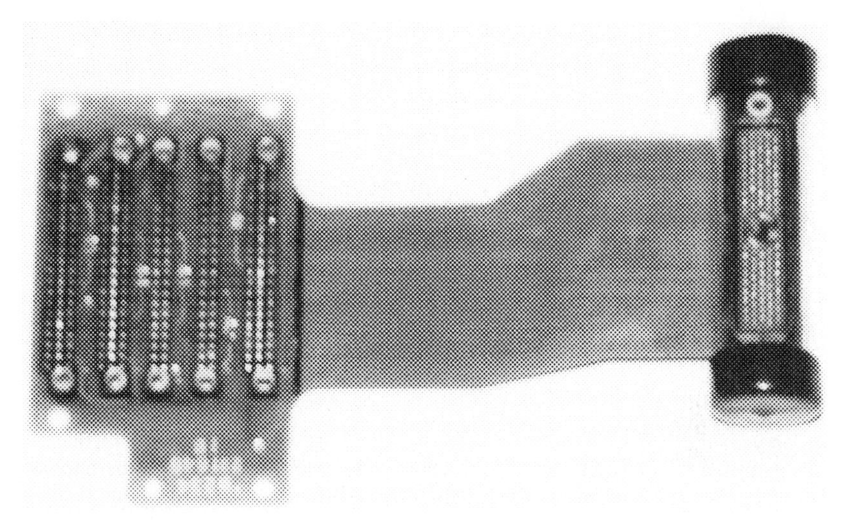

FIGURE 2.5 Typical rigid-flexible PWB. (*Courtesy of Teledyne Electromechanisms, Nashua, NH.*)

Installation and assembly benefits include elimination of miswiring and indexing errors encountered in discrete wired systems. A reduction in the installer skill and training, increased speed of installation, and mounting simplifications leading to reduced hardware requirements are other benefits.

Quality control benefits are achieved through the adaptability of the product to automated inspection, a simplification of error-cause analysis, and more effective error-cause correction.

Products are enhanced through reduced weight, volume, and cost. Fewer inter-connections are usually required, leading to increased system reliability. Reductions in product inventory, maintenance, and field service time and expense are also realized.

2.3 MATERIALS FOR PRINTED-WIRING BOARDS

As delineated in Sec. 2.2, PWBs are designed in various sizes and shapes, use a variety of processes and materials, and perform a variety of electrical, structural, and sometimes thermal functions. Paramount to achieving a PWB that performs its intended function reliably, is producible, and is fabricated for the lowest cost possible, the designer must have a fundamental knowledge of the materials used in the end item. However, the knowledge must not be limited to the electrical, mechanical, thermal, and chemical properties of the end-item material. Materials knowledge must include environmental effects (thermal, mechanical, and humidity) on properties and their impacts on the material's performance in a given design. In addition, imposed manufacturing-process-related stresses must also be considered. The following sections address these aspects in materials used in PWBs.

2.3.1 Rigid PWB Materials

As discussed in Sec. 2.2.1, organic rigid PWBs consist of a dielectric material onto which is patterned some form of metallization, which creates the actual circuit.

Fiber-reinforced resin dielectric materials, referred to as laminates, clad with copper sheets, are most commonly used for rigid PWB applications. The following subsections describe the typical laminate anatomy, the materials that comprise typical laminates, the variety of laminates in use today, the specifications that govern their fabrication and end-item properties, and the problems one may encounter with laminate materials.

Anatomy of a PWB Laminate. The laminate properties are related to the constituents in the composite laminated structure, that is, to the anatomy of the laminate. In this section the nature of the PWB laminate construction is explained and the rules of the various constituents are discussed.

An organic rigid PWB laminate consists of three major elements and some auxiliary elements. The major elements are the fabric, the resin (which combined comprise the dielectric), and the metal foil (usually copper). The auxiliary elements are the adhesion promoters or treatments which are applied to assure maximum adhesion of the resin to the fabric and to the foil.

The manufacture of a copper-clad PWB laminate begins in a machine called a treater. Fundamentally, the operational sequence is that the fabric is fed off the fabric roll and through a dip pan containing resin. The resin-curing agent mixture in the dip pan is called *A stage,* a term used to describe totally unreacted resin.

The resin impregnates the fabric, is passed through a set of metering rolls to control the thickness, and then passes through a treating oven for partial cure (polymerization) of the resin onto the fabric. After the resin-soaked fabric is partially cured in the treater oven, the fabric-resin combination is called *B stage,* or *prepreg.* These two terms are used to describe the partially cured resin. Finally, the B-stage coated fabric is cut into predetermined sizes for laminating (as big as 4 ft by 8 ft).

The B stage is especially critical since it can be undercured (understaged) or overcured (overstaged). Ideally, the B stage will be dry to the touch, but capable of reflow and optimized bonding in the laminating press. If the B-stage sheets are undercured, they may not be dry to the touch, and the resin will flow out of the laminate during processing under heat and pressure in the laminating press. Dry spots will then exist in the laminate. On the other hand, if the B-stage sheets are overcured, poor bonding will occur in the lamination processing. This will result in poor bond strength, both between layers of fabric and to the copper foil. Delamination and poor copper peel strength will occur as problems in the final PWB laminate. Hence the B stage must be closely controlled for optimum PWB laminates to be produced.

Two other important factors in handling B-stage material are (1) resin aging and (2) moisture layers on the B-stage sheets. The resins used to impregnate the fabric are an organic polymer. The nature of polymer reactions is such that resin curing, or polymerization, will slowly continue at all times, the reaction rate being a function of temperature. Therefore since an optimum B stage is only partly cured, the curing will continue toward overcure, especially in warm or hot conditions, such as summer shipping and storage.

Under any given set of storage temperature conditions, a specific useful life, or shelf life, will exist for any given B-stage sheet. Thus cool shipment and storage conditions are usually recommended for B-stage stacks. The B-stage sheets must also be shipped and stored in dry conditions, since moisture films can be condensed onto cool sheets. When laminated together into a PWB laminate, the B-stage sheets with invisible moisture films will result in moisture entrapment between layers in the cured laminate. During subsequent soldering operations on the PWB, this entrapped moisture will explode into small entrapped delamination spots. These white spots, known as blisters, can be sufficiently large or dense to affect PWB performance and reliability. Often it is necessary to dry boards in an oven before assembly and soldering to drive out any moisture and to thereby prevent blisters.

The overall important points, then, are that B-stage sheets should be kept cool and dry (perhaps in sealed plastic bags) during shipping and storage. The laminate or B-stage sheet supplier can advise on optimum storage and shelf-life conditions. The end-item prepreg critical factors are maintenance of the ratio of fabric to resin, final material thickness, and degree of cure (polymerization) of the resin. The critical machine parameters to be monitored are metering roll thickness, speed of the fabric through the treater, and air velocity and oven temperature within the treater.

The final cured laminate is referred to as *C-stage* laminate and is achieved by pressure and heating in a laminating press. A final copper-clad C-stage laminate of a given thickness is made up of a number of thin B-stage laminates. This complete stack, including the copper foil, would be pressed together and heated between flat plates, or platens, in a heated laminating press for the time required to completely polymerize the epoxy resin at the selected press temperature. Laminates used in the manufacture of single- and double-sided PWBs are usually thick laminates (>0.030 in) and are made up of a number of thin unclad laminate B-stage plies. Multilayer PWBs use thin clad C-stage laminate plies bonded together with thin B-stage plies. Each ply thickness is typically in the range of 0.004 to 0.008 in. As an example, 0.062-in-thick double-sided NEMA grade FR-4 epoxy-glass laminate would contain enough layers of B-stage epoxy-resin-soaked glass sheets to achieve the desired final pressed thickness, with copper foil on both the top and the bottom of the stack. The final product would be processed to conform to NEMA or military specification standards.

Prior to discussing the various laminates, their properties, and uses, the major elements—fabric, resin, and metal—will be highlighted.

Fabric Materials and Construction. There are two materials which usually constitute the base fabric of the PWB laminate: paper and E-glass. Specialized laminate materials using S2-glass, quartz, and aramid (aromatic polyamide polymer) fibers and ceramic sphere fillers have been developed over the last 20 years. Table 2.1 shows the properties of most of these materials. In some instances a hybrid mixture of these materials has been investigated by laminate suppliers to try to optimize end-item properties.

Paper-based materials are used with flame-retardant resins. They are used in low-cost PWBs where laminate dimensional stability is not critical and where holes are punched in the material. Their use is mainly limited to single- and double-sided laminates for consumer electronics such as toys, calculators, and radios.

E-glass is a borosilicate-type glass that is the most widely used fabric material in PWB applications. IPC-EG-140 documents the specifications for finished fabrics woven from E-glass materials. Both randomly orientated glass fiber matts and woven glass fiber fabrics are produced. Its material properties satisfy the electrical and mechanical needs of most applications. It is used in single-sided, double-sided, and multilayer applications for aerospace, computer, communication, and industrial control products.

TABLE 2.1 PWB Fiber Material Properties

	E-glass	S-glass	Quartz	Aramid
Specific gravity, g/cm^3	2.54	2.49	2.20	1.40
Tensile strength, kg/mm	350	475	200	400
Young's modulus, kg/mm	7400	8600	7450	13,000
Maximum elongation, %	4.8	5.5	5.0	4.5
Specific heat, cal/(g · °C)	0.197	0.175	0.230	0.260
Thermal conductivity, W/(m · °C)	0.89	0.9	1.1	0.5
CTE, ppm/°C	5.0	2.8	0.54	−5.0*
Softening point, °C	840	975	1420	300
Dielectric constant at 1 MHz	5.8	4.52	3.5	4.0
Dissipation factor at 1 MHz	0.0011	0.0026	0.0002	0.001

*CTE given is along fiber axis; radial CTE = 60 ppm/°C.
Source: From Senese.[3]

S2-glass-, quartz-, and Kevlar-based fabrics have recently been introduced in laminates mainly for use in PWB SMT applications requiring laminates with CTEs approaching alumina (6 ppm/°C). They are also being considered in PWBs requiring lower dielectric constant end-item properties. Their use in applications to date has been limited to high-reliability military and aerospace products.

S2-glass is a structural-grade glass. IPC-SG-141 documents the specifications for finished fabrics woven from S-glass materials. Interest in this material for use in laminates lies in the fact that it has a lower CTE than E-glass. In addition it has a slightly lower dielectric constant than E-glass. Laminate materials made from this material process similar to E-glass laminates. S2-glass is a more expensive material than conventional E-glass.

Quartz is actually a fused silica. IPC-QF-143 documents the specifications for finished fabrics woven from quartz material. Once again the interest in quartz lies in its very low CTE for use in SMT laminates. The quartz material is very expensive when compared to glass due to a limited supply of fibers and the fact that it is produced by a foreign source. Laminates manufactured with quartz fabric exhibit occasional voiding due to the low fiber-to-resin ratio needed to maintain a low-CTE laminate. In addition, quartz is very tough on drill bits, making the drilling process of the PWB plated-through holes a more expensive operation. A second important property in quartz is the fact that it has a lower dielectric constant than E- or S2-glass or the aramid fibers and thus would lend itself to be used in laminates requiring a lower dielectric constant than typical FR-4 or polyimide-glass laminates.

The aramid fiber is manufactured by E.I. duPont de Nemours & Co. under the trade name Kevlar. IPC-A-142 documents the specifications for finished fabrics woven from aramid material. A second aramid type fiber has begun to be investigated for SMT laminate applications. It is being manufactured in Japan by Tejin Ltd. under the trade name Technora. The aramid fiber with its negative CTE along the fiber axis will produce a laminate material with the lowest CTE values when compared to the other available fiber materials, all other factors being equal, such as equivalent fiber-to-resin ratios. The high radial CTE can lead to some laminate resin problems (see page 2.29). Drilling of aramid laminates leaves frayed ends of the aramid fiber in the hole barrel, requiring a plasma etch to smooth the hole barrel, prior to plating. The material is also more expensive than conventional E-glass materials.

Specialized microwave applications require substrates with the physical properties of plastic, but with a dielectric constant near that of alumina, namely, 10. Specialized materials have been developed using alumina spheres as the filler material in a polytetrafluoroethylene (PTFE)/E-glass composite laminate.

Numerous types of base fabrics made from these materials are used, such as woven continuous fiber (Table 2.2), short randomly oriented glass fibers known as glass-matte, electrical-grade papers, and others. The woven continuous-fiber fabric is the most common type used in PWB applications. As Table 2.2 shows, cloths are designated by three- or four-digit "style" numbers indicating weight, thickness, and thread count, with an increase in the number indicating an increase in the fabric thickness. The choice of fabric used in a particular laminate is dependent on electrical requirements, PWB mechanical properties (CTE in particular), dielectric thickness and tolerance, and circuit filling needs during processing.

Some discussion of fabric treatments will be useful for a better understanding of the laminates and laminate problems, as presented in Table 2.24. Many fabrics, especially glass and Kevlar, do not bond well to the resins that are used to impregnate them. Thus to assure a strong laminate, it is necessary to apply to the fabric an adhesion-promoting treatment. The optimum treatment will vary, depending on the fabric-resin combination being used. Inadequate bonding of the fabric-resin interface can result in migration of water, plating solutions, copper etchants, and other liquids into the laminate during PWB fabrication processes.

A second glass treatment of importance should also be mentioned. This treatment, known as a sizing, is applied by glass fabric weavers to the glass during the weaving operations. Sizing is often waxlike and provides lubricity to the fibers during the mechanical weaving operation. This treatment must be removed from the woven glass fabric, otherwise poor adhesion of the resin to the glass fabric will occur. This can result in the same type of problems mentioned above. Removal of this sizing treatment is especially difficult at the fiber intersections in the woven fabric. Failure to remove the sizing agent from these intersections can result in poor adhesion of the resin at the intersections, usually observable as small white crosses, known as measling or crazing, in the final cured laminate.

Resin Systems. A wide range of resin materials is in use in today's organic PWBs, with new formulations being brought to market continually. Chapter 1 gives more in-depth presentation of many of the resin systems discussed in this section. As mentioned previously, most resin systems used in organic PWB laminates are thermosetting plastics, with thermoplastic materials used primarily in microwave applications and in molded PWB applications. Thermoplastics are those plastics that will remelt upon heating to some melting temperature. Thermosetting plastics will not remelt upon heating. Thermosetting plastics will soften upon application of heat, however, as will thermoplastics. Both will soften dramatically at the glass transition temperature of the plastic.

The glass transition temperature is the temperature at which the plastic changes from a rigid or harder material to a softer or glassy type material. It is a definite characteristic of all plastic materials, but is not a property of the resin system where molecular bonds are broken. The glass transition temperature T_g is the point where the physical properties of the resin change due to a weakening of the resin system's molecular bonds. The physical properties (CTE, flexural strength, tensile strength, and so on) undergo significant changes with temperature in the slope of the curve of the property versus the temperature above and below T_g (Fig. 2.6). There is a transition region approximately 25°C below and above T_g where gradual changes in the material properties occur. In general T_g is defined by the intersection of the property curves above and below this transition region. The glass transition temperature is normally determined

TABLE 2.2 PWB Fabric Cloth Makeup

Fabric range	Style: plain weave	Thickness,*† in	Thread count per inch ±3* ($W \times F$)	Weight ±5%,* oz/yd²
G1 (glass)	104	0.0010	60 × 52	0.56
	106	0.0012	56 × 56	0.73
	1070	0.0014	60 × 35	1.05
	107	0.0015	60 × 35	1.05
	1080	0.0020	60 × 47	1.40
	108	0.0022	60 × 47	1.40
G2 (glass)	2112	0.0030	40 × 39	2.05
	112	0.0036	40 × 39	2.10
	2113	0.0029	60 × 56	2.30
	2313	0.0030	60 × 64	2.40
	113	0.0032	60 × 64	2.43
	2125	0.0035	40 × 39	2.60
	1125	0.0039	40 × 39	2.60
	2116	0.0036	60 × 58	3.10
	116	0.0038	60 × 58	3.10
	1675	0.0040	40 × 32	2.90
	2119	0.0034	60 × 46	2.80
	119	0.0038	54 × 50	2.80
	2165	0.0040	60 × 52	3.55
	1165	0.0042	60 × 52	3.55
S-glass	6106	0.0013	56 × 56	0.72
	6080	0.0020	60 × 47	1.45
	6180‡	0.0022	55 × 55	1.20
	6212	0.0032	40 × 39	2.04
	6313	0.0033	60 × 64	2.40
	6120‡	0.0035	60 × 60	3.10
	6216	0.0035	60 × 58	3.09
	6116	0.0036	60 × 58	3.10
	6628	0.0067	44 × 32	6.03
G3 (glass)	7628	0.0067	44 × 32	6.00
	1528	0.0065	42 × 32	5.95
	7637	0.0089	44 × 22	7.00
	7642	0.0099	42 × 20	6.70
	6628	0.0067	44 × 32	6.03
	6628	0.0067	44 × 32	6.03
A (aramid)	120	0.0040	34 × 34	1.70
	108	0.0020	60 × 60	0.80
	177	0.0030	70 × 70	0.93
Q (quartz)	503	0.0050	50 × 40	3.30
	525	0.0030	50 × 50	2.00

*Based on finished-goods states in which heat cleaning and finishing have been applied.

†These values should not be used for computation of dielectric thickness in board design or layout. Tolerance is 20% on fabric ranges G1 and G2 and 10% on all others.

‡4- or 8-harness satin weave.

Source: From MIL-S-13949 and Senese.[3]

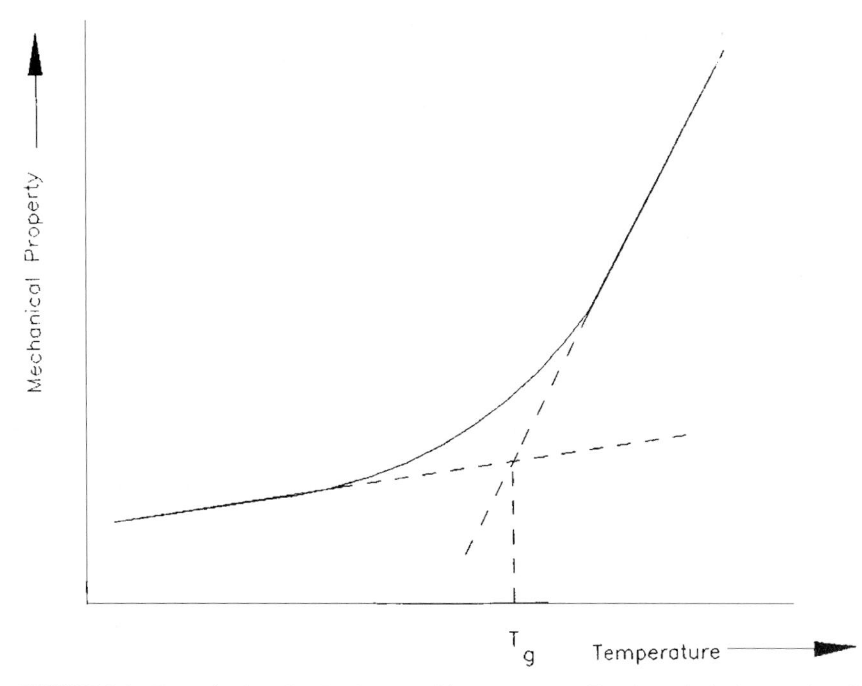

FIGURE 2.6 Determination of resin glass transition temperature T_g, where physical properties of resin exhibit significant change.

by measuring the change in the slope of the volume expansion of the resin using differential scanning calorimetry (DSC) or thermomechanical analysis (TMA).

There are three base resin systems that are most often used in today's laminates: epoxies, polyimides, and polytetrafluoroethylene (PTFE). Recent advanced material laminates are beginning to use cyanate ester resins as a fourth base system.

The most common resin systems in use today are the standard epoxies. These are used in NEMA grade G-10 and FR-4 laminates and have a relatively low T_g (105 to 125°C). The G-10 epoxy is a general-purpose bisphenol A difunctional epoxy, while the FR-4 epoxy is a brominated bisphenol A difunctional epoxy. The bromines make the FR-4 epoxy flame-retardant. These epoxies are easy to process and B-stage and have excellent adhesion to copper at room temperature. The low T_g causes a large percent expansion, which weakens the copper adhesion during soldering temperatures.

Manufacturers perform modifications to the base G-10 and FR-4 epoxies using smaller amounts of many materials, including bismaleimide triazine (BT), polyimide, or tetrafunctional epoxy, to create a wide variety of difunctional epoxy materials. These epoxies are created mainly to increase the base epoxy resin T_g and improve the chemical and thermal stress resistance of the bisphenol A epoxies. The type of resin added and the percentage can cause T_g to vary between 125 and 200°C. Raising the T_g of the epoxy resin usually leads to a resin system that is more brittle than the base epoxy. This can make the new resin more difficult to process and can lead to laminate reliability problems in harsh service environments. A resin based on fluorine amines and epoxides has recently been developed by 3M which has a high T_g and is fracture-resistant (ductile).

Polyimides are the second major type of resin system in use today for organic PWB laminates. Up until the 1980s the polyimide used in PWB applications was Kerimid 601 from a European source. The 601 polyimide results from the polymerization of bismaleimide and methylene dianiline (MDA) with a subsequent high-temperature cross-linking to develop its physical properties. Concerns about carcinogenic effects of MDA have lead to the development of variations of this resin.

Polyimide exhibits many important attributes which make it more attractive for use in laminates. The polyimide resins exhibit T_g values greater than 200°C. Due to this high-temperature heat resistance, drill smear, which may be encountered with epoxy-based laminates during the plated-through-hole drilling operation, is nonexistent. In addition to the higher T_g, they exhibit superior adhesion to copper at soldering temperatures and have a lower CTE than epoxies. They have excellent Z-axis dimensional stability, stable electrical values, and superior chemical resistance when compared to epoxies.

Their disadvantages are that they are quite brittle, and hence more care needs to be followed during their processing. They cost more and have a higher moisture absorption and a lower flammability rating. Modifications to polyimides are usually done with epoxies to improve their processability, reduce laminate moisture absorption, and improve their adhesion characteristics.

Many new polyimide formulations are coming to market today due to their use in multichip modules (MCMs) as a thin-film dielectric. The formulations strive to decrease the material dielectric constant, improve the resin ductility, lower the resin moisture absorption, and improve the processability. See Chap. 11 for more details on this subject.

To improve electrical performance (lower signal propagation delay) in high-speed applications, the laminate dielectric constant must be reduced. Although S2-glass, quartz, and Kevlar fibers do have lower dielectric properties than E-glass (see Table 2.1), the lowering of the resin dielectric constant is the main driver in achieving lower-dielectric-constant laminate materials. In that regard, most low-dielectric-constant laminates on the market today use PTFE as the resin system.

PTFE is an ultrahigh-molecular-weight molecule that consists of symmetrically oriented fluorine atoms around a carbon center, giving it a nonpolar nature. It has strong primary chemical bonds, making it thermally stable. The nonpolar nature and the unique molecular nature of PTFE are responsible for the following:

PTFE resins are nonhydroscopic

Unlike with other plastics, there are no reactive groups in the molecule which would cause further cross-linking to occur during the lamination process.

The glass transition temperature does not strictly apply to PTFE. Its properties are maintained throughout its useful temperature range of well below 0 to 327°C.

These three unique PTFE characteristics lead to PTFE resins exhibiting stable dielectric-constant and dissipation-factor properties under varying frequency, temperature, and humidity conditions.

A number of manufacturers have laminate materials on the market today using PTFE resins. The main drawback to PTFE-based laminates is their low modulus, which is due to the low PTFE modulus and the high lamination processing temperatures required compared to conventional epoxies or polyimides. Applications of PTFE resin system laminates tend to be limited to relatively specialized small microwave substrates.

Cyanate ester resin systems have a dielectric constant of 2.8, approaching that of PTFE (2.2). Cyanate ester systems have been available for the past 15 years and have

TABLE 2.3 Typical Resin System Physical Properties

Type	T_g, °C	Elastic modulus, 10^6 lb/in^2	Poisson's ratio	CTE, ppm/°C	Thermal conductivity, Btu/) (h • ft • °F)	Dielectric constant at 1 MHz
Epoxy	130	0.5	0.35	58	0.14	4.5
Polyimide	260	0.6	0.33	49	0.14	4.3
Cyanate ester	260	0.5	0.35	55	0.14	3.8
PTFE	—	0.05	0.46	99	0.14	2.6

been known as triazine. In the past they have been added as modifiers to epoxy resins. Recent advances by Dow Chemical and others have improved the cycloaliphatic backbone structure, dicyclopentadiene. This has improved the dielectric properties and moisture resistance while maintaining standard processes in both the laminate and the PWB operations. This has lead to the recent development of low-dielectric-constant and low-CTE laminates when the cyanate ester resin system is used with S2-glass or quartz fabrics.

Table 2.3 lists some of the major physical properties for various resin systems. They have been grouped into epoxy, polyimide, PTFE, and cyanate ester. The actual properties vary somewhat about these norms due to the differences in manufacturers' formulations. The user should consult the manufacturer for the exact physical properties of the resin being used.

Metals in Laminates. The third major component of a PWB laminate is the metal used to create the interconnection circuitry. Copper is used almost exclusively. In thin-film MCM and microwave applications, aluminum is sometimes used as the interconnect metal. This section will not discuss aluminum. Certain SMT applications use a clad metal, copper-Invar-copper (CIC) or copper-molybdenum-copper (CMC), as a constraining core. This will be briefly discussed upon completing the discussion of copper.

Copper foils used in PWB applications are covered in IPC-MF-150. Two types of copper foil are used: electrodeposited (E), the most prevalent, or wrought (W). The wrought foils are usually limited to special applications, such as flexible printed circuitry where high ductility is essential. The electrodeposited and wrought types are further subdivided into classes to reflect functional performance and testing properties. Table 2.4 covers the eight IPC class descriptions, Table 2.5 gives the minimum mechanical properties for each class, and Table 2.6 is an applications guide for copper foil usage.

Class 1, 2, 3, and 4 electrodeposited foils are used predominantly in laminates. The class 1 and 2 foils are a more brittle foil and are not generally used in high-performance laminate applications where substantial thermal stress ranges are to be incurred. In class 1 foils, fracture without deformation will occur under relatively low stress levels. This fracturing occurs under a combination of thermal stress imposed on a copper with degraded grain-boundary strength. This degradation is due to such factors as codeposited organics, which can cause easy separation of grain boundaries at elevated temperature. Class 3 and 4 foils are much more ductile at elevated temperature and as such are used in the higher-thermal-stress environments. If one studies Table 2.5, it is apparent that at room temperature the mechanical proper-

TABLE 2.4 Metal Foil Designations

Class (IPC-MF-150)	Designation (MIL-S-13949)	Description
1	C	Copper, standard electrodeposited (STD, type E)
2	G	Copper, high ductility, electrodeposited (HD, type E)
3	H	Copper, high temperature, elongation electrodeposited (HTE, type E)
4	J	Copper, annealed, electrodeposited (ANN, type E)
5	A	Copper, as rolled, wrought (AR, type W)
6	K	Copper, light cold-rolled, wrought (LCR, type W)
7	L	Copper, annealed, wrought (ANN, type W)
8	M	Copper, as rolled, wrought, low-temperature annealable (ARLT, type W)
—	B	Copper, rolled (treated)
—	D	Copper, drum side out (double treated), electrodeposited
—	N	Nickel
—	U	Aluminum
—	Y	Copper-Invar-copper
	O	Unclad

Source: From IPC-MF-150 and MIL-S-13949.

ties of the foils overlap, and it would appear that they are equivalent. At elevated temperature, however, the ductilities of the classes are significantly different. Testing the mechanical properties of foils at elevated temperature to distinguish between classes can be done using a hot rupture test methodology developed by Zakraysek of General Electric.[4]

As mentioned previously, copper foil thicknesses are described in terms of area weight (oz/ft^2). The various common foil thicknesses are listed in Table 2.7. The side of the copper foil which is to be bonded to the base laminate may undergo a treatment to promote better adhesion. The type of treatment is designated by a letter, as shown in Table 2.8.

The copper foil treatment is an adhesion promoter that is selective for the resin being used. It is usually some form of black oxide treatment. Oxidized materials are generally known to provide optimized bond strength for most bonding systems. Failure to achieve an optimum bond of the copper foil results in poor bond strength of the etched copper circuits—a very important factor in PWB performance and reliability. In the resin treatment and lamination operation, great care must be taken to optimize resin bond strength to the copper, to the fabric, and between the thin sheets of resin-soaked fabric which make up the laminate. Failure in any of these can result in end-product problems. Even with the very best adhesion between sheets (known as interlayer bond strength), there is always a weakness between layers. This is clearly indicated by the fact that voltage breakdown occurs 20 to 25 times faster parallel to these layers than perpendicular to them.

SMT applications requiring controlled CTEs have led to the use of clad metals in PWB applications. These metals are either Invar or molybdenum clad on their outer surface with copper. The thickness ratio of the Invar or molybdenum to copper con-

TABLE 2.5 Copper Foil Minimum Mechanical Properties*

Class	Weight,‡ oz	Tensile strength at room temperature (23°C)		Tensile strength at elevated temperature (180°C)	
		10^3 lb/in^2	MPa	10^3 lb/in^2	MPa
		Type E			
1	1/2	30	207	N/A	N/A
	1	40	276	N/A	N/A
	2+	40	276	N/A	N/A
2	1/2	15	103	N/A	N/A
	1	30	207	N/A	N/A
	2+	30	207	N/A	N/A
3	1/2	30	207	15	103
	1	40	276	20	138
	2+	40	276	20	138
4	1	20	138	15	103
	2+	20	138	15	103
		Type W			
5	1/2	50	345	—	—
	1	50	345	20	138
	2+	50	345	40	376
6	1	25–50	177–345	N/A	N/A
	2+	25–50	177–345	N/A	N/A
7	1/2	15	103	—	—
	1	20	138	14	97
	2+	25	172	22	152
8†	1/2	15	103	N/A	N/A
	1	20	138	N/A	N/A
	2	20	138		

*The copper foil shall conform to the tensile requirements when tested in both longitudinal and transverse directions.

†Properties given are following a time-temperature exposure of 15 min at 177°C (350°F).

‡Minimum properties for testing copper weights less than 1/2 oz shall be agreed to between user and vendor.

Source: From IPC-MF-150.

trols the CTE of the clad metal. The ratio of the thickness of the clad metal to the thickness of the laminate dielectric, their respective CTEs, and their elastic moduli determine the CTE of the overall PWB.

Laminate manufacturers have developed CIC and CMC clad dielectrics as cores for PWBs. These cores usually consist of an epoxy/E-glass or polyimide/E-glass thin dielectric (0.004 to 0.006 in thick) sandwiched between thin CIC or CMC layers (usually 0.006 in thick). One or two cores may be used, depending on the application. In addition, the CIC or CMC layers can serve as power or ground planes.

The CIC and CMC foil requirements are currently defined in IPC-CF-152. Further discussion and the material properties of CIC and CMC are contained in Sec. 2.3 and in Table 2.26.

TABLE 2.6 Applications Guide for Copper Foil

				Maximum strain range (%)/minimum bend diameter (mil) accommodated by 1-oz foil[a]					
				Flex to install		Continuous flexing,	Flex to install		Continuous flexing,
				Single bend	Low cycle fatigue[b]	high cycle fatigue[b]	Single bend	Low cycle fatigue[b]	high cycle fatigue[b]
Type	Grade	Class	Handling	At room temperature			At elevated temperature (180°C)[c]		
E	1	All	Good	Not recommended for applications requiring foil flexing or bending					
E	2	All	Good	30/3.3	7.1/38	0.19/1470	Not applicable		
E	3	All	Good	20/5.6	5.3/52	0.18/1560	15/7.9	4.2/63	0.17/1610
E	4	All	Caution[h]	50/1.4	10.3/26	0.21/1340	30/3.3	6.9/39	0.15/1230
W	5[d]	All	Good	30/3.3	7.5/36	0.32/870	15/7.9	4.2/63	0.17/1610
W	5[e]	All	Good	65/0.8	13.1/20	0.32/870	TBD	TBD	TBD
W	6[f]	All							
W	7	All	Caution[h]	65/0.8	12.5/21	0.32/870	45/1.7	9.5/28	0.20/1370
W	8[g]	All	Good	25/4.0	6.2/44	0.15/1890	TBD	TBD	TBD

[a]Larger maximum strain range and smaller minimum bend diameter values indicate superior performance for a given strain mode.

[b]Low cycle fatigue < 500 cycles to failure; high cycle fatigue > 10^4 cycles to failure. The values given here have been calculated for 20 and 10^5 fatigue cycles for low and high cycle fatigue, respectively. For the calculations the minimum mechanical properties for 34-μm [1-oz] copper foil given in the respective slash sheets have been used. Typical property values can be considerably higher. (See IPC-TR-484, "Results of Copper Foil Ductility Round Robin Study" April 1986.)

[c]The values for elevated-temperature applications should primarily be used for qualitative purposes, since unproven assumptions were necessary for their calculation.

[d]W5 foil is highly anisotropic due to the rolling process; in the rolling direction the performance values can surpass those given for W7 foil.

[e]The values given apply in the foil rolling direction only. For cross-machine direction values see previous line.

[f]Choice of temper allows tradeoff in handling/high cycle fatigue versus high strain/low cycle fatigue properties. Grades 5 and 7 are the limiting tempers.

[g]Handling characterization is for as-received foil; maximum strain range/minimum bend diameter values apply to W8 foil after low-temperature annealing at 177°C (350°F) for 15 min.

[h]Handling difficulties normally occur only with thinner foils.

Source: From IPC-MF-150.

TABLE 2.7 Copper Foil Thickness Requirements

Designator	Thickness by weight*		Thickness by gauge (for reference only)		Tolerance	
	oz/ft²	g/m²	in nominal	mm nominal	Type E	Type W
E	0.146 (1/8)	44.57	0.0002	0.005	N/A	N/A
Q	0.263 (1/4)	80.18	0.0004	0.009	N/A	N/A
T	0.350 (3/8)	106.9	0.0005	0.012	N/A	N/A
H	0.500 (1/2)	153	0.0007	0.018	↑	↑
M	0.750 (3/4)	229	0.0010	0.025	+10%	+5%
1	1	305	0.0014	0.035	of	of
2	2	610	0.0028	0.071	thickness	thickness
3	3	915	0.0042	0.106	or weight	or
4	4	1221	0.0056	0.142		weight
5	5	1526	0.0070	0.178	↓	↓
6	6	1831	0.0084	0.213	↓	↓
7	7	2136	0.0096	0.249	↓	↓
10	10	3052	0.0140	0.355	↓	↓
14	14	4272	0.0196	0.496	↓	↓

*Fractions in parentheses represent common industry terminology.
Source: From IPC-MF-150.

TABLE 2.8 Copper Foil Treatment Designations

Designation	Description
N	No treatment, no stain proofing
P	No treatment, with stain proofing
S	Single-sided treatment, with stain proofing
D	Double-sided treatment, with stain proofing

Source: From IPC-MF-150.

Laminate Types and Specifications. As was mentioned, organic PWB laminates are by far the largest group of materials used for PWB applications in the electronics industry. Combining materials from the three major groups discussed—fabric, resin, and metal—leads to a wide variety of available laminates whose properties are tailored to meet specific PWB application requirements. There are strong industry standards for the laminates and the PWBs made using these laminates, as explained hereafter.

The major groups which issue these standards are the National Electrical Manufacturers Association (NEMA), the Department of Defense (DOD) for military (MIL) specifications, and the Institute for Interconnecting and Packaging Electronic Circuits (IPC). Laminates defined by NEMA are covered in document LI-1-89. This document is currently undergoing extensive revision. Laminates meeting DOD and MIL requirements are specified in MIL-P-13949, currently at revision G. This document is currently being revised to MIL-S-13949, soon to be at revision H. It contains specifications for both clad-metal C-stage as well as unclad B-stage laminate materials. The laminate materials meeting IPC specifications are divided across five documents. ANSI/IPC-L-108 covers the requirements for metal-clad thin (>0.001-in) base material laminates for multilayer PWBs. ANSI/IPC-L-109 covers the requirements for B-stage laminate materials used in bonding the layers of high-temperature multilayer PWBs. ANSI/IPC-L-112 covers the requirements for metal-clad laminated polymeric sheets containing two or more types of reinforcing materials for PWBs. In this document, the basic laminate thicknesses are 0.031 in and greater. ANSI/IPC-L-115 defines the laminate materials used for manufacturing single- and double-sided PWBs 0.020 in and greater in thickness. ANSI/IPC-L-125 covers the requirements for high-speed high-frequency performance laminates, both clad and unclad (<0.002 in thick), for use in microwave and other high-speed electrical applications.

Basically, PWB laminators manufacture PWB laminates to NEMA and/or IPC standards. Many of these laminates are qualified to the more rigid DOD requirements for military electronic systems. IPC is an important major industry association whose documents include test standards, workmanship standards, PWB operations standards, and much more. There are, of course, numerous other important industry groups for specific objectives, specific geographical technology transfer, and other purposes. It should be understood that the members of the industry, both producers and suppliers, make up the bodies of both NEMA and IPC and are responsible for inputting the technical data within these material specifications that are issued through these industry groups. The DOD uses the inputs from the industry in updating and modifying MIL-S-13949, but it is the DOD that controls and issues this document.

The major NEMA and military standard grades for PWB laminates are shown in Tables 2.9, 2.10, and 2.11. These standards are based on the type of resin and the type of fabric used in the laminate, and they address copper-clad laminates as well as prepreg.

The NEMA standard grades are used in a variety of commercial applications. FR-2, FR-3, CEM-1, CEM-3, FR-4, and FR-5 are used most widely. The paper-base FR-2 laminates are used in low-cost consumer products such as toys, TV games, and calculators. The FR-3 laminate with its higher electrical and physical properties is used in televisions, computers, and communications equipment. CEM-1 laminate has punching properties similar to FR-2 and FR-3, but with electrical properties approaching FR-4. It is used in industrial electronics, automobiles, and smoke detectors. CEM-3 is higher in cost than CEM-1 and is more suited to plated-through-hole applications. It is used in automobiles and home computers. FR-4 is the most widely used laminate material due to its excellent physical, electrical, and processing properties. It is used in aerospace, computers, automotive, and industrial control applications. FR-5 is used in applications requiring higher heat resistance than attainable with FR-4.

In advanced electronic packaging and interconnecting, many trends require higher-performance PWB laminate properties than are available with the industry standard NEMA laminates just discussed. Many of the newer high-performance laminates are covered in MIL-S-13949, specification sheets 6 to 26. Other laminates are continually being developed and will no doubt be added to these documents in the near future.

Three major advanced PWB laminate requirements predominate:

1. Higher-thermal-stability laminates (higher glass transition temperature in particular) to withstand increasing system thermal densities and to stabilize critical physical properties, particularly thermal expansion (CTE)
2. Lower-thermal-expansion laminates, especially x–y-axis CTE, to eliminate or minimize solder joint cracking in surface-mount packages
3. Lower-dielectric-constant laminates to minimize signal propagation delay and losses in high-speed digital or microwave applications

TABLE 2.9 NEMA Copper-Clad Laminate Requirements*

LI-1-1989 grades	XXXP	XXXPC	FR-1	FR-2	FR-3	FR-4	FR-5	FR-6	CEM-1	CEM-3	G-10	G-11
Fabric description	Paper base	Paper base	Paper base	Paper base	Paper base	E-glass fabric base	E-glass fabric base	E-glass matte	E-glass cloth surf cellulose paper core	E-glass cloth surf nonwoven glass core	E-glass fabric base	E-glass fabric base
Resin description	Phenolic	Phenolic	Phenolic	Phenolic	Epoxy, flame-resistant	Epoxy, flame-resistant	Epoxy, flame-resistant	Polyester, flame-resistant	Epoxy, flame-resistant	Epoxy, flame-resistant	Epoxy, general-purpose	Epoxy, temperature-resistant
Copper peel, min. lb/in 1 oz after solder float	6.00	6.00	6.00	6.00	8.00	8.00	8.00	7.00	7.00	8.00	8.00	8.00
Volume resistivity, min. MΩ • cm	10^4	10^4	10^3	10^4	10^4	10^6	10^6	10^6	10^6	10^6	10^6	10^6
Surface resistivity, min. MΩ	10^3	10^3	10^2	10^3	10^3	10^4	10^4	10^4	10^4	10^4	10^4	10^4
Water absorption, max. %												
0.031 in thick	1.30	1.30	5.60	1.30	1.00	0.50	0.50	—	0.50	0.50	0.50	0.50
0.062 in thick	1.00	0.75	3.60	0.75	0.65	0.25	0.25	0.40	0.30	0.25	0.25	0.25
Dielectric breakdown, min. kV parallel to lam.	15.00	15.00	5.00	15.00	30.00	40.00	40.00	30.00	40.00	40.00	40.00	40.00
Permittivity (dielectric constant) at 1 MHz, max. average	4.80	4.80	6.00	4.80	4.80	5.40	5.40	4.30	5.20	5.40	5.40	5.40
Loss tangent at 1 MHz max. average	0.040	0.040	0.060	0.040	0.040	0.035	0.035	0.030	0.035	0.035	0.035	0.035
Flexural strength, 10^3 lb/in^2												
Lengthwise	12.00	12.00	12.00	12.00	20.00	60.00	60.00	15.00	50.00	50.00	60.00	60.00
Crosswise	10.50	10.50	10.00	10.50	16.00	50.00	50.00	—	40.00	40.00	50.00	50.00
Flammability, min.	N/A	N/A	UL94 V–1	UL94 V–1	UL94 V–1	UL94 V–1	UL94 V–1	— —	UL94 V–0	UL94 V–0	UL94 V–0	UL94 V–0

*All values are for 0.031-in-thick material except for: (1) FR-6 properties are for 0.062 in thick; (2) volume and surface resistivities are for 0.062-in thick; (3) water absorption gives both 0.031- and 0.062-in thicknesses.

Source: From NEMA LI-1-1989.

TABLE 2.10 Copper-Clad Laminate Requirements

MIL-S-13949, sheet #	2	4	5	6	7	8	9
Type designation	GB	GF	GH	GP	GR	GT	GX
Fabric description	Woven E-glass	Woven E-glass	Woven E-glass	Nonwoven E-glass	Nonwoven E-glass	Woven E-glass	Woven E-glass
Resin description	Polyfunct. epoxy	Difunct. epoxy	Polyfunct. epoxy	PTFE	PTFE	PTFE	PTFE
Copper peel, min. lb/in							
1 oz as received	8.0	6.0	6.0	5.0–7.0	5.0–7.0	5.0	5.0
1 oz at elevated temperature	3.0	5.0	5.5	4.0–5.0	4.0–5.0	2.0	2.0
Volume resistivity, min. MΩ • cm	1.00E + 06	1.00E + 06	1.00E + 07	1.00E + 06	1.00E + 06	1.00E + 06	1.00E + 06
	1.00E + 05	1.00E + 03	1.00E + 05	1.00E + 05	1.00E + 05	1.00E + 05	1.00E + 05
Surface resistivity, min. MΩ	1.00E + 04	1.00E + 04	1.00E + 08	1.00E + 04	1.00E + 04	1.00E + 04	1.00E + 04
	1.00E + 03	1.00E + 03	1.00E + 05	1.00E + 03	1.00E + 03	1.00E + 03	1.00E + 03
Water absorption, max. %							
0.031 in thick	0.80	0.80	0.80	0.30	0.10	0.20	0.20
0.062 in thick	0.35	0.35	0.35	0.20	0.10	0.10	0.10
Dielectric breakdown, min. kV parallel to laminate	40	40	40	30	30	20	20
Electric strength, min. v/mil	—	750	—	—	—	—	—
Permittivity (dielectric constant)							
at 1 MHz	5.4	5.4	4.5	2.4	2.20–2.40	2.8	2.40–2.60
at 50 MHz	5.4	—	4.5	—	at X band	—	at X band
Loss tangent							
at 1 MHz	0.03	0.03	0.2	0.001	0.0015 max	0.005	0.0022
at 50 MHz	0.03	—	0.2	—	at X band	—	at X band
Flexural strength, 10^3 lb/in^2							
Lengthwise	60	60	60	8	3.5	12	12
Crosswise	50	50	50	6	3.5	10	10
Arc resistance, min. seconds	60	60	90	120	180	180	180
Flammability, max. burn time, seconds	N/A	15	15	N/A	N/A	N/A	N/A

*Sheets 1 and 3 were canceled June 1988.

Source: From MIL-S-13949.

TABLE 2.10 Copper-Clad Laminate Requirements (*Continued*)

10	14	15	17	19	22	24	25	27	29
GI	GY	AF	AI	QI	BF	GM	CF	SC	GC
Woven E-glass	Woven E-glass	Woven aramid	Woven aramid	Woven quartz	Nonwoven aramid	Woven E-glass	Nonwoven poly-glass	Woven S2-glass	Woven E-glass
Polyimide	PTFE	Polyfunct. epoxy	Polyimide	Polyimide	Epoxy	Triazene or BT epoxy	Polyfunct. epoxy	Cyanate ester	Cyanate ester
5.0	5.0	5.0	5.0	5.0	3.0	6.0	10.0	4.5	4.5
5.0	2.0	4.0	4.0	5.0	2.5	3.0	1.0	4.5	4.5
6.00E + 04	1.00E + 06	1.00E + 06	6.00E + 04	6.00E + 04	1.00E + 06	1.00E + 06	1.00E + 06	1.00E + 06	1.00E + 06
6.00E + 04	1.00E + 05	1.00E + 03	1.00E + 03	6.00E + 04	1.00E + 03	1.00E + 05	1.00E + 03	1.00E + 04	1.00E + 04
1.00E + 04	1.00E + 04	1.00E + 04	1.00E + 04	1.00E + 04	1.00E + 04	1.00E + 06	1.00E + 05	1.00E + 06	1.00E + 06
6.00E + 04	1.00E + 03	1.00E + 03	1.00E + 03	6.00E + 04	1.00E + 03	1.00E + 05	1.00E + 03	1.00E + 04	1.00E + 04
1.00	0.20	2.00	2.00	1.00	2.00	0.80	—	1.00	1.00
0.50	0.10	2.00	2.00	0.50	2.00	0.35	—	1.00	1.00
40	20	40	40	40	40	40	40	40	40
750	—	750	750	750	1500	750	750	750	750
5.4 —	2.15–2.40 at X band	4.5	4.2	3.4	4.0 max.	4.8	5.4	4.3	4.5
0.025 —	.0015 at X band	0.035	0.025	0.01	—	0.02	0.03	0.015	0.015
60	6	50	50	60	50	60	6	50	50
45	5	40	40	45	40	50	5.5	40	40
120	120	60	60	120	60	60	60	120	120
—	N/A	15	15	—	15	15	15	25–30	25–30

TABLE 2.11 Prepreg Requirements

MIL-S-13949, sheet #	11	12	13	16	18	20	21	23	26	28	30
Type designation	GE	GF	GI	AF	GH	QI	AI	BF	GM	SC	GC
Fabric description	Woven E-glass	Woven E-glass	Woven E-glass	Woven aramid	Woven E-glass	Woven quartz	Woven aramid	Nonwoven aramid	Woven E-glass	Woven S2-glass	Woven E-glass
Resin description	Difunct. epoxy	Difunct. epoxy	Polyimide	Polyfunct. epoxy	Polyfunct. epoxy	Polyimide	Polyimide	Epoxy	Triazene or BT epoxy	Cyanate ester	Cyanate ester
Volatiles content, %	0.75	0.75	2	1.5	0.75	4	1.5	1.5	2	2	2
Electric strength, min. V/mil	750	750	750	750	750	750	750	1500	750	750	750
Permittivity at 1 MHz	5.4	5.4	5.4	4.5	5.4	3.6	4.2	4	4.3	4.3	4.5
Loss tangent at 1 MHz	0.035	0.035	0.025	0.035	0.035	0.01	0.035	0.025	0.02	0.015	0.015
Flammability, seconds	—	15	—	15	15	—	15	15	15	15	15

Source: From MIL-S-13949.

For varying electronic system requirements, any one or any combination of these requirements might exist. Most of the advanced PWB laminates have been developed to achieve the higher glass transition requirement, many have been developed to meet the lower *x–y*-axis CTE requirement, and a smaller number have been developed to meet the lower-dielectric-constant requirement.

Basically, in the anatomy of a PWB laminate, the glass transition requirement is largely controlled by the resin component of the laminate, the *x–y*-axis (in-plane) CTE is largely controlled by the fabric constituent of the laminate, the *z*-axis (out-of-plane) CTE is controlled by both the fabric and the resin, and the dielectric constant is largely controlled by the resin component, although selection of the fabric can have a substantial influence on the dielectric constant.

To achieve higher glass transition temperature PWB laminates, several major higher-temperature resin systems have been developed. The resin systems are discussed on page 2.15. Laminates made using these resin systems include NEMA grades G-11 and FR-5 and MIL-S-13949 grades GI, GH, GM, AI, QI, and CF.

To achieve lower *x–y*-axis PWB CTE, two fibers have predominated, mainly, aramid fiber (such as duPont's Kevlar) and quartz fiber. These fibers were discussed on page 2.13. Laminates using these fibers have been made and qualified to military grades AE, AI, and QI (see Table 2.10).

Finished PWBs using Kevlar-based laminates have two problems not associated with glass-based laminates. First, a phenomenon known as resin microcracking will occur during subsequent environmental thermal cycling. Microcracking consists of small cracks in the resin, usually at the PWB surface and sometimes within the finished laminate cross-section. These cracks are caused by high stresses built up in the resin due to the high radial CTE of the Kevlar fabric. They occur at the crossover points of the fabric. The severity of microcracking is a function of the resin system used. Microcracking is most severe in brittle polyimide systems. Modified polyimide and epoxy systems can limit its occurrence somewhat. Microcracking can cause failure in copper traces, especially when low-ductility (class 1 and 2) copper in combination with fine-line (<0.008 in wide) copper traces are used in a design. Figure 2.7 shows typical surface resin microcracks.

The second problem that may develop in PWBs manufactured with laminates using Kevlar fabrics is premature failure in the barrel of copper plated-through holes during subsequent environmental thermal cycling. This is once again due to the high radial CTE of the Kevlar fabric, which imposes a high *z*-axis (out-of-plane) CTE in the PWB. It should be pointed out that this phenomenon is not limited to Kevlar applications, but will occur whenever a material is used to constrain the *x–y*-axis (in-plane) PWB CTE. This usually will cause an increase in the *z*-axis CTE. The onset of cracking is determined by the thermal cycling temperature extremes, the copper ductility in the barrel of the hole, the aspect ratio of the plated-through hole (the aspect ratio is the ratio of the board thickness to the finished-hole diameter), and the copper plating thickness. Barrel cracking can be controlled through the use of ductile copper plating (class 3), limiting the plated-through-hole aspect ratio (<3.5:1), increasing the copper thickness in the hole barrel from 1 mil minimum to <1.5 mil minimum, or applying an electroless or electrolytic nickel plate of 0.3 to 0.5 mil over the copper in the barrel of the hole.

Microwave applications in aerospace and telecommunications equipment would require the use of NEMA type GT and GX or military grades GP, GR, GT, GX, or GY materials. Most of these laminates consist of PTFE/E-glass laminate composites.

The type GP and GR laminates are produced from a mixture of randomly oriented E-glass fibers and PTFE resin yielding a low-loss printed-circuit material with many unique properties. The fibers used are either a nominal 0.5-μm-diameter E-glass

FIGURE 2.7 Typical surface resin microcracking in PWB manufactured from Kevlar-based laminate.

microfiber in short lengths or a long-length E-glass fiber. These materials resist water and chemical absorption, resulting in excellent maintenance of their electrical properties. Holes may be easily machined or punched due to the very fine, randomly dispersed E-glass fibers. The product is very flexible and can be bent and used in conformable PWB or antenna configurations. It maintains excellent dimensional stability and physical strength. Applications for this product are in high-frequency analog and digital communications systems, missile and radar components, electronic warfare and countermeasures, and satellite communication equipment.

Type GT, GX, and GY laminates are produced from a mixture of woven-fabric E-glass and PTFE resin. They are manufactured to closely controlled thickness tolerances and precise dielectric constant values. MIL-S-13949 designates that type GT materials have dielectric constants <2.8 at 1 MHz, type GX materials have dielectric constants from 2.4 to 2.6 at x band, and type GY materials have dielectric constants from 2.15 to 2.40 at x band. Applications for these products exist in many areas of microwave technology, including ILS and MLS radar systems, phased array antennas, satellite communication equipment, TVRO and DBS low-loss circuits for LNA, LNB, and BDC circuits, cellular communication systems, microwave transmission devices, and a variety of RF componentry, including power dividers, mixers, amplifiers, couplers, and antennas.

Specialized woven E-glass PTFE laminate materials filled with an alumina particle to produce a high-dielectric-constant (10) laminate as a substitute for alumina ceramic substrates are also manufactured. They are used in applications which may see high vibration or stress and are more rugged than conventional alumina ceramic substrates.

Electrical Properties of PWB Laminates. There are several laminate electrical properties the designer must be concerned with. These are permittivity (dielectric constant) and dissipation factor, current-carrying capacity, dielectric strength and breakdown, and insulation resistance.

1. *Permittivity and Dissipation Factor.* These two material properties are commonly grouped together when discussing laminate material properties. High-frequency electrical applications such as microwave, computer, and radar signal processing applications often require PWB laminates or substrates with a lower permittivity or dissipation factor than found in standard FR-4 laminates. These are often referred to as low-loss laminates.

The permittivity is more commonly known as the dielectric constant. The dielectric constant of a material is a dimensionless value which is defined as the ratio of the capacitance of a capacitor with a given dielectric to the capacitance of the same capacitor with air as the dielectric. It is the electrostatic energy storage capability of an insulating material. The dielectric constant of air is 1.

A given laminate's dielectric constant is determined by the dielectric constants of the constitutive fabric and resin materials and the laminate fabric-to-resin ratio. To illustrate this, Fig. 2.8 presents data on the variation of the dielectric constant for a cyanate ester resin system as a function of fabric material and percent resin content. It will also vary with frequency, temperature, and humidity. The dielectric constant is usually measured at 1 MHz using either ASTM-D-150 or IPC-TM-650, method 2.5.5.3.

The dielectric constant plays a major role in high-frequency digital and microwave applications in determining the electrical propagation delay as well as the impedance and capacitance associated with the PWB. For high-frequency applications the designer prefers a low-dielectric-constant laminate to decrease the signal propagation delay.

The dissipation factor (sometimes referred to as loss tangent) is the second dimensionless property. It is defined as the ratio of the total power loss in a dielectric mater-

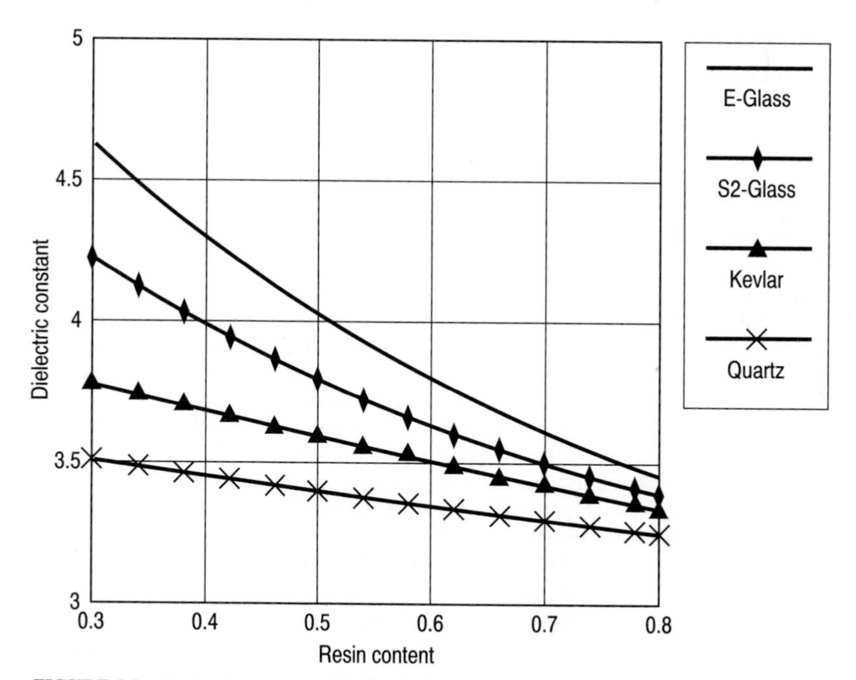

FIGURE 2.8 Dielectric constant of cyanate ester resin system.

TABLE 2.12 Dielectric Constants and Dissipation Factors of Typical Laminates

Materials	Permittivity at 1 MHz, condition D 24/23	Dissipation factor at 1 MHz	
		Condition A	Condition D 24/23
Ordinary applications			
XXXP	—	0.028	0.03
XXXPC	4.1	0.028	0.03
FR-2	4.5	0.024	0.026
FR-3	4.3	0.024	0.026
CEM-1	4.4	0.027	0.028
CEM-3	4.6	0.020	0.022
FR-6	4.1	0.020	0.028
G-10	4.6	0.018	0.019
FR-4	4.6	0.018	0.020
G-11	4.5	0.019	0.020
FR-5	4.3	0.019	0.028
GI	4.8	0.020	0.030
High-frequency applications			
GT	2.8	0.005	0.006
GX	2.8	0.002	0.002
Polystyrene	2.5	0.00012–0.00025*	0.00012–0.00066*
Cross-linked polystyrene	2.6	0.0004–0.0005†	0.0004–0.0005

*At 10 MHz.

†At 10 GHz.

Source: Fιom Coombs.[5] Reprinted with permission.

ial to the product of the voltage and current in a capacitor in which the material is a dielectric. It varies as a function of frequency. It is usually measured at 1 MHz per ASTM D-150 or IPC-TM-650, method 2.5.5.3.

The major NEMA and military grades for low-loss materials used in microwave applications are GP, GR, GY, GT, and GX. Other often used ungraded low-loss laminates include polystyrene and cross-linked polystyrene, polyethylene and polypropylene, polyphenylene oxide, polysulfones, and other low-loss high-thermal-performance thermoplastics. Table 2.12 lists typical values for the permittivity and the dissipation factor of some important laminates and substrates. Chapter 1 provides further explanations of these and other electrical properties of plastic and insulating materials.

2. *Current-Carrying Capacity.* The current-carrying capacity of a copper conductor (in amperes) in a PWB is limited by either the allowable temperature rise in the conductor or allowable voltage drop. This is a function of copper thickness, trace width, environmental conditions, conductor density, laminate glass transition temperature, current applied, and whether the trace is within the PWB or on the PWB external surface. Figure 2.9 can be used to estimate temperature increases above ambient versus current. The curves include a 10 percent derating factor on a current basis. It is also best to further derate the values another 15 percent when the PWB is less than 0.032 in thick or the conductors are of greater than 2-oz thickness.

3. *Dielectric Strength and Breakdown.* The dielectric strength is the ability of an insulating material to resist the passage of a disruptive electric discharge produced by an electric stress. It is given in volts per mil and is measured through the thickness of a laminate per ASTM D-149 or IPC-TM-650, method 2.5.1.

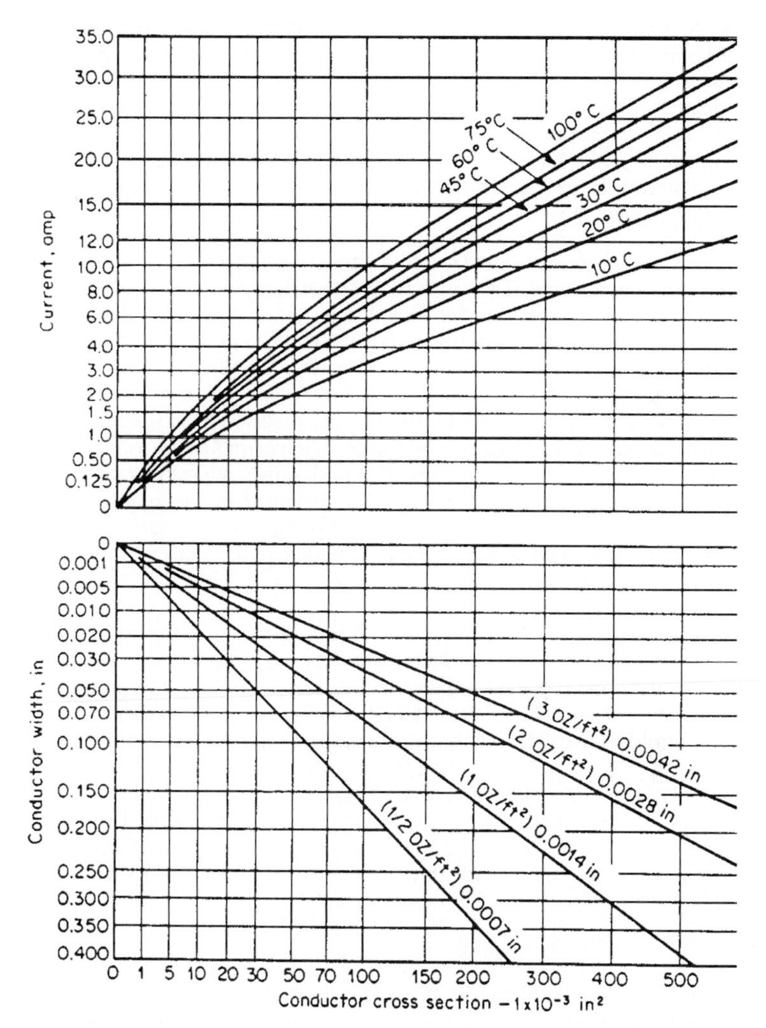

FIGURE 2.9 Conductor cross section and current-carrying capacity. (*From MIL-STD-275.*)

Dielectric breakdown is the value of the disruptive charge (in kilovolts) measured between two Pratt and Whitney no. 3 taper pins inserted in the laminate on 1-in centers perpendicular to the lamination. It is also measured using ASTM D-149 or IPC-TM-650, method 2.5.6.

Fundamentally, high voltages can cause severe damage to laminates by voltage breakdown (which is especially rapid for voltage applied in the *x–y*-axis, parallel to the laminations), high-voltage puncture, corona, and carbon tracking. Most laminates are highly subject to all of these high-voltage failure modes, but there are numerous ways to minimize all of these destructive problems.

4. *Insulation Resistance.* The insulation resistance between two holes or conductors in a laminate or PWB are composed of two components: surface resistivity and volume resistivity. The surface resistivity (usually given in megohms) is the ratio of the dc potential applied between two conductive points on an insulating material surface to the total current between the points. The volume resistivity (usually given in ohm-centimeters) is the ratio of the dc potential applied to electrodes embedded in a dielectric material to the current between them. They may be determined by using IPC-TM-650, method 2.5.17.1.

These properties are usually measured at specific temperature-humidity conditions to evaluate their degradation. The tests are of greatest importance when the test condition matches the end-use environmental conditions.

The volume resistivity is very high, above 10^{15} $\Omega \bullet$ cm at room temperature and low humidity for most laminate materials. However, the resistivity values drop very rapidly as temperature or humidity, or both, are increased. It is not uncommon for volume resistivity values to decrease to 10^6 $\Omega \bullet$ cm or lower in difficult temperature-humidity environments. For high performance it is desirable that volume resistivity values remain above 10^9 or 10^{10} $\Omega \bullet$ cm. Reduced values will recover when the temperature or humidity condition is removed. Improvement in retaining high resistivity in humid environments is achieved by conformal coatings or low-surface-energy treatments being applied to the completed PWB assembly. This is discussed in Sec. 2.3.5, which covers PWB coatings.

Physical and Mechanical Properties of PWB Laminates. Certain physical and mechanical properties of PWB laminates are very important in the design of PWB assemblies. Some important data on these factors are presented at this point.

Laminate Thickness. Laminate material thicknesses and tolerances are set per IPC-L-108 and MIL-S-13949 for thin clad laminates used in multilayer applications and in IPC-L-115 for single- and double-sided applications. Thickness is measured after removal of the copper cladding using a mechanical micrometer or by microsection using the closest point between metal claddings for the value. Data from these publications are given in Tables 2.13 and 2.14. The military specification does not allow double-sided laminates with thicknesses of less than 0.0035 in between the clad metal.

TABLE 2.13 Laminate Thicknesses and Tolerances, in mil

Base thickness	Class 1*	Class 2*	Grade A†	Grade B†	Grade C†
1– 4.5	±1	±0.75	—	—	—
4.6– 6.0	±1.5	±1	—	—	—
6.1– 12.0	±2	±1.5	—	—	—
12.1– 19.9	±2.5	±2	—	—	—
20.0– 30.9	—	—	±3	±2.5	±1.5
31.0– 40.9	—	—	±6.5	±4	±2
41.0– 65.9	—	—	±7.5	±5	±2
66.0–100.9	—	—	±9.0	±7	±3
101.0–140.9	—	—	±12.0	±9	±3.5
141.0–250.0	—	—	±22.0	±12	±4

*From IPC-L-108.

†From IPC-L-115.

TABLE 2.14 Laminate Thicknesses and Tolerances,* in mil

Nominal thickness of base laminate without cladding	Class 1 PX, paper base only	Class 1 reinforced	Class 2 reinforced	Class 3† reinforced	Class 4 for microwave application GR, GX, and GY	Class 5† reinforced	
						–	+
1.0– 4.5	—	± 1.0	± 0.7	±0.5	—	0.5	1.0
4.6– 6.5	—	± 1.5	± 1.0	±0.7	—	0.7	1.2
6.6– 12.0	—	± 2.0	± 1.5	±1.0	±0.75‡	1.0	1.5
12.1– 19.9	—	± 2.5	± 2.0	±1.5	±1.0	1.5	2.0
20.0– 30.9	—	± 3.0	± 2.5	±2.0	±1.5	2.0	2.5
31.0– 40.9	± 4.5	± 6.5	± 4.0	±3.0	±2.0	3.0	3.5
41.0– 65.9	± 6.0	± 7.5	± 5.0	±3.0	±2.0	3.0	3.5
66.0–100.9	± 7.5	± 9.0	± 7.0	±4.0	±3.0	4.0	4.5
101.0–140.9	± 9.0	±12.0	± 9.0	±5.0	±3.5	5.0	5.5
141.0–250.0	±12.0	±22.0	±12.0	±6.0	±4.0	6.0	6.5

*Tolerance value is determined by nominal base thickness (less cladding). Tolerance is applied over base plus cladding, with no additional tolerance for cladding thickness allowed. Tolerance of class 5 materials is applied to base thickness.

†These tighter tolerances are available only through product selection on most material types.

‡For some base materials, materials below a certain base thickness are not covered by this specification. For example, types GT, GX, and GY are not covered for less than 10-mil core thickness, and core thicknesses below 3.5 mil are not currently covered for any double-sided laminate.

Source: From MIL-S-13949.

TABLE 2.15 Laminate Bow and Twist, Maximum Total Variation (on basis of 36-in dimension),* in percent

	Class A			Class B		
Thickness,† in	All types, all weights, metal (one side)	All types, all weights, metal (two sides) Glass	All types, all weights, metal (two sides) Paper	All types, all weights, metal (one side)	All types, all weights, metal (two sides) Glass	All types, all weights, metal (two sides) Paper
0.20 and over	—	5	—	—	2	—
0.030 or 0.031	12	5	6	10	2	5
0.060 or 0.062	10	5	6	5	1	2.5
0.090 or 0.093	8	3	3	5	1	2.5
0.120 or 0.125	8	3	3	5	1	2.5
0.240 or 0.250	5	1.5	1.5	5	1	1.5

*Values apply only to sheet sizes as manufactured and to cut pieces having either dimension not less than 18 in.

†For nominal thicknesses not shown in this table, the bow or twist for the next lower thickness shown shall apply.

Source: From MIL-S-13949.

Bow and Twist. Bow and twist of laminate panels is a measure of the deviation of the planarity of the laminate or PWB panel expressed in inches per inch. Acceptable limits of bow and twist per MIL-S-13949 are given in Table 2.15. Laminate materials specified in IPC-L-115 typically have values of bow and twist between 0.010 and 0.025 in/in. These values differ by material, panel size, and thickness. The reader should consult the applicable IPC material specification sheet. Panel bow and twist can be measured per IPC-TM-650, method 2.4.22.

Flexural Strength. The laminate flexural strength is a measure (in pounds per square inch) of the load that a beam will stand without fracturing when loaded in the center and supported on the ends. The test methods are covered in ASTM D-790 and IPC-TM-650, method 2.4.4. Flexural strengths of typical PWB materials are given in Table 2.16.

Coefficient of Thermal Expansion. The CTE of a material is the change in linear length per unit length per degree change in temperature. Both x–y-axis (in-plane) and z-axis (out-of-plane) CTEs are important parameters. CTE test methods are covered in ASTM D-696 as well as in IPC-TM-650, method 2.4.41 or 2.4.41.1.

Laminate CTEs vary due to the CTEs of the constitutive fiber and resin materials, the fiber weave and orientation, and the ratio of fiber to resin in the laminate. In a finished PWB, the percent copper loading in a multilayer PWB and the mounting method of the PWB to any carrier plates or heat sinks will affect the CTE of the overall laminate. CTEs of some typical laminate materials are given in Table 2.17.

Regarding the CTEs listed in Table 2.17, most PWB laminates have CTE values of 7 to 25 $\times 10^{-6}$ cm/(cm • °C) in the x and y axes (parallel to laminations); z-axis CTE values are much higher, often 40 to 110 $\times 10^{-6}$ cm/(cm • °C). The reason for this great difference is that the x–y-axis expansion is limited by the low CTE values of the fabric in the laminate (see Table 2.1), while the z-axis expansion is largely controlled by the resin component in the laminate (except when Kevlar fabric is used) (see Table 2.3).

Thermal Conductivity. Regarding the thermal conductivity of PWB laminates,

TABLE 2.16 Typical Laminate Flexural Strengths, Minimum Average, in lb/in^2

Material	Lengthwise	Crosswise
XXXPC	12,000	10,500
FR-2	12,000	10,500
FR-3	20,000	16,000
FR-4	60,000	50,000
FR-5	60,000	50,000
FR-6	15,000	15,000
G-10	60,000	50,000
G-11	60,000	50,000
CEM-1	35,000	28,000
CEM-3	40,000	32,000
GT	15,000	10,000
GX	15,000	10,000
GI	50,000	40,000

Source: From Coombs.[5] Reprinted with permission.

TABLE 2.17 Typical Laminate CTEs, in ppm/°C

Material	*x, y* axes	*z* axis
Polyimide E-glass	15–18	45–60
Epoxy E-glass	15–18	45–60
Modified epoxy/aramid	6.5–7.5	95–110
Modified epoxy/quartz	11.0–14.0	55–65
FR-2	12–25	—
FR-3	12–25	—
XXP	12–17	—
Polystyrene	70	—
Cross-linked polystyrene	57	—

Source: From Harper.[1]

these values are all very low [about 0.14 Btu/(h • ft^2 • °F)], as is the case for nearly all electrically insulating materials. Thus heat cannot be removed efficiently from PWB assemblies through the laminate materials. Other thermal management methods must be used. The reader is referred to Harper[1] for further information on this subject.

Flammability. With respect to flammability, the ratings of Underwriters Laboratories are most commonly used. These ratings for various laminates are shown in Table 2.18. The requirements for the various UL ratings are as follows:

94V–0. Specimens must extinguish within 10 s after each flame application and a total combustion of less than 50 s after 10 flame applications. No samples are to drip flaming particles or have glowing combustion lasting beyond 30 s after the second flame test.

94V–1. Specimens must extinguish within 30 s after each flame application and a total combustion of less than 250 s after 10 flame applications. No samples are to drip flaming particles or have glowing combustion lasting beyond 60 s after the second flame test.

94V–2. Specimens must extinguish within 30 s after each flame application and a total combustion of less than 250 s after 10 flame applications. Samples may drip flame particles, burning briefly, and no specimen will have glowing combustion beyond 60 s after the second flame test.

94HB. Specimens are to be horizontal and have a burning rate less than 1.5 in/min over a 3.0-in span. Samples must cease to burn before the flame reaches the 4-in mark.

TABLE 2.18 UL Flammability Ratings for Typical Laminates

Grade	UL classification	Grade	UL classification
XXXPC	94HB	G-10	94HB
FR-2	94V–1	FR-4	94V–0
FR-3	94V–0	G-11	94HB
CEM-1	94V–0	FR-5	94V–0
CEM-3	94V–0	FR-6	94V–0

Source: From Coombs.[5] Reprinted with permission.

Copper Peel Strength. The peel strength of the copper circuitry is very important in PWB assemblies. The basic peel strength test pattern is shown in Fig. 2.10. The wide end of each copper strip is peeled back from the edge of the specimen and gripped in a machine in which the force required to pull the copper away from the laminate may be measured. The foil is pulled perpendicular to the specimen surface at a rate of 2 in/s minimum and the force recorded. The force is converted to pounds per inch width of peel. Initially, peel strengths of most laminates are very good, with peel values of 5 to 12 lb/in of conductor width, as measured using test patterns of NEMA, IPC, or MIL-S-13949 standards. Additional peel tests are usually done on laminate copper traces to simulate peel during soldering, after soldering, and after plating. It is not uncommon for the initial peel values to be reduced by 50 percent after or during these operations.

Copper Surface. Copper surface standards in laminates are controlled by IPC-L-108 and MIL-P-13949 for thin clad laminates used in multilayer applications and IPC-L-115 for single- and double-sided applications. Foil indentations, wrinkles, and scratches are the three criteria delineated in these documents. Foil indentations are specified on the basis of a point count per square foot of surface area. Wrinkles visible to the naked eye are not allowed. Scratches longer than 4 in and deeper than 5 percent of the nominal foil thickness are not allowed.

Thermal Stress. Laminates are subjected to an elevated thermal stress test to ensure good adhesion of copper and fabric to the resin as well as to ensure that an excessive level of volatiles is not trapped in the laminate. The test involves floating a specimen (usually 2 by 2 in) in a solder pot for 10 s. IPC-TM-650, method 2.6.8.1, is used. Any signs of surface blistering, interlayer delamination, blistering, or crazing is cause for rejection.

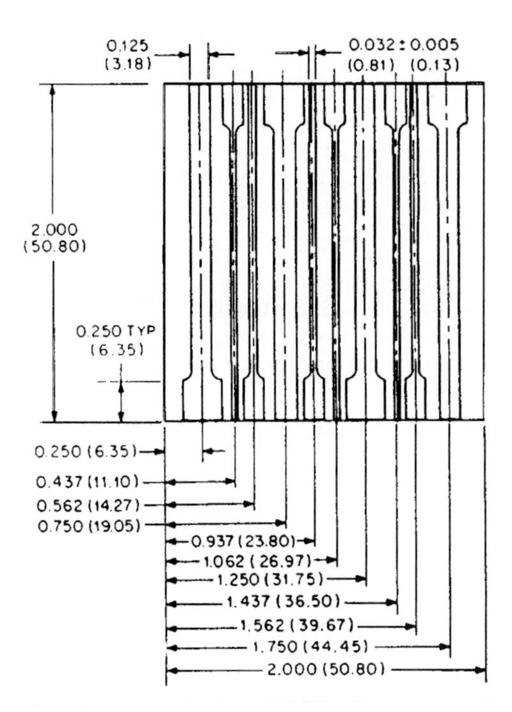

FIGURE 2.10 Typical peel strength test pattern.

Water Absorption. Water absorption is defined as the ratio of the weight of water absorbed by the laminate material to the weight of the dry material expressed as a percent. IPC-TM-650, method 2.6.2.1, can be used to measure this property. Cross-linked polystyrenes and GX and GT materials will have values of 0.01 to 0.2 percent, paper-based XXXP and XXPC products range from 0.65 to 1.3 percent, FR-3, FR-4, and G-10 epoxy laminates range from 0.2 to 0.5 percent, and GI materials range from 1 to 1.5 percent.

Chemical Resistance. The processing to which a laminate material is subjected during PWB fabrication incorporates a wide variety of solvents and aqueous solutions. These solutions are absorbed to some degree by the laminate. This absorption can damage the laminate in a number of ways. The copper bond strength to the resin may be degraded, and wicking of the solution along the fiber bundle can lead to material discoloration and a degradation of dielectric strength and volumetric resistance. IPC-TM-650, method 2.3.4.2, is used to test a laminate's resistance to various chemicals used in PWB processing.

Fungus Resistance. Regarding fungus resistance, epoxy laminates are generally considered fungus-resistant, while phenolic laminates are not.

Dimensional Stability. Dimensional stability is the resistance of a laminate to planar dimensional changes through processing. The changes may be either expansion or shrinkage. It is one of the critical material parameters in multilayer PWB processing. Its value is given in inches per inch of laminate length, and it is measured using IPC-TM-650, method 2.4.39.

Laminate properties which determine the dimensional stability are type of resin, resin content, fabric construction including fabric type, weave, and thickness, and the lamination cycle used in fabricating the laminate. Improvements in dimensional stability occur by lowering resin contents, lowering lamination pressures, and using heavier-fabric materials.

Laminate Problems. The complexity of the overall PWB life cycle is such that complete freedom from laminate problems is usually unavoidable. Problems can occur in laminate manufacture or in PWB manufacturing processes. Certain types of problems are common, and some of these problems and the corrective actions are listed in Tables 2.19 to 2.23.

TABLE 2.19 Measling and Blistering

Cause	Corrective action
Entrapped moisture	Check drill hole quality; check for delamination during drilling
Excessive exposure to heat during fusing and hot-air leveling	Monitor equipment for proper voltage regulation and temperature
Laminate weave exposure	Ensure that there is sufficient butter coat; that all wet processes are checked when materials are changed
Measles related to stress on boards with heavy ground planes	Postbake panels prior to drilling; ensure that laminate has balanced construction; use optimum preheat conditioning before IR fusing or hot-air leveling
Handling when laminate temperature exceeds its glass transition temperature	Allow boards to cool to ambient temperature before handling
Measling when large components or terminals are tight enough to cause excessive stress when heated	Check tooling for undersized holes; loosen tight terminals

Source: From Coombs.[5] Reprinted with permission.

TABLE 2.20 Dimensional Stability

Cause	Corrective action
Undercured laminate	Check laminate vendor for glass transition temperature
Distorted glass fabric (FR-4)	Examine etched panels for yarns parallel to warp and fill direction
Dense hole patterns with fine lines and spaces	Prebake laminate in panel form before drilling
Dimensional change parallel to grain differs from that of cross grain	Have laminator identify grain direction.

Source: From Coombs.[5] Reprinted with permission.

TABLE 2.21 Warp and Twist

Cause	Corrective action
Improper packing	Ensure that packing skids are flat and have sufficient support
Storage	Stack material horizontally rather than vertically
Distorted glass fabric (FR-4)	Examine etched panels for yarns parallel to warp and fill direction
Excessive exposure to heat	Maintain proper temperature and exposure times in all heat-related fabricating processes (IR fusing, hot-air leveling, and solder mask applications)
Improper handling after exposure to heat applications	Cure solder mask horizontally; use proper cooling techniques after IR fusing and hot-air leveling
Unbalanced laminate construction	Work with laminator to get balanced construction

Source: From Coombs.[5] Reprinted with permission.

TABLE 2.22 Copper Foil

Cause	Corrective action
Fingerprints due to improper handling	Handle copper cladding with gloves
Oils from punching, blanking, or drilling	Degrease with proper solvent
Poor solderability on print-and-etch boards	Degrease to remove contaminants; use highly activated flux; check laminator for procedure to remove excessive antistaining compound

Source: From Coombs.[5] Reprinted with permission.

TABLE 2.23 Copper Bond Strength

Cause	Corrective action
Pad or trace lifting during wave soldering	Check for undercut to overexposure during etch; check with laminator that solvents used in circuit manufacturing process are compatible with laminate
Pad or trace lifting during hand soldering	Hand-soldering device is too hot or wattage is too high for application; device operator is applying heat too long to soldered pad
Pad or trace lifting during processing	Check laminate supplier for bond strength properties of lot of material in process; check for undercut during etch; if pads lift after solder leveling, check fuser voltage or hot-air leveling equipment temperature

Source: From Coombs.[5] Reprinted with permission.

TABLE 2.24 Laminate Problems—Causes and Corrective Actions

Problems and causes	Corrective action
Resin starvation	Burn laminate to determine glass content; control laminating conditions
Voids in laminate from volatile components	Expose prepreg to vacuum; mold laminate in vacuum press
Surface dents	Clean laminating fixtures; do lay-up in controlled atmosphere
Voids in thick laminates	Control pressures and temperatures; use thermocouple in lay-up
Movement of multiple laminates in one pressing	Ensure that each lay-up is of the same thickness
Voids from insufficient prepreg plies	Use at least two plies of 108 glass and add up to four if the inner layer is 2-oz copper
Warp and twist	Ground and supply planes must be symmetrical; check warp direction in prepreg

Troubleshooting Material Problems in PWB Laminates. In the manufacture of the base laminate, and in the processing of the laminate into a finished PWB assembly, there are certain common problems which often occur. A listing of an important group of these problems, their probable sources, and the possible corrective actions is given in Table 2.24.

Structural Core Materials in PWAs. In many PWA designs, in particular SMT applications, the finished PWB is mounted to another material, usually for one of three reasons:

1. The second material functions as a structural stiffener to provide sufficient rigidity to the PWB in order to survive environmental shock and vibration inputs.
2. The material is used to reduce the CTE of the PWB in order to match the CTE of components such as ceramic leadless chip carriers. This in turn reduces the stresses imposed on solder joints during any environmental thermal cycling the PWA may experience during its functioning lifetime.
3. The material is used to conduct heat that is being dissipated by the electrical components away from the PWA and into a heat exchanger in order to provide component reliability.

In many instances, particularly in aerospace and military applications, this material performs all three functions. The following discusses the variety of material options available to the PWA designer today, compares their material properties, provides manufacturing insights, and presents lists of applications which may require a particular material use.

Table 2.25 presents basic material properties of the constitutive materials, which either by themselves or in combination constitute the PWA cores in use in today's electronics. Table 2.26 lists the properties of the core materials that are combinations of the basis materials given in Table 2.25.

Aluminum. Aluminum materials have been used for many years in both military and commercial products for SMT as well as for through-hole applications. Aluminum material properties are the baseline properties that generally are used to compare the advantages and disadvantages of new material products.

TABLE 2.25 Typical Mechanical Properties of PWA Core Constitutive Material

Material	Density, g/cm^3	CTE, 10^{-6} ppm/°C	Thermal conductivity, W/(m · °C)	Specific conductivity [W/(m · °C)]/(g/cm^3)	Young's modulus, 10^6 lb/in^2	Yield strength, 10^3 lb/in^2	Ultimate strength, 10^3 lb/in^2	Elongation, %
Aluminum 1100H18			218		10.0			
Aluminum 6063T6	2.68	23.4	201	74.86	10.0	31.0	35.00	
Aluminum 6061T6	2.71	23.6	156	57.51	10.0	40.0	45.00	17
Copper	8.95	16.7	390	43.58		30.0		
Invar	8.03	1.7	12	1.49	20.0			
Molybdenum	10.28	4.5	142	13.81	47.0			
Beryllium S-200F	1.85	11.4	216	116.76	44.0	35.0	47.00	2
Beryllium oxide	3.06	6.4	250	81.70	56.0			
Graphite IG-610	1.85	4.5	128	69.02	1.7	4.5		
Graphite P100	2.20	−1.6	520	236.36				
Graphite P120	2.10	−1.6	640	304.76	120.0			
Graphite K1100	2.15	−1.6	1050	488.37	130.0	400.0		

TABLE 2.26 Typical Mechanical Properties of PWA Core Final Material

Material	Density, g/cm^3	CTE, 10^{-6} ppm/°C	Thermal conductivity, W/(m · °C)	Specific conductivity [W/(m · °C)]/(g/cm^3)	Young's modulus, 10^6 lb/in^2	Yield strength, 10^3 lb/in^2	Ultimate strength, 10^3 lb/in^2	Elongation, %
Aluminum 6061T6	2.72	23.6	156	57.35	10.0	40.0		
Cu/Invar/Cu, 12.5/75/12.5	8.34	*x–y* 45.–4.7	*x–y* 100 *z* 19	11.99	20.3	55–70		
Cu/Invar/Cu, 20/60/20	8.45	*x–y* 5.1–5.3	*x–y* 167 *z* 24	19.76	19.6	45–60		
Cu/moly/Cu, 5/90/5	10.08	*x–y* 5.1	*x–y* 96 *z*	9.52	44.0	100.0		
Cu/moly/Cu, 13/74/13	9.89	*x–y* 5.7	*x–y* 120 *z*	12.13	39.0	87.0		
Cu/moly/Cu, 20/60/20	9.70	*x–y* 6.5	*x–y* 140 *z*	14.44	35.0	75.0		
Cu/moly/Cu, 25/50/25	9.56	*x–y* 7.9	*x–y* 155 *z*	16.22	32.0	62.0		
Moly/graphite/moly, 10/80/10	3.54	*x–y* 4.8	*x–y* 130 *z*	36.77	—	>2.0		
Beryllium S-200F	1.85	11.4	216	116.76	44.0	35.0	47	2
Be-BeO, E20	2.07	8.7	205	110.14	44.0	44.0	40	0.2
Be-BeO, E40	2.29	7.5	215	101.31	46.0	46.0	35	0.1
Be-BeO, E60	2.55	6.1	220	94.12	48.0	48.0	30	0.05
Beryllium-aluminum, AlBeMet 140	2.26	16.3	200	88.50	20.0	30.0	40	15
Beryllium-aluminum, AlBeMet 162	2.10	13.9	210	100.00	26.0	45.0	60	5

Two series of aluminums are used predominantly: 1100, usually H18, and 6061, usually T6, although 6063 T6 grades with their higher thermal conductivities are being considered. The 1100 series has a slightly higher thermal conductivity than the 6061 series. The 6061 series has a higher structural modulus. The 1100 series is most easily manufactured by punching while the 6061 series is easily machinable; both can be cast. Aluminum materials are the least expensive as far as raw material cost is concerned, when compared to all other materials presented in this chapter. The 1100 series predominates in through-hole applications while the 6061 series predominates in SMT applications.

Aluminums have a relatively low density and are used in many applications where weight is critical. Their structural modulus provides a relatively stiff assembly, though the thickness of the material must be increased to provide very high structural natural frequencies. The increased thickness tends to raise the assembly weight in very-critical-weight applications to unacceptably high levels. This drives the designer to use materials with greater stiffness-to-weight ratios.

Aluminum has a very high CTE, 24×10^{-6} ppm/°C, when compared to the CTE of alumina components, 6×10^{-6} ppm/°C. PWBs bonded to aluminum structures must use one of two methods to prevent solder joint failures during environmental thermal cycling. The first method is to use components with leads of sufficient structural compliance to reduce solder joint strain. In the second method, the PWB would be soft-bonded to the aluminum using a silicone or gum rubber adhesive of sufficient thickness to uncouple the PWB structurally from the aluminum.

Copper. Copper may be used in applications requiring the removal of high power from a PWA. It has a thermal conductivity 1.8 times that of aluminum. Its biggest drawback is its weight density, which is 3.3 times that of aluminum.

Copper is easy to machine. To prevent corrosion of the copper, protective coatings of nickel plate are provided. Uses for copper have been in high-power-density through-hole applications and SMT applications where weight is not a consideration.

Clad Metal and Graphite Materials. Two clad metal materials predominate, copper-Invar-copper (CIC) and copper-molybdenum-copper (CMC). Both metal systems use outer copper surfaces clad to either Invar or molybdenum on the interior. The thickness ratio of the Invar or molybdenum to copper controls the CTE of the clad metal. Both of these materials are used when SMT applications require constraining materials to lower the CTE of the PWB in order to match that of the alumina chip carrier components.

A detailed explanation of CIC and CMC foils used within the PWB is given on page 2.20. While some designs use internal PWB CIC or CMC foils, other designers have hard-bonded completed PWB structures to either CIC or CMC structural members. The choice of which method to use normally depends on two factors. In order to use thin clad metal cores integral to the PWB, the manufacturer's PWB facility must have the technological expertise in PWB processing with these materials. PWBs using clad metal internal foils require expertise in etching, plating, laminating, and drilling, which vary from standard FR-4/copper-clad processes. The second factor is weight. Both materials have very high weight densities, which limits their use in many applications. A lighter-weight assembly could be made with thin clad metal inner cores bonded to a lightweight structure rather than a thin conventional PWB bonded to a thick CIC or CMC structural core.

In comparing these two clad metal systems, CIC has seen the most industry usage, particularly in military applications. CIC is about 21 percent lighter, slightly less expensive, and easier to machine than CMC. CMC has about double the strength of CIC, is better able to resist warpage, and remains relatively flat and stable compared to CIC. CMC is slightly more expensive than CIC.

Texas Instruments and Metalwerk Plansee of Austria have recently developed a new clad heat-sink core material consisting of 0.010-in molybdenum sheets bonded by a high-temperature vacuum brazing process to a lightweight central graphite plate. This molybdenum-graphite-molybdenum (MGM) composite, when compared to conventional CIC or CMC clad metals, offers a material that has less than 50 percent of the density (although the density is still 16 percent greater than that of aluminum), has a lower CTE, and has better structural dampening characteristics.

Existing plate sizes are limited to 12 by 18 in due to brazing furnace limitations. It is being produced in thicknesses ranging from 0.040 to 0.125 in with a tolerance of ±0.003 in. Flatness can be held to 0.002 in/in with a 5-μm surface finish.

Finished products can be machined by means of carbide drilling tools for hole drilling and profiling, using carbide end mills, or with an Nd:Yag laser.

Beryllium (Be). For applications where weight is extremely critical and high structural rigidity is required, beryllium is the material of choice based on technical merit. Pure beryllium has 68 percent the density of aluminum, has a structural modulus 4.4 times greater than aluminum, a CTE 47 percent lower than aluminum, and a thermal conductivity 23 percent greater than aluminum.

Beryllium has three major drawbacks: lack of multiple major suppliers, extremely high cost, and potential health hazards. Brush Wellman, Inc., is the major supplier of beryllium, with proven ore reserves of 73 years using 1992 consumption rates. Its high raw-material cost and difficulty in machining, joining, and forming make beryllium structures very expensive. It should be noted that 2 to 6 percent of the world's population can develop allergic reactions to inhaled beryllium particles. This is known as berylliosis. Three factors are required for the person to contract the disease: they must be allergic, they must inhale particle sizes finer than 10 μm in diameter, and they must be exposed to concentrations >2 $\mu m/m^3$ over an 8-h period. For this reason, manufacturing operations using beryllium-based materials are carried out in well-ventilated facilities with strict OSHA policies regarding the disposal of this material. Solid pieces of beryllium are not a health hazard and can be handled without extreme safety precautions.

Be-BeO Composites. One of the latest developments in electronic composite materials are metal-matrix composites (MMCs) consisting of single-crystal beryllium oxide (BeO) particles dispersed in a metallic beryllium matrix. The Be-BeO composites exploit the high thermal conductivity and chemical stability of BeO and the light weight and high modulus of beryllium. The resultant composites exhibit a high modulus, good thermal conductivity, low density, and a low CTE relative to aluminum.

The composites are manufactured using powder metallurgy techniques. Machining of the composite material into a final product is difficult, although advances in ultrasonics, cubic boron nitride, and diamond tooling are being used today to produce end-item products from powder metallurgy blanks. Dry and cold pressing procedures to produce near-net-shape and net-shape products are being developed. Under development are also injection molding processes. Plating, alodining, and anodizing can be used to prepare the surface for further processing or as a protective measure.

Three forms are in use today, depending on the specific material properties needed: E20, E40, and E60. The E designation refers to "electronic grade" products; the number refers to the volume percent loading of BeO in the product, 20 being 20 percent BeO. Under development are heat-sink products for SMT applications using these alloys. Most applications for this material revolve around housings for components such as transmit-receive modules for advanced radar applications.

Aluminum-Beryllium Alloys. An aluminum-beryllium alloy called Lockalloy was developed by Lockheed Aircraft Corporation some 25 years ago as a lightweight structural material used in the folding ventral fin of the SR-71 Blackbird aircraft. Recently Brush Wellman has developed a new group of aluminum-beryllium alloys based on the Lockalloy process, using the trademark AlBeMet.

The AlBeMet products are made by incorporating refinements into the basic powder metallurgical processing techniques used to produce Lockalloy. Current products are made from atomized powder consolidated by extrusion or hot isostatic press (HIP). AlBeMet consists of finely dispersed beryllium particles in an aluminum matrix. The beryllium particles act as reinforcing agents. The alloys developed have the stiffness of steel with a density 25 percent less than aluminum. AlBeMet product designations are AlBeMet-XXX, with the first X referring to the aluminum matrix, 1 being pure 1100 aluminum.

A wide variety of manufacturing techniques can be used to produce end-item products with AlBeMet materials. Health and safety procedures associated with beryllium-containing materials must be observed. The material machines much like magnesium, with fully annealed material being easier to machine than as-extruded stock. It does not require an etching treatment after machining as does beryllium in order to remove machine marks and damage. Electron-beam and TIG welding techniques can be used and the material can be welded to some aluminum grades. AlBeMet can be brazed, and epoxy bonding can be used. Platings such as alodine, nickel, and paints adhere well. The cost of the raw material is still very high compared to aluminum. Heat sinks for SMT applications are under development using this material as well as structural housings and covers for small electronic modules.

Graphite Matrix Composites. These composites using high-strength graphite fibers impregnated into a metal or plastic matrix have been under development for a number of years. Several major manufacturers, such as DWA, Sparta Inc., Americom, Courtalds Inc., AMOCO, and Advanced Composite Products to name a few, have been involved in various research projects to develop these composites for electronics applications. The main thrust areas for this development revolve around reducing the weight and the CTE of the resultant composites while increasing their thermal conductivity. For PWA applications, the most work has been done in developing thermal planes for SMT applications.

A number of high-strength pitch graphite fibers have been used (P75, P100, P120, and K1100X). The graphite fibers are produced in sheets with the fiber oriented in one direction (x). Along the x axis the sheet exhibits a high value of thermal conductivity and excellent x–z rigidity. Thermal conductivity in the y and z directions is relatively poor, as is the y–z rigidity.

The two most prevalent metals in use today as base materials are aluminum and copper. Aluminum serves as the base material for structures requiring light weight. Copper serves as the base material for structures requiring superior thermal conductivity.

Epoxies have served for the most part as the plastic base matrix. The graphite-plastic plates have been shown to provide a structure 30 percent lighter in weight than aluminum, provide a structure with much superior rigidity compared to aluminum, provide an in-plane CTE nearly matching that of ceramic, and provide superior in-plane thermal conductivity.

These composite products are made by laying up alternating sheets of the base material (aluminum, copper, or plastic) and graphite. Varying the percentage loading of graphite to base material and the ply orientation of the graphite sheets allows a wide variety of composite properties to be tailored to a particular application. For example, in order to maintain near isotropic in-plane properties, sheets of graphite fiber need to be laid up or plied in alternating x–y orientations. Most products under development today use fiber ply orientations varying between 15 and 90° and percentage loading of graphite to base material by volume of 20 to 60 percent. Isostatic pressure and heating of the laid up stack of materials fuse the graphite and base material together into the composite structure.

For the most part plates of the composite material are easily machinable. It should be noted, however, that machining exposes the graphite fibers, which must then be protected in order to prevent corrosion of the composite. The most common means of protection for graphite-aluminum composites is ion vapor deposition of 0.002 to 0.003 in of aluminum, and for graphite-copper or graphite-plastic composites, 0.001 in of nickel plating.

It should be noted that graphite-plastic structures are susceptible to cracking of the plastic base material due to the extreme mismatch of the CTEs of the graphite fiber and the plastic material. Therefore plastic materials have to be chosen which are somewhat compliant. Brittle polyimides would be more susceptible to the cracking than many epoxies. Care needs to be taken in developing the composite manufacturing process and in particular the heat-up–cool-down cycle. Cracking may be induced during the initial processing by a nonoptimal heat-up–cool-down cycle through the development of high stresses in the plastic. Subsequent thermal cycling of the product has shown that the initial cracks will grow in length, and in extreme cases they can cause fracture of the graphite fibers within the structure. This could lead to a degradation in the initial CTE, thermal conductivity, or structural moduli.

2.3.2 Flexible and Rigid-Flexible PWB Materials

The materials and the anatomy of flexible printed wiring (FPW) are quite different from those of the various rigid PWB constructions discussed earlier in this chapter. As was mentioned, these flexible circuitry forms are used in the many product designs and applications where their physical flexibility and formability offer advantages. These applications and their advantages are many, and they are increasing constantly. FPW is etched or formed to the desired circuit pattern, and thus is similar to rigid PWBs, except that it is flexible in form. The anatomy of the flexible conductor system in both flexible and rigid-flexible materials is similar in cross section, however, and these cross-sections are shown in Figs. 2.3 and 2.4.

As Figs. 2.3 and 2.4 illustrate, the flexible circuits in both flexible and rigid-flexible systems consist of three important materials: the base film, the conductor, and the adhesive. The metal foil is bonded and subsequently etched to the desired pattern on the adhesive-coated base film. A second adhesive-coated film is bonded on top of the formed circuit pattern. This top adhesive-coated film is usually of the same material construction as the bottom, or base, adhesive-coated film. The top adhesive-coated film is known as a cover coat film. It protects the circuitry much as conformal coatings protect circuitry on rigid PWB circuits. This cover coat is usually laminated to the base film under heat and pressure, after the circuit pattern has been formed and any cleaning operations are completed. Although many types of films and adhesives can be used, thermal stability must be considered in their selection, especially where parts will be soldered to the circuitry or when vapor degreasing operations are used for cleaning.

Completed FPW products are covered by a number of government and industry documents. The military delineates the requirements for the design of FPW and rigid FPW in MIL-STD-2118 and the fabrication of FPW and rigid FPW in MIL-P-50884. IPC covers FPW in three documents: single- and double-sided FPW design standards in IPC-D-249, performance specifications for single- and double-sided FPW in IPC-

FC-250A, and performance specifications for rigid FPW in IPC-RF-245. The following sections discuss in detail the constitutive materials used in the fabrication of FPW.

Dielectric Films. There are four IPC documents covering the dielectric films used in the fabrication of FPW. IPC-FC-231B covers the basis dielectric film material. IPC-FC-232B covers adhesive-coated dielectric films used as cover sheets in FPW. IPC-FC-241B is the specification covering metal-clad FPW dielectrics, and IPC-FC-FLX is a listing of specification sheets for the various basis dielectric film and adhesive materials used in FPW.

For the reasons mentioned, and for the general good stability and other good material characteristics which they possess, the films used for FPW are usually either polyimide or polyester. While there are other sources of these materials, perhaps the most widely used are the duPont polyimides (Kapton H) and the duPont polyesters (Mylar).

Polyimides are dark yellow films with thermal stabilities of over 200 to 250°C. Polyimide films can withstand soldering temperatures. They are not attacked by any known organic solvent and cannot be fused. Standard film thicknesses are 0.0005, 0.001, 0.002, 0.003, and 0.005 in. There are variations of the basic Kapton H film to develop special properties. Kapton XT is a polyimide film which has been developed with alumina incorporated into the polymer to increase the thermal conductivity. Kapton XC has been developed with increased electrical conductivity for special applications. Polyimides are considerably higher in cost than polyesters (about 20 times).

Polyesters are transparent films containing no plasticizers. They do not become brittle under normal conditions. Their low service temperature range of −70 to 150°C limits their use in applications requiring soldering. Thus polyesters, sometimes identified by their chemical class (polyethylene terephthalate), are best used when mechanical rather than soldered interconnections are employed for component attachment.

Other types of films which are sometimes used are aramids (duPont Nomex), polyester epoxies (Rogers' Bend-Flex), Upilex S polyimide film (W. L. Gore's Gore-Clad), fluorocarbons (duPont Teflons), and polyetherimides (General Electric Ultem).

Nomex is a high-temperature paper material made from random-fiber aramid materials. It is very hydroscopic and tends to absorb processing chemicals readily. It has fairly low tear strength and a dielectric constant about half that of Kapton H.

Bend-Flex is manufactured using a nonwoven mat of Dacron polyester and glass fibers saturated in a B-stage epoxy. Copper foils are directly clad to the B-stage material. The dielectric material is available in 0.005- to 0.030-in thicknesses.

Gore-Clad is a new material developed by W. L. Gore and Associates for use in rigid-flexible applications. This system uses Upilex S polyimide film. This film exhibits superior dimensional stability, chemical resistance, and moisture resistance when compared to traditional Kapton H material systems. The adhesive system is a thin cyanate ester modified with PTFE. The material is not recommended for dynamic flexible applications. Material properties of the various dielectric films are given in Table 2.27.

Adhesives. Two IPC documents delineate the requirements for adhesive films used in the fabrication of FPW products. IPC-FC-233A covers the adhesive materials and IPC-FC-FLX is a specification sheet listing the materials. Table 2.28 lists adhesive properties.

The adhesive types used to bond the metal foil to the dielectric film are most commonly epoxies, polyesters, or acrylics. New product developments have resulted in adhesiveless material systems for bonding between the copper and the dielectric. In addition, an electrically conductive adhesive system has been introduced to eliminate plated-through-hole interconnections.

TABLE 2.27 Material Properties of Flexible Printed-Wiring Dielectric Film

IPC-FC-231, specification sheet	Polyimide 1	FEP 2	Polyester 3	Epoxy polyester 4	Aramid paper 9
Tensile strength, min. lb/in^2	24000	2500	20000	5000	4000
Elongation, min. %	40	200	90	15	4
Initial tear strength, min. grams	500	200	800	1700	—
Density, g/cm^3	1.40	2.15	1.38	1.53	0.65
Dimensional stability, max. %	0.15	0.3	0.25	0.2	0.30
Flammability, min. % oxygen	30	30	22	28	24
Dielectric constant, max., at 1 MHz	4.00	2.30	3.40	—	3.00
Dissipation factor at 1 MHz	0.0120	0.0007	0.0070	—	0.01
Volume resistivity (damp heat), Ω • cm	10^6	10^7	—	10^5	10^6
Surface resistivity (damp heat), MΩ	10^4	10^7	—	10^3	10^5
Dielectric strength, min. V/mil	2000	2000	2000	150	390
Moisture absorption, %	4.00	0.10	0.80	1.00	13.00
Fungus resistance	Non-nutrient	Non-nutrient	Non-nutrient	Non-nutrient	Non-nutrient

Source: From IPC-FC-231B, class 3 properties.

TABLE 2.28 Material Properties of Flexible Adhesive Bonding Films

IPC-FC-233, specification sheet #	1 Acrylic	2 Epoxy	3 Polyester
Peel strength, lb/in width			
As received	8.00	8.00	5.00
After solder float	7.00	7.00	N/A
After temperature cycling	8.00	8.00	5.00
Flow max, squeeze out mil/mil	5.00	5.00	10.00
Volatiles content, %	2.00	2.00	1.50
Flammability, min. % oxygen	15.00	20.00	20.00
Solder float, (IPC-TM-650, method 2.4.13 B)	Pass	Pass	N/A
Chemical resistance, % (IPC-TM-650, method 2.3.2 B)	80	80	90
Dielectric constant, max., at 1 MHz	4.00	4.00	4.60
Dissipation factor, max., at 1 MHz	0.05	0.06	0.13
Volume resistivity, min. MΩ · cm	10^6	10^8	10^6
Surface resistance, min. MΩ	10^5	10^4	10^5
Dielectric strength, V/mil	1000	500	1000
Insulation resistance, min. MΩ	10^4	10^4	10^5
Moisture absorption, max. %	6	4	2
Fungus resistance (IPC-TM-650, method 2.6.1)	Nonnutrient	Nonnutrient	Nonnutrient

Epoxy systems include modified epoxies known as phenolic butyrals and nitrile phenolics. These systems are used widely; they are generally lower in cost than acrylics but higher in cost than polyesters. Epoxies have good high-temperature resistance, a lower moisture absorption rate than acrylics, a lower CTE than acrylics, and remain in good condition in all approved soldering systems. They also have very long-term stability at elevated temperatures in environmental conditions up to 250°F.

Acrylic systems are most often used in high-temperature soldering applications. They exhibit excellent adhesion to both the copper interconnect as well as the polyimide films. Typical bond strengths to copper are 8 to 15 lb/in at room temperature. These adhesives have very controlled flow when bonding cover sheets to the base laminate. They are easy to process and exhibit excellent batch-to-batch process consistency. There are a number of properties that will affect multilayer products adversely. Their high moisture absorption rate of 3 percent makes products not properly dried prior to soldering delaminate. They have a low T_g (90°F) and a high CTE (400 to 600 ppm/°F). In multilayer applications these properties can cause delamination and barrel cracking of plated-through holes when the product is subjected to extreme thermal environments.

Polyesters are the lowest-cost adhesives used and the only adhesives that can be used properly with polyester films for base laminate and polyester cover film. The major drawback of this system is their low heat resistance.

New material systems have been developed recently to provide designers with thinner films and laminates through the use of adhesiveless flexible circuits. Flex Products' UltraFlex is an example. This product uses advanced vacuum deposition technology to bond copper to polyimide. Parlex has developed PALFlex as an adhesiveless flexible laminate. This process uses specially treated polyimide film to which an ultrathin coating of a barrier metal is applied to promote copper adhesion. The copper is then electrodeposited onto the film. Rogers uses a cast-on process to apply copper to the base dielectric film in its Flex-I-Mid 3000 material. Sheldahl produces a product called Novaclad, which uses a semiadditive process to deposit copper onto the base film. These material systems provide for applications which may use thinner copper for finer line widths. They have improved dimensional stability and improved flame retardance, and they have extended the dynamic flexible life of the completed FPW product.

A new adhesive system recently introduced by Sheldahl is Z-link. Z-link is an anisotropic adhesive system that is impregnated with conductive particles dispersed uniformly throughout the adhesive layer. The material system has two important attributes that determine its performance. First, a solder alloy is used as the conductive spheres. When laminated, the spheres fuse to the mating surfaces, creating a low-resistance high-current-carrying interface. Second, the adhesive thermosetting formulation gives it the mechanical strength to withstand severe thermal conditions. This material is being used in multilayer applications to replace plated-through-hole interconnections.

Foils. The metal foils used for conductors in flexible circuitry most commonly consist of electrodeposited copper or rolled annealed copper, as defined in IPC-MF-150. The electrodeposited foil is economical. It has a rougher base bonding surface, which enhances bonding to the base dielectric, and a vertical grain structure. Rolled annealed copper foil is normally supplied with a bonding enhancing treatment due to its smoother surface and has a horizontal grain structure. This grain structure makes it better suited for flexing applications, making it the material of choice in dynamic applications. Other conductors may be used for special cases. For copper and other foils important features to specify are hardness, flexibility, and bondability to the adhesive being used.

2.3.4 Solder Masks

In PWB assemblies there are many instances when it is desired to protect patterned conductor surfaces on the laminate, such as etched copper circuitry, from being coated by solder during either soldering or tinning processes. In these instances a solder resist, more commonly known as a solder mask, is applied to the surface of the PWB laminate, usually after the conductor pattern has been formed and the PWB panel cleaned. After the application of the solder mask, solder can then be selectively applied to the areas of PWB holes, pads, and conductor line areas to which component parts are to be attached in soldering operations. The absence of solder under the PWB coating not only saves solder, but eliminates the many problems that can occur when solder is entrapped under coatings.

Functions and Design Considerations. While the primary function of solder masks is to prevent solder movement from the solder land areas to attached conductors and plated-through holes, solder masks actually serve many other functions. These are summarized as follows:

Reducing solder bridging and electrical short circuits

Reducing the volume of solder pickup to obtain cost and weight savings

Reducing solder pot contamination (copper and gold)

Protecting PWB circuitry from handling damage such as dirt or fingerprints

Providing an environmental barrier

Filling the space between conductor lines and pads with material of known dielectric characteristics

Providing an electromigration barrier for dendritic growth

Providing an insulation or dielectric barrier between electrical components and conductor lines or via interconnections when components are mounted directly on top of these features

The following design factors influence decisions to use solder masks:

Criticality of system performance and reliability

Physical size of the PWB

PWB metallizations (tin-lead, copper, gold)

Density of line widths and spacings

Amount and uniformity of metallization

Size and number of drilled plated-through holes

Annular ring tolerance for plated-through holes

Component placement on one or both sides of the PWB

Need to have components mounted directly on top of a surface conductor

Need to tent via holes in order to keep molten solder out of selected holes

Need to prevent the flow of solder up via holes that have components situated on top of them

Likelihood of field repair or replacement of components

Need to contain solder on the lands in order to maintain the proper volume of solder required for the component lead solder joint

Choice of specifications and performance class that will give the solder mask the properties necessary to achieve the design goals

It should be pointed out that solder masks, which do not cover the entire board and component assembly, do not serve the same functions as conformal coatings discussed in Sec. 2.3.5.

Specifications and Selections. The use of solder masks is covered in specification IPC-SM-840, which calls out three classes of performance that the designer can specify:

Class 1—Consumer: Noncritical industrial and consumer control devices and entertainment electronics

Class 2—General Industrial: Computers, telecommunication equipment, business machines, instruments, and certain noncritical military applications

Class 3—High Reliability: Equipment where continued performance is critical; aerospace and military electronic equipment

In addition to calling out the performance classes, the specification assigns responsibility for the quality of the solder mask to the materials supplier, the PWB fabricator, and, finally, the PWB user.

Solder masks may be temporary or permanent. The following paragraphs will concentrate on permanent solder mask technology. The various types are illustrated in the technology chart in Fig. 2.11. It can be seen that various chemistries, uncured forms, application technologies, and curing methodologies exist. A selection guide for choosing among the permanent solder masks is given in Table 2.29. Permanent solder masks become a functional part of the PWB assembly, and hence their performance in the system life cycle must be considered.

The demand for permanent coatings on PWBs has greatly increased as the trend toward surface mounting and higher circuit density has increased. When the conductor line density was low, there was little concern about solder bridging, but as the density increased, the number and the complexity of the components increased also. At the same time the soldering defects, such as line and component short circuits, increased greatly. Inspection, testing, and rework costs accelerated as efforts went into locating and repairing the offending solder defects. The additional cost of the mask on one or both sides of the PWB was viewed as a cost-effective means to offset the higher inspection, testing, and rework costs. The addition of a mask also had the added value for the designer of providing an environmental barrier on the PWB. This feature was important and dictated that the materials considered for a permanent solder mask should have similar physical, thermal, electrical, and environmental performance properties as the laminate material.

There are four major types of permanent solder masks available on the market today: screen-printable thermally cured liquids, photo-imageable liquids, photo-imageable dry films, and photo-imageable liquid/dry films.

The screen-printable thermally cured solder masks have been in existence the longest. These masks, which are usually epoxy-based, are applied mainly by means of a screen printer using either a stainless-steel or a polyester mesh. The screen has openings which are aligned to the PWB features and define the mask pattern to be applied to the PWB. Curing of the mask occurs at elevated temperature (usually about 125°C).

These masks have strong hard surfaces and excellent adhesion to the PWB surface. They are used on PWBs having relatively coarse feature patterns (>0.025 in). Fine mask features (<0.025 in) are difficult to maintain in older epoxy formulations due to limitations in screen-printer technology, smearing of the liquid, and bleeding of sol-

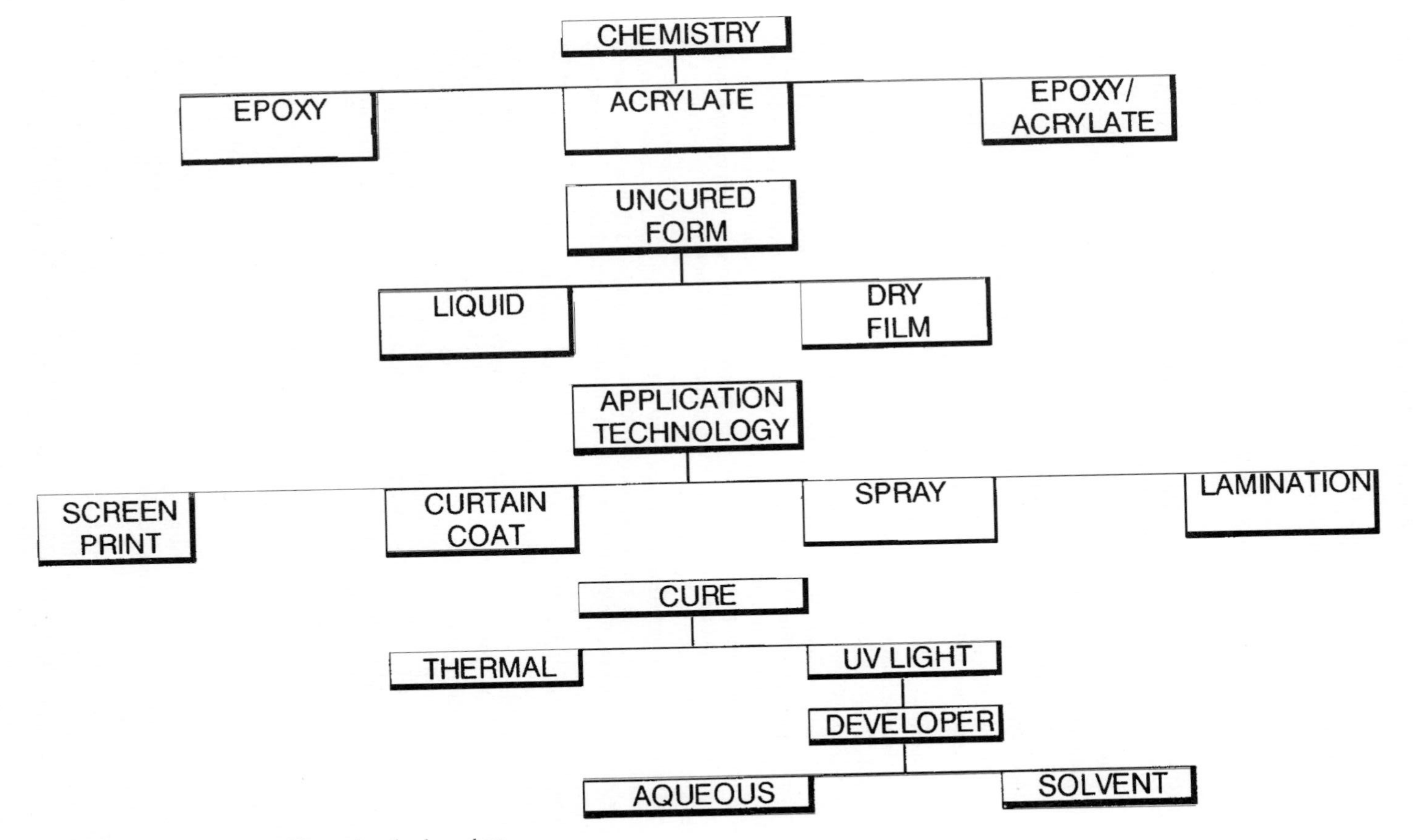

FIGURE 2.11 Permanent solder mask technology chart.

TABLE 2.29 Selection Guide for Permanent Solder Masks*

Feature	Screen print		Dry film		Liquid photoresist solvent
	Thermal	UV	Aqueous	Solvent	
Soldering performance	1	1	1	1	1
Ease of application	1	1	2	2	2
Operator skill level	2	2	2	2	2
Turnaround time	2–3	2–3	2	2	2
Inspectability	2–3	2–3	3	3	3
Feature resolution	3	3	1	1	1
Adhesion to Sn-Pb	1	3	1–2	1–2	1
Adhesion to laminate	1	1–2	1	1	1
Thickness over conduct or lines	3–4	3–4	1–2	1–2	2
Bleed or residues on pads	3–4	3	1	1	1
Tenting or plugging of selected holes	4	4	1	1	4
Handling of large panel size with good accuracy	3	3	1	1	1
Meeting IPC SM-840 class 3 specification	3–4	3–4	1–2	1	1
Two-sided application	4	4	1	1	3–4
Capital equipment cost	4	4	1–2	1–2	1

*1 = good or high; 2 = moderate; 3 = fair; 4 = poor or low.

Source: From Coombs.[5] Reprinted with permission.

vents from the mask during cure. A further limitation is the tendency of the mask to retain residues during further processing steps.

The advent of new screen-printing technologies coupled with a new generation of liquid polymers is causing a reevaluation of this solder mask technology for finer-pitch SMT applications. Many of the new thermal cure liquids on the market today are based on polyfunctional epoxies which cure to a high cross-link density, providing a low-porosity coating. In addition, advances have been made to reduce their tendency to smear during printing and to bleed during cure.

With regard to the photo-imageable solder masks, three types are used: liquid, dry film, and liquid/dry film. The liquid photo-imageable solder masks are formulated from either epoxy, acrylate, or epoxy-acrylate chemistries. The dry films are formulated from either acrylate or epoxy-acrylate chemistries. The liquid/dry-film masks use a combination of both mask technologies. The mask chemistry can be either solvent or aqueous developable. Two different polymerization formulations are available, bulk and surface.

Bulk polymerization, the most widely used, provides fairly straight sidewalls after development. Liquid bulk-polymerized masks yield the narrowest finished feature sizes, 0.001 to 0.003 in. The smallest dry-film bulk-polymerized mask feature size is limited to 0.006 in. Disadvantages to bulk polymerization are shoot-through on double-sided PWBs and light scattering. On shoot-through, the ultraviolet light travels through the PWB and partially cures the mask on the far side of the PWB, with the image being developed on the near side. The scattering of the ultraviolet light source may cause exposure of the mask inside the barrel of the plated-through hole or partial exposure on the solder lands.

Surface polymerization masks exhibit undercutting of the mask during development of about 0.0005 to 0.003 in. This prevents feature sizes of less than 0.004 to 0.007 in in size. Pin holes may develop in these masks when foreign material is present between the solder mask surface and the ultraviolet light source.

Dry-film solder masks are supplied by their manufacturers in rolls. The thickness is usually 0.002 to 0.004 in. The major advantage in using a dry-film mask over a liquid mask is its ability to tent holes up to 0.030 to 0.040 in in diameter. Liquid solder masks, while lacking the hole-tenting characteristics of the dry films, tend to give lower finished mask thicknesses than the dry film—usually 0.001 to 0.002 in thick over the laminate and 0.0007 to 0.001 in over metallized traces.

Pencil hardness is a value assigned solder masks which defines their ability to resist abrasion or their scratch resistance. Current solder mask formulations have pencil hardness values of 3H to 8H, with 8H being the most scratch-resistant. Dry-film masks tend to have values in the 3H to 4H range, while liquid photo-imageable masks have values from 5H to 6H. Table 2.30 presents typical material property data on photo-imageable dry-film solder masks and Table 2.31 lists material property data on photo-imageable liquid solder masks.

Processing Considerations. The major processing steps for solder mask applications are cleaning and preparation of the PWB surface, solder mask application, and solder mask development and cure.

Cleaning and Preparation of the PWB Surface. Optimal solder mask appearance and adhesion will not occur unless the PWB surface has been carefully prepared. The main objectives in this process step are to remove all surface contaminants, roughen up the surface, and dry the prepared surface thoroughly to promote mask adhesion.

If only bare copper circuitry or other hard metallizations such as nickel are present on the surface conductors prior to mask application, the following steps are usually employed. An acid preclean is used first to remove oils and oxidation on the surface. This is followed by a mechanical scrubbing to remove further residues and roughen the surface. Bristle brushes, composite brushes, or the combination of a pumice slurry and bristle brushes are used in this step.

The critical parameter to be maintained in this process is the brush pressure. Too little pressure, and contaminants will remain trapped on the PWB, or the surface will not be roughened sufficiently to promote adhesion of the mask. Too high a brush pressure, and fine circuit features may be broken or grooves may be made in the surface under the solder mask, which may promote entrapment of developing solutions.

Where tin-lead is present on the PWB prior to solder mask application, mechanical scrubbing and pumice scrubbing are not normally used. This is due to smearing of the malleable tin-lead or imbedding of the pumice material in the tin-lead during this operation. Usually the precleaning of PWBs with tin-lead features involves a solvent degreasing in Freon or 1,1,1-trichloroethane followed by a light chemical cleaning.

The final steps in this process are a rinse in deionized water to remove any remnants from the scrubbing or cleaning processes, followed by a thorough drying of the PWB surface. In some instances, particularly when dry-film solder masks are used, an oven bake of the PWB prior to solder mask application is recommended to remove any entrapped moisture in the laminate.

Solder Mask Application. Application of the solder mask is usually accomplished through one of four procedures: screen printing, curtain coating, spray coating, or dry-film application. Screen printing of oven-cured and some liquid photo-imageable

TABLE 2.30 Material Property Data for Typical Dry-Film Photo-Imageable Solder Masks*

Type	Dynamask KM Epoxy	Conformask 1000	Conformask 2000 Epoxy	Laminar DM	Vacrel 8000	Vacrel 8100	Valu 8000†	Valu 8200†
Solids content, %	100	100	100	100	100	100	100	100
Cured film thickness, in	0.003 or 0.004	0.0023	0.0023	0.003 or 0.004	0.003 or 0.004	0.003 or 0.004		
UL rating	94V–0	94V–0	94V–0	94V–0	94V–0	94V–0	94V–0	94V–0
IPC-SM-840A rating	Class III	Class III	Class III	Class III	Class III	Class III	Class III	Class III
Pencil hardness	3H	2H	3H	2H–3H 2H–3H	4H–5H	4H	4H–6H	4H–6H
Insulation resistance, Ω	4.0×10^{12}	3.3×10^{12}	1.1×10^{11}	2.0×10^{14} 1.0×10^{14}	5.0×10^{14}	5.0×10^{14}	3.0×10^{14}	3.0×10^{14}
Dielectric strength, V/mil	3434 3386	1952	2969	1366 1600	3506	2829	3517	4310
Dielectric constant at 1 MHz	4.23 3.41	—	—	—	3.6	3.8	3.5	3.2
Dissipation factor at 1 MHz	0.031 3.41	—	—	—	0.033	0.042	0.027	0.027
Volume resistivity, Ω • cm	6.0×10^{15} 6.7×10^{15}	—	—	—	3.0×10^{14}	9.0×10^{14}	3.0×10^{16}	1.5×10^{16}
Surface resistivity, Ω	1.0×10^{17} 3.0×10^{17}	—	—	—	10^{15}	3.6×10^{13}	8.0×10^{13}	7.0×10^{13}
Solvent resistance	Pass	Pass	Pass	Pass	Pass	Pass	Pass	Pass

*Where two values are given, the upper value is for 0.003-in-thick film, the lower value for 0.004-in-thick film.

†Value system by duPont is a combination liquid and dry-film system.

TABLE 2.31 Material Property Data for Typical Liquid Photo-Imageable Solder Masks

Type	Lea Ronal OP SR 5500 Epoxy/acrylate	Coates Imagecure AQ Epoxy/acrylate	Chemline Photobond 3000 Acrylate	WR Grace AM-300 Epoxy/acrylate	Lackwerke Peters SD-2461-AE Epoxy	M and T Photomet 1001 Acrylate	Ciba-Geigy Probimer 52 Epoxy	Dynachem EPIC SP-100 Epoxy
Solids contents, %	70	68	65	100	80	100	38	76
Cured film thickness, in	0.0005–0.0015	0.0007–0.002	0.0008–0.0015	0.001–0.0055	0.0007–0.002	0.002–0.006	0.0005–0.0017	0.0005–0.0017
UL rating	94V–0	94V–0	94V–0	94V–0	94V–0	94V–0	94V–0	94V–0
IPC-SM-840A rating	Class III	Class III	Class III	Class III	Class III	Class III	Class III	Class III
Pencil hardness	5H	7H	4H	4H	6H	5H	4H	6H
Insulation resistance, Ω	9.3×10^{12}	1.0×10^{12}	4.0×10^{13}	1.4×10^{10}	—	2.0×10^{14}	2.0×10^{11}	1.2×10^{12}
Dielectric strength, V/mil	2500	2080	1703	3500	1778	3000	3048	3018
Dielectric constant at 1 MHz	3.5	2.8	3.6	3.5	3.6 at 0.8 MHz	4.3	3.7	4.07
Dissipation factor at 1 MHz	0.042	—	0.015 at 0.5 MHz	0.011	0.040 at 0.5 MHz	0.007	0.002	0.035
Volume resistivity, Ω · cm	10^{13}	10^{15}	—	10^{15}	10^{13}	9.0×10^{14}	10^{13}	2.7×10^{15}
Surface resistivity, Ω	2.0×10^{13}	10^{13}	4.0×10^{12}	10^{10}	10^{13}	4.0×10^{13}	10^{10}	3.7×10^{15}
Solvent resistance	Pass	Pass	Pass	Pass	Pass	Pass	Pass	Pass

masks uses stainless-steel or polyester-mesh screens which define the finished solder mask features. A manual or semiautomated screen-printing machine mounts the mesh screen and is used to align the screen to the PWB. The liquid mask material is squeegeed through the openings in the screen onto the PWB.

Parameters to be controlled in order to optimize this printing process include choosing the optimal screen mesh size (110 to 200 mesh polyester monofilament is typical), optimal screen mesh tension and emulsion, maintaining the proper mask liquid viscosity, and using a proper durometer rubber squeegee (60 to 70 durometer, well sharpened, and 10 to 20° from the vertical is typical).

Photo-imageable liquid solder masks can be applied by screen printing, using a screen to coat the PWB completely with the mask material, but are also applied through the use of either curtain-coating or spray-coating equipment. Curtain-coating equipment applies the mask material on one side of the PWB at a time by passing it horizontally through a flowing curtain of the liquid mask material. The material is then oven-dried to allow application of the material to the second side of the PWB.

Critical parameters to be controlled in this process to ensure a consistent mask deposit and coverage are maintenance of a consistent solder mask viscosity and temperature and using a proper conveyer speed through the equipment to ensure that a thin uniform coat is applied. Some formulations of liquid photo-imageable materials require the PWB to be preheated prior to application. The oven-drying cycle must remove a sufficient amount of the solvents from the mask to allow direct contact with the equipment conveyer system. Typically 85 to 90 percent of the solvents must be removed during this step.

Spray-coating equipment has the advantage of applying the mask material to both sides of the PWB prior to oven drying. The spray-coating process uses an electrostatic spray system for application. In this system the PWB is grounded and processed vertically between a grounded back panel and the spray head. The spray head atomizes negatively charged masked material, which is attracted by the grounded PWB. The equipment automatically sprays one side of the PWB and then flips the PWB to coat the second side.

There are a few drawbacks to this process. Some waste of mask material occurs. Solder mask "wrap around," a buildup of material on the panel edge, can occur due to charge buildups. Solder mask "skipping," a lack of mask coverage, can occur due to like charges building up on adjacent traces which repel the mask material.

Dry-film mask materials are supplied as 0.002- to 0.004-in-thick films sandwiched between protective film layers. The mask is cut to size according to the size of the PWB panel to be masked. One side of the protective film is removed and the mask is applied to the panel through the use of specialized lamination equipment. The second protective film coat is removed just prior to film development. The lamination equipment process uses heat and a vacuum to cause the solder mask to flow and conform to the panel surface with no entrapped air pockets.

Temperature is the important process parameter to be maintained. Too low a temperature results in improper flow and air entrapment, while too high a temperature causes thin coverages on circuitry and flow of material into plated-through holes, leading to difficulties in later development. Fine-line closely spaced circuitry can be difficult for dry-film masks to conform to.

The liquid/dry-film systems use a specialized machine which applies a thin coating of the liquid solder mask to the PWB and floods closed the PWB plated-through holes. Then a thin dry-film mask is applied immediately over the liquid to trap the liquid in place.

Solder Mask Development and Cure. The development of both liquid and dry-film photo-imageable masks occurs under ultraviolet light in specially designed equipment. In the ultraviolet curing process the mask contains components that undergo polymerization and cross-linking by photo-generated radical and cationic species which occur during the exposure to an ultraviolet light source. This ultraviolet light exposure causes almost instantaneous hardening of the material into a cured film. A phototool is used with the final solder mask pattern in a negative ultraviolet-light-blocking image. (The areas to keep the solder mask are exposed.) This tool is aligned to the PWB in a frame which is then drawn down on the panel in a vacuum to eliminate space between the panel and the mask surface. It is then exposed to an ultraviolet light source, which produces an initial cure of the exposed mask areas.

Removal of the phototool occurs next, and the final image is developed in either an aqueous or a solvent chemistry. It should be noted that aqueous chemistries are being favored by more users due to the imposition of stricter EPA guidelines on the storage, use, and disposal of toxic solvents. This development removes the unwanted, non-cured mask film.

Developing critical parameters include the chemistry temperature and concentration, spray pressures, conveyer speed, rinsing, and drying. Fine-line mask features may not adhere well to the surface if exposed to spray pressures that are too high.

Acrylate and epoxy-acrylate masks usually require a secondary ultraviolet "bump" to complete the cure. This bump completes the cross-linking and makes the mask more robust. Epoxy-based masks usually require a thermal cure to complete the cross-linking.

Solder Mask over Bare Copper (SMOBC). One major solder mask application technology is called SMOBC. This term stands for the application of "solder mask over bare copper." A problem for conventional copper-tin-lead electroplated PWBs is the flow of tin-lead solder under the solder mask during the wave or vapor phase or infrared soldering process. This flow of molten metal underneath the mask can prevent the mask from adhering to metal or laminate. If the mask fractures because of this hydraulic force, the surface integrity is lost and the effectiveness of the mask as an environmental or dielectric barrier can be severely impaired. In fact, such breaks can actually trap moisture, dirt, and soldering flux and serve as a conduit to direct liquids down to the mask-laminate interface. This situation could lead to serious reliability or performance concerns.

The SMOBC process addresses the tin-lead flow problem by eliminating the use of tin-lead electroplating on the conductor lines under the solder mask. An all-copper plated-through-hole PWB is often produced by a "tent-and-etch" process. This process is one in which the PWB is drilled and plated with electroless copper. This is immediately followed by copper panel plating with the full-thickness copper required for the plated-through holes. A dry-film resist process is used with a negative phototool (clear conductor lines and pads) to polymerize the dry-film resist in only those clear areas of the phototool. This polymerized resist will protect the lines and plated-through holes during a copper-etching process, which will remove all the unwanted background copper. The photoresist is then stripped off, and the solder mask material is applied and processed through curing. Tin-lead is next added to the open component pads and the plated-through holes by the hot-air leveling process.

An alternative process uses conventional procedures to create a pattern-plated board. After etching, however, the metal etch resist is removed chemically, leaving the underlying copper bare.

The primary function of tin-lead in the plated-through holes and on the component pads is to improve solderability and appearance. It is important to demonstrate the solderability of the holes and pads on the SMOBC panel. One method of accomplishing this is by a hot-air leveling process which places a thin coating of molten tin-lead on only those copper areas of the plated-through hole that have not been covered by the solder resist. This hot-air leveling process improves the ability of the copper surface to be soldered. It also improves the appearance and solderability after longer-term storage of the plated-through holes and pad surfaces. Since there is no flowable metal under the solder mask during the hot-air leveling step or later during component soldering, the mask maintains its adhesion and integrity.

The lower metallization height of the conductor lines allows the use of a thinner dry-film resist and also makes the liquid and screen-printing applications somewhat easier. One variation on the basic SMOBC process is to produce the PWB by the conventional pattern-plate copper and tin-lead process followed by etching of the background copper. Then another photoresist step is used to tent the holes and pads so that the tin-lead can be stripped selectively from the conductor lines. This would be followed by infrared or oil reflow, cleaning, and application of a solder mask.

A second process variation strips off the tin-lead plating completely and is followed by cleaning, solder mask application, and hot-air leveling. There are still other PWB fabricators who do not like either of these processes and are opting to use a nonflowable copper-etchant-resistant metal-like tin-nickel under the solder mask. The major shortcoming to tin-nickel is that it is considerably more difficult to solder with low-activity soldering fluxes.

2.3.5 Conformal Coatings

Systems of synthetic resins usually dissolved in solvent vehicles are applied to completed PWAs in many applications. When cured they form a secure plastic-film encapsulation around PWAs, called a conformal coating. Their purpose is to seal out environmental contaminants to which the assembly may be exposed during its lifetime. However, they may also seal in contamination products from the manufacturing process not removed by a proper precleaning.

Conformal coatings were originally developed to protect sensitive electronic assemblies from the hostile environments experienced in the military, aerospace, and marine domains. As the level of integration has increased in the electronics industry with the use of SMT and the inherent finer lead pitches associated with VLSI circuitry, the use of conformal coatings has spread into the commercial arena.

Conformal coating usage occurs for a variety of reasons. In humid or marine salt-laden atmospheres the insulation resistance of PWB laminates often drops very significantly due to ionic contaminants that form conductive solution paths between circuits. Conformal coatings provide a major slowing of this degradation and in addition can prevent fungal attack. Abrasion-related failures due to mishandling and environmental dirt can be averted with conformal coatings. Corrosion of metallizations due to chemicals such as jet fuel and engine fluid contaminants in the environment can also be reduced. Coatings can provide added insulation protection between conductors against increased voltage levels. If the coating is formulated to contain an alumina trihydrate powdered filler, then improved resistance to carbon tracking results in high-voltage arcing conditions.

Since most conformal coatings were developed in response to the harsh environments experienced in aerospace and military applications, many coatings on the market today are identified in military specification MIL-I-46058. Further, coatings selected from the qualified products list (QPL) for this specification will have already been qualified to high-performance standards.

Two other organizations have documentation on conformal coatings. IPC covers the qualification and performance of PWA conformal coatings in IPC-CC-830.

Underwriters Laboratory has qualified three groupings of conformal coatings: acrylics, solvent-based and water-based polyurethanes, and high-temperature silicones. These have passed rigorous UL746C and 94V test qualifications and are being used by the consumer electronics manufacturers to improve the reliability of their products.

The five classes of coatings listed in MIL-I-46058 are acrylic (type AR), polyurethane (type UR), epoxy (type ER), silicone (type SR), and polyparaxylylene (type XY). Some comments on each of these coating types, along with polyimides and diallyl phthalates, are presented in the following sections. Chapter 1 covers the chemistry of each of the coating types in greater depth.

Acrylic Coatings. Acrylics are excellent coating systems from a production standpoint because they are relatively easy to apply. Furthermore, application mistakes can be corrected readily, because the cured film can be removed by soaking the PWA in a chlorinated solvent such as trichloroethane or methylene chloride. Spot removal of the coating from isolated areas to replace a component can also be accomplished easily by saturating a cloth with a chlorinated solvent and gently soaking the area until the cured film is dissolved.

Since most acrylic films are formed by solvent evaporation, their application is simple and is easily adaptable to manufacturing processes. Also, they reach optimum physical characteristics during cure in minutes instead of hours.

Acrylic films have desirable electrical and physical properties, and they are fungus-resistant. Further advantages include long pot life, which permits a wide choice of application procedures; low or no exotherm during cure, which avoids damage to heat-sensitive components; and no shrinkage during cure. The most obvious disadvantage of the acrylics is poor solvent resistance, especially to chlorinated solvents.

Polyurethane Coatings. Polyurethane coatings are available as either single- or two-component systems. They offer excellent humidity and chemical resistance and good dielectric properties for extended periods of time.

In some instances the chemical resistance property is a major drawback because rework becomes more costly and difficult. To repair or replace a component, a stripper compound must be used to remove effectively all traces of the film. Extreme caution must be exercised when the strippers are used, because any residue from the stripper may corrode metallic surfaces.

In addition to the rework problem, possible instability or reversion of the cured film to a liquid state under high humidity and temperature is another phenomenon that might be a consideration. However, polyurethane compounds are available to eliminate that problem.

Although polyurethane coating systems can be soldered through, the end result usually involves a slightly brownish residue, which could affect the aesthetics of the board. Care in surface preparation is most important, because a minute quantity of moisture on the substrate could produce severe blistering under humid conditions. Blisters, in turn, lead to electrical failures and make costly rework mandatory.

Single-component polyurethanes, although fairly easy to apply, require anywhere from 3 to 10 days at room temperature to reach optimum properties. Two-component polyurethanes, on the other hand, provide optimum cure at elevated temperatures within 1 to 3 h and usually have working pot lives of 30 min to 3 h.

Epoxy Coatings. Epoxy systems are available as two-component compounds only for coating electronic systems. Epoxy coatings provide good humidity resistance and high abrasive and chemical resistance. They are virtually impossible to remove chemically for rework, because any stripper that will attack the coating will also attack epoxy-coated or potted components as well as the epoxy-glass PWB. That means that the only effective way to repair a board or replace a component is to burn through the epoxy coating with a knife or soldering iron.

When epoxy is applied, a buffer material must be used around fragile components to prevent fracturing from shrinkage during the polymerization process. Curing of epoxy systems can be accomplished either in 1 to 3 h at elevated temperature or in 4 to 7 days at room temperature. Since epoxies are two-component materials, a short pot life creates an additional limitation in their use.

Silicone Coatings. Silicone coatings are especially useful for high-temperature service (approximately 200°C). The coatings provide high humidity and corrosion resistance along with good thermal endurance, which makes them highly desirable for PWAs that contain high heat-dissipating components, such as power resistors.

Repairability, which is a prime prerequisite in conformal coating, is difficult with silicones. Because silicone resins are not soluble and do not vaporize with the heat of a soldering iron, mechanical removal is the only effective way to approach spot repair. That means the cured film must be cut away to remove or rework a component or assembly. In spite of some limitations, silicone coatings fill a real need because they are among the few coating systems capable of withstanding temperatures of 200°C.

In general silicones have other disadvantages. Their pot life is usually limited, bonding to certain surfaces may require a priming agent, and they have very high CTEs. A high CTE can lead to the failure of fragile solder joints under environmental thermal excursions if the application procedure is not controlled to prevent a thick nonuniform coating.

Polyparaxylylene Coatings. Polyparaxylylene coatings are commonly known as Parylene coatings. This is the tradename given to these coatings by its developer, Union Carbide Company.

Parylenes are unique in several important ways. First, they are applied in a vapor-deposition process under a 0.1-torr vacuum. The deposition occurs at room ambient in a solvent-free atmosphere, which places no thermal, mechanical, or chemical stress on the PWA. This vacuum chamber process does require special equipment.

A second important unique characteristic is that the highly active gaseous monomer molecules within the vacuum chamber are not hindered by line-of-sight application procedures such as spray deposition. A thin uniform coating occurs on all exposed surfaces, including sharp corners and edges of components and solder joints, in tiny crevices, and under components. Typical coating thicknesses are 0.5 to 2 mil. This outstanding feature produces a coated PWA which offers better protection against humid and contaminated environments than most other coatings. Tests have shown Parylenes to provide the most effective protective barrier in nuclear biological chemical (NBC) warfare environments. They have also been shown to withstand and protect PWAs from the damaging effects of decontamination with highly caustic agents such as DS_2.

The requirement for special vacuum-deposition equipment is often a major disadvantage, but there are numerous companies that will provide coating services. Since Parylene coats all exposed surfaces, the masking procedures of parts such as connector pins require greater care and process development. The tenacious nature of the film causes a more difficult repairability procedure to be implemented. Removal is usually accomplished through mechanical means or by burning off the film using a solder iron. Reapplication of Parylene to the repaired area must occur in the vacuum chamber. In many instances a repair of a Parylene-coated assembly may be through the use of another type of coating material.

Polyimide Coatings. Polyimide coating compounds provide high-temperature resistance and also excellent humidity and chemical resistance over extended periods of time. Their superior humidity resistance and thermal range qualities are offset by the need for high-temperature cure (from 1 to 3 h at 200 to 250°C). High cure temperatures limit the use of these coating systems on most PWAs. Because the polyimides were designed for high-temperature and chemical resistance, chemical removal and burn-through soldering cannot be successful.

Diallyl Phthalate Coatings. Diallyl phthalate (DAP) varnishes also require high-temperature cure (approximately 150°C), which limits their use on PWAs. Furthermore, their removal with solvents or with a soldering iron is difficult, owing to their excellent resistance to chemicals and high temperatures (350°F).

Coating Selection. A designer has many considerations to be aware of in selecting the optimal conformal coating for a particular application. There is not one universal conformal coating that is right for all applications.

Manufacturers have developed a vast variety of coatings under each of the coating types mentioned. The QPL for MIL-I-46058, dated January 21, 1992, lists 21 type AR, 10 type ER, 19 type SR, 34 type UR, and 5 type XR coating materials distributed by 21 different manufacturers. These various formulations are developed through the addition of accessory chemicals and solvent vehicles to the basic resin systems to optimize the cured films' desired properties. As an example, dyes are added as an aid in identification and fluorescins are added to aid inspection for coating uniformity under ultraviolet light. Increased film flexibility occurs through the addition of plasticizers, and adhesion enhancements occur through the use of wetting agents.

There are a number of important attributes that are desirable when selecting a conformal coating. When uncured, it should have a long pot life to allow adaquate manufacturing application procedures. It should be easy to apply, adhere to the PWA securely, and be environmentally safe. When cured, it should not shrink excessively and induce damaging mechanical stress on component solder joints. The coating should be a long-term effective barrier to moisture and chemical contaminants, provide adequate electrical insulation resistance, not crack under its intended thermal environment, and be repairable. Table 2.32 gives a listing of coating selection attributes; Table 2.33 lists important physical characteristics of the various coating materials.

TABLE 2.32 Conformal Coating Selection Chart

	Acrylic	Urethane	Epoxy	Silicone	Parylene	Polyimide	DAP
Application	A	B	C	C	C	C	C
Removal (chemically)	A	B	—	C	—	—	—
Removal (burn through)	A	B	C	—	C	—	—
Abrasion resistance	C	B	A	B	A	A	B
Mechanical strength	C	B	A	B	A	B	B
Temperature resistance	D	D	D	B	C	A	C
Humidity resistance	B	A	C	B	A	A	A
Potlife	A	B	D	D	—	C	C
Optimum cure	A	B	B	C	A	C	C
Room temperature	A	B	B	C	A	—	—
Elevated temperature	A	B	B	C	—	C	C

Property ratings (A–D) are in descending order, A being optimum.

TABLE 2.33 Conformal Coating Physical Characteristics

Properties	Acrylic	Urethane	Epoxy	Silicone	Parylene	Polyimide	DAP
Volume resistivity, $\Omega \cdot cm$	10^{15}	2×10^{11}	10^{12}	2×10^{15}	6×10^{16}	10^{16}	1.8×10^{16}
(50% RH, 23°C)		11×10^{14}	10^{17}				
Dielectric strength, V/mil					5600		
Dielectric constant							
60 Hz	3–4	5.4–7.6	3.5–5.0	2.7–3.1	3.15	3.4	3.6
1 kHz	2.5–3.5	5.5–7.6	3.5–4.5	—	3.10	3.4	3.6
1 MHz	2.2–3.2	4.2–5.1	3.3–4.0	2.6–2.7	2.95	3.4	3.4
Dissipation factor							
60 Hz	0.02–0.04	0.015–0.048	0.002–0.010	0.007–0.010	0.020	—	0.010
1 kHz	0.02–0.04	0.043–0.060	0.002–0.020	—	0.019	0.002	0.009
1 MHz	0.010–0.040	0.050–0.070	0.030–0.050	0.001–0.002	0.013	0.005	0.011
Thermal conductivity, 10^{-4} cal/(s · cm · °C)	3–6	1.7–7.4	4–5	3.5–7.5	—	—	4–5
Thermal expansion, 10^{-5}/°C	5–9	10–20	4.5–6.5	6–9	3.5	4–5	
Tensile strength, lb/in^2					10,000		
Yield strength, lb/in^2					8000		
Elongation to break, %					200		
Yield elongation, %					2.9		
Resistance to heat, °F continuous	250	250	250	400	—	500	350
Effect of weak acids	None	Slight to dissolve	None	Little or none	None	Resistant	None
Effect of weak alkalides	None	Slight to dissolve	None	Little or none	None	Slow attack	None

Coating Process. Proper preparation, application, and curing procedures must be followed to ensure that conformal coatings will perform their intended function.

First a thorough cleaning of completed PWAs is essential. Conformal coatings, once applied and cured, will do as good a job of sealing in contaminants and causing later reliability problems if the PWA is improperly cleaned, as preventing contamination from outside sources.

The following are typical procedures used after the PWA has undergone the visual, physical, and electrical testing to prepare for the coating process:

1. Vapor degrease or ultrasonic clean in a solvent to remove flux residues and other contaminants from the previous processes.
2. Remove ionic salts and other contaminants not soluble in solvents by rinsing in deionized water or alcohol.
3. Dry for 2 h at 66°C to remove any last traces of cleaners.
4. Handle with gloves and store in environmentally controlled areas prior to coating.

The coating process itself is normally accomplished in one of five different ways: spraying, dipping, flow coating, brushing, and vapor depositing.

Spray Coating. Spray-coat processes are very amenable to automated high-volume conformal coating applications. The process variables, which must be optimized to ensure repeatable results, include the material viscosity and solids content, spray pressure, nozzle design (hole sizes as well as patterns within the nozzle), distance of the nozzle from the PWA, and spray pattern across the PWA (velocity of the spray nozzle across the PWA, number of passes used, and spray angles used during the process). Advanced computer-controlled machinery is available from manufacturers to accomplish this. For low volumes manual spraying is used.

The main drawback to the spray process is that shadowing due to varying component heights and component leads makes applying a uniform coat difficult. Applying the coating under components is very difficult with this process.

Dipping. Dipping is the second method. This process can produce even uniform coatings across a PWA. To ensure that all voids are properly coated, the main process variables that must be controlled are the coating material viscosity and the immersion and withdrawal speeds of the equipment used. Evaporation of the solvents in the coating bath during this process is an important contributor to changing material viscosities and hence must be monitored. Typical immersion speeds are 2 to 12 in/min. Automated equipment for this process is available in the industry.

Masking of areas in which coating material is not wanted makes this part of the process more critical to control than when using the spraying process.

Flow Coating. Flow coating, often referred to as curtain coating, applies the material to the PWA by passing it through a "curtain" of flowing material. Material viscosity, curtain flow rate, and the velocity of the PWA as it passes through the material are important process variables. Programmable deposition equipment is available for this process. Some equipment can be programmed to vary or even stop the flow when areas that are to be left uncoated pass within the coating curtain window.

Brushing. This operation, which is manual in nature, is normally used only when repair of conformal coatings is required. This can be due to either reworking of components on the PWA or improper manufacturing processes that have left small void areas in the coating.

Vapor Deposition. This process is normally done only with the Parylene coatings. Curing of most material coating systems is either through letting the coated assemblies remain at room temperature for significant periods of time (usually 24 h, but sometimes days) or through cure of the material at an elevated temperature (usually from 15 min to 2 h at temperatures of about 170°F).

A recent development in conformal coating technology has been the implementation of ultraviolet-curable coatings. There are currently 10 different ultraviolet-curable coatings recognized in MIL-I-46058, three type AR, three type ER, three type UR, and one type SR from six different manufacturers. All but one material consist of 100 percent solids.

The ultraviolet curing materials offer many advantages. Conventional thermoplastic and thermoset chemistry material properties can be formulated through the use of the ultraviolet-curable coatings. During the manufacturing process, shorter cure cycles are needed; for the most part the cure occurs in less than 1 min. Little or no solvent emissions occur due to the 100 percent solids content of most of the systems. Reduced space requirements and reduced energy consumption are other achievable advantages.

In the ultraviolet curing process the liquid ultraviolet monomer, oligomer, and polymer components undergo polymerization and cross-linking by photogenerated radical and cationic species, which occur during the exposure to an ultraviolet light source. This exposure causes almost instantaneous hardening of the material into a cured film. The most critical item which must be optimized in this process is the development of a correct coating formulation matched to a particular ultraviolet light source.

There are a few less than optimal features of the ultraviolet systems. Shadowing of the ultraviolet light source due to component variations and component leads causes a secondary cure to be imposed on most of the ultraviolet systems on the market today. This cure ranges from 24 h at room temperature to as little as 10 to 15 min at temperatures elevated to 250 to 300°F. The materials are also difficult to repair and rework.

2.4 FABRICATION PROCESSES FOR PRINTED-WIRING BOARDS

The fabrication of PWBs that meet the performance requirements of today's electronic systems presents a considerable challenge to the "would-be" fabricator. Although fabrication processes have evolved and improved over the years to accommodate changing design requirements, most of the basic subtractive methods for producing rigid and flexible PWBs have remained the same for the last 20 years.

As mentioned in Sec. 2.2, three types of PWBs are in broad use today: rigid, flexible, and rigid-flexible. This section will focus primarily on one class of rigid PWBs, multilayer. Fabrication processes up to electroplating will be covered. Subsequent operations are discussed in detail in other sections of this handbook or in Coombs.[5] The growing worldwide demand for multilayer technology has driven more and more fabricators to develop the manufacturing technologies required to produce these sometimes complicated multilevel electronic structures. The fabrication processes associated with multilayer PWBs do, however, encompass the manufacturing requirements for single- and double-sided boards. Therefore this singular discussion of the technology is also appropriate for rigid single- and double-sided PWBs.

2.4.1 Data Preparation and Preproduction Tooling

The first step in the fabrication of PWBs is the preproduction tooling phase. This process converts the computer-generated PWB design information into the tools and data needed to produce the working PWBs. Some of the most significant improvements in the overall fabrication scenario have come as a result of the proliferation of computer technology. Computer-aided manufacturing (CAM) and computer-aided design (CAD) equipment such as that shown in Figs. 2.12 and 2.13 is now found in most engineering departments of PWB fabricators. This equipment is used to manipulate and check the myriad of information needed to produce PWBs. CAD data files contain information which is ultimately processed to produce the tooling needed to transfer circuit images onto the PWB material, drill holes for component placement, and electrically test the completed board.

One of the most important steps in the preproduction process is *panelization.* The panelization process creates the optimum panel layout for the PWB to be produced. The number of images on the panel is optimized based on material size. Test coupons, plating balancing patterns, tooling hole locations, and tab connectors are all added at the panelization step. Most modern CAM workstations are equipped with software capable of handling this function.

In addition to accurate and speedy manipulation of CAD data, some CAM workstations are equipped with software that literally checks the integrity of the design data before manufacturing begins. This design-data check identifies problems such as improper circuit locations, circuit and spacing width violations, pad and hole size compatibility, and feature distortions. This early detection of potential defects prevents the subsequent manufacture of noncompliant PWBs.

2.4.2 Phototool Generation

Many of the more veteran PWB fabricators remember the days of the camera and a steady hand with lithographer's tape as the preferred method of producing phototools. Those days have given way to the age of the photoplotter. It is important to note here that the quality of the phototool is becoming increasingly important because of requirements for ever smaller circuits, greater detail, and increased density. The integrity of these tools, which are used to transfer images to the PWB material, is extremely important in minimizing the production of defective parts. The production of quality phototools is directly dependent on the quality of the CAD/CAM data and plotting equipment used for their manufacture.

Photoplotters in use today are of two general types: vector and raster. Vector plotters have been around for many years and produce very precise plots. They require less computer storage capacity because plotting is performed by locating and positioning x–y coordinates and thus need only store those coordinates. The problem with vector plotters is speed: they operate very slowly. A complicated high-density circuit pattern with many lines and features could take several hours to complete using this type of plotter.

Raster plotters, which usually employ lasers, digitize the entire plot area into pixels, or what may be described as individual image components. As the pixel size decreases, resolution increases, thus allowing sharper line and feature definition. Resolution in the range of 0.00025 mil (1 mil = 0.001 in) is required for some of today's high-density designs. The tradeoff here, however, is a requirement for much more computer capacity than is needed for vector plotters. In addition, the processing of large numbers of pixels requires some increase in plot time, although still far less than vector plot times.

Despite the expensive requirement for computer memory and the increased time required to plot complicated patterns, raster plotters are the plotters of choice in most

FIGURE 2.12 CAD-CAM equipment.

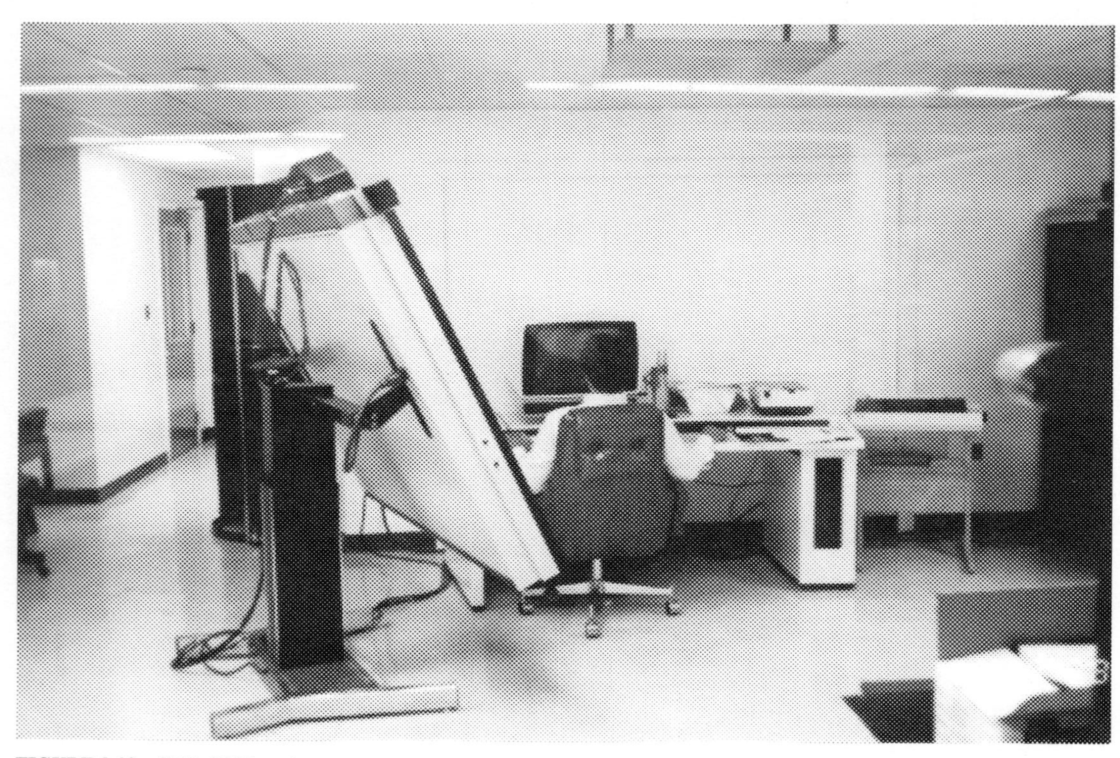

FIGURE 2.13 CAD-CAM equipment.

PWB fabrication shops. Their ability to produce very-high-quality phototools in generally shorter periods of time makes them ideal for fabricators that focus on both volume and high quality.

The overall accuracy of the image transfer process has been greatly increased as a result of the laser photoplotter's ability to produce multiple image phototools. The elimination of copying steps, which were once required to produce "step and repeat" phototools from original single-image films, is one of the most significant contributions to this increased accuracy. First-generation phototools directly from the platens of photoplotters are the most accurate mechanisms for producing circuit images.

The selection of the most appropriate photoplotter is difficult and should be accomplished based on the complexity and volume of the PWBs to be built by the fabricator. As detailed in Table 2.34, many manufacturers have entered the photoplotter market. Prices range from $85,000 to $390,000, depending on the number of features and the level of accuracy desired. These prices are not precise and are intended to provide general price ranges for the various classes of plotters.

Phototools. Two primary types of phototools are used in PWB fabrication today: diazo and photographic material (usually polyester-based). Although glass phototools are very dimensionally stable and are used in some shops, problems with storage, handling, and cost make this approach unattractive, especially for higher-volume fabricators.

One of the most important considerations in choosing the type of medium for phototooling is dimensional stability. Dimensional changes in polyester-based phototools will have a direct impact on the quality of the board produced. It is important to note that the dimensional integrity of these materials is directly proportional to the level of environmental control in the production and storage area. In order to limit dimensional changes to 1 mil or less over 10 in, rigid control of temperature and relative humidity is essential.

A key to producing phototools with good dimensional stability is the control of the film exposure process. One of the ways to avoid dimensional changes during this operation is to allow the unexposed film to reach equilibrium in its use area prior to processing. This is especially true of materials that have been refrigerated for some time. The rule of thumb is to allow 1 h of equilibration for each mil of base thickness. The point to remember is that the longer the period of time the film has to reach equilibrium with its use environment, the more stable the product.

Many shops have opted to use photographic films as opposed to diazos in order to eliminate a copying step. Photographic films are imaged directly on the photoplotter and subsequently processed for use as a phototool. Diazo phototools must be made from these "originals" and are used as "shop" film because of their durability and versatility.

2.4.3 Artwork Registration and Material Preparation

The choice of materials for PWBs is part of the design requirements and as such is not left to the discretion of the fabricator. However, the methods by which the material is handled and prepared for subsequent processing are extremely important considerations when developing the fabrication process. This is especially true in the case of multilayer PWBs since the alignment of layers is critical. The success of the image transfer process hinges on the ability of the fabricator to manage the steps of the registration scenario.

TABLE 2.34 Photoplotter Comparison Matrix

Manufacturer and model	Platen type	Image area	Accuracy, mil	Repeatability, mil	Minimum pixel, mil	Min. line	Pixel rate, millions pixels/sec	Imaging speed, in^2/min	Glass	Auto-load	Develop convey	Base price, US$
Lavenir Technology Inc.												
RPG-1622/si	Flatbed	16.0 × 22.0	1.5	0.75	2.0	5.0	0.1	25	No	No	No	44,000
Mivatec Inc.												
Miva 1450	Flatbed	14.0 × 22.0	0.5[b]	0.5	0.5	2.0	1.3	20	Yes	No	No	40,000
Miva 1651	Flatbed	16.0 × 22.0	0.5[b]	0.5	0.5	2.0	1.3	20	Yes	No	No	49,000
Miva 2201	Flatbed	22.0 × 26.0	0.5	0.3	0.5	2.0	1.7	25	Yes	No	No	79,000
Miva 2201H	Flatbed	22.0 × 26.0	0.5	0.3	0.25	0.1	1.3	5	Yes	No	No	82,000
Optronics												
FilmMaster 2000	Drum	19.6 × 25.0	0.5	0.2	0.5	1.0[b]	16.3	245	No	Optional	Optional	165,000
Optrotech												
LP5008A	Drum	25.0 × 37.0[e]	0.4	0.2	0.25	1.5	13.1	49a,f	No	Optional	Optional	170,000
LP5008E	Drum	25.0 × 37.0	0.4	0.2	0.125a	0.75	34.5	37	No	Optional	Optional	230,000
LP5008G	Drum	25.0 × 37.0	0.3	0.15	0.125a	0.75	34.5	37	No	Optional	Optional	400,000
Orbot												
FP-2125	Drum	20.0 × 25.0	0.8	0.2	0.2[a]	0.5	12.0	29[a]	No	Optional	No	180,000
FP-2125M	Drum	22.0 × 25.0	0.8	0.2	0.2	0.5	13.3	32	No	Optional	No	190,000
FP-2143	Drum	31.0 × 42.8	0.8	0.2	0.2	0.5	17.1	41	No	Optional	Optional	220,000
FP-2225	Drum	20.0 × 25.0	0.8	0.2	0.1	0.5	66.6	40	No	Optional	No	230,000
FP-2225M	Drum	22.0 × 25.0	0.8	0.2	0.1	0.5	73.3	44	No	Optional	No	240,000
FP-2243	Drum	31.0 × 42.8	0.8	0.2	0.1	0.5	93.3	56	No	Optional	Optional	280,000

SECMA Inc.												
SWIFTLine 2026	Drum	20.5 × 25.5	0.2[b]	0.08[b]	0.125[a]	0.8	61.9	58[a]	No	Optional	Optional	100,000
SWIFTLine 2636	Drum	26.0 × 36.0	0.2	0.08	0.125	0.8	65.1	61	No	Optional	Optional	230,000
SWIFTLine 3242	Drum	32.0 × 42.0	0.28	0.12	0.125	0.8	57.6	54	No	Optional	Optional	330,000
SWIFTLine 4247	Drum	42.0 × 47.0	0.28	0.12	0.125	0.8	64.0	60	No	Optional	Optional	360,000
SWIFTLine 4763	Drum	47.0 × 63.0	0.4	0.16	0.125	0.8	60.8	57	No	Optional	Optional	390,000

[a] Indicates that the pixel size is user-settable. The best resolution (smallest pixel size) is shown, with the plotting speed at that resolution. In all cases, setting the machine to a lower resolution dramatically increases the plotting speed.

[b] Indicates that the manufacturer's specification is qualified. Consult the applicable notes in the manufacturer's published specifications for further details.

[c] Plotting speed varies depending on plot complexity.

[d] A manual cartridge handling system is used to carry film to the developer.

[e] Other size configurations (up to 34.0 × 36.0 in) are optional.

[f] The 5008 plotter exhibits a geometric increase in speed at decreased resolutions. Thus the plot speed with a 0.25-mil pixel is 148 in^2/min.

Artwork Registration. There are many elements in the overall registration or layer or feature alignment process for multilayer boards. The process starts with the alignment of the internal artwork layers. These phototools must first be aligned front to back. As mentioned, the stability of the artwork is a variable that is, to some extent, controlled by the fabricator.

The environment in which the phototool is stored contributes to dimensional integrity. Mylar-based artwork is affected by temperature changes (0.000015 in/in per 1°C change in temperature) and relative humidity (0.000009 in/in per 1 percent change in relative humidity). The importance of controlling these environmental parameters becomes very clear when one considers that a 24-in phototool could experience growth of 2 to 3 mil as a result of a 15 percent change in relative humidity or a 5° change in temperature. Artwork distortion of this magnitude manifests itself in misalignment of features throughout the finished multilayer board.

In the days before fabricators had the capability to produce their own artwork, they were forced to use films supplied by their customers. The built-in inaccuracies inherent in the phototool production process were passed on to the fabricator. Now that fabricators are able to plot film, fewer variables are introduced into the registration and image-transfer process. Much more freedom is given to board manufacturers, allowing them to add targets or registration symbols to each layer of artwork. These features on the working artwork allow the use of both manual and automated alignment processes. The choice of registration system is driven by many factors, including the degree of accuracy required by the PWB design. In the case of multilayer boards used in sensitive military and medical applications, registration tolerances must be minimized in all operations. Several approaches exist for the preparation of artwork for the image-transfer process. They range from fully manual processes greatly dependent on the skill level of the operator to sophisticated automated systems employing computers and numerical control.

Many fabricators choose a system that requires the punching or drilling of the unexposed photographic film with holes or slots corresponding to pin locations in the photoplotter platen. The film is pinned to the plotter and exposed. This approach contains inherent inaccuracies associated with the punching or drilling process as well as the location of the pins in the platen. As much as 0.003 in of misalignment is commonplace using this technique. Since registration and subsequent processing tolerances are cumulative, a misalignment of this amount at the very beginning of the fabrication cycle could be disastrous to the finished product.

Another more recent approach to artwork registration is postimage punching of the artwork. Film is exposed on the photoplotter without pinning to the platen. This approach requires the plotting of registration targets in the same location on all layers of the PWB. These targets are usually a set of concentric circles with cross hairs denoting the center point. They are located diagonally along the longest dimension of the image. Plotter pinning inaccuracies are virtually eliminated using this method. Once the film is plotted and developed, the job of punching the alignment holes or slots becomes critical to the accuracy of the registered artwork package.

One of the most accurate methods of punching the processed artwork employs an automated computer-controlled system with a vacuum bed and a series of closed-circuit television cameras. The registration symbols are aligned to preset camera positions. The film is held in place by vacuum while punching takes place. A computer stores the punch location data, which aids in compensating for any dimensional shifts that may have occurred. This method of producing punched phototools has reduced tolerance buildup in multilayer packages by as much as 65 percent.

Material Registration. Alignment holes or slots corresponding to the location of holes or slots in the artwork must be punched or drilled into the inner-layer material so that alignment for image transfer is possible. Location accuracy is critical to the multilayer process because the overall board quality depends on layer-to-layer alignment. Inner-layer pads must be aligned with a great degree of accuracy so that subsequently drilled holes fall as close to the center of each pad as possible. Layer-to-layer misalignment results in hole "break-out," which could cause short circuits or open circuits in the completed multilayer board.

Coombs[5] describes a registration system based on the use of orthogonal slots. The system makes use of four slots, with those of the x axis being at right angles to those of the y axis (Fig. 2.14). The advantage of this configuration is based on the contention that movement of the panel during processing occurs from the center. Therefore expansion and contraction error is spread over the entire panel instead of accumulating at one side or along one edge. Many fabricators have experimented with and adopted this approach and found it to reduce layer-to-layer misalignment. Automatic punches are available employing the slot configuration.

Material Preparation. Inner-layer material must be prepared for the image-transfer process not only by providing tooling holes for subsequent processing but also by conditioning the copper surface for the application of photoresist. Conventional methods of surface treatment and cleaning include both chemical and mechanical processes. These processes are designed to remove oxides, fingerprints, coatings, and other contaminants that could impede resist adhesion to the copper surface.

The use of mechanical scrubbing machines is very commonplace in the industry. This method is used by fabricators producing medium to large volumes of PWBs. Scrubbing machines are usually conveyorized, employing top and bottom rotating brushes so that both sides of the material are cleaned simultaneously. Water is usually

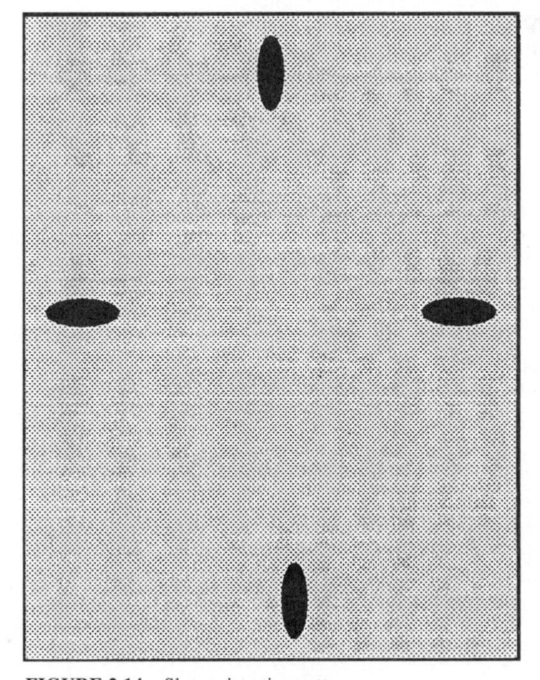

FIGURE 2.14 Slot registration pattern.

introduced to the copper surface via a series of high-pressure nozzles to aid cleaning and provide a medium for removed copper and oxidation. Treatment of this "slurry" in order to render it benign to the environment is an extremely important consideration. Filtering and recycling are essential.

Chemical cleaning usually involves the use of hydrochloric acid in an immersion and rinse process. Although this method of cleaning copper surfaces is fairly effective, it does involve the use of hazardous chemicals and requires elaborate handling and disposal systems. Some low-volume high-reliability fabricators have used this method successfully because it imparts fewer stresses to thin material, thereby reducing dimensional shifts.

Pumice scrubbing has been used by a number of fabricators and has proven to cause problems related to pumice particles entrapped in the surface topography of the copper cladding. This contamination prevents proper adhesion of photoresist to copper surfaces. In addition, dried pumice particles are easily transferred to surfaces that will come in contact with plating baths, causing contamination.

2.4.4 Image Transfer

The transfer of circuit images to the PWB material is one of the most critical processes of the fabrication cycle. The integrity of the etched or plated feature is directly related to how well the image-transfer steps are optimized. These circuit features are defined by the resist material, which determines the pattern to be plated or etched into the copper cladding. We will concentrate here on inner-layer print and etch processing.

The two methods of image transfer are photo printing and screen printing. Photo printing in PWB fabrication usually involves liquid or dry-film resists while screen printing makes use of only screenable liquid resists.

The choice of resist materials and related processes for their application should be made with several factors in mind. They include:

1. Line width and spacing requirements
2. Capital equipment requirements
3. Lot size and overall volume
4. Floor space and plant layout

Since the introduction of dry-film resists by duPont in the late 1960s many fabricators have opted to use this method of image transfer. The transfer method of choice had been screen printing up to that point, although liquid resists were used by some fabricators. The limitations of both liquid and screenable resists are well documented. In the case of liquid resists the major drawbacks include handling, consistency of the coating (especially when dip techniques are used), contamination, and the inability to "tent" holes effectively so that pattern plating may be performed. Screen printing was used extensively in the early days of commercial-grade PWB fabrication because it was capable of producing the rather simple designs of that era. Circuit widths and spaces were commonly in the 0.015-in to 0.025-in range. With the introduction of more complex circuit patterns and smaller circuit widths, screen printing became less attractive because of its inability to produce circuit widths below 0.012 in consistently. The limitations on the thickness of the resist coating also proved to be an obstacle in some pattern plating applications.

In addition to the inherent limitations of the resulting screened coating, the number of screen-printing process variables is very high. They include screen material, mesh size, mesh material, squeegee hardness, squeegee pressure, squeegee speed, ink viscosity, substrate condition, and curing time.

Dry-film resists may be either positive or negative working, depending on how they react to light energy. The most widely used dry film, by far, is negative working. These resists are polymerized by exposure to ultraviolet energy and become insoluble in the developing chemistry. Dry-film resists may also be categorized by the nature of the chemicals needed to develop and strip the unexposed and exposed resist.

Resist Exposure and Development. Photoresist exposure, or photo polymerization, is a complicated series of chemical reactions. The process begins, however, with the accurate alignment of the phototool to the resist-coated inner layer (in the case of multilayer PWBs). The two are brought into intimate contact and the resist is exposed to ultraviolet energy in equipment similar to that shown in Fig. 2.15. This exposure

FIGURE 2.15 Resist exposure equipment.

process is critical and must be tightly controlled. Underexposed resist is soft and may be washed away inadvertently in the developing process. Overexposure, on the other hand, causes ultraviolet energy to bleed underneath protected photoresist and results in the plugging of spaces between tracks. Exposure control mechanisms include variable-density step tablets, which measure the change in the level of resist polymerization with varying exposure times. Methods of measuring the exposure energy by means of photometers are also used to optimize the exposure and development process.

Removal of the unexposed photoresist (development) is usually accomplished in horizontal, conveyorized systems similar to that shown in Fig. 2.16. Equipment should be designed with fan nozzles that flood the panel with solution at the very front end of the developing chamber. The nozzles, which should be high-volume and low-pressure, are then able to dislodge the softened resist. Acid dips and warm-water rinsing are sometimes used following development to aid in hardening the polymerized resist and removing developer residues.

Etching. The removal of copper from PWB material (etching) to achieve the desired circuit pattern is an extremely complex topic. This technology is covered in great detail in Coombs[5] and is also discussed in Chap. 9. The most commonly used etchants are cupric chloride, hydrogen peroxide–sulfuric acid, ferric chloride, ammonium persulfate, chromic-sulfuric acid, and alkaline ammonia. Some others are also used, but this partial list covers most of the major processes. Waste disposal and pollution control have historically been all-consuming issues related to the etch process.

Resist Stripping. Removal of the polymerized photoresist after etching is the final step in the image-transfer process. Most dry-film resists have been formulated such

FIGURE 2.16 Resist development equipment.

that they are easily removed in aqueous-alkaline or solvent strippers. Horizontal conveyorized equipment, very similar to developing equipment, is used extensively.

2.4.5 Lamination

Once internal layers have been produced and verified, the next step in the fabrication process is lamination, where all internal layers of the multilayer board are bonded together under heat and pressure. This process distinguishes double- and single-sided boards from multilayer ones. It is important to note that the lamination process is very interdependent on tooling, materials, equipment, and the lamination cycle. It is also important to note here that fabrication errors at this point cause the scrapping of more than an individual layer. The cost of errors becomes much more significant.

Oxide Treatment. In order to increase the bondability of the copper surface once the inner layer has been fabricated, many PWB manufacturers "grow" an adhesion-promoting oxide coating on the surface of the copper. These coatings are usually referred to as black oxide (low temperature) or red oxide (high temperature). Typical oxide coatings are sodium chlorite–based and are stabilized with sodium hydroxide or trisodium phosphate. Once the coating has been deposited, a drying step is required. The coated inner layers should be laminated not more than 8 hrs. after coating. In addition they should be handled with lint-free gloves.

Tooling. The most common material for lamination fixtures is steel. Steel is the material of choice because it more closely approximates the thermal expansion characteristics of the epoxy-glass-copper matrix of the PWB below the glass transition temperature of the resin system. Above that temperature, the CTE increases significantly. A good thermal match is required to minimize stresses induced during the lamination cycle. The fixtures should be checked frequently for dents, contamination, and scratches because at high temperature and pressure, plate defects are tansferred to the PWB. At least two pins should be used to align the layers to be laminated. As mentioned, slots have become a very popular mechanism for registration. The orthogonal slot system described previously is also the preferred method of achieving quality alignment of layers during the lamination process.

Materials. The bonding material used in the lamination process is the semicured equivalent of the inner-layer material and is referred to as B-stage or prepreg. This material is not fully polymerized or cross-linked. It is interwoven between the completed inner layers and under heat and pressure becomes fully cured. B-stage material should be stored in sealed bags at 45 to 50°F and 50 to 55 percent relative humidity. The material should be allowed to come to room temperature before use. Gloves should be used to handle the prepreg B-stage material as well as the inner-layer C-stage material to avoid contamination.

Monitoring of prepreg properties is an important element of a successful lamination process. The following are the most critical of these properties:

1. *Resin flow* describes the tendency of the resin to flow once its minimum viscosity is achieved during the lamination cycle. Too much flow causes resin starvation, while too little flow causes too thick PWBs.
2. *Resin content* describes the percentage of resin in the resin-fabric matrix. Again, too little resin results in resin-starved PWBs and too much resin results in too thick boards.

3. *Gel time* refers to the time required for the resin to begin to change phases under heat and begin to cross-link into a cured system.
4. *Volatile content* provides a measure of the amount of residual volatiles remaining in the resin which could, upon heating, form bubbles in the PWB.

One of the most popular methods of testing incoming prepreg is *scaled flow*. This test allows the fabricator to simulate press cycles and optimize the lamination process by defining pressed thickness and flow properties of the prepreg in a setup that is very similar to actual laminating. Scaled-flow testing has essentially replaced percent resin flow as the primary method of establishing press parameters.

Equipment. The most important single piece of equipment in the lamination process is the press. Typical laminating presses are electric or steam heated and are usually multiple opening. They may be either conventional or vacuum assisted. Today's presses are designed to produce temperatures in excess of 500°F and pressures up to 500 lb/in^2. This wide range of heat and pressure provides fabricators the ability to laminate today's myriad of PWB materials. Temperature and pressure gauges are required so that monitoring these parameters is possible during the lamination cycle.

Press maintenance is critical to any successful lamination operation. Surface temperature profiling must be accomplished regularly to assure uniform heating on all regions of both platens. Thermocouples sandwiched between two sheets of release or insulating sheets should be placed on each platen to determine whether temperature variations exist across the surface of the platen. Uneven heating (hot or cold spots) causes voids and nonuniform resin distribution in the finished PWB. Platen parallelism and flatness are also critical press parameters. Platen parallelism should be maintained within 0.003 to 0.005 in in order to assure uniform resin flow and consistent thickness across the entire PWB.

The most recent advance in lamination technology is the use of vacuum-laminating processes. Vacuum laminating removes air, moisture, and residual volatiles from the laminating package, greatly reducing the incidence of blistering between layers of the multilayer PWB. The application of a vacuum equivalent to 30 to 40 mm Hg greatly reduces the amount of pressure required to achieve the same level of volatile and moisture removal. Because the vacuum removes these trapped gases, the lamination package requires far less "squeezing." This lowered pressure (reduced in some cases from 350 to 150 lb/in^2) commensurately reduces the amount of misregistration resulting from inner-layer movement during the lamination cycle. In addition, the vacuum removal of residual gases provides for a much more homogeneous resin system, resulting in a higher-quality PWB.

Lamination Lay-up and Press Cycle. Figure 2.17 shows a conventional multilayer package inside a press. The lay-up process should occur in an environmentally controlled area with filtered air. Multiple boards may be placed in the packages if separated by a good thermal conductor which is sufficiently thick (0.025 in) to avoid transfer of circuit discontinuities from board to board.

Press cycles are driven by the PWB material used and the thermal properties of the press and lay-up materials. The most important aspects of the cycle are related to the changes in resin viscosity during heat up, pressure application, and cool down in the lamination process. Laminate and prepreg producers provide guidelines for developing optimum cycles. Most of today's prepregs possess very predictable properties so that supplier data sheets are usually very good starting points for the fabricator to develop tailored press cycles.

Figure 2.18 shows viscosity curves for typical electric and steam presses operating

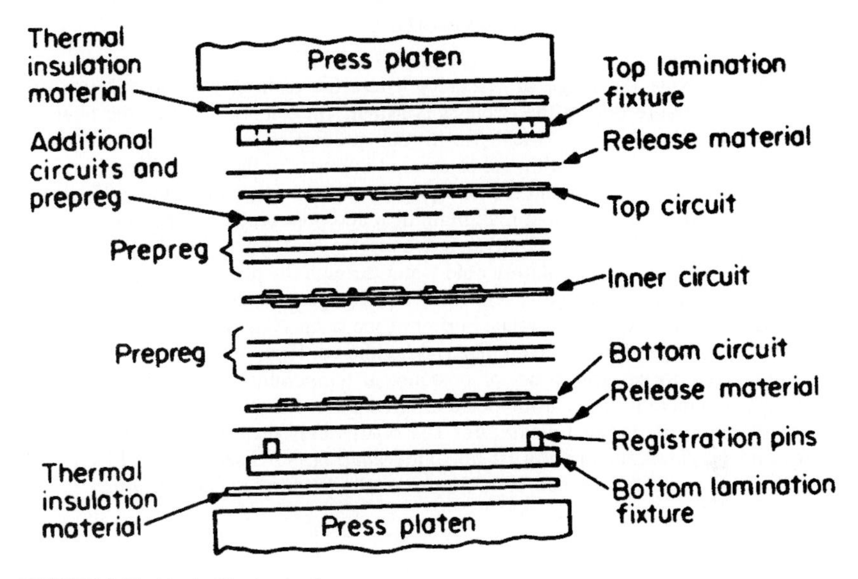

FIGURE 2.17 Typical lamination lay-up.

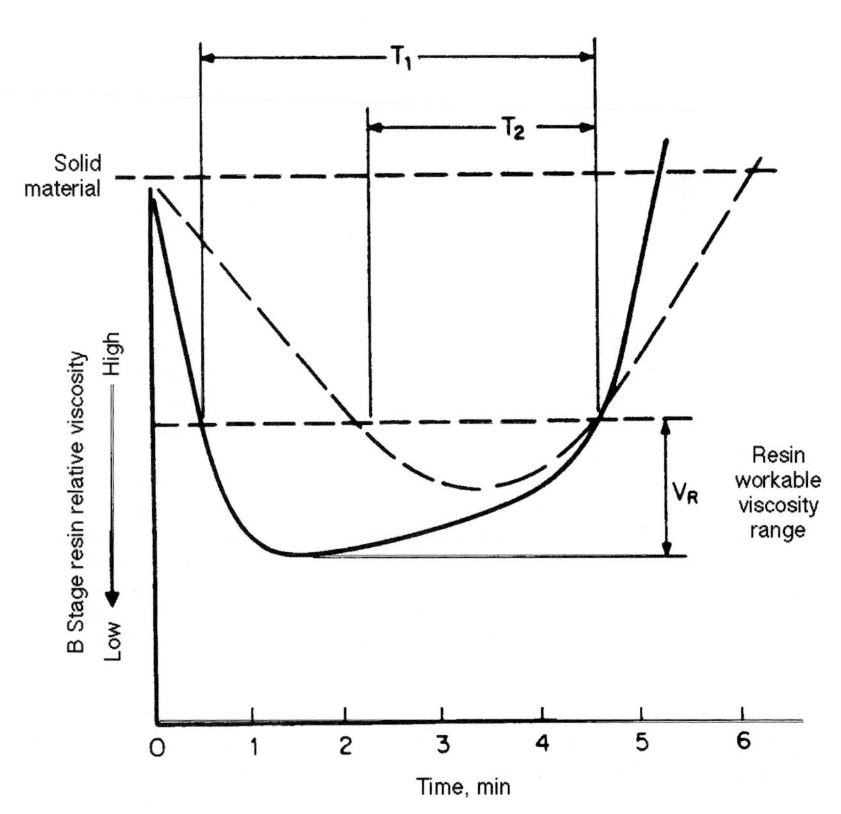

FIGURE 2.18 Resin viscosity curves.

at about 350°F. Note the range of viscosity where the resin is workable. Since the application of pressure during the cycle is constant, close monitoring of the heat-up characteristics of the press is critical. Resin melt properties are also important since these properties determine the ultimate functional integrity of the PWB. Depending on the characteristics of the resin and the heat profile of the press, lamination cycle times usually range from 45 to 90 min at maximum heat and pressure. The laminate package is cooled to around 100 to 115°F, either in a separate cool-down portion of the press or in the hot press itself by circulating cold water through the platens.

Once the lamination package is removed from the press, a postbake period is recommended in order to assure complete curing of the resin system. Postlamination curing temperatures depend on the resin material laminated. High-temperature resins such as polyimide require several hours of postcure at temperatures in excess of 400°F. Most fabricators have found that the postcuring step is more effective if accomplished with the PWBs in the laminating fixture. The boards may then be removed from the fixtures and prepared for drilling. If an oxide coating was used, it should now be removed with a 5- to 10-s dip in a 10 to 20 percent hydrochloric acid solution.

2.4.6 Drilling

Holes are drilled in the laminated PWB in order to provide a method of interconnection between top, bottom, and internal layers. These holes are subsequently plated with copper in order to provide electrical continuity where required. The ability of the drilled hole to accept this plating is therefore critical to the ultimate performance of the PWB. In addition to providing the electrical interconnecting mechanism, drilled holes also provide the sites for the placement of components. Therefore location accuracy and structural integrity are also critical.

The drilling operation comprises many very interdependent elements which must be tightly controlled and optimized. These elements are of two major categories: (1) materials (drills, entry materials, and back-up materials), and (2) equipment (operating parameters).

Materials.

Drills. Most PWB drills are made of tungsten carbide because of some of the unique properties of this material. It may be procured at reasonable costs, is extremely hard yet machinable, and is easily handled. This combination of properties is ideal for producing small (0.004-in) and larger (0.125-in) holes in most of today's PWB materials. Although some drill manufacturers have attempted to improve hardness by surface treatments and ion implantation, conventional drills have proven to provide high-quality holes as long as the critical elements mentioned previously are optimized.

Common shank drills are the most widely used because of their compatibility with spindles used on today's drilling equipment. Figure 2.19 shows the geometry of the typical common shank drill. Bit geometry is extremely important for minimizing drilling temperatures and maximizing chip-removal efficiency. Point angle (115 to 130°) and helix angle (30 to 40°) are two of the most important measured features. Drilling temperatures are also lowered by reducing the amount of bit surface area which comes in contact with the hole wall. This is accomplished by removing some of the material directly behind the cutting area.

The condition of the primary cutting surface of the drill tip is the most important aspect of the drill's contribution to quality holes. The cutting lip (Fig. 2.19) must be as sharp and free from gouges and discontinuities as possible. It is therefore important to handle drills with extreme care and institute a rigid inspection of incoming drills.

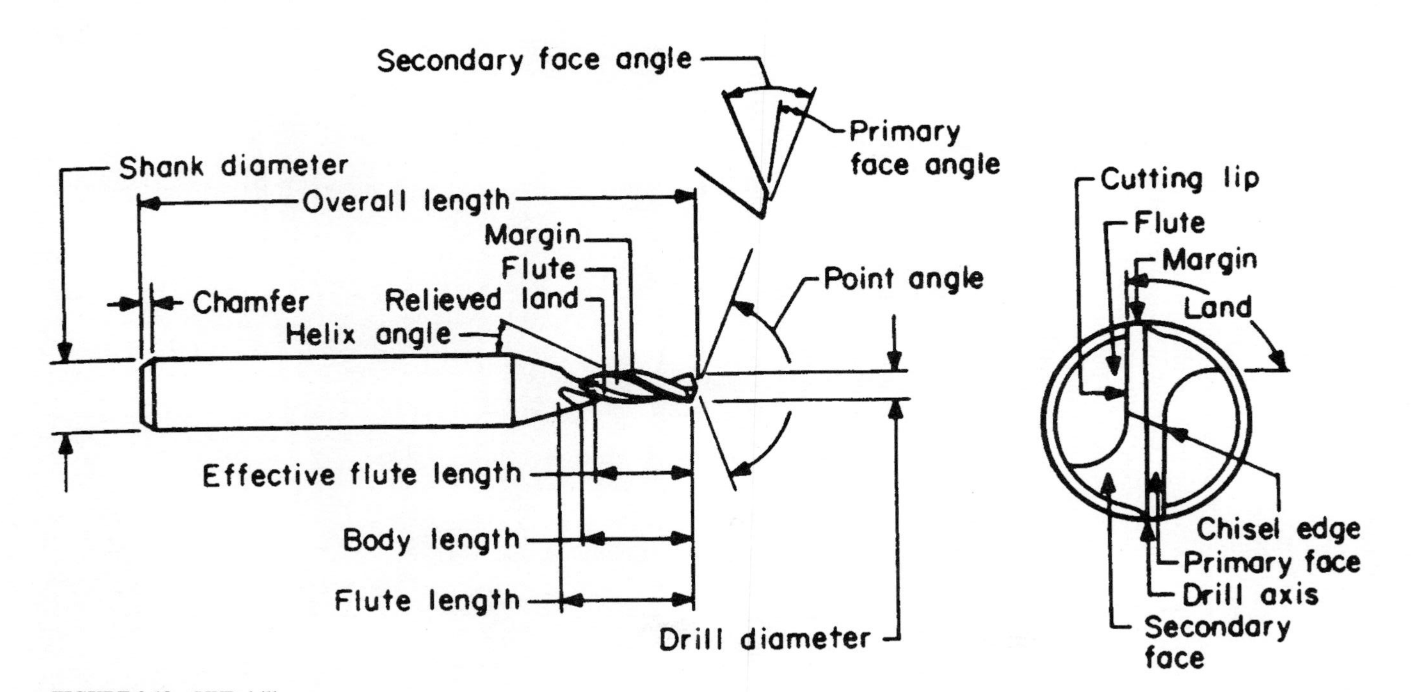

FIGURE 2.19 PWB drill.

Entry and Back-up Materials. The purpose of entry materials is manyfold. They should minimize entry burrs, protect the copper cladding of the PWB from the spindle pressure foot, and minimize drill wander. In addition the material chosen should not contribute to hole contamination, excessive drill wear, or excessive drilling temperatures. The most commonly used materials are solid aluminum, aluminum-clad phenolics, and paper-based phenolics.

Back-up materials, which are placed underneath the copper-clad PWB, should be selected to do much more than prevent exit burrs. Once the hole has been made in the PWB, the drill then enters the back-up material and, upon retracting, pulls with it debris lodged within the flutes. This debris causes a myriad of problems, including hole-wall gouging, contaminated drill bits, and excessive drill wear. The back-up material chosen must be soft enough to produce chips that will not damage the hole walls, yet it must have a surface hard enough to prevent exit burring. The most effective and commonly used materials are hardboard, aluminum-clad phenolic or hardboard, and vented aluminum.

Equipment. Most drilling machines used in today's multilayer PWB fabrication shops are multiple-spindle CNC with automatic tool changers. The cost of these machines ranges from $80,000 to $500,000, depending on the number and capability of spindles, level of accuracy and repeatability, level of automation, and the number of sensors that detect such things as broken drill bits. Drilling equipment such as the five-spindle model shown in Fig. 2.20 should be installed in well-lighted dust-free temperature- and humidity-controlled environments.

The capabilities of drilling machines are varied. Table travel speeds may be as fast as 500 in/min. Spindle speeds on more expensive machines exceed 100,000 r/min,

FIGURE 2.20 Five-spindle drilling machine.

while in-feed rates exceed 200 in/min. These features allow this equipment to produce 300 to 400 holes per minute, accommodating both moderate and high-volume PWB fabricators.

The implementation of a well-disciplined preventive maintenance program is essential for the upkeep of these very sophisticated machines. The most important elements of the program should be the following:

1. Spindle and collet maintenance should include regularly scheduled checks for run out (which should not exceed 0.005 in), adequate cooling via a well-maintained heat exchanger, and thorough cleaning with a noncorrosive solvent.

2. Tachometers should be used to verify and calibrate spindle speed and in-feed rates.

3. Chip-removal (vacuum) systems should be checked often to assure maximum efficiency.

4. All pneumatic systems should be checked regularly for efficient operation.

5. Filters should be changed regularly.

6. Debris generated from the drilling operation should be removed from the table and other surfaces during each shift. This debris should never be blown from these surfaces using compressed air. Vacuum cleaners should always be used.

Operating Parameters. Most PWB fabricators will agree that the most important drill-machine parameters are spindle speed and in-feed rate. These two settings, along with the quality of the drill cutting surfaces, more than any others, determine the impact of the drill on the quality of the hole produced. Spindle speed, measured in revolutions per minute (r/min), and in-feed rate, measured in inches per minute (in/min), are interrelated by a term called *chip load.* This term is derived by dividing in-feed rate by spindle speed:

$$\text{Chip load (in/r)} = \frac{\text{in-feed rate (in/min)}}{\text{spindle speed (r/min)}}$$

Chip load, therefore, is defined in terms of inches per revolution, that is, the distance the drill travels into the PWB per each revolution, or the rate of penetration. The choice of chip load is very much dependent on the size of the drill used, the PWB material, and the ratio of copper to material in the PWB.

The surface speed of the drill, measured in surface feet per minute (SFM), is also a very important parameter. SFM is usually limited to under 250 for drill diameters of less than 0.012 in. The surface speed may be increased to 600 SFM for larger drills (up to 0.050 in). Generally in multilayer applications average chip loads range from 0.002 to 0.0035 in/r, while surface speed is maintained at 400 to 500 SFM. Chip loads are lowered when drilling small holes, under 0.006 in, to avoid excessive drill breakage. The choice of chip load and surface speed should be made only after a thorough evaluation, which includes the drilling of a statistically significant number of holes with varying parameters and hole sizes. Cross-sectional evaluations should be made to adequately examine the quality of the hole.

Another important operating parameter frequently overlooked is the pressure exerted on the PWB stack by the pressure foot assembly. The entry, PWB, and back-up must be held firmly in place during the drill stroke. If this does not occur, severe burrs are generated. Delamination of the PWB and drill breakage are also caused by insufficient clamping force on the PWB package.

A "maximum hit count," or maximum number of holes drilled with each drill, should be established in order to avoid the possibility of using a dull drill. The maximum number should be determined based on the type of material drilled, the number of boards in the PWB drill stack (although it is generally recommended to drill multilayers only one high), and the entry and back-up material used. Most modern drill machines are equipped with software that allows a maximum hit count to be programmed.

Drilled-Hole Evaluation. Cross-sectional evaluations should be performed both before and after the plating of drilled holes. Evaluations performed before plating allow assessments of the level of loose debris present in the hole, hole-wall topography, and ploughing or riffling. Postplating examinations reveal the presence of resin smear. Smear is generated as the resin melts over the inner layer during the drilling operation. It prevents electrical continuity, which leads to intermittent or catastrophic failure of the finished PWB. Figure 2.21 shows the presence of smear in a plated-through hole. Resin smear appears as a dark line between the plated barrel copper and the inner-layer copper track or pad. The formation of smear may be eliminated or greatly reduced by optimizing the drilling process.

2.4.7 Preplating Operations

The first operation following drilling is deburring of the top and bottom surfaces of the PWB. This process is usually performed in a conveyorized machine which employs rotating oscillating compacted brushes with a high-pressure water spray. It should be noted here that the discharge (effluent) from this operation cannot be directed to the public drainage system, and thus the deburring equipment must be a closed system. Some low-volume shops may use manual sanding tools, especially if large burrs are present. This method of deburring is not recommended because of the inher-

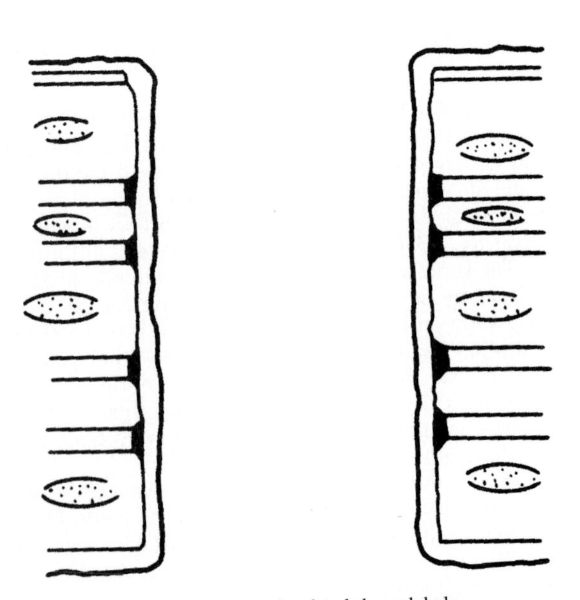

FIGURE 2.21 Resin smear in plated-through hole.

ent nonuniformity of the process as well as the need for ventilation and respirators to protect operators from the copper dust generated.

Treatment of the hole prior to plating is critical to the success of all subsequent plating steps. Two methods are widely used: chemical and plasma. Chemical cleaning makes use of either concentrated sulfuric acid, sometimes followed by a permanganate neutralizer, or chromic acid. These chemicals dissolve resin debris from the hole walls as well as removing resin smear from inner-layer copper.

In addition to cleaning holes prior to plating, these chemicals are also used for the "etch-back" process. Etch-back refers to the removal of resin surrounding the inner layer at the wall of the hole, as indicated in Fig. 2.22. This procedure is accomplished by allowing the drilled PWB to be exposed to the cleaning chemicals for a longer period of time than is required for cleaning only. The amount of time varies with the amount of etch back required. Generally etch back of 0.001 in to 0.003 in is acceptable and will not cause copper-foil cracks. Etch back greatly enhances the structural integrity and electrical reliability of the plated-through hole by providing additional inner-layer copper surface area for subsequent metallization. The glass constituent of the resin-glass matrix is not removed by this process, thus requiring the PWB to be submersed in a solution of ammonium bifluoride and hydrochloric acid as a glass etch.

Plasma hole treatment employs a vacuum chamber into which gases capable of forming a plasma (when an RF voltage is applied) are introduced (Fig. 2.23). Oxygen-carbon tetrafluoride plasmas are used widely. Desmearing as well as etch back may be accomplished with this method. It is extremely attractive to today's fabricators because it eliminates the need for hazardous acids and the related handling and disposal issues. Gas inlet position, rack configuration, and level of vacuum are all extremely important parameters that must be tightly controlled.

2.4.8 Plating Operations

The initial plating step for PWBs is electroless copper deposition. As the name suggests, it is a chemical deposition, as opposed to an electroplating process. It provides

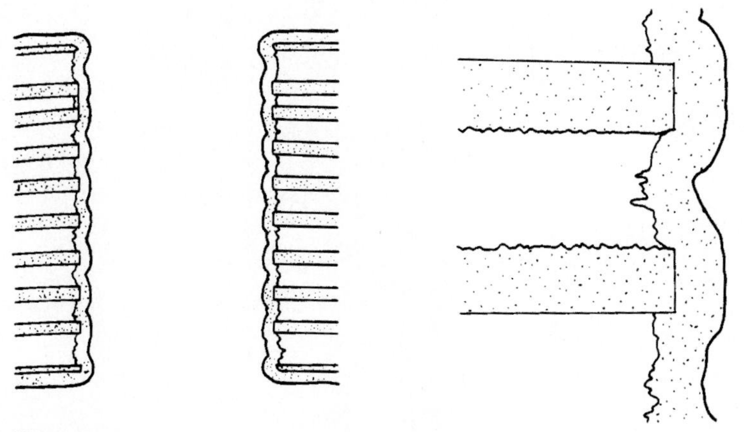

FIGURE 2.22 Etch back in drilled holes.

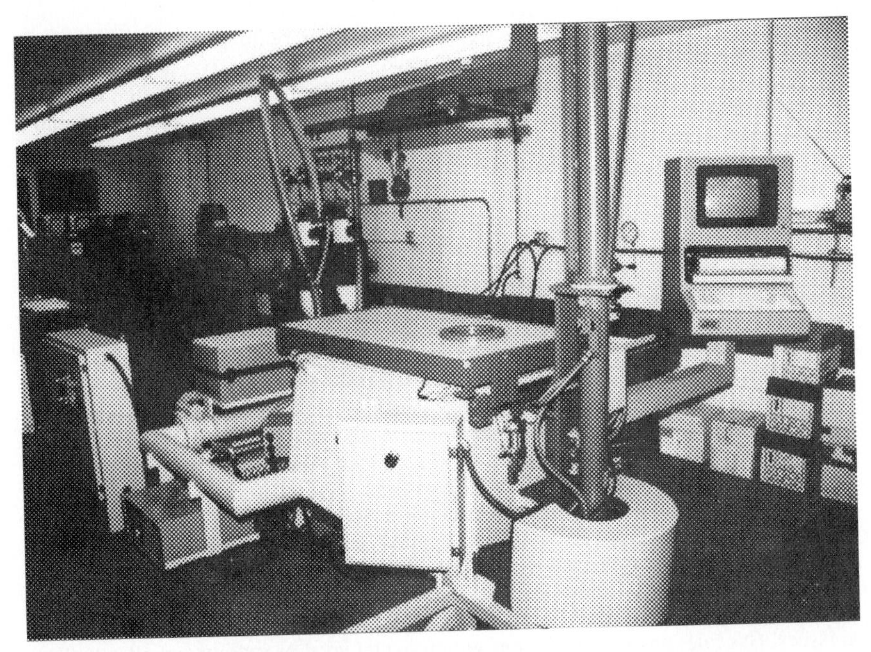

FIGURE 2.23 Plasma etch equipment.

FIGURE 2.24 Electroless copper deposition line.

sites for subsequent electroplating by depositing a thin layer of copper (20 to 100 μin) in the holes and on all surfaces exposed to the solution. The process is performed in a series of tanks (Fig. 2.24) that clean, microetch, and catalyze the surfaces for the copper-reducing step. Copper and tin-lead plating are exhaustively discussed in other sections of this handbook, as is panel and pattern plating. Detailed discussions may be found in Coombs.[5]

BIBLIOGRAPHY

References

1. C. A. Harper, *Electronic Packaging and Interconnection Handbook,* McGraw-Hill, New York, 1991.
2. M. B. Miller, *Dictionary of Electronic Packaging, Microelectronic, and Interconnection Terms,* Technology Seminars, Lutherville, Md., 1990.
3. T. Senese, "N-8000/S-Glass for CTE Matching to Leadless Ceramic Chip Carriers, NELCO Laminates," presented at the Smart VI Conf., Jan. 16, 1990.
4. L. Zakraysek, "Rupture Testing for the Quality Control of Electrodeposited Copper Interconnections in High-Speed, High-Density Circuits," publication unknown.
5. C. Coombs, *Printed Circuits Handbook,* 3d ed., McGraw-Hill, New York, 1988.

Further Reading

"Al-Be Alloy Technology," Tech. Viewgraph presentation, Brush Wellman, Elmore, Ohio, Apr. 15, 1992.

Angelo, M.: "Post Processing: Multilayer Registration for the 1990s," *Printed Circuit Fab.,* Apr. 1992.

Ashhurst, A.: "Structure and Properties of AlBeMet Alloys," Tech. Presentation, AeroMat, Brush Wellman, Elmore, Ohio, 1991.

"Beryllium and Beryllium Composites as Electronic Substrate Materials," Tech. Viewgraph presentation, Brush Wellman, Elmore, Ohio, 1991.

"Beryllium-Beryllium Oxide Metal Matrix Composites: Lightweight-Electronic Materials," Tech. Viewgraph presentation, Brush Wellman, Elmore, Ohio, 1991.

Bouska, G.: "High Speed Low Dielectric Substrate Material," Tech. Paper, Norplex Oak, La Crosse, Wis.

Bullock, K.: "Photoimageable Solder Masks Revisited," *Electron. Packag. Prod.,* pp. 40–44, June 1992.

Chamberlain, G.: "Materials Put the Flex in Flexible Circuits," *Des. News,* pp. 92–98, Oct. 26, 1992.

Chamblin, J.: "Cyanate Ester Offers Light Weight, Low Cost CTE PWBs for Advanced Surface Mount Technology Applications," *Proc.,* pp. 677–684.

Chouinard, D.: "Liquid Photoimageable Solder Masks," in *Proc. NEPCON West* (1991), pp. 898.

Clark, E.: "Status Report: The Global Multilayer Industry," *Printed Circuit Fab.,* pp. 34–39, Apr. 1992.

"Copper-Invar-Copper Technical Data Sheets," Texas Instruments, Attleboro, Mass.

"Copper-Moly-Copper Technical Data Sheets," Climax Speciality Metals, Cleveland, Ohio.

Eurich, J.: "A New Flex Material System for Rigid-Flex Applications," in *Proc. NEPCON West* (1991), pp. 1094–1103.

"Flexible Circuit Design Guide," Teledyne Electro-Mechanisms, 1991.

Galvin, T.: "Liquid Photoimageable Solder Mask Technologies," *Electron. Packag. Prod.,* pp. 54–56, June 1991.

Grensing, F.: "Thermal and Mechanical Properties of Beryllium-Beryllium Oxide Composites," in *Developments in Advanced Aerospace Materials,* TMS Fall Meeting (Cincinnati, Ohio, Oct. 21, 1991).

Guiles, C.: "CTE Materials for PCBs," *Surface Mount Technol.,* July 1991.

Guiles, C.: "High Performance Materials for Printed Wiring Boards," in *Proc. NEPCON West* (1991), pp. 349–359.

Gurley, S.: *Flexible Circuits, Design and Application,* Marcel Dekker, New York, 1984.

Hardesty, R., and F. Grensing: "AlBeMet for Avionics," Tech. Presentation, Brush Wellman, Elmore, Ohio, Jan. 7, 1992.

Henningsgard, R. C.: "Photoplotters: Creating the Critical Shadows," *Printed Circ. Fab.,* July 1992.

Herber, R.: "The Parallel Movement in Solder Mask Technology," *Electron. Packag. Prod.,* pp. 113–134, Sept. 1991.

"High Performance Laminates Technical Data Sheets," Arlon Electronic Substrates Div., Rancho Cucamonga, Calif.

Hnojewyj, O.: "Study of MIL-Spec UV Conformal Coatings: Phase I," in *Proc. NEPCON West* (1991), pp. 1680–1708.

Markstein, H.: "Low TCE Metals and Fibers Prove Viable for SMT Substrates," *Electron. Packag. Prod.,* Jan. 1985.

McKeever, T.: "Selecting Solder Masks for SMT," *Electron. Packag. Prod.,* Suppl., pp. 26–27, Nov. 1990.

Olson, R.: "The Application of Paralene Conformal Coating Technology to Chemical Agent Protection," in *Proc. NEPCON West* (1991), pp. 1655–1668.

Olson, R.: "Method of Quantifying Relative Stress in Conformal Coatings," in *Proc. NEPCON West* (1991), pp. 2202–2206.

Olson, R.: "The Performance of Various Military Classes of Conformal Coatings as Department of Defense Requires in the Chemical Warfare Area," *IPC Tech. Rev.,* pp. 13–19.

Ozmat, B. and N. Reheis: "A New Composite Core Material for Surface Mount Technology Applications," in *Proc. Technical Program of Surface Mount International* (Aug. 1991), pp. 569–592.

"Paralene Technical Data Sheets," Paratronix, Melbourne, Fla.

"Product Bulletins for PWB Laminate Materials," Norplex Oak, La Crosse, Wis.

"PTFE Woven Glass Laminates Technical Data Sheets," Arlon Microwave Materials Div., Bear, Dela.

Ringling, W.: *Rigid-Flex Printed Wiring Design for Production Readiness,* Marcel Dekker, New York, 1988.

Samsami, D.: "Solder Mask Materials," *Electron. Packag. Prod.,* Eng. Fact File, p. 39m Feb, 1991.

Seraphim, D. P., R. C. Lasky, and C. Li: *Principles of Electronic Packaging,* McGraw-Hill, New York, 1989.

Stopperan, J.: "The Changing Technology of Rigid-Flexible Circuits," *Electron. Packag. Prod.,* pp. 34–37, Nov. 1992.

Waryold, J., and J. Lawrence: "Selection Criteria for the Use of Conformal Coatings," in *Proc. NEPCON West* (1991), pp. 2174–2182.

Industry Specifications

ASTM D-149, "Test Method for Dielectric Breakdown Voltage and Dielectric Strength of Solid Electrical Insulating Materials at Commercial Power Frequencies."

ASTM D-150, "Test Methods for AC Loss Characteristic and Permittivity Dielectric Constant of Solid Electrical Insulating Materials."

ASTM D-696, "Test Method for Coefficient of Linear Thermal Expansion of Plastics."

ASTM D-790, "Test Method for Flexural Properties of Unreinforced and Reinforced Plastics and Electrical Insulating Materials."

IPC-T-50, "Terms and Definitions for Interconnecting and Packaging Electronic Circuits."

IPC-L-108, "Specification for Thin Metal Clad Base Materials for Multilayer Printed Boards."

IPC-L-109, "Specification for Resin Preimpregnated Fabric (Prepreg) for Multilayer Printed Boards."

IPC-L-112, "Standard for Foil Clad, Composite Laminate."

IPC-L-115, "Specification for Rigid Metal-Clad Base Materials for Printed Boards."

IPC-L-125, "Specification for Plastic Substrates, Clad or Unclad, for High Speed/High Frequency Interconnections."

IPC-EG-140, "Specification for Finished Fabric Woven from `E' Glass for Printed Boards."

IPC-SG-141, "Specification for Finished Fabric Woven From `S' Glass for Printed Boards."

IPC-A-142, "Specification for Finished Fabric Woven from Aramid for Printed Boards."

IPC-QF-143, "Specification for Finished Fabric Woven from Quartz (Pure Fused Silica) for Printed Boards."

IPC-CF-148, "Resin Coated Metal for Printed Boards."

IPC-MF-150, "Metal Foil for Printed Wiring Applications."

IPC-CF-152, "Composite Metallic Material Specification for Printed Wiring Boards."

IPC-FC-231, "Flexible Bare Dielectrics for Use in Flexible Printed Wiring."

IPC-FC-232, "Specifications for Flexible Adhesive Bonding Films."

IPC-FC-FLX, "Specifications Sheets for IPC-FC-231, IPC-FC-232, IPC-FC-233, and IPC-FC-241."

IPC-FC-232, "Specifications for Adhesive Coated Dielectric Films for Use as Adhesive Coated Cover Sheets for Flexible Printed Wiring."

IPC-FC-241, "Metal-Clad Flexible Dielectrics for Use in Flexible Printed Wiring."

IPC-RF-245, "Performance Specification for Rigid-Flex Printed Boards."

IPC-D-249, "Design Standard for Flexible Single- and Double-Sided Flexible Printed Wiring."

IPC-FC-250, "Specification for Single- and Double-Sided Flexible Printed Wiring."

IPC-D-275, "Design Standard for Rigid Printed Boards and Rigid Printed Board Assemblies."

IPC-RB-276, "Qualification and Performance Specification for Rigid Printed Boards."

IPC-D-300G, "Printed Board Dimensions and Tolerances."

IPC-HF-318, "Microwave End Product Board Inspection and Test."

IPC-D-322, "Guidelines for Selecting Printed Wiring Board Sizes Using Standard Panel Sizes."

IPC-MC-324, "Performance Specification for Metal Core Boards."

IPC-D-325, "Documentation Requirements for Printed Boards."

IPC-TM-650, "Test Methods Manual."

IPC-CC-830, "Qualification and Performance of Electrical Insulating Compound for Printed Board Assemblies."

IPC-SM-840, "Qualification and Performance of Permanent Polymer Coating (Solder Mask) for Printed Boards."

UL796, "Safety Standard for Printed Wiring Boards."

Military Specifications

MIL-STD-275, "Design Standard for Rigid Printed Boards and Rigid Printed Board Assemblies."

MIL-P-13949, "General Specification for Plastic Sheet, Laminated, Metal-Clad (for Printed Wiring)."

MIL-STD-2000, "Soldering Requirements for Soldered Electrical and Electronic Assemblies."

MIL-I-46058, "Insulating Compound, Electrical (for Coating Printed Circuit Assemblies)."

Mil-P-50884, "Printed Wiring, Flexible."

Mil-P-55110, "General Specification for Printed Wiring Boards."

MIL-STD-2118, "Flexible and Rigid Flexible Printed Wiring for Electronic Equipment, Design Requirements For."

MIL-P-50884, "Printed Wiring, Flexible and Rigid Flex."

CHAPTER 3
CERAMICS, GLASSES, AND DIAMOND

Douglas M. Mattox

3.1 INTRODUCTION

In the earlier edition of this handbook, the chapter "Ceramics, Glasses, and Micas" gave a broad treatment of these materials, including both passive and active devices. Thus in addition to insulators, capacitors, piezoelectrics, and ferromagnetic materials were also included. For a treatment of comparable scope in the present edition, this list would have to be expanded to include electrooptic materials, high-temperature superconductors, varistors, chemisorption sensors, and ionic conducting sensors. In addition to scope, a wealth of development has occurred with each of these materials, which would thus make even a superficial treatment of all of the topics impossible within the constraints of this chapter and would be inconsistent with the objectives of a handbook. Because of the depth of treatment required for the most basic understanding of the electronically active materials, they are being omitted from this chapter. Readers interested in these topics will find excellent surveys elsewhere.[1–6]

This chapter will be confined to passive insulating ceramics, glasses, and diamond films. Diamond films are included because they have emerged as an electrical insulating material of growing significance in advanced electronic packaging and will undoubtedly displace traditional ceramic insulating layers in many applications when heat dissipation is a premium concern.

3.2 PHYSICAL AND ELECTRICAL TESTS AND MEASUREMENTS

The properties of ceramics and glasses which are of most interest to the electrical or electronics design engineer and their ASTM test procedures are listed in Table 3.1.

Ceramics and glasses historically serve as electrical insulators in many electrical and electronics applications. Selection is usually based on secondary mechanical, thermal, and chemical properties, assembly characteristics, and availability. Because ceramics and glasses differ so often in assembly procedures, they will be treated separately.

TABLE 3.1 ASTM Property Test Methods

Properties	ASTM test methods
Physical	
Specific gravity	C-329, C-20
Porosity	C-373
Flexure strength	C-674
Hardness	C-730, E-18
Thermal	
Thermal expansion	F-228, C-539
Thermal conductivity	C-408, C-177
Thermal shock resistance	D-116
Electrical	
Dielectric constant and dissipation factor	D-2149, D-150, D-2520
Dielectric strength	D-149
Electrical resistivity	D-257, D-1829

3.3 CERAMICS

The great majority of traditional ceramic materials are electrical insulators. A survey of ceramic insulators can be found in Buchanan.[7] In describing the traditional materials, terminology frequently is used which relates to the method of densification. An introduction to some of this terminology follows.

Traditional ceramics formulations are rooted in the unique properties of clay. Clay [$Al_2Si_2O_5(OH)_4$] has the familiar quality of being plastic when wet. This forming quality made it an important component in the earliest ceramic materials. Over the centuries additives of quartz, or flint (SiO_2), and feldspar ($KAlSi_3O_8$) to the local clays brought benefits in making thermal densification (firing) easier. Because of the three major ingredients, these formulations became known as *triaxials.*

During thermal densification, a series of complex chemical reactions occur between these materials. Clay first gives off the OH groups, leaving a very-fine-particle reactive material. This is followed by several further reactions of the clay transformation products toward the very-fine-particle end product, mullite ($SiO_2 \cdot Al_2O_3$). During baking above 900°C, glasses begin to form from the feldspar and flint components. During firing, the fluid glass lubricates rearrangement of particles, acts as a solvent for material transport of the more refractory phases, and leads to grain growth and particle bonding (sintering). The resulting material is a mixture of a glassy matrix phase interpenetrated with small needlelike mullite crystals and quartz or flint particles, which have become smaller because of dissolution. This produces very strong microstructures, which from antiquity to today, have dependably served an exceptionally wide range of applications from dinnerware to sanitary ware to electrical insulation.

The turn of the century inventions of the automobile[8] and radio placed new demands on the performance of ceramic materials, which led to a rapid evolution in the composition and quality of these materials. In particular the cyclic thermal environment of the combustion engine and the high-temperature corrosion led empiricists to make radical compositional shifts in the basic triaxial compositions toward pure aluminum oxide. Radio transmission demanded less electrically lossy insulators; therefore the alkali-bearing feldspars were eliminated from the high-alumina triaxials. Not all applications demand the premium materials, and so many of the earlier materi-

TABLE 3.2 Ceramic Insulator Candidates for Electronic Manufacture

Material	Dielectric constant at 1 MHz	Thermal conductivity W/(m · K)	Flexure strength, 10^3 lb/in^2	Thermal expansion coefficient to 200°C, ppm/°C	Density, g/cm^3
Alpha quartz (SiO_2)	4.1	2	20	13.0	2.2
Aluminum nitride (AlN)	10.0	200	43	4.4	3.2
Beryllium oxide (BeO)	6.8	300	24	7.0	2.9
Boron nitride (hexagonal)	4.1	60	16	3.8	2.2
Boron nitride (cubic)	4.1	1300	16	3.8	2.2
Cordierite ($2MgO \cdot 2Al_2O_3 \cdot 5SiO_2$)	4.5	2.5	10	2.5	2.2
Diamond	5.5	2000	200	1.1	3.5
Forsterite ($2MgO \cdot SiO_2$)	6.2	3.3	24	9.8	2.9
Fused quartz (SiO_2)	3.8	1.6	8	0.5	2.2
Mullite ($3Al_2O_3 \cdot 2SiO_2$)	6.8	6.7	20	4.0	3.1
Polyimide board	3.6	<1	15	15.0	1.3
Porcelain, electrical	6.5	2.0	15	5.0	2.4
Silicon nitride (Si_3N_4)	8.1	21	100	2.0	3.3
Spinel ($MgO \cdot Al_2O_3$)	8.3	15	24	7.0	3.6
Steatite ($MgO \cdot SiO_2$)	5.7	2.5	24	7.2	2.7

als persist in use today. Thus economic considerations frequently dictate the choice of material for electrical applications.

Today high-frequency use of insulators, especially in computers, has placed a premium on lowering the dielectric constant and metallizing the insulators with high-conductivity metals. As is frequently the case, no material seems to possess all the desired properties designers wish to incorporate in a single design. Thus tradeoffs continue to be the requirement for designers. The following treatment of the ceramic and glass materials, while not exhaustive, is especially aimed at providing the data and guidance for making the best compromises now and in the foreseeable future.

The realistic scope of ceramic insulators for electronic manufacturing, packaging, and assembly is listed in Table 3.2 along with some materials properties. All of the materials in the table are commercially available. The most important of all the listed materials is aluminum oxide.

3.3.1 Electrical Porcelains

Electrical porcelains are triaxial formulations (clay, flint, and feldspar), which are closely related in composition to the compositions of fine china. Because the formulations are greatly affected by the regional clays and feldspars, there are no standard compositions (Table 3.3). Typical formulations contain about 45 weight percent of a mixture of clays, 25 to 35 percent feldspar, 0 to 20 percent flint, and 0 to 40 percent aluminum oxide. One common feature of the triaxial formulations and a number of other early compositions still used for electrical applications is that they are densified

TABLE 3.3A Typical Physical Properties of Traditional Ceramic Dielectrics—Vitrified Products

Material	1 High-voltage porcelain	2 Alumina porcelain	3 Steatite	4 Forsterite	5 Zircon porcelain	6 Lithia porcelain	7 Titania, titanate ceramics
Typical applications	Power-line insulation	Spark-plug cores, thermocouple insulation, protection tubes	High-frequency insulation, electrical appliance insulation	High-frequency insulation, ceramic-to-metal seals	Spark plug cores, high-voltage high-temperature insulation	Temperature-stable inductances, heat-resistant insulation	Ceramic capacitors, piezoelectric ceramics
Specific gravity, g/cm^3	2.3–2.5	3.1–3.9	2.5–2.7	2.7–2.9	3.5–3.8	2.34	3.5–5.5
Water absorption, %	0.0	0.0	0.0	0.0	0.0	0.0	0.0
Coefficient of linear thermal expansion, at 20–700°C, 10^{-6} in/(in) · (°C)	5.0–6.8	5.5–8.1	8.6–10.5	11	3.5–5.5	1	7.0–10.0
Safe operating temperature, °C	1,000	1,350–1,500	1,000–1,100	1,000–1,100	1,000–1,200	1,000	
Thermal conductivity, cal/(s · cm · °C)	0.002–0.005	0.007–0.05	0.005–0.006	0.005–0.010	0.010–0.015	—	0.008–0.01
Tensile strength, lb/in^2	3,000–8,000	8,000–30,000	8,000–10,000	8,000–10,000	10,000–15,000	—	4,000–10,000
Compressive strength, lb/in^2	25,000–50,000	80,000–250,000	65,000–130,000	60,000–100,000	80,000–150,000	60,000	40,000–120,000
Flexural strength, lb/in^2	9,000–15,000	20,000–45,000	16,000–24,000	18,000–20,000	20,000–35,000	8,000	10,000–22,000
Impact strength ($^1/_2$-in rod), ft · lb	0.2–0.3	0.5–0.7	0.3–0.4	0.03–0.04	0.4–0.5	0.3	0.3–0.5
Modulus of elasticity, 10^{-6} lb/in^2	7–14	15–52	13–15	13–15	20–30	—	10–15
Thermal shock resistance	Moderately good	Excellent	Moderate	Poor	Good	Excellent	Poor
Dielectric strength ($^1/_4$-in-thick specimen), V/mil	250–400	250–400	200–350	200–300	250–350	200–300	50–300
Resistivity at room temperature, Ω/cm^3	10^{12}–10^{14}	10^{14}–10^{15}	10^{13}–10^{15}	10^{13}–10^{15}	10^{13}–10^{15}	—	10^{3}–10^{15}
T_e value,* °C	200–500	500–800	450–1,000	above 1,000	700–900	—	200–400
Power factor at 1 MHz	0.006–0.010	0.001–0.002	0.0008–0.0035	0.0003	0.0006–0.0020	0.05	0.0002–0.050
Dielectric constant	6.0–7.0	8–9	5.5–7.5	6.2	8.0–9.0	5.6	15–10,000
L grade (JAN Spec. T-10)	L-2	L-2–L-5	L-3–L-5	L-6	L-4	L-3	—

*Temperature at which resistivity is 1 MΩ · cm.

Source: From Von Hippel.[9]

TABLE 3.3B Typical Physical Properties of Traditional Ceramic Dielectrics —Semivitreous and Refractory Products

	8	9	10	11
Material	Low-voltage porcelain	Cordierite refractories	Alumina, aluminum silicate refractories	Massive fired talc, pyrophyllite
Typical applications	Switch bases, low-voltage wire holders, light receptacles	Resistor supports burner tips, heat insulation, arc chambers	Vacuum spacers, high-temperature insulation	High-frequency insulation, vacuum-tube spacers, ceramic models
Specific gravity, g/cm^3	2.2–2.4	1.6–2.1	2.2–2.4	2.3–2.8
Water absorption, %	0.5–2.0	5.0–15.0	10.0–20.0	1.0–3.0
Coefficient of linear thermal expansion, at 20–700°C, 10^{-6} in/(in) · (°C)	5.0–6.5	2.5–3.0	5.0–7.0	11.5
Safe operating temperature, °C	900	1,250	1,300–1,700	1,200
Thermal conductivity, cal/(s · cm · °C)	0.004–0.005	0.003–0.004	0.004–0.005	0.003–0.005
Tensile strength, lb/in^2	1,500–2,500	1,000–3,500	700–3,000	2,500
Compressive strength, lb/in^2	25,000–50,000	20,000–45,000	15,000–60,000	20,000–30,000
Flexural strength, lb/in^2	3,500–6,000	1,500–7,000	1,500–6,000	7,000–9,000
Impact strength ($^1/_2$-in rod), ft · lb	0.2–0.3	0.2–0.25	0.17–0.25	0.2–0.3
Modulus of elasticity, 10^{-6} lb/in^2	7–10	2–5	2–5	4–5
Thermal shock resistance	Moderate	Excellent	Excellent	Good
Dielectric strength ($^1/_4$-in-thick specimen), V/mil	40–100	40–100	40–100	80–100
Resistivity at room temperature,	10^{12}–10^{14}	10^{12}–10^{14}	10^{12}–10^{14}	10^{12}–10^{15}
T_e value,* °C Ω/cm^3	300–400	400–700	400–700	600–900
Power factor at 1 MHz	0.010–0.020	0.004–0.010	0.0002–0.010	0.0008–0.016
Dielectric constant	6.0–7.0	4.5–5.5	4.5–6.5	5.0–6.0
L grade (JAN Spec. T-10)	—	—	—	—

*Temperature at which resistivity is 1 MΩ · cm.

Source: From Von Hippel.[9]

with the aid of a glassy phase. These products have therefore been called vitrified bodies. Until the 1930s ceramic densification relied on densification in the presence of a glassy phase. In addition, since many of these products are used in outdoor-service transmission and distribution line insulation, the insulators have a glass coating, or glaze, bonded to the surface. Its purpose is to inhibit the absorption of airborne contaminants in the body, which would provide arc paths.

The electrical porcelains are the oldest of the reliable ceramic insulators and exist in a very wide range of compositions, particularly with regard to alumina content. Their development was slowed until the early part of the twentieth century by the inability to make higher-temperature kilns. Thus while properties will be given for these materials, there is such a range of compositions that no general properties can be relied on, but must be determined from the manufacturer. Table 3.3 lists properties for representatives of these general classes of vitrified materials. The difference between high-voltage porcelain and the alumina porcelain is simply in the alumina content and therefore the firing temperature. It is to be noted that these materials, especially the high-voltage porcelain, are lossier than some of the other vitrified ceramics. This is due to the substantial alkali ions present from the feldspar and their mobility in the glassy phase. These materials are widely used today in power distribution. Figure 3.1 shows the dependence of the dielectric constant and tan δ on temperature and frequency.

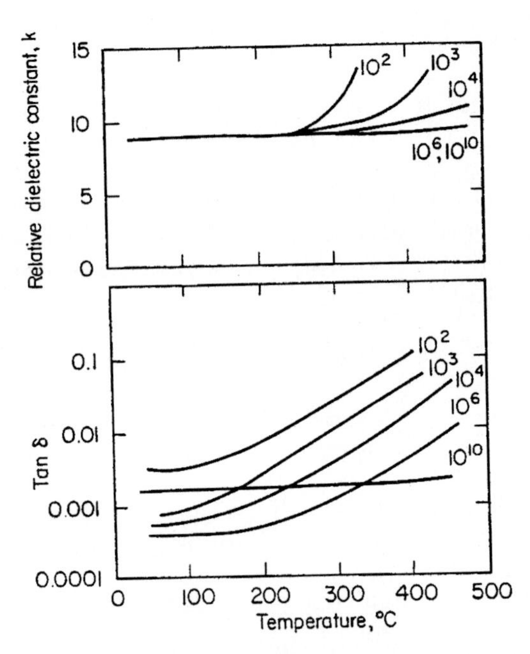

FIGURE 3.1 Dependence of dielectric constant and tan δ on temperature and frequency. (*From Harrison and Moratis,*[10] *p. 6-9.*)

3.3.2 Steatite

Steatite porcelains came into prominence in the 1920s because of their excellent application to radio transmission, which required low dielectric loss (about one-third to one-fourth that of porcelain[11]). The basic material was derived from a naturally occurring magnesium silicate soap stone to which some clay and feldspar might be added. As Table 3.3 shows, steatite has many of the virtues of the electrical porcelains, but with a much lower dissipation factor. A grading system was specified to describe the dissipation factor as follows:

Grade	L-1	L-2	L-3	L-4	L-5	L-6
Maximum dissipation factor	0.15	0.070	0.035	0.016	0.008	0.004

This grading system was applied to most insulators at the time, as noted in Table 3.3. One advantage of the steatite compositions was in the natural lubrication of the talc powders. Prior to the development of inorganic additives, this allowed low-cost high-tolerance pressing of the insulator shapes. Because of the absence of significant vitreous phase in the best materials, the firing range was quite narrow, however. The lowest-loss materials are formed using additional MgO to combine with excess silica and barium oxide as a flux. The fired body consists of enstatite ($MgSiO_3$) crystals bonded together by a glassy matrix.[12] The properties of a typical steatite appear in Table 3.3. Figure 3.2 shows the dependence of steatite's resistivity on temperature, and Fig. 3.3 shows the dependence of its dielectric constant and tan δ on temperature and frequency.

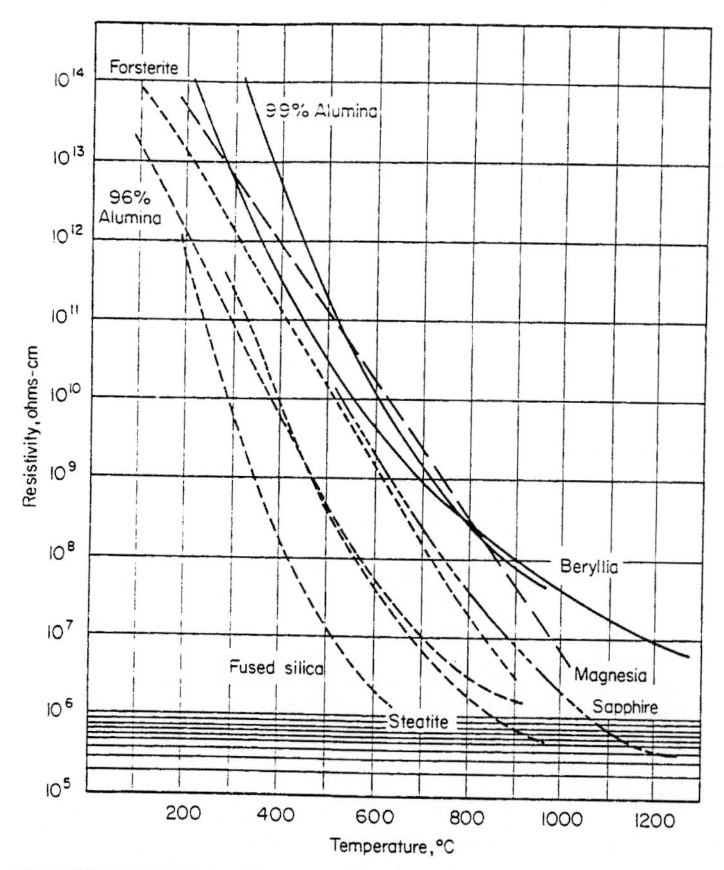

FIGURE 3.2 Change in volume resistivity for various ceramics as a function of temperature. (*From Harrison and Moratis,*[10] *p. 6-10.*)

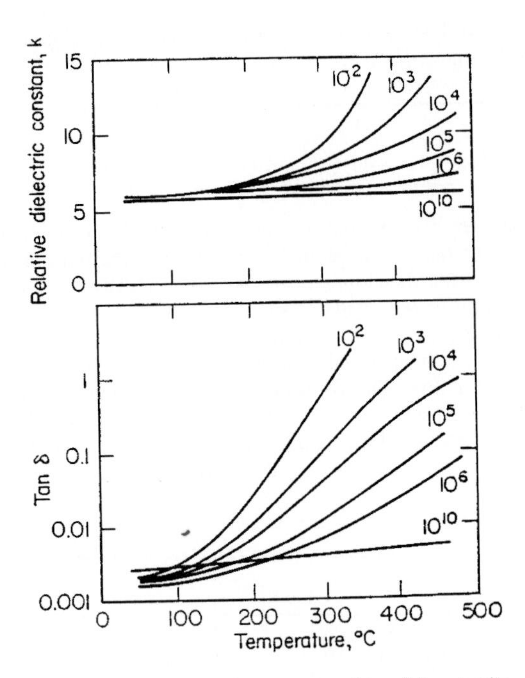

FIGURE 3.3 Dielectric constant and tan δ for steatite ceramic over a range of temperatures and frequencies. *(From Harrison and Moratis,*[10] *p. 6-9.)*

A handy form of steatite-based ceramic is readily available for small-lot manufacture and laboratory applications. Massive fired talc or pyrophyllite, called lava talc, is available for custom fabrication.[13] It is obtained in the unfired state, where it is readily machinable, and then fired with little net dimensional change.

3.3.3 Cordierite

Cordierite ($Mg_2Al_4Si_5O_{18}$) evolved as an important ceramic material where thermal-shock resistance was at a premium. Its low thermal expansion has resulted in its application in electric heater plates, resistor cores, thermocouple insulation, burner nozzles, catalytic substrates, radomes, and computer substrates. The properties of a typical cordierite appear in Table 3.3. Electrically it is not as low in loss as steatite or forsterite, but it is preferred for thermal-shock applications or when minimal expansion mismatch is of concern, as when attached to low-expansion silicon in large chip electronic package designs. Cordierite bodies are more difficult to make than porcelains and alumina because of their narrow firing range.

3.3.4 Forsterite

Forsterite developed as an alternative to steatite and cordierite because of its ease of firing. The electrical and dielectric properties of forsterite materials are quite attractive. The major drawback of forsterite-based materials is its high thermal expansion (Table 3.3) and attendant poor thermal-shock resistance. The relatively high thermal expansion is occasionally advantageous when an insulator must be bonded to higher-expansion materials such as common glasses and some metals. A forsterite funnel has been developed and commercialized for cathode-ray tubes. The funnel is joined to a conventional glass faceplate and neck.

Forsterite is one of a very few relatively high expansion ceramics. Because of an absence of alkali in the vitreous phase, the material remains a good electrical insulator at high temperatures. A comparison of resistivities at elevated temperatures is provided in Fig. 3.2.

3.3.5 Aluminum Oxide (Al_2O_3)—*Alumina*

Alumina ceramics, or high-alumina ceramics, refers to those ceramics that have alumina contents greater than 85 percent and usually 90 percent. They evolved by the 1940s from the spark plug compositions which were developed to obtain higher strength and resistance to the fluxing action of lead compounds contained in antiknock fuels. As a by-product of these developments, the resulting materials found wide favor in electrical applications because of their excellent mechanical strength (350 MPa or 5×10^4 lb/in^2), electrical resistivity (>10^{14} $\Omega \cdot$ cm), dissipation factor (<0.005), and dielectric strength (>200 V/mil).[14]

The aluminum oxide–based ceramic insulators (aluminas) are a natural extension of traditional, triaxial, porcelain (clay, feldspar, flint) ceramics. Starting from the compositions of fine-grade dinnerware, electrical porcelains were formulated with higher levels of alumina because of the improved strength (Table 3.4). The trend toward higher alumina contents brought the additional benefit of high thermal conductivity, which made high-alumina ceramics superior to the alternatives of the day. Aluminum oxide is 20 times higher in thermal conductivity than most oxides. The flexure strength of commercial high-alumina ceramics is two to four times greater than that of most oxide ceramics. Aluminum oxide is an expensive ingredient, particularly in that it adds to the cost of firing because of its more refractory nature. As a simple rule of thumb, however, the fired body's strength tends to be proportional to the aluminum oxide content (see Tables 3.3 and 3.4).

Aluminas are fabricated from aluminum oxide powders with various percentages of sintering promoters[15] which produce the so-called glassy phase. The latter additives reduce the densification temperatures to between 1500 and 1600°C. The minor additives and impurities in high-alumina ceramics have a substantial impact on the final electrical properties. Dielectric constant and dissipation factor are directly affected by impurities, such as Si, Ti, Mg, and Ca,[16] and indirectly by the phase distributions at the grain boundaries.[17]

Based on the final application, the powders may be pressed, extruded, or prepared in slurries for slip casting or tape casting, as is practiced for electronic substrate production. In the tape casting process, a slurry of powders is prepared, usually incorporating organic-based binders, dispersants, deflocculants, and solvents.[18] By drawing the "milk-shake"-like slurry under a knife-edge, called a doctor blade, onto a moving easy-release coated polymer or glass surface, a thin coating is formed which becomes a flexible sheet after drying. This "green" (unfired) tape may be peeled from the supporting surface, cut or punched to shape, and subjected to further processing. The resulting sheets of material are densified at elevated temperatures in an atmosphere reflective of the subsequent processing of the green tape. If the alumina is not metallized, air firing is used. Without any further processing the resulting product is the common aluminum oxide substrate.

TABLE 3.4 Typical Properties of High-Alumina Ceramics

Property	Aluminum oxide content, %							
	85	90	90*	94	96	99.5	99.9	99.9
General								
Specific gravity	3.41	3.60	3.69	3.62	3.72	3.89	3.96	3.99
Color	White	White	Black	White	White	Ivory	Ivory	Transparent white
Surface finish,† μin	63, 39, 8	63, 20, 3.9	63, 39, 3.9	63, 51, 12	63, 51, 12	35, 20, 3.9	20, 35, <1.2	24, 35, <1.2
Water absorption	None	None	None	None	None	None	None	None
Gas permeability‡	None	None	None	None	None	None	None	None
Mechanical								
Hardness, R45N	73	79	75	78	78	83	90	85
Tensile strength, lb/in^2								
25°C	22,000	32,000	33,000	28,000	28,000	38,000	45,000	30,000
1000°C	—	15,000	—	15,000	14,000	—	32,000	15,000
Compressive strain, lb/in^2								
25°C	280,000	360,000	350,000	305,000	300,000	380,000	550,000	370,000
1000°C	—	75,000	—	50,000	—	—	280,000	70,000
Flexural strength, lb/in^2								
25°C	43,000	49,000	53,000	51,000	52,000	55,000	80,000	41,000
1000°C	25,000	—	—	20,000	25,000	—	60,000	25,000
Modulus of elasticity, lb/in^2	32×10^6	40×10^6	45×10^6	41×10^6	44×10^6	54×10^6	56×10^6	57×10^6
Shear modulus, lb/in^2	14×10^6	17×10^6	18×10^6	17×10^6	18×10^6	22×10^6	23×10^6	23.5×10^6
Thermal								
Specific heat, cal/(g · °C)	0.22	0.22	0.25	0.21	0.21	0.21	0.21	0.21
Coefficient of expansion, 10 in/(in · °C)								
25–200°C	5.3	6.1	7.2	6.3	6.0	7.1	6.5	6.5
25–800°C	6.9	7.7	8.1	7.6	8.0	8.0	7.8	7.8
25–1200°C	7.5	8.4	—	8.1	8.4	—	8.3	8.3

Property								
Conductivity, cgs								
20°C	0.035	0.040	0.030	0.043	0.059	0.085	0.093	0.095
400°C	0.016	0.017	0.018	0.017	0.024	0.028	0.032	0.032
800°C	0.010	0.010	—	0.010	0.013	0.015	0.015	0.015
Maximum service temperature,¶ °C	1400	1500	1500	1700	1700	1750	1900	1900
Electrical								
Dielectric strength, V/mil								
0.25 in	240	235	135	220	210	220	240	230
0.050 in	440	450	415	425	370	430	460	510
0.01 in	720	760	720	720	580	840	800	—
Dielectric constant								
1 kHz	8.2	8.8	22.0	8.9	9.0	9.8	9.9	10.1
1 MHz	8.2	8.8	9.8	8.9	9.0	9.7	9.8	10.1
100 MHz	8.2	8.8	—	8.9	9.0	—	—	10.1
Dissipation factor								
1 kHz	0.0014	0.0006	0.3000	0.0002	0.0011	0.0002	0.0020	0.0005
1 MHz	0.0009	0.0004	0.0200	0.0001	0.0001	0.0003	0.0002	0.00004
100 MHz	0.0009	0.0004	—	0.0005	0.0002	—	—	0.00006
Loss factor								
1 kHz	0.011	0.005	6.6	0.002	0.010	0.002	0.020	0.0050
1 MHz	0.007	0.004	0.200	0.001	0.001	0.003	0.002	0.0004
100 MHz	0.007	0.004	—	0.004	0.002	—	—	0.0006
Volume resistivity, Ω · cm								
25°C	$>10^{14}$	$>10^{14}$	$>10^{14}$	$>10^{14}$	$>10^{14}$	$>10^{14}$	$>10^{15}$	—
1000°C	—	8.6×10^5	4.0×10^4	5.0×10^5	1.0×10^6	—	1.1×10^7	—
T_e value, § °C	850	960	—	950	1000	—	1170	—

*Opaque.

†Values pertain to as-fired, ground, and polished conditions, respectively.

‡No helium leak through plate 0.25 mm thick by 25.4 mm diameter at 3×10^7 torr vacuum versus about 1 atm helium pressure for 15 s at room temperature.

¶No load.

§Temperature at which resistivity is 1 MΩ · cm.

Source: Coors Porcelain Company. From Sampson and Mattox,[14] pp. 1.60–1.61.

The surface finish of ceramics used to be quite limited by the method of manufacture. Considerable refinement in powder development and processing has made excellent surface finishes possible without further processing. Generally aluminas may have surface finishes of 3 to 25 μm/in, resulting from normal processing. For very smooth finishes (<2 μm/in) the surfaces have to be lapped or polished. Glazing offered a low-cost improvement over processed surfaces. Today there are fabrication techniques which can make further improvements in surface finish.[19] Only time will tell whether these processes can be price-competitive with lapping and polishing.

Aluminum oxide ceramics, like most ceramic materials, are rarely used apart from being bonded to metals. The means by which this is accomplished frequently dictate the processing technique. Metallization of aluminas is usually accomplished by either high-temperature firing or low-temperature thick-film processing (see Sec. 3.7.1 and Chap. 8).

For many reasons the use and amount of glassy phase in high-alumina ceramics vary considerably, particularly in substrates. Normally high-alumina ceramics are identified by the percent of aluminum oxide they contain. The quoted figures generally range between 90 and 100 percent.[20] This designation is misleading in the sense that it does not refer to the percentage of crystalline Al_2O_3 but to the weight percent of chemically analyzable Al_2O_3. Often one-third of the glassy phase is composed of alumina and is included in the total analysis. This means that the proportion of crystalline Al_2O_3 is usually less than the reference percent. This becomes important when selecting the grade of high-alumina ceramic because of thermal conductivity. Although the composition of the glassy phase and its volume fraction are usually considered proprietary, the effect of glassy phase on conductivity in high-alumina ceramics is very pronounced (Fig. 3.4). The sharp transition has been attributed to the crystalline Al_2O_3 grains losing their connectivity[12] (p. 637) with the resultant rapid deterioration in thermal conductivity.[21] The effect on strength and the dielectric constant is much smaller. Porosity exerts a big effect on thermal conductivity (Fig. 3.5) as does temperature (Fig. 3.6).

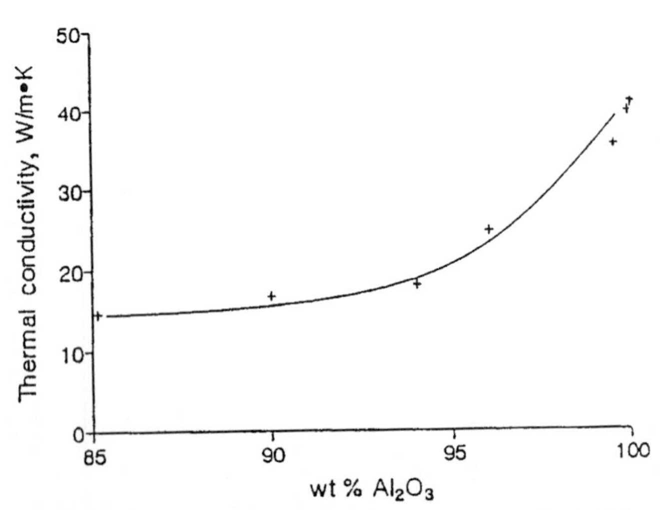

FIGURE 3.4 Effect of weight percent Al_2O_3 on thermal conductivity. *(From Sampson and Mattox,[14] p. 1.62.)*

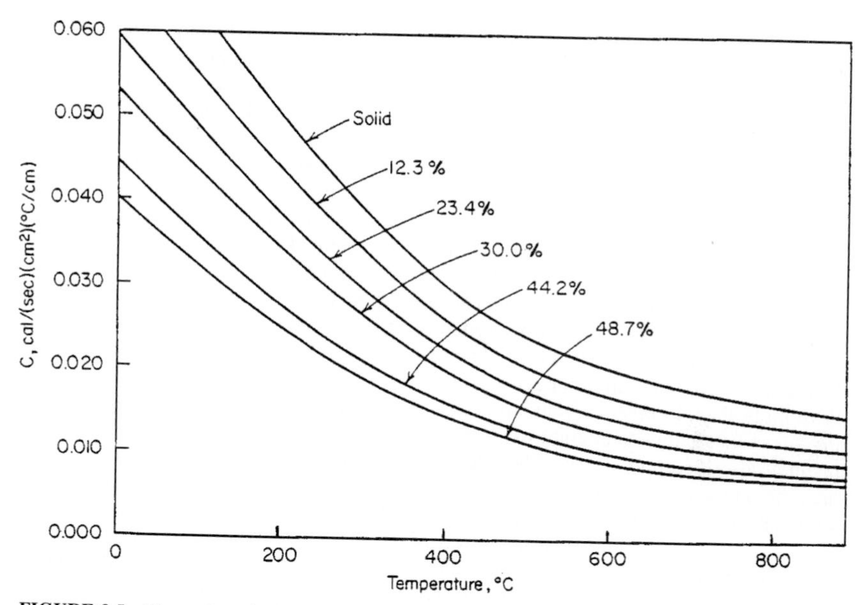

FIGURE 3.5 Thermal conductivity of Al_2O_3 with various amounts of porosity. (*From Harrison and Moratis,*[10] *p. 6-11.*)

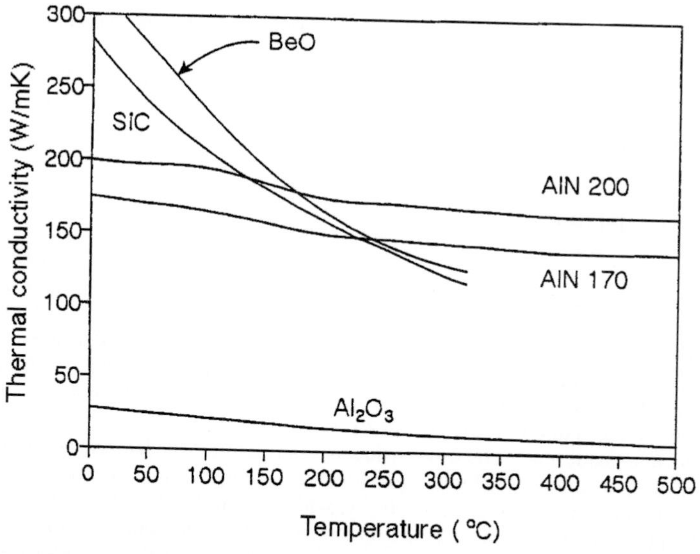

FIGURE 3.6 Thermal conductivity of ceramic substrate materials with temperature. (*From Sampson and Mattox,*[14] *p. 1.64.*)

Ultra-low-loss alumina can be made by eliminating the glassy phase. This is accomplished by sintering pure fine-grained alumina powder at high temperatures (1600°C). To maintain strength through controlled small grain size, a small amount of MgO (≈$^1/_4$ percent) is added. The resulting density is so high that the material is transparent enough to be used for arc tubes in high-pressure sodium-vapor lamps.[22]

A final comment on high-alumina manufacturing relates to the manufacture of cofired multilayer ceramics.[23] Because of areal and resolution limitations in two-dimensional circuitry, high-density packages, chips, and discrete devices are frequently attached to the planar lead termini of three-dimensional circuitry formed by a layering and interconnecting strategy called multilayer (see Chap. 2). In the cofired version, green tapes have feed-through (via) holes punched or laser-bored in them. Then conductor lead traces are carefully applied by silk screening. The various circuit layers are applied to separate green sheets. The resulting sheets are stacked after sequentially filling the via holes with conductor paste and careful registering. The assembly may have more than 40 layers which, after stacking, are consolidated initially in a laminating press using low temperatures (≈90°C). The laminated stack is then fired in the manner described previously. When the green tape is alumina-based, the metallization must be a refractory metal such as tungsten or molybdenum because of the high firing temperature. This results in higher lead resistivities than thick-film metallization.

Until recently only alumina-based cofired multilayer substrates were available. An alternative to cofired circuitry was produced by building up multiple thick-film applications and firing to alumina substrates. Recently, however, a number of low-fire tape systems[24] based on extensions of thick-film compositions have been commercialized, which permit cofired multilayer processing in thick-film equipment. Low-fired substrates matching the expansions of alumina as well as lower-expansion lower-dielectric-constant systems have been developed (see Sec. 3.6). The lower dielectric constant and expansion, contrasted with alumina-based cofired multilayer materials, come at the expense of poor thermal conductivity.

Efforts to find alternatives to alumina arise from a number of concerns. Logic applications find speed limited by the high dielectric constant of the substrate and the high resistivity of the lead traces. Power applications find the thermal conductivity of alumina inadequate for thermal management. Increases in chip size find the thermal expansion mismatch between substrate and die difficult to manage. Of the alternatives which are emerging for electronic packaging applications it is clear that no single material on the realistic horizon will answer all of alumina's deficiencies.

3.3.6 Beryllia (BeO)

Beryllium oxide–based ceramics are in many ways superior to alumina-based ceramics.[25] The major drawback is beryllia's potential toxicity. Beryllium and its compounds are a group of materials which are potentially hazardous and must be handled in such a way that proper precautions are taken to prevent potential accidents. Using proper care, the material has been widely used in electronics applications from automotive ignitions to defense systems. To understand the safe handling of this material the reader is referred to Powers.[26]

Beryllium oxide materials are particularly attractive for electronics applications because of their unique combination of electrical, physical, and chemical properties, particularly their high thermal conductivity, which is roughly 10 times higher than that of the commonly used alumina-based materials. Figure 3.6 compares beryllia's thermal conductivity to that of alumina and some emerging alternative materials; Fig. 3.7 compares beryllia's thermal conductivity to some common metals used in electronics

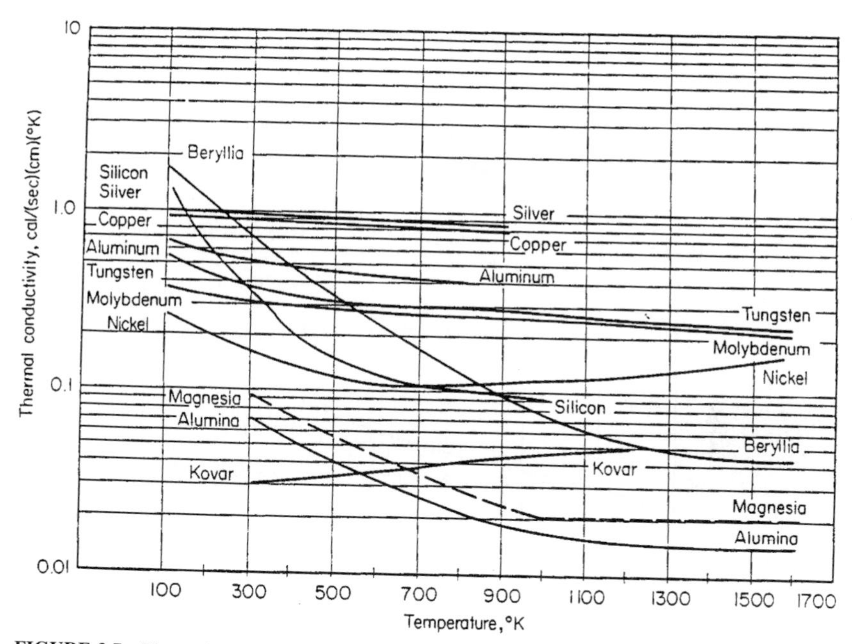

FIGURE 3.7 Thermal conductivity as a function of temperature for various materials. (W/m · K = 418 × cal/(s · cm · K). (*From Harrison and Moratis,*[10] *p. 6-22.*)

applications. Table 3.2 shows that beryllia is of lower dielectric constant and thermal expansion than alumina, but slightly lower in strength. Figure 3.8 shows the temperature dependence of the dielectric constant for beryllia versus that of alumina ceramics and fused silica.

Beryllia materials are fabricated in much the same way as alumina materials, although the toxic properties of the powders mandate that they be processed in laboratories equipped to handle the powders safely. This results in there being a very limited number of suppliers. Simple cutting, drilling, and postmetallization are also handled by especially equipped vendors. Reliable thick-film systems may be applied to beryllia substrates. Such coatings may require less elaborate safety precautions, and are often applied by the device fabricators. The general safety requirements for the source substrate and package materials impact design and prototype scheduling, but not unreasonably. Prices are considerably higher than for alumina-based materials; thus use is generally restricted to applications where thermal management considerations cannot be served in any other way.

3.3.7 Nonoxide Ceramics

Aluminum Nitride (AlN). Aluminum nitride–based substrates[27] and packages began to appear in the late 1980s as an alternative to the toxicity concerns of beryllia. As Fig. 3.4 shows, the thermal conductivity of aluminum nitride is comparable to that of beryllia, but deteriorates less with temperature. Aluminum nitride's dielectric constant

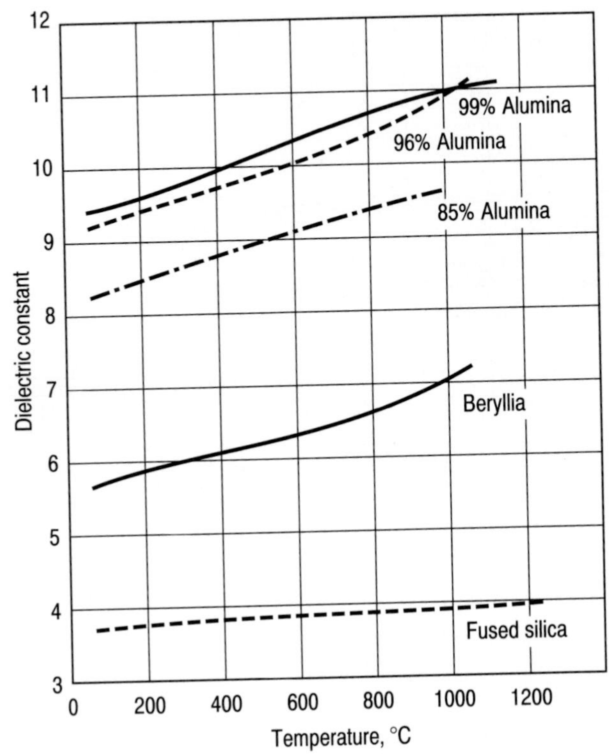

FIGURE 3.8 Dielectric constant as a function of temperature measured at a test frequency of 4 GHz. (*From Harrison and Moratis,*[10] *p. 6-23.*)

is comparable to that of alumina, and thus a liability in high-speed device applications, but its thermal expansion is low (4 ppm/°C) and comparable to that of silicon. Because of its expansion, it is being used in conjunction with large silicon die or substrate attachments, commonly called multichip designs.

The fabrication of aluminum nitride–based materials is very different than that of alumina- and beryllia-based materials and must be factored into application considerations. The high thermal conductivity is attributable to the presence of aluminum nitride and prevention of its oxidation to alumina. Thus aluminum nitride powders are sintered at very high temperatures in N_2 atmospheres. Because the powders are costly and sintering is difficult, the product costs are comparable to those of beryllia materials. As Fig. 3.6 indicates, thermal conductivity is available in two ranges, the highest being the most difficult and costly to fabricate. The lowest conductivity level is associated with cofire metallization. Because tungsten undergoes unfavorable reaction with nitrogen, cofired aluminum nitride is fired in vacuum. This inhibits sintering, resulting in lower thermal conductivity, about twice that of alumina. New developments as of this writing indicate that this problem may be solved in the near future.[28]

Because aluminum nitride is a relatively new material, and because it is a nitride and not an oxide, glass-bonding characteristics are not well defined and conventional thick-film alumina-based pastes are thermodynamically incompatible with aluminum nitride.[29] Therefore thick-film coating systems are in the early stages of development and reliability is not firmly established (Chap. 8). Nevertheless, the major thick-film manufacturers are active in the development of compatible systems.[30] Aluminum nitride reacts with the alkaline environments frequently used in cleaning and degreasing operations, and they must be avoided. To summarize, while alumina- and beryllia-based materials and their thick-film companions are technically mature materials, at the beginning of the 1990s aluminum nitride is only emerging.

Silicon Carbide (SiC). Silicon carbide gained prominence as a heating element. It is made by pressing into rods granular silicon carbide with a temporary binder and firing at 2000 to 2500°C. After sintering, the bulk resistance of the body is essentially the same as that of the individual grains. A typical resistivity of these rods at room temperature is 0.2 Ω · cm, decreasing to 0.1 Ω · cm at 1000°C. Nonlinear silicon carbide is used for lightning arrestors and in voltage-limiting resistor applications. Practically the entire voltage drop in silicon carbide occurs at the interface between the grains and is described by the relation $V = KI^n$, where K is a constant depending on the geometry of the body and n is a constant determined by the manufacturing process. The exponent n generally varies from 0.1 to 0.35. The high resistance of nonlinear silicon carbide is attributed to a grain-boundary barrier.

Beginning in the 1980s, because of its high thermal conductivity, silicon carbide has found some special application as a substrate material because of its special processing chemistry[31] (Table 3.2). While normally an electrical conductor, silicon carbide has been able to function as an insulator because of a beryllia-based intergranular insulating phase which interrupts continuity between the individual grains of the polycrystalline sintered material. Conductivity can be observed within the individual grain feature size of 40 μm. A major drawback of silicon carbide vis-à-vis aluminum nitride, beryllia, and alumina is its high dielectric constant (≈40). The material is more costly than alumina-based materials, but comparable to beryllia and aluminum nitride at this time. There is no known development effort of thick-film systems comparable to that known for aluminum nitride. It would appear that this material should be viewed as a selective alternative to beryllia and aluminum nitride.

Boron Nitride (BN). Boron nitride ceramics[32] are used occasionally when concerns of thermal conductivity or machinability are prominent. Hexagonal boron nitride is formed by pyrolytic deposition or by hot pressing boron nitride powders in the presence of a borate glass phase. The resulting material is soft and easily machined. Of chief interest is the thermal conductivity of the material, which is about twice that of the better aluminas, that is, 60 W/(m · K). It cannot be metallized nor sealed to, so its use is very limited.

Cubic boron nitride would be an excellent substrate material, with a thermal conductivity of 1300 (W/m · K), which is only exceeded by diamond. Unfortunately it is even harder to manufacture than diamond; so it is not an option in the foreseeable future.

3.3.8 Zirconium Oxide (ZrO_2)–Based Materials

While not strictly used in electronics applications, zirconium oxide–based ceramic materials deserve some recognition as the premier representatives of a new class of ceramic materials that may be regarded as nonbrittle. Ceramic materials are normally characterized by their brittle fracture, and as such must only be used in structural applications in very careful designs. Without compositional modification, zirconium oxide undergoes a structural inversion at elevated temperature, which leads to extensive internal cracking and failure. This inversion may be substantially retarded by adding a second phase, usually a CaO or Y_2O_3, which suppresses the phase that undergoes the inversion. The preferred compositions are partially stabilized and known as partially stabilized zirconias (PSZ).

It has been learned how to make PSZ[33] with a particular crystalline structure which is transformable under mechanical loading. This represents an alternative energy dissipation mechanism in place of the creation of new surface, that is, brittle fracture. It renders the material nonbrittle, or "tough," to use the correct terminology. PSZs are commonly manufactured at a toughness level approaching that of cast iron. As such it is used for industrial shears, scissors, knives, medical prostheses, and the like. It represents an ideal insulating material for electrical or electronics application when mechanical abuse is a consideration.

3.3.9 Mica

Mica is a natural insulating material, once widely used as a dielectric in capacitors and motor insulation, electron tubes, and appliances. Commercially important mica is a naturally occurring form of a potassium-alumino-silicate material. Muscovite or ruby mica [$KAl_3Si_3O_{10}(OH)_2$] is the most desirable form for capacitor use because of its low dissipation factor and high dielectric strength. Phlogopite or amber mica [$KMg_3AlSi_3O_{10}(OH)_2$] has better resistance at high temperatures than muscovite, but its electrical properties are not so well suited for use as capacitor dielectrics. Phlogopite mica is used to insulate motor and generator commutator segments. A synthetic mica, fluorophlogopite [$KMg_3AlSi_3O_{10}F_2$], has been prepared for special applications. Another synthetic mica is reconstituted mica, a sheet material made from thin flakes of natural or synthetic mica that have been ground up, pressed into sheet form, and heated under pressure to bond the mica flakes together. The properties are highly reproducible in sheet form.

Mica has a unique combination of electrical properties not available in other materials. The relatively stable dielectric constant, the low dissipation factor, and the high dielectric strength of mica are used to form stable and reliable capacitors for critical circuit applications. Disadvantages of mica are its small capacitance-to-volume ratio and its high cost in the form of large natural sheets. A listing of ASTM specifications and methods of testing mica and mica products is given in Table 3.5; the properties of mica are listed in Table 3.6.

TABLE 3.5 ASTM Specifications and Test Methods for Mica and Mica Products

D-748-59	Natural block mica and mica films suitable for use in fixed mica-dielectric capacitors
D-351	Natural muscovite mica based on visual quality
D-2131	Natural muscovite mica splittings
D-352	Pasted mica used in electrical insulation
D-1082	Power factor and dielectric constant of natural mica
D-1677	Untreated mica paper used for electrical insulation sampling and testing
D-374	Thickness of solid electrical insulation
D-1039	Testing glass-bonded mica used as electrical insulation
F-12	Mica bridges for electron tubes
F-48	Dimensioning mica bridges
F-652	Measuring mica stampings or substitutes used in electron devices and lamps

TABLE 3.6 Properties of Mica

Property	Natural Mica		Synthetic mica, fluorophlogopite
	Muscovite	Phlogopite	
Volume resistivity, Ω · cm			
20°C	10^{16}–10^{14}	Lower than muscovite at low temperature	10^{16}–10^{17}
500°C	10^{8}–10^{10}	Higher at high temperatures	
Dielectric strength (1–3-mil specimen), V/mil	3000–6000	3000–4000	2000–3000
Dielectric constant			
100 Hz	5.4		
1 MHz	5.4	6	5–6
100 MHz	5.4		
Dissipation factor			
100 Hz	0.0025		
1 MHz	0.0003	0.001–0.01	0.0004–0.0007
100 MHz	0.002		
Safe operating temperature in vacuo, °C	350–450	—	1000
Water absorption	Practically zero	Practically zero	Practically zero
Compressive strength, lb/in^2	25,000	15,000	

Source: From Harrison and Moratis,[10] p. 6-37.

3.4 GLASSES

Glasses are important materials in electronics applications. They serve in more varied roles than the ceramic materials just described.[34] Historically they were used in tube envelopes, capacitor dielectrics, and substrates because of their reproducible electrical properties, ease of fabrication into complex shapes, and low cost. They were valued for their insulating properties, their surface smoothness, their controllable thermal deformability, or their ability to serve as a bonding phase. Today they are used as substrates, delay lines, passivation layers, capacitors, resistor and conductor bonding phases, package sealants, and insulating bushings. Some discussion of the broad properties of glasses follows.

Silicate glasses exhibit dielectric constants ranging from 3.8 for pure silica to 15 for high-lead glasses. Dissipation factors vary from tan $\delta < 0.001$ for vitreous silica to 0.01 for soda-borosilicate glasses at 60 Hz. At room temperature most glasses are excellent electrical insulators (resistivity $>10^{14}$ Ω · cm), although some are semiconductors. Electrical resistivity is inversely controlled by the alkali present. In order to reduce alkali levels, PbO is frequently used to reduce processing temperatures without elevating conductivity.

Thermal conductivity for most glasses is very similar and very poor [<2 W/(m · K)], as is flexure strength, which varies very little between compositions (≈70 MPa or 10^4 lb/in^2). The thermal expansion of glasses varies between ≈0 ppm/°C for pure SiO_2 glass and ≈12 ppm/°C for some compositions. There is one feature of thermal expansion which is peculiar to glass and is important for designers to understand. In addition to an absence of structure, glasses are distinguished by their nonlinear thermal expansion. Glasses undergo an abrupt increase in thermal expansion within 20 to 50°C of their softening points (Fig. 3.9). The temperature at which the effect begins is called the glass transformation point T_g. The commonly quoted thermal expansion coefficients for glass describe the expansion *below* T_g. Bonding to metals and ceramics occurs at temperatures *above* T_g, so the selection of matching thermal expansions must take this into account. Data this complete are normally only obtained by special request from the manufacturers. Compatibility selection must consider this detail.[35]

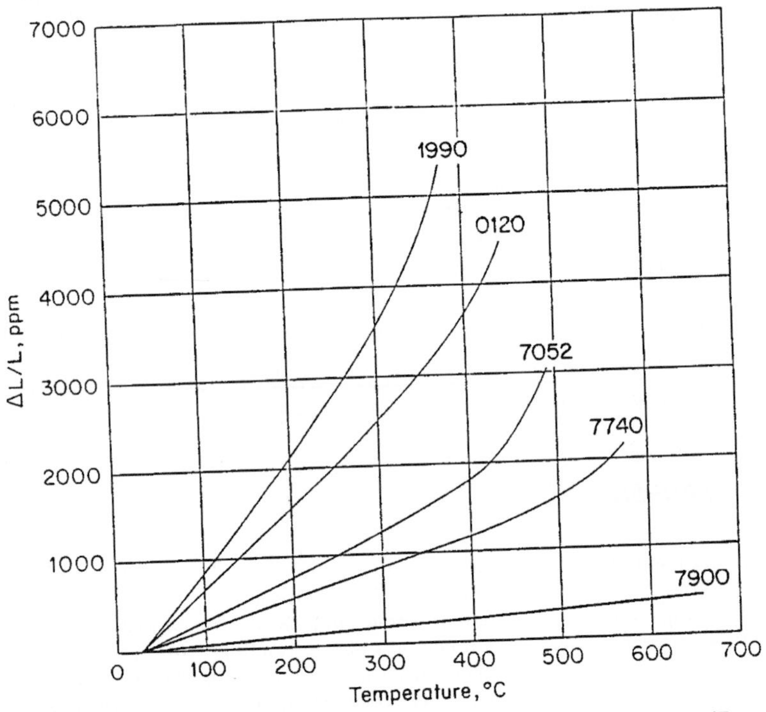

FIGURE 3.9 Thermal expansion characteristics of some commercial glasses. (*From Sampson and Mattox,*[14] *p. 1.69.*)

Glasses are noncrystalline solids. They have the random atomic structure of a liquid which gets frozen in as the melted glass is cooled. The feature which permits glasses to remain amorphous rather than crystallizing is the very high viscosity of the melted material. This phenomenon is frequently used to advantage when glasses are employed. At suitable temperatures, glasses will slowly deform and fill cavities for insulator feed-throughs. Due to their high surface tension, glasses will also fill and be retained in the cavities of a filler powder which may be an insulator (capacitors), a semiconductor (resistors), or a conductor in thick-film applications while simultaneously wetting the associated substrate.

Since glasses are not definite crystals but are closer to chemical mixtures, they may have an infinite range of compositions. In spite of this, almost all glasses are based on four glass-forming oxides: SiO_2, B_2O_3, P_2O_5, and occasionally GeO_2. The first three are often intermixed. To these basic oxides are added a number of softeners, or fluxes (the oxides of Li, Na, K, Rb, Cs, and Pb) and extenders (the oxides of Mg, Ca, Sr, and Ba). Another common group of nonfunctional additives are colorants (the oxides of Co, Mn, Fe, and Cr).

It is best to think of glass as a liquid which has been cooled to a temperature where it has become so stiff that for all practical purposes it is rigid. This viewpoint helps in understanding the various viscosity points which are frequently reported for glasses (Fig. 3.10). The different points refer to temperatures where the glass has become less stiff than at room temperature. These points are defined in some cases by how long it takes for strain to be relieved by viscous flow at the temperature, that is, the strain point (hours), or the annealing point (minutes), or how long it takes for a fiber to deform under its own weight, the softening point. The working point is where the glass may be "pushed around." These temperatures are used to determine where particular glasses may be sealed or bonded. Some brief discussion of common glasses is presented next.

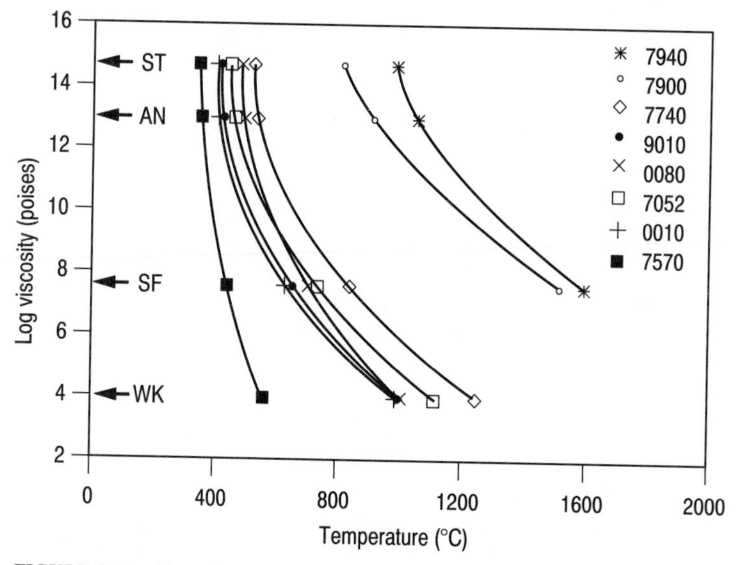

FIGURE 3.10 Glass viscosity as a function of temperature. ST—strain point; AN—annealing point; SF—softening point; WK—working point. (*From Sampson and Mattox,*[14] *p. 1.68.*)

3.4.1 Commercial Glasses

Table 3.7 lists some common glasses used in electronics along with some of their properties; Table 3.8 shows the composition of some commercial glasses. The additives in glass have a great effect on the viscosity points, thermal expansions, and electrical properties, as Figs. 3.9, 3.10, and 3.11 illustrate.[36,37] Table 3.9 gives the dielectric properties as a function of frequency for most of the glasses of Table 3.7. These are the major design parameters used in glass selection.

TABLE 3.7 Properties of Common Electronic Packaging Glasses

Glass* code	Type	Thermal expansion 0–300°C, 10^{-7}/°C	Viscosity data, °C				Density, g/cm^3	Young's modulus, 10^6 lb/in^2	Poisson's ratio	Log_{10} of volume resistivity			Dielectric properties at 1 MHz and 20°C		
			Strain point	Annealing point	Softening point	Working point				25°C	250°C	350°C	Power factor, %	Dielectric constant	Loss factor, %
0010	Potash soda lead	93	395	435	625	985	2.86	8	0.21	17.+	8.9	7.0	0.16	6.7	1
0080	Soda lime	92	470	510	695	1005	2.47	10.0	0.24	12.4	6.4	5.1	0.9	7.2	6.5
7052	Borosilicate	46	435	480	710	1115	2.28	8.2	0.22	17.0	9.2	7.4	0.26	4.9	1.3
7570	High lead	84	340	365	440	560	5.42	8.0	0.28	—	10.6	8.7	0.22	15.0	3.3
7740	Borosilicate	33	515	565	820	1245	2.23	9.1	0.20	15.0	8.1	6.6	0.50	4.6	2.6
7900	96% silica	8	820	910	1500	—	2.18	10.0	0.19	17.0	9.7	8.1	0.05	3.8	0.19
7940	Fused silica	5.5	990	1050	1580	—	2.20	10.5	0.16	—	11.8	10.2	0.001	3.8	0.0038
9010	Potash soda barium	89	405	445	650	1010	2.64	9.8	0.21	—	8.9	7.0	0.17	6.3	1.1

*Corning Glass Works.

Source: From Sampson and Mattox,[14] p. 1.67.

TABLE 3.8 Estimated Commercial Glass Compositions

Glass code*	Type	SiO_2	Na_2O	K_2O	CaO	MgO	Li_2O	PbO	B_2O_3	Al_2O_3
0010	Electrical	63	8	6	0.3	—	—	22	0.2	0.5
0080	Lamp	74	16	1	5	3	—	—	—	1.0
0120	Lamp	56	5	9	—	—	—	30	—	—
7050	Sealing	67	5	1.0	—	0.3	—	—	25	1.7
7070	Low loss	70	—	0.5	0.1	0.2	1–2	—	28	1.1
7570	Solder seal	3	—	—	—	—	—	75	11	11
7720	Electrical	73	2	2	—	—	—	6	17	—
7740	Labware	80	3.5	0.5	—	—	—	—	14	2
7900	96% SiO_2	96	0.2	0.2	—	—	—	—	3	0.6
7910	Fused SiO_2	99.5	—	—	—	—	—	—	—	—
8870	Sealing	35	—	7	—	—	—	58	—	—

*Corning Glass Works.

Source: From Sampson and Mattox,[14] p. 1.68.

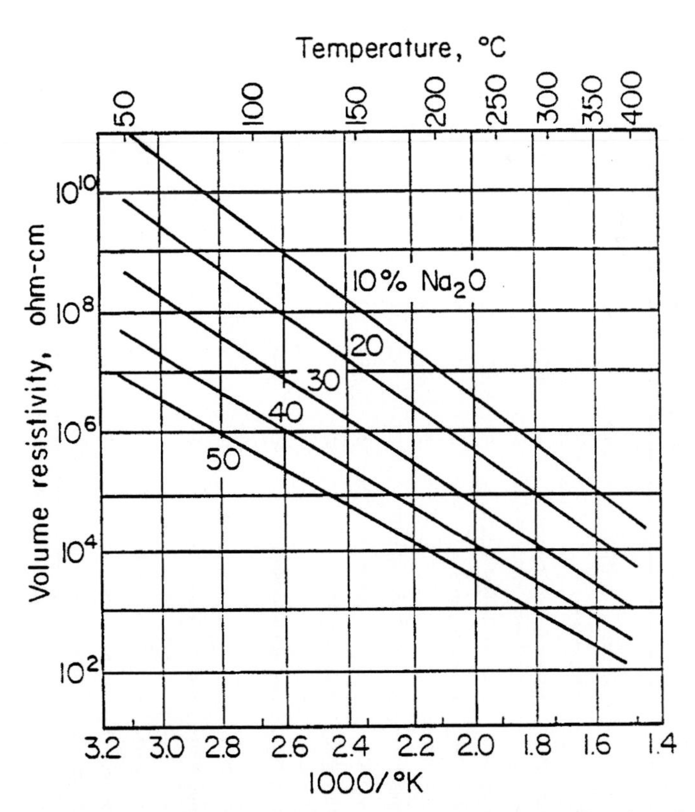

FIGURE 3.11 Effect of increasing the amount of Na_2O in soda-silica glass on volume resistivity. (*From Harrison and Moratis,*[10] *p. 6-28.*)

TABLE 3.9 Dielectric Properties of Several Commercial Glasses

Glass	Composition	Temper-ature	Dielectric constant						Dissipation factor					
		°C	60 Hz	1 kHz	1 MHz	100 MHz	3 GHz	25 GHz	60 Hz	1 kHz	1 MHz	100 MHz	3 GHz	25 GHz
Corning 0010	Soda-potash-lead silicate ~20% lead oxide	24	6.70	6.63	6.43	6.33	6.10	5.87	0.0084	0.00535	0.00165	0.0023	0.0060	0.0110
Corning 0120	Soda-potash-lead-silicate	23	6.76	6.70	6.65	6.65	6.64	6.51	0.0050	0.0030	0.0012	0.0018	0.0041	0.0127
Corning 1990	Iron-sealing glass	24	8.41	8.38	8.30	8.20	7.99	7.84	—	0.0004	0.0005	0.0009	0.00199	0.0112
Corning 1991		24	8.10	8.10	8.08	8.00	7.92	—	0.0027	0.0009	0.0005	0.0012	0.0038	
Corning 7040	Soda-potash-borosilicate	25	4.85	4.82	4.73	4.68	4.67	4.52	0.0055	0.0034	0.0019	0.0027	0.0044	0.0073
Corning 7050	Soda-borosilicate	25	4.90	4.84	4.78	4.75	4.74	4.64	0.0093	0.0056	0.0027	0.0035	0.0052	0.0083
Corning 7060 (Pyrex)	Soda-borosilicate	25	—	4.97	4.84	4.84	4.82	4.65	—	0.0055	0.0036	0.0030	0.0054	0.0090
Corning 7070	Low-alkali potash-lithia-borosilicate	23	4.00	4.00	4.00	4.00	4.00	3.9	0.0006	0.0005	0.0008	0.0012	0.0012	0.0031
Corning 7720	Soda-lead borosilicate	24	4.75	4.70	4.62	—	4.60	—	0.0093	0.0042	0.0020			
Corning 7750	Soda-borosilicate ~80% silicon dioxide	25	—	4.42	4.38	4.38	4.38	—	—	0.0033	0.0018	—	0.0043	
Corning 7900	96% silicon dioxide	20	3.85	3.85	3.85	3.85	3.84	3.82	0.0006	0.0006	0.0006	0.0006	0.00068	0.0013
Fused silica 915c	Silicon dioxide	25	—	3.78	3.78	3.78	3.78	—	—	0.00026	0.00001	0.00003	0.0001	
Quartz (fused)	100% silicon dioxide	25	3.78	3.78	3.78	3.78	3.78	3.78	0.0009	0.00075	0.0001	0.0002	0.00006	0.00025

Source: From Harrison and Moratis,[10] p. 6-33.

Pure SiO_2 Glass. Vitreous silica or fused silica, often inaccurately called fused quartz, is used only rarely, and often where thermal-shock considerations are primary. Its thermal expansion coefficient is nearly zero. Coated and uncoated windows are typical applications, including tungsten, halogen, mercury, and metal halide lamp envelopes. Pure silica glass is costly because of its exceptional melting temperature requirements (>1900°C). Pure silica glass has the lowest dielectric constant of any ceramic material (3.8).

96% SiO_2 Glass. Where the properties of pure silica are wanted at lower cost, an ingenious synthesis technique[38] permits fabrication of 96 percent silica glass with very similar properties to 100 percent silica. This glass is easily shaped and is thus appealing where complicated shapes, impossible to form from pure silica glasses, are wanted. The glass is formed by melting a lower-temperature glass which, after forming and heat treatment, separates into two intertwined glass phases. One of the phases is easily digested in acid, leaving a skeleton of 96 percent silica. This skeleton sinters densely at temperatures below 900°C, leaving a 96 percent silica glass.

Soda-Lime Glass. These compositions are based on Na_2O, CaO, SiO_2, and lesser amounts of K_2O, MgO, Al_2O_3, and so on. They define the bulk of common glasses, windows, tumblers, and the like. They are chemically durable and of moderately high thermal expansion but are not good in thermal-shock applications. The alkali varies between 12 and 15 weight percent, typically resulting in dielectric constants of 6 to 9. These glasses, with and without small levels of PbO, are frequently used as the compositions of feed-through bushing insulators, that is, hermetic metal packages. Their proven chemical durability is attested by their application in food processing.

Borosilicate. These are the common laboratory and cookware glass compositions. They are low in thermal expansion and show excellent chemical durability. They have high electrical resistivities and low dielectric constants. By themselves they are not common electronics materials, but they are the glassy binder phase in some of the newer low-temperature firing substrate systems.

Lead Alkali Borosilicate (Solder/Sealing Frits). These glasses are often based on PbO, B_2O_3, Al_2O_3, BiO, ZnO, and SiO_2 and are widely used in sealing and adhesive applications because of their low melting and softening temperatures.[39] They are usually used as powder additives by the vendors of thick-film pastes. Their compositions are considered proprietary but are compounded with a view to the ranges of materials compatibility requirements in semiconductor processing. PbO and BiO are additives which exhibit various compatibility problems and are included or not, as the material requires.

The low melting glasses are also used as "glues" in applying ceramic or metal lids for hermetically sealed packages. In some cases these glasses are formulated to crystallize (devitrify). The advantage is that the crystallized products frequently have much different properties than the glasses. This is a means of getting low or very high expansion.

3.4.2 Commercial Glass Properties

Electrical Resistivity. Because of its high resistivity, glass is widely used for electrical insulation (Fig. 3.12). While resistivity is largely determined by the composition

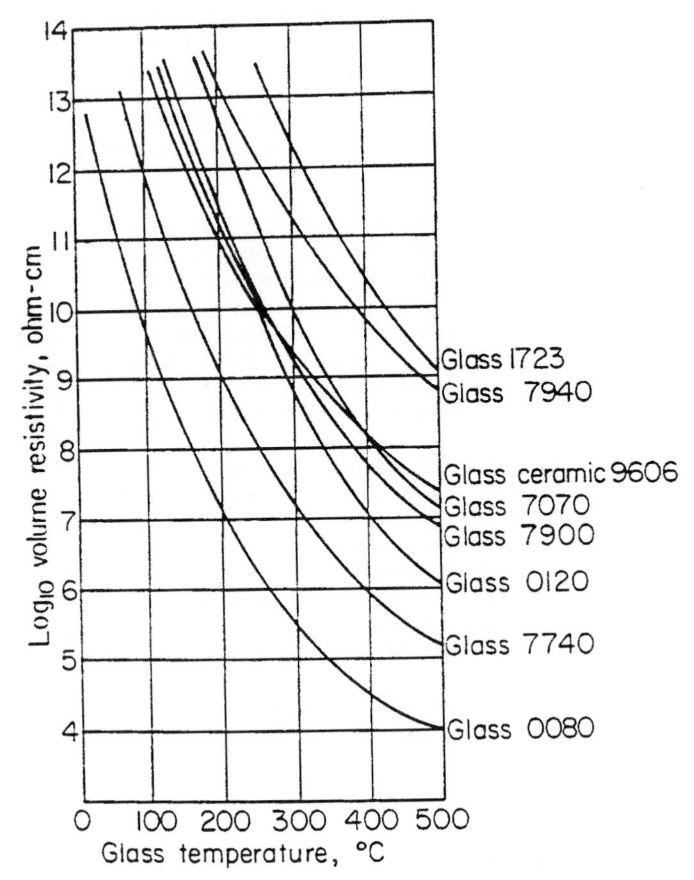

FIGURE 3.12 Volume resistivity as a function of temperature for several Corning glasses. (*From Harrison and Moratis,*[10] *p. 6-28.*)

of the glass, it is affected by temperature, moisture, and structural defects. Since the factors that affect volume and resistivity are somewhat different, these properties are treated separately.

Volume Resistivity. Electrical conductivity in most oxide glasses occurs by ion transport under the influence of an electric field. Small ions such as Li^+ and Na^+ can easily move through the glass structure. The K^+, Mg^{2+}, Ca^{2+}, Pb^{2+} and Ba^{2+} ions, however, are larger and less mobile and, therefore, are less able to contribute to the electrical conductivity. Increasing alkali content disrupts the degree of cross-linking in the network polyhedra. Consequently the volume resistivity is decreased in proportion to the alkali content (Fig. 3.12). Anomalous behavior[40] is exhibited by glasses of mixed alkali where the resistivity of the admixtures is higher than that of the end points (Fig. 3.13).

Semiconducting glasses, exhibiting electronic rather than ionic conductivity, form the exception to this discussion. Glasses with resistivities in the range of 10^2 to 10^9 $\Omega \cdot$ cm have been reported in the Ge-P-V oxide system[41,42] and in the elemental glasses

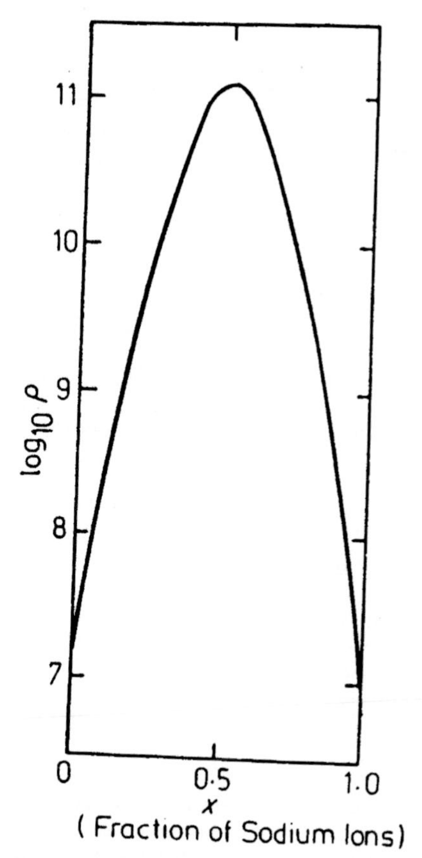

FIGURE 3.13 Electrical resistivity of (26-x) $Li_2O \cdot xNa_2O \cdot SiO_2$ glasses.[40]

of S, Se, or Te combined with one or more of the elements Si, Ge, P, As, The, and Pb.[43] These latter glasses are referred to as chalcogenide glasses. Electronic devices are described that use semiconducting glasses as the solid-state switch or memory component.[44,45]

With increasing temperature the volume resistivity of nonsemiconducting oxide glasses decreases (Fig. 3.12) because of the thermal weakening or breakdown of the bonds between the individual polyhedra. This causes the volume of the interstices to increase, which in turn permits higher mobility of the modifying cations. At the melting temperature, the resistivity of the glass decreases to as low as 10 to 100 Ω • cm, thereby making it possible to use the glass itself as the electric conductor. This behavior is used for primary melting of glasses and boosting of gas-fired furnaces, as well as for glass-to-glass sealing in energy-conserving, double-pane windows in some cases.

Surface Resistivity. The relatively low surface resistivity exhibited by glass under certain conditions is usually caused by either adsorbed moisture or accompanying electrolytic processes at the surface. The effect of atmospheric moisture becomes pronounced above 50 percent relative humidity (Fig. 3.14). Below 50 percent the sur-

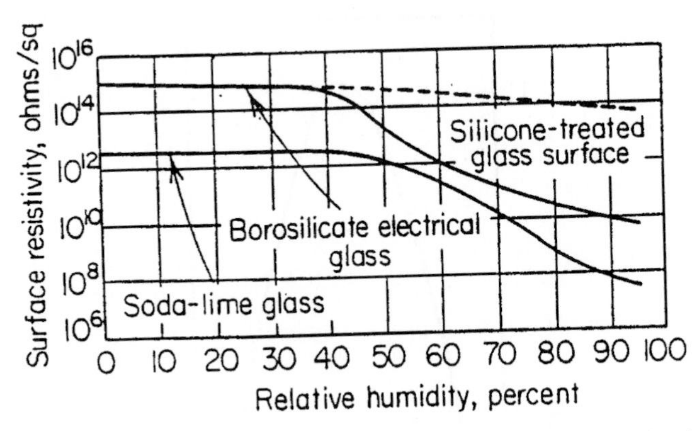

FIGURE 3.14 Effect of relative humidity on surface resistivity of several glasses. (*From Harrison and Moratis,*[10] *p. 6-29.*)

face water does not form a coherent monolayer; therefore it does not provide an electrical path for conduction.

Surface conductivity can be a more serious factor in the confined environments of hermetic packages which are designed to operate for long periods of time. Because these glasses are low melting glasses, they do not have the chemical durability and structural integrity of the high-silicate glasses and are prone to field-assisted electrolytic surface decomposition. These processes must be factors in both materials selection and design.[46]

Dielectric Strength. A practical requirement of a dielectric is that it not break down under applied voltage stress. Dielectric breakdown can occur by thermal and intrinsic processes. Thermal breakdown is due to localized heating caused by inhomogeneities in the electric field and in the dielectric itself. With increasing temperature, the dielectric strength of typical alumina ceramics decreases by about 40 to 50 V/mil per 100°C. If thermal heating is permitted to increase without limit, chemical breakdown of the dielectric will result in the passage of even higher currents, which in turn leads to fusion, vaporization, and finally puncture of the dielectric.

Intrinsic or electronic breakdown occurs when conduction electrons are accelerated to sufficiently high energies by local field gradients to liberate valence electrons by collision. This avalanche effect continues at an accelerating rate until finally dielectric breakdown results. The dielectric strength of various ceramic materials at several frequencies is given in Table 3.10.

The effect of the testing medium on the dielectric strength is shown in Fig. 3.15. The intrinsic breakdown strengths of borosilicate glass (curve *A*) and soda-lime-silicate glass (curve *B*) are both very high and linearly dependent on the specimen thickness. When these glasses are tested in various media, such as insulating and semiconducting oils, the breakdown strengths are much lower and nonlinearly related to specimen thickness. A carefully designed glass insulator can have an impulse-voltage strength of 1.7×10^6 V/cm in air.[47]

The dielectric strength of glass decreases with increasing temperature because of the greater ionic conductivity at the higher temperatures (Fig. 3.16). Regions of instability due to the channeling of the current and the formation of hot spots result in dielectric breakdown by the puncture of the dielectric material.

TABLE 3.10 Dielectric Strengths of Various Insulating Materials at Frequencies from 60 Hz to 100 MHz, in rms V/mil

Material	Thick-ness, mil	60 Hz	1z kHz	38 kHz	180 kHz	2 MHz	18 MHz	100 MHz
Polystyrene (unpigmented)	30	3174	2400	1250	977	725	335	220
Polyethylene (unpigmented)	30	1091	965	500	460	343	180	132
Polytetrafluoroethylene (Teflon*)	30	850	808	540	500	375	210	143
Monochlorotrifluoroethylene (Kel-F)‡	20	2007	1478	1054	600	354	129	29†
Glass-bonded mica	32	712	643	—	360	207	121	76
Soda-lime glass	32	1532	1158	—	230	90	55	20†
Dry-process porcelain	32	232	226	—	90	83	71	60†
Steatite	32	523	427	—	300	80	58	56†
Forsterite (AlSiMag-243)	65	499	461	455	365	210	112	74
Alumina, 85% (AlSiMag-576§)	55	298	298	253	253	178	112	69

*Trademark of E. I. duPont de Nemours & Co.

†Puncture with attendant volume heating effect.

‡Trademark of Minnesota Mining and Manufacturing Co.

§Trademark of American Lava Corp.

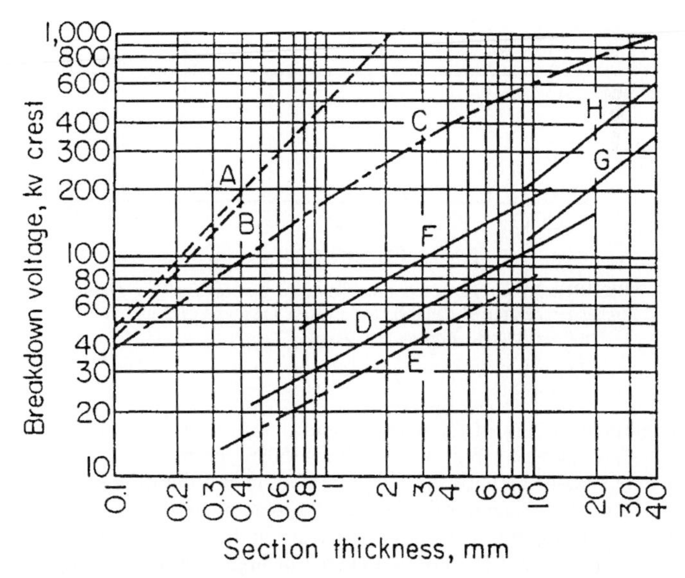

FIGURE 3.15 Breakdown voltage versus thickness of glass for different conditions at room temperature, 60-Hz voltage raised continuously. *A*—intrinsic dielectric strength of borosilicate glass; *B*—intrinsic dielectric strength of soda-lime glass; *C*—highest test values available for borosilicate glass; *D*—borosilicate glass plate immersed in insulating oil; *E*—soda-lime glass plate immersed in insulating oil; *F*—borosilicate glass plate immersed in semiconducting oil; *G*—borosilicate glass power-line insulator immersed in insulating oil; *H*—borosilicate glass power-line insulator immersed in semiconducting oil. (*From Harrison and Moratis,*[10] *p. 6-31.*)

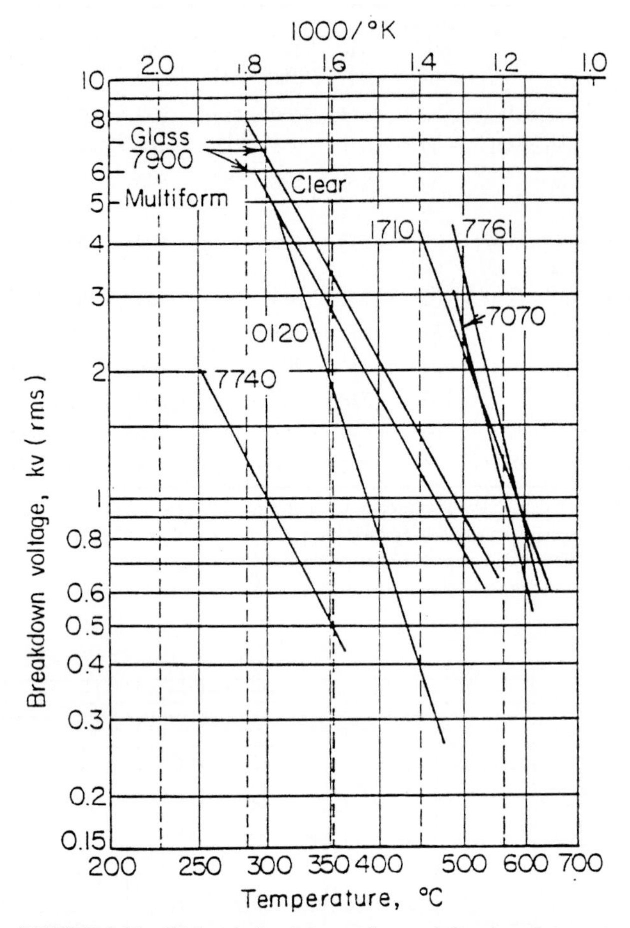

FIGURE 3.16 Dielectric breakdown of several Corning glasses at higher temperatures, 1 min breakdown for thickness of 2 mm at 60 Hz. (*From Harrison and Moratis,*[10] *p. 6-31.*)

The combined effects of temperature and duration of applied voltage stress on Pyrex* glass are shown in Fig. 3.17. Below −50°C the breakdown strength is quite high and independent of both the temperature and the duration of applied voltage stress. Above room temperature, however, the combined effects of higher temperatures and longer testing times do cause a decrease in the dielectric strength.

The dielectric strength of glass is lowered at higher frequencies by dipole relaxation processes (Fig. 3.18). Data listed in Table 3.11 indicate that the dielectric strength of ordinary glass at 100 MHz is 2 to 5 percent of the corresponding value measured at 60 Hz.

---

*Pyrex is a registered trademark of Corning Glass Works, Inc.

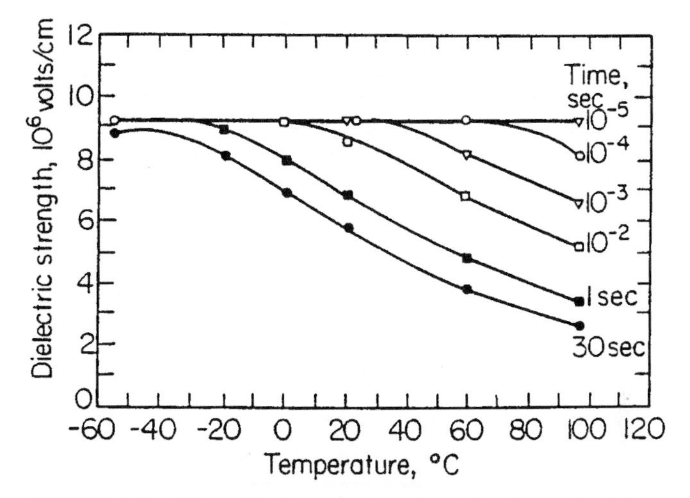

FIGURE 3.17 Effect of test temperature and test duration on breakdown strength of Pyrex glass. (*From Harrison and Moratis,*[10] *p. 6-31.*)

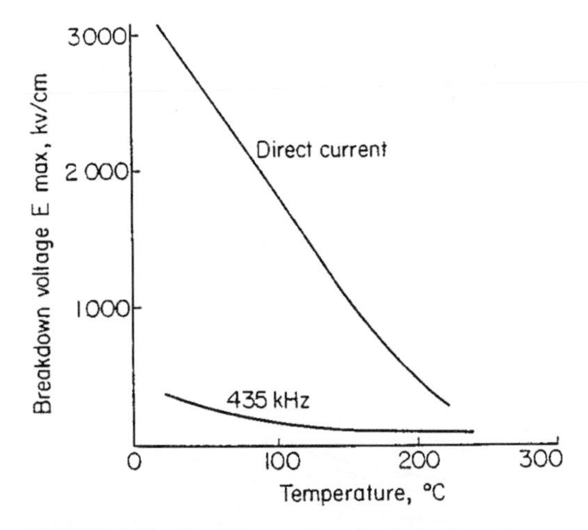

FIGURE 3.18 Breakdown voltage in air for lime glass as a function of temperature for direct current and for 435-kHz alternating current. (*From Harrison and Moratis,*[10] *p. 6-31.*)

Dielectric Constant. The dielectric constants of the more common types of glass (silicates, phosphates, borosilicates, and aluminosilicates) vary from a low of 3.8 for fused silica to 8 to 10 for high-lead glasses (Fig. 3.19). The high polarizability of the Pb^{2+} ion gives rise to the high dielectric constant of high-lead glasses, whereas the absence of such ions in fused quartz accounts for the low permittivity.

TABLE 3.11 Dielectric Breakdown of Ordinary Glass as a Function of Frequency

	Breakdown voltage, rms kV	
Frequency	0.030 in thick	0.125 in thick
60 Hz	38.2	38.4
1 kHz	28.3	
14 kHz	21.0	
150 kHz	7.5	
2 MHz	2.65	9.25
4.2 MHz	2.0	
9.8 MHz	1.7	
18 MHz	1.7	3.4
100 MHz	0.63	1.9

Source: From Chapman and Frisco.[37]

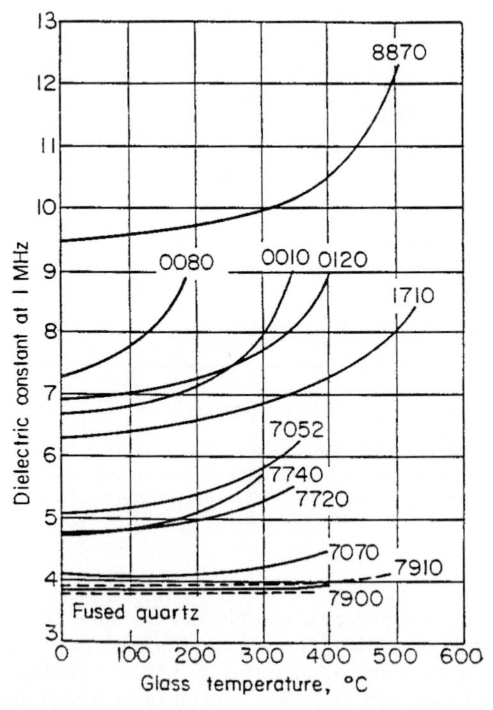

FIGURE 3.19 Dielectric constants of several Corning glasses at 1 MHz as a function of temperature. (*From Harrison and Moratis,*[10] *p. 6-32.*)

The dielectric constant of glass increases with rising temperature in proportion to the polarizability of the medium. When stability of the dielectric properties is required, glasses with rigid structures, such as fused silica and high-silica glasses, are used.

The dielectric constant of glass undergoes steplike reductions at higher frequencies as various polarizing mechanisms lose their ability to keep up with the field, generally following the formula: $\epsilon' = \epsilon_{inf} + (\epsilon_s - \epsilon_{inf})/(1 + \omega^2\tau^2)$, where ϵ' is the real part of the permittivity, ϵ_s the dc permittivity, ϵ_{inf} the permittivity at infinite frequency, ω the angular frequency (2πf), and τ the appropriate mechanism's relaxation time. Advanced computers are approaching the frequency where one of these drop-offs occurs. Designers must be aware of the potential instability as this region is traversed.

Dissipation Factor. The total dielectric loss in glass is the sum of four different loss mechanisms, namely, space charge polarization, dipole relaxation, atomic polarization, and electronic polarization losses, as shown in Fig. 3.20. The magnitude of each of these loss mechanisms varies with frequency and temperature. Space charge polarization and dipole relaxation losses (curves 1 and 2 in Fig. 3.20) are the predominant loss mechanisms at low frequencies; whereas atomic polarization and electronic polarization losses (3 and 4) are the controlling mechanisms at high frequencies. These losses generally increase with rising temperatures. The total losses combined constitute the dissipation factor.

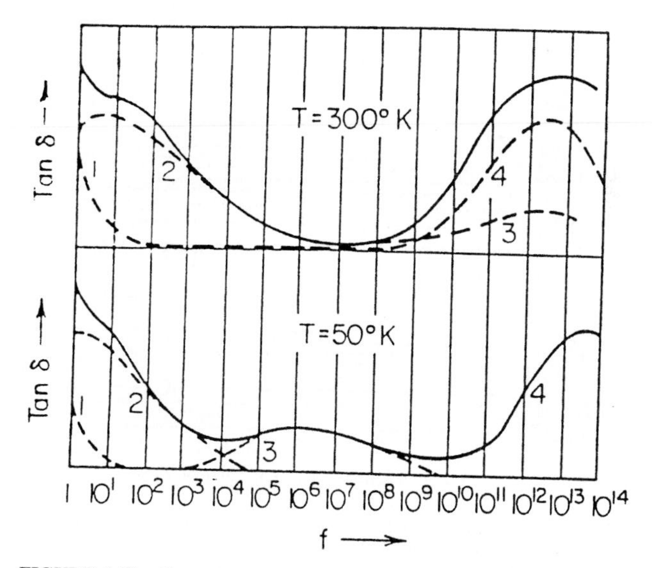

FIGURE 3.20 General shape of tan δ as a function of the frequency at 300 and 50 K. Solid curves give total losses, that is, the sum of four different contributions: 1—space charge polarization losses; 2—dipole relaxation losses; 3—atomic polarization losses; 4—deformation losses. (*From Harrison and Moratis,*[10] *p. 6-34.*)

Space charge polarization losses are important only at frequencies below about 50 Hz, where they are described by the relation tan $\delta = (2\pi f \rho \epsilon' \epsilon_0)^{-1/2}$, where ϵ_0 is the permittivity of free space. Since the conductivity $1/\epsilon$ of glass increases with rising temperatures, so also does the space charge polarization loss. Likewise, alkali-rich glasses exhibit greater space charge polarization losses than high-silica glasses.

Dipole relaxation losses are experienced up to about 10^6 Hz. They result from the energy that is given up by mobile ions as they jump over small distances in the network. Dipole losses are greatest when the frequency of the applied field is equal to the relaxation time of the dipole. At frequencies either far above or far below the resonant frequency, the losses are the least. Because of its random structure, glass has more than one relaxation time. This results in a broad distribution of dipole relaxation losses with frequency, rather than one discrete loss peak (curve 2 in Fig. 3.20).

The dissipation factor at 1 MHz and 250°C for various commercial glasses (Fig. 3.21) reveals a correlation between high resistivity and low dielectric losses. This is further illustrated in Fig. 3.22 by glasses with low alkali content, which not only have lower losses but also show greater temperature stability. Above 200°C losses become excessive in practically all glasses because of increased ionic conduction.

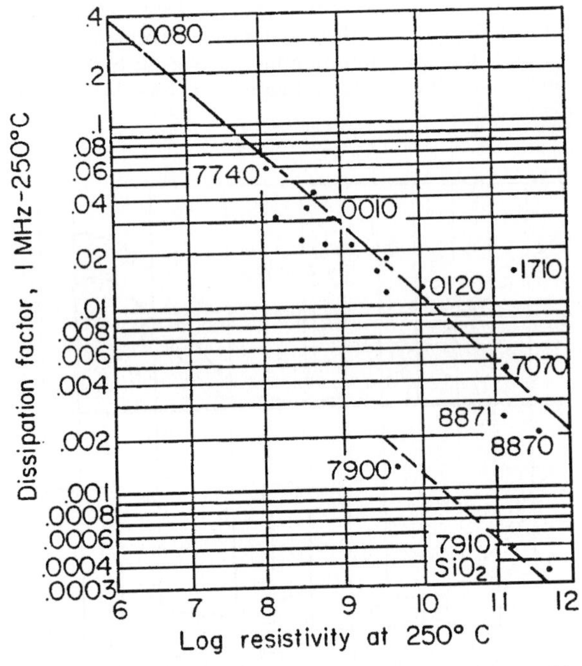

FIGURE 3.21 Relationship between log dissipation factor and log resistivity of several Corning glasses at 250°C. (*From Harrison and Moratis,*[10] *p. 6-35.*)

Atomic polarization losses occur by a resonance phenomenon involving both the network-forming ions and the network-modifying ions. As shown in Fig. 3.20, curve 3, vibration losses take place over a broad frequency range because of the variation in both the mass and the location of the different ions in the glass network. The resonant frequency of this loss mechanism is given by $f_{res} = (A/M)^{1/2}$, where A is a constant relating the displacement and the restoring force, and M is the mass of the ion.

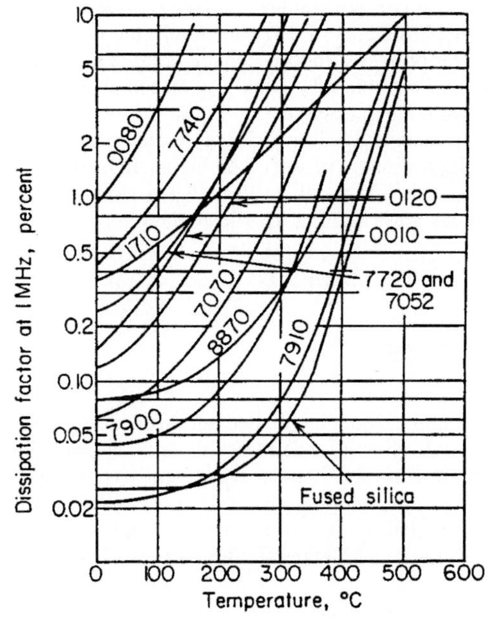

FIGURE 3.22 Dissipation factors of several Corning glasses at 1 MHz as a function of temperature. (*From Harrison and Moratis,*[10] *p. 6-35.*)

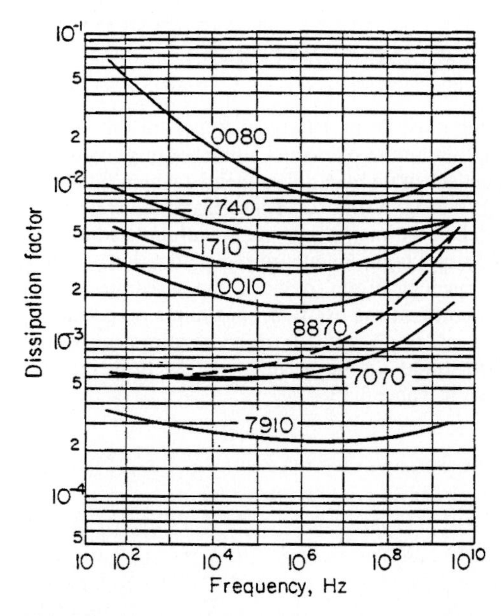

FIGURE 3.23 Dissipation factors at room temperature over a range of frequencies for several Corning glasses. (*From Harrison and Moratis,*[10] *p. 6-36.*)

Electronic polarization losses are similar to dipole relaxation losses in that they are associated with small displacements of electron clouds and their nuclei, the glass network. Electronic polarization losses involve smaller charge movements than do either space charge polarization or dipole relaxation loss mechanisms, and they occur in the region of 10^{13} Hz at room temperature (curve 4 in Fig. 3.20).

In general heavy ions such as Pb^{2+} or Ba^{2+} will decrease the resonant frequency of the network, whereas lighter ions will shift it to higher frequencies. This is shown in Fig. 3.23, where glass 7910, a high-silica glass, has a slowly increasing dissipation factor at 10^{10} Hz whereas glass 8870, a high-lead glass, has a rapidly increasing loss factor at this frequency.

3.5 GLASS-CERAMICS

Glass-ceramics[48] are ceramics which are melted and formed as glasses and which have been designed to be crystallizable after forming, at temperatures below the glass melting and deformation temperatures. The process frequently combines the advantages of low-cost forming of glass with the properties of the crystallized phase. The resulting material is dominantly crystalline and offers special properties. Corningware* is the most familiar example. Table 3.12 gives the properties for some commercial glass-ceramics, glass, and ceramics. These materials may be twice as strong as common glasses, and some may be machinable.[49] This latter property has put the material in demand for small-production-volume ceramic insulator parts since it may be machined as a metal. They are used in electronics in specialty applications.

3.6 LOW-FIRE CERAMIC SUBSTRATES

From the 1960s through most of the 1980s substrates for thin- and thick-film circuitry were mostly limited to aluminum oxide and a number of glasses. Beginning in the late 1980s a number of new substrate materials appeared which fired at the temperatures of thick-film processing, that is, <1000°C. These materials were aimed at a number of targets, including matched thermal expansion to alumina and to silicon, low-dielectric constant materials (<5), and high-conductivity metallizations, gold, silver, and copper. As circuit speed has risen in digital devices, the substrate and package dielectrics become part of the circuit and impose new constraints on packaging materials. Since no single material can satisfy all of the property requirements of advanced packaging applications,[50] different tradeoffs, have emerged. The properties of two commercialized systems are shown in Table 3.13.

*Corningware is a registered trademark of Corning Glass Works, Inc.

TABLE 3.12 Comparison of Properties of Pyroceram, Glass, and Ceramic

	Pyroceram		Glass				Ceramic		
Property	9606	9608	Fused silica 7940	Vycor 7900	Pyrex 7740	Lime glass 0080	High-purity alumina (93%+)	Steatites ($MgO-SiO_2$)	Forsterite ($2MgO-SiO_2$)
Specific gravity at 25°C	2.61	2.50	2.20	2.18	2.23	2.47	3.6	2.65–2.92	2.9
Water absorption, %	0.00	0.00	0.00	0.00	0.00	0.00	0.00	0–0.03	0–0.01
Gas permeability	0	0	0	0	0	0	0	—	0
Thermal									
Softening temperature, °C†	—	—	1584	1500	820	696	1700	1349	1349
Specific heat (25°C)	0.185	0.190	0.176	0.178	0.186	0.200	0.181		
Specific heat mean (25–400°C)	0.230	0.235	0.223	0.224	0.233	0.235	0.241		
Thermal conductivity, cgs, at 25°C mean temperature	0.0087	0.0047	0.0032	0.0036	0.0026	0.0025	0.042–0.086¶	0.0062–0.0065	0.010
Linear coefficient of thermal expansion at 25–300°C, 10^{-7} in/(in • °C)	57	4–20‡	5.5	8	32	92	73 (20–500°C)	81.5–99 (20–500°C)	99 (20–500°C)
Mechanical									
Modulus of elasticity, 10^{-6} lb/in^2	17.1	12.5	10.5	10.0	9.1	10.2	40–50	15	20
Poisson's ratio	0.25	0.25	0.17	0.19	0.20	0.22	0.21		
Modulus of rupture (abraded), 10^{-3} lb/in^2	18–20	12–14	5–9	5–9	6–10	6–10	50§	20§	19§
Knoop hardness									
100 g	657	593	—	463	418	—	1850		
500 g									

TABLE 3.12 Comparison of Properties of Pyroceram, Glass, and Ceramic *(Continued)*

	Pyroceram		Glass				Ceramic		
Property	9606	9608	Fused silica 7940	Vycor 7900	Pyrex 7740	Lime glass 0080	High-purity alumina (93%+)	Steatites ($MgO-SiO_2$)	Forsterite ($2MgO-SiO_2$)
Electrical									
Dielectric constant									
At 1 MHz									
25°C	5.58	6.78	3.78	3.8	4.6	7.2	8.81	5.9	6.3
300°C	5.60	—	—	3.9	5.9				
500°C	8.80	—	—	—	—	—	9.03		
At 10 GHz									
25°C	5.45	6.54	3.78	3.8	4.5	6.71	8.79	5.8	5.8
300°C	5.51	6.65	3.78						
500°C	5.53	6.78	3.78	—	—	—	9.03		

Source: From Harrison and Moratis,[10] p. 6-46.

TABLE 3.13 Alumina Compatible Materials

Property	Al_2O_3 (90%)	DuPont (851AT)*	IBM†
Dielectric constant	9.0	8.0	4.0
Dissipation factor	0.0014	—	
Inner-layer metallization	W or Mo	Au	Cu
Flexural strength, MPa	460	84	200
Thermal conductivity, W/(m • K)	16	2	<4
Coefficient of thermal expansion, ppm/°C	7.2	7.9	3.0

*From Eustice et al.[51]
†From Kumar and Tummala.[52]

3.7 CERAMIC AND GLASS SEALING

Ceramics and glasses are rarely used apart from being joined to metals. This is particularly true for electronics applications, where the glass or ceramic is usually a hermetically sealed insulator, and a metal penetration must be made through it. Ceramic-to-metal sealing is a very broad subject in its own right and cannot be treated exhaustively in this modest space. For applications in electronics it is important to know a few of the most common metallization and joining techniques.

3.7.1 Ceramic Metallization

The simplest means of producing a solderable metallization on common ceramics is to use a fired metal organic resinate coating.[53] These are typically based on the precious metals gold, rhodium, and platinum and are most commonly encountered as the lustrous decorative coatings on fine china and goblets. The resinate is merely painted on the ceramic and given an oxidizing bake between 550 and 850°C. The resulting coatings can be solderable as they are for metal flange attachments on distribution-line insulators.

Ceramics and glasses can be metallized by another relatively simple technique, flame and plasma spraying. This technique is most often used for mechanical connection of metals to ceramics. Basically, molten metal droplets are sprayed onto the substrate of interest, ceramic or glass. The molten droplets are made by passing metal powder through either a flame or an arc to melt them into discrete spherical droplets. The droplets are carried by an inert gas to the substrate to be coated. Adherence is not chemical but mechanical; therefore it is common to roughen the surface mechanically by grit blasting prior to coating. The resulting coatings are 5 to 20 percent porous and are thus not hermetic. This coating technique is very adaptable to small-volume large specimens. The reverse process, that is, coating ceramic or glass powders on metals, is frequently practiced as well, with the same limitations. Because molten metal is sprayed on the ceramic or glass substrate, considerable heating occurs, which would damage chip materials. While considerable substrate heating occurs, the temperatures are nowhere near the conventional temperatures of densification of the ceramic materials.

A more common technique for electronic application is that used for high-aluminum oxide ceramics and beryllium oxide ceramics. The fired ceramic is coated with tungsten or Mo-Mn powder and the assembly is baked in a moist reducing atmosphere at temperatures approaching 1600°C.[54] At these temperatures a glassy phase in the alumina or beryllia is drawn by capillary action into the interstices of the coated powder, and upon cooling, the glassy phase glues the metal powder to the surface.[55] The process is commonly referred to as the moly-manganese process. In ceramics with insufficient glassy phase, a glass is provided in the form of a powder along with the metal powder. The resulting metallization is usually electrochemically plated with nickel or copper, or both.

The process has been adapted to the manufacture of multilayer embedded lead-trace substrate materials, commonly called cofired. The final product is a hermetically sealed and buried three-dimensional circuit. Its drawback is in the high-lead trace conductivities compared to the precious metals and copper.

Aluminum oxide substrates may be metallized by thick-film processing.[56] Thick-film processing has traditionally been performed in oxidizing atmospheres at moderate temperatures, 800 to 1000°C. The metals are the precious metals gold, silver, platinum, palladium, and their combinations. A glassy phase is usually incorporated in the thick-film paste.

Each metallization system has advantages and disadvantages. The advantage of the high-temperature metallization schemes with molybdenum or tungsten is their moderate cost. The disadvantage is the high resistivity of the resultant films. These relative merits are reversed for thick-film materials. Another advantage of thick-film metallization is that the metallization is often applied by the device fabricator and not the ceramic vendor. This can allow for greater design modification efficiency.

3.7.2 Glass-to-Metal Joining

For high-reliability devices electrical connection is made by metal penetrations through insulating glass bushings. The glass is directly bonded by fusion to the feed-through pin and package wall, or bonded ring. Because glasses are so much weaker than ceramics, expansion matching is more critical. Remembering that glasses undergo a sharp nonlinearity in their thermal expansions, matching is not accomplished by matching published low-temperature coefficients. To join a metal and a glass, the total contractions are matched from a temperature referred to as the set point. This temperature is defined by the temperature where glass and metal, joined at elevated temperature, experience mutual stress upon further cooling. Above this temperature the glass dissipates stress by viscoelastic relaxation. Below this point the stress cannot be dissipated.

A matching contraction seal normally requires the following relation between published thermal expansion: $\alpha_g = \alpha_p = \alpha_r$ (Fig. 3.24). A limited series of alloys have been formulated which undergo an anomalous thermal expansion nonlinearity reminiscent of glass. These are based on 40 to 50 percent Ni-Fe alloys.[57] These metals, when combined with appropriate glasses, form the only truly "matching seals" and are used in the severest thermal-shock applications. They are frequently used in feed-through applications depicted in Fig. 3.24.

Another approach to glass-to-metal compatibility is often employed in glass-to-metal feed-throughs and is known as the compression seal. In this technique a true expansion-matching pin and glass insulator combination is selected. The ring is selected from a higher-expansion alloy, typically 1010 steel. The expansion relationships are shown in Fig. 3.24. Upon cooling from the sealing temperature, the metal is attempting to contract more than the pin and glass, resulting in the glass being thrust in compression. Ceramics are extremely strong in compression, and the seal is rendered much stronger by this strategy. Full potential is not realized because of tendencies for the sealing glass to form mensicuses with the pin and ring, which expose a

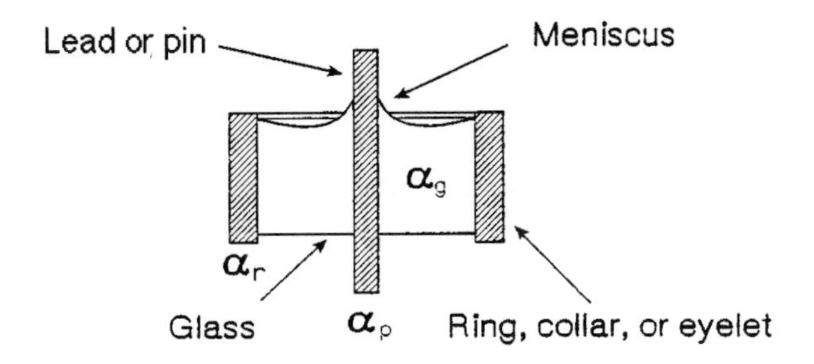

FIGURE 3.24 Cross section of glass-to-metal feed-through seal. (*From Sampson and Mattox,*[14] *p. 1.70.*)

region of tension. Properly traded off, a more reliable seal is achieved. The feed-throughs described are brazed or soldered to metal packages.

Housekeeper Seals. Housekeeper seals, which are named for their inventor,[58] are glass-to-metal seals which make no allowance for contraction matching. Instead, the metal is shaped to a feather edge and embedded in the softened glass (Fig. 3.25). The malleability of the metal is used to absorb the mutual strain between glass and metal. Such seals are used with copper and Kovar, for example. This is the only reliable metal joining scheme for the zero-expansion 96+ percent glasses.

These seals are found in high-power vacuum tubes, arc tubes, and lamps. This is a seal of last resort, and because of the required feathering, probably is ruled out of most feed-through applications because of the high electrical resistance.

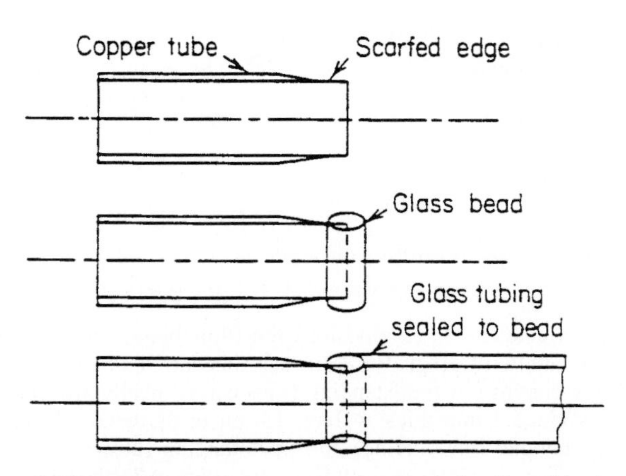

FIGURE 3.25 Housekeeper type seal and procedure for making it. (*From Harrison and Moratis,*[10] *p. 6-49.*)

3.8 GLAZES

Glazes are often applied to ceramics to improve their smoothness and reduce the physical absorption of contaminants. These glass coatings are very similar to the glass compositions already described and may be applied from powder slurries at the same time as the original firing or in postfiring.

3.9 DIAMOND

As Table 3.2 indicates, diamond is a material of exceptional properties for electronics applications. Its outstanding potential benefit is its heat transmission and heat-spreading capability, while being an excellent low-expansion insulator. Diamond is the best thermal conductor known. Its heat-conducting ability is five times better than that of copper. Natural diamond is a prohibitively expensive material for most applications, the price scaling exponentially with volume. Synthetic high-pressure formed diamond has a much more attractive price structure, but is not a good insulator because of densification-promoting phases added during processing. Until the early 1970s the production of diamond by physical vapor-deposition techniques was a matter of controversy and speculation. Thin-film diamond coatings and free-standing films are now an established science and an emerging technology.

The elements of the current deposition practices are credited to Russian workers[59] who identified critical conditions necessary for true diamond film deposition, that is, gas activation and the presence of atomic hydrogen to suppress graphite deposition. Since that time their results have been duplicated and improved upon in many laboratories around the world.[60] It has been learned that the growth of true diamond films is quite easy in that activation has been accomplished by a seemingly inexhaustible range of techniques, including thermal activation,[61] microwave,[62] dc discharge,[63] and combustion.[64] Figure 3.26 illustrates the basic elements of diamond deposition. The source of carbon is usually a dilute concentration of a hydrocarbon in a hydrogen carrier gas, that is, 0.2 to 5 percent CH_4 in H_2.

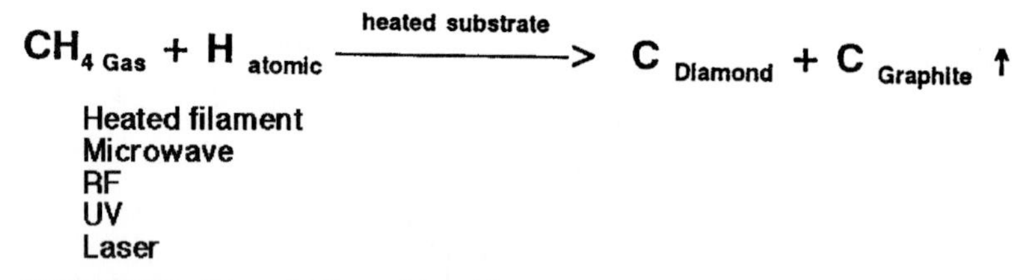

FIGURE 3.26 Schematic diamond deposition process.

Applications-oriented research envisions the films being used for heat spreaders, optical windows, and tool facings. Research has concentrated on increasing deposition rates to make the coatings or free-standing films economically attractive. Research is under way to produce 1-mm-thick wafers, 15 cm in diameter, at very competitive costs by 1995. Diamond film coatings and free-standing substrates are commercially available today.[65] The properties of such films are given in Table 3.14.

TABLE 3.14 Diamond Film Properties

Property	Value
Thermal conductivity	1000–1300 W/(m • K)
Log electrical resistivity	>12 Ω • cm
Dielectric constant at 45 MHz to 20 GHz	5.6
Loss tangent	<0.002
Coefficient of thermal expansion	2.0 ppm/°C
Mechanical strength	>1000 MPa
Transmission, 500-nm thick	
At 500 nm	56%
At 3 μm	36%
At 10 μm	71%
Absorption coefficient at 10 μm	0.1 cm^{-1}
Index of refraction at 10 μm	2.34

Source: From Norton Diamond Film.[65]

Diamond films are being evaluated as heat spreaders in microelectronic packaging, that is, lateral heat sinks. Research shows that beyond the thickness of several hundred micrometers, no further gain in heat extraction is realized.[66] Diamond can be grown on a wide range of materials, including metals, carbides, and oxides. The best adherence is presently realized on carbide forming metals of low expansion such as molybdenum. Silicon is also coated effectively. It appears that the expansion match is quite important. There is also a need to seed the deposition. For example, diamond polishing or grinding enhances deposition. One drawback of the technology is that temperatures approaching 900°C are required for deposition. This technology is very much an evolving technology, which will reward those who follow the diamond-film deposition literature.

REFERENCES

1. I. Burn, "Ceramic Capacitor Dielectrics," in *Engineered Materials Handbook,* vol. 4: *Ceramics and Glasses,* ASM International, 1991, pp. 1112–1118.
2. K. Uchino, "Piezoelectric Ceramics," in *Engineered Materials Handbook,* vol. 4: *Ceramics and Glasses,* ASM International, 1991, pp. 1119–1123.
3. A. Goldman, "Magnetic Ceramics (Ferrites)," in *Engineered Materials Handbook,* vol. 4: *Ceramics and Glasses,* ASM International, 1991, pp. 1160–1165.
4. G. H. Haertling, "Electrooptic Ceramics and Devices," in *Engineered Materials Handbook,* vol. 4: *Ceramics and Glasses,* ASM International, 1991, pp. 1124–1130.
5. T. K. Gupta, "Varistors," in *Engineered Materials Handbook,* vol. 4: *Ceramics and Glasses,* ASM International, 1991, pp. 1150–1155.
6. R. B. Poeppel, M. T. Lanagan, U. Balachandran, S. E. Dorris, J. P. Singh, and K. C. Goretta, "High-Temperature Superconductors," in *Engineered Materials Handbook,* vol. 4: *Ceramics and Glasses,* ASM International, 1991, pp. 1156–1160.

7. R. C. Buchanan, *Ceramic Materials for Electronics,* Marcel Dekker, New York, 1986, pp. 1–66.
8. T. G. McDougal, "History of AC Spark Plug Division, General Motors Corporation," *Ceram. Bull.,* vol. 28, 1949, pp. 445–448.
9. A. R. Von Hippel, *Dielectric Materials and Applications,* MIT Technology Press and Wiley, New York, 1954.
10. D. E. Harrison and C. J. Moratis, in *Electronics Materials and Processes Handbook,* C. A. Harper (ed.), McGraw-Hill, New York, 1970.
11. H. Thurnauer, "Development and Use of Electronic Ceramics Prior to 1945," *Ceram. Bull.,* vol. 56, 1977, pp. 219–224.
12. W. D. Kingery, H. K. Bowen, and D. R. Uhlmann, *Introduction to Ceramics,* Wiley, New York, 1976, pp. 308–309.
13. G. S. Brady and H. R. Clauser, *Materials Handbook,* 11th ed., McGraw-Hill, New York, 1977, p. 776.
14. R. Sampson and D. M. Mattox, "Materials for Electronic Packaging," in *Electronic Packaging and Interconnection Handbook,* C. A. Harper (ed.), McGraw-Hill, New York, 1991, pp. 1.60–1.61.
15. J. R. Floyd, "How to Tailor High-Alumina Ceramics for Electrical Applications," *Ceram. Ind.,* pp. 44–47, Feb. 1969; pp. 46–49, Mar. 1969.
16. N. M. Atlas and H. H. Nakamura, "Control of Dielectric Constant and Loss in Alumina Ceramics," *J. Am. Ceram. Soc.,* vol. 45, 1962, pp. 467–471.
17. J. R. Floyd, "Effect of Secondary Crystalline Phases on Dielectric Losses in High-Alumina Bodies," *J. Am. Ceram. Soc.,* vol. 47, 1964, pp. 539–543.
18. R. E. Mistler, D. J. Shanefield, and R. B. Runk, "Tape Casting of Ceramics," in *Ceramic Processing before Firing,* G. Y. Onoda, Jr., and L. L. Hench (eds.), Wiley, New York, 1978, pp. 411–448.
19. R. Block, "CPS Microengineers New Breed of Materials," *Ceram. Ind.,* Apr. 1988, pp. 51–53.
20. "Coors Ceramics—Materials for Tough Jobs," Coors Data Sheet, K.P.G.-2500-2/87 6429.
21. D. M. Mattox, "Glassy Phase Limited Thermal Conductivity for Alumina Ceramics," in *Proc. ISHM '91 Conf.* (Orlando, Fla., Oct. 1991).
22. D. W. Richerson, *Modern Ceramic Engineering,* Marcel Dekker, New York, 1992, pp. 528–531.
23. W. S. Young and S. H. Knickerbocker, *Ceramic Materials for Electronics,* Marcel Dekker, New York, 1986, pp. 489–526.
24. H. T. Sawhill, A. L. Eustice, S. J. Horowitz, J. Gar-El, and A. R. Travis, "Low Temperature Co-Fireable Ceramics with Co-Fired Resistors," in *Proc. Int. Symp. on Microelectronics* (ISHM, Oct. 6–8, 1986), pp. 173–180.
25. "Ceramic Products," Brush Wellman, Cleveland, Ohio.
26. M. B. Powers, "Potential Beryllium Exposure while Processing Beryllia Ceramics for Electronics Applications," Brush Wellman, Engineered Materials, Ceramics Div., Cleveland, Ohio.
27. N. Iwase, K. Anzai, and K. Shinozaki, "Aluminum Nitride Substrates Having High Thermal Conductivity," *Solid State Technol.,* Oct. 1986, pp. 135–137.
28. L. E. Dolhert, A. L. Kovacs, J. W. Lau, J. H. Enloe, E. Y. Luh, and J. Stephan, "Performance and Reliability of Metallized Aluminum Nitride for Multichip Module Applications," *Int. J. Hybrid Microelectron.,* vol. 10, no. 4, pp. 113–120, 1991.
29. T. Yamaguchi, and M. Kageyama, CHMT, *IEEE Transactions,* vol. 12, no. 3, pp. 402–405, 1989.
30. E. S. Dettmer, H. K. Charles, Jr., S. J. Mobley, and B. M. Romenesko, "Hybrid Design and Processing Using Aluminum Nitride Substrates," *ISHM 88 Proc.,* pp. 545–553.

31. E. S. Dettmer and H. K. Charles, Jr., "AlN and SiC Substrate Properties and Processing Characteristics," in *Advances in Ceramics,* vol. 31, American Ceramic Society, Columbus, Ohio, 1989, pp. 87–106.
32. "Combat Boron Nitride, Solids, Powders, Coatings," Carborundum Product Literature, Form A-14, 011 Effective 9/84.
33. R. Kamo, "Adiabatic Diesel Engines," in *Engineered Materials Handbook,* vol. 4: *Ceramics and Glasses,* ASM International, 1991, pp. 987–994.
34. G. W. McLellan and E. B. Shand, *Glass Engineering Handbook,* McGraw-Hill, New York, 1984.
35. W. H. Kohl, *Materials and Techniques for Electron Tubes,* Reinhold, New York, 1960, pp. 394–469.
36. L. J. Frisco, "Frequency Dependence of Electric Strength," *Electro-Technol.,* vol. 68, Aug. 1991, pp. 110–116.
37. J. J. Chapman and L. J. Frisco, *Elec. Mfg.,* vol. 53, p. 136, May 1959; E. B. Shand, *Glass Engineering Handbook,* 2d ed., McGraw-Hill, New York, 1958, p. 79.
38. W. P. Hood, U.S. patent 2,106,744, Feb. 1, 1938.
39. *Ceramic Source '90,* vol. 5, American Ceramic Society, Columbus, Ohio, 1990, p. 356.
40. R. H. Doremus, *Glass Science,* Wiley, New York, 1973, p. 162.
41. B. H. V. Janakirama-Rao, "Structure and Mechanism of Semiconductor Glasses," *J. Am. Ceram. Soc.,* vol. 48, pp. 311–319, 1965.
42. J. D. Mackenzie, "Glasses for Electronic Applications—Why and What," in *Ceramic Transactions,* vol. 20: *Glasses for Electronic Applications,* K. M. Nair (ed.), American Ceramic Society, Columbus, Oh., 1990, pp. 341–354.
43. J. D. Mackenzie, "Semiconducting Glasses," in *Physics of Electronic Ceramics,* L. L. Hench and D. B. Dove (eds.), Marcel Dekker, New York, 1971, p. 157.
44. N. F. Mott, "Conduction in Glasses Containing Transition Metal Ions," *J. Non-Cryst. Solids,* vol. 1, 1969, p. 1.
45. S. R. Ovshinsky and H. Fritzsche, "Amorphous Semiconductors for Switching, Memory, and Imaging Applications," *IEEE Trans. Electron Dev.,* vol. ED-20, 1973, p. 91.
46. C. R. S. Needes, "The Accelerated Life Testing of Copper Thick Film Multilayer Materials," in *Proc. 1986 Int. Symp. on Microelectronics* (Oct. 1986), pp. 840–847.
47. J. B. Birks, *Modern Dielectric Materials,* Heywood, London, 1960, p. 19.
48. P. W. McMillan, *Glass-Ceramics,* Academic Press, London, 1979.
49. "Machineable Glass Ceramic," Product Literature, Corning Glass Works, Corning, N.Y.
50. M. Mattox, "New Choices for Electronic Packaging Materials," in *Proc. 6th Int. SAMPE Electronic Materials and Processes Conf.* (Baltimore, Md., June 1992).
51. A. L. Eustice et al., "Low Temperature Co-Fireable Ceramics: A New Approach for Electronic Packaging," in *Proc. 36th Electronics Components Conf.* (Seattle, Wash., May 1986), pp. 37–47.
52. A. H. Kumar and R. R. Tummala, "State-of-the-Art Glass-Ceramic/Copper, Multilayer Substrate for High Performance Computers," *Int. J. Hybrid Microelectron,* vol. 14, no. 4, pp. 137–150, 1991.
53. T. T. Hopper, "How to Apply Noble Metals to Ceramics," *Ceram. Ind.,* June 1963.
54. H. Mizuhara and T. Oyama, "Ceramic/Metal Seals," in *Engineered Materials Handbook,* vol. 4: *Ceramics and Glasses,* ASM International, 1991, pp. 502–510.
55. D. M. Mattox and H. D. Smith, "The Role of Manganese in the Metallization of High Alumina Ceramics," *Ceram. Bull.,* vol. 64, 1985, pp. 1363–1369.
56. R. E. Cote and R. J. Bouchard, "Thick Film Technology," in *Electronic Ceramics,* L. M. Levinson (ed.), Marcel Dekker, New York, 1988, pp. 307–370.

57. A. P. Tomsia and J. A. Pask, "Glass/Metal and Ceramic/Metal Seals," in *Engineered Materials Handbook,* vol. 4: *Ceramics and Glasses,* ASM International, 1991, pp. 493–501.

58. W. G. Housekeeper, U.S. patent 1,294,466, Feb. 18, 1919.

59. V. P. Varnin, I. G. Teremetskaya, D. V. Fedoseev, and B. V. Deryagin, "Growth of Diamond from Highly Nonequilibrium Gas Media," *Sov. Phys. Dokl.,* vol. 29, 1984, pp. 419–421.

60. W. Zhu, B. R. Stoner, B. E. Williams, and J. T. Glass, "Growth and Characterization of Diamond Films on Nondiamond Substrates for Electronic Application," *Proc. IEEE,* vol. 79, 1991, pp. 621–646.

61. Y. H. Lee, P. D. Richard, K. J. Bachmann, and J. T. Glass, "Bias-Controlled Chemical Vapor Deposition of Diamond Thin Films," *Appl. Phys. Lett.,* vol. 56, 1990, pp. 620–622.

62. Y. A. Liou, A. Inspektor, R. Weimer, and R. Messier, "Low-Temperature Diamond Deposition by Microwave Plasma-Enhanced Chemical Vapor Deposition," *Appl. Phys. Lett.,* vol. 55, 1989, pp. 631–633.

63. K. Suzuki, A. Sawabe, H. Yasuda, and T. Inuzuka, "Growth of Diamond Thin Films by DC Plasma Chemical Vapor Deposition," *Appl. Phys. Lett.,* vol. 50, 1987, pp. 728–729.

64. L. M. Hanssen, W. A. Carrington, J. E. Butler, and K. A. Snail, "Diamond Synthesis Using an Oxygen-Acetylene Torch," *Mater. Lett.,* vol. 7, 1988, pp. 289–292.

65. Norton Diamond Film, Northboro, Mass.

66. D. J. Pickrell and D. S. Hoover, "Chemical Vapor Deposition of Diamond for Electronic Packaging Applications," *INSIDE ISHM,* July/Aug. 1991, pp. 11–15.

CHAPTER 4
SEMICONDUCTOR MATERIALS

R. Noel Thomas and Albert A. Burk, Jr.

4.1 PROPERTIES OF SEMICONDUCTORS

Semiconductors are the raw materials of the electronics industry, and thin single-crystal slices or wafers are used as substrates for fabricating electronic or optoelectronic devices and integrated circuits. They are frequently referred to as elemental, such as silicon, or compound semiconductors, such as gallium arsenide, from the column placement of the element or the constituent elements in the periodic table. The important semiconductors consist of the elements between columns IIA and VIA, as shown in Table 4.1. This chapter deals mainly with the properties and preparation of those semiconductors which are of the most significance to our world today—silicon (and for historical reasons, germanium), GaAs, and InP of group III-V and CdTe and HgCdTe of group II-VI. In addition SiC and diamond are discussed briefly because of their potential to significantly advance solid-state device technology and the current intensive materials development of these "new" semiconductors. Information on other semiconductors is, however, included and referenced where possible. The focus is on semiconductor materials technology, which serves the needs (in terms of starting substrates and active layer fabrication) for successful device fabrication. The intent of this chapter is to provide useful information to the semiconductor device community for broader understanding of the materials issues and device-materials interactions, as well as to others requiring broad coverage of this topic.

4.1.1 Semiconductor Principles

Physics of Semiconductors.[1,2] From the application of quantum mechanics, electrons in a solid are described as having allowed energy levels or energy bands. These bands are separated by regions or forbidden gaps, which the electrons cannot occupy. Electrons in the outermost shell of the atoms in the solid, the valence electrons, are in their lowest energy states, or the valence band. A set of energy levels—the conduction band—exists above the forbidden gap, to which an electron can be excited by imparting kinetic energy, thereby causing electrical conduction. The magnitude of conduction in the three different classes of solids—metals, insulators, and semiconductors—is described in terms of both an atomistic representation and an energy-band representation, as illustrated in Fig. 4.1. In metals, such as aluminum, the valence electrons

TABLE 4.1 Elements in the Periodic Table That Form Common Semiconductors

IIB	IIIA	IVA	VA	VIA
	5 **B** Boron	6 **C** Carbon	7 **N** Nitrogen	8 **O** Oxygen
	13 **Al** Aluminum	14 **Si** Silicon	15 **P** Phosphorus	16 **S** Sulfur
30 **Zn** Zinc	31 **Ga** Gallium	32 **Ge** Germanium	33 **As** Arsenic	34 **Se** Selenium
48 **Cd** Cadmium	49 **In** Indium	50 **Sn** Tin	51 **Sb** Antimony	52 **Te** Tellurium
80 **Hg** Mercury	81 **Tl** Thallium	82 **Pb** Lead	83 **Bi** Bismuth	84 **Po** Polonium

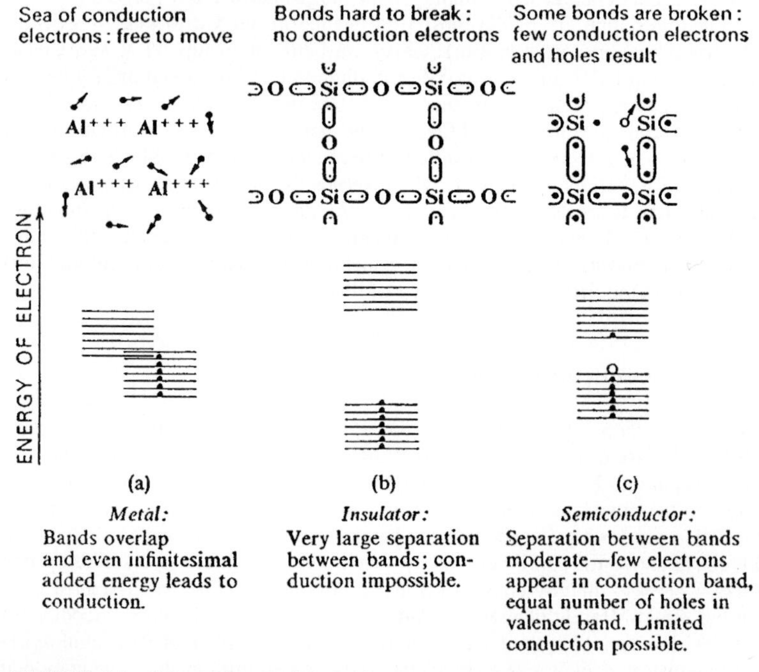

FIGURE 4.1 Schematic atomistic and energy-band representation. (*a*) Conductor. (*b*) Insulator. (*c*) Intrinsic semiconductor.[1]

constitute a sea of free electrons which can be excited by either an electric field or temperature. Since the energy bands overlap (Fig. 4.1*a*), very little energy is required to excite electrons to the conduction band and achieve motion of the electrons. Hence metals typically have high electrical conduction. In insulators, such as silicon dioxide, where the valence electrons form strong bonds between neighboring atoms, there are no free electrons to participate in conduction. This means that in our energy-band picture (Fig. 4.1*b*) the forbidden gap is large. All the electrons occupy the valence band while the conduction band remains empty. It is impossible to impart energy to the valence electrons except at very high electric fields. For this reason, insulators have low electrical conduction.

A semiconductor such as silicon represents the intermediate case shown in Fig. 4.1*c*. Thermal excitation causes some of the silicon bonds to be broken, resulting in free electrons in the valence band. The smaller band gap between valence and conduction bands, compared to an insulator, enables electrons to jump up into the conduction band. Under an applied electric field, electrons in the conduction band and "holes" (a charge deficit corresponding to the bonding electron before the bond was broken) in the valence band gain kinetic energy and conduct electricity.

The magnitude of the energy gap accounts for the different properties of semiconductors. In absolutely pure semiconductors (which cannot be realized in practice) the electron and hole concentrations are equal and are called the intrinsic carrier concentration n_i. Since the intrinsic carrier concentration depends only on the breakage of silicon bonds, we would therefore expect n_i to be a function of the vibrational energy of the lattice (hence the temperature) and the bonding energy, which in terms of the band diagram corresponds to the energy gap or band gap E_g. Both these dependencies are borne out by experimental data,[3–5] as shown in Fig. 4.2. Here the intrinsic carrier concentrations n_i for germanium, silicon, and GaAs are shown as a function of temperature. It is evident that (1) n_i increases very rapidly with increasing temperature, and (2) at a given temperature n_i decreases very sharply with increasing band gap E_g. The data can be summarized by an exponential temperature dependence, $n_i \propto e^{-E_a/kT}$, where the activation energy E_a is approximately $E_g/2$.

In the real world, impurities are introduced intentionally (for doping) as well as unintentionally (as contamination during crystal growth) into a single-crystal semiconductor at concentrations which are much larger than n_i. In the case where the impurity phosphorus, which has five valence electrons, is added to silicon, which has four electrons, the extra electron on the phosphorus atom cannot be accommodated in the regular bonding of the silicon lattice. Thus since it is out of place, it is easily removed or ionized. The ionization energy of phosphorus in silicon is only about 0.05 eV and is much less than the band gap of 1.1 eV of silicon. At room temperature, lattice vibrational energy is sufficient to completely ionize phosphorus and other column V impurities in silicon, providing an equal concentration of conduction electrons except at very high phosphorus concentrations. Under complete ionization, the electron concentration $n = N_d$, where N_d denotes the concentration of donor impurities. The column V impurities in silicon are the donor impurities in silicon, because they donate an electron to the conduction band. The band diagram for extrinsically doped silicon is shown in Fig. 4.3. For *n*-type silicon (Fig. 4.3*a*), the phosphorus donor ions are denoted by positive charges 0.044 eV below the conduction-band edge.

The analogous case where an impurity that has three valence electrons, such as boron (a column III element), is introduced into the silicon lattice as shown in Fig. 4.3*b*. The column III impurities having a deficit of one electron compared to silicon can be considered to carry a hole. This hole can be removed relatively easily at room temperature since its ionization energy is about 0.05 eV. The hole concentration $p = N_a$, where N_a denotes the concentration of acceptor impurities. The boron acceptor ions

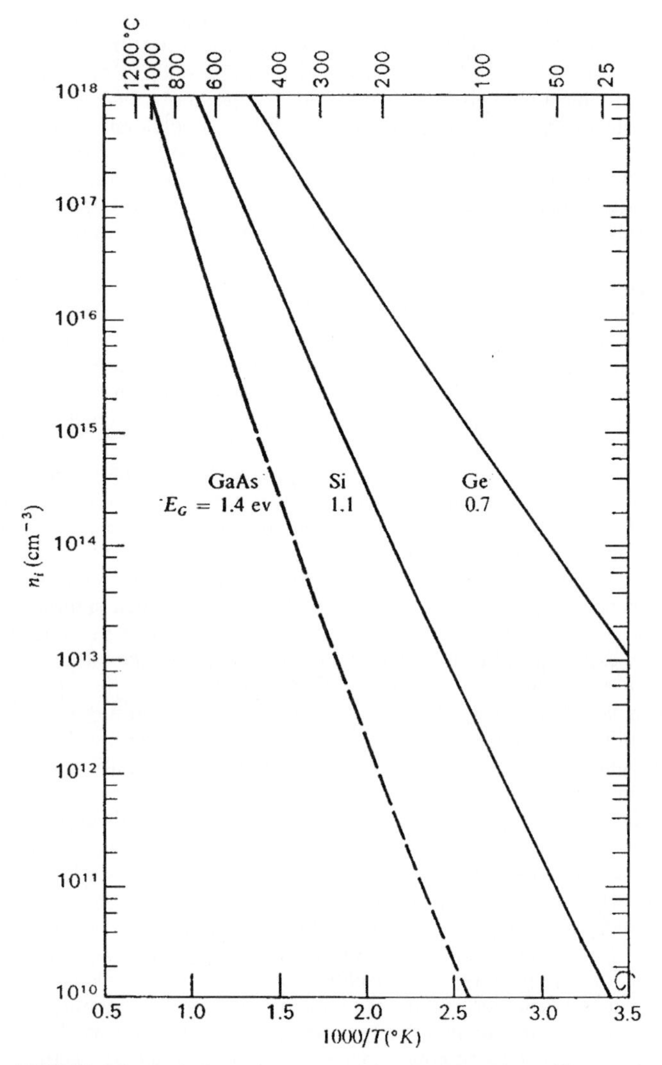

FIGURE 4.2 Intrinsic carrier concentration for germanium, silicon, and GaAs (calculated).[1]

(and other column III elements in silicon) are shown as negative charges located 0.045 eV above the valence band edge in Fig. 4.3*b*.

In general, both acceptor- and donor-type impurities can be present simultaneously. The conductivity type of the semiconductor is then determined by that impurity which is present in the greater concentration. The corresponding majority carrier concentration will then be given by

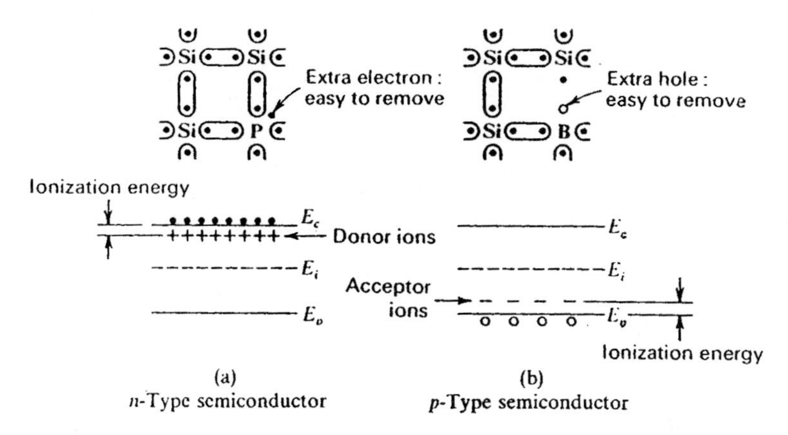

FIGURE 4.3 Schematic atomistic and energy-band representations of extrinsic semiconductors.[1]

$$n = N_d - N_a, \qquad \text{when } N_d > N_a \tag{4.1}$$

or

$$p = N_a - N_d, \qquad \text{when } N_a > N_d \tag{4.2}$$

The importance of the band-gap energy of different semiconductors and of impurity ionization energies in determining the semiconducting properties has resulted in the data[6,7] shown in Fig. 4.4, which maps the energy levels (ionization energies) of different impurities in germanium, silicon, and GaAs.

Drift Velocity and Mobility. Charge transport in semiconductors under the influence of an electric field is related to the drift velocity and the mobility. In a uniform *n*-type semiconductor, electrons are in continual thermal motion interrupted by collisions. There is no net displacement until an electric field is applied, when an additional velocity component is superimposed upon the random motion, giving rise to the drift velocity v_{drift}, with a direction opposite to the applied field for electrons. The drift velocity is given by the average acceleration between collisions $qE/2m^*$, and the time interval between collisions t_{coll},

$$v_{\text{drift}} = \frac{qE}{2m^*} t_{\text{coll}} \tag{4.3}$$

or

$$v_{\text{drift}} = \mu E \tag{4.4}$$

where μ is the mobility of the electrons, q the electronic charge, E the applied electric field, and m^* the effective mass of the electrons (which replaces the mass of the free electron by correcting for the effects of crystalline lattice).

Figure 4.5 shows measurements[8] of the drift velocities of electrons and holes in silicon as a function of the applied electric field. The linear relationship $v_{\text{drift}} = \mu E$ holds only at low electric fields, where the drift velocity is small in comparison to the thermal velocity of carriers in silicon at room temperature, about 10^7 cm/s. At large enough fields a maximum drift or saturation velocity is reached. The carrier mobility

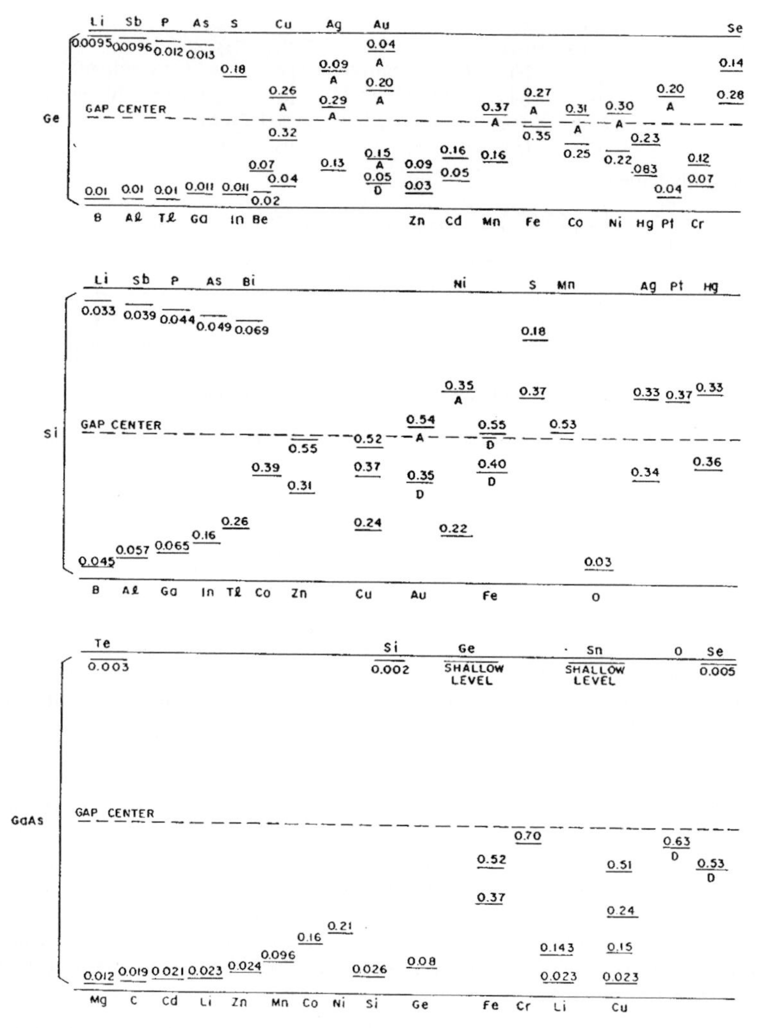

FIGURE 4.4 Measured ionization energies for various impurities in germanium, silicon, and GaAs. Band gaps at 300 K are 0.803, 1.12, and 1.42 eV, respectively.[2]

reflects scattering mechanisms during charge transport, with impurity and lattice scattering being the dominant processes. Mobility can be expressed as

$$\frac{1}{\mu} = \frac{1}{\mu_i} + \frac{1}{\mu_l} \tag{4.5}$$

Impurity scattering is caused by collisions with fixed charges, such as ionized donors or acceptors, and depends on the total concentration of negatively and positively

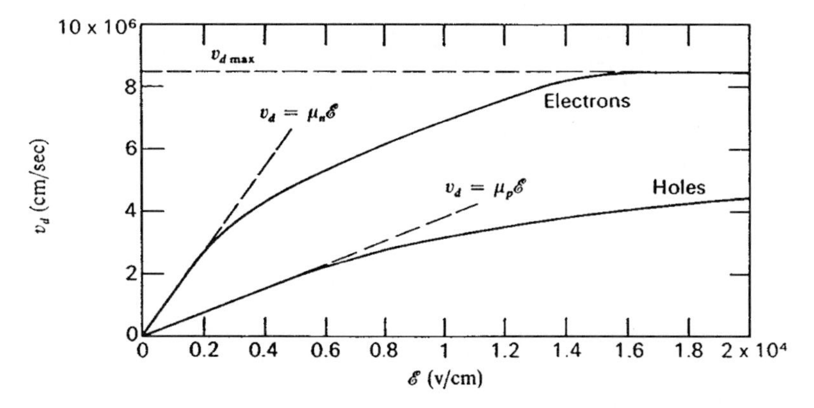

FIGURE 4.5 Drift velocity of carriers in silicon as a function of electric field.[1]

charged ions. The mobility dominated by impurity scattering μ_i can be shown theoretically to be proportional to $T^{3/2}/N_T$. Lattice scattering is due to the thermal vibrations of the atoms, and lattice scattering dominated mobility μ_l is theoretically proportional to $1/T^{3/2}$. Figure 4.6 shows measurements[6,9,10] of electron and hole mobilities in germanium, silicon, and GaAs versus the total ionized impurity concentration N_T at 300 K. Two features are of note and common to all semiconductors: (1) mobility reaches a maximum value at low impurity concentration corresponding to lattice scattering, and (2) the mobility of electrons is higher than that of holes. The influence of temperature on hole mobility in p-type silicon with two different doping levels is illustrated[11] in Fig. 4.7. Two regions can be distinguished. At low temperatures, impurity scattering dominates, and separate curves are seen for the different doping concentrations. At high temperatures, lattice scattering dominates, and the doping concentration has little effect. The mobility decreases with temperature in this region, and experimentally mobilities are found to follow a $T^{-2.5}$ dependence rather than the theoretically predicted $T^{-1.5}$ dependence for lattice scattering.

The saturation drift velocity in all semiconductors is limited by impurity scattering to about 10^7 cm/s at high electric fields. The maximum electric field, or the breakdown field of a semiconductor, is determined by the process of impact ionization. As the electric field is increased, a thermally generated electron-hole pair can gain sufficient kinetic energy before colliding with the lattice so that silicon-silicon bonds can be fractured, leading to the formation of additional electron-hole pairs. These in turn are accelerated, forming additional electron-hole pairs by this avalanche gain process. The critical electrical breakdown field for the avalanche process to occur depends on the doping density for a given semiconductor and increases with E_g for different semiconductors. In p-n junctions with high doping levels, breakdown due to tunneling of carriers across the band gap (zener breakdown) occurs at very high electric fields.

Resistivity. The resistivity of n-type semiconductors is given by

$$\rho = \frac{1}{q\mu_n n} \simeq \frac{1}{q\mu_n N_d} \tag{4.6}$$

and similarly for p-type semiconductors by

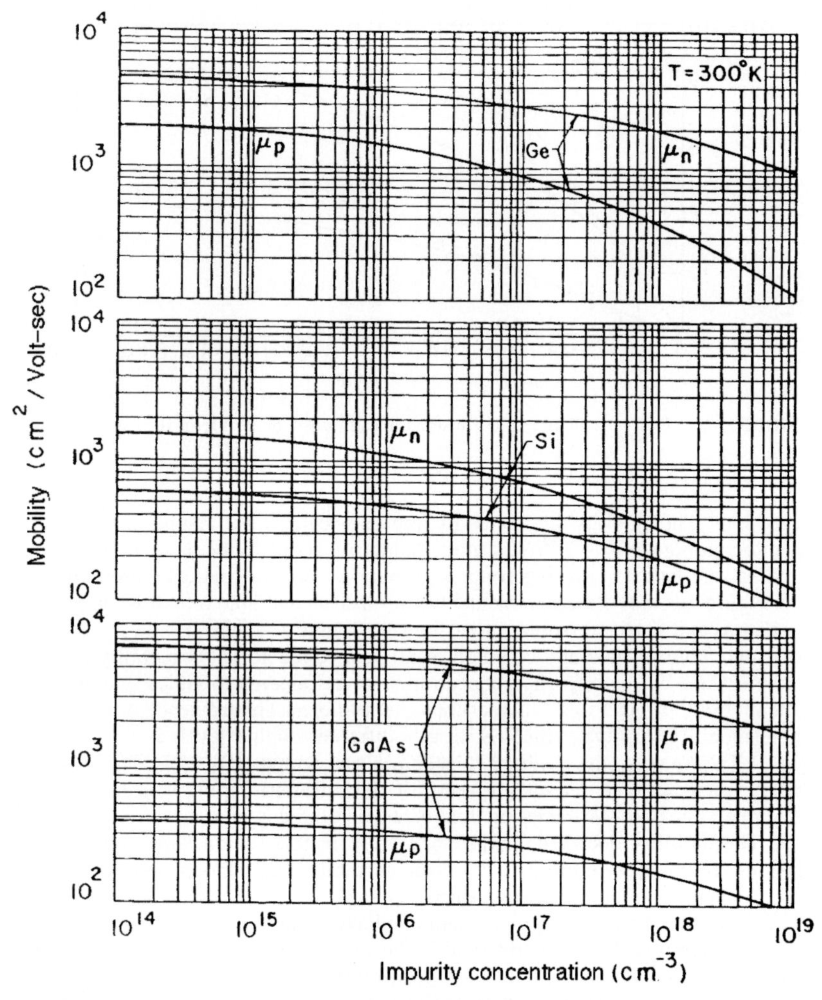

FIGURE 4.6 Drift mobility of germanium and silicon and Hall mobility of GaAs at 300 K versus impurity concentration N_T.[1]

$$\rho = \frac{1}{q\mu_p p} \simeq \frac{1}{q\mu_p N_a} \tag{4.7}$$

where μ_n and μ_p are the electron and hole mobilities, and n and p are the carrier concentrations, respectively.

In general, when both carrier types are taken into consideration, the resistivity is given by

$$\rho = \frac{1}{q(\mu_n n + \mu_p p)} \simeq \frac{1}{q(\mu_n N_d + \mu_p N_a)} \tag{4.8}$$

Table 4.2 lists some of the most useful properties of important semiconductors.

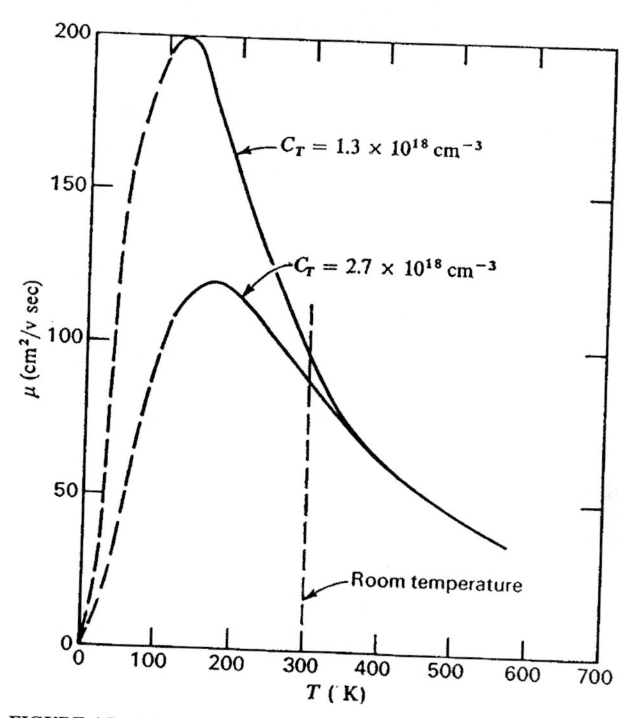

FIGURE 4.7 Effect of temperature on mobility of carriers in silicon.[1]

TABLE 4.2 Properties of Important Semiconductors

	Semiconductor	Melting point, °C	Density, g/cm³	Band gap, eV	Mobility, cm²/V • s		Dielectric constant
					Electrons	Holes	
Element	C (diamond II)		3.51	5.47	1800	1600	5.5
	Si	1417	2.33	1.12	1500	600	11.8
	Ge	937	5.33	0.80	3900	1900	16
	Sn/(grey)			≈0.09			
IV-IV	α-SiC		3.21	3.1	400	50	10
	ß-SiC		3.21	3.00	250	50	10
III-V	BN			≈7.5			
	BP			6			
	GaSb	712	5.62	0.67	4000	1400	15
	GaAs	1237	5.31	1.43	8500	400	10.9
	GaP	1465		2.24	110	75	10
	InSb	525		0.16	78000	750	17
	InAs	942	5.7	0.33	33000	460	14.5
	InP	1062	4.79	1.29	4600	150	14
II-VI	CdS		4.82	2.42	300	50	10
	CdSe	1350	5.81	1.7	800		10
	CdTe	1092	6.06	1.58	1050	100	10.2
	ZnS		4.1	3.6	165		8
IV-VI	PbS	1062	5.42	0.41	600	700	17
	PBTe	904		0.32	6000	4000	30

Band Structure. The electronic band structure of silicon (and germanium) as a function of momentum differs from that of GaAs and many other compound semiconductors in one important manner. GaAs is a direct band-gap material, since the valence-band minima are aligned with the conduction-band minima, whereas silicon is an indirect band-gap semiconductor, as shown in Fig. 4.8. In GaAs, transitions between bands are easier, since only an energy change, and not a momentum change also, is required. These direct transitions between the valence band and the conduction band lead to the most unique properties of compound semiconductors—the ability to emit or absorb light. Events involving light emission are unlikely in silicon due to the changes in both energy and momenta which are required. Other compound semiconductors with GaAs-like band structures include AlSb, GaSb, InSb, InAs, CdS, CdSe, ZnO, and ZnS. These properties make compound semiconductors uniquely suitable for optical devices. Solid-state lasers and light-emitting diodes (LEDs) as well as detectors and solar cells utilize this phenomenon. The wavelength of the light emitted or detected is determined by the band gap of the compound semiconductor.

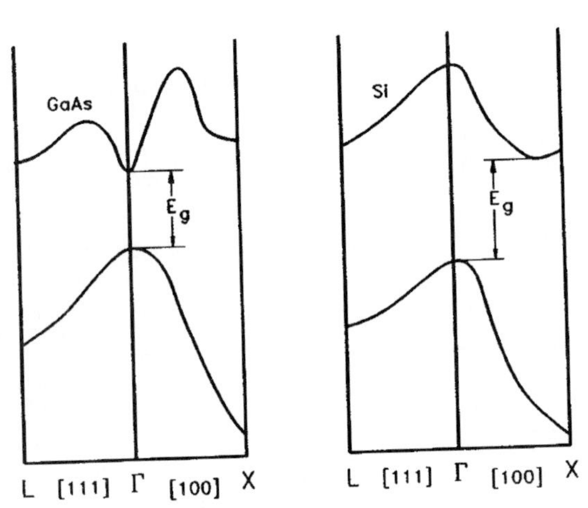

FIGURE 4.8 Electronic band structures of GaAs (direct band gap) and silicon (indirect band gap).[2]

4.1.2 Crystal Structure

Semiconductor devices are almost always fabricated on single-crystal substrates because grain boundaries adversely influence the transport of carriers and, therefore, device performance. The electronic properties of semiconductors are intimately related to their lattice structures and to the degree of crystalline perfection which can be realized in practice.

Diamond, Zinc Blende, and Wurtzite. Figure 4.9 illustrates the cubic lattice structure in its various forms. Many of the important semiconductors are of the diamond or zinc-blende lattice, where each atom is surrounded by four equidistant nearest neighbors which lie at the corners of a tetrahedron. The diamond and zinc-blende lattices can be considered as two interpenetrating face-centered cubic lattices. The diamond lattice represents the structure of elemental column IV silicon and germanium, the zinc-blende lattice that of GaAs, where one sublattice is gallium and the other is

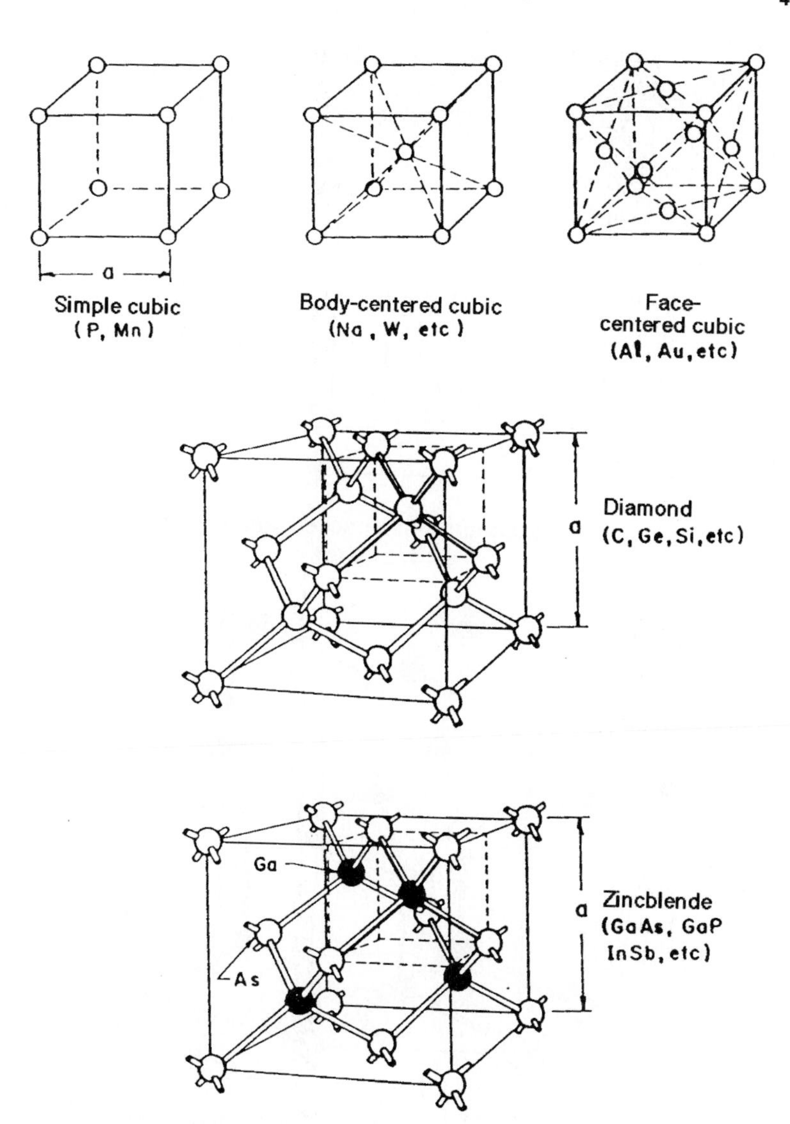

FIGURE 4.9 Cubic, diamond, and zinc-blende unit-cell crystal structures.[2]

arsenic. Most of the column III-V compound semiconductors crystallize in the zinc-blende structure. However, other semiconductors (including some III-V compounds) crystallize in the wurtzite structure, which also has a tetrahedral arrangement of nearest neighbors. Figure 4.10 shows the hexagonal close-packed lattice as well as the wurtzite lattice, which corresponds to two interpenetrating hexagonal close-packed lattices. SiC crystallizes in the latter structure with the existence of many polytypes. Some compound semiconductors, such as ZnS and CdS, can crystallize in both the zinc-blende and the wurtzite structures.

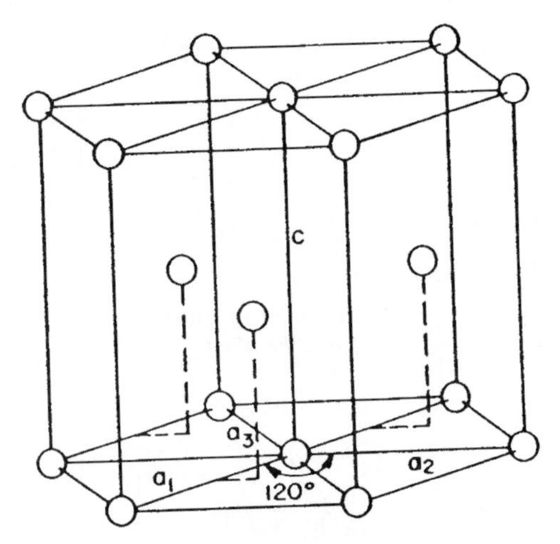

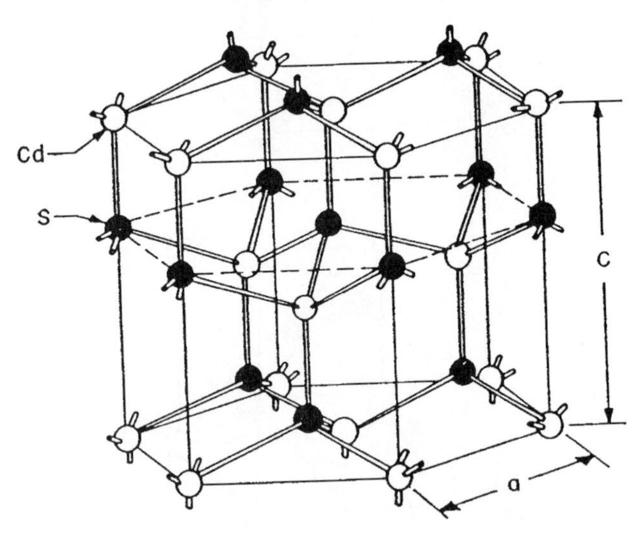

FIGURE 4.10 Hexagonal close-packed and wurtzite unit-cell structures.[2]

Miller Indices. Crystal planes and directions are normally defined by the use of Miller indices, a convention which allows a single plane, or a set of planes, through a cubic lattice structure to be denoted as $\{h, k, l\}$. The Miller indices of some of the important crystal planes in the cubic lattice structure are illustrated in Fig. 4.11. The convention of, for example, <100> denotes the growth direction of a crystal, while wafers cut perpendicularly to the length of the crystal are (100)-oriented. Hexagonal or wurtzite structure crystals are normally characterized in terms of the basal lattice constant a and the unit-cell repeat distance c. Table 4.3 lists the lattice constants and structures of various semiconductors.[12,13]

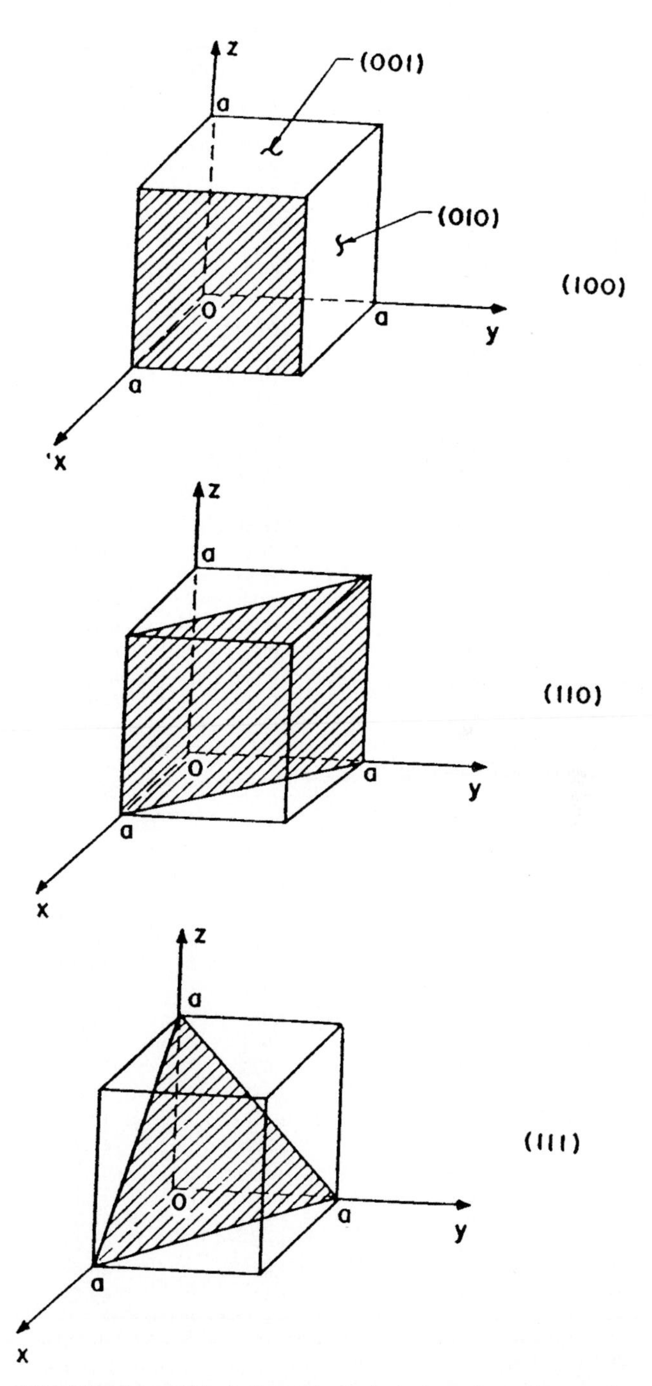

FIGURE 4.11 Miller indices of principal planes in cubic crystal; *a*—lattice constant.[2]

TABLE 4.3 Lattice Spacing Constant for Different Semiconductors

*a. Cubic lattice spacing**

Alphabetical		Numerical	
Material	*a*, Å	Material	*a*, Å
AlSb	6.1355	C (diamond)	3.5668
C (diamond)	3.5668	SiC	4.3596
CdS	5.820	ZnS	5.4093
CdSe	6.05	Si	5.4307
CdTe	6.481	GaP	5.4504
GaAs	5.6533	GaAs	5.6533
GaP	5.4504	Ge	5.6575
GaSb	6.0961	ZnSe	5.6687
Ge	5.6575	CdS	5.820
InAs	6.0584	HgS	5.851
InP	5.8687	InP	5.8687
InSb	6.4788	PbS	5.935
HgS	5.851	CdSe	6.05
HgSe	6.084	InAs	6.0584
HgTe	6.460	HgSe	6.084
PbS	5.935	GaSb	6.0961
PbSe	6.122	ZnTe	6.1037
PbTe	6.460	PbSe	6.122
Si	5.4307	AlSb	6.1355
SiC	4.3596	HgTe	6.460
Sn	6.5041	PbTe	6.460
ZnS	5.4093	InSb	6.4788
ZnSe	5.6687	CdTe	6.481
ZnTe	6.1037	Sn	6.5041

b. Hexagonal lattice spacing†

		Angstroms	
Material	Polytype	*a*, Å	*c*, Å
CdSe		4.2985	7.0150
CdS		4.1368	6.7163
SiC	4H	3.073	10.053
	6H	3.073	15.079
	15R	3.073	37.70
	21R	3.073	52.78
	33R	3.073	82.94
	51R	3.073	128.18
ZnS	2H	3.819	6.246
	4H	3.814	12.46
	6H	3.821	18.73
	8H	3.82	24.96
	10H	3.824	31.20
	9R	3.82	28.08
	12R	3.82	37.44
	15R	3.83	46.88
	21R	3.82	65.52

*W. B. Pearson, *Handbook of Lattice Spacings and Structures of Metals and Alloys,* vol. 2, Pergamon, New York, 1967.

†L. V. Azaroff, *Introduction to Solids,* McGraw-Hill, New York, 1960.

Crystalline Defects. Although the highest crystalline lattice perfection is sought for all semiconductor devices, in reality a variety of crystal defects plague melt-grown crystals, ranging from twin formation to the generation of dislocations, lineage, and, in the worst case, degeneration into polycrystalline regions. On the microscopic scale, vacancies, interstitials, impurity-vacancy clusters, and, in the case of compound semiconductors, stoichiometric defects are cause of concern in the operation, yields, and reliability of present-day sophisticated devices and integrated circuits (ICs). Active-layer, thin-film growth brings its own set of defects in addition to the above, including microtwins, stacking faults, and hillock-type growth, which detract from the specularly smooth surface desired for device processing. A brief description of our understanding of some of the principal lattice defects is given in this section.

Point Defects: Vacancies and Interstitials. The simplest defect arises when an atom is missing from its site in a lattice (vacancy) or when an extra atom, for which there is no lattice site available, is forced into the lattice (interstitial). When the interstitial originally came from a lattice site, the interstitial vacancy is called a Frenkel defect. The number of vacancies increases exponentially with temperature, especially near the melting point of a semiconductor, and can be "quenched" in a crystal (for example, during crystal growth) by rapid cooling from near the melting point. The technological importance of vacancies and interstitials lies in the enhanced diffusion effects (through the migration of vacancies in the opposite direction) and impurity point defect complex formation which can occur. In the melt growth of semiconductor crystals, point defects play an important but a poorly understood role and are influenced by the thermal history of the grown crystal (slow cooling or quenching) and whether it is subjected to postgrowth annealing.

Stacking Faults. A face-centered cubic structure may be built up by laying down close-packed planes in a sequence *ABCABCA*.... A stacking fault occurs when the sequence goes wrong, as in *ABCBCABC*..., where an *A* layer is missing. This is a common fault in epitaxial single-crystal layer growth, and the ease of stacking fault generation has been correlated to polytypism in compound semiconductors such as SiC, and to twin-plane formation in bulk crystals.

Twin Boundaries. Melt-grown III-V and II-IV crystals often exhibit twinning, a defect in which one region of the crystal is the mirror image of another. In commercial semiconductor growth, this defect can have a devastating effect since, for example, the desired <100>-oriented crystals take on a <211> orientation with deleterious consequences to (100) wafer yields. A pictorial representation of twin formation is given in Fig. 4.12 where, after twinning, the lattice is a mirror image of the original lattice structure. Twinning has been related[14] to the covalent/ionicity of III-V and II-VI compound semiconductors. In particular, the probability of twin formation has been empirically found to relate to the stacking fault energy, as shown in Fig. 4.13, suggesting that when the extra energy required for a small atomic mismatch decreases, the incidence of twinning increases dramatically. Silicon, with a high stacking fault energy, rarely shows any tendency to twin during crystal growth, while CdTe and InP, with a low stacking fault energy, typically can only be grown as crystals with numerous twin planes. While the cause of twinning in the compound semiconductors is still poorly understood, the maintenance of a contamination-free melt-solid interface, utilization of very low thermal gradients, and reduction of melt turbulence during growth have empirically been found to impact the production of twin-free III-V crystals (GaAs and InP). Twin formation remains a major problem in CdTe and other II-VI crystals.

Dislocations. In its simplest form, an edge dislocation[15] occurs when an extra layer of atoms is introduced into the lattice, as shown in Fig. 4.14*a*. In the diamond (germanium and silicon) and zinc blende (GaAs and other III-V crystals), edge dislocations lie in the (111) planes and the <110> directions. Screw dislocation is another type

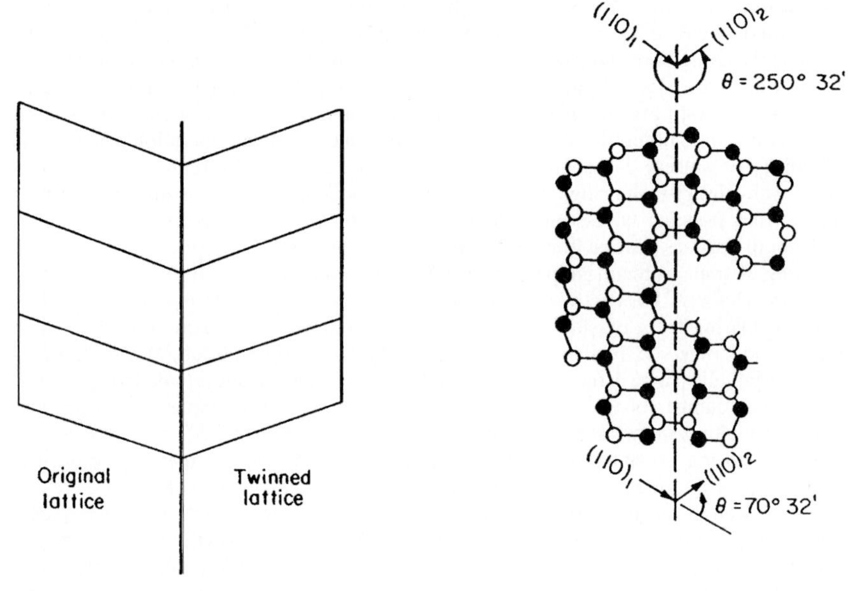

FIGURE 4.12 Twin-plane formation. (*a*) Lattice structure reverts to mirror image. (*b*) Twinning in zinc-blende structure.

of dislocation, which is more difficult to visualize and consists of atoms along a line which rearrange themselves in a helix whose axis is at the center of the discontinuity (Fig. 4.14*b*). During melt growth, dislocations can be propagated from the seed crystal or result from large thermal gradients, which cause the critical resolved shear stress of the solid semiconductor to be exceeded at a particular temperature. Unfortunately both large and small impurity atoms are readily accommodated and cluster at dislocation lines and networks, with deleterious effects usually in minority carrier devices. Dislocation generation often occurs also in active layer formation (such as in heteroepitaxy, to accommodate lattice mismatch) and during device fabrication processes (because of temperature cycling and stress caused by doping and metallization).

4.1.3 Electrical

Some of the more important properties of silicon have already been discussed in familiarizing readers with some of the basic semiconductor concepts. Additional data for other semiconductors are discussed here, including specific properties relating to GaAs and InP in response to emerging device technologies for high-speed digital (for computing) and analog power generation at high frequencies (for radar and communications), and properties of semiconducting SiC and diamond, where intensive materials and device efforts are currently under way.

Resistivity. While much of the pioneering crystal growth and materials development work occurred in germanium, its importance in device fabrication was quickly eclipsed by silicon because of the near-perfect properties of the silicon–silicon dioxide interface (in terms of achieving devices with passivated, low surface leakage currents), which forms the basis of today's highly developed silicon IC industry.

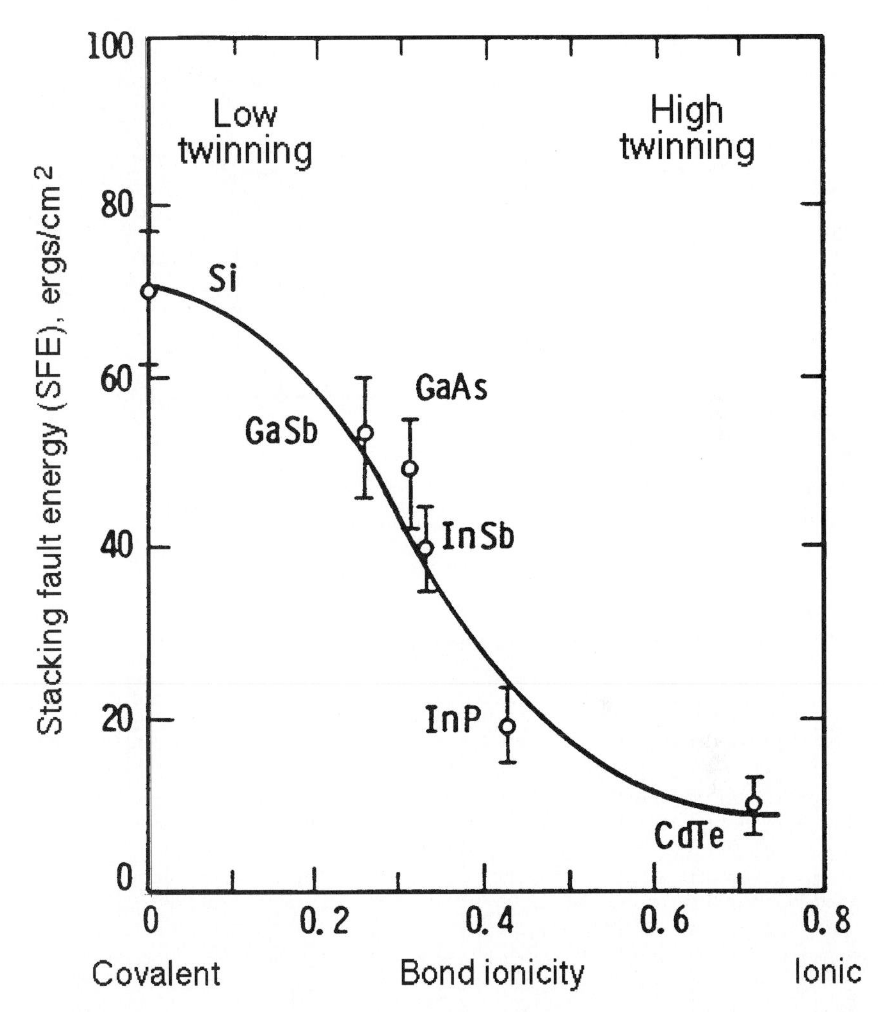

FIGURE 4.13 Stacking-fault energy (SFE) for various semiconductors and relationship to twin formation.[14]

Unfortunately germanium oxide is unsuitable for passivation and modern germanium usage addresses specialized applications which take advantage of its remarkable purity and optical properties, for example, as radiation or long-wavelength infrared detectors or as high-transmission windows.

The variation of resistivity of germanium, silicon, and GaAs as a function of impurity concentration has been determined experimentally[6,16,17] and is shown in Fig. 4.15. Increasing interest over the past decade in very-high-resistivity semiconductors has been dictated by specific device technologies. Nuclear particle radiation and infrared detectors, which require high-resistivity germanium and silicon, have resulted in single crystals of germanium exceeding 30 $\Omega \cdot$ cm and silicon exceeding $10^5\ \Omega \cdot$ cm by reducing unwanted contamination in the growth of crystals. In the case of GaAs and

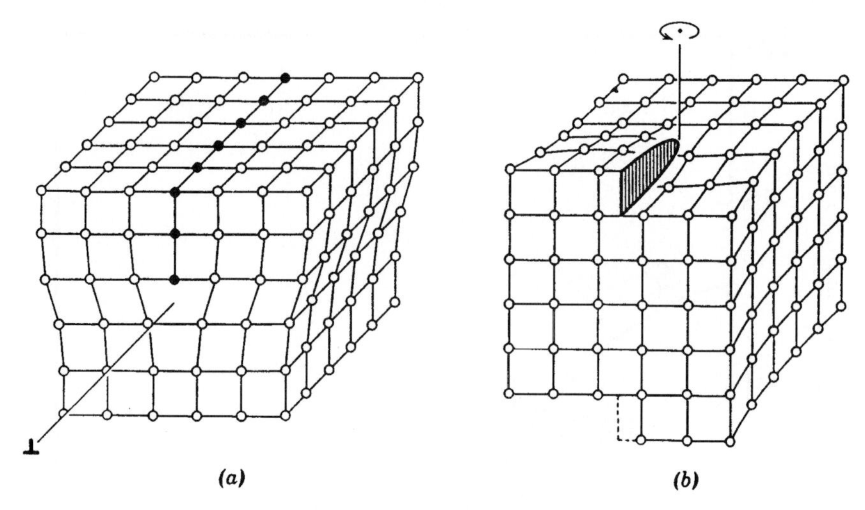

FIGURE 4.14 (*a*) Edge and (*b*) screw dislocation formation in cubic crystals.

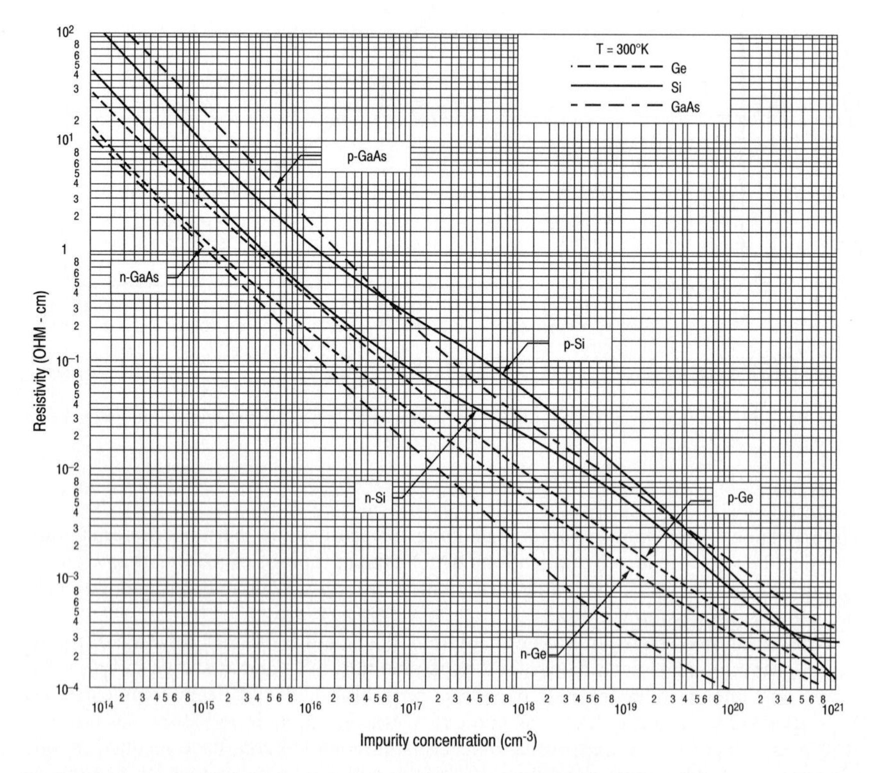

FIGURE 4.15 Resistivity versus impurity concentration for germanium, silicon, and GaAs at 300 K.[2]

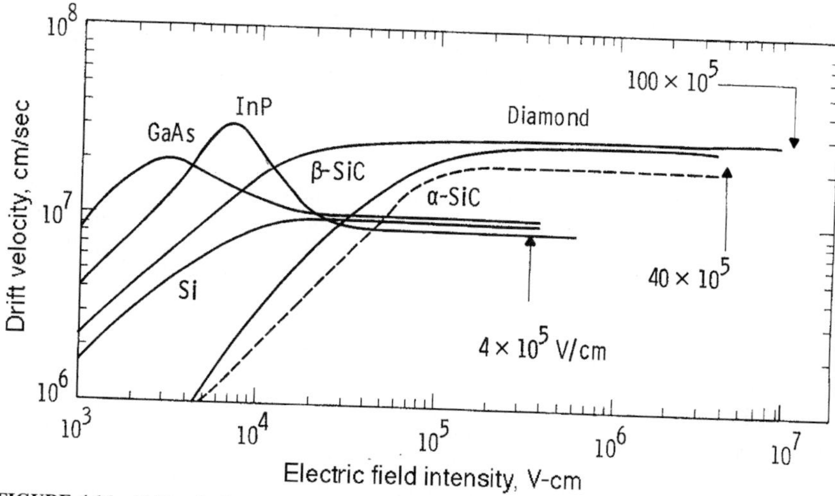

FIGURE 4.16 Drift velocity at high electric fields for different semiconductors. Breakdown electric fields are indicated.[19]

InP, their higher electron mobility and the velocity overshoot at low electric fields (compared to silicon) makes them suitable for high-frequency devices. This has directed considerable attention to the preparation and properties of very-high-resistivity or semi-insulating III-V compounds. Semi-insulating substrates prevent unwanted parasitic resistive and capacitative effects at high frequencies, especially in monolithic IC technology. Resistivities in the 10^7- to 10^8-Ω • cm are attainable.

Saturation Velocity and Breakdown Voltage. Silicon device and IC technology is highly developed and ubiquitous in its applications. Other semiconductor technologies thrive only in applications where limitations are clearly indicated by the basic properties of silicon. Optoelectronic applications are dominated by the III-V compounds (because of the direct band gap of GaAs and InP, whereas silicon is an indirect bandgap semiconductor). Power generation and high-speed device applications are other areas where silicon is being challenged by not only the III-V compounds, but by older well-known semiconductors where dramatic improvements in the materials technology are being achieved. Two contenders which are undergoing intensive development at present are silicon carbide and diamond. The promise of SiC and diamond is illustrated in Fig. 4.16, where the saturation velocity (speed capability) and breakdown voltage (power capability) are shown. Both these device requirements are limited by the ability to remove heat dissipation efficiently, thus requiring semiconductors of higher thermal conductivities. Table 4.4 lists the thermal conductivity, the saturation

TABLE 4.4 Comparison of Silicon, GaAs, SiC, and Diamond Properties at Melting Point

Semiconductor	E_g, eV	ϑ_{sat}, cm/s cm/s	κ W/(cm • K)	ϵ_r	E_{BD}, V/cm	K(FoM) rel. Si
Si	1.1	1×10^7	1.5	12	3×10^5	1
GaAs	1.4	2×10^7	0.5	12.8	3.5×10^5	4.6
SiC	2.9	2×10^7	5.0	10	4×10^6	10
Diamond	5.5	2.7×10^7	20.0	5.5	$\approx 1 \times 10^7$	324

velocity, and the dielectric constants for silicon, GaAs, SiC, and diamond. A figure of merit proposed by Keyes[18] considers both switching speed and power dissipation in monolithic circuit design and is given by

$$K(\text{FoM}) \propto \kappa \left(\frac{cv}{4\pi\epsilon_r} \right)^{1/2} \tag{4.9}$$

where κ is the thermal conductivity in W/(cm • K) and ϵ_r the dielectric constant. The Keyes figure of merit is also listed in Table 4.4 and shows the potential advantages of an SiC-based technology in the near term and the promise of semiconducting diamond if the difficulties[19] of producing single-crystal diamond larger than today's gem quality and size can be overcome.

Minority Carrier Lifetime. All the electrical properties discussed so far are related to majority carrier devices. Mainstream silicon IC technology (high density, signal processors, digital memories) and III-V optoelectronic devices rely on minority carriers. Minority carrier lifetime refers to the lifetime of an electron in *p*-type or a hole in *n*-type semiconductors. It is an exceedingly important parameter in device technology since p-n junction and MOSFET leakage current decreases with increasing lifetime. The simplest concept of carrier lifetime involves band-to-band (auger) recombination, where the lifetime of a minority carrier is determined by the recombination of an electron in *p*-type material by the excitation of holes from the valence band by thermal or optical means. Band-to-band recombination in *p*-type material would therefore be expected to be proportional to the majority carrier concentration *p,* or the net acceptor concentration. Generally the band-to-band recombination is

$$\tau \propto \frac{1}{N_{a,d}} \tag{4.10}$$

where τ is the carrier lifetime of electrons or holes in a *p*- or *n*-semiconductor and $N_{a,d}$ the corresponding acceptor or donor concentration.

Carrier lifetimes limited by the band-to-band recombination process are rarely encountered in practice. Instead, the carrier lifetime is almost always determined by recombination generation through intermediate centers, and is related to the presence and concentration of near midgap recombination centers. The carrier lifetime τ is related to the density of midgap centers N_t by the expression

$$\tau = \frac{1}{\alpha v_{\text{th}} N_t} \tag{4.11}$$

where α is the capture cross section of the mid-band-gap impurity or defect level, v_{th} is the thermal velocity, and N_t the concentration of mid-band-gap impurities.

Carrier lifetimes of 10 ms and greater can be achieved[20,21] in very high-purity, dislocation-free silicon, but 100 μs and smaller lifetimes are more typical in processed devices due to the unwanted influx of fast diffusing transition metals (such as copper, gold, and iron) during thermal processing at concentrations of 10^{14} cm^{-3} and higher. Figure 4.17 shows measurements of the photoconductive decay lifetime for as-grown silicon. The auger recombination lifetime is also shown. Minority carrier lifetimes in III-V compounds are significantly lower (typically in the nanosecond range) because of lower purity levels, the higher concentration of mid-band-gap impurities, and the influence of dislocations which serve as sinks for these lifetime-killing impurities and electrically active defects. Maximizing minority carrier lifetime through improved purity, stoichiometry control, and inhibiting dislocation generation is an important preoccupation in the materials technology for solid-state LEDs, lasers, and optoelectronic circuits.

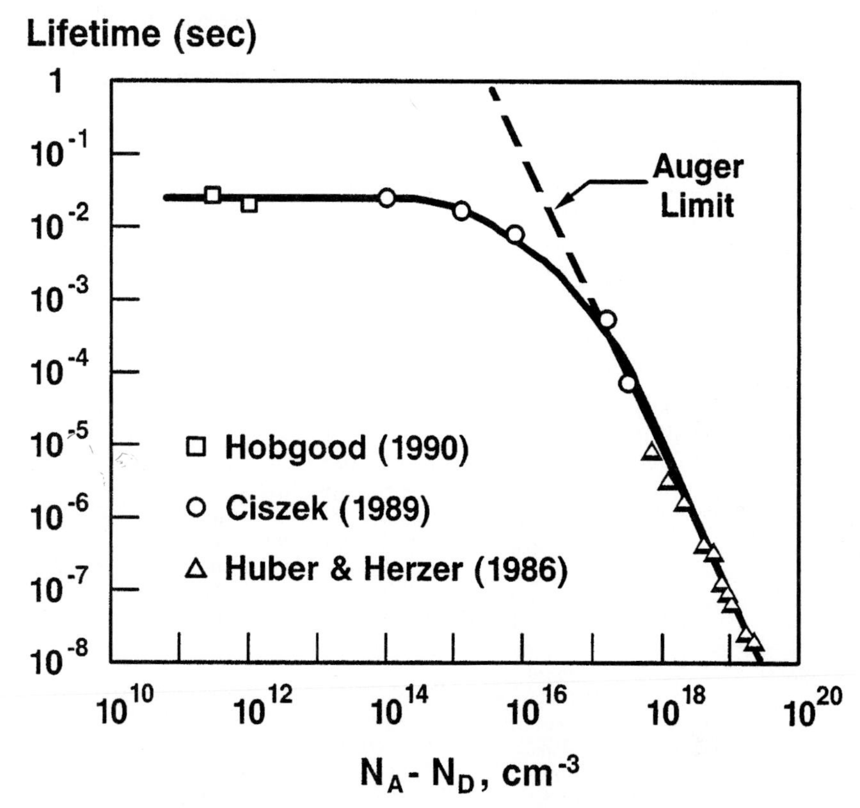

FIGURE 4.17 Measured photoconductive decay lifetime in silicon compared to calculated lifetime for auger recombination.[20,21]

4.1.4 Impurities in Semiconductors

While the goal in semiconductor crystal growth is to produce crystals of known and uniform dopant concentrations from wafer to wafer along a crystal, impurity segregation, solid solubilities, and the diffusion of impurities in different semiconductors generally impose fundamental limitations to achieving this ideal. These principles, and how they affect impurity distribution in melt-grown crystals, are discussed here.

Impurity Distribution in Melt-Grown Crystals.[22] When a single crystal is grown from a melt doped with a specific impurity, the axial and radial dopant distribution is determined by the segregation (distribution) coefficient, while the maximum dopant concentration is limited by the solid solubility. As a semiconductor freezes during melt growth, the concentration of impurities incorporated in the solid N_s, is usually different from the concentration in the liquid at the interface N_l. The equilibrium segregation coefficient k_0 is defined as

$$N_s = k_0 N_1 \tag{4.12}$$

TABLE 4.5 Segregation Coefficients of Elements in Germanium and Silicon

Group	Element	Germanium	Silicon
IA	Lithium	0.002	0.01
IB	Copper	1.5×10^{-5}	4×10^{-4}
	Silver	4×10^{-7}	
	Gold	1.3×10^{-5}	2.5×10^{-5}
IIB	Zinc	4×10^{-4}	$\sim 1 \times 10^{-5}$
	Cadmium	$>1 \times 10^{-5}$	
IIIA	Boron	17	0.80
	Aluminum	0.073	0.0020
	Gallium	0.037	0.0080
	Indium	0.001	4×10^{-4}
	Thallium	4×10^{-5}	
IVA	Silicon	5.5	1
	Germanium	1	0.33
	Tin	0.020	0.016
	Lead	1.7×10^{-4}	
VA	Nitrogen	—	$<10^{-7}$(?)
	Phosphorus	0.080	0.35
	Arsenic	0.02	0.3
	Antimony	0.0030	0.023
	Bismuth	4.5×10^{-5}	7×10^{-4}
VIA	Oxygen	—	0.5
	Sulfur	—	10^{-5}
	Tellurium	$\sim 10^{-6}$	
Transition elements	Vanadium	$<3 \times 10^{-7}$	
	Manganese	$\sim 10^{-6}$	$\sim 10^{-5}$
	Iron	$\sim 3 \times 10^{-5}$	8×10^{-6}
	Cobalt	$\sim 10^{-6}$	8×10^{-6}
	Nickel	3×10^{-6}	
	Tantalum		10^{-7}
	Platinum	$\sim 5 \times 10^{-6}$	

Source: *Electronic Materials and Processes Handbook,* First Edition, McGraw-Hill, Inc., New York.

An equilibrium segregation coefficient k_0 of unity would enable uniform doping of semiconductors prepared by melt growth. Unfortunately k_0 is usually less than unity, so that impurity concentrations vary along the length of a grown crystal. Data in Tables 4.5 and 4.6 give k_0 for major semiconductors and their common impurities.

The segregation coefficient can be calculated from the phase diagram if known, but k_0 is usually determined experimentally. In practical melt-growth systems growth rarely occurs at rates slow enough for the true equilibrium state under which phase diagrams are determined. As a result (when k_0 is less than unity), an impurity concentration enhancement occurs in the liquid at the solid-liquid interface, since the impurity rejection rate is higher than the impurities which can be transported by diffusion. An impurity pileup in the melt at the melt-crystal interface occurs, as shown in Fig. 4.18. An impurity depletion would be the result for an impurity with k_0 greater than unity. The diffusion boundary-layer thickness depends on the growth rate and melt stirring, and is determined by crystal rotation, melt rotation, and thermal convection.

TABLE 4.6 Segregation Coefficients of Elements in Group III-V Compound Semiconductors

Doping agent		Compound					
Group	Impurity element	AlSb	GaSb	GaAs	InAs	InP	InSb
IB	Copper	0.01	—	<0.002	—	—	6.6×10^{-4}
	Silver	—	—	0.1	—	—	4.9×10^{-5}
	Gold	—	—	—	—	—	1.9×10^{-6}
IIA	Magnesium	—	—	0.3	0.7		
	Calcium	—	—	<0.02			
IIB	Zinc	0.02	0.2–0.3	0.1–0.9	0.77	—	4
	Cadmium	0.002	0.02	<0.02	0.13	—	0.26
IIIA	Boron	0.01–0.02					
	Aluminum	—	—	3			
	Gallium	—	—	—	—	—	2.4
	Indium	~1	—	0.1			
IVA	Carbon	0.6	—	0.8			
	Silicon	0.4	~1	0.1	0.4		
	Germanium	0.3	0.3	0.2–0.3	0.07	0.05	0.045
	Tin	2×10^{-4} 8×10^{-4}	0.01	0.03	0.09	0.03	0.057
	Lead	0.01	—	<0.02	<0.01		
VA	Phosphorus	—	—	2	—	—	0.16
	Arsenic	—	2–4	—	—	—	5.4
	Antimony	—	—	<0.02			
VIA	Sulfur	0.003	0.06	0.3	1.0	~0.8	0.1
	Selenium	0.003	0.2–0.4	0.44–0.55	0.93	~0.6	0.2–0.5
	Tellurium	0.01	0.4	0.3	0.44	—	0.5–3.5
Transition elements	Vanadium	0.01					
	Manganese	0.01	—	0.05			
	Iron	0.02	—	0.003			
	Cobalt	0.002					
	Nickel	0.01	—	<0.02	—	—	6×10^{-5}

Source: *Electronic Materials and Process Handbook,* First Edition, McGraw-Hill, Inc., New York.

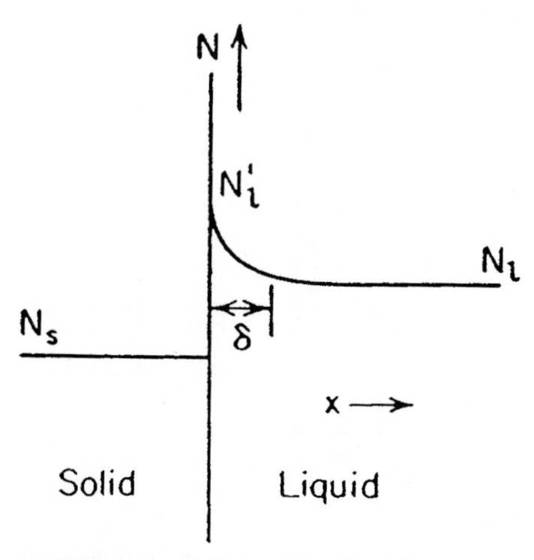

FIGURE 4.18 Impurity pileup at crystal-melt interface illustrating importance of the diffusion boundary layer in equilibrium melt growth.[22]

The effective segregation coefficient k_{eff} is therefore highly dependent on the growth conditions used and is usually less than the equilibrium segregation coefficient k_0. The variation in impurity concentration along a crystal grown from a melt is given by

$$N_s(x) = N_0 k_{\text{eff}}(1 - x)k_{\text{eff}}\ 1/k_{\text{eff}} \tag{4.13}$$

where x is the fraction of the original melt solidified.

When k_{eff} approaches unity ($k_{\text{eff}} = 0.8$ for boron in silicon), the concentration is relatively uniform throughout the crystal, whereas when k_{eff} is much less than unity, the axial impurity distribution increases significantly from the crystal seed to tang end and approximates

$$N_s(x) = \frac{N_0 k_{\text{eff}}}{1 - x} \tag{4.14}$$

The preceding theory describes impurity distributions in crystals grown from a melt, such as Czochralski or Bridgman growth. For cases when only a zone is molten at a given time, such as float zone or zone leveling,

$$N_s(x) = 1 - (1 - k_{\text{eff}})\, e^{-k_{\text{eff}} X/L} \tag{4.15}$$

where X is the total length of the crystal and L the zone length.

The impurity distribution for both normal freezing and zone passing is shown[23] in Fig. 4.19 for comparison. Two points of note are that (1) impurity segregation behavior in melt growth permits purification of a semiconductor by rejection of the last to freeze portion of the crystal, and (2) since k_{eff} depends on the diffusion layer boundary thickness (which is determined by the microscopic growth velocity across the growth interface and therefore influenced by melt motion due to thermal convection or

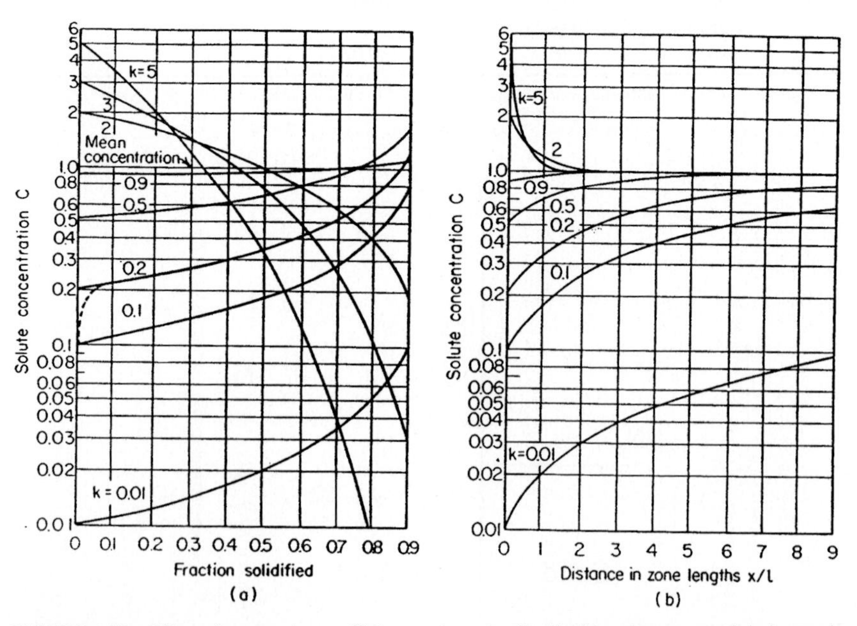

FIGURE 4.19 Effect of segregation coefficient on impurity distributions along a crystal during equilibrium melt growth. (*a*) Czochralski or Bridgman growth. (*b*) Float zone or zone leveling.[23]

imposed rotations), radial inhomogeneities in impurity incorporation are predicted and observed experimentally.

The maximum concentration of any impurity which can be accommodated in a semiconductor at any given temperature is given by the equilibrium solid solubilities, which are shown[24] in Fig. 4.20 for different impurities in germanium and silicon. However, other limitations may determine the highest impurity concentrations that can be attained in single crystals, such as constitutional supercooling, or the impurity incorporation may be electrically inactive.

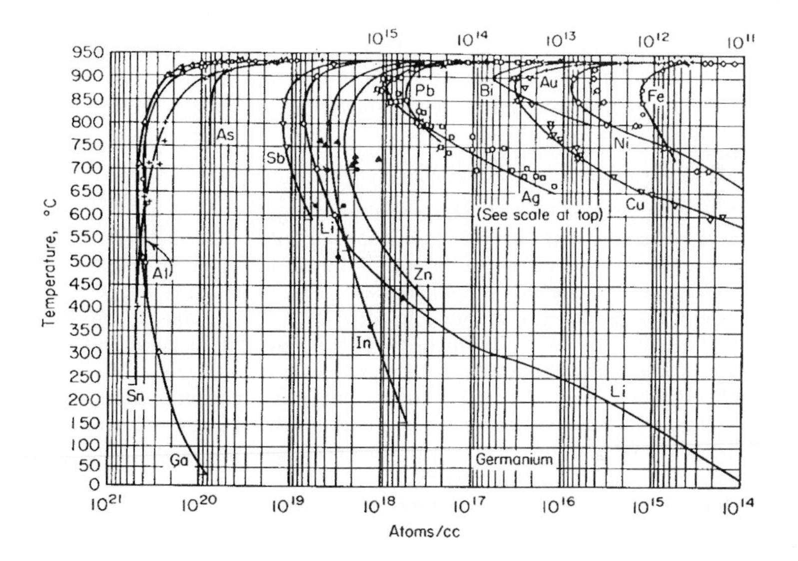

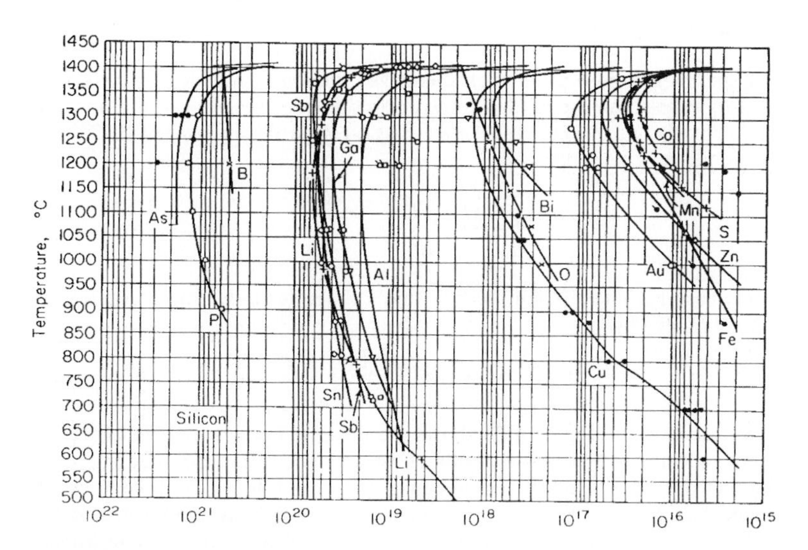

FIGURE 4.20 Solid solubility of various elements in germanium and silicon as a function of temperature.[24]

Solid-State Diffusion. The introduction of impurities into a semiconductor to form diffused layers is commonly used in the silicon planar process. Diffusion theory yields solutions for the variation of impurity concentration with depth from the surface, involving the complementary error function[25]

$$C(x, t) = C(s)\ \mathrm{erfc}\left(\frac{x}{2Dt}\right) \tag{4.16}$$

where $C(s)$ is the surface concentration, $C(x, t)$ the concentration at a depth x after diffusion at a given temperature for time t, and D the diffusivity.

From a materials technology point of view our interest lies more in the physical meaning and implications of the diffusivities of different impurities than in the formation of diffused layers. The diffusivities of different elements in germanium, silicon, and GaAs versus reciprocal temperatures are shown[26,27] in Fig. 4.21. The diffusivity can represented by

$$D = D_0\, e^{-E_a/kT} \tag{4.17}$$

where k is Boltzmann's constant and E_a the activation energy.

For silicon E_a values for both acceptor and donor impurities range between 3 and 4 eV, and the atomistic mechanism is one of substitutional diffusion, that is, diffusion proceeds by the impurities jumping into silicon lattice vacancies. Thus the activation energy corresponds to the energy required to form vacancies in the silicon lattice rather than that required to move the impurity. Diffusivities of acceptor and donor impurities in germanium are similar, but with E_a in the 2- to 3-eV range. The transition metals (Ni, Fe, Au, Zn, and Mn) in particular move by interstitial diffusion and much more rapidly than the substitutional impurities. Tables 4.7 and 4.8 summarize our knowledge of the diffusivities of impurities in germanium and silicon, and in the III-V compound semiconductors, respectively.

Note that the mobility μ and the diffusivity D are related by

$$D = \left(\frac{kT}{q}\right)\mu \tag{4.18}$$

4.1.5 Other Properties

Optical. Experimental absorption coefficients as function of wavelength for several semiconductors at or above the fundamental absorption edge (band-to-band transition) are shown[28–31] in Fig. 4.22. The optical transmission T through a layer of thickness x for normal incidence is given by

$$T = \frac{(1 - R^2)\mathrm{e}^{-ax}}{1 - R^2\, e^{-2ax}} \tag{4.19}$$

where a is the absorption coefficient in cm^{-1} and R the reflectivity. Photon energy and wavelength are related by

$$E_\lambda = h\nu \quad \text{or} \quad \frac{1.24}{\lambda} \tag{4.20}$$

for E_λ in eV and λ in μm. Figure 4.23 shows the measured optical transmittance[32] of various semiconductors out to wavelengths of 14 μm. The refractive indices of most semiconductors over this spectral region range between 3 and 4, with some wavelength dependence.[32] At wavelengths beyond the band edge, absorption due to free carriers and lattice vibrations occurs, which, although small compared to band-to-band absorption, can be important in infrared-device technology.

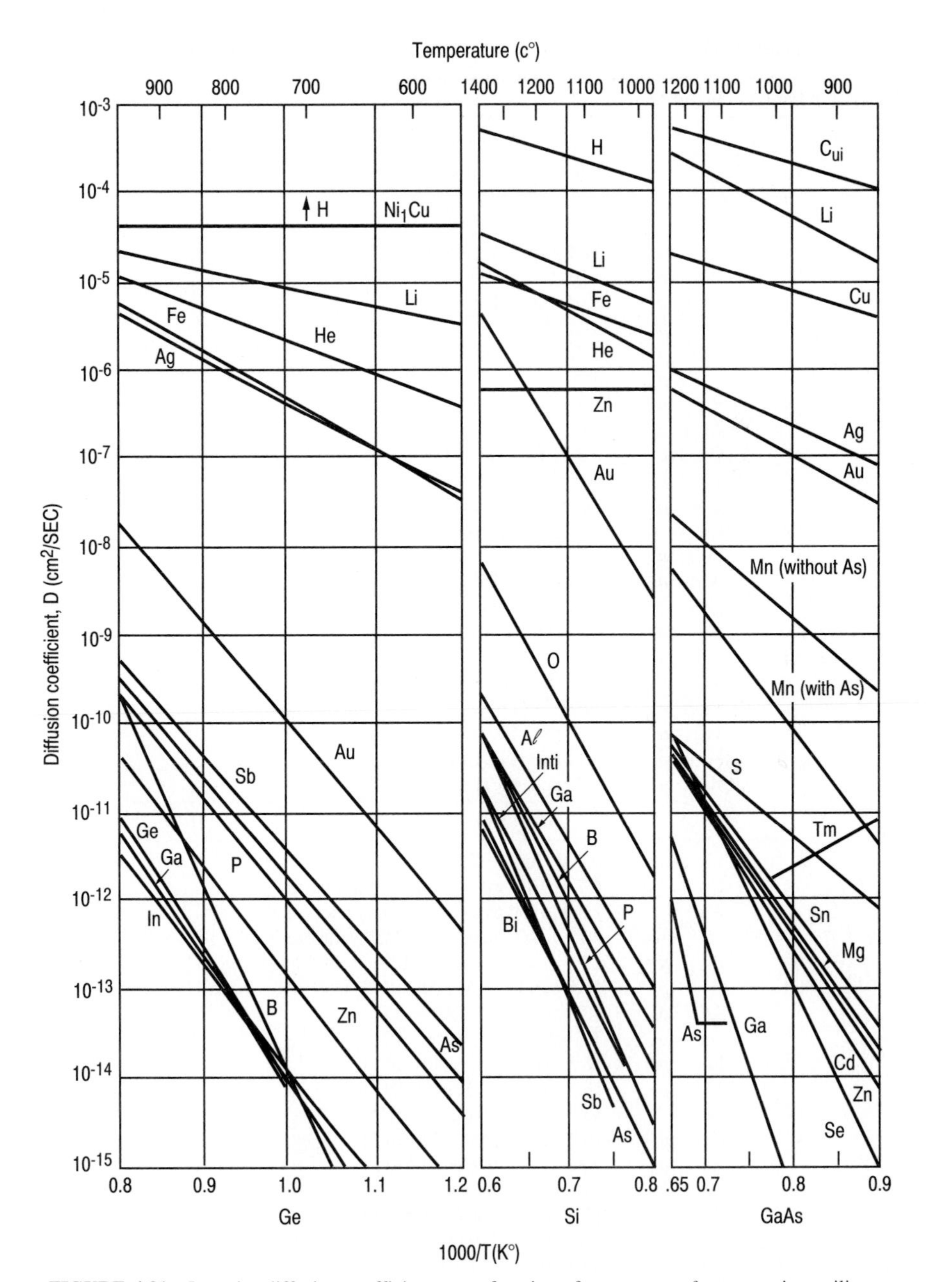

FIGURE 4.21 Impurity diffusion coefficients as a function of temperature for germanium, silicon, and GaAs.[26,27]

TABLE 4.7 Diffusion Coefficients of Elements in Germanium and Silicon

Group	Impurity		D
			Germanium
I	Li	n	$0.0025 \exp(-0.51/kT)$
	Cu	p	3×10^{-5} (T1173)
	Ag	p	$0.044 \exp(-1.0/kT)$
	Au	n	$12.6 \exp(-2.25/kT)$
II	Be	p	$<10^{-11}$ (T1123)
	Zn	p	$0.65 \exp(-2.5/kT)$
		p	—
	Cd	p	$1.75 \times 10^{9} \exp(-4.4/kT)$
		p	—
III	B	p	$1.8 \times 10^{9} \exp(-4.55/kT)$
	Al	p	—
	Ga	p	3.5×10^{-13} (T1123)
	Tl	p	7.0×10^{-14} (T1123)
	In	p	$20 \exp(-3.0/kT)$
IV	Ge		$(7.8 \pm 3.4) \exp(-2.98/kT)$
V	P	n	2.4×10^{-11} (T1123)
	As	n	$2.12 \exp(-2.4/kT)$
	Sb	n	$(1.3 \pm 1.0) \exp[(-2.26 \pm 0.07)/kT]$
	Bi	n	$4.7 (\exp -2.4/kT)$
VI	O	n	$\sim 10^{-8}$
	S	n	1×10^{-9} (T1193)
	Se	n	$\sim 10^{-10}$ (T1193)
	Te	n	$\sim 10^{-11}$ (T1193)
VII	Fe	p	$0.13 \exp(-1.1/kT)$
	Ni	p	$0.8 \exp(-0.91/kT)$
			Silicon
I	Li	n	$0.0023 \exp(-0.66/kT)$
	Cu		$4.7 \times 10^{-3} \exp(-0.43/kT)$ interstitial
	Au		$2.4 \times 10^{-4} \exp(-0.387/kT)$ interstitial
			$2.75 \times 10^{-3} \exp(-2.04/kT)$ substitutional
II	Zn	p	$1 \times 10^{-6} - 1 \times 10^{-7}$ $(1273 < T < 1573)$
III	B	p	$10.5 \exp[(-3.6 \pm 40\%)/kT]$
	Al	p	$8.0 \exp[(-3.4 \pm 40\%)/kT]$
	Ga	p	$3.6 \exp[(-3.5 \pm 40\%)/kT]$
	In	p	$16.5 \exp[(-4.0 \pm 40\%)/kT]$
	Tl	p	$16.5 \exp[(-4.0 \pm 40\%)/kT]$
IV	Ge		$6.26 \times 10^{-5} \exp(-5.3/kT)$
V	P	n	$10.5 \exp[(-3.6 \pm 40\%)/kT]$
	As	n	$0.32 \exp[(-3.6 \pm 40\%)/kT]$
	Sb	n	$5.6 \exp[(-3.9 \pm 40\%)/kT]$
	Bi	n	$10.30 \exp[(-4.7 \pm 40\%)/kT]$
VI	S	n	$0.92 \exp(-2.2/kT)$
VII	Mn	n	»2×10^{-7} (T1473)
VIII	Fe	n	$0.0062 \exp(-0.87/kT)$

Source: *Electronic Materials and Processes Handbook,* McGraw-Hill, Inc., New York.

TABLE 4.8 Diffusion Coefficients of Elements in Group III-V Compound Semiconductor

	D_0, cm^2/s					
Element	AlSb	GaP	GaAs	GaSb	InAs	InSb
S			2×10^{-5}		7×10^{0}	
Se					1×10^{1}	
Te			4×10^{-2}	4×10^{-4}	3×10^{-5}	2×10^{-7}
Zn	3×10^{-1}	1×10^{0}	5×10^{-7}		3×10^{-3}	
Cd			5×10^{-2}		4×10^{-4}	
Ge					4×10^{-6}	
Sn				2×10^{-5}	2×10^{-6}	6×10^{-8}
Cu	3×10^{-3}		3×10^{-2}			
Ag			7×10^{-3}			7×10^{-4}
Au			1×10^{-3}			1×10^{-7}
Mn			8×10^{-3}			

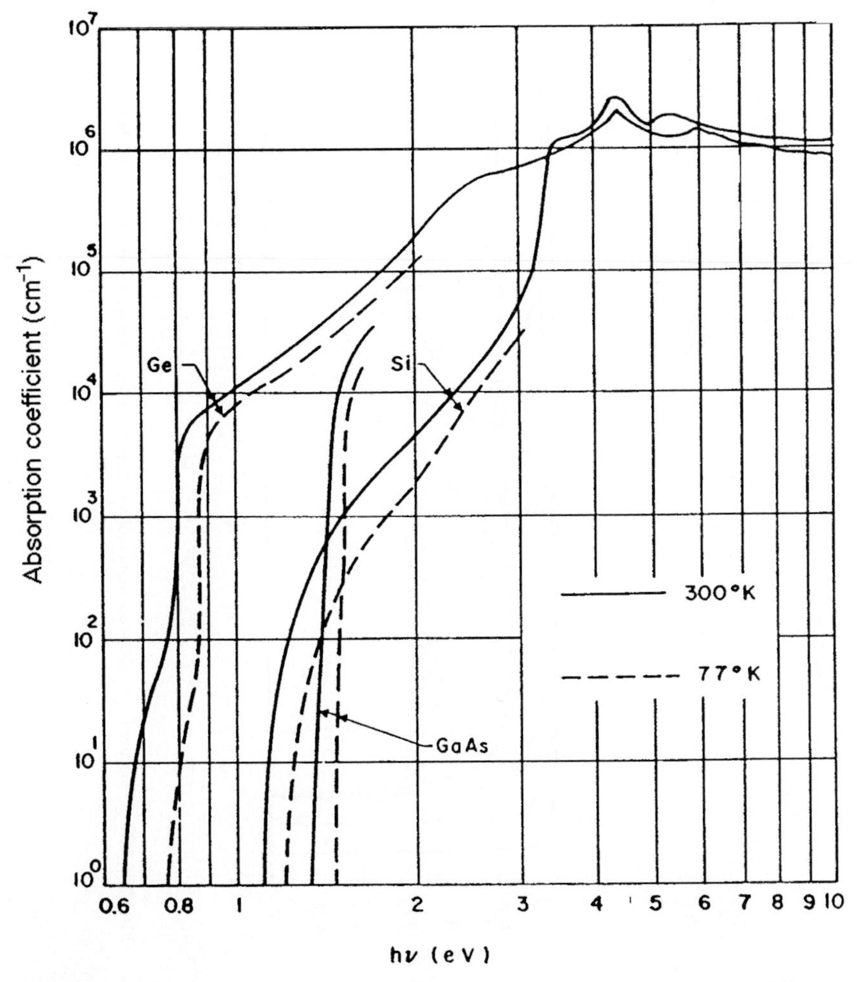

FIGURE 4.22 Measured optical absorption coefficients for germanium, silicon, and GaAs.[28–31]

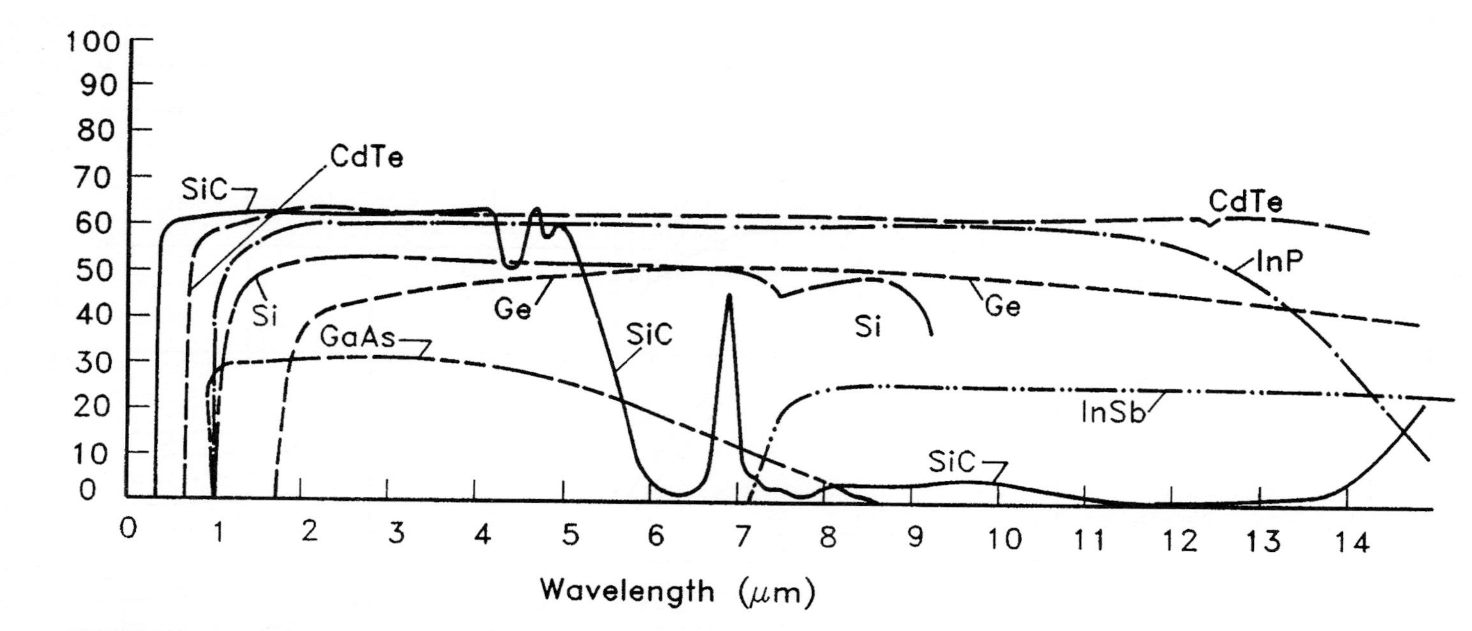

FIGURE 4.23 Measured optical transmittance of various semiconductors as a function of wavelengths.[32]

Thermal. The thermal conductivities and expansion coefficients of several semiconductors at room temperature are tabulated in Table 4.9 and the variations with temperature are shown[33–35] in Fig. 4.24. The thermal as well as the mechanical properties of semiconductors at or near their melting points, which are of significant importance in crystal growth, are unfortunately not well documented.

TABLE 4.9 Thermal Conductivity and Expansion Coefficients of Semiconductors at Room Temperature

Semiconductor	Thermal conductivity, W/(cm • K)	Thermal expansion coefficient, K^{-1}
Diamond	20.0	1.2×10^{-6}
Si	1.5	2.3
Ge	0.06	5.7
SiC	5.0	4.5
GaP	0.08	5.3
GaAs	0.5	5.7
GaSb	0.03	6.9
InP	0.07	4.4
InAs	0.02	5.3
InSb	0.02	5.0

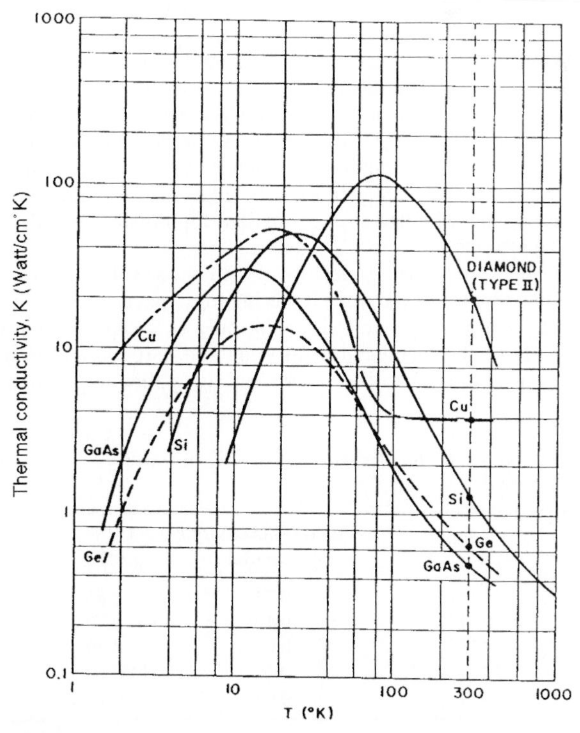

FIGURE 4.24 Variation of measured thermal conductivity with temperature for several semiconductors.[33–35]

4.2 BULK SEMICONDUCTOR CRYSTAL GROWTH

Modern semiconductor devices and IC fabrication technology require (1) defect-free single-crystal substrates, (2) preferably round cross-section wafers (because of the industrial investment in processing equipment for handling such wafers), and (3) ever-larger-diameter wafers (since device yields and therefore manufacturing costs decrease dramatically with increasing wafer size). Because of its importance to solid-state device technology, the growth of single-crystal semiconductors has been studied extensively. Growth from a semiconductor melt has proven to be highly successful for almost all the important semiconductors.

Historically melt growth was first pioneered by Czochralski[36] (1918 in Germany), Obreimov and Schubnikov[37] (1924 in Russia), and Bridgman[38] (1925 in the United States) as a means of producing single-crystal metal and semimetal samples for determining some of their fundamental properties. Subsequently these techniques were applied to semiconductors. The Czochralski process was first applied to germanium growth by Teal and Little[39] (1950), while Bridgman growth has become a generic description of both the Obreimov-Schubnikov and the Bridgman-Stockbarger[40] methods, and was applied to compound semiconductors. The Czochralski and Bridgman techniques and their variants are the principal growth methods employed in semiconductor growth.

The ability to produce near-perfect large-diameter crystals of any semiconductor depends on the thermophysical properties of the semiconductor, which determine how heat and mass transport can be managed for a particular materials system. The growth of large semiconductor single crystals with high compositional purity, low concentration of stoichiometric and point defects, and high crystalline perfection becomes increasingly difficult as one progresses from elemental germanium and silicon through the III-V and II-VI compound semiconductors. Table 4.10 shows the thermal conductivity κ, the critical resolved shear stress (CRSS), extrapolated to near the melting point, and the stacking fault energy (SFE) of several important semiconductors. These fundamental properties provide important guidelines to determine which technique is most suitable for a particular semiconductor, whether the large-diameter crystals can indeed be produced, and whether twin formation, high-dislocation densities, and so on, are likely to be problems.

The success of the Czochralski process with silicon, and the overwhelming importance of this semiconductor device and IC technology, have made it the most highly developed crystal-growth technique today. Not only is it capable of producing very large single crystals of the elemental semiconductors (silicon and germanium) but, through liquid encapsulation, it has been successfully applied to producing large com-

TABLE 4.10 Thermophysical and Mechanical Properties of Selected Semiconductor Materials at Their Respective Melting Points

Element or compound	Melting point, °C	Thermal conductivity κ W/(cm • K)	CRSS, MPa	SFE, ergs/cm^2
Si	1420	0.21	1.85	70
Ge	960	0.17	0.70	63
GaAs	1238	0.07	0.40	48
InP	1070	0.10	0.36	20
CdTe	1092	0.01	0.11	10

pound III-V semiconductor single crystals of GaAs, InP, and GaP. Unfortunately the Czochralski technique suffers from several deficiencies, including contamination of high-purity melts as a result of dissolution of the crucible material, inhomogeneous incorporation of unwanted impurities and dopants in the growing crystal (as a result of buoyancy-driven melt convection), and the detrimental effects of the high thermal gradients which are necessary for good diameter control in the Czochralski method.

For these reasons Bridgman growth and its variants continue to be important, especially for the III-V and II-VI compounds, but they suffer the disadvantage of producing crystals with the D-shaped cross section of the containing boat. The growth techniques described here are commonly used for producing single crystals of elemental silicon, III-V (GaAs, InP, and GaP), and II-VI (CdTe and HgCdTe) compounds. Recent progress in silicon carbide crystal growth is included because of its importance to future advanced solid-state device technology.

4.2.1 Silicon Crystal Growth

Polycrystalline Silicon.[41] Semiconductor-grade polycrystalline silicon is manufactured from metallurgical-grade silicon (98 to 99 percent pure) which is produced from the reduction of quartzite (SiO_2) with carbon. Purification to the very high levels of semiconductor-grade silicon (fractional parts per billion) is accomplished in four basic process steps.

1. Submerged-electrode arc furnace reduction of quartzite and carbon-reducing agents (coal, coke, and wood chips) to metallurgical-grade silicon,

$$SiO_2 + 2C = Si + 2CO \tag{4.21}$$

2. Conversion of metallurgical silicon by chlorination into trichlorosilane via a fluidization bed reaction with anhydrous HCl,

$$Si + 3HCl = SiHCl_3 + H_2 \tag{4.22}$$

3. Purification of $SiHCl_3$ by multiple distillation
4. Deposition of semiconductor-grade silicon by chemical-vapor deposition from $SiHCl_3$ with hydrogen,

$$SiHCl_3 + 2H_2 = 6HCl + 2Si \tag{4.23}$$

The conversion of metallurgical silicon to silane (SiH_4) and its pyrolytic decomposition of silicon is also used (Komatsu process), although not widely because of the instability of silane.

The silicon is deposited on thin single-crystal rods (called slim rods), having a quality comparable to the produced silicon, which are heated by passing a current through them in large vacuum reactors (Siemens' process). The resulting polycrystalline silicon rods are of remarkable purity. Today large-diameter polysilicon of very high purity is produced by both the Siemens trichlorosilane process and the Komatsu silane process. Table 4.11 lists the boron, phosphorus, and carbon concentrations typical of these sources of polycrystalline silicon.

Float-Zone Growth. The Siemens group[42] in Munich, together with Bell Laboratories,[23] pioneered much of the science and technology of float zoning during the 1950s and 1960s. The Siemens researchers, however, were responsible for much

TABLE 4.11 Impurity Content of Commercially Available Polysilicon

Impurity	Silane-based,* ppba	Trichlorosilane,† ppba	
		Standard	UHP
B	0.01	0.02	0.007
P	0.06	0.10	0.024
C	200	200	<200

*Union-Carbide.

†Hemlock Semiconductor.

of the evolutionary engineering of production-sized large silicon-crystal zoning equipment, to the point where today 4- and 5-in-diameter crystal growth capability is available commercially. Up to 6-in silicon crystals are being developed by merchant suppliers such as Wacker Chemie in Germany and SEH and Komatsu in Japan.

Float zoning, illustrated in Fig. 4.25, consists in melting a zone of molten silicon in a vertical polysilicon rod by means of an RF heating coil. The molten zone is freely

FIGURE 4.25 Silicon float zoning showing liquid zone molten by RF heater suspended between polycrystalline (upper) and recrystallized silicon (lower). (*Courtesy Siemens A.G.*)

suspended by surface tension and RF levitation forces. As a result no contamination of the polysilicon occurs from a containing crucible (as in other growth methods) and exceptional high-purity levels can be maintained. The floating-zone process consists of two main steps. First under vacuum the molten zone is formed at one end of the silicon rod and slowly moved through its length to further purify the silicon rod by the segregation of impurities into the molten zone and by evaporation of volatile impurities. The second step, which is carried out under a partial pressure of an inert gas, usually argon, is to attach a seed crystal and slowly move the molten zone through the seed crystal into the attached polysilicon rod, thereby converting it into single-crystal silicon. The upper and lower drive shafts holding the polysilicon rod are rotated independently for uniform thermal distribution.

Modern commercial float-zone equipment permits growing dislocation-free <111>- and <100>-oriented silicon crystals at 4-in diameters with exceptional purity. For example, the Siemens VZA-9 zoner, which is well engineered to decouple zoning from unwanted external mechanical, thermal, and electrical perturbations, accommodates a 1.5-m-long up to 100-mm-diameter polysilicon rod in an 80-cm-diameter working chamber and has a demonstrated capability for zone purification in vacuum and the conversion (usually in an argon atmosphere) to dislocation-free 4-in-diameter silicon crystals up to 15 kg in weight. Figure 4.26 illustrates some of the principal steps in the growth of 4-in-diameter high-purity silicon crystals. Figure 4.26*a* shows an 80-mm high-purity polysilicon rod after grinding, coning, and etching before introduction into the chamber. The as-grown surface texture of polysilicon rods is always too rough to be used directly. Figure 4.26*b* shows the rod after attachment to the seed

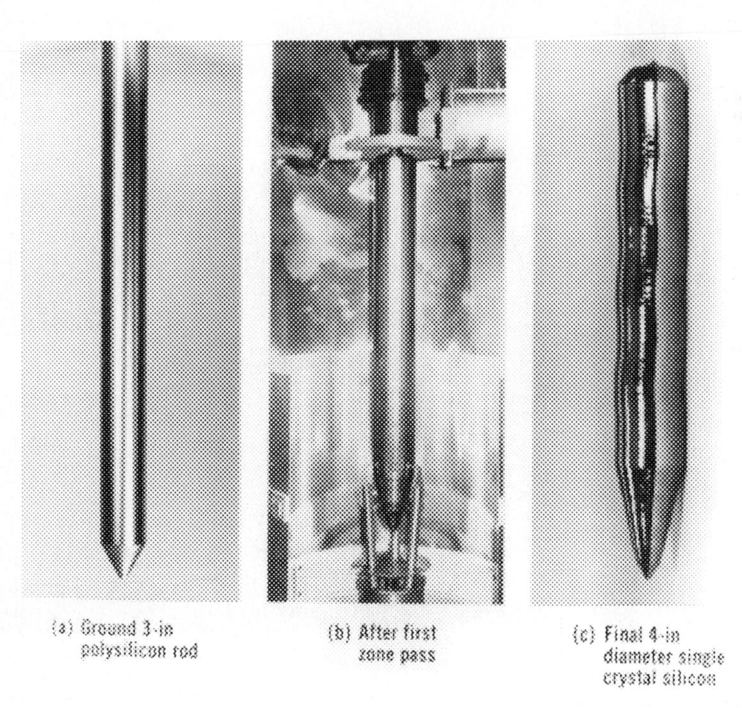

FIGURE 4.26 Stages in large-diameter silicon float zoning. (*a*) 3-in-diameter ground and etched polysilicon rod. (*b*) After first zone pass. (*c*) 4-in-diameter <100>-oriented single crystal.[47]

crystal and after the first zone purification (usually performed in an argon atmosphere to avoid the formation of whiskers). Note the three-point support of the ingot mass at the bottom and the fixed pancake "needle-eye" RF heater coil at the top of the photograph. Figure 4.26*c* shows a 4-in-diameter <100>-oriented silicon single crystal approximately 30 in long, grown after three vacuum passes and a single-crystal conversion step carried out in argon.

Dislocation-free seeding, first developed by Dash[43] for Czochralski growth, was adapted to float zoning. The steep axial thermal gradients (100 to 200°C/cm) of zoning large-diameter crystals make the system absolutely intolerant of any dislocations, either propagated from the seed or subsequently introduced by either an electrical perturbation or premature engagement of the crystal support mechanism. Crystals are either completely free of dislocations or degenerate quickly into large-grain polycrystals. The Dash method consists of necking down the cross section of the single-crystal seed to about 3 mm diameter over a 1-in length so that dislocations will grow out to the surface, and growing several centimeters at sufficiently high speed so that high vacancy densities are generated to annihilate edge dislocations. The X-ray topograph in Fig. 4.27*a* shows the successful removal of dislocations in <111> seeding, and in Fig. 4.27*b* the subsequent cone growth. The rotational striations in this dislocation-free crystal are made visible through copper decoration of the etched section and show the development of the central (111) facet.

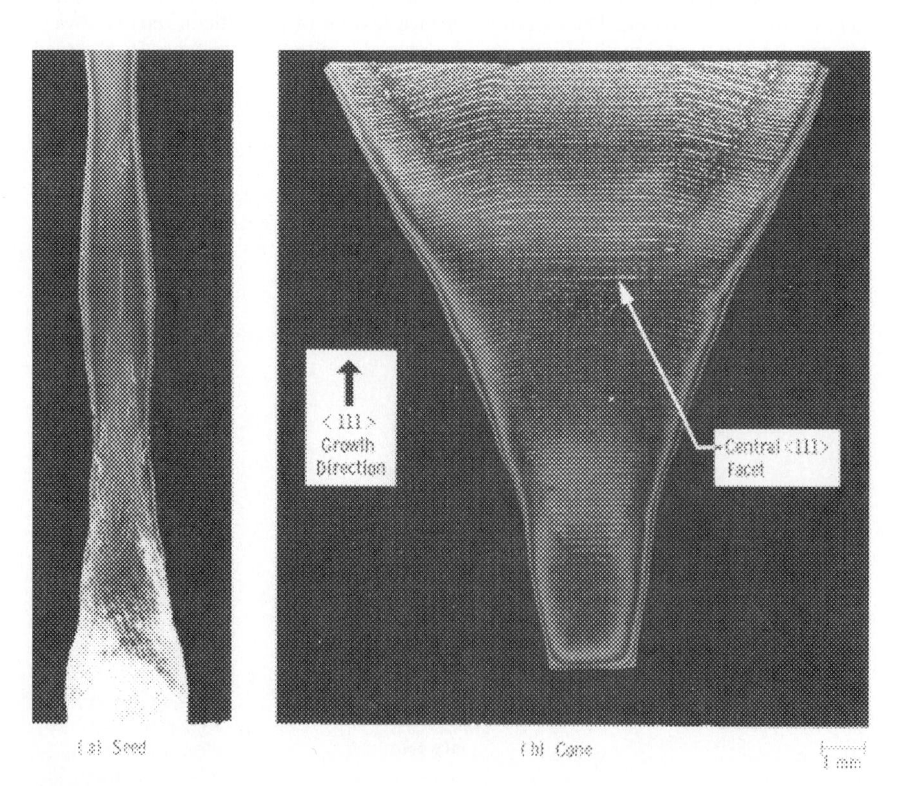

FIGURE 4.27 X-ray topographs. (*a*) Removal of dislocations during seeding. (*b*) Development of central (111) facet in float-zone growth.[47]

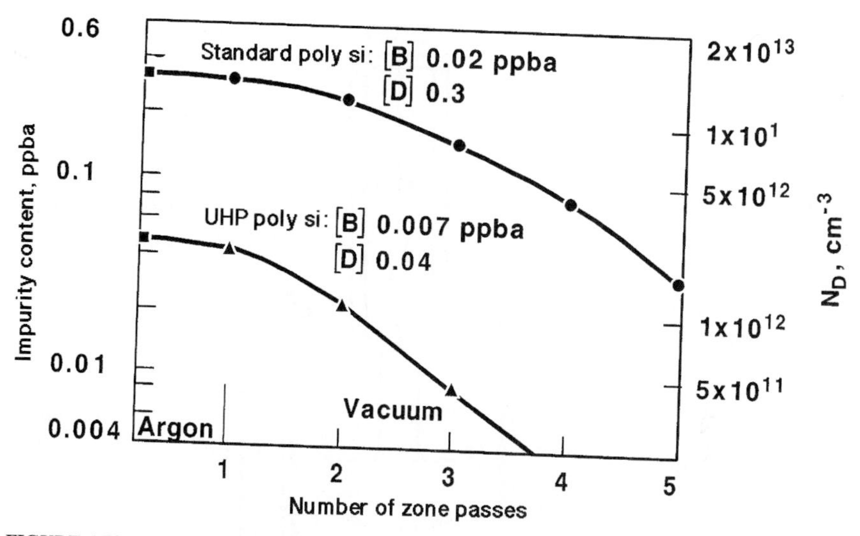

FIGURE 4.28 Purification with number of zone passes for commercial and ultrahigh-purity polysilicon. Single-crystal 3-in-diameter silicon with p-type resistivities approaching 100,000 Ω • cm have been achieved.[47]

Multipass zoning in vacuum is used to remove the phosphorus impurity, as illustrated by the typical data shown in Fig. 4.28. Typically three to four zone passes in vacuum appear optimal to reduce the donor concentration below the inherent low boron content of the polysilicon (and yield p-type resistivities in excess of 10,000 Ω • cm). Ultra-high-purity polysilicon enabled single crystals of exceptionally high resistivities (100,000 Ω • cm, p-type) and photoconductive lifetime (10 ms) to be grown. The intrinsic room-temperature resistivity of silicon is 210,000 Ω • cm. Photoluminescence measurements indicated a boron concentration of about 3×10^{11} cm^{-3} and donor concentrations below the detection limit (about 1×10^{11} cm^{-3}).

Czochralski Growth. Silicon Czochralski growth is shown schematically in Fig. 4.29. The starting charge of broken pieces of polycrystalline silicon is held in a crucible made of high-purity fused silica, which is held in a graphite susceptor. This assembly is seated within a concentric graphite "picket-fence" resistive heater (or a water-cooled RF heater) with various heat shieldings. The charge is molten down under an argon partial pressure, and growth is initiated by dipping a seed crystal into the melt, waiting for thermal equilibrium, and reducing the input power while slowly withdrawing the solidifying solid crystal. Both crystal and crucible rotations and lifts are utilized. Dash seeding is used to eliminate dislocations from the seed and to prevent their propagation. Modern silicon Czochralski crystal pullers have precision temperature control and diameter control, and are highly automated (to the point that all the processes from heat up to cool down can be carried out almost without human intervention). Pullers used in manufacturing are also generally well characterized in terms of the effects of pull rate and rotation rates (to control melt stirring), and designed to exercise some control over thermal gradients within the melt-crystal system. This follows from the need to control principally the oxygen concentration (inadvertently incorporated in the silicon by crucible dissolution) which is important in IC processing. Growth rates of 50 mm/h are

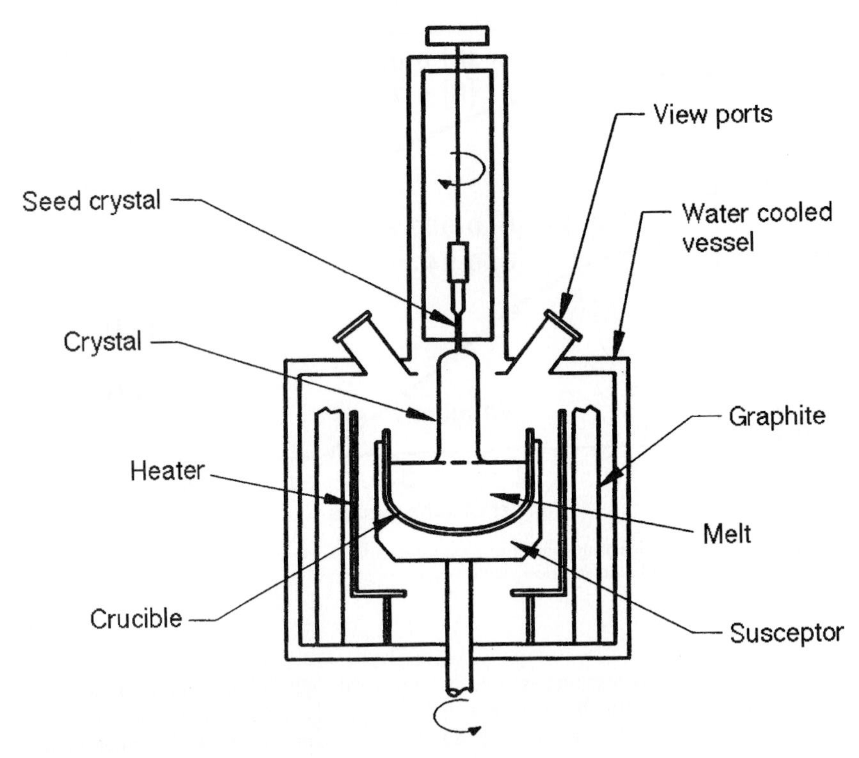

FIGURE 4.29 Schematic of Czochralski growth process.[76]

typical, and silicon crystals with diameters up to 200-mm and weighing 80 kg are typical of modern production. Figure 4.30 illustrates 150-mm and 200-mm-diameter silicon crystals produced by the Czochralski process.

Magnetic Czochralski Growth. Buoyancy-driven thermal convection causes impurity striations in all melt-grown crystals and is the major driver for today's push to grow crystals in the microgravity environment of space. The effect of a magnetic field on a large-volume conductive semiconductor melt is similar. The thermal convection which normally takes place is suppressed by the inductive magnetic drag or the "magnetic viscosity" of the melt. The important advantages to producing better-quality silicon and other semiconductor crystals using magnetic field stabilization were predicted to include:

- Improved macro- and microscale uniformity because of quiescent melts free of temperature fluctuations
- Improved purity due to the reduced crucible contamination of melts
- Increased boundary-layer thickness and, therefore, an increase in the effective segregation coefficient of impurities

Magnetic fields were first applied to semiconductor crystal growth to eliminate impurity striations in small InSb crystals grown by the horizontal Bridgman[44] method

FIGURE 4.30 Czochralski pulled silicon crystals, 150-mm (left) and 200-mm diameter. (*Courtesy Kayex, Hamco.*)

in the mid-1960s and by the Czochralski technique[45] in the United States in 1970. However, the extension of this "magnetic melt stabilization" to large commercially significant Czochralski silicon pulling occurred at Sony[46] in Japan.

The effects of axial and transverse magnetic fields on melt temperature fluctuations, the distribution of low k_{eff} dopants, and the oxygen content in silicon Czochralski growth have been investigated.[46,47] Silicon crystals of 3 and 4 in in diameter were pulled from 20-kg melts contained in 30-cm-diameter fused silicon crucibles at magnetic fields up to 5000 G. The suppression of natural convection again resulted in good diameter control and excellent surface texture. The normal shimmer observed on the melt surface in the absence of a field was not evident with the use of either field configuration.

Thermal Effects. Immersing a thermocouple just below the surface of such a melt as the magnetic field is slowly increased reveals two main observations: the predicted reduction in the temperature fluctuations to less than 0.1°C at fields of about 1500 gauss and a progressive cooling of the center of the melt with increased field strength (Fig. 4.31). Cooling occurs as the convective component to the transfer of heat from

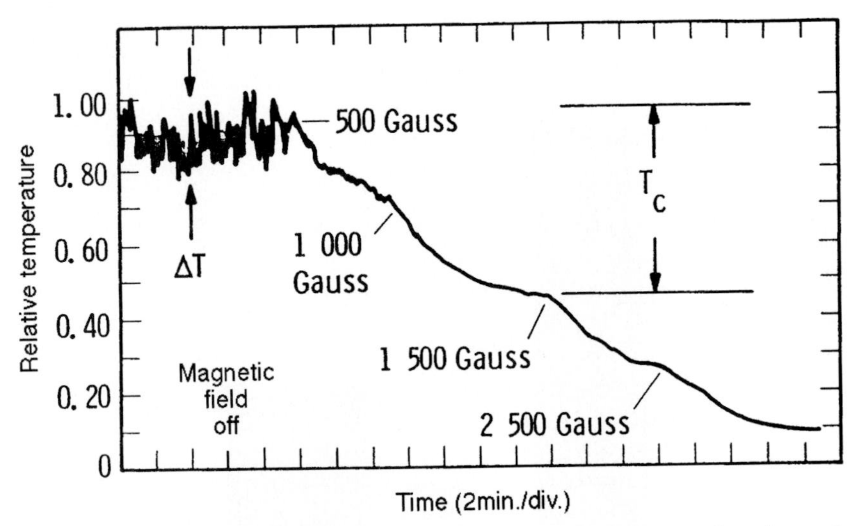

FIGURE 4.31 Thermocouple probing at center of 20-kg silicon melt, illustrating effect of magnetic field on temperature fluctuations and melt cooling.[47]

the crucible walls is reduced and becomes dominated by conductive transfer. One of the most significant differences between axial and transverse magnetic-field melt stabilization is in the relative magnitude of the melt cooling observed in our experimental equipment. About a 10°C temperature decrease was measured (just below the surface of the melt) as the axial field was increased up to 1500 G, but only a 2 to 3°C drop was observed when a 1500-G transverse field was imposed across the melt. During crystal growth, when the melt interface must naturally be maintained at the freezing temperature of silicon, the inference is that the crucible walls must be maintained at a significantly higher temperature (and radial gradients are higher) in axial magnetic field stabilization than in a transverse field or in the absence of a magnetic field.

Impurity Distribution. Both magnetic field configurations are effective in improving both the micro- and the macrouniformity of impurity incorporation in crystals through the elimination of temperature fluctuations caused by buoyancy-driven thermal convection. Magnetic melt stabilization improves the microresistivity of low-segregation-coefficient k_{eff} dopants such as antimony or gallium in silicon dramatically, and the effective segregation coefficient increases with applied field strength, as predicted theoretically. Figure 4.32 shows spreading resistance measurements along a longitudinal section of antimony-doped crystals grown with and without transverse magnetic field stabilization. Figure 4.33 shows measurements of the increase in k_{eff} for gallium in silicon with increasing field strength for 4-in-diameter crystals grown in an axial magnetic field. The increased k_{eff} reflects an increased thickness of the diffusion boundary layer as the melt motion is suppressed.[48] Recent studies[49] of small gallium-doped germanium crystals grown by the vertical Bridgman method demonstrate that k_{eff} approaches unity, as predicted under very strong fields (about 50,000 G). Macrouniformity is also improved through the elimination of striations associated with turbulent convection, but unfortunately magnetic melt stabilization does not remove the ever-present rotational striations which result from small-temperature asymmetries of the hot zone, which interact with the crystal rotation in all practical large-diameter crystal-growth systems.

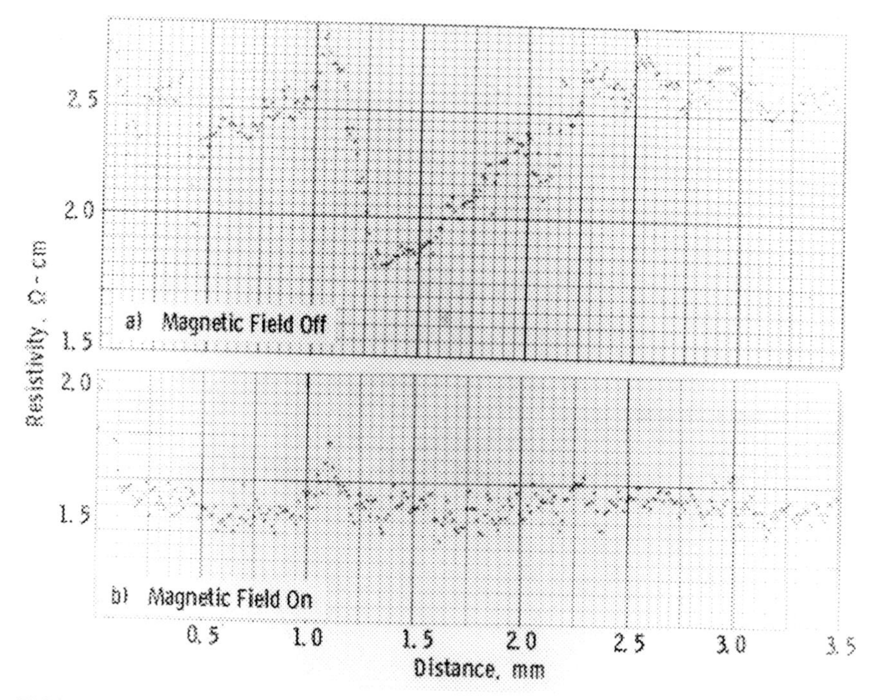

FIGURE 4.32 Longitudinal spreading resistance profiles for antimony-doped silicon crystals grown (*a*) without and (*b*) with a 1500-G transverse field.[47]

Oxygen in silicon. The magnetic field configuration utilized for melt stabilization has a major effect on the incorporation of oxygen in silicon. It is also of major technological importance because of the use of intrinsic oxygen gettering in VLSI processing. The effect of crystal growth parameters and field strength on oxygen concentration and distribution is dramatically different in the two field configurations (Fig. 4.34*a*). In the case of an axial field, increasing the field strength increases the oxygen concentration, whereas transverse magnetic fields reduce the oxygen content to as low as 4 ppma. The data[47] shown in Fig. 4.34*b* indicate that the oxygen content changes with both crucible and crystal rotation. Transverse magnetic field stabilization is clearly preferable for obtaining controlled oxygen concentrations and uniform radial distributions in Czochralski silicon over a wide range of concentrations of interest to VLSI processing. At fields between 1500 and 3500 G the oxygen content can be conveniently controlled by the crucible rotation and is relatively independent of the crystal rotation rate.

Ribbon Growth. The intense interest in photovoltaic solar cells in the 1970s led to the development of techniques for producing high-quality silicon in thin ribbon form. It was recognized that at least an order of magnitude cost reduction in producing silicon was essential if photovoltaic technology was to be competitive with other forms of electrical generation. Ribbon growth holds this promise, bypassing the expense of converting silicon ingots into polished wafers.

Edge-defined film-fed growth (EFG)[50] and the dendritic web process[51] are two techniques which can produce thin (typically 0.1- to 0.2-mm-thick) ribbons up to 7 cm

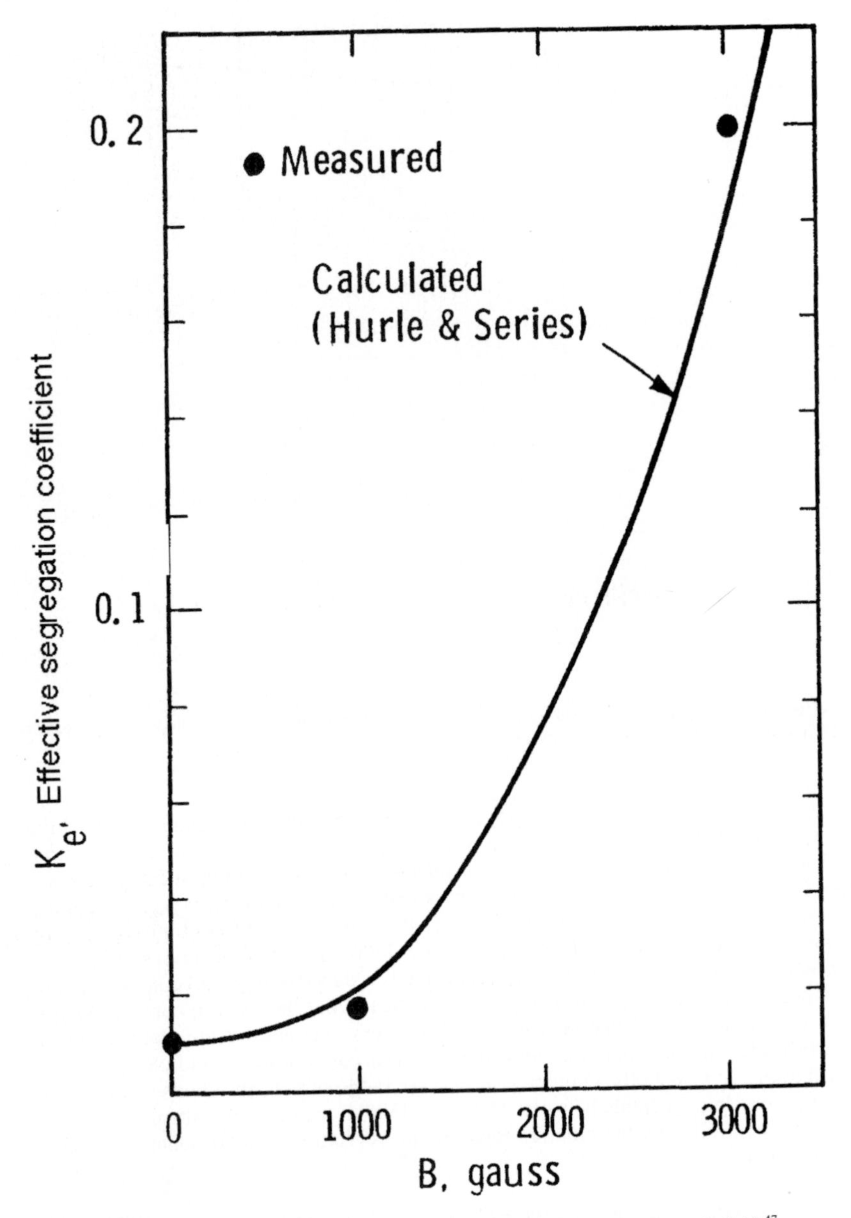

FIGURE 4.33 Increase in k_{eff} of gallium in silicon crystals grown in axial magnetic field.[47]

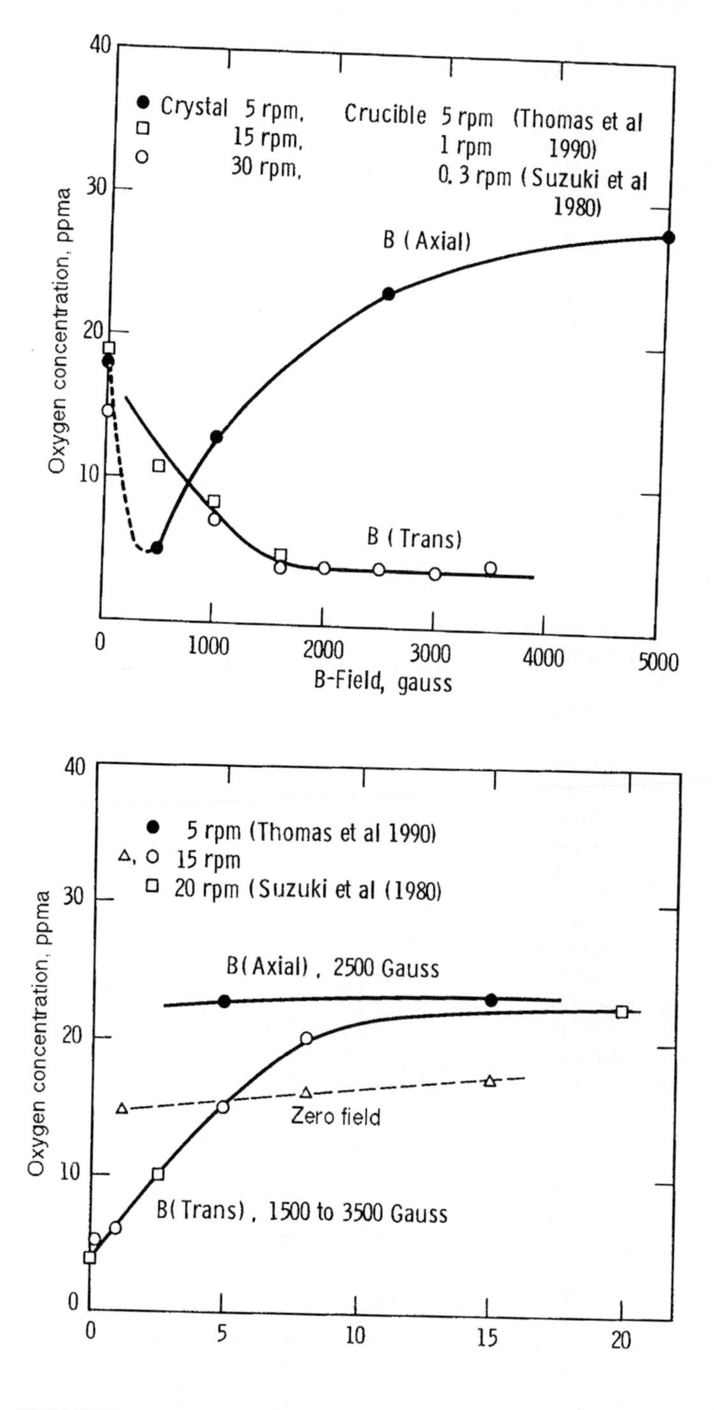

FIGURE 4.34 Effect of axial and transverse magnetic fields upon interstitial oxygen content in 4-in-diameter silicon crystals. (*a*) Variation with field strength. (*b*) Variation with crystal rotation.[46,47]

wide and many meters in length. EFG relies on pulling the silicon ribbon through a shape-defining die made of high-purity graphite, while the dendritic web process produces a ribbon crystal whose form is controlled by crystallography and surface tension forces. Dendritic web silicon is of higher purity than the EFG grown silicon, which suffers from carbon contamination (about 10 ppma) introduced from the graphite orifice. Silicon dendritic web always grows with a central twin plane throughout its length and has mirrorlike flat surfaces. Both EFG and the dendritic web contain dislocation densities of about 10^3 to 10^4 cm^{-2}.

The dendritic web method shown in Fig. 4.35 consists of dipping a thin seed crystal (a dendrite cut from a ribbon crystal) into an undercooled silicon melt contained in a shallow rectangularly shaped crucible. As the melt temperature falls, the seed first spreads laterally to form a button, and as it is raised, two secondary dendrites propagate from each end of the button into the melt. The button and the dendrites form a frame supporting a liquid film which crystallizes, forming a silicon web. Since the melt is supercooled a few degrees, removal of the latent heat up the growing ribbon crystal is no longer the growth-limiting step. Dendritic growth proceeds at high speed (typically 5 cm/min). The dendrites grow in the <211> direction and the silicon web crystallizes as a (111)-oriented atomically flat surface with a central twin plane through the entire web. Dendritic web silicon has a similar compositional purity to Czochralski-grown silicon because of the fused silica crucible. Significant engineering advances have been made to control web thickness and to increase the widths (up to 7 cm) and lengths (up to 22 m) produced, as well as improving melt replenishment techniques. Figure 4.36 illustrates the progress made in increasing the ribbon width in recent years.

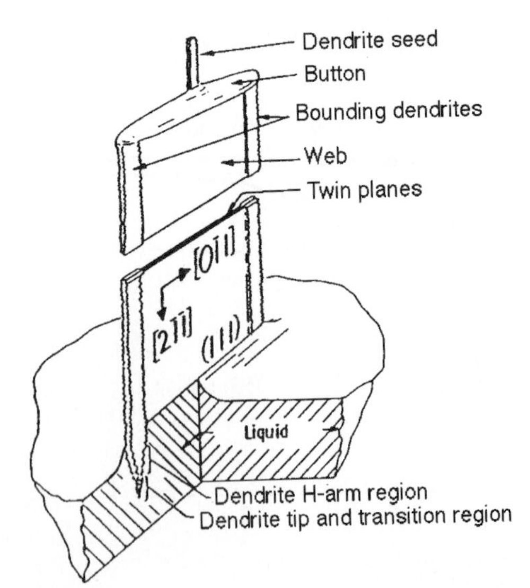

FIGURE 4.35 Schematic depiction of silicon dendritic web growth. Seed button, bounding dendrites, and crystallographic orientation of ribbon are shown.[51]

FIGURE 4.36 Increase in web width with improved control of furnace temperature profile. Maximum width of almost 7 cm has been achieved.[51]

4.2.2 III-V Compound Semiconductors

Melt-growth methods have also been successful for producing high-quality bulk crystals of III-V semiconductors. The basic difference between the growth of elemental and compound semiconductors is twofold, (1) the difficulty of maintaining stoichiometry—the correct chemical ratio of the constituent species, since one component is usually highly volatile—and (2) achieving high crystalline perfection for semiconductors with lower thermal conductivities, critical resolved shear stress, and stacking fault energies than elemental silicon. The growth techniques utilized to produce crystals of the important III-V compounds InSb, GaAs, InP, and GaP are discussed here.

Purification and Synthesis. The constituents of the III-V compounds are amenable to high purification in their elemental form by vacuum zone refining (column III) and by volatilization and distillation (column V). Purity levels of six (99.9999 percent) or seven nines (99.99999 percent) can typically be attained, but these are several orders of magnitude worse than the fractional parts per billion purity of elemental silicon and germanium. Polycrystalline III-V compound starting charges are prepared from slowly reacting the two components in a thick-walled quartz tube (to withstand the significant vapor pressure which the volatile column V will develop) in a horizontal furnace designed to maintain a high/low temperature gradient as in Bridgman growth. In the case of InSb, for example, which melts congruently at 525°C and 50 percent composition, and has a low vapor pressure over the melt, the following process is used. Both indium and antimony are purified by zone leveling in vacuum, synthesized into polycrystalline InSb in a quartz or fused silica boat, and followed by multiple zone-leveling passes for purification. Single crystals can be prepared by Bridgman or conventional Czochralski growth. However, InSb crystals exhibit unique growth and twinning habits, which make the <211> orientation the preferred growth direction. The (111) and (100) wafers normally used for device work are formed by slicing the ingots at the appropriate angle to the <211> growth direction, resulting in D-shaped wafers.

Similar purification and synthesis techniques are adapted to the constraints imposed by the particular constituent elements of other III-V compounds.

Horizontal Bridgman/Gradient Freeze Techniques.[52] Horizontal Bridgman (HB) is the least expensive means of producing bulk compound semiconductor crystals and is still used for approximately 75 percent of the GaAs single crystals produced today. The method is shown schematically in Fig. 4.37. Crystal growth is usually a two-step process. First is the synthesis of polycrystalline compound semiconductor by the exothermic reaction of high-purity (six or seven nines purity) gallium and arsenic starting materials in a sealed heavy-walled quartz tube. The second step consists of loading the polycrystalline GaAs in a quartz boat with a seed crystal at one end. The boat with an excess arsenic is contained within a sealed heavy-walled quartz tube within a furnace. Crystal growth consists of establishing the temperature profile shown in Fig. 4.37 and either moving the boat horizontally through this temperature gradient, freezing the melt to the "template" of the seed crystal (HB), or slowly reducing the temper-

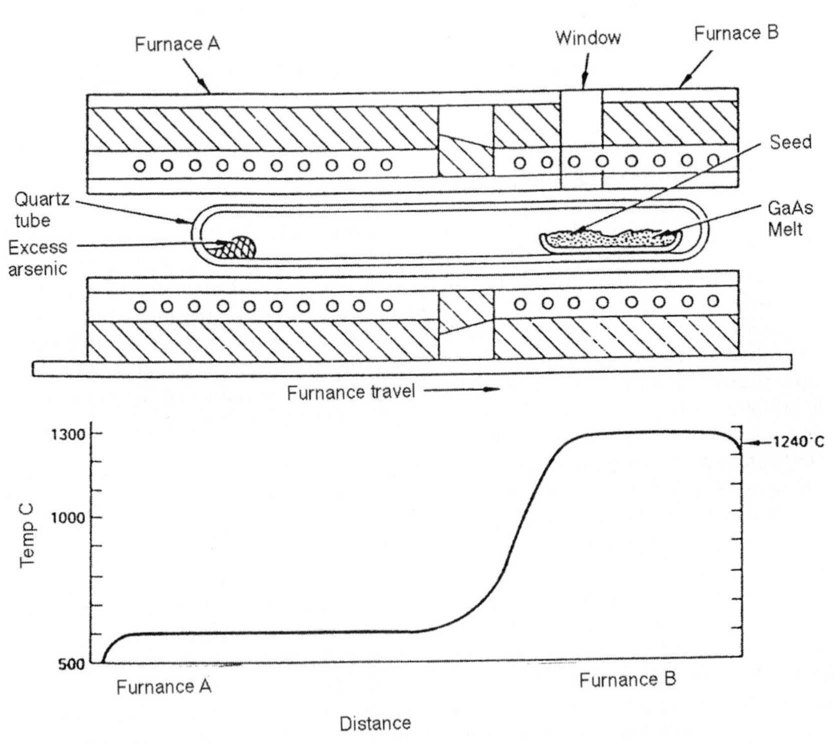

FIGURE 4.37 Schematic of furnace and temperature profile for horizontal Bridgman/gradient freeze growth of GaAs.[52]

ature profile through the stationary seed-crystal–melt system (gradient freeze). These techniques are favored for the industrial production of *n*-type, GaAs, InP, and GaP crystals for optoelectronics applications and have been developed to produce in GaAs large 3-in-diameter <111>-oriented crystals approaching 10 kg in weight

For preparing high-purity GaAs crystals, the horizontal Bridgman/gradient freeze technology suffers from several problems. First melt-container interactions cause twinning and polycrystalline growth (which has given rise to proprietary surface preparations, such as roughening of the quartz boat surface or the use of carbon coatings to overcome this difficulty); second, melt-container interaction results in the introduction of *n*-type silicon contamination; and finally, boat-grown crystals favor the <111> orientation with a characteristic D-shaped cross section of (111) or (100) wafers. Figure 4.38 shows a Bridgman-grown GaAs crystal. Nevertheless the Bridgman/gradient freeze technology produces III-V single crystals with low dislocation densities since the thermal gradients are low (5 to 10°C/cm), which is a significant advantage in optoelectronic device technology as dislocations are generally sinks for minority carrier recombination centers.

Vertical Gradient Freeze. Vertical Bridgman and Gradient Freeze (VGF) growth methods on a small scale were first applied to III-V compounds in the 1970s with the goal of producing low-dislocation, cylindrically-shaped crystals. Overpressure of the volatile column V component and encapsulation techniques were used to maintain crystal stoichiometry. In the mid-1980s, Gault and other researchers[53] at AT&T evolved the first reproducible VGF technology for producing large-diameter crystals of GaAs, InP, and GaP. Their method maintains stoichiometry by column V compo-

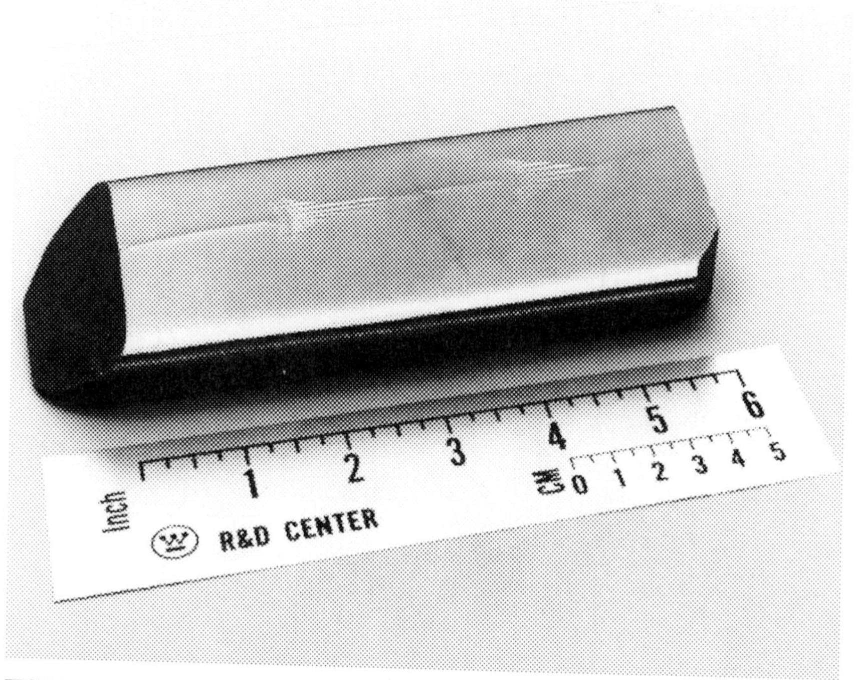

FIGURE 4.38 Silicon-doped <111> GaAs crystal grown by Bridgman process.

nent overpressure, and low thermal gradients and controlled postsolidification cooling rates to minimize stress-induced dislocation generation.

The VGF crystal growth equipment as configured for GaP is shown in Fig. 4.39*a*. It is surrounded by two "picket-fence" coaxial graphite heaters and contained within a water-cooled stainless-steel pressure vessel. Various modifications are required for GaAs and InP crystal growth. Presynthesized polycrystalline GaP is contained in a pyrolytic boron nitride (pBN) crucible with 50-mm diameter at the top, which tapers down to a 6-mm-diameter seed well containing a single-crystal GaP seed. A reservoir containing red phosphorus is located below the support pedestal to provide a partial pressure of phosphorus over the melt. A top cap of pBN provides a leak path for phosphorus vapor during growth, but maintains the desired partial pressure during the length of the growth run. The furnace is fitted with appropriate translation and rotation shafts and various monitor and control thermocouples.

The axial temperature profiles of the hot zone are set empirically by the furnace control shown in Fig. 39*b*. Over the seed-well region the gradient is 40°C/cm for meltback purposes and 8°C/cm over the 50-mm region of the crystal. The two profiles shown correspond to temperatures of 1465 and 1300°C in the seed well, the difference providing the amount of cooling necessary to solidify a 1-kg charge of GaP. During growth the system is pressurized to 54 atm, and the growth rate is 3 mm/h. The growth cycle, which is automated and includes a controlled cool down at 100°C/h for 5 h, is approximately 40 h. When the system is at ambient temperature, the crystal slips free of the tapered pBN crucible, which is reusable. This VGF technique has been successful in producing low-dislocation, <111>- and <100>-oriented crystals of GaAs and InP

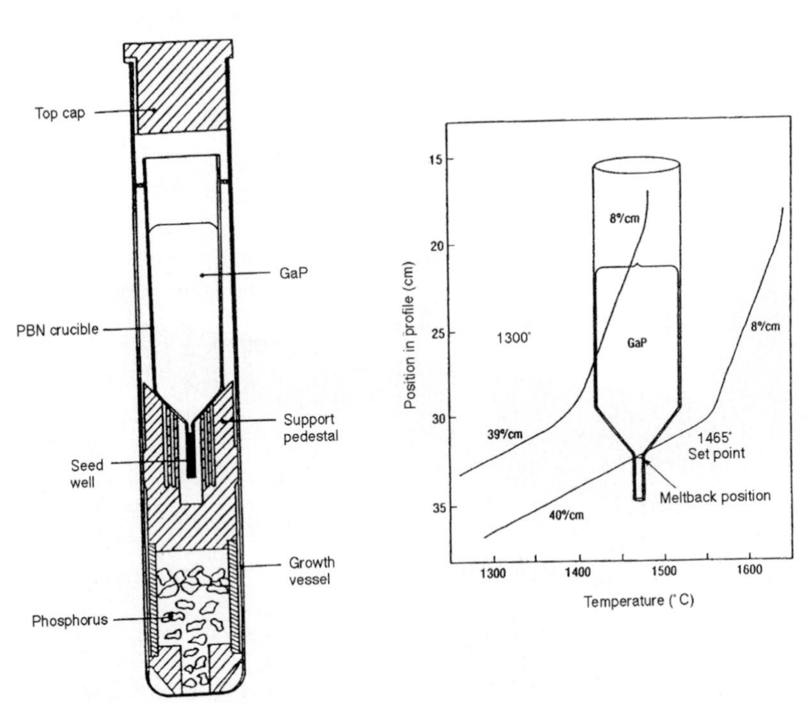

FIGURE 4.39 Vertical Bridgman growth of III-V compounds. (*a*) Growth vessel and contents for GaP. (*b*) Axial temperature profiles.[53]

(of semiconducting and semi-insulating properties) and GaP. Twin formation, however, remains a problem in the growth of <111> and especially <100> InP crystals.

Liquid-Encapsulated Czochralski Technique.[52] The feasibility of using liquid encapsulation to overcome the volatility of binary compounds was first demonstrated in 1963 by Metz et al.[54] in the Bridgman growth of PbTe crystals. Two years later the technique was successfully applied to the Czochralski process[55] for the growth of III-V compounds, from which the important high-pressure, liquid-encapsulated Czochralski (HP-LEC) growth technology has emerged. A well-developed commercial equipment base now exists as a family of Cambridge HP-LEC pullers capable of producing large-diameter crystals of GaAs, InP, and GaP.

Liquid encapsulation (Fig. 4.40*a*) prevents dissociation of the III-V melt during crystal growth, so that a stoichiometric crystal can be pulled. It relies on enclosing a liquid melt of a III-V compound with an inert liquid and applying an overpressure of an inert gas greater than the vapor pressure of the volatile constituent at the melting point of the compound, to minimize loss of the volatile column V element. For GaAs the vapor pressure of arsenic at the melting point is 0.8 atm, and for InP and GaP the corresponding vapor pressures of phosphorus are 27 and 54 atm, respectively. GaAs can therefore be grown with 2 atm overpressure (low-pressure LEC technology), whereas the much higher overpressure capability of the Cambridge HP-LEC pullers (150 atm maximum) is essential for InP and GaP. An additional important advantage was introduced with the advent of pBN crucibles. Traditionally semi-insulating GaAs crystals were grown by melt doping with high levels of chromium (a deep-level acceptor in GaAs) to counteract the 10^{17}-cm^{-3} silicon contamination introduced from the quartz boat or crucible. With pBN crucibles[56] silicon contamination of the melt is significantly reduced and large-diameter undoped semi-insulating GaAs crystals can be produced.[57]

Gallium Arsenide Crystals. The HP-LEC process allows in-situ synthesis of the GaAs melt starting from elemental gallium and arsenic, followed by the growth of the crystal without opening up the puller. The process sequence typically consists of the following steps.

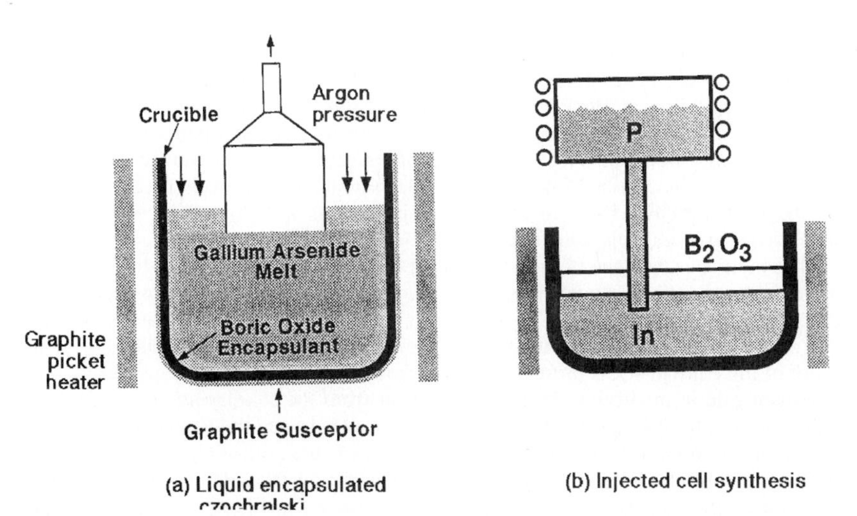

FIGURE 4.40 (*a*) Liquid encapsulated Czochralski and (*b*) injection cell synthesis techniques used in growth of GaAs and InP crystals.

1. In situ synthesis of a GaAs melt of known composition is achieved by slowly heating up accurately weighed amounts of 7N purity gallium and arsenic, which are covered with a tightly fitting solid boric oxide disk in a pBN crucible. A slight excess of arsenic (10 to 20 g for a 4-kg total charge) is added for the small arsenic loss which occurs before the B_2O_3 becomes molten and completely encapsulates the charge above 400°C. An exothermic reaction begins at about 800°C and eventually raises the melt temperature to the 1238°C melting point of GaAs. During this synthesis phase, argon is maintained at a pressure of 40 atm and then slowly depressurized to 20 atm.

2. Growth of monocrystalline GaAs proceeds in the manner used in silicon Czochralski growth, by dipping a seed crystal through the transparent boric oxide layer while the melt temperature is slowly ramped down. Seeding and slowly increasing to the crystal diameter requires slowly increasing the pull rate up to between 9 and 12 mm/h. Once the desired crystal diameter is reached, the system is "handed off" to a differential-weight automatic diameter control (the crystal is weighed continuously by a sensitive weighing cell in the pull rod and the error signal controls the heater power). Crystal and crucible rotations are used along with pull rates to minimize melt thermal instabilities. The latter is important for avoiding twin formation in large <100> GaAs crystals. The operator monitors the growth via a closed-circuit TV image of melt/boric oxide/crystal and from continuous printouts of the melt temperature, differential weight.

3. Growth is completed with a controlled annealing and cool down for stress relief and to improve the uniformity of electrical properties. The starting melt composition is determined from weighing the crucible and its contents (weigh-in) and the crucible, remnant contents, and the crystal at the end of the process (weigh-out).

This technology has now become a reliable method for producing undoped semi-insulating GaAs crystals up to 4-in in diameter and weighing 10 to 12 kg. The semi-insulating GaAs has the important virtue of being thermally stable during device and IC processing. The technology is also applicable for producing semiconducting GaAs for optoelectronic applications. Typical crystals yield 50 to 150 wafers, in contrast to mature silicon crystal technology, which produces crystals containing many hundreds of wafers. Figure 4.41 shows 3- and 4-in-diameter semi-insulating GaAs crystals grown by the HP-LEC method.

These improvements in GaAs crystal growth technology were brought about through achieving better compositional purity and understanding and controlling the role of stoichiometry-related defects. Figure 4.42 shows the effect of controlling melt stoichiometry[58,59] on the resistivity and the mobility of undoped GaAs crystals. Undoped semi-insulating GaAs is the result of compensation of shallow carbon acceptors (the dominant residual impurity) by a deep donor defect level (EL2). The latter has been identified[60] as due to arsenic antisite defects As(Ga), whose concentration increased sharply in stoichiometric or arsenic-rich melts.

The main disadvantage of the Czochralski process comes about from the inherently high axial and radial gradients (about 100°C/cm), which are necessary for good diameter control. LEC-grown III-V crystals contain dislocation densities usually between 10^4 and 10^5 cm^{-2}, which result from the crystallographic glide associated with the relief of the thermal stress present during growth. The boric oxide encapsulant plays a dominant role in modifying the transfer of heat from the solidifying crystal. A large temperature difference occurs between the hotter interior and the cooler surface of the crystal. Therefore, as the thermoelastic stress exceeds the critical resolved shear stress (CRSS), plastic deformation and dislocation generation take place. Moreover, since this stress relief mechanism occurs at temperatures only slightly below the melting point, dislocation propagation leads to polygonization into cell-like networks and lin-

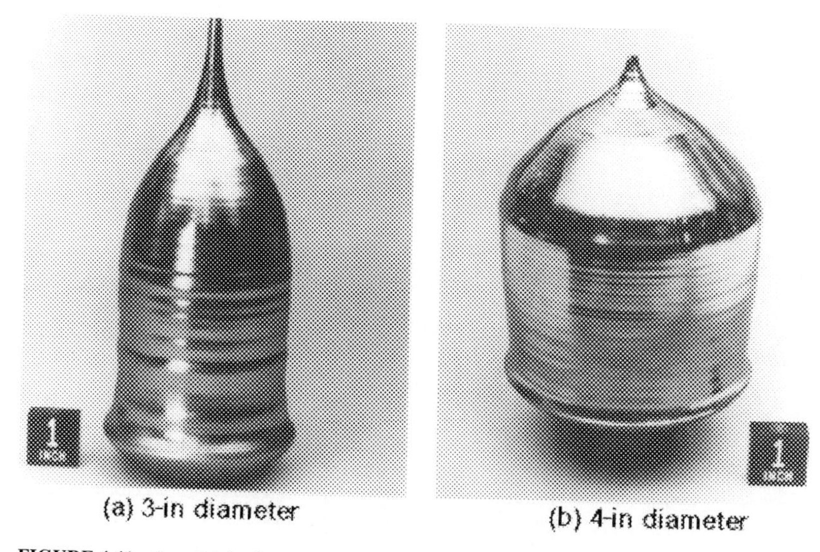

FIGURE 4.41 3- and 4-in-diameter semi-insulating GaAs crystals pulled by HP-LEC method.[47]

eage. A thermoelastic stress model[61] predicts qualitatively the fourfold symmetry of dislocation maps of (100) wafers, as shown in Fig. 4.43. The mechanism of solid solution hardening is intended to increase the CRSS of GaAs (and InP) and produce 3-in-diameter GaAs crystals which are nearly dislocation-free. Heavy *n*-type doping produces dislocation-free semiconducting GaAs and InP, while heavy indium doping or alloying (1 to 2 percent) has yielded[57,62] dislocation-free semi-insulating GaAs. Unfortunately this method increases slip, wafer brittleness, and breakage.

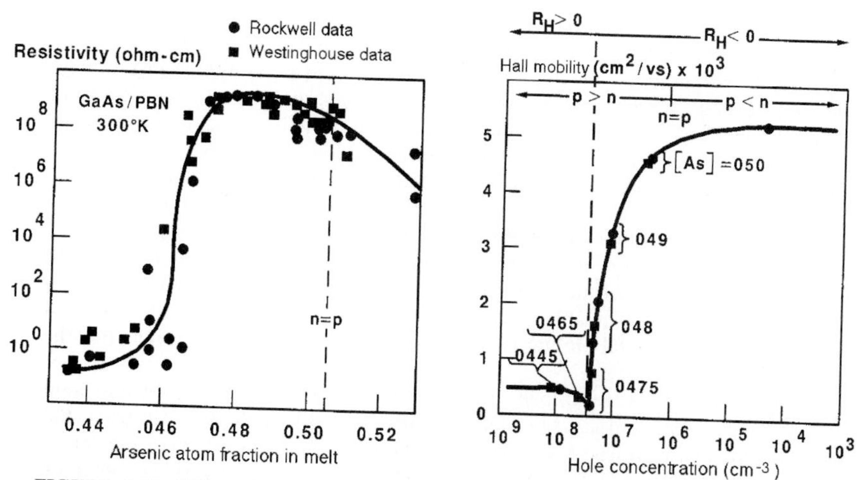

FIGURE 4.42 Effect of melt composition on (*a*) resistivity and (*b*) mobility of undoped GaAs crystals.[57–59]

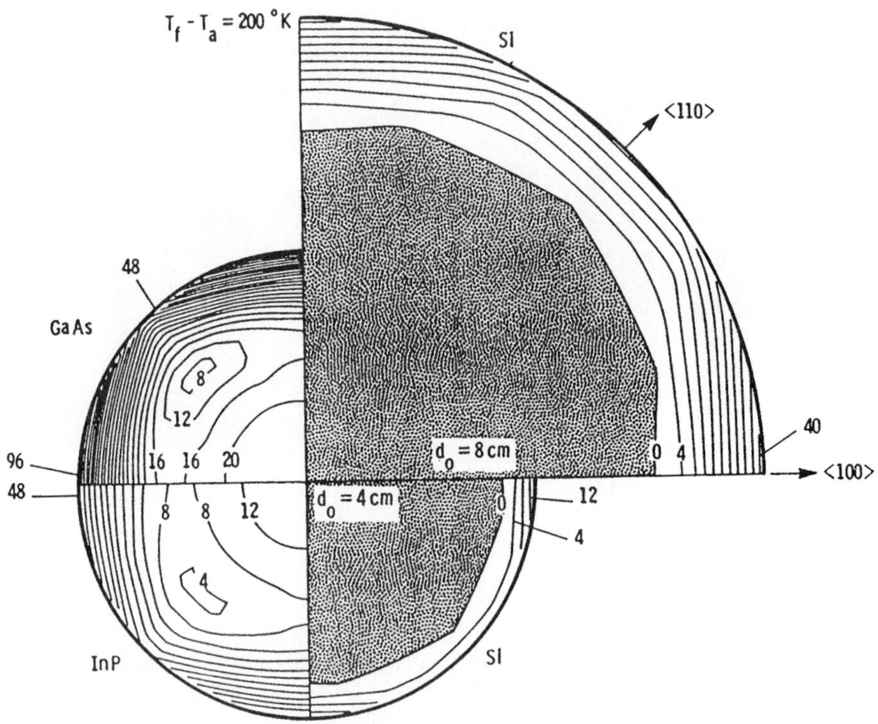

FIGURE 4.43 Comparison of calculated thermal stress (dislocation density) for LEC-grown <100> GaAs, InP, and silicon crystals.[61]

Indium Phosphide Crystals. HP-LEC growth of InP is more complex because of the higher pressures required, the ease with which twinning occurs, and the fact that iron doping must be employed to achieve semi-insulating behavior. Polycrystalline InP is usually prepared by reacting indium with phosphorus, as shown in Fig. 4.44, using thick-walled quartz capable of withstanding the high vapor pressure of phosphorus. In-situ synthesis by injection of phosphorus into an indium melt by means of an injection cell (Fig. 4.40*b*) is practiced successfully at a few laboratories.

InP crystals grown using presynthesized polycrystalline InP charge follow the methodology developed for GaAs, except at higher argon overpressures of 27 atm. An important technique,[63] which avoids the strong prevalence of twinning in large-diameter <100>-oriented InP crystals, has recently emerged. It consists in initiating crystal growth by allowing the melt to undercool and permitting the seed crystal to grow out as a flat crown (under presumably a very low radial thermal gradient) until the desired diameter is reached. Crystal pulling is started at this point and usually results in twin-free InP crystals (Fig. 4.45).

For unknown reasons, deep-level defects related to stoichiometric effects do not appear to form in InP, and semi-insulating resistivities still rely on melt doping with transition elements, with iron being the best candidate, in spite of its low segregation coefficient and the attendant problems of undercompensation of residual donors if doping is too light (10^{16}-cm^{-3} range), and FeP_2 precipitation at heavy doping levels.

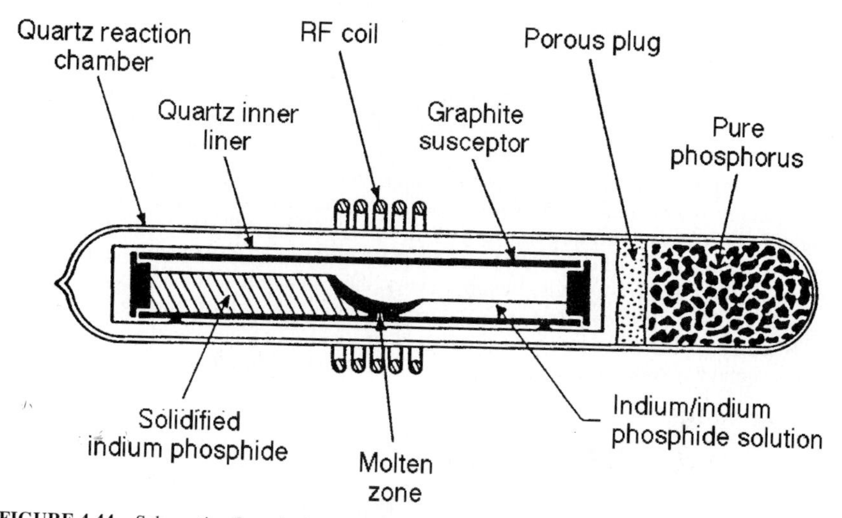

FIGURE 4.44 Schematic of synthesis of high-purity polycrystalline InP by zone leveling.

FIGURE 4.45 3-in-diameter <100> InP crystal pulled from 4-kg melt by HP-LEC technique. (*Courtesy D. H. Hobgood, Westinghouse STC.*)

4.2.3 II-VI Compounds

Mercury cadmium telluride (HgCdTe) has the desirable feature of possessing an approximately linear variation of its band gap with composition over a range which makes it an excellent infrared detector material whose wavelength sensitivity can be controlled by its composition (Fig. 4.46). However, the cutoff wavelength of a HgCdTe detector changes rapidly with composition, so that very high compositional homogeneity is essential for useful detector array fabrication. Because of the importance of HgCdTe for infrared imaging systems, much effort has been devoted to the growth of high-quality bulk crystals of this difficult ternary II-VI compound, as well as CdTe and CdZnTe crystals, which serve as host substrates for epitaxially grown HgCdTe layers. In the case of CdTe and CdZnTe, melt growth by horizontal or vertical Bridgman[64] methods has been developed to yield <111>-oriented single crystals which contain several twin-plane boundaries (horizontal Bridgman) or 2-in-diameter

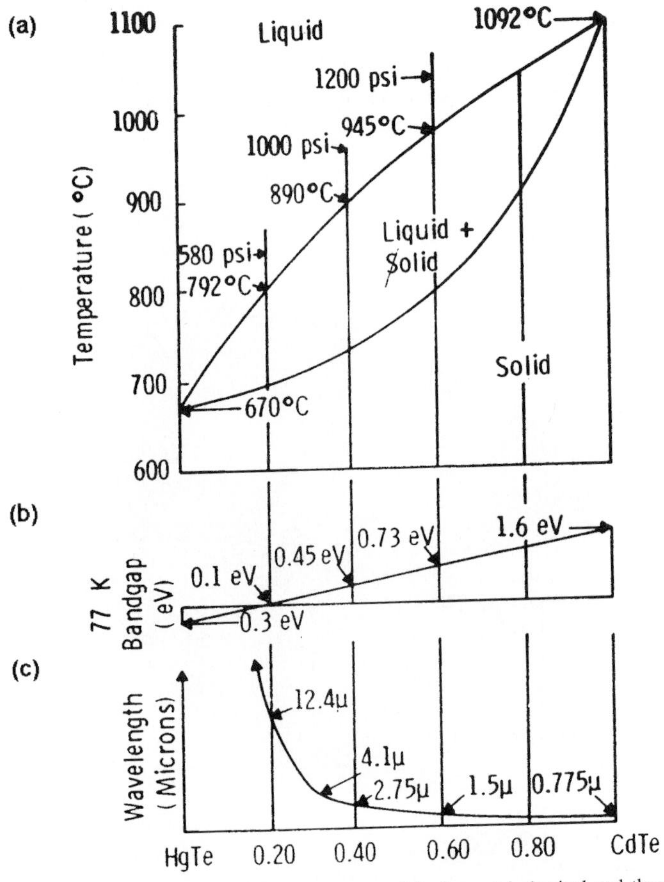

FIGURE 4.46 Variation with composition of fundamental physical and thermodynamical properties of $Hg_{1-x}Cd_xTe$. (a) Pseudobinary phase diagram. (*b*) Band gap. (*c*) Variation of cutoff wavelength with wavelength.[65]

ingots consisting of large polycrystalline regions of random orientation with respect to each other (vertical Bridgman). Irregularly shaped (111) substrates are "mined" by slicing either selected polygrains or slicing between twin-plane boundaries. Alloying of zinc produces crystals of improved crystalline quality and better lattice match to HgCdTe.

The intent here is to review briefly the production of bulk crystals of the ternary HgCdTe with the challenges imposed by the toxicity of the constituents and the explosion potential which results from the high mercury-vapor pressure involved in growth.

HgCdTe Growth. Ternary III-V and II-VI compounds can in principle be grown by most melt-growth techniques, especially if there is complete miscibility of the two compounds in the solid state and a relatively simple phase diagram. The principal difficulty is one of maintaining constant composition throughout the crystal because of the usually wide separation between the liquidus and solidus temperatures. The pseudobinary phase diagram from HgTe to CdTe for $Hg_{1-x}Cd_xTe$ (Fig. 4.46a) illustrates this difficulty. Directional freezing of the liquid alloy will result in the formation of solid alloy solutions varying continuously in composition along the growth axis. Figure 4.46*c* illustrates the strong dependence of λ_{co} on the composition of the $Hg_{1-x}Cd_xTe$ over the 8- to 12-μm spectral band, which is of great importance to infrared detectors used for terrestrial imaging. Other major concerns include the very high vapor pressure of mercury (about 70 atm) at the 800 to 900°C temperatures required for the synthesis and growth of HgCdTe crystals, and the onset of constitutional supercooling (a phenomenon whereby a breakdown in crystalline structure occurs and one or more of the constituents precipitate out of solution).

The growth of single-crystal HgCdTe of constant composition thus requires (1) rapid freezing to avoid severe segregation of HgTe and CdTe (requiring high thermal gradients to suppress constitutional supercooling, but resulting in polycrystalline growth), (2) handling the very high mercury overpressures during synthesis and growth, and (3) protecting the operator from the toxicity of these constituents, which is compounded by the high mercury overpressures.

A unique method[65] involving synthesis/quench/solid-state recrystallization has been developed, which yields single crystals of HgCdTe of random orientation and of high lattice defect densities. The technique has been developed and refined in North America as the production source for excellent photoconductive detectors operating in the 8- to 12-μm atmospheric window for military infrared imaging systems.

HgCdTe Growth by Quench and Solid-State Recrystallization.[65] Furnace requirements for the growth of HgCdTe are dictated by the need to contain a very corrosive melt of mercury, cadmium, and tellurium at temperatures up to 800 to 900°C and very high mercury overpressures without endangering the operator. Thick-walled quartz ampoules meet these requirements. This technique consists of first synthesizing HgCdTe of the required composition from carefully weighed constituent charges, and carefully slowly ramping the temperature of the ampoule while being cognizant of exothermic reactions, which can lead to explosive mercury overpressures. Typically rocking of the ampoule/furnace is used to enhance mixing. The next step involves the rapid quenching of the HgCdTe charge to room temperature (made difficult by the thick-walled ampoule) to produce a fine-grain polycrystalline structure. The final step, that of solid-state recrystallization, requires annealing at temperatures below the solidus temperature (650 to 700°C) for periods of 5 to 10 days. A successful growth typically results in a single-crystal HgCdTe ingot of 2-cm diameter and up to 5 cm long, and having a compositional uniformity better than ±0.005, which satisfies the needs of the infrared detectors.

Melt Growth of HgCdTe. Melt-growth methodology, including Bridgman, zone melting, liquid encapsulated Czochralski, traveling solvent zone melting, and slush growth, have been applied to the growth of HgCdTe crystals.[65] However, the slow-moving interface of these equilibrium processes invariably results in a steady variation of the crystal composition. A limited number of wafers of the desired compositional homogeneity is yielded from each crystal. The vertical Bridgman method, adapted and engineered for the severe demands imposed by the vagaries of HgCdTe growth, has been the most successful,[66] especially when growth rates as low as 0.25 mm/h are employed. Figure 4.47 is representative of the compositional change, as measured optically from the cutoff wavelength, of wafers cut from the leading to the trailing edge of a crystal. Although the quality of Bridgman-grown HgCdTe equals or surpasses that produced by the quench/recrystallization method, the inherent low wafer yields make it unattractive for production.

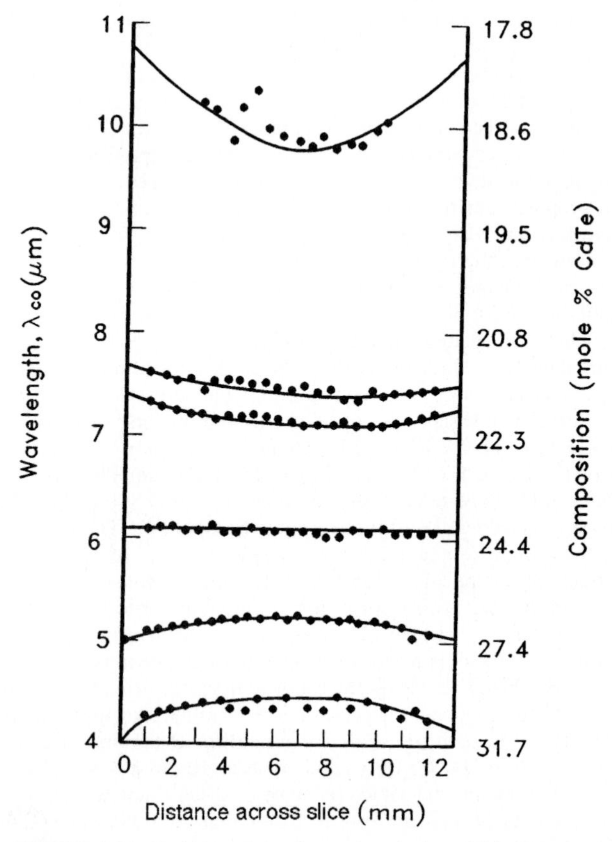

FIGURE 4.47 Variation of cutoff wavelength along Bridgman grown HgCdTe crystal.[66]

4.2.4 Silicon Carbide

Silicon carbide and semiconducting diamond possess properties which offer the potential of vastly superior device technologies in the future. The basic materials technology base is however still in its infancy. Diamond growth is still very immature but can now be prepared as self-standing polycrystalline layers or gem-stone size semiconducting diamond single crystals. The development of single-crystal SiC ingots and of single polytype, while also in its formative development stages, has made significant progress in the past few years.

Small (up to 5 by 7 mm) single-crystal platelets of SiC were produced by the Lely[67] vapor transport growth technique. Modern SiC crystal growth was pioneered in 1978 by Tairov and Tsvetkov[68] by the high-temperature vapor transport or sublimation of SiC onto a seed crystal. SiC does not exist in the liquid state, so that growth from the melt, as with silicon or III-V compounds, cannot be considered. Crystal growth proceeds by sublimation of a SiC source at between 2000 and 2500°C under near vacuum conditions, and the transport of the SiC vapor species onto a seed crystal held at a temperature some 100°C cooler than the source. The technique offers round cross-section 6H-polytype SiC single crystals capable of being scaled up to produce wafers of the diameters required today for conventional device processing.

A physical vapor transport system[69] designed for the growth of high-purity SiC crystals is shown in Fig. 4.48. A water-cooled double-quartz-wall induction-heated construction is used. Because nitrogen is a shallow *n*-type donor impurity, and a pernicious impurity source in SiC, high-temperature vacuum degasing techniques are employed. The method consists of sublimation at about 2400°C from a reservoir of

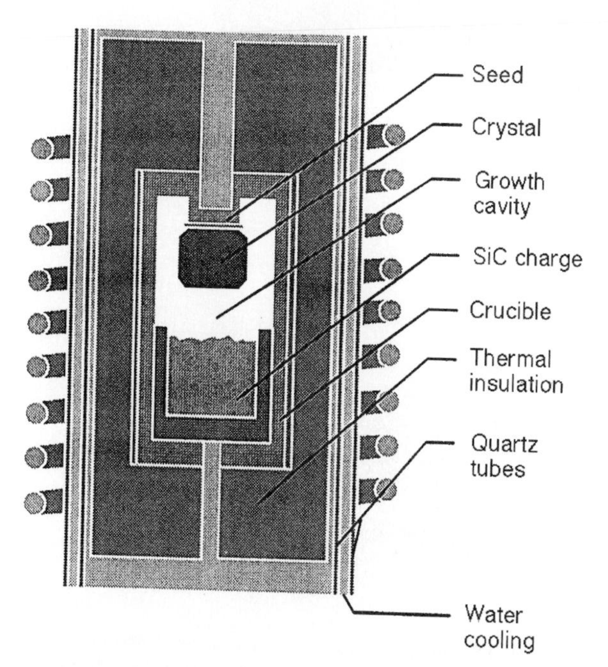

FIGURE 4.48 Schematic of physical vapor transport growth of SiC crystals.[69]

SiC (usually very small Lely platelets) onto a seed crystal mounted above the SiC charge. Lely platelets are used as seed crystals as these platelets exhibit nearly perfect crystallographic surfaces with <0001> basal planes. A temperature difference of about 100°C is maintained between the seed and the source. The induction coil can be adjusted horizontally and vertically for control of the thermal gradients.

Figure 4.49*a* illustrates a $1^1/_2$-in-diameter SiC crystal grown at a growth rate of 2 mm/h. The single crystal was grown along the <0001> orientation (*c* axis of 6H-SiC) and is single 6H polytype. Often polycrystalline growth occurs at the crystal periphery due to random nucleation of SiC crystallites from the graphite container. Figure 4.49*b* is representative of 1-in-diameter SiC polished wafers produced from such crystals by diamond internal diameter sawing and polishing techniques. Impurity concentrations (around 10^{16} cm^{-3}), microscopic voids, high dislocation densities and lineage, and mixed polymorphism are currently characteristic of this developing technology.

FIGURE 4.49 (*a*) 1.5-in-diameter 6H-SiC single crystal. (*b*) (0001) SiC wafer cut and polished from such crystals. Lely-grown platelets are shown for comparison. (*Courtesy D. L. Barrett, Westinghouse STC.*)

4.3 WAFER PREPARATION

Semiconductor crystals or ingots must be converted to thin wafers with flat mirrorlike featureless surfaces, which are free from any residual polishing damage, for electronic device processing. In silicon and GaAs IC production, stringent wafer specifications as to flatness, taper, and bow are dictated by the increasing use of submicrometer lithography to define fine device features. The standards to which silicon and GaAs wafers are produced have been developed by the American Society for Testing and Materials (ASTM)[70] and the Semiconductor Equipment and Materials Institute (SEMI),[71] and encompass specifications of wafer diameters, thickness, primary and secondary orientation, flats location, orientation (on-orientation or up to several degrees off-orientation), wafer flatness, and so on, as well as electrical and chemical measurement standards. Residual surface and subsurface damage remaining from the polishing process and chemical contamination of surfaces are less easily specified.

4.3.1 Wafer Preparation Technology

The process[72] of fabricating wafers from ingots is similar for all semiconductors, with individual process steps modified to account for the differing chemistry and the hardness and brittle nature of all semiconductors. The typical process flow diagram for silicon wafer preparation is shown in Table 4.12. The steps typically involved in the fabrication of silicon or GaAs wafers from cylindrical ingots consist in the following:

1. Cropping the crown and taper ends of the ingot using an outer diameter diamond cropping saw, so that the ingot has flat ends and can be conveniently mounted in a lathelike centered horizontal grinder, where both ends of the ingot are held rigidly.

2. Grinding the ingot to constant diameter and forming primary and secondary orientation flats on a horizontal grinder. The ingot is rotated slowly, while material is removed by a high-speed diamond grinding wheel, which moves back and forth along the ingot length. As much as 0.2 mm is removed per pass, and all grinding and cutting operations require copious water cooling. In the case of GaAs or InP ingots, precautionary measures must be taken to enclose and efficiently vent the grinder in order to safeguard the operator from dust and the possibility of arsine or phosphine generation. This operation is followed by chemical etching to remove work damage to the surface. Placement of the orientation flats is determined from Laue X-ray measurements.

3. Slicing the ingot using an internal-diameter (I/D) diamond saw into wafers, as illustrated in Fig. 4.50 for a GaAs ingot. The I/D saw allows the use of very thin blades which are attached to a heavy flange under tension and stretched in the manner of a drumhead to prevent flexing. The I/D periphery of the blade is coated with microscopic diamonds in a rigid matrix. The thickness of the diamond-coated stainless-steel blade is between 0.006 and 0.009 in, and the materials loss in producing a wafer is referred to as the kerf loss. Automated I/D diamond saws are available from several manufacturers and permit wafer indexing, so that a large number of wafers of a fixed thickness can be sawn without operator attention. Prior to sawing, the blade is "dressed," that is, tensioned by mechanical adjustments to cut parallel-sided wafers with smooth surfaces free of saw blade marks. The consequences of blade deflection during wafer sawing can result in wafers with unacceptable bow, taper, or warp, as shown schematically in Fig. 4.51. For this reason, sawing is continuously monitored with a position sensor to maintain correct operation. The crystallographic orientation of the ingot surface is again measured by Laue X-ray measurement, with the ingot mounted on a goniometer stage. This enables the wafer orientation [for example, exactly on (100) orientation, or several degrees off-orientation toward a specified crystal direction] to be established accurately.

4. Removing the saw damage and producing a flat wafer with the minimum bow and taper by mechanical grinding or lapping using usually Al_2O_3 grit of specified size in an aqueous suspension. Figure 4.52 illustrates typical multiwafer double-sided equipment in use today for lapping and polishing semiconductor wafers.

5. Edge rounding (or contouring) the lapped wafers, so that there are no sharp edges which can break during device processing, causing dust contamination of the wafer surfaces. This is accomplished by an automated multiwafer edge grinder.

6. Chemical-mechanical polishing of the wafers using either double-sided or single-sided polishing techniques or a combination of both. Chemical-mechanical polishing generally uses an oxidizing reagent (such as NaOH for silicon or NaOCl for GaAs and InP) combined with a colloidal SiO_2 suspension which acts in combination with polishing pads (judged most suitable from manufacturers' specifications and recommendations) as a microscopic abrasive to continuously remove the oxide formed.

7. Polished wafers are scrubbed, cleaned in detergent, deionized water, and solvents (such as methanol or proprietary commercial solvents), and finally spun dry. Wafer cleaning is also undertaken with automated scrubber equipment using light pressures to avoid introducing new surface damage. Figure 4.53 illustrates an automated multiwafer scrubber. All the polishing procedures, testing, and packaging operations are carried out under clean-room conditions.

Water polishing technology in silicon and GaAs has advanced to the stage where large-diameter (8-in-diameter in silicon and 4-in-diameter GaAs) wafers can be produced with featureless damage-free surfaces, of flatness within 5 μm or less, and with fewer than 5 particulates over the wafer. Automated laser scanning instruments have been developed to map the flatness, tape, and bow (Fig. 4.54) and the particle contamination on wafers.

4.3.2 Surface and Subsurface Evaluation

Remnant surface and subsurface damage has been a major concern in wafer preparation. The development of wafer preparation procedures for producing GaAs wafers free of residual polishing damage is shown in Fig. 4.55. During grinding, sawing, and lapping material is removed on a macroscopic scale by a harder abrasive material. On a microscopic scale the surface is rough with microcracks and dislocation networks penetrating several tens of micrometers into the bulk wafer. The depth of residual damage after sawing, lapping, and polishing of GaAs was measured by cross-sectional transmission electron microscopy (XTEM).

Free chemical etching, which removes material on a microscopic scale, removes work damage, but in an uncontrollable manner. Most chemical etches preferentially attack damaged or dislocated regions or become locally depleted, resulting in unacceptable flatness and surface morphology. A detailed investigation[57] of chemical-mechanical polishing has been conducted by delineation etching,[73] transmission electron microscopy, and laser backscattering mapping.[74] Delineation etching refers to the use of a sensitive defect-revealing etch for evaluating damage in polished wafers. The etch consists of a dilute (1 percent) bromine-methanol solution to which a small quantity of a surfactant (Ethoquad C/25, manufactured by Akzo Chemie America) has been added. Defects such as microscopic residual scratches become decorated and can be

TABLE 4.12 Process Flow Diagram for Silicon Wafer Preparation

Category	Process step	Measurement
	Cropping	Oxygen
		Carbon
		Resistivity
Crystal shaping	O.D. grinding	Diameter
	Flat grinding	Flat width
	Etching	Surface roughness
		Thermal slip
	Wafering	Orientation
		Thickness
		Flatness, taper
		Bow, damage
Wafer shaping	Heat treatment*	Resistivity
	Edge contouring	Edge Profile
	Lapping*	Thickness
		Flatness
		Taper, damage
	Wafer etching	Surface, roughness
	Polishing	Visual inspection
	Cleaning	Thickness
		Flatness
		Visual inspection
Wafer finishing	Back side damage*	Surface roughness
	Marking*	
	Cleaning	Visual inspection
	Packaging	

*Signifies optional process.

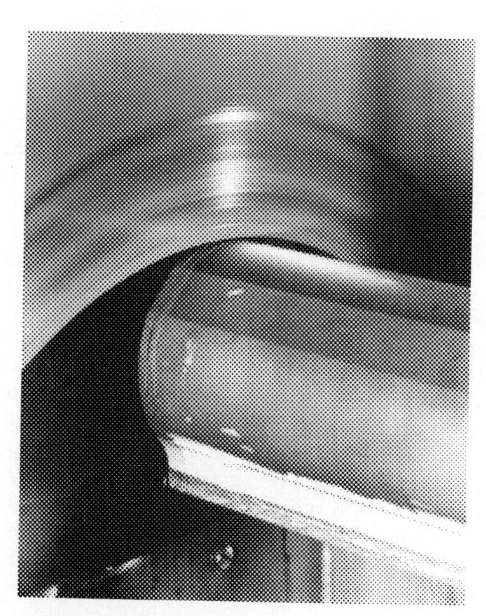

FIGURE 4.50 $5\frac{1}{2}$-in I/D diamond saw slicing wafers from ground GaAs ingot.

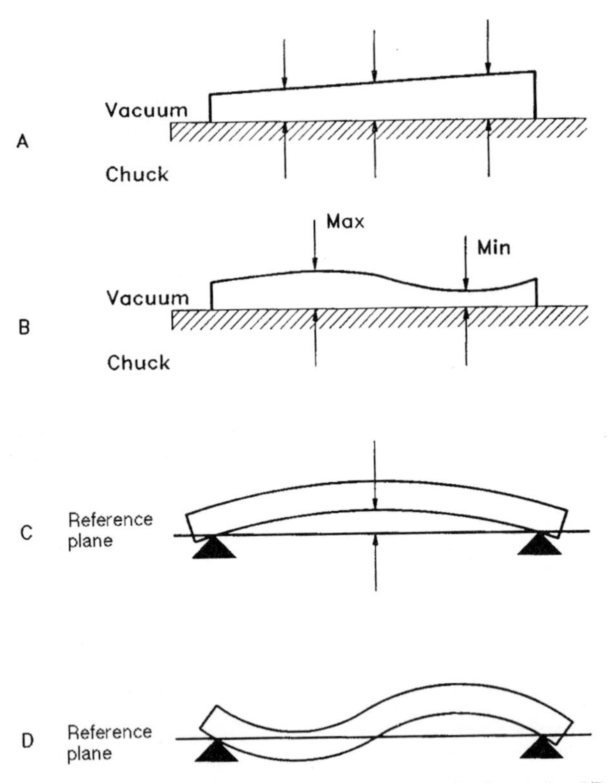

FIGURE 4.51 Possible consequences of blade deflection during I/D diamond sawing causing wafer bow, taper, and warp.[72]

viewed using Normarski or dark-field microscopy. Figure 4.56 illustrates the use of delineation etching to monitor damage removal in a two-stage polishing process developed[75] for GaAs wafer polishing. The methodology employed is that wafer flatness and geometric integrity of the wafer are determined by the sawing and lapping processes, while the polishing processes were tailored to produce flat, featureless surfaces, free of damage, without loss of the desired wafer flatness. Two-sided polishing is first utilized to remove most of the residual damage while maintaining the geometrical integrity of the wafers. Subsequently a single-side polish (which removes less than 10 μm) is applied with very light polishing pressures to produce wafers free of residual damage. Damage-free 3-in-diameter GaAs wafers produced by this process are consistently flat (TIR) to better than 3 μm. Polished GaAs wafers damage readily and subsequent wafer cleaning processes (such as wafer scrubbing) must be closely monitored to ensure that the wafers remain damage-free.

The laser scanning technique developed to assess residual wafer damage consists of scanning a wafer surface in an x–y raster with an He:Ne laser. The measured backscattering of the laser light is utilized to detect and quantify the presence of surface and subsurface damage. This photon backscattering system (PBS)[74] produces wafer maps indicative of the residual polishing damage: wafers free of damage produce uniformly "blue" maps whereas highly damaged wafers are "red." (The colors

FIGURE 4.52 Two-sided multiwafer lapper and polisher.

refer to the digitized color maps of the wafers.) GaAs wafers with "blue" maps (and low backscattering values) consistently showed[75] better device performance and yields compared to "red" wafers. Figure 4.57 illustrates the difference between the uniformity of radial sheet conductance of GaAs wafers implanted uniformly with silicon ions to produce an n-type layer in damaged and undamaged GaAs wafers.

For less developed semiconductors (which are not available as cylindrical ingots) test slices are usually prepared by wire sawing (using fine tantalum or molybdenum wire in an abrasive slurry) and hand lapped. Wafer polishing consists of free etching using a rotation wheel to achieve approximately laminar flow conditions of the etchant over the wafer surface. Bromine-in-methanol is commonly used in laboratory polishing of III-V and II-VI compounds, but is highly corrosive and a major health hazard, so that appropriate precautions must be employed.

FIGURE 4.53 Automated cassette-to-cassette wafer scrubber which utilizes either brush scrubbing or solvent jet cleaning.

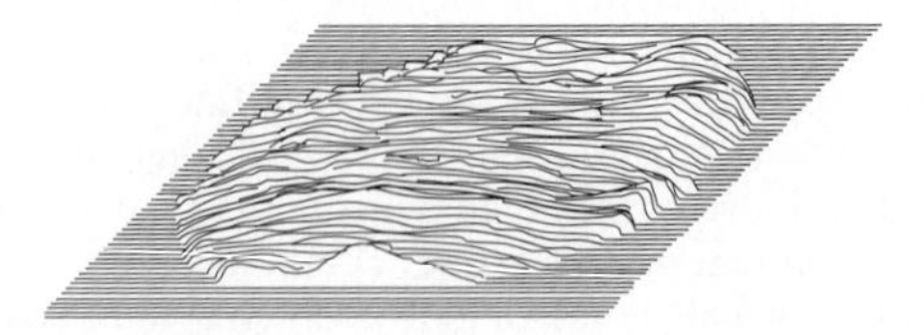

FIGURE 4.54 Laser interferometric mapping of wafer flatness.

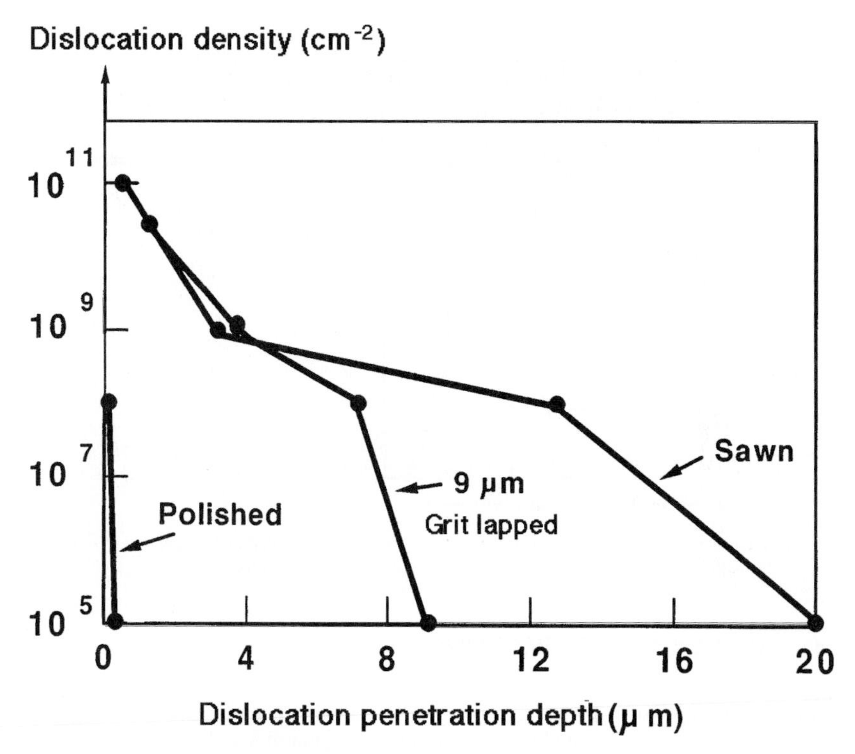

FIGURE 4.55 Cross-sectional TEM of damage depth caused by (*a*) diamond I/D sawing, (*b*) lapping, and (*c*) conventional polishing of GaAs wafers.[57,75]

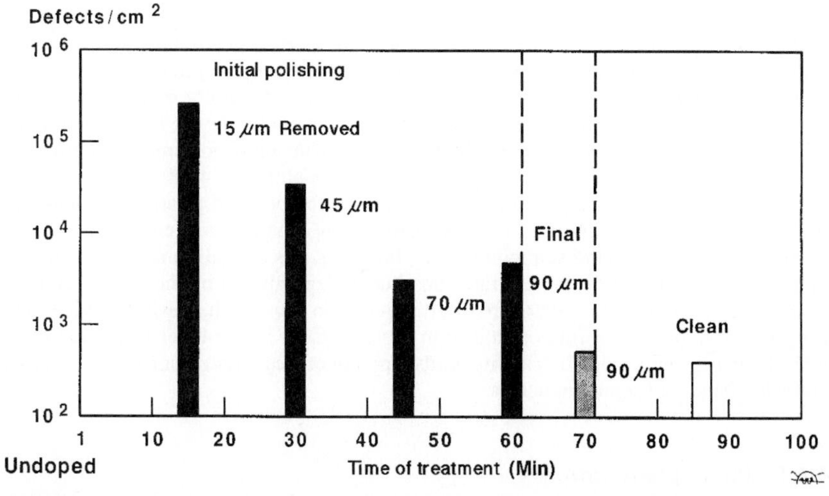

FIGURE 4.56 Reduction in residual surface and subsurface damage in polishing 3-in-diameter GaAs wafers while maintaining less than 3-µm TIR flatness. Damage measured in defects per square centimeter as revealed by delineation etching.[75]

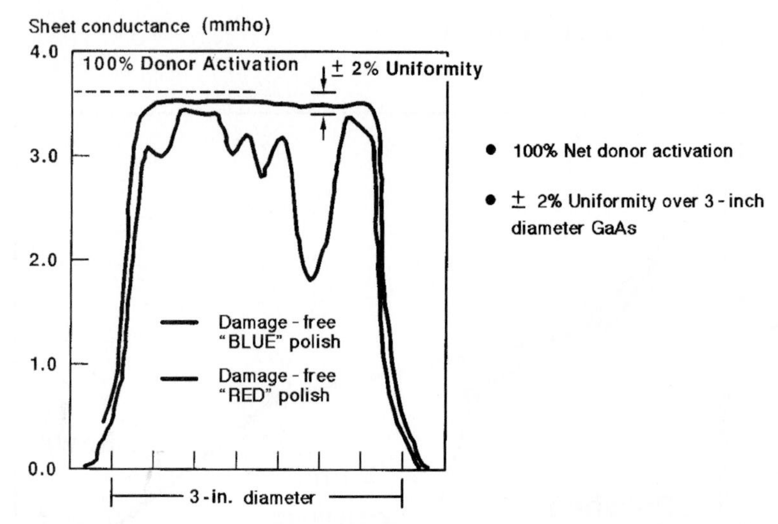

FIGURE 4.57 Correlation of uniformity of *n*-type layer implanted in 3-in-diameter GaAs wafers with low-damage ("blue") and high-damage ("red") PBS characteristics.[57,75]

4.4 ACTIVE-LAYER FORMATION

In almost all solid-state device technologies, especially ICs, the active devices and other components are fabricated in the near-surface region of a wafer within a micrometer or less of the surface. The electrical role of the wafer bulk material is generally a passive one, although it may be either highly conductive to act as a common ground contact, or it may be insulating or semi-insulating to isolate high-frequency devices or circuits. High-voltage power device technology is a notable exception, where the high-resistivity wafer must block the high voltage in the off-condition and contribute to current conduction in the on-condition. Also, other properties of the bulk wafer may be utilized. For example, intrinsic gettering by oxygen-related precipitates in Czochralski silicon is an important phenomenon in IC production.[76] A variety of techniques have been developed for preparing these so-called active layers for device fabrication, either within the surface of a wafer by methods such as diffusion or ion implantation, or by the growth of an epitaxial layer on the surface of the wafer (Fig. 4.58).

In silicon, active-layer techniques are highly developed and permit excellent control of layer quality and purity, thickness, and doping type and concentrations. Active-layer growth of compound semiconductors, however, has an additional dimension and complexity introduced by the separate constituent elements and by the fact that ternary compound semiconductor layers are often required to provide the desired device function. This section is therefore organized to review briefly active-layer formation in silicon, focusing mainly on III-V compounds, and concludes with recent work in narrow and wide band-gap semiconductors.

4.4.1 Ion Implantation

Diffusion and ion implantation enable more heavily doped layers to be formed[76] in the surface of semiconductor wafers. Diffusion processes are well established in silicon,

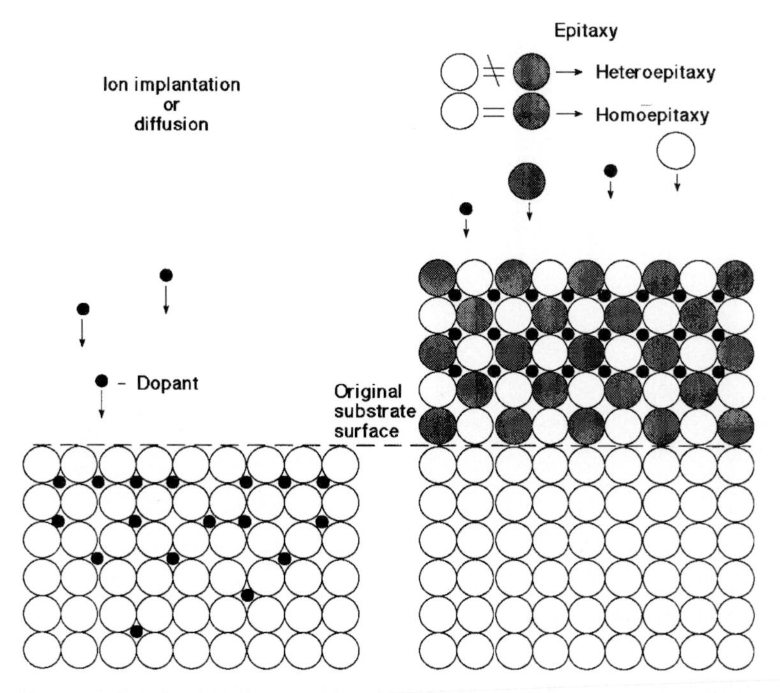

FIGURE 4.58 Active-layer formation. (*a*) Diffusion or ion implantation of dopant atoms below original substrate surface. (*b*) Epitaxy where substrate crystal lattice is replicated by additional semiconductor layers having the same (homoepitaxy) or differing (heteroepitaxy) composition and controlled doping.

while ion implantation has been extended to other semiconductors because of its versatility to precisely add impurity atoms (or, more correctly, add charge which governs device operation) in the formation of active layers.

Ion implantation[77,78] enables positive ions of almost any element to be introduced into a semiconductor substrate. After a series of collisions with the substrate nuclei, usually resulting in considerable disruption to the host, the atom comes to rest at a random depth (and random crystallographic location), which is determined primarily by the kinetic energy, mass, and charge of the incoming ion and the average atomic number and density of the atoms in the solid. Ion implantation can be used to modify surfaces for diverse purposes, including enhanced hardness and corrosion resistance of metals,[79,80] but was first envisioned by Shockley in 1957[81] as a means of doping semiconductors. It took another full decade before numerous laboratories were actively developing ion-implanted silicon and germanium semiconductor devices.[82] It was not until the early 1970s, however, that reasonable success was obtained in the ion implantation of dopants in GaAs.[83,84] This was due to the difficulty in annealing a compound semiconductor without altering its stoichiometry and also, by today's standards, due to the poor quality of the GaAs substrates generally available at the time.

Figure 4.59 contains a schematic of a typical production-type ion implantation system for semiconductors. It includes an ion source and an analyzer magnet which allows the selection of a particular ion by its charge-to-mass ratio. The selected ion is then accelerated to the desired energy (typically 50 to 400 keV), focused into a beam,

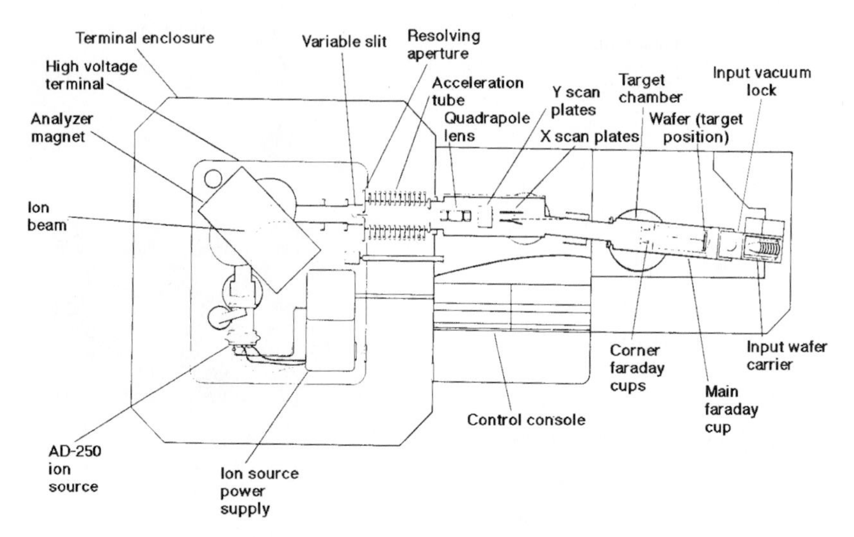

FIGURE 4.59 Schematic of typical production ion implantation system for semiconductors. (*Courtesy Varian Associates.*)

and rastered electrostatically across the semiconductor substrate in a uniform pattern. The target position is mounted at an angle from the source of the ion beam so that any neutral species generated by the ion beam will not impinge upon the substrate. The dose (the number of ions per unit area which are implanted into the wafer surface), typically 10^{12} to 10^{14} ions/cm^2 is measured by integrating the current entering a Faraday cup. Commercial ion implanters are automated so that a cassette of wafers can be implanted with ions having a series of energies and doses. Throughputs are typically quite high, exceeding tens of wafers implanted per hour.

Implanted ions lose energy and slow down as they enter a substrate by two mechanisms, elastic scattering by the substrate nuclei and inelastic scattering by the electrons, involving energy loss. The resulting doping profile or distribution of dopant atoms as a function of depth in the semiconductor for a single implant is approximately gaussian,

$$N(x) = \frac{N_t}{\sigma\sqrt{2\pi}}\, e^{-(x-R)/2\sigma^2} \tag{4.24}$$

where $N(x)$ is the doping density at depth x, N_t the total dose, σ the standard deviation of the dopant distribution, and R the projected range perpendicular to the wafer surface. In the gaussian limit, the peak implant doping density in atoms/cm^2 is simply obtained by the equation

$$N_p = \frac{N_t}{\sigma\sqrt{2\pi}} \tag{4.25}$$

The projected range and standard deviation of an implant can be determined from tables[81] resulting from theoretical calculations assuming implantation into amorphous solids based on the work of Lindhard, Scharff, and Schiott (LSS), or by means of curves such as those presented in Fig. 4.60 which more correctly take into account the

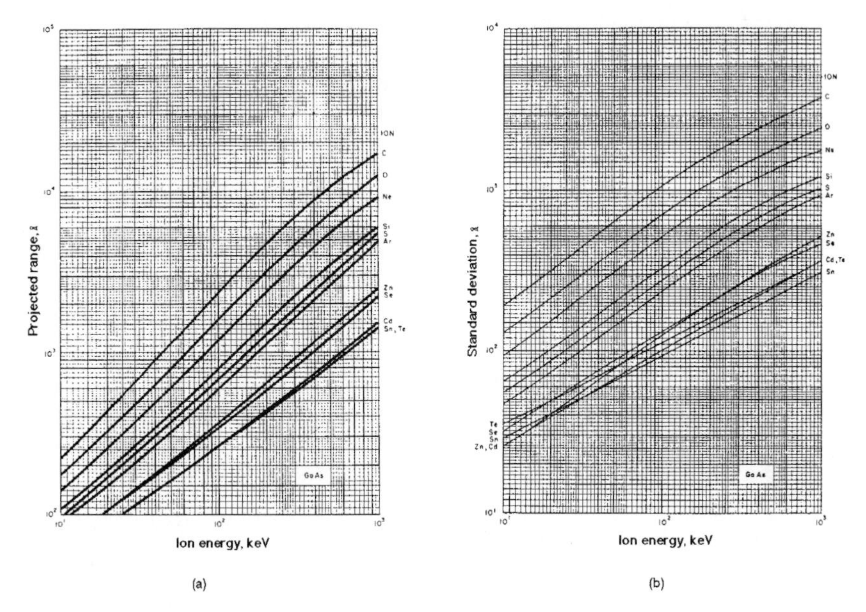

FIGURE 4.60 (*a*) Projected ranges and (*b*) standard deviations as a function of ion energy for various elements implanted into GaAs.[81,85]

electronic stopping mechanism in ion implantation.[85] While any single implant profile will be limited to an approximately gaussian shape, several implants of tailored ion energy and doses can be combined to yield more complex profiles. A unique feature of ion implantation is that when combined with photolithographic patterning steps, it can produce different active-layer doping profiles adjacent to one another on the same wafer while maintaining the planar surface that facilitates subsequent processing steps.

One complication, which can significantly alter the actual doping profile from simple gaussian, is known as channeling. Semiconductors are monocrystalline solids, so that depending on the orientation of the crystal, channels of unoccupied spaces exist for the impinging ion beam, which can result in dramatic increases in the depth the ions reach inside the crystal. To minimize the effects of channeling, wafers are tilted 6 to 9° and rotated more than 30° off axis to present an amorphouslike target.[86] This is successful at eliminating channeling for high-energy ions, but less so for low-energy ions. Even if the ion beam is originally off-axis with regard to the crystal channeling direction, scattering will result in some ions being redirected into a channel. These factors result in undesirable exponential tails of dopant atoms extending significantly farther into the semiconductor than predicted by a simple gaussian distribution, as shown in Fig. 4.61.

After implantation, a wafer must be annealed to eliminate the significant damage induced by the ion beam and to allow the dopant atoms to diffuse to the appropriate lattice sites where they will be electrically active as donors or acceptors. The two most common techniques for annealing implanted wafers are furnace anneals and rapid thermal annealing (RTA). For silicon, an elemental semiconductor, the anneal step poses little difficulty. GaAs, however, needs to be protected from the loss of arsenic, which is thermodynamically favored at the 800°C temperatures required to anneal GaAs.[87,88]

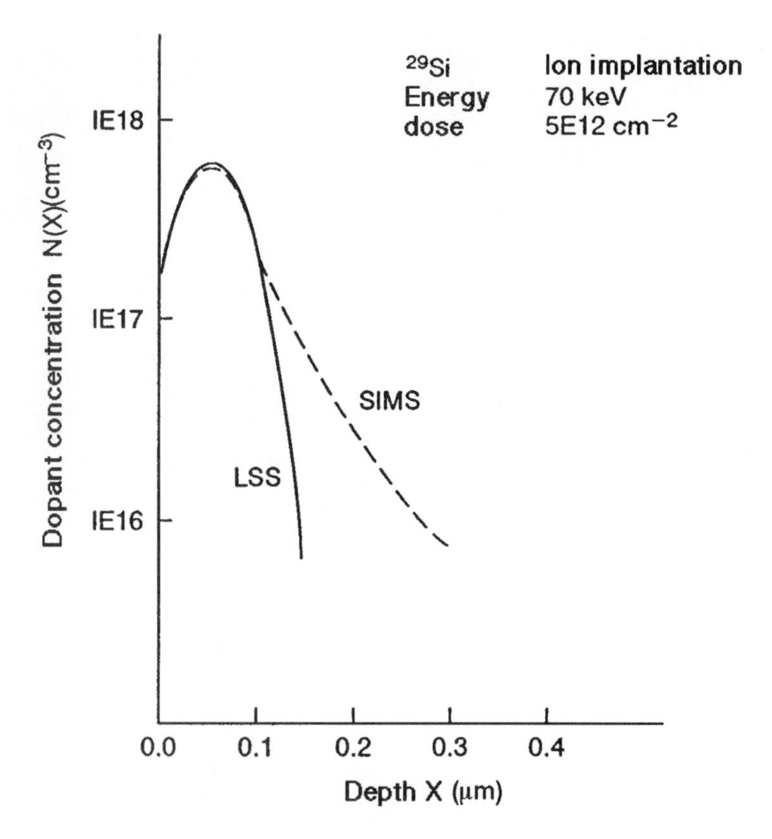

FIGURE 4.61 Comparison of LSS predictions of ion implantation profile for ^{29}Si into GaAs and experimental SIMS measurements demonstrating exponential dopant tail caused by channeling.[78]

The final distribution of dopant atoms in an ion-implanted active layer is the result of both ion implantation scattering processes and the diffusion which occurs to a greater or lesser extent during the annealing step. The use of RTA, however, can minimize the contribution of diffusion of the implanted species. Electrical activation in ion-implanted active layers depends on the implanted species occupying usually a substitutional site during the annealing, and a 100 percent efficiency is rarely achieved. In addition, background impurities in semiconductor substrates have a major impact on the free carrier distribution in the low doped tails of the doping profiles, causing a major impact in device performance.[89] For ion-implanted *n*-type active layers intended for GaAs metal semiconductor field effect transistors, coimplants of beryllium have been employed successfully to both sharpen and control the free-carrier distribution on the substrate side of the active channel in order to enhance device pinchoff characteristics, tighten threshold voltage distribution, and increase gain.[90]

4.4.2 Epitaxy—General Considerations

The alternate method for producing active layers is to grow additional semiconductor layers having the desired electrical and optical properties on the host substrate (Fig. 4.58). The requirements for epitaxial growth were determined by Royer[91] in 1928 with the use of X-ray diffraction measurements. In addition to reproducing the substrate's crystalline symmetry, the difference in the bulk lattice constants of the substrate a and the overgrowth b could not exceed about 15 percent for epitaxial growth to occur,

$$\epsilon = \frac{100(b - a)}{a} \tag{4.26}$$

In epitaxial growth, if the semiconductor film has the same composition as the substrate, the process is referred to as homoepitaxy. If the growing layer has a different composition from the substrate material, the process is referred to as heteroepitaxy. In contrast to the vast majority of deposition processes where the growing film structure does not depend on the crystal structure of the substrate, epitaxy requires that the growing layer reproduce the planar spacing and periodicity of the semiconductor substrate exactly.

The earliest model of epitaxial growth was developed by Frank and van der Merwe (FM), who postulated a two-dimensional layer-by-layer growth mode (Fig. 4.62). Alternate growth modes also shown in Fig. 4.62 are Volmer-Weber growth, where three-dimensional nuclei are formed immediately upon the bare substrate, and Stanski-Krastanov growth, where the initial growth is two-dimensional, followed by three-dimensional growth.[92] Upon coalescence of the growth nuclei all these modes have been shown to yield epitaxial layers. The operative growth mode is determined by numerous crystallographic, thermodynamic, and kinetic factors, including degree of lattice mismatch, crystallographic orientation of the substrate, relative surface energy of the substrate and epilayer, and the sticking coefficients and diffusion rates of adatoms. These factors determine whether it is more likely for growth to occur by nucleation on a surface, or by lateral growth of an existing epitaxial monolayer island at a kink or an edge (step) site.

While not the only epitaxial growth mode, the theory of Frank and van der Merwe does provide a means of rationalizing the characteristics of epitaxial layers in general, and an important class of epitaxial materials in particular, known as pseudomorphic or strained layer semiconductors. In pseudomorphic layers, the misfit of the bulk values of the substrate and the epitaxial layer are large, but the epitaxial layer still conforms to the substrates in plane lattice spacing. In these strained layers, as shown in Fig. 4.63, the epitaxial overgrowth takes place by matching with the in-plane lattice spacing of the substrate, while the approximate bulk density is maintained by a tetragonal dislocation. In the case shown, the lattice spacing perpendicular to the crystal face is elongated if the epitaxial layers in plane bulk lattice constant are larger than those of the substrates, and is compressed if the epitaxial layers in plane bulk lattice constant are smaller than those of the substrate. The epitaxial layer is elastically strained in this fashion up to a critical thickness h_c, after which dislocations in the epitaxial layer are created to relieve the strain. Epitaxial layers are then said to be "relaxed" as their lattice constants approach their bulk values (strictly speaking, the relaxed layers should no longer be called epitaxial layers). The ability to grow strained layers increases greatly the number of heterostructure layer combinations available for active layer formation, particularly given the limited types of commercially available semiconductor substrates. Importantly the strain present in these layers also alters their optical and electrical properties to values not available in natural-

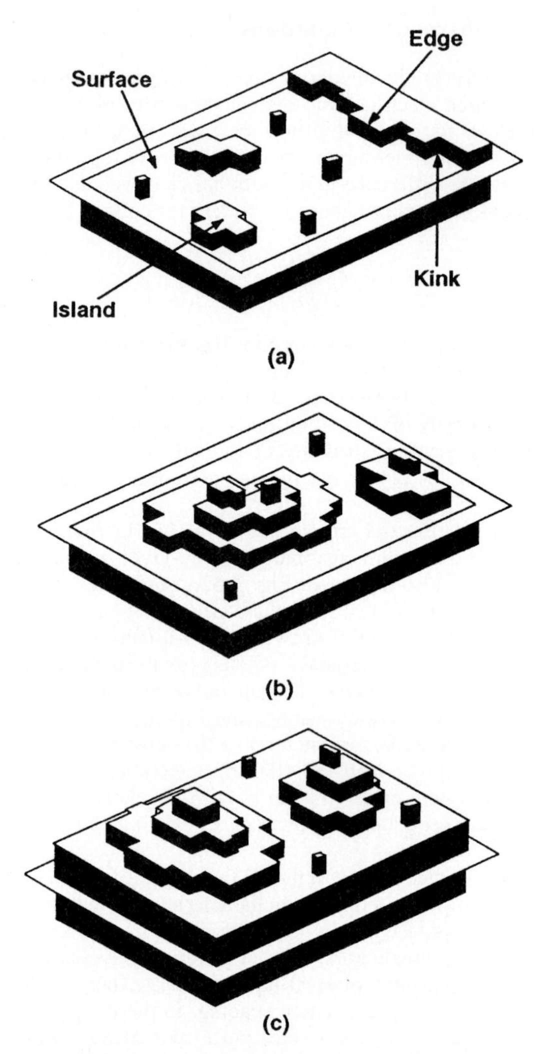

FIGURE 4.62 Three epitaxial growth modes of overlayers on a semiconductor substrate. (*a*) Frank–van der Merwe (FM) or two-dimensional growth. (*b*) Volmer-Weber (VW) or three-dimensional growth. (*c*) Stanski-Krastanov (SK) or two-dimensional followed by three-dimensional growth. Horizontal plane indicates substrate-epitaxial layer interface.[92]

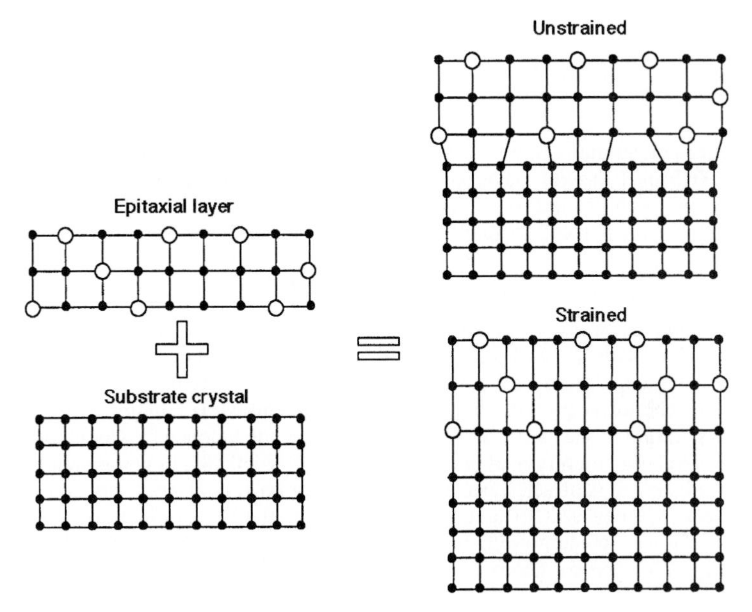

FIGURE 4.63 Schematic representation of strained layer or pseudomorphic epitaxy.[101]

ly occurring semiconductor compounds, offering even more variety in the selection of active-layer characteristics.

4.4.3 Silicon Epitaxy[93]

The commonly used techniques of silicon epitaxial growth today consist of chemical-vapor deposition and molecular-beam epitaxy. Solid-phase regrowth is another technique that is finding increasing use for high-frequency devices.

Chemical-Vapor Deposition.[94,95] The chemical vapor deposition (CVD) of silicon is carried out by reacting silicon-containing gases over heated silicon wafers in a reaction chamber. High-quality homoepitaxy occurs if the silicon surfaces are clean and the temperature is high enough to impart surface mobility to the depositing atoms. Commercial silicon epitaxy employs either the reduction of chlorosilanes (usually trichlorosilane $SiHCl_3$, because of its lower cost), or the pyrolytic decomposition of silane SiH_4.

All the available chlorosilanes can be employed for silicon epitaxy. Dichlorosilane decomposes according to

$$SiH_2Cl_2(v) \rightarrow Si(s) + 2HCl(v) \tag{4.27}$$

and epitaxy occurs at 1050 to 1150°C. Prior to growth, native oxide and water vapor is desorbed by heating the reactor and the silicon wafers in hydrogen at these temperatures. Trichlorosilane and silicon tetrachloride also use a hydrogen carrier gas to undergo reduction according to

$$SiHCl_3(v) + {}^3/_2\, H_2 \rightarrow Si(s) + 3HCl(v) \tag{4.28}$$

and

$$SiCl_4(v) + 2H_2 \rightarrow Si(s) + 4HCl(v) \tag{4.29}$$

The higher temperatures required with the higher chlorinated silanes, 1100 to 1250°C for epitaxial-layer growth, have become a major disadvantage in forming the abrupt, narrow doping profiles necessary for silicon VLSI circuits.

In contrast, the pyrolytic reduction of silane occurs at lower temperatures,

$$SiH_4(v) \rightarrow Si(s) + 2H_2(v) \tag{4.30}$$

At 850 to 900°C, epitaxial silicon forms, while at temperatures below 600°C, silicon is deposited as polysilicon. Epitaxy is carried out at atmospheric pressure to $^1/_{10}$ atm pressure using a hydrogen-reducing ambient to remove and avoid the regrowth of surface oxide. The lower growth temperature of silane epitaxy is a major advantage, and laboratory ultrahigh vacuum CVD systems have been developed,[96] which are capable of defect-free silicon epitaxy at temperatures as low as 600°C. In commercial production this advantage is offset, however, by the higher cost of silane, the need for higher reactor leak integrity, and difficulties associated with the formation of powdery silicon deposits in lower temperature zones of the reactor. Figure 4.64 shows schematically some of the CVD reactor geometries in use today.

Molecular-Beam Epitaxy.[94,95] In molecular-beam epitaxy (MBE) single-crystal layers are formed in ultrahigh vacuum by directing a stream of atoms from an evaporating source at the heated substrate. The condensing atoms will diffuse about on the surface until they reach a low-energy site where the "adatoms" bond and extend the underlying crystalline lattice by a vapor-to-solid-phase layer growth. In the case of silicon, temperatures[97] above 350°C are sufficient to allow the growth of high-quality epitaxial layers.

Silicon MBE has grown in importance over CVD, especially for VLSI processing, for this very reason. The typical 600 to 700°C growth temperatures that are used in silicon MBE reduce out-diffusion of, for example, buried diffused layers into the epitaxial layer or between differently doped layers. The result is that MBE is particularly suited for precise submicrometer layered structures requiring ultra-abrupt doping transitions.

A silicon MBE system shown schematically in Fig. 4.65 consists of an ultrahigh vacuum chamber (10^{-9} torr pressure or lower), an electron-gun silicon source, and Knudsen effusion cells to provide dopants. The evaporated silicon and dopant beam are directed at a heated silicon wafer. The dopant beams generated by the Knudsen cell usually employ boron, aluminum, or gallium for p-type doping, and antimony for n-type. The sticking coefficients of phosphorus and arsenic are too low and their evaporation rates too rapid for controlled n-doping. The crystalline perfection of silicon MBE layers is highly dependent on achieving silicon surfaces completely free of oxide prior to the start of film growth. In-situ cleaning can be accomplished by high-temperature annealing at 1100 to 1250°C for several minutes or by low-energy inert-gas ion sputtering, followed by a short anneal at 800 to 900°C to reorder the surface damage caused by the sputter clean. Lower-temperature cleaning methods are preferred in device processing, particularly if the silicon wafers are oxide patterned or have buried diffused layer structures in them. Ex-situ methods of "surface passivating" have been developed[98] (by high-frequency etching immediately prior to introducing the wafers into the chamber), which yield atomically clean surfaces with more modest annealing temperatures of 600 to 800°C in ultrahigh vacuum.

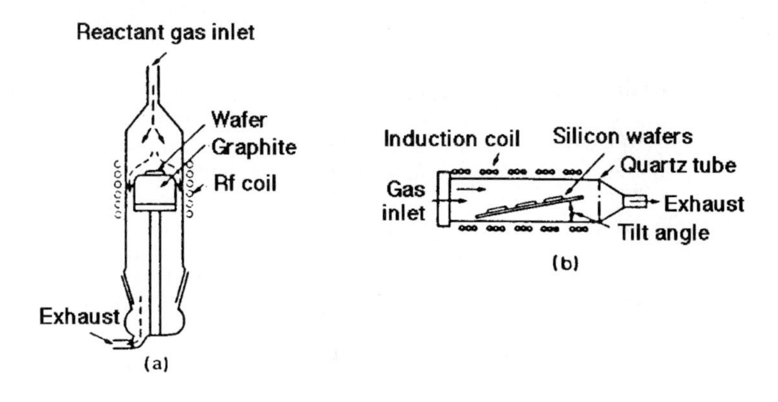

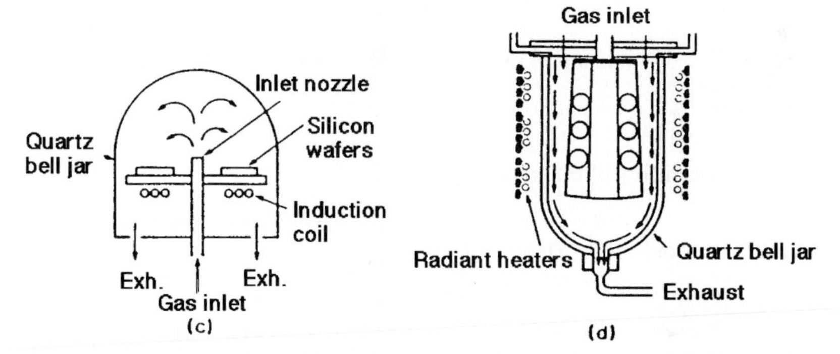

FIGURE 4.64 Schematic silicon CVD reactor geometries. (*a*) Vertical flow. (*b*) Horizontal flow. (*c*) Modified vertical (or pancake) reactor. (*d*) Down-flow cylindrical reactor.[76]

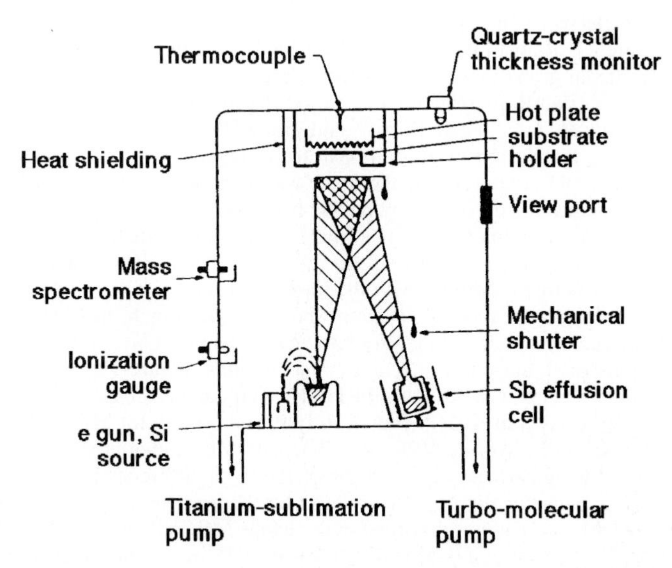

FIGURE 4.65 Typical arrangement of silicon MBE growth system.[76]

Commercial silicon MBE equipment today can handle multiple 4-in wafers and typically comprises multiple chambers, one or more for film growth and another for in-situ preparation and surface analysis. Cassette entry locks and interchamber wafer transfer mechanisms are available to maintain vacuum at all times.

Solid-Phase Regrowth. When silicon is subjected to high-dose ion implants, the surface layer becomes amorphous by the heavy ion damage to the crystalline lattice. Annealing above 600°C will, however, cause the amorphous layer to recrystallize from the interface with the underlying substrate. The reordered interface moves toward the surface, resulting in solid-phase epitaxial regrowth.

A particular application of this technique is the so-called SIMOX process.[99] Silicon wafers are implanted with oxygen at doses of 2×10^{18} cm^{-2} at 600°C and subsequent annealing at 1300°C for 4 to 10 h, to form a thin (several 0.1-μm) epitaxial silicon layer which is separated from the host silicon substrate by a 0.4-μm-thick buried oxide layer. This is useful for electrically isolating the active layer in high-frequency devices. By using high-resistivity silicon wafers for the SIMOX process, silicon CMOS devices and circuits operating at frequencies up to 22 GHz have been demonstrated recently.[100]

4.4.4 III-V Compound Epitaxy

Various epitaxial growth techniques have also become highly developed for III-V compound semiconductors. These include liquid-phase epitaxy (LPE), two common inorganic vapor-phase epitaxies (Cl-VPE and hydride-VPE), metalorganic chemical-vapor deposition (MOCVD), molecular-beam epitaxy (MBE), and various hybrid techniques combining some of the best features of MOCVD and MBE.

Figure 4.66 contains a plot of the band-gap and lattice constants for the III-V compound semiconductors. For the sake of clarity, only the III-V-based tertiary and quaternary compounds are shown by the connecting lines. Also included in Fig. 4.66 are the analogous elemental column IV, column IV-VI, and column II-VI compounds. The options in electrical and optical properties represented by the semiconductors listed in this figure are almost limitless.[101,102] Table 4.13 contains a preview of the attributes of the epitaxial growth technologies which are utilized in forming semiconductor active layers.[103] Excellent reviews of epitaxial growth have been published by Pashley,[104] Peercy et al.,[92] Stringfellow,[103] and Tsang.[105]

In considering the design of an epitaxial active layer beyond the previously discussed crystallographic requirements there are also thermodynamic and kinetic considerations as to whether the desired semiconductor can be fabricated by a certain growth technique.[106] Thermodynamics serve as a guide in predicting whether a particular epitaxial layer of a given alloy composition can be produced by a given reaction as long as the growing layer is in equilibrium with the immediate source of reactants. Under normal growth conditions used for III-V compound epitaxial growth, this basic condition is satisfied. Thermodynamics can therefore predict the epitaxial layer alloy composition given the vapor- or liquid-phase reactant concentrations. Figure 4.67 shows the aluminum and indium content in AlGaAs, InGaAs, InGaP, and AlInAs, and the phosphorus and antimony content in AlAsP, GaAsP, InAsP, and InAsSb for MOCVD.[107]

While LPE, Cl-VPE, hydride-VPE, MOCVD, and MBE are approximately at equilibrium at the growing surface, Fig. 4.68 shows the large difference in the net thermodynamic driving force for the growth of GaAs by these processes. LPE and Cl-VPE have very low thermodynamic driving forces, while MOCVD and MBE have much higher impetus to deposit epitaxial layers. Hydride-VPE is an intermediate case. The

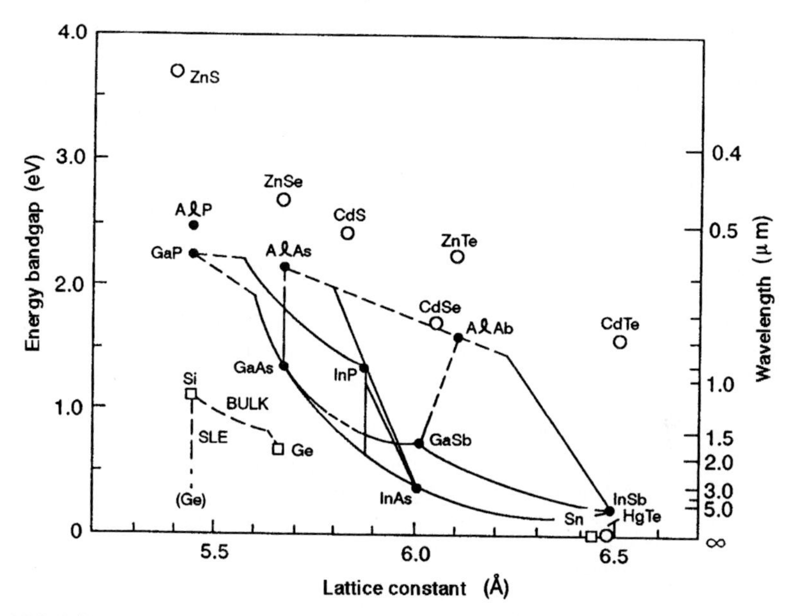

FIGURE 4.66 Band-gap energy and wavelength versus lattice constant for elemental group IV semiconductors and some of the analogous IV-IV, III-V, and II-VI compound semiconductors. Connecting lines (shown only for group IV and III-V) denote alloy semiconductors; solid and dashed lines indicate direct and indirect band gaps, respectively.[101,102]

large driving force of both MOCVD and MBE allows atomically abrupt heterostructure interfaces, and under altered growth conditions (low V/III ratios) the production of interesting thermodynamically disallowed alloys such as $GaAs_{1-x}Sb_x$, where $0.8 > x > 0.2$. The much smaller driving force in LPE and Cl-VPE can also be advantageous. By simply increasing the temperature of the substrate (or altering the gas composition in VPE) one can reverse the deposition reaction, allowing for in-situ etching of substrates or for epitaxially grown layers to obtain improved interfacial properties.[108] The low thermodynamic driving forces in LPE and VPE are also conducive to truly selective epitaxy, where deposition occurs only on the exposed semiconductor surface and not on the masking material. The addition of a halogen to the chemistry of MOCVD either as $AsBr_3$[109] or as part of a novel gallium source, diethylgallium chloride,[110] can impart these same characteristics to MOCVD.

Liquid-Phase Epitaxy. Liquid-phase epitaxy (LPE)[111] is a simple, safe, and inexpensive way of growing small samples of epitaxial active layers. In LPE a semiconductor substrate is brought into contact with a slightly supercooled liquid (1°C) of the semiconductor to be deposited. The composition of deposited layers is predicted by thermodynamic phase diagrams.[112] Figure 4.69 contains a schematic of a typical LPE apparatus.[113] Due to its simplicity and low levels of oxygen contamination, LPE was the first successful means of producing high-quality GaAs/AlGaAs heterostructures. GaAs/AlGaAs was the archetypal heterostructure because of a near perfect lattice match between GaAs and AlAs (5.6532 Å and 5.0607Å, respectively), resulting in

TABLE 4.13 Comparison of Relative Capabilities Demonstrated by Epitaxial Growth Techniques

Technique	Strengths	Weaknesses
LPE	Simple; high purity	Scale economics; inflexible; nonuniformity; ≥20-Å interface widths
Cl-VPE	Simple; high purity	No Al alloys; Sb alloys difficult; ≥20-Å interface widths
HVPE	Well-developed; large scale	No Al alloys; Sb alloys difficult; hazardous sources; graded interfaces; complex process/reactor; control difficult
MBE	Simple process, uniform; high-purity GaAs; abrupt interfaces; in-situ monitoring; moderate scale	As/P alloys difficult; oval defects; poor InP purity
MOCVD	Most flexible; high purity; abrupt interfaces; low interfacial recombination velocity; selective growth; uniform; large scale; simple reactor	Expensive reactants; most parameters difficult to control accurately; hazardous precursors (especially using hydrides)
MOCVD/MBE	Combines flexibility of room-temperature vapor sources of MOCVD with high uniformity, simple process, and in-situ monitoring of MBE	Still under development; some require hydrides

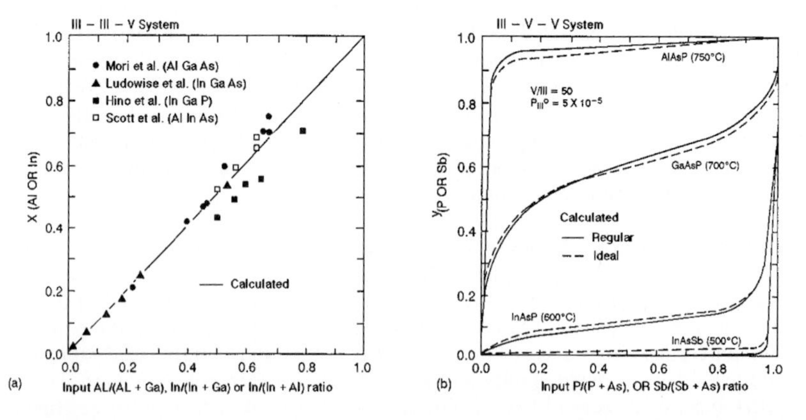

FIGURE 4.67 Comparison of experimental and calculated compositions. (*a*) III-III-V systems as a function of input group III mole ratio. (*b*) Calculated compositions of III-V-V systems as a function of input group V mole ratio.[107]

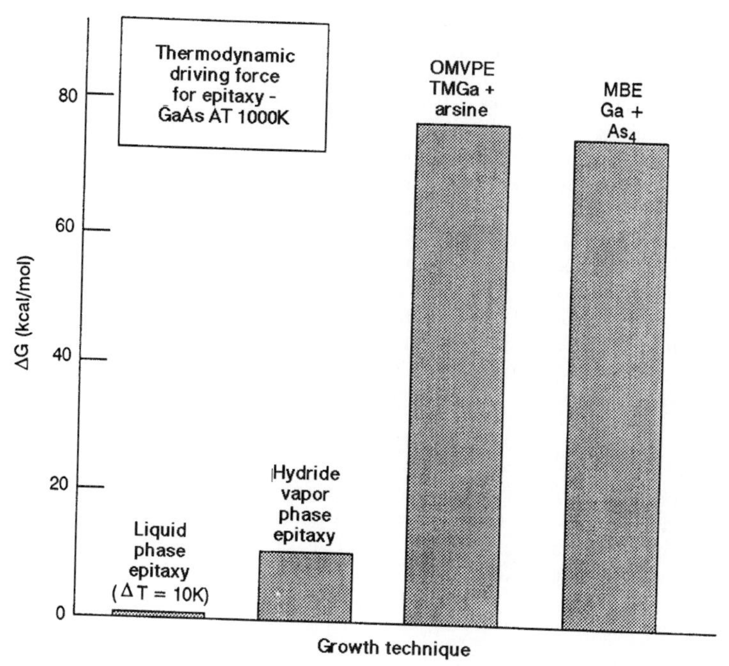

FIGURE 4.68 Estimated thermodynamic driving force, Gibbs free-energy difference between reactants and products, for LPE, HVPE, OMVPE (MOCVD), and MBE growth of GaAs at 1000 K.[103]

near ideal interfaces, and because the alloy system has a direct band gap ranging over the useful range of 1.4 to 2.0 eV.

In order to produce multiple layers by LPE it is necessary to move the substrate into contact with a series of melts. Despite some very creative attempts, the difficulty in dealing with large substrates and multiple melts has limited LPE layer control to several 100Ås with poor uniformities. Furthermore LPE active layers suffer from morphological defects known as meniscus lines, a consequence of the liquid-solid contact. LPE growth, however, is capable of producing layers of high minority carrier lifetimes

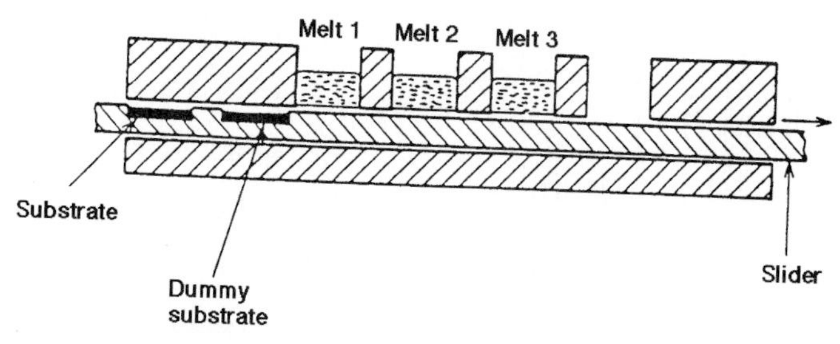

FIGURE 4.69 Typical LPE growth apparatus with three separate wells for differing epitaxial layers. Assembly is placed in multizoned furnace and blanketed with hydrogen gas flow.[113]

and has until recently been the commercial standard for the production of discrete GaAs/AlGaAs and InGaAsP/InP double heterostructure lasers and LEDs.[114,115]

Chloride-Vapor-Phase Epitaxy. Chloride-vapor-phase epitaxy (Cl-VPE) is historically the first vapor-phase growth technique able to grow ultrahigh-purity GaAs, and as a result has had wide commercial application in microwave devices where purity is essential. Originally described by Knight[108] in 1964, this growth technique, as depicted in Fig. 4.70, consists of a hot-wall quartz reactor containing a boat of gallium liquid or solid GaAs in an 820°C source region and a GaAs substrate (seed crystal) held at a lower 720°C temperature. High-purity hydrogen carrier gas from a palladium cell diffuser is bubbled through $AsCl_3$, which has a vapor pressure approximately equal to that of water. The carrier gas now saturated with $AsCl_3$ vapor is admitted to the quartz reactor with a large excess of additional hydrogen carrier gas. A portion of the $AsCl_3$ vapor (referred to as the growth $AsCl_3$) is passed over the gallium or GaAs solid source, where it is thermally cracked into As_4 and HCl according to

$$2AsCl_3 + 3H_2 \rightarrow 6HCl + {}^1/_2\, As_4 \tag{4.31}$$

If the source contains liquid gallium the As_4 dissolves into the gallium until saturated in dissolved arsenic. A thin GaAs crust forms on the surface of the liquid gallium. In the case of the solid GaAs source, this step in the process is not required as the source is already saturated in arsenic.

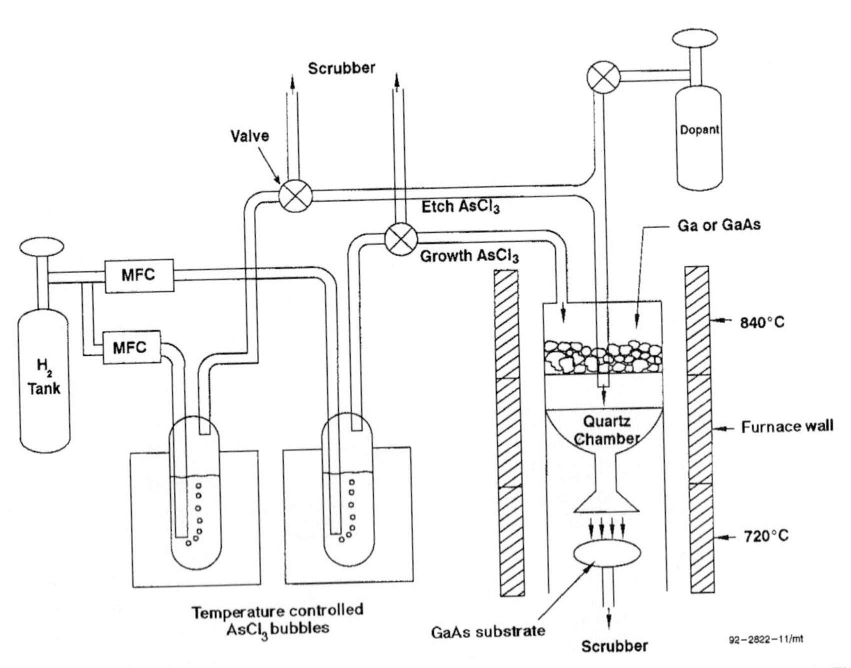

FIGURE 4.70 Schematic of chloride-VPE reactor for GaAs growth. In hydride-VPE reactor $AsCl_3$ would be replaced by arsine and hydrochloric acid.[108]

Once GaAs is available in the source zone, the HCl formed is then able to react, forming volatile GaCl and As_4 according to

$$GaAs(s) + HCl(g) \rightarrow GaCl(g) + \tfrac{1}{2} H_2(g) + \tfrac{1}{4} As_4(g) \qquad (4.32)$$

A second source of $AsCl_3$ vapor, (referred to as etch $AsCl_3$), is allowed to bypass the source material, providing an additional source of HCl and As_4, which allows in-situ wafer etching, growth rate control, and reduced unintentional doping levels in the resulting GaAs epitaxial layers.[116] GaAs materials resulting from Cl-VPE have always been recognized as having the ultimate purity, with 77 K Hall mobilities exceeding 200,000 $cm^2/V \cdot s$ and net carrier concentrations of less than 10^{14} cm^{-3}.[119]

Uniform epitaxial growth across large-area wafers is difficult to obtain with Cl-VPE (and hydride VPE) because the growth rate as a function of temperature offers only a narrow temperature window where growth rate is relatively constant,[117] and because typically VPE is performed at atmospheric pressure. Even in the region of relative temperature insensitivity, where the deposition reaction is diffusion limited, this results in increased sensitivity to the difficult-to-control gas flow.[118] Another difficulty can be eliminated by replacing the liquid gallium source with its fluctuating GaAs crust surface area with a solid GaAs source. The implementation of solid sources has also been successful in the much more demanding growth of GaInAsP layers lattice matched to InP.[119]

Hydride-Vapor-Phase Epitaxy. Hydride-vapor-phase epitaxy (H-VPE) is chemically very similar to Cl-VPE with one major distinction. Instead of generating the needed column V element and HCl from the same source ($AsCl_3$ or PCl_3), the column V element is delivered as a hydride (AsH_3 or PH_3) and HCl is supplied directly from a tank source. The HCl is passed over boats containing the column III gallium or indium to form the needed volatile GaCl or InCl.

The hydride technique results in one major improvement over the chloride technique, that being an independent control of the column III and V vapor concentrations facilitating the growth of GaAsP and GaInAsP. In general the hydride technique results in lower-purity material than the chloride technique due to the variable purity of compressed HCl and AsH_3 source gases. (MOCVD is still impacted by variable AsH_3 purity.) Given careful control and purified sources, however, H-VPE can match Cl-VPE's purity levels.[120] H-VPE's principal application has been in the commercial production of discrete GaAsP LEDs, where background doping levels are not as critical as they are in microwave devices.[121]

The principal drawback of these vapor-phase epitaxy methods is their chemical inability of growing AlGaAs layers. This is due primarily to the disparate temperatures needed to grow AlAs >1000°C and GaAs 700°C using Cl transport and the destructive chemical reaction of the quartz reactor walls with the highly reactive AlCl.[122,123] It was this inability to grow AlGaAs layers that led to the development of MOCVD and MBE active-layer growth techniques.

Metalorganic Chemical-Vapor Deposition. Metalorganic chemical-vapor deposition (MOCVD)[124] together with MBE are the dominant active-layer growth technologies for advanced heterostructure active-layer devices. MOCVD is also referred to as organometallic vapor-phase epitaxy (OMVPE) and metallorganic vapor-phase epitaxy (MOVPE). It describes the pyrolytic decomposition of an organometallic compound and a metal hydride over a heated semiconductor substrate to form an epitaxial film and volatile organic by-products. A schematic of a typical MOCVD reactor is contained in Fig. 4.71. In MOCVD, as in the hydride VPE technique, a large excess of the

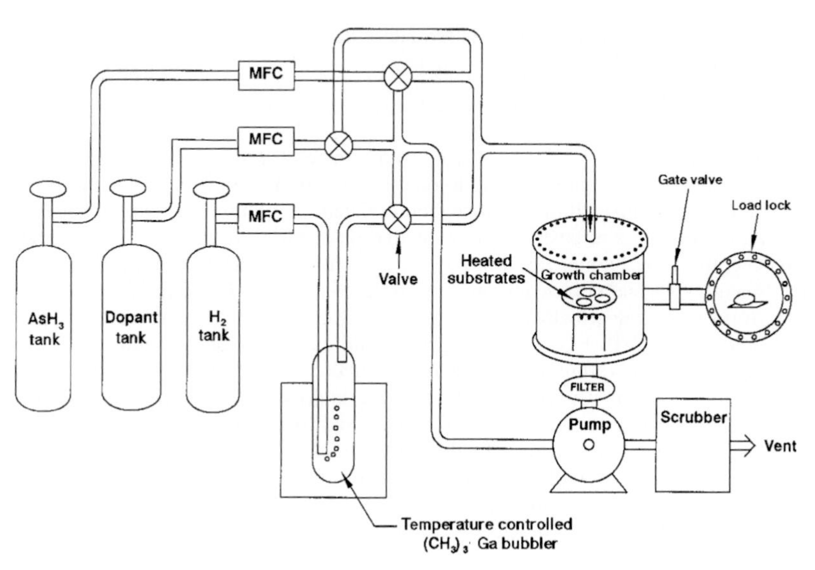

FIGURE 4.71 Schematic of an MOCVD reactor.

column V element is delivered as a hydride, for example, as arsine, AsH_3, or phosphine, PH_3. The column III species, however, is always delivered as an organometallic vapor via saturated hydrogen carrier gas bubbled through the liquid organometallic. The organometallic, typically trimethyl gallium $(CH_3)_3Ga$ for GaAs growth, is maintained at a constant temperature to assure that a consistent and controllable amount of the reagent is delivered to the growth chamber. Due to the irreversible nature of the net MOCVD reaction, the substrate holder is heated inductively (in which case the holder is called a susceptor) or resistively or radiatively, with the heating element being in contact with or in close proximity to the substrate holder. This minimizes unwanted decomposition and deposition on the chamber walls. After the gases flow over the substrate, they typically pass through filters to remove gross particulates and condensates and then into a pump. While MOCVD can be performed at atmospheric pressure, pumps are often employed to reduce growth-chamber pressure to as low as 20 torr, with 76 torr being common. Reduced-pressure growth results in numerous advantages, including faster gas velocities, shorter residence times in the growth chamber, and the inhibition of unwanted convective flow, all contributing to approximate monolayer abruptness in layer composition and doping, and reduced-gas-phase prereaction of the often moderately stable chemical precursors. The extreme toxicity of arsine and phosphine gas require their complete removal from the MOCVD effluent by chemical scrubbing, activated charcoal absorption, or combustion.

The greatest strengths of MOCVD are its extreme versatility in the number of active-layer materials that it can deposit and its relatively low cost. Table 4.14 contains a compilation of many of the organometallic precursors and the semiconducting materials grown by MOCVD.[125] GaAs deposition by MOCVD using starting materials of trimethyl gallium, $(CH_3)_3Ga$, and arsine, AsH_3, proceeds according to

$$(CH_3)_3Ga(g) + AsH_3(g) \rightarrow GaAs(s) + 3CH_4(g) \tag{4.33}$$

TABLE 4.14 Semiconductors Grown by MOCVD[125]

Compound	Reactants	Compound	Reactants
AlAs	TMA1-AsH_3	$Ga_{1-x}In_xAs$	TMGa-TEIn-AsH_3
AlN	TMA1-NH_3	$AlGa_yIn_{1-x-y}P$	TMA1-TMGa-TMIn-PH_3
GaAs	TMGa-AsH_3		TMA1-TMGa-TEIn-PH_3
	TMG-TBA	$InAs_{1-x}P_x$	TEIn-AsH_3-PH_3
	TEGa-AsH_3	$Ga_{1-x}In_xAs_{1-y}P_y$	TEIn-TMGa-PH_3-AsH_3
	DEGaCl-AsH_3	ZnS	DEZn-H_2S
GaN	TMGa-NH_3	ZnSe	DEZn-H_2Se
GaP	TMGa-PH_3	ZnTe	DEZn-DMTe
	TEGa-PH_3	CdS	DMCd-H_2S
GaSb	TEGa-TMSb	CdSe	DMCd-H_2Se
	TMGa-TMSb	CdTe	DMCd-DMTe
InAs	TEIn-AsH_3	HgTe	Hg-DMTe
InP	TEIn-PH_3	HgCdTe	Hg-DMCd-DMTe
	TMIn-PH_3	SnTe	TESn-DMTe
InSb	TMIn-SbH_3	PbTe	TMPb-DMTe
$GaAs_{1-x}P_x$	TMGa-AsH_3-PH_3		TEPb-DMTe
$GaAs_{1-x}Sb_x$	TMGa-AsH_3-SbH_3	$Pb_{1-x}Sn_xTe$	TMPb-TESn-DMTe
	TMGa-TMSb-AsH_3	PbS	TMPb-H_2S
$Ga_{1-x}Al_xAs$	TMGa-TMA1-AsH_3	PbSe	TMPb-H_2S
	TeGa-TMA1-AsH_3		

Figure 4.72 contains a simplified view of the GaAs deposition process.[126] As a trimethyl gallium molecule approaches the heated GaAs substrate, it partially decomposes, losing one or two methyl radicals to form dimethyl or monomethyl gallium intermediates. The gallium-containing species then diffuse toward the surface through a region largely depleted of trimethyl gallium by the facile surface deposition reaction. Arsine being present in excess (typically V/III ratios exceed 30:1), it is not appreciably depleted. The gallium intermediates adsorb on the growing GaAs surface, where they migrate before loosing the final methyl radical. It is generally accepted that arsine decomposes primarily heterogeneously under typical MOCVD growth conditions, adsorbing on the GaAs surface by forming arsenic hydride AsH. In the ideal case then the surface-adsorbed monomethyl gallium and AsH react to deposit GaAs and release gaseous methane. That this is usually indeed the case is consistent with the less than $10^{16}cm^{-3}$ background carbon doping levels that can be obtained in MOCVD deposited GaAs.[127]

As shown in Fig. 4.73, there exists a wide range of temperatures from 550 to 750°C where the growth rate of GaAs is only a weak function of temperature. The relatively temperature-insensitive rate of diffusion across the depleted layer by the galli-

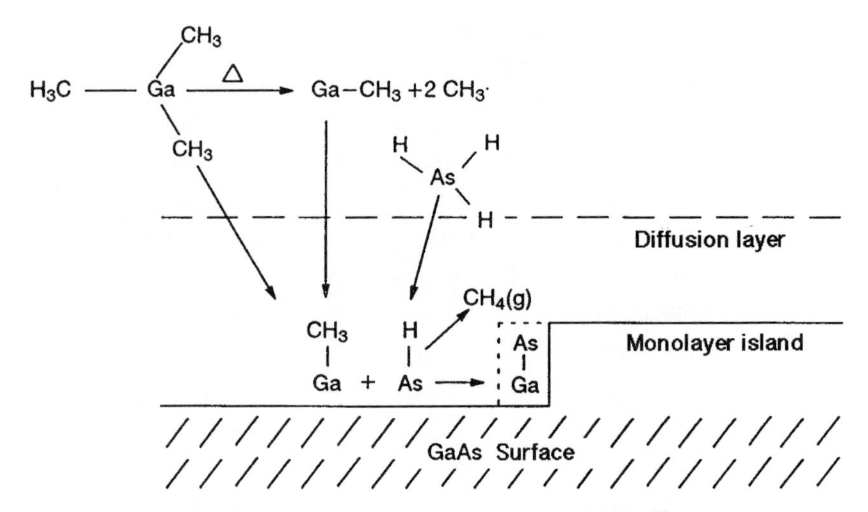

FIGURE 4.72 Simplified view of the GaAs deposition process in MOCVD.[126]

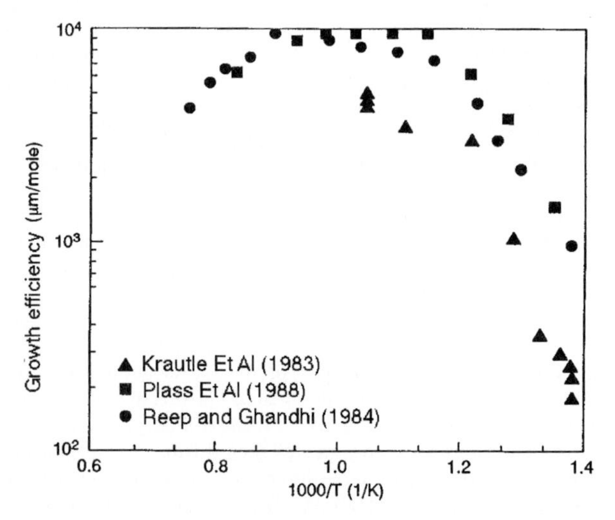

FIGURE 4.73 Growth efficiency versus reciprocal temperature for MOCVD growth of GaAs from trimethylgallium and arsine.[126]

um precursors is the rate-determining step for the reaction, thus yielding the desired broad constant growth region. Importantly the diffusion limitation allows the final GaAs growth step, from the surface-adsorbed arsenic and gallium containing species, to reach thermodynamic equilibrium, that is, to provide time for the species to diffuse across the crystal surface and come to rest in the appropriate crystallographic position, thereby enabling high-quality epitaxial GaAs growth. The region of constant growth rate is significantly wider than that for inorganic VPE, so that in principle it is somewhat easier to obtain good intrawafer thickness uniformity in MOCVD. Even with this temperature insensitivity, however, uniform GaAs growth is difficult to achieve in MOCVD due to several factors related to the difficulty in predicting and controlling the flow of heated gases in the reactors. In addition the dopant incorporation efficiency of the most common GaAs *n*-type dopant silicon from silane and the most desirable *p*-type dopant carbon from CCl_4 or $(CH_3)_3As$ are strongly temperature dependent. Under the proper growth conditions an alternate source of silicon from disilane is relatively temperature-insensitive.[127] As of yet, however, a temperature-insensitive source of carbon has not been identified. Further, compositional uniformity of arsenic- and phosphorus-containing alloys such as GaInAsP is very temperature-sensitive because of the incomplete cracking of arsine, and even more so phosphine, at typical growth temperatures.[128] Attempts at modeling the complex MOCVD process while providing valuable insights into trends of growth rate, carbon incorporation, and intentional doping as a function of temperature pressure and V/III ratio have also shown an extreme sensitivity to difficult-to-control reactor conditions, such as reactor-wall temperature.[126] As currently practiced, MOCVD reactor design is still a very empirical process.

Molecular-Beam Epitaxy.[114,129,130] III-V MBE equipment differs from that of silicon MBE (Fig. 4.65) in that resistively heated effusion cells containing elemental column III and column V sources replace the e-beam silicon sources. In GaAs MBE growth, the basic components are adsorption, surface diffusion, dissociation, and incorporation. Gallium has a sticking coefficient near unity at temperatures typically employed for GaAs growth, about 580°C. As_2 and As_4, however, will not condense on a bare GaAs surface, although they do have finite surface lifetimes in a weakly adsorbed precursor state. On a gallium-coated GaAs surface, however, As_2 and As_4 will adsorb with sticking coefficients of unity and 0.5, respectively. Consistent with this observation, the mechanism of MBE GaAs growth is initiated by gallium adsorption on the GaAs surface. The gallium diffuses on the surface until it reacts with As_2 or As_4 and is effectively immobilized. If given sufficient time to diffuse before coming in contact with As_2 or As_4, gallium adatoms preferentially bond to kink or step sites on the edge of a growing GaAs monolayer island. As was the case for MOCVD growth of III-V compounds, MBE is performed with an excess column V flux under which conditions the growth rate is determined by the rate of column III atom arrival at the surface. For typical III-V growths[114] at 1-μm/h growth rates of group V fluxes are 10^{15} to 10^{16} molecules/cm^2 • s. Column III element fluxes are 10^{14} to 10^{15} atoms/cm^2 • s, and dopant species fluxes are 10^8 to 10^{11} atoms/cm^2 • s.

Due to the precise layer control and abruptness inherent in the MBE technique, its versatility in the number of evaporable sources, and the in-situ measures of layer quality, composition, and thickness, MBE has been an excellent tool for heterostructure device development, not just in the III-V family but for numerous other semiconductors. In fact all of the compounds in Table 4.14 can also be grown by MBE. Like any growth technology, MBE has had its difficulties, including high cost, high levels of oval defects, and difficulty in growing phosphorus-containing compounds, particularly those with two column V components.

There are now, however, commercial MBE reactors with the capability of growing simultaneously on five or more 3-in-diameter wafers with multiple platters and load locks with preheaters to speed equipment throughput. For device profiles of 1 μm or less thickness the cost is now comparable for MBE and MOCVD wafers at approximately $1000 per 3-in wafer. That MBE is no longer just a research and development tool is demonstrated by its commercial use in MMICs, AlGaAs/GaAs lasers, and surface emitting lasers. Careful control of reactor cleanliness substrate preparation, and particularly the redesign of gallium effusion cells to include heated necks in order to eliminate gallium spitting, has dramatically reduced the density of oval defects from over 1000 cm^{-2} to less than 100 cm^{-2}, in some cases approaching the defect densities available in VPE and MOCVD of fewer than 10 cm^{-2}.

The difficulty experienced in growing phosphorus-containing compounds, particularly those with two group V components such as GaAsP or GaInAsP, is due to the fact that arsenic and phosphorus, being solid sources, have changing surface areas, and hence flux varies with usage. As group III-V growth is not sensitive to column V flux, as long as there is an excess, this does not represent a difficulty for single column V containing compounds. In order to lattice match GaInAsP to InP, however, the composition of arsenic and phosphorus must be tightly controlled. An added difficulty with phosphorus usage is that it exists in a mixture of allotropic forms, each with a different vapor pressure, and with thermal cycling the amounts of the different allotropes vary. As a result of this difficulty, MOCVD has dominated in mixed column V semiconductor growth. The recent successful advent of cracked arsine and phosphine sources is described in the next section.

Hybrid MOCVD/MBE Growths. In order to overcome some of the shortcomings inherent in both MOCVD and MBE growth, several researchers have developed various hybrid techniques, all attempting to incorporate the best features of both growth technologies. All of the hybrid techniques shown in Fig. 4.74, combine the low-pressure ($<<10^{-2}$ torr) molecular flow environment of the standard solid-source MBE growth technique either with column III or V, or with both column III and V nonelemental vapor sources. These hybrid techniques eliminate the complex and difficult to

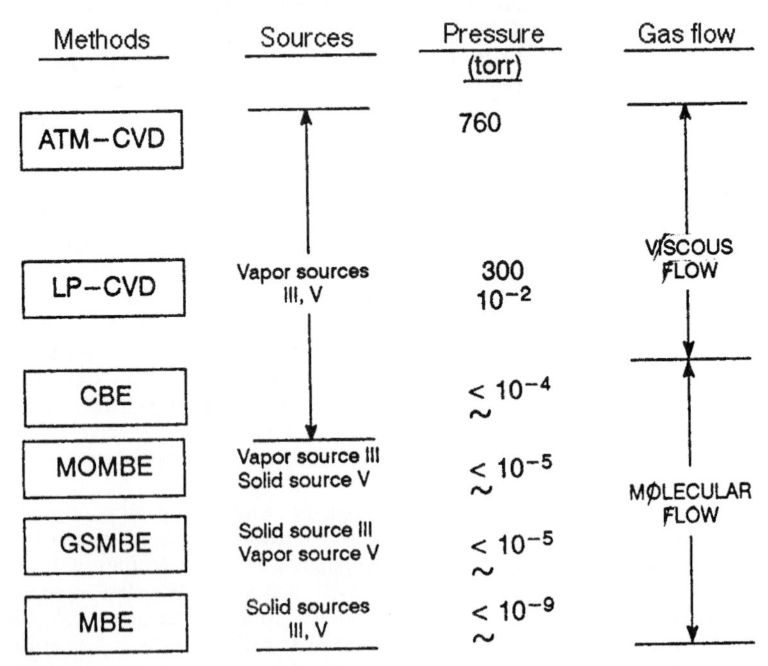

FIGURE 4.74 Names and growth conditions for several hybrid MOCVD/MBE growth techniques.

control viscous flows of MOCVD while taking advantage, to varying degrees, of the convenience of the room-temperature vapor sources and mass-flow controllers used in MOCVD. Unlike the case for MOCVD, each of the hybrid technologies is distinct, with active-layer growth mechanisms ranging from MOCVD-like to MBE-like.

Gas-Source MBE. This is the most MBE-like growth technique and the most mature of the hybrid technologies. In gas-source MBE (GSMBE) the solid column V effusion cells are replaced with arsine and phosphine gas sources, where controlled leaks of the hydrides are cracked into As_2 and P_2 beams in a high-temperature 1000°C alumina bed.[131] Single-crystal growth of GaAs and InP was first achieved by Panish using this method in 1980.[132] There are now commercial gas cracker cells and growing acceptance in the MBE community. The only alterations in equipment and procedure required to use column V gas sources are an increased pumping capacity to remove the H_2 generated by the gas cell, the requisite safety precautions and equipment for arsine and phosphine, and precautions for dealing with solid phosphorus deposits during chamber cleaning. The application of group V gas sources to MBE has enabled the growth of the otherwise problematic InP lattice-matched GaInAsP lasers.[133]

Metalorganic MBE. The vapor from an organometallic column III source through a controlled leak is used in place of the standard column III elemental effusion cells in metallorganic MBE (MOMBE). This technique allows the study of heterogeneous metal alkyl reactions at a growing III-V surface without the complication of homogeneous gas-phase reactions, with the obvious benefit of understanding the MOCVD and CBE processes.

Chemical-Beam Epitaxy. Chemical-beam epitaxy (CBE) was originally performed by Tsang[134] in 1984. Here the column III and V sources are vapors at room temperature. In this hybrid growth technology the simplicity inherent in the molecular flow regime of MBE along with all the in-situ characterization capabilities is combined with the convenience of mass-flow controller delivery. A side benefit of CBE, although somewhat mitigated by the similar success in solid-source MBE, is a great reduction in oval defect sources. Like GSMBE, CBE has resulted in the capability of growing quaternary arsenic- and phosphorus-containing compounds. Perhaps the greatest advantage portended by CBE is the precise compositional control of both group V and III precursors, enabling very well lattice-matched active layers. Tsang has suggested that standard elemental effusion cells would require constant temperature control to ±0.1°C to match the compositional control offered by CBE. Similarly the difficult-to-eliminate pressure transients upon layer switching, which are manifested by unwanted compositional variations in MOCVD, are not present in CBE.

While CBE has eliminated the difficult-to-control flows inherent in MOCVD, it has a significantly different growth mechanism from the well-understood MBE or GSMBE. This is a consequence of metal alkyls instead of group III atoms impinging upon the substrate. Also unlike MOCVD, these alkyls must decompose heterogeneously without the benefit of homogeneous gas-phase reactions. The commonly employed MOCVD group III sources $(CH_3)_3Ga$ and $(CH_3)_3Al$ have resulted in unavoidably high levels of carbon incorporation, in the 10^{15}- and 10^{16}-cm^{-3} range, respectively. This, however, has been turned into a benefit in CBE. Electrically active carbon doping concentrations as high as 10^{21} cm^{-3} have been claimed[107] and $(CH_3)_3Ga$ has been used to produce carbon-doped GaAs bases in HBTs. To reduce unwanted carbon incorporation in CBE-grown GaAs, and carbon and oxygen contamination in AlGaAs, new chemically tailored gallium-, aluminum-, and arsenic-containing precursors such as trimethylamine gallane (TMAG), $(CH_3)_3N\text{-}GaH_3$, trimethylamine alane (TMAA), $(CH_3)_3N\text{-}AlH_3$, and phenylarsine, $PhAsH_2$, are being investigated.[135–137]

Atomic-Layer Epitaxy.[137–140] With atomic-layer epitaxy (ALE), first proposed by Suntola and Autson,[141] the layers are grown one monolayer at a time by successive exposure of the substrate to one component of a compound semiconductor and then to the other component. The monolayer nature of ALE growth is the consequence of monolayer self-limiting adsorption of at least one of the constituents of a compound semiconductor, for example, that of monoalkyl gallium shown in Fig. 4.75. Note the important distinction from the two-dimensional growth that can occur in standard MBE (or MOCVD) growth. MBE two-dimensional growth occurs under the simultaneous presence of both components of a compound semiconductor and is the result of the diffusion of the cation adsorbed on the GaAs surface to a preferred but not required low-energy step-edge or step-kink site in a growing film. Once the gallium reaches a favorable site, the subsequent incorporation and heterogeneous incorporation reaction with the loosely adsorbed anion source (As_2 or As_4 for GaAs) are rapid. Even starting with an atomically flat layer, this form of two-dimensional growth typically does not continue for more than a few atomic layers, as evidenced by RHEED studies.[142] This is due to the restricted diffusion length of gallium adatoms on GaAs when they are exposed to As_2 or As_4. The gallium may react with As_2 or As_4 before it reaches a site on the edge of the growing GaAs monolayer island, thus starting a new layer before the subsequent one is complete. In migration-enhanced epitaxy (MEE), a special case of MBE,[142] the cation diffusion length is greatly increased as the cation is allowed to diffuse in the absence of the anion by alternating the supply of the anion and cation source. This procedure allows two-dimensional growth to continue indefi-

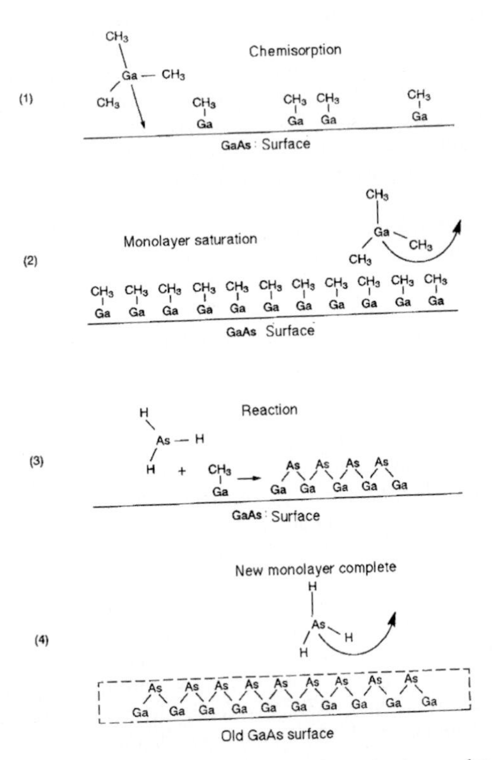

FIGURE 4.75 GaAs ALE growth mechanism under MOCVD-like conditions using trimethylgallium and arsine.[141]

nitely as long as the cycle times result in an alternating supply of roughly a monolayer of both cations and anions. An advantage of ALE in comparison to MEE is that the timing requirement is relaxed as the adsorption of the cation source molecule is self-limiting at one monolayer.

The potential advantages of ALE over other growth techniques are numerous, including atomic-layer compositional and doping transition abruptness, ultimate active-layer thickness uniformity and control, multiple-wafer capacity reactors (up to 30 3-in wafer systems have been demonstrated), and identical growth rates on all exposed faces of nonplanar substrates.[138] A very significant difficulty with ALE, however, is an extremely low growth rate, 10 to 100 times slower than the typical 1 μm/h found in conventional MBE. As of yet there are no commercial applications of ALE active layers.

4.4.5 Narrow-Band-Gap Materials

Si:Ge/Si Strained Layers. The success of pseudomorphic growth in the III-V compounds showed the way to the $Si_{1-x}Ge_x/Si$ strained-layer heterostructures. The first topical conference[143] on these heterostructures was held in 1990. Already, however, $Si_{1-x}Ge_x/Si$ HBTs have demonstrated 75-GHz operation,[144] rivaling that of III-V HBTs. The first successful strained-layer $Si_{1-x}Ge_x/Si$ growths were performed in the 1980s using MBE.[101] The primary challenges in $Si_{1-x}Ge_x/Si$ MBE is that, due to the low vapor pressure of the group IV elements, electron-beam evaporation is required to supply a sufficient growth flux, and that it is much more difficult to dope growing silicon MBE films than was the case for the III-V compounds. The difficulty with doping in silicon MBE is that dopants tend to segregate on the growth surfaces, resulting in low incorporation efficiencies and very nonabrupt doping transitions. Abrupt *n*-type doping can still be achieved, however, with antimony by using in-situ low-energy ion implantation where the substrate is biased, and Si^+ ions from the silicon source drive antimony adatoms into the growing silicon film.[145] Abrupt *p*-type doping can be made by using high-temperature boron sources.[146]

Device-quality $Si_{1-x}Ge_x/Si$ heterostructures have also been grown very recently at 550°C by UHV-CVD at 1 m torr employing silane, SiH_4, and germane, GeH_4, as reagents.[147] This low temperature is required to allow approximately two-dimensional growth of pseudomorphic layers and to assure abrupt doped junctions. The CVD-based system is more amenable to the multiple-wafer growths required in silicon processing.

Infrared Detector Materials. The dominant semiconductor active layer used for infrared detectors is the well lattice-matched II-VI semiconductor alloy system, mercury cadmium telluride, $Hg_xCd_{1-x}Te$.[148–152] A plot of the band-gap energy and lattice constant of HgCdTe and other II-VI compounds is contained in Fig. 4.76. By varying the cadmium mole fraction from 0 to 1 the band gap of HgCdTe can be tuned all the way from 0.8 to 30 μm. Mole fractions of ≈0.3 and ≈0.2 correspond to the infrared atmospheric spectral windows of 3 to 5 and 8 to 12 μm, respectively. As a direct band-gap semiconductor, HgCdTe exhibits a high optical absorption, allowing 10- to 15-μm-thick layers to have 100 percent internal quantum efficiencies.

Despite being worked on for some 32 years, HgCdTe materials technology is less mature than that of the III-V semiconductors. Standard lattice-matched substrates for HgCdTe, CdZnTe, and CdTe cannot be obtained as single-crystal boules, and mercury out-diffusion from HgCdTe is a problem at temperatures much above 200°C, resulting in vacancy *p*-type doping. The majority of HgCdTe active layers for photovoltaic detectors are produced by LPE, which is well suited for these thick structures.

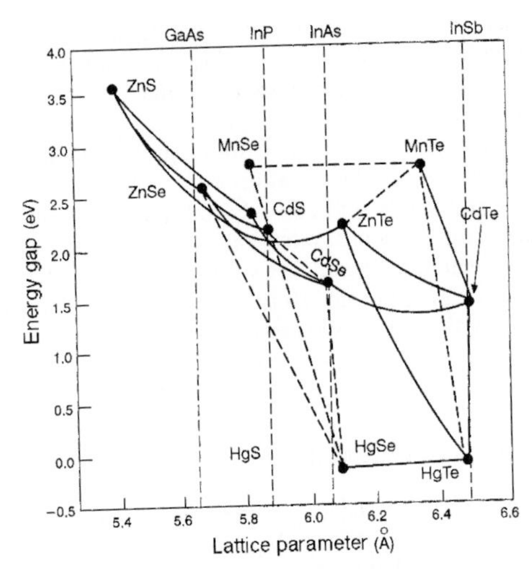

FIGURE 4.76 A plot of band-gap energy and lattice constant for II-VI family of semiconductors.[152]

Because low-temperature growth can be achieved (150 to 200°C), MBE has been the most successful vapor-phase technique with intentional *n*- and *p*-type doping possible for the alloy semiconductor,[145] MOCVD typically requires growth temperatures greater than 300°C with resulting complexities associated with mercury out-diffusion. Despite these difficulties, active HgCdTe layers are being developed by both MBE and MOCVD, because of the possibility of achieving high-quality HgCdTe on large-area lattice-mismatched semiconductor substrates such as silicon and GaAs.

In the 3- to 5-μm band the III-V semiconductor InSb can compete directly with HgCdTe. The previously described high level of development of III-V growth and processing weighs heavily in favor of InSb for 3- to 5-μm applications. $In_xGa_{1-x}Sb$/InAs strained-layer superlattices allow small band gaps to be achieved, however, demonstrating absorption out to 19 μm.[153] Even specialized superlattices based on relatively wide-band-gap GaAs/AlGaAs have been fabricated which show optical absorption out to 10 μm.[154]

4.4.6 Wide-Band-Gap Materials

Large-band-gap materials have physical properties that make them both challenging to grow and uniquely suited for certain device applications. The large band gap is a consequence of the strong bond strengths between the constituent atoms requiring increased temperatures for equilibrium crystal growth but in return providing enhanced chemical, mechanical, and thermal stability. The large band gap, while at the same time making ohmic contact formation difficult, also results in desirable high breakdown voltages and provides the potential for blue to ultraviolet light emission and detection.

Chemical-Vapor Deposition of Silicon Carbide. CVD of epitaxial silicon carbide originally was undertaken on foreign substrates such as silicon or titanium carbide. In this process, temperature cycling of the substrate in a carbon-rich atmosphere is employed to generate nucleation sites, and ß-silicon carbide (3C-polytype) epitaxial layers are then deposited using a mixture of silane and propane, or silane and acetylene. This process was used in the 1980s to achieve single-crystal growth of ß-silicon carbide on (001) silicon substrates.[155,156] However, the 20 percent lattice mismatch between these two crystal structures gives rise to a great number of crystalline defects, such as dislocations, microtwins, stacking faults, and inversion domain boundaries.[157] Consequently this technique has not been very useful in the fabrication of electronic devices. A better lattice (1 percent) match can be achieved[158] using titanium carbide substrates, but as yet no large-area defect-free substrates have been made available for silicon carbide deposition. The recent availability of 6H-SiC wafers has focused CVD efforts on SiC homoepitaxy with considerable success.

Initial growth of homoepitaxial SiC layers used Lely-grown SiC platelet crystals as substrates. The layers were grown on the basal (0001) SiC plane at temperatures above 1550°C and showed very high densities of defects, including stacking faults, dislocations, and double-position boundaries (the coalescence of regions of inverting stacking sequence). With the availability of 6H-SiC wafers prepared from single SiC crystals, produced by physical-vapor transport, device-quality epitaxial layers of both α-SiC and ß-SiC have been demonstrated. The tilt angle of the substrate [that is, the amount off-axis from the (0001) basal plane], the polarity of the surface (carbon or silicon face), and the pregrowth surface treatment of the substrate were found[159] to be important factors which affected the structure and morphology of the epitaxial layers. Typically SiC CVD is carried out in a horizontal reaction chamber using a high-purity silicon carbide–coated graphite susceptor to support the wafers. The susceptor is heated inductively to the growth temperature of 1450°C. A water jacket surrounds the fused-silica growth chamber to keep the reactor walls cold and minimize the transport of impurities to the growing film. Propane and silane gases in hydrogen at atmospheric pressure are then used to deposit epitaxial SiC layers. Single 6H-polytype epitaxial SiC layers with excellent surface morphology and background impurity levels in the mid-10^{15}-cm^{-3} range have been achieved. Through the addition of dopants these films may be doped *n*-type (nitrogen) or *p*-type (aluminum) for the creation of a host of electronic devices, such as bipolar transistors, *p-n* junction diodes, MOS devices, and MESFETs.

Gallium Nitride Growth.[160] Of the III-V mononitrides, GaN, with a band gap of 3.19 eV (wurtzite structure) and 4.54 eV (zinc blende), is the most highly developed. It is of considerable interest as an optoelectronic material for the fabrication of sources and detectors in the solar blind ultraviolet spectrum and of electrooptic, piezoelectric, and acoustooptic modulators.

Recent research[160] on GaN has been concerned with the growth of ß-GaN layers by various techniques at temperatures below about 1000°C, using AlN buffer layers between the substrate (usually sapphire) and the GaN layer, and with the use of electron-beam irradiation to achieve *p*-type conduction for *p-n* junction formation.

Modified MBE techniques (including an ion source and an electron cyclotron resonance plasma source to produce activated and ionized nitrogen),[161,162] plasma techniques,[163] and MOCVD,[164–166] have been used to achieve ß-GaN layers on various other substrates. Numerous substrate materials have been used for the deposition of GaN, the most common being (0001) sapphire. Yoshida et al.[163] initially showed that the surface morphology of the GaN layers improved significantly when an AlN buffer layer was initially deposited on the sapphire. Other researchers have since achieved high-quality GaN epitaxial layers using 50-µm-thick AlN buffer layers by MOCVD. Electron mobilities greater than 300 $cm^2/V \cdot s$ at room temperature and electron concentrations as low as 1×10^{17} cm^{-3} were reported.

In-situ *n*-type doping of CVD-grown GaN layers with zinc and magnesium and other impurities has been achieved, but not the formation of *p*-type conductivity. However, low-energy electron-beam irradiation of GaN films (LEEBE) has recently been shown to produce[166] *p*-type GaN layers with hole concentrations of 2×10^{16} cm^{-3} and a mobility of 8 $cm^2/V \cdot s$ at 293 K. This work has led to the fabrication of *p-n* junction LEDs. GaN are, however, highly defective because of the lattice mismatch with all host substrates tried to date. The future development of a GaN-based device technology will face many obstacles, such as the lack of a stable oxide and the formation of ohmic contacts.

REFERENCES

1. A. S. Grove, *Physics and Technology of Semiconductor Devices,* Wiley, New York, 1987.
2. S. M. Sze, *Physics of Semiconductor Devices,* Wiley, New York, 1969.
3. R. N. Hall and J. H. Racette, *J. Appl. Phys.,* vol. 35, p. 379, 1964.
4. F. J. Morin and J. P. Maita, *Phys. Rev.,* vol. 96, p. 28, 1954.
5. F. J. Morin and J. P. Maita, *Phys. Rev.,* vol. 94, p. 1525, 1954.
6. S. M. Sze and J. C. Irvin, *Solid-State Electron.,* vol. 11, p. 599, 1968.
7. E. M. Conwell, *Proc. IRE,* vol. 46, p. 1281, 1958.
8. E. J. Ryder, *Phys. Rev.,* vol. 90, p. 766, 1953.
9. M. B. Prince, *Phys. Rev.,* vol. 92, p. 681, 1953.
10. K. B. Wolfstirn, *J. Phys. Chem. Solids,* vol. 16, p. 279, 1960.
11. G. L. Pearson and J. Bardeen, *Phys. Rev.,* vol. 75, p. 865, 1949.
12. W. B. Pearson, *Handbook of Lattice Spacings and Structure of Metals and Alloys,* Pergamon, New York, 1967.
13. L. A. Azaroff, *Introduction to Solids,* McGraw-Hill, New York, 1960.
14. H. Gottschalk, G. Balzer, and H. Alexander, *Phys. Stat. Sol.,* vol. (a)45, p. 207, 1978.
15. H. W. Hayden, W. G. Moffatt, and J. Wulff, *The Structure and Properties of Materials: Mechanical Behavior,* vol. 3, Wiley, New York, 1965.
16. D. B. Cuttriss, *Bell Sys. Tech. J.,* vol. 40, p. 509, 1991.
17. J. C. Irvin, *Bell Sys. Tech. J.,* vol. 41, p. 387, 1962.
18. R. W. Keyes, *Proc. IEEE,* vol. 60, p. 255, 1972.
19. M. W. Geis, H. I. Smith, A. Argoitia, and J. Angus, G. H. M. Ma, J. H. Glass, J. Butler, C. J. Robinson, and R. Pryor, *Appl. Phys. Lett.,* vol. 58, p. 2485, 1991.
20. T. F. Cisek, T. Wang, and T. Schuyler, *J. Electrochem. Soc.,* vol. 136, p. 230, 1989.
21. H. M. Hobgood, P. Ravishankar, and R. N. Thomas, *Semiconductor Silicon 1990,* Electrochem. Soc., Princeton, N.J., p. 58, 1990.
22. W. R. Runyan, *Silicon Semiconductor Technology,* TI Electronics ser., McGraw-Hill, New York, 1965.
23. W. G. Pfann, *Zone Melting,* Wiley, New York, 1966.
24. F. A. Trumbore, *Bell Sys. Tech. J.,* vol. 39, pp. 208, 210, 1960.
25. H. S. Carslaw and L. Derick, *J. Electrochem. Soc.,* vol. 104, p. 547, 1957.
26. R. M. Burger and R. P. Donovan, [eds.), *Fundamentals of Silicon Integrated Device Technology,* vol. 1, Prentice-Hall, Englewood Cliffs, N.J., 1967.
27. D. L. Kendall, Rep. 65-29, Dept. of Material Science, Stanford University, Stanford, Calif., Aug. 1965.

28. W. C. Dash and R. Newman, *Phys. Rev.,* vol. 99, p. 1151, 1955.
29. H. R. Philipp and E. A. Taft, *Phys. Rev.,* vol. 113, p. 1002, 1959; *Phys. Rev. Lett.,* vol. 8, p. 13, 1962.
30. D. E. Hill, *Phys. Rev.,* vol. 133, p. A866, 1964.
31. J. A. Carruthers, T. H. Geballe, H. M. Rosenberg, and J. M. Ziman, *Proc. R. Soc. (London),* vol. 238, p. 502, 1957.
32. W. L. Wolfe and G. J. Zissis (eds.), *The Infrared Handbook,* Infrared Information Analysis (IRIA) Center, Environmental Research Inst. of Michigan, 1989.
33. M. G. Holland, *Phys. Rev.,* vol.134, p. A471, 1964.
34. G. K. White, *Austr. J. Phys.,* vol. 6, p. 397, 1953.
35. R. Berman, *Proc. R. Soc. (London),* ser. A, vol. 200, p. 171, 1953.
36. J. Czochralski, *Z. Phys. Chem.,* vol. 92, p. 219, 1918.
37. J. Obreimov and L. Schubnikov, *Z. Physik,* vol. 25, p. 31, 1924.
38. P. W. Bridgman, *Proc. Am. Acad. Arts Sci.,* vol. 60, p. 303, 1925.
39. G. K. Teal and J. B. Little, *Phys. Rev.,* vol. 78, p. 647(A), 1950.
40. D. C. Stockbarger, *Rev. Sci. Instr.,* vol. 7, p. 133, 1936; *Trans. Faraday Soc.,* no. 5, p. 294, 1949.
41. L. D. Crossman and J. A. Baker, in *Semiconductor Silicon 1977,* H. R. Huff and E. Sirtl (eds.), Electrochemical Soc., Princeton, N.J., 1977.
42. W. Keller and A Muhlbauer, *Preparation and Properties of Solid State Materials,* vol. 5: *Floating-Zone Silicon,* Marcel Dekker, New York, 1981.
43. W. C. Dash, *J. Appl. Phys.,* vol. 29, p. 736, 1958.
44. H. P. Utech and M. C. Flemmings, *J. Appl. Phys.,* vol. 37, p. 202, 1966; H. A. Chedzey and D. T. J. Hurle, *Nature,* vol. 210, p. 933, 1966.
45. A. F. Witt, C. J. Herman, and H. C. Gatos, *J. Mater. Sci.,* vol. 5, p. 822, 1970.
46. T. Suzuki, N. Isawa, Y. Okubo, and K. Hoshi, *Semiconductor Silicon 1981,* H. R. Huff, R. Krieger, and Y. Tekeishi (eds.), Electrochemical Soc., Princeton, N.J., 1981, p. 90.
47. R. N. Thomas, H. M. Hobgood, P. S. Ravishankar, and T. T. Braggins, *Solid-State Technol.,* p. 163, Apr. 1990; *J. Crystal Growth,* vol. 104, p. 617, 1990.
48. D. T. J. Hurle and R. W. J. Series, *J. Crystal Growth,* vol. 73, p. 1, 1985.
49. D. H. Matthiesen, J. J. Wargo, S. Motakef, D. J. Carlson, J. S. Nakos, and A. F. Witt, *J. Crystal Growth,* vol. 85, p. 557, 1987.
50. K. V. Ravi, *J. Crystal Growth,* vol. 39, p. 1, 1977.
51. R. G. Seidensticker, A. M. Stewart, and R. H. Hopkins, *J. Crystal Growth,* vol. 46, p. 51, 1979.
52. R. K. Willardson and A. C. Beer (eds.), *Semiconductors and Semimetals: Semi-Insulating GaAs*, vol. 20, Academic Press, New York, 1984.
53. W. A. Gault, E. M. Monberg, and J. E. Clemans, *J. Crystal Growth,* vol. 74, p. 491, 1986.
54. E. P. A. Metz, R. C. Miller, and R. Mazelsky, *J. Appl. Phys.,* vol. 33, p. 2016, 1962.
55. J. B. Mullin, R. J. Heritage, C. H. Holiday, and B. W. Straughan, *J. Crystal Growth,* vol. 34, p. 281, 1968.
56. E. M. Swiggard, S. H. Lee, and F. W. Von Batchelder, *Conf. Ser. Inst. Phys.,* no. 33b, p. 23, 1977.
57. R. N. Thomas, S. McGuigan, G. W. Eldridge, and D. L. Barrett, *Proc. IEEE,* vol. 76, p. 778, 1988.
58. D. E. Holmes, R. T. Chen, K. R. Elliott, and G. G. Kirkpatrick, *J. Appl. Phys. Lett.,* vol. 40, p. 46, 1982.

59. L. B. Ta, H. M. Hobgood, A. Rohatgi, and R. N. Thomas, *J. Appl. Phys.*, vol. 58, p. 5771, 1982.
60. J. Lagowski and H. C. Gatos, in *Proc. 13th Int. Conf. on Defects in Semiconductors*, (AIME, 1985), p. 73.
61. A. S. Jordan, R. Caruso, and A. R. Von Neida, *Bell Sys. Tech. J.*, vol. 59, p. 593, 1980.
62. G. Jacob, in *Proc. Semi-Insulating III-V Materials,* Shiva Publ., Nantwich, U. K., 1982, p. 402.
63. S. Yoshida, S. Ozawa, T. Kijima, J. Suzuki, and T. Kikuta, *J. Crystal Growth,* vol. 113, p. 221, 1991.
64. K. Zanio, in *Semiconductors and Semimetals: Cadmium Telluride,* vol. 13, R. K. Willardson and A. C. Beer (eds.), Academic Press, New York, 1978.
65. W. F. H. Micklewaite, in *Semiconductors and Semimetals,* vol. 18, R. K. Willardson and A. C. Beer (eds.), Academic Press, New York, 1981, chap. 3.
66. P. Capper and J. E. Harris, *J. Crystal Growth,* vol. 46, p. 575, 1979.
67. J. A. Lely, *Deut. Keram. Ges.,* Berlin, vol. 32, p. 229, 1955.
68. Yu. M. Tairov and V. F. Tsvetkov, *J. Crystal Growth,* vol. 43, p. 209, 1978.
69. D. L. Barrett, G. G. Seidensticker, W. Gaida, and R. H. Hopkins, *J. Crystal Growth,* vol. 109, p. 17, 1991.
70. W. M. Bullis and R. I. Scace, *Proc. IEEE,* vol. 57, p. 1639, 1969.
71. J. W. Carlson, *Solid-State Technol.,* vol. 16, p. 49, 1973.
72. R. L. Lane, "Silicon Wafer Preparation," in *Handbook of Semiconductor Silicon Technology,* W. C. O'Mara, R. B. Herring, and L. P. Hunt (eds.), Noyes Publ., Park Ridge, N.J., 1990, chap. 4.
73. T. J. Magee and C. Leung, Tech. Rep. SBIR Phase I, ONR Contract N00014-84-C-0057, 1984.
74. R. M. Silva, F. D. Orazio, and J. M. Bennett, *Opt. News,* vol. 12a, p. 10, 1986.
75. D. L. Barrett, G. W. Eldridge, R. C. Clarke, and R. N. Thomas, in *Proc. 1988 GaAs ManTech. Conf.* (Nashville, Tenn., 1988).
76. R. B. Herring, in *Handbook of Semiconductor Silicon Technology,* W. C. O'Mara, R. B. Herring, and L. P. Hunt (eds.), Noyes Publ., Park Ridge, N.J., 1990, chap. 5, p. 259.
77. B. J. Sealy, *Mater. Sci. Technol.,* vol. 4, p. 500, 1988.
78. B. L. Sharma, *Def. Sci. J.,* vol. 39, p. 353, 1989.
79. S. T. Picraux, *Phys. Today,* p. 38, Nov. 1984.
80. V. Ashworth, R. P. M. Proctor, and W. A. Grant, in *Treatise on Materials Science and Technology,* vol. 18, J. K. Hirvonen (Ed.), Academic Press, New York, 1980, p. 176.
81. J. F. Gibbons, W. S. Johnson, and S. W. Mylroie, *Projected Range Statistics, Semiconductors and Related Materials,* 2d ed., Dowden, Hutchinson and Ross, Stroudsburg, Pa., 1975.
82. J. M. Mayer, L. Eriksson, and J. A. Davies, *Ion Implantation in Semiconductors, Silicon and Germanium,* Academic Press, New York, 1970.
83. J. M. Woodcock and D. J. Clark, in *Gallium Arsenide and Related Compounds, Proc. Inst. Phys. Conf.,* ser. 24 (1975), p. 331.
84. R. K. Surridge, B. J. Sealy, A. D. E. D'Cruz, and K. G. Stephens, in *Gallium Arsenide and Related Compounds, Proc. Inst. Phys. Conf.,* ser. 33a, (1977), p. 161.
85. R. G. Wilson and G. R. Brewer, *Ion Beams with Applications to Ion Implantation,* Robert E. Krieger, Huntington, N.Y., 1979, p. 357.
86. D. H. Rosenblatt, W. R. Hitchens, R. E. Anholt, and T. W. Sigmon, *IEEE Electron Device Lett.,* vol. 9, no. 3, p. 139, 1988.

87. T. Sato, M. Tajima, and K. Ishida, *Appl. Phys. Lett.,* vol. 51, p. 755, 1987.

88. M. Arai, K. Nishiyama, and N. Watanabe, *Jpn. J. Appl. Phys.,* vol. 19, p. L563, 1980.

89. R. Anholt and T. W. Sigmon, *J. Electron. Mater.,* vol. 17, p. 5, 1988.

90. K. T. Short and S. J. Pearton, *J. Appl. Phys.,* vol. 64, p. 1206, 1988.

91. L. Royer, *Bull. Soc. Fr. Mineral. Crist.,* vol. 51, p. 7, 1928.

92. P. S. Peercy et al., *J. Mater. Res.,* vol. 5, p. 852, 1990.

93. H. M. Liaw and J. W. Rose, in *Epitaxial Silicon Technology,* B. J. Bagila (ed.), Academic Press, New York, 1986, chap. 1, p. 1.

94. S. S. Iyer, in *Epitaxial Silicon Technology,* B. J. Bagila (ed.), Academic Press, New York, 1986, p. 91.

95. J. C. Bean, in *Impurity Doping Processes in Silicon,* F. Wang (ed.), North-Holland, Amsterdam, 1981.

96. D. W. Grieve and M. Racanelli, *J. Vac. Sci. Technol.,* vol. B8, p. 511, 1990.

97. R. N. Thomas and M. H. Francombe, *Solid-State Electron.,* vol. 12, p. 799, 1969.

98. P. G. Grunthaner, F. J. Grunthaner, R. W. Fathauer, T. L. Lin, M. H. Hecht, L. D. Bell, W. J. Kaiser, F. D. Schowengerdt, and J. H. Mazur, *Thin Solid Films,* vol. 183, p. 197, 1989.

99. J. P. Colinge, *Silicon-on-Insulator Technology: Materials of VLSI,* Kluwer Academic Publ., 1991.

100. A. K. Agarwal, M. C. Driver, M. H. Hanes, H. M. Hobgood, P. G. McMullin, H. C. Nathanson, T. W. O'Keefe, T. J. Smith, J. R. Szedon, and R. N. Thomas, *Tech. Digest IEDM,* p. 687, 1991.

101. J. C. Bean, in *Silicon-Molecular Beam Epitaxy,* vol. 2, E. Kasper and J. C. Bean (eds.), CRC Press, Boca Raton, Fla., 1988.

102. Crystal Data Determinate Tables, vol. 4, 3d ed., in *Inorganic Compounds,* U. S. Dept. of Commerce and National Bureau of Standards, Washington, D.C., 1978.

103. G. B. Stringfellow, *J. Crystal Growth,* vol. 115, p. 1, 1991.

104. D. W. Pashley, in *Materials Science and Technology, Epitaxial Growth,* pt. A, J. W. Matthews (ed.), Academic Press, New York, 1975, p. 1.

105. W. T. Tsang, *J. Crystal Growth,* vol. 95, p. 121, 1989; vol. 120, p. 1, 1992.

106. G. B. Stringfellow, *J. Crystal Growth,* vol. 68, p. 111, 1984.

107. H. Seki and A. Konkitu, *J. Crystal Growth,* vol. 98, p. 118, 1989.

108. J. R. Knight, D. Effer, and P. R. Evans, *Solid-State Electronics,* vol. 8, Pergamon, Great Britain, 1995.

109. T. F. Kuech, E. Marshall, G. J. Scilla, R. Potemski, C. M. Ransom, and M. Y. Hung, *J. Crystal Growth,* vol. 77, p. 539, 1986.

110. T. F. Keuch, M. A. Tischler, and R. Potemski, *Appl. Phys. Lett.,* vol. 54, p. 910, 1989.

111. H. Nelson, *RCA Rev.,* vol. 24, p. 603, 1963.

112. K. Ishida, H. Tokunaga, H. Ohtani, and T. Nishizawa, *J. Crystal Growth,* vol. 98, p. 140, 1989.

113. I. A. Dorrity, J. D. Grange, and P. K. Wickenen, in *Gallium Arsenide: Materials, Devices and Circuits,* M. J. Morgan and D. V. Morgan (eds.), Wiley, New York, 1985, p. 98.

114. W. T. Tsang, in *Semiconductors and Semimetals,* vol. 22A, R. K. Willardson and A. C. Beer (eds.), Academic Press, New York, 1985, p. 95.

115. V. Swaminathan and A. T. Macrender, *Materials Aspects of GaAs and InP Based Structures,* Prentice-Hall, Englewood Cliffs, N.J., 1991, p. 78.

116. J. V. Dilorenzo, *J. Crystal Growth,* vol. 17, p. 189, 1972.

117. D. W. Shaw, *J. Crystal Growth,* vol. 8, p. 117, 1971.

118. J. P. Chane, *J. Electrochem. Soc.,* vol. 127, p. 913, 1980.

119. P. Vohl, *J. Crystal Growth,* vol. 54, p. 101, 1981.

120. J. K. Abokwah, T. N. Peck, R. A. Walterson, G. E. Stillman, T. S. Low, and B. Skromme, *J. Electron. Mater.,* vol. 12, p. 681, 1983.

121. M. G. Craford and W. O. Groves, *Proc. IEEE,* vol. 61, p. 862, 1973.

122. W. D. Johnston, Jr., and W. M. Callahan, in *Gallium Arsenide and Related Compounds, Inst. Phys. Conf.,* ser. 33a, 1977, p. 311.

123. G. B. Stringfellow and H. T. Hall, *J. Crystal Growth,* vol. 43, p. 47, 1978.

124. G. B. Stringfellow, *Organometallic Vapor-Phase Epitaxy, Theory and Practice,* Academic Press, San Diego, Calif., 1989.

125. J. J. Coleman and P. D. Dapkus, in *Gallium Arsenide Technology,* D. K. Ferry (ed.), SAMS, Inc., Indianapolis, Ind., 1985, p. 86.

126. K. F. Jensen, D. I. Fotiadis, and T. J. Mountziaris, *J. Crystal Growth,* vol. 107, p. 1, 1991.

127. P. R. Hageman, M. H. J. M. de Croon, J. N. K. Reek, and L. J. Giling, *J. Crystal Growth,* vol. 116, p. 169, 1992.

128. C. H. Chen, D. S. Cao, and G. B. Stringfellow, *J. Electron. Mater.,* vol. 17, p. 67, 1988.

129. A. Y. Cho, *J. Crystal Growth,* vol. 161, p. 1, 1991.

130. B. A. Joyce, J. Zhang, T. Shitara, J. H. Neave, A. Taylor, S. Armstrong, M. E. Pemble, and C. T. Foxon, *J. Crystal Growth,* vol. 115, p. 338, 1991.

131. M. B. Panish, H. Temkin, and S. Sumski, *J. Vac. Sci. Technol.,* vol. B3, p. 657, 1985.

132. M. B. Panish, *J. Electrochem. Soc.,* vol. 127, p. 2729, 1980.

133. L. Goldstein, *J. Crystal Growth,* vol. 105, p. 93, 1990.

134. W. T. Tsang, *Appl. Phys. Lett.,* vol. 45, p. 1234, 1984.

135. A. C. Jones, C. R. Whitehouse, T. Martin, and P. A. Lane, in *Conf. Dig., 6th Int. Conf. on MOVPE,* (Cambridge, Mass., June, 1992).

136. C. R. Abernathy, A. S. Jordan, S. J. Pearton, W. S. Hobson, D. A. Bohling, and G. T. Muhr, *Appl. Phys. Lett.,* vol. 56, p. 2654, 1990.

137. M. Weyers, *J. Crystal Growth,* vol. 107, p. 1021, 1991.

138. A. Usui and H. Watanabe, *Ann. Rev. Mater. Sci.,* vol. 21, p. 185, 1991.

139. K. Adomi, J. I. Chyi, S. F. Fang, T. C. Shen, S. Strite, and H. Morkoc, *Thin Solid Films,* vol. 205, p. 182, 1991.

140. M. A. Herman, *Vacuum,* vol. 42, p. 61, 1991.

141. T. Suntola and J. Autson, U. S. Patent no. 4058430, 1977.

142. Y. Horikoshi, M. Kawashima, and H. Yamaguchi, *Appl. Phys. Lett.,* vol. 50, p. 1686, 1987; *Jpn. J. Appl. Phys.,* vol. 27, p. 169, 1988.

143. *Proc. Topical Conf. on Silicon Based Heterostructures, J. Vac. Sci. Technol.,* vol. B9, p. 2011, 1991.

144. G. L. Patton, J. H. Comfort, B. S. Meyerson, E. F. Crabbe, G. J. Scilla, E. de Fresart, J. M. C. Stork, J. Y. C. Sun, D. L. Harame, and J. N. Burghartz, *Electron Device Lett.,* vol. 11, p. 171, 1990.

145. Y. Ota, *J. Vac. Sci. Technol. A,* vol. 2, p. 393, 1989.

146. S. S. Rhoc, R. P. G. Karunasiri, C. H. Chern, J. S. Park, and K. L. Wang, *J. Vac. Sci. Technol. B,* vol. 7, p. 327, 1989.

147. B. S. Meyerson, K. J. Uram, and F. K. LeGoues, *Appl. Phys. Lett.,* vol. 53, p. 2555, 1988.

148. *Proc. 1991 U.S. Workshop on the Physics and Chemistry of Mercury Cadmium Telluride and Other II-VI Compounds, J. Vac. Sci. Technol. B,* vol. 10, p. 1352, 1992.

149. O. K. Wu and G. S. Kamath, *Semicond. Sci. Technol.,* vol. 6, p. C6, 1991.
150. L. M. Smith and J. Thompson, *GEC J. Res.,* vol. 7, p. 72, 1989.
151. J. Mullin, D. J. Cole-Hamilton, S. J. C. Irvine, J. E. Hails, J. Gross, and J. S. Gough, *J. Crystal Growth,* vol. 101, p. 1, 1990.
152. D. W. Kisker, *J. Crystal Growth,* vol. 98, p. 127, 1989.
153. T. D. Golding, H. D. Shih, J. T. Zborowski, W. C. Fan, C. C. Horton, P. C. Chow, A. Vigliante, B. C. Covington, A. Chi, J. M. Anthony, and H. F. Schacke, *J. Vac. Sci. Technol. B,* vol. 10, p. 880, 1992.
154. K. K. Choi, B. F. Levine, C. G. Bethea, J. Walker, and R. J. Malik, *Appl. Phys. Lett.,* vol. 50, p. 1814, 1987.
155. S. Nishino, J. A. Powell, and H. A. Will, *Appl. Phys. Lett.,* vol. 42, p. 460, 1983.
156. P. Liaw and R. F. Davis, *J. Electrochem. Soc.,* vol. 132, p. 642, 1985.
157. P. Pirouz, C. M. Chorey, and J. A. Powell, *Appl. Phys. Lett.,* vol. 50, p. 221, 1987.
158. J. D. Parsons, in *Proc. Mater. Res. Soc. Symp.,* vol. 97, p. 271, 1987.
159. J. A. Powell, D. J. Larkin, I. G. Matus, W. J. Choyke, J. L. Bradshaw, L. Henderson, M. Yoginathan, J. Yang, and P. Pirouz, *Appl. Phys. Lett.,* vol. 56, p. 1442, 1990.
160. R. F. Davis, *Proc. IEEE,* vol. 79, p. 702, 1991.
161. R. C. Powell, G. A. Tomasch, Y. W. Kim, J. A. Thornton, and J. E. Greene, in *Proc. Mater. Res. Soc. Symp.,* vol. 162, p. 525, 1990.
162. T. P. Humphreys, C. A. Sukow, R. J. Nemanichy, J. B. Posthill, R. A. Rudder, S. V. Hattangady, and R. J. Markunas, in *Proc. Mater. Res. Soc. Symp.,* p. 531, 1990.
163. S. Yoshida, S. Misawa, and S. Gonda, *Appl. Phys. Lett.,* vol. 42, p. 427, 1983; *J. Vac. Sci. Technol. B,* vol. 1, p. 250, 1983.
164. H. Amato, N. Sawaki, I. Akasaki, and Y. Toyoda, *Appl. Phys. Lett.,* vol. 48, p. 353, 1986.
165. M. A. Khan, J. N. Kuznia, J. M. Van Hove, D. T. Olsen, S. Krishnankutty, and R. M. Kolbas, *Appl. Phys. Lett.,* vol. 58, p. 526, 1991.
166. H. Amamo, M. Kito, K. Hiramatsu, and I. Akasaki, *Jpn. J. Appl. Phys.,* vol. 28, p. L2112, 1989.

CHAPTER 5
METALS

Don E. Harrison and John D. Harrison

5.1 INTRODUCTION

The aim of this chapter is to provide the electronics engineer with information on the properties of metals that may influence the design or affect the performance of electronic devices. A summary is given on the composition, fabrication, and properties of metals to indicate the range of alloys available for structural applications. In subsequent sections more detailed information is given on the properties of metals of importance in electronics applications.

5.2 CHARACTERISTICS OF METALS

5.2.1. Composition

Chemical composition is the most commonly specified criterion for metals. Composition determines the basic mechanical and physical properties of a metal or alloy. Other factors, such as fabrication method and heat treatment, will influence these properties, but the composition determines the fundamental characteristics. Many metals are used in the "pure" state, free of all but impurity elements. Addition of alloying elements is made to alter the physical or mechanical properties of pure metal.

5.2.2 Material Manufacturing Variables

Alloys having the same composition can differ in their physical and mechanical properties because of the fabrication methods used to produce the desired shape. Metals used in the manufacture of electronic equipment are usually either cast or wrought materials.

Wrought Metals. The majority of metals and alloys are furnished in a wrought form, that is, the metal was subjected to some working operation that deformed it while it was in the solid state. Among these working operations are rolling, drawing, extruding, and forging. They affect the grain size, crystallographic orientation, homogeneity, soundness, size and shape of inclusions, size and shape of metallurgical phases, and so on. These microstructural factors affect the strength, ductility, formability, and the directional physical and mechanical properties of a metal as well as its behavior during subsequent heat treatment.

Cast Metals. Cast metal objects are made by pouring molten metal into a mold of the desired shape. No metal deformation occurs, and therefore the properties of the metal are dependent on the casting process, soundness of the casting, chemical composition, and heat treatment. A major factor in the properties of cast metals is their soundness. Nearly all castings have some defects, such as porosity, gas holes, and shrinkage voids. The grain size of a casting decreases as the cooling rate increases, and small grain size generally means increased strength. Thus a small thin-wall casting will be of higher strength than a thicker-wall massive casting of equal soundness. Not all metals can be satisfactorily cast into equally complex shapes, although practically all metals are cast into ingots for working into shapes by deformation methods. In the iron, copper, aluminum, magnesium, and zinc alloy families, specific alloys have been developed for casting and for wrought alloys. Some alloys are available in both wrought and cast forms, but most are available in one form or the other.

5.2.3 Strengthening Mechanisms

Strain Hardening. Cold working increases the strength and hardness of metals. Practically all metals strain-harden to some extent, and this is the only means available for strengthening some alloys. Care must be taken to ensure that manufacturing processes (for example, brazing and welding) do not exceed the recrystallization temperature. Otherwise the metal will revert back to the annealed state, with the loss of the strain-hardening effect.

Precipitation Hardening (Age Hardening). In many alloy compositions the solid solubility of one constituent in the other increases with temperature. By heating and quenching, solute can be retained in solution. This process is known as solution treating. Subsequent aging at a suitable temperature precipitates out the solute uniformly throughout the crystallographic structure, causing lattice distortions, which result in strengthening. Alloys of this type offer the advantage that forming can be done in the annealed or soft condition, with subsequent heat treatment being used to achieve higher strength. Parts may be welded, brazed, or deformed during fabrication, and then brought to high strength after these operations by solution treating and aging. Many metals can be solution-treated and then cold-worked and aged to achieve higher strength than could be attained by either heat treatment or cold working alone.

Phase-Transformation Hardening. Steel, cast iron, and certain titanium alloys can be hardened by manipulation of phase transformations. This mechanism takes advantage of the fact that a normally two-phase alloy can be converted to a single-phase alloy by heat treatment at elevated temperatures. This single phase is retained by quenching to room temperature. Subsequent tempering at moderately elevated temperature causes the alloy to revert to the two-phase structure, but one with much finer particles of each phase and having greater phase dispersion than before. This fine structure increases both strength and hardness.

5.2.4 Mechanical Properties

After chemical composition the most commonly specified property is mechanical strength. Usually this includes ultimate strength, yield strength, and percentage elongation as determined on a standard tensile test bar. These properties are often all that needs to be known in order to make a material selection for mechanical design.

However, these values represent data taken on machined test bars at room temperature under one short-time loading. A number of factors can lower the design-allowable strength to values below the published levels. They include the following.

Repeated Loads (Fatigue). When loads are repeated, the allowable stress is decreased to well below the tensile strength of the material. In fact, with most metals subjected to 10 million stress cycles, the breaking stress can be one-third of the tensile strength. In considering fatigue data, attention should be paid to the type of specimen, the type of loading, and the surface of the test specimen. Notches on the surface can further reduce fatigue strength to as low as one-fifth of the tensile strength. Improvement in fatigue life is brought about by improving surface finish, by working the surface by shot peening, and by other methods of introducing a compressive stress in the surface.

Sustained Loads. Failure under long-time sustained loads may occur as a result of creep or fracture at stresses below the yield strength as determined by the standard tensile test. Although this type of failure is associated with relatively high temperatures, creep rupture can occur at room temperature.

Temperature Effects. The strength of metals decreases as temperatures increase. Ductility usually increases until temperatures near the melting point are reached. The reverse occurs with decreasing temperatures, that is, strength increases and ductility decreases. With the exception of steel, decreasing ductility with decreasing temperature is not particularly severe in most alloys. No drastic embrittlement occurs even at cryogenic temperatures. Therefore room-temperature tensile test values are useful measures for selecting materials. However, service conditions and type of loading can be important factors in determining a safe design stress.

5.2.5 Effect of Heat Treatment on Physical Properties

Heat treatment of a metal will affect electrical conductivity, thermal conductivity, density, thermal expansion, and sometimes the elastic modulus. Most of these changes are insignificant for engineering purposes. However, electrical conductivity differences can be quite dramatic. Precipitation-hardenable alloys are especially prone to changes in electrical conductivity with changes in heat treatment. The solution-treated condition usually has the lowest conductivity, since the alloying elements are in solution, where they increase electron scatter. Annealing results in agglomeration of the alloying elements, essentially clearing the matrix of scattering centers, and thereby increasing the electrical conductivity. Aging of solution-treated materials will usually result in some increase in electrical conductivity, but not to the degree achieved by annealing. Owing to atomic rearrangement, some metals change in density when heat-treated. Beryllium-copper is one such material. Beryllium-copper is also an example of a material that can undergo a change in elastic modulus when heat-treated.

5.2.6 Manufacturing Considerations

Economics of fabrication as well as engineering performance may determine materials selection. Awareness of the physical and mechanical properties that most affect fabricating operations aids in materials selection.

Formability. Tensile strength values provide a reasonable guide for selecting a formable material. Percentage elongation is a good indicator of how much a material can be deformed. Also, a large difference between the ultimate tensile strength and the yield strength indicates good formability. Minimum bend radii are also published for most materials and are sometimes included in specifications. For drastic forming operations such as deep drawing, more sophisticated specifications of material properties are required. Grain-size control is one of these. Data on optimum grain size along with minimum elongation exist for most materials.

Machinability. Most alloys can be machined without great difficulty. Many alloys are made with specific alloy additions which enhance machinability. The additions usually have the advantage that small chips are formed during machining. Many of these alloys have the disadvantage of being less formable, more brittle, and less weldable because of these additions. Care should be taken to ensure that the free-machining additions do not interfere with other fabrication operations.

Joining. Correct alloy selection for metals joining is important. Often the alloy selected determines the joining method. Bolting, riveting, and other fastening methods present few problems for most metals, but the metallurgical processes of welding, brazing, and soldering require considerable thought.

Weldability ratings are qualitative and are generally based on the ease of welding by conventional techniques. A metal or alloy may get a poor weldability rating for a number of reasons, including (1) tendency to crack during welding, (2) poor corrosion resistance of the weld, or (3) brittleness of the weld. Weldability is a function of the process used for welding. Much of the success in welding reactive metals such as aluminum, magnesium, and titanium is due to the use of the inert-gas shielded-arc method, in which an inert gas is used to shield the molten metal from the oxidizing effects of air. Electron-beam and laser-beam welding methods that give narrow deep-penetration welds permit successful welding of alloys and combinations of alloys formerly considered unweldable.

Brazing alloy selection must consider the melting point of the alloy to be brazed in comparison to the melting point of the brazing alloy. The effect of the brazing temperature on producing further heat treatment of the part should also be considered. If it is desired to have the final properties of a part in a heat-treated condition, the brazing alloy must have a melting point above the temperature of the required heat treatment. The method used for brazing is also important. Torch brazing locally heats the joint and may result in distortion. Furnace brazing requires heating of the whole part, which results in less distortion, but requires fixturing in most situations. Induction brazing gives fast, localized heating. Dip brazing involves immersing the area to be brazed into a molten flux bath.

Soldering is a low-temperature process that can be used on a wide variety of metals. Some metals such as aluminum and magnesium alloys are soldered with great difficulty. These alloys may be plated with solderable coatings and then soldered like other metals. Solders are generally galvanically different from the base metal. Consequently, solder joints may have poor corrosion resistance.

5.3 COMPARATIVE DATA ON METALS

Data on the range of properties available in pure metals and alloys are presented in Figs. 5.1 to 5.8.

5.3.1 Yield-Strength Comparisons

The yield strength of a metal is generally used as the maximum permissible stress to which a part may be subjected in service. Figure 5.1 compares the yield strengths of a number of alloys. There is a wide overlap of yield strengths among commonly used alloys. At stress levels below 40,000 lb/in^2 most alloy groups have members which are adequately strong. Therefore selection is made either on the basis of some other property or on the basis of economics.

5.3.2 Modulus of Elasticity and Stiffness

Figure 5.2 compares the elastic moduli of a number of alloys. Since this is the property that indicates the stiffness of a part, it plays an important role in design. There is little variation in elastic moduli within an alloy group, even though yield strengths may vary as much as 2000 percent. The specific stiffness values (the quotient of elastic modulus divided by density) of various alloys are compared in Fig. 5.3. Commonly used structural materials such as steel, aluminum, magnesium, and titanium have similar stiffnesses, whereas nickel and copper alloys have somewhat lower values. The outstanding material in this property is beryllium, with a value of 6 to 12 times that of the common structural materials.

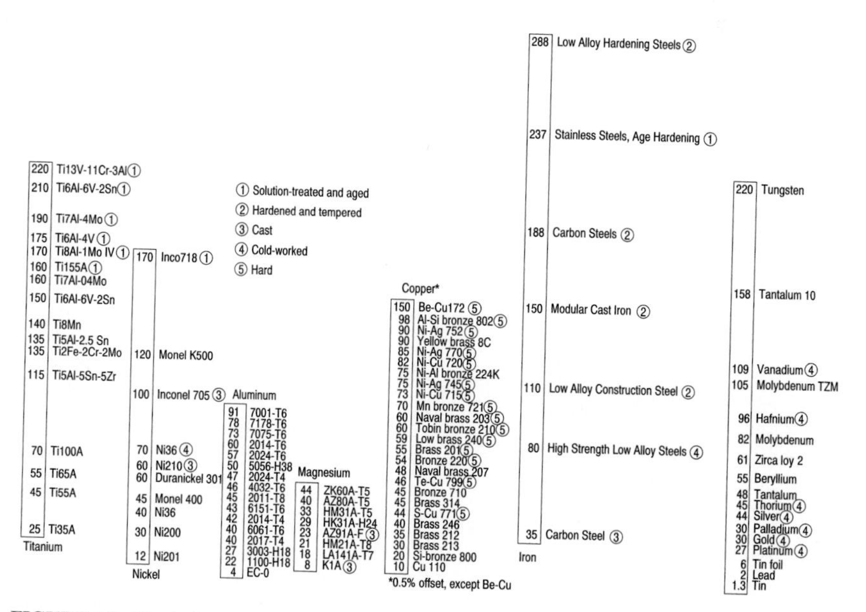

FIGURE 5.1 Typical values of tensile yield strength of some metals at room temperature (10^3 lb/in^2).

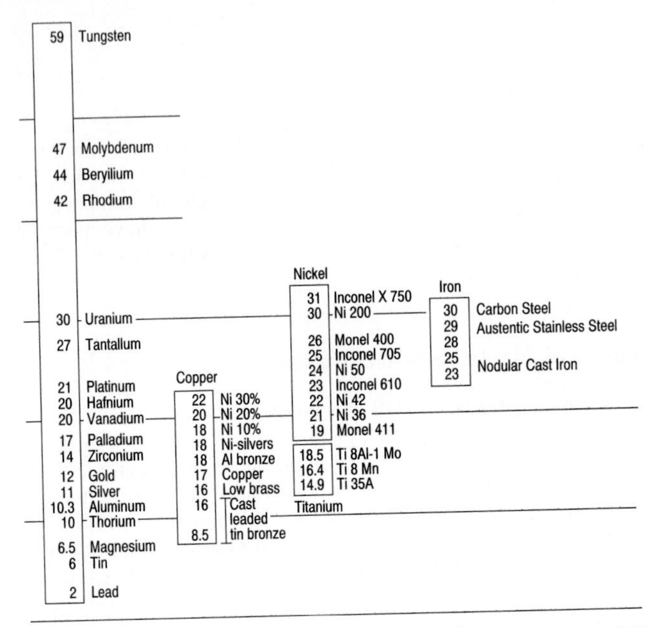

FIGURE 5.2 Modulus of elasticity of some metals at room temperature (10^{-6} lb/in^2).

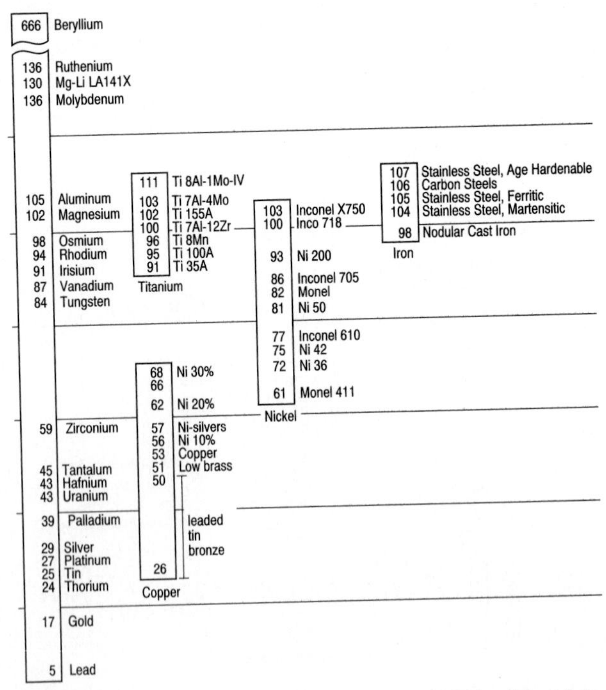

FIGURE 5.3 Specific stiffness of some metals at 70°F (modulus of elasticity/density × 10^{-6} in).

5.3.3 Strength-Density Ratios

In Fig. 5.4, the strength-to-weight ratios of a number of materials are compared. This figure shows that, when compared on a weight basis, some of the weaker low-density materials are superior to the high-strength high-density materials. Figure 5.5 compares the effect of temperature on the yield strength-to-density ratios of several alloys. This figure illustrates that room-temperature comparisons do not necessarily hold at elevated temperatures.

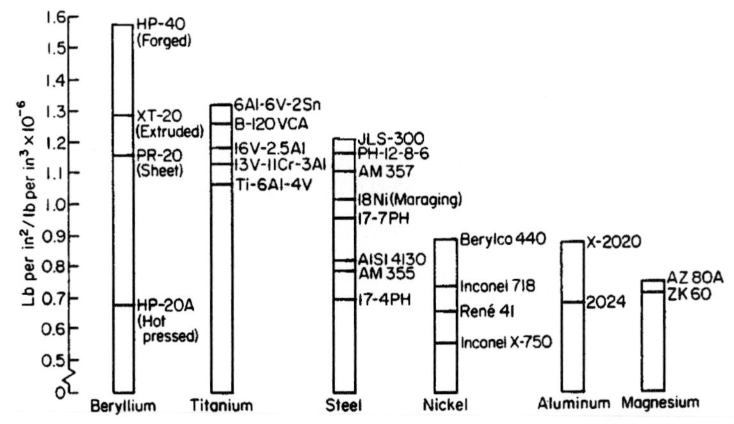

FIGURE 5.4 Ultimate tensile strength-to-density ratios for various structural materials at room temperature.

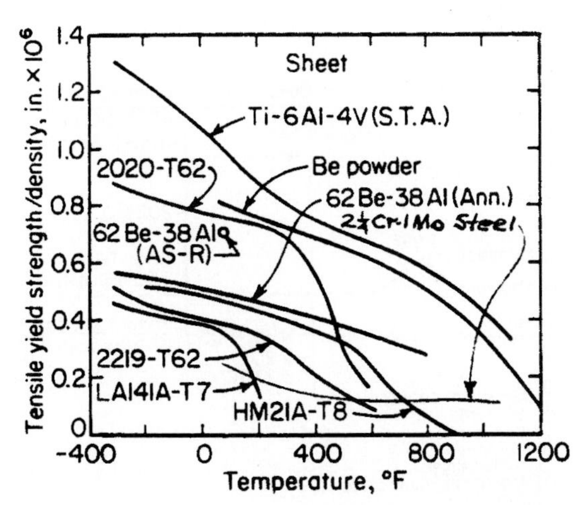

FIGURE 5.5 Comparison of tensile yield-to-density ratios for some structural metals at various temperatures.

5.3.4 Thermal Conductivity

In Fig. 5.6 thermal conductivities are compared. Metals exhibit the highest conductivity in the pure state. Alloying lowers the conductivity, sometimes drastically. Silver, copper, gold, and aluminum, among the metals, and diamond (C^{13}) have the highest thermal conductivities.

5.3.5 Electrical Resistivity

Figure 5.7 graphs the comparative electrical resistivities of certain metals. Alloying will drastically affect electrical resistivity.

5.3.6 Heat Absorption

Figure 5.8 gives a comparison of the ability of metals to absorb heat. These values are for pure metals (with the exception of steel). In Table 5.1 the thermal properties of a number of pure metals are listed.

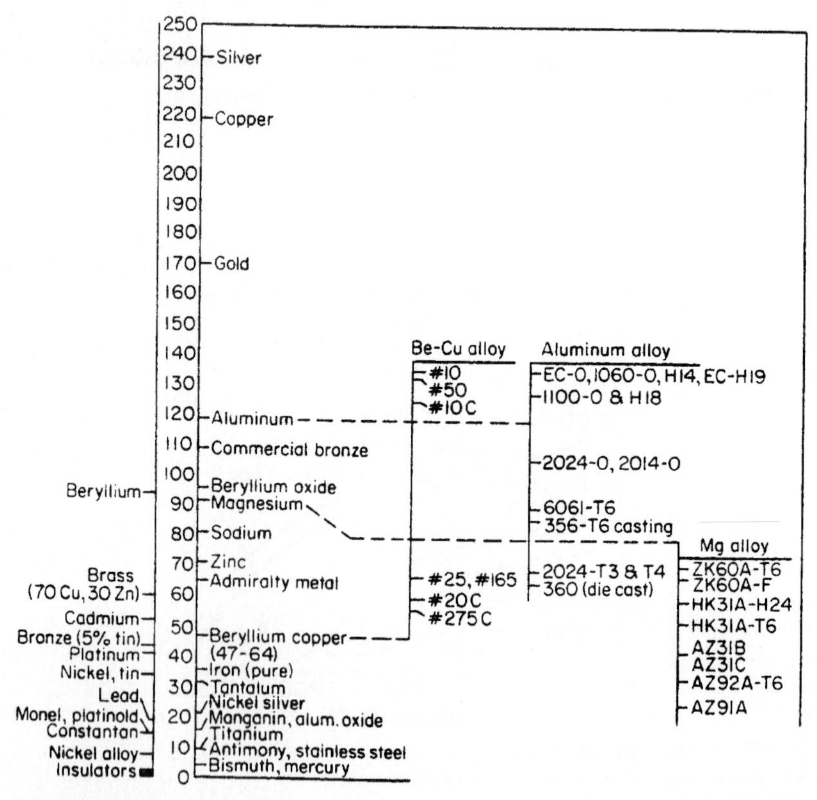

FIGURE 5.6 Thermal conductivity at 212°F for some materials used in electronic equipment.

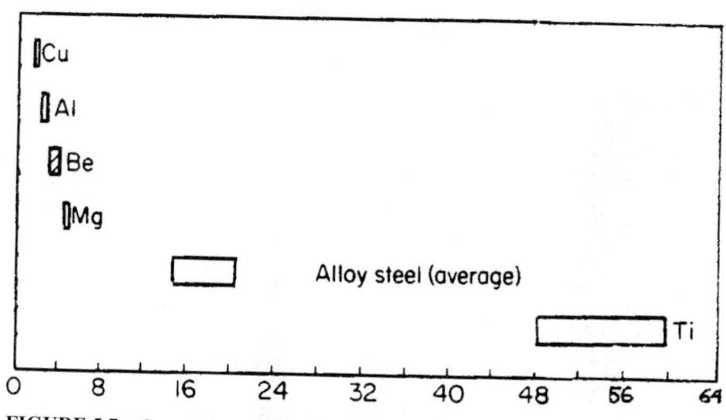

FIGURE 5.7 Comparison of electrical resistivities for several pure metals.

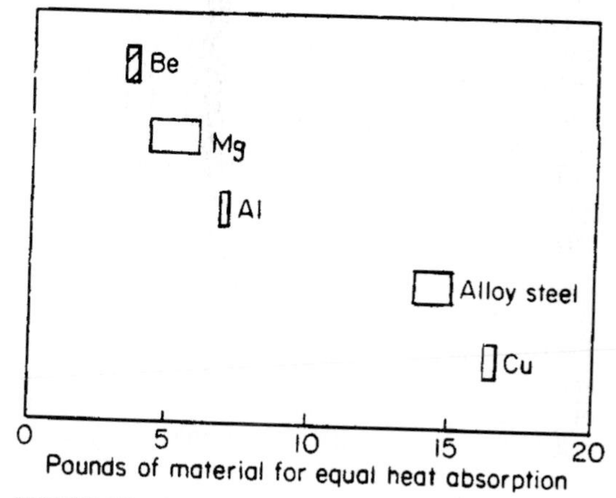

FIGURE 5.8 Comparison of heat-absorption abilities for several pure metals.

TABLE 5.1 Thermal Properties of Several Pure Metals

Property	Be	Mg	Al	Ti	Fe	Cu
Melting temperature						
°C	1277	650	660	1668	1536	1083
°F	2332	1202	1220	3035	2797	1981
Boiling temperature						
°C	2770	1107	2450	3260	3000	2595
°F	5020	2025	4442	5900	5430	4703
Thermal conductivity, cal/(s • cm • °C)	0.35	0.367	0.53	0.41	0.18	0.94
Linear thermal expansion						
μin/in • °C	11.6	27.1	23.6	8.41	11.7	16.5
μin/in • °F	6.4	15.05	13.1	4.67	6.5	9.2
Specific heat, cal/g • °C	0.45	0.25	0.215	0.124	0.11	0.092
Heat of fusion, cal/g	260	88	94.5	104*	65	50.6

*Estimated

Source: From *Electronic Materials & Processes Handbook, Edition 1.*

5.4 CORROSION

Corrosion is the inevitable result of exposing unprotected metals, such as plain carbon steel, zinc, copper, or aluminum, to an environment that is humid (70 to 80 percent relative humidity), contaminated with chlorides or sulfates, and which undergoes a diurnal temperature drop of 15 to 20°F. (Boyer and Gall,[1] pp. 29.51–29.54; 32.20–32.29)

5.4.1 Fundamentals of Electrochemical Corrosion

Electrochemical corrosion of metals in a natural environment, whether in the atmosphere, in water, or underground, is caused by a flow of electricity from one metal to another, or from one part of a metal surface to another part of the same surface where conditions permit the flow of electricity. For the flow of energy to take place, either a moist conductor or an electrolyte must be present. An electrolyte is an electricity-conducting solution containing ions, which are atomic particles or radicals bearing an electrical charge. Ions are present in solutions of acids, alkalis, and salts. Water, especially salt water, is an excellent electrolyte. Ions pass from a negative area to a positive area through the electrolyte. For electrochemical corrosion to occur in metals, there must be (1) an electrolyte, (2) an area or region on a metallic surface with a negative charge, (3) a second area with a positive charge, and (4) an electrically conductive path between (2) and (3). These components are arranged to form a closed electric circuit. In the simplest case, the anode would be one metal, such as iron, the cathode would be another, perhaps copper, and the electrolyte might or might not have the same composition at both anode and cathode. Furthermore, the anode and the cathode could be of the same metal.

The cell shown in Fig. 5.9 illustrates the corrosion process in its simplest form. This cell includes the following essential components: (1) an electrolyte in contact with the anode and the cathode, (2) a metal anode, (3) a metal cathode, and (4) a

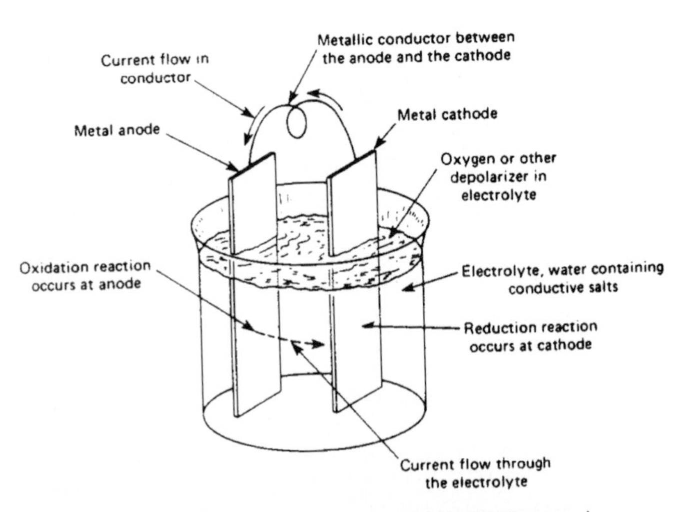

FIGURE 5.9 Simple cell showing components necessary for corrosion.

metallic conductor between the anode and the cathode. If this cell were constructed and allowed to function, an electric current would flow through the metallic conductor and the electrolyte, and if the conductor were replaced by a voltmeter, a potential difference between the anode and the cathode could be measured. The anode would corrode. Chemically this is an oxidation reaction. During metallic corrosion, the rate of oxidation equals the rate of reduction. Thus a nondestructive chemical reaction, reduction, proceeds simultaneously at the cathode. In most cases hydrogen gas is produced on the cathode. When the gas layer insulates the cathode from the electrolyte, current flow stops, and the cell is polarized. However, oxygen or some other depolarizing agent is usually present to react with the hydrogen, which decreases this effect and allows the cell to continue to function. Contact between dissimilar metallic conductors or differences in the concentration of the solution cause the difference in potential that results in electric current. Any lack of homogeneity on the metal surface or in its environment may initiate attack by causing a difference in potential, and this results in localized corrosion. The metal undergoing electrochemical corrosion need not be immersed in a liquid, but may be in contact with moist soil, or may have moist areas on its surface. Consult Evans[2] and Uhlig[3] for a more comprehensive understanding of corrosion phenomena.

5.4.2 Types of Corrosion

The various types of corrosion may be classified into nine categories: uniform corrosion, pitting corrosion, selective leaching, intergranular corrosion, crevice corrosion, galvanic corrosion, erosion corrosion, stress corrosion, and corrosion fatigue.

Uniform Corrosion. Metals most commonly attacked by uniform, or general, corrosion are plain carbon steels, high-strength low-alloy weathering-type steels, zinc and galvanized steel, cadmium-coated steel, and, to a more limited extent, copper and copper alloys. Uniform corrosion of steels takes place as the result of electrochemical action between closely spaced microanodes and microcathodes on the metal surface. In the case of zinc and galvanized steel exposed to the atmosphere, the acidity of a uniform film of condensed dew is neutralized by reacting with zinc carbonate on the surface of the zinc. A similar reaction results in the formation of a patina on copper and its alloys. However, in this instance the corrosion products are insoluble toward rain water and seal the surface against further attack. The same phenomenon occurs in weathering-type steels, where a tightly adherent impervious oxide film develops that protects the substrate from further atmospheric attack. In contrast, the rust or corrosion product on plain carbon steel is not protective and merely slows the rate of attack. The corrosion products that form on zinc and cadmium in industrial environments are not completely protective. They result in limited service life, because they form water-soluble cadmium and zinc sulfates. However, in a rural environment the basic zinc carbonate film that forms offers a long service life. In a marine atmosphere, a protective zinc chloride film forms on the zinc coating and extends its service life.

Pitting Corrosion. Pitting is a type of localized cell corrosion. Pitting develops when the anodic or corroding area is small in relation to the cathodic or protected area. For example, pitting can occur where large areas of the surface are covered by applied coatings, or deposits of various kinds, and breaks exist in the continuity of the protective coating. Pitting may also develop on bare, clean metal surfaces because of irregularities in the physical or chemical structure of the metal. Electrical contact between dissimilar materials or concentration cells (areas of the same metal where oxygen or conductive salt concentrations in water differ) accelerates the rate of pitting. Severe pitting is often caused by concentration differences in the electrolyte, especially dissolved oxygen. When part of the metal is in contact with water relatively low in dissolved oxygen, it is anodic to adjoining areas in contact with water higher in dissolved oxygen. This lack of oxygen may be caused by exhaustion of dissolved oxygen within the confines of a crevice (Fig. 5.10). Pits require a long incubation period before they are made evident by sudden failure. Pits grow downward and rarely originate on vertical walls, except in the case of a break in a protective coating under conditions of total immersion. Pit growth is autocatalytic. Once begun, the rate of penetration by such pits is accelerated proportionately as the bottom of the pit becomes more anodic as a result of oxygen depletion and increased metal ion concentration in the electrolyte. In the case of stainless steel and aluminum, the presence of chloride ions in a slightly acidic solution can result in widely scattered pit formation. The chloride ion is thought to be responsible for disrupting the passive or protective films on stainless steel and aluminum. Grains of sand lying on a damp stainless-steel surface can stimulate pit formation beneath the grains by shutting off access to oxygen. Corrosion inhibitors can stimulate attack if present in lower-than-optimum concentrations in aqueous solutions. Other stimuli for pitting are the bubbling of air against a submerged surface, residual stresses along a line of bending in fabrication, and certain types of inclusions inherent in the metal processing.

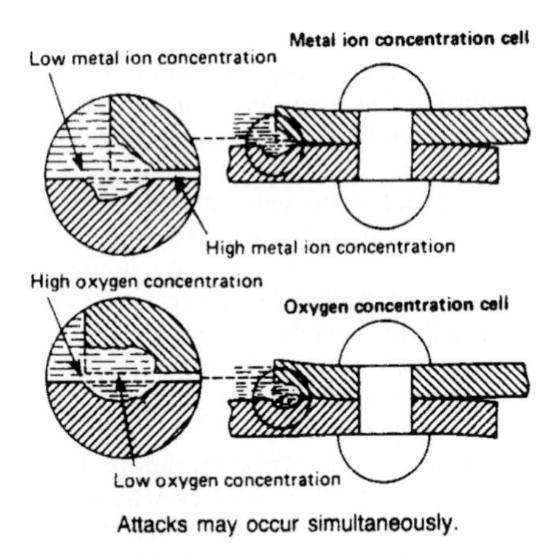

FIGURE 5.10 Corrosion caused at crevices by concentration cells.

Selective Leaching. Selective leaching is the removal of an element from an alloy by corrosion. Two mechanisms can result in selective leaching: (1) one metal is selectively dissolved, leaving the other metals behind, and (2) two metals in an alloy are dissolved, and one redeposits on the surface. The loss of molybdenum from nickel alloys in molten sodium hydroxide occurs by the first mechanism while dezincification of brasses occurs by the second. Dezincification occurs in brasses containing less than 85 percent copper. Zinc corrodes preferentially, leaving a porous residue of copper and corrosion products. Alpha brass containing 70 percent copper and 30 percent zinc (C 26000) is particularly susceptible to dezincification when exposed in an aqueous electrolyte at elevated temperatures. Dezincification can proceed in the absence of oxygen, as evidenced by the fact that zinc corrodes slowly in pure water. However, dissolved oxygen increases the rate of attack. Analyses of dezincified areas usually show 90 to 95 percent copper, with some of it present as copper oxide. With many alloys, selective leaching is not readily detected by visual examination. A copper flash is usually visible on dezincified copper alloys, but this is not positive evidence of dezincification because a copper flash can deposit from even small amounts of copper salts in aqueous solution without the occurrence of dezincification.

Intergranular Corrosion. Intergranular corrosion of stainless steels is encountered most frequently in welded assemblies. Although the mechanical properties of a welded steel structure are hardly affected, exposure to an aggressive environment, such as seawater spray, can cause extensive damage by selective corrosion of the grain-boundary region. During welding, carbon diffuses to the grain boundaries, where it forms carbides. This carbide formation can deplete the region adjacent to the grain boundary of chromium to below the 12 percent level needed to maintain a passive state. Chromium depletion results in the creation of a strong active-passive cell between a large cathodic grain and a small anodic grain boundary. Under load, mechanical failure occurs along these weakened grain boundaries. This phenomenon is called *heat sensitization.* For austenitic stainless steels the sensitizing temperature range is between 1550 and 750°F. Slow cooling through this range results in a vulnerable situation, whereas rapid cooling avoids any serious damage. The extent of damage is based on the length of time the metal is held at either the top or the bottom of the sensitizing temperature range. The degree of sensitization is a function of the carbon content of the metal. An 18-8 stainless steel containing 0.1 percent carbon or more will be extensively sensitized after only a 5-min exposure at 1080°F, whereas a steel containing 0.03 percent carbon sustains very little damage. Heating the welded structure to about 2000°F to dissolve the carbides, then cooling rapidly through the sensitizing temperature range, will avoid the problem. Alternatively, alloy additions of titanium or niobium that tie up carbon preferentially can be used to avoid heat sensitization.

Crevice Corrosion. In the presence of an electrolyte, disproportionate access of air to a crevice results in the formation of a differential aeration cell. Even mud deposits can create conditions conducive to crevice corrosion. More recognizable crevices occur at back-to-back angles on transmission towers, bolted joints of steel and aluminum, the fretting surfaces in bolted joints, and accidental or design arrangements involving contact of metal with wood, rubber, plastic, glass, wax, asbestos, concrete, and so on. The contact area is generally of a capillary nature and thus draws in a small amount of liquid from which the air is quickly exhausted in comparison with the liquid film outside the capillary, which is constantly saturated with air. Aluminum and stainless steels are notoriously susceptible to attack in oxygen-deficient situations. Bolted joints, unless they are checked for tightness or involve the use of high-strength bolts, can undergo such attack beneath the bolt head or beneath the washer as well as within the fretting surfaces of the joint (Fig. 5.10).

Galvanic Corrosion. The potential for electrochemical corrosion between dissimilar metals can be estimated from the galvanic series that lists metals and alloys arranged according to their tendency to corrode when in galvanic contact (Table 5.2). Metals close to one another in the table generally do not have a strong effect on each other. However, the more any two metals are separated, the stronger is the corroding effect on the one higher in the list. The galvanic series should not be confused with a similar electromotive force series, which gives exact potentials based on highly standardized conditions which rarely exist in nature. A galvanic series based on immersion in seawater is a useful indicator of the rate of corrosion between different metals or alloys when they are in contact in an electrolyte (Table 5.2).

TABLE 5.2 Galvanic Series in Seawater

Corroded end (anodic, or least noble)
Magnesium Magnesium alloys
Zinc Galvanized steel or galvanized wrought iron
Aluminum alloys 5052, 3004, 3003, 1100, 6053, in this order
Cadmium
Aluminum alloys 2117, 2017, 2024, in this order
High-strength low-alloy steel Plain carbon steel Cast iron
Ni-Resist (high-nickel cast iron)
Type 410 stainless steel (active)
50-50 lead-tin solder
Type 304 stainless steel (active) Type 316 stainless steel (active)
Lead Tin
Copper alloy 280 (Muntz metal, 60%) Copper alloy 675 (manganese bronze A) Copper alloys 464, 465, 466, 467 (naval brass)
Nickel 200 (active) Inconel alloy 600 (active)
Hastelloy B Chlorimet 2
Copper alloy 270 (yellow brass, 65%) Copper alloys 443, 444, 445 (admiralty brass) Copper alloys 608, 614 (aluminum bronze) Copper alloy 230 (red brass, 85%) Copper 110 (ETP copper) Copper alloys 651, 655 (silicon bronze) Copper alloy 715 (copper, nickel, 30%) Copper alloy 923, cast (leaded tin bronze G) Copper alloy 922, cast (leaded tin bronze M)
Nickel 200 (passive) Inconel alloy 600 (passive)
Monel alloy 400
Type 410 stainless steel (passive) Type 304 stainless steel (passive) Type 316 stainless steel (passive) Incoloy alloy 825
Inconel alloy 625 Hastelloy C Chlorimet 3
Silver
Titanium
Graphite
Gold
Platinum
Protected end (cathodic, or most noble)

Source: From Boyer and Gall.[1]

Galvanic action between two dissimilar metals may generate a noise voltage in the signal path of low-level circuitry. A galvanic couple may be the cause of premature failure in metal components of water-related structures or it may be advantageously exploited. Galvanizing iron sheet is an example of useful application of galvanic action or cathodic protection. Iron is the cathode and is protected against corrosion at the expense of the sacrificial zinc anode. Alternatively, a zinc or magnesium anode may be located in the electrolyte close to the structure and connected electrically to the iron or steel. This method is referred to as cathodic protection of the structure. Iron or steel becomes the sacrificial anode when in electrical contact with copper, brass, or bronzes when both metals are submerged in an electrolyte. Weld metal may be anodic to the basis metal, creating a corrosion cell when immersed (Fig. 5.11).

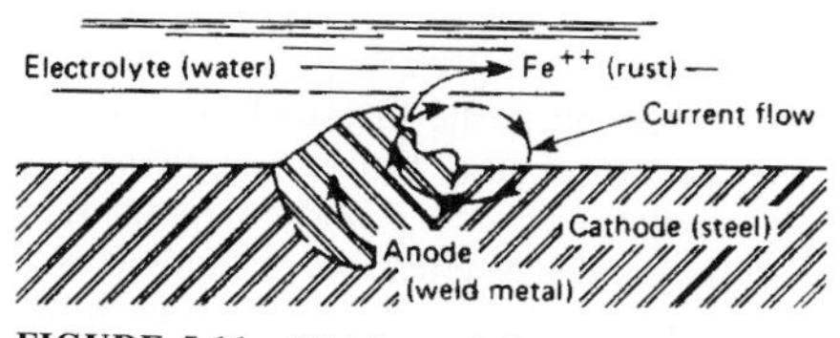

FIGURE 5.11 Weld metal forming a corrosion cell on steel.

Although the galvanic series represents the potential available to promote a corrosive reaction, the actual corrosion is difficult to predict. Electrolytes may be poor conductors, or long distances may introduce large resistance into the corrosion-cell circuit. More frequently, corrosion residues form a partially insulating layer over the anode. A cathode may become inactive because of adsorbed gas bubbles. The effect of such conditions is to reduce the theoretical consumption of metal by corrosion. The area relationship between the anode and the cathode may also strongly affect the corrosion rate; a high ratio of cathode area to anode area produces more rapid corrosion of the anode. In the reverse case, the cathode polarizes, and the corrosion rate soon drops to a negligible level. The passivity of stainless steels is attributed to either the presence of a corrosion-resistant oxide film or an oxygen-caused polarizing effect, durable only as long as there is sufficient oxygen to maintain the effect over the surfaces. In most natural environments, stainless steels will remain in a passive state and thus tend to be cathodic to ordinary iron and steel. Change to an active state usually occurs only where chloride concentrations are high, as in seawater or in reducing solutions. Oxygen starvation also produces a change to an active state. This occurs where the oxygen supply is limited, as in crevices and beneath contamination of partially fouled surfaces.

Erosion Corrosion. The erosion-corrosion phenomena show up in a variety of ways. One of the most common is that occurring in pipes through which a variety of corrosive and noncorrosive fluids flow at differing velocities. Protective films may develop such as the thin oxide films on aluminum and stainless-steel pipes, or the mixed lead oxide–lead sulfate films that form in lead pipes used to carry sulfuric acid. Generally simple lamellar or straight-line flow causes no problem until an obstruction appears and turbulent flow develops. Such turbulence can physically disrupt a protective film and expose bare metal to subsequent corrosion. This cycle repeats itself so that thinning and perforation occur at an accelerated rate. Disruption of the surface film can result in a form of galvanic corrosion in which the small exposed surface acts as an anode surrounded by a large (coated) cathode. Accelerated attack occurs at the anode. Protective films can be abraded by the force of cavitation created by the turbulence that is generated by changes in the diameter of piping or by bends and elbows. Water flowing at 39 ft/s can be aggressive at pH 3 to 5 and 7 to 9. In contrast, no damage occurs at pH 6 and 10 to 13. Corrosion inhibitors can minimize the corrosive effects of turbulent fluid action. The use of smooth interiors with few bends and protective coatings are means for mitigating the effects of fluid flow.

Stress Corrosion. Stress-corrosion cracking is produced by the synergistic action of tensile stresses, a specific environment, and a susceptible alloy, and it results in failure where one would not necessarily occur if only two of these conditions were met. The process of stress-corrosion cracking generally involves crack initiation, subcritical crack growth, and final failure when the stress-corrosion crack reaches a critical size such that the tensile strength or fracture toughness of the remaining material is exceeded. It is the subcritical crack growth phase of this process that is meant by stress-corrosion cracking. Although the physical manifestation of stress-corrosion cracking is obvious (that is, a cracked or broken sample or part), the mechanisms and atomic-level processes involved are many and complex. As a result, the effects of environmental and metallurgical variables are difficult if not impossible to predict. Despite many years of research and experience, stress-corrosion cracking remains the predominant cause of unexpected failures in many segments of industry, such as the petrochemical and chemical process industries, and is a continuing concern to the aerospace industry.

Corrosion Fatigue. Repeated or fluctuating stresses in a corrosive environment can produce progressive cracking known as corrosion fatigue. Once fatigue cracks have formed, corrosion at the crack tip may accelerate the rate of crack growth. The cracking process is affected by stress-cycle frequency, stress wave shape, stress ratio, and time-dependent environmental effects.

5.5 APPLICATIONS OF METALS IN ELECTRONICS

5.5.1 Electrical Contact Materials

Commercial materials for electrical contacts are divided into two categories based on the method used to form them, (1) wrought materials, which include both pure metals and alloys, and (2) composite materials, which include powder metallurgy products and internally oxidized silver alloys (Boyer and Gall,[1] pp. 20.16–20.20). Hunt[4] gives additional information on electrical contact materials.

Copper Metals. Copper is widely used in electrical contacts because of its high electrical and thermal conductivity, low cost, and ease of fabrication. The main disadvantages of copper contacts are low resistance to oxidation and corrosion. In many applications, the voltage drop resulting from the film developed by normal oxidation and corrosion is acceptable. In air, copper has moderate resistance to arcing, welding, or sticking. When these characteristics are important, copper-tungsten or copper-graphite mixtures are used. However, in a helium atmosphere, a Cu-CdO contact performs similarly to an Ag-CdO contact. Copper alloy contacts are used to carry high currents in vacuum interrupters. Pure copper is relatively soft, anneals at low temperatures, and

TABLE 5.3 Properties of Copper Metals Used for Electrical Contacts

UNS number	Solidus temperature °C	Solidus temperature °F	Electrical conductivity, % IACS	Hardness OS035 temper	Hardness H02 temper	Tensile strength OS035 temper MPa	Tensile strength OS035 temper ksi	Tensile strength H02 temper MPa	Tensile strength H02 temper ksi
C11000	1065	1950	100	40 HRF	40 HRB	220	32	290	42
C16200	1030	1886	90	54 HRF	64 HRB*	240	35	415*	60*
C17200	865	1590	15 to 33†	60 HRB‡	93 HRB¶	495‡	72‡	655¶	95¶
C23000	990	1810	37	63 HRF	65 HRB	285	41	395	57
C24000	965	1770	32	66 HRF	70 HRB	315	46	420	61
C27000	905	1660	27	68 HRF	70 HRB	340	49	420	61
C50500	1035	1900	48	60 HRF	59 HRB	276	40	365	53
C51000	975	1785	20	28 HRB	78 HRB	340	49	470	68
C52100	880	1620	13	80 HRF	84 HRB	400	58	525	76

* H04 temper.
† Depends on heat treatment.
‡ TB00 temper.
¶ TD02 temper.
Source: From Boyer and Gall.[1]

lacks the spring properties sometimes desired. Some copper alloys which are harder than pure copper and have better spring properties are listed in Table 5.3. However, improved spring properties are obtained only at the expense of electrical conductivity. Precipitation-hardened alloys, dispersion-hardened alloys, and powder metal mixtures can provide a wide range of mechanical and electrical properties.

Applications. Copper-base metals are commonly used in plugs, jacks, sockets, connectors, and sliding contacts. Contact force and length of slide must be kept high to overcome surface tarnish and avoid excess contact resistance and high levels of electrical noise. Yellow brass (C27000) is preferred for plugs and terminals because of its machinability. Phosphor bronze (C50500 or C51000) is preferred for thin socket and connector springs and for wiper switch blades because of its strength and wear resistance. Nickel silver is sometimes preferred over yellow brass for relay and jack springs because of its high modulus of elasticity and strength, and also for its resistance to tarnishing and better appearance. Sometimes copper alloy parts are nickel plated to improve surface hardness, reduce corrosion, and improve appearance. However, nickel forms a thin, hard oxide film that has high contact resistance; very high contact force and long slides are necessary to rupture this film. To maintain low levels of resistance and noise, copper contacts should be plated or overlaid with a precious metal.

Silver Metals. Silver, in pure or alloyed form, is the most widely used material for make-and-break contacts (1 to 600 A). The mechanical properties and hardness of pure silver are improved by alloying, but its thermal and electrical conductivity are adversely affected. Figure 5.12 shows the effect of different alloying elements on the hardness and electrical resistivity of silver. Properties of the principal silver metals used for electrical contacts are given in Table 5.4. Silver is widely used in contacts that remain closed for long periods of time, and in the form of electroplate it is widely

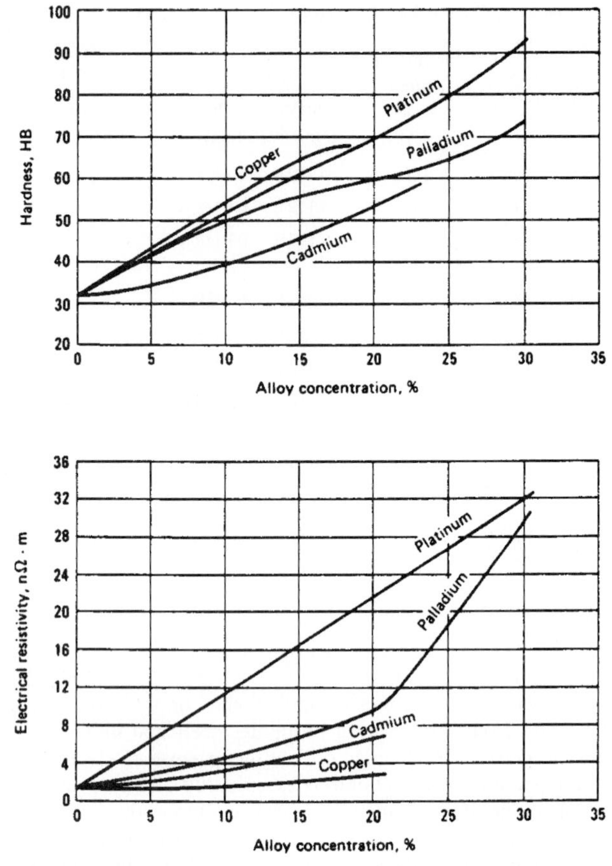

FIGURE 5.12 Hardness and electrical resistivity versus alloy content for silver alloy contacts.

used as a coating for connection plugs and sockets. It is also used on contacts subject to occasional sliding, such as in rotary switches, and to a limited extent for low-resistance sliding contacts such as slip rings.

The electrical and thermal conductivities of silver are the highest of all metals at room temperature, and as a result, silver will carry high currents without excessive heating, even when the dimensions of the contacts are only moderate. Although good thermal conductivity is desired once the contact is in service, such conductivity increases the difficulty of assembly welding. In contact with certain materials, such as phenol fiber, and under an electric potential, silver can migrate through the insulation and degrade its dielectric strength.

Oxidation resistance in air is the reason why silver contacts are used instead of copper. Although silver oxide has a high resistivity, it dissociates slowly upon heating at about 350°F, and rapidly at about 650°F. Arcing removes silver oxide. Silver is vulnerable to attack by sulfur or sulfide gases in the presence of moisture. The resulting film may produce significant contact resistance, particularly where contact force, voltage, or current is low. Direct current induces the migration of silver ions from the

TABLE 5.4 Properties of Silver Metals Used for Electrical Contacts

Alloy	Solidus temperature		Electrical conductivity	Hardness, HR15T		Tensile strength				Density,	Elongation in 50 mm or 2 in, %	
						Annealed		Cold-worked				
	°C	°F	% IACS	Annealed	Cold-worked	MPA	ksi	MPa	ksi	δ-g/cm^3	Annealed	Cold-worked
99.9Ag	960	1760	104	30	75	172	25	310	45	10.51	55	5
99.55Ag-0.25Mg-0.2Ni			70	61	77	207	30	345	50	10.34	35	6
99.47Ag-0.18Mg-0.2Ni-0.15Cu			75	64	84	—	—	—	—	10.38	—	—
99Ag-1Pd	—	—	79	44	76	179	26	324	47	10.14	42	3
97Ag-3Pd	977	1790	58	45	77	186	27	331	48	10.53	37	3
97Ag-3Pt	982	1800	45	45	77	172	25	324	47	10.17	37	3
92.5Ag-7.5Cu	821	1510	88	65	81	269	39	455	66	10.34	35	5
90Ag-10Au	971	1780	40	57	76	200	29	317	46	11.03	28	3
90AG-10Cu	777	1430	85	70	83	276	40	517	75	10.31	32	4
90Ag-10Pd	999	1830	27	63	80	234	34	365	53	10.57	31	3
86.8Ag-5.5Cd-0.2Ni-7.5Cu	—	—	43	72	85	276	40	517	75	10.10	43	3
85Ag-15Cd	877	1610	35	51	83	193	28	400	58	10.17	55	5
77Ag-22.6Cd-0.4Ni	—	—	31	50	85	241	35	469	68	10.31	55	4
75Ag-24.5Cu-0.5Ni	—	—	75	78	85	310	45	552	80	10.00	32	4
72Ag-28Cu	777	1430	84	79	85	365	53	552	80	9.95	20	5
60Ag-23Pd-12Cu-5Ni	—	—	11	86	93	538	78	758	110	10.51	22	3

Source: From Boyer and Gall.[1]

matrix into the sulfide film. Consequently, high currents cause the film to become somewhat conducting. The resistance of a silver sulfide film decreases with increasing temperature, and it begins to decompose at 680°F.

Limitations of pure silver contacts are due to the transfer of metal from one electrode to the other. This condition leads to the formation of buildups on one contact surface and holes in the other. When used in dc circuits, silver contacts are subject to ultimate failure by mechanical sticking as a direct result of metal transfer. The direction of transfer is generally from the positive contact to the negative, but under the influence of arcing, the direction may be reversed. With high currents or inductive loads, it may be desirable to shunt the load with a resistance-capacitance (*RC*) protection network to reduce erosion. When arcings produce a 5-A glow discharge in air, the rate of erosion of silver is unusually high because of a chemical interaction with air to form $AgNO_2$. For low resistance and low noise levels, the design of the device must provide sufficient force and slide to break through any silver sulfide film and maintain film-free metal-to-metal contact at the interface. Connectors should have high slide force and a pound or so of normal force. Rotary switches that have up to 0.11 lb normal force and considerable slide should have a protective coating of grease to reduce sulfiding and to remove abrasive particles. In low-noise transmission circuits silver should not be used on relay and other butting contacts that have less than 0.044 lb force; other precious-metal coatings, such as gold or palladium, should be used instead of silver. A silver sulfide film has a characteristic voltage drop of several tenths of a volt. Where this drop is tolerable, silver contacts will provide reliable contact closure. Failure to close, however, may be greater than with other precious metal contacts because of impacted dirt, with the sulfide film acting as a dirt catcher. For many applications silver is too soft to give acceptable mechanical wear. Alloying additions of copper, cadmium, platinum, palladium, gold, and other elements are effective in increasing the hardness and modifying the contact behavior of silver.

Gold Metals. Pure gold has unsurpassed resistance to oxidation and sulfidation, but a low melting point and susceptibility to erosion limit its use in electrical contacts to situations where the current is not more than 0.5 A. Although oxide and sulfide films do not form on gold, a carbonaceous deposit sometimes develops when a gold contact is operated in the presence of organic vapors. The resistance of this film may be several ohms. Gold in contact with palladium or rhodium has very low contact resistance. The low hardness of gold can be increased by alloying with copper, silver, palladium, or platinum, but usage is necessarily restricted to low-current applications because of the low melting point. Properties of gold and its alloys are listed in Table 5.5. If low tarnish rates and low contact resistance are to be preserved, the gold content should not be less than about 70 percent.

Platinum Group Metals. Platinum and palladium are the two most important metals of the platinum group. These metals have a high resistance to tarnishing, and therefore provide reliable contact closure for relays and other devices having contact forces of less than 0.11 lb. High melting points, low vapor pressures, and resistance to arcing make platinum and palladium suitable for contacts that close and open a load, particularly in the range up to 1 A. Low electrical and thermal conductivities and high cost generally exclude them from use for currents above about 5 A. Palladium has an arcing limit only slightly less than that of platinum, and it gives comparable performance in relays for telephones and similar services handling 1 A or less. Palladium is a satisfactory substitute for platinum in these applications. Properties of platinum and palladium are presented in Table 5.6.

TABLE 5.5 Properties of Gold Metals Used for Electrical Contacts

	Solidus temperature		Electrical conductivity,	Hardness, HR15T		Tensile strength				Density,
						Annealed		Cold-worked		
Alloy	°C	°F	% IACS	Annealed	Cold-worked	MPa	ksi	MPa	ksi	Mg/m^3
99Au	1085	1985	74	40	65	—	—	—	—	19.36
90Au-10Cu	932	1710	16	76	91	400	58	705	102	17.18
75Au-25Ag	1029	1885	17	50	77	—	—	—	—	15.96
72.5Au-14Cu-8.5Pt-4Ag-1Zn	954	1750	10	88	96	—	—	—	—	16.11
72Au-26.2Ag-1.8Ni	—	—	14	61	81	230	33	345	50	15.56
71Au-5Ag-9Pt-15Cu	—	—	8	88.5	75*	700	101	1170	170	16.02
69Au-25Ag-6Pt	1029	1885	10	70	84	275	40	415	60	15.92
50Au-50Ag	—	—	—	—	—	—	—	—	—	13.59

*Rockwell 15N.

Source: From Boyer and Gall.[1]

TABLE 5.6 Properties of Platinum and Palladium Metals Used for Electrical Contacts

Alloy	Solidus temperature		Electrical conductivity,*	Hardness, HR15T		Tensile strength				Density,	Elongation in 50 mm or
						Annealed		Cold-worked			
	°C	°F	% IACS	Annealed	Cold-worked	MPa	ksi	MPa	ksi	Mg/m^3	2 in, %
99.9Pt	1770	3220	15	60	73	138	20	241	35	21.45	35
95Pt-5Ru	1775	3230	5	84	89	414	60	793	115	20.57	18
92Pt-8Ru	—	—	4	86	91	483	70	896	130	20.27	15
90Pt-10Ir	1780	3240	7	87	92	379	55	620	90	21.52	12
89Pt-11Ru	1815	3300	4	91	96	586	85	1034	150	19.96	12
86Pt-14Ru	1843	3350	3	93	99	655	95	1172	170	19.06	10
85Pt-15Ir	1787	3250	6	90	95	517	75	827	120	21.52	12
80Pt-20Ir	1808	3290	5	93	97	689	100	1000	145	21.63	12
75Pt-25Ir	1819	3310	5	95	98	862	125	1172	170	21.68	10
73.4Pt-18.4Pd-8.2Ru	—	—	4	90	92	517	75	862	125	17.77	12
65Pt-35Ir	1899	3450	4	97	99	965	140	1344	195	21.80	8
99.9Pd	1554	2830	16	62	78	193	28	324	47	12.17	28
95Pd-5Ru	1593	2900	8	79	89	372	54	517	75	12.00	15
89Pd-11Ru	1649	3000	6	85	92	483	70	689	100	12.03	13
72Pd-26Ag-2Ni	1382	2520	4	82	90	469	68	689	100	11.52	13
60Pd-40Ag	1338	2440	4	65	91	372	54	689	100	11.30	28
60Pd-40Cu	1199	2190	8	82	92	565	82	1331	193	10.67	20
35Pd-9.5Pt-9Au-14Cu-32.5Ag	1085	1985	5	90	94	689	100	1034	150	11.63	18

*For material in annealed condition.

Source: From Boyer and Gall.[1]

Chemical properties of the platinum group metals include high resistance to oxidation, sulfidation, and salt water. Platinum will not form a stable oxide at any temperature. Palladium is resistant to oxidation at ordinary temperatures. If heated above 660°F, it will oxidize slowly to form an oxide that is stable at room temperature. However, palladium oxide decomposes rapidly when heated to 1470°F or by arcing. Palladium oxide is an insignificant factor in the reliability of telephone-type relays. However, the presence of organic vapors in the contact area can seriously influence the life and reliability of electrical contacts, especially in the case of low-force precious metal contacts universally employed in high-reliability low-noise circuits. The damaging organic vapors may arise from coil forms, wire coatings, insulation, soldering flux, potting and sealing compounds, and other organics in associated electrical equipment, as well as from external sources. Organic contamination may produce two distinctly different forms of contact damage: activation and polymer formation. Activation is the development of a carbon deposit on the contacting surfaces formed by the decomposition of the organic contaminant in the arc. These deposits increase arc erosion markedly. Carbon deposits decrease the current needed to sustain an arc, thereby prolonging arcing time. A 95 percent reduction in contact life may result from activation caused by the presence of organic vapors. Activation can be reduced or eliminated by using insulating materials that are not sources of organic vapors, by adequate ventilation, or by absorbing the organic vapors. Polymer formation is the development of a polymerlike insulating brown powder on contacts in dry circuits (those not carrying current on make or break), and may lead to transient open circuits. The insulating brown powder is believed to result from the adsorption of the organic vapor on the contact surface, followed by its polymerization by the friction associated with contact operation. The sliding motion both forms the polymer and pushes it outside the slide area, where it builds up as a brown powder. A transient open circuit occurs when some of the built-up powder falls into the contact area. Controlled experiments have shown that the type of contact metal influences the amount of polymer formed. The greatest amount of polymer is formed on the platinum metals; lesser amounts are formed on gold and some base metals; polymer does not form on silver. Elimination of materials that lead to organic vapors is a possible solution to polymer formation, but one that is difficult to carry out. The polymer formation problem was solved in telephone circuits by cladding one of a mating pair of palladium contacts with a very thin layer of gold. In dry circuits one gold surface significantly reduces polymer formation, although in working circuits the gold soon wears off and exposes the palladium base.

Erosion and sticking of platinum group metal contacts will occur if the arcing current limit of about 1 A is exceeded. Contact life is long if the current is kept below this value. With currents higher than 1 A, or with inductive loads, it may be desirable to shunt the load or contact with a resistance-capacitance network to reduce erosion and failures caused by snagging of pits and buildups. For equal volumes of metal, platinum or palladium contacts have about 10 times the life of silver contacts.

Resistance and noise reduction in relays and relay-type switch contacts for talking circuit transmission favors the use of platinum group metals. Palladium contacts are essentially noise-free, and are used in preference to platinum or gold alloys because of lower cost. For these reasons, palladium is used almost exclusively for telephone systems in the United States. In a few isolated instances, where the palladium talking circuit contacts were subjected to vibration in service, noise troubles developed because of polymer formation. The difficulty was usually overcome by using gold alloys to minimize polymer production.

Tungsten and Molybdenum. Most tungsten and molybdenum contacts are made in the form of composites, with silver or copper as the other principal component.

Tungsten has the highest boiling point (10,700°F) and melting point (5625°F) of all metals; it also has very high hardness at both room and elevated temperatures. Therefore, as a contact material it offers excellent resistance to mechanical wear and electrical erosion. The main disadvantages of tungsten are low corrosion resistance and high electrical resistance. After a short period of operation, an oxide film will build up on tungsten contacts, resulting in very high contact resistance. Considerable force is required to break through this film, but high pressure and considerable impact cause little damage to the underlying metal because of its high hardness.

The high boiling and melting points of molybdenum at 10,040°F and 4730°F, respectively, are second only to those of tungsten and rhenium. Molybdenum is not used as widely as tungsten because it oxidizes more readily and erodes faster on arcing than tungsten. Because the density of molybdenum (0.369 lb/in^3) is about half that of tungsten (0.697 lb/in^3), molybdenum contacts are advantageous where mass is important. The cost of molybdenum contacts by volume is also lower.

Applications of tungsten contacts in switching devices with closing forces of more than 4.41 lb and in circuits with high voltages and current of not more than 5 A include automotive ignitions, vibrators, horns, voltage regulators, magnetos, and electric razors. In low-voltage dc devices, tungsten is always used as the negative contact, and is paired with a positive contact made of precious metal. Tungsten rods or strips that are consolidated by swaging or rolling from sintered powder compacts have very poor ductility. They cannot be cold-worked, in contrast to other contact materials. Tungsten disks are usually cut from rods or punched from strips and then brazed directly to functional parts such as circuit breaker arms, brackets, or springs.

Molybdenum is widely used for mercury switches because it is not attacked, but only wetted, by mercury. It is also used in make-and-break contacts. Molybdenum strips and sheets are made by swaging or rolling sintered powder compacts. Disks made from rods or sheets are brazed to blanks or other structural components. Table 5.7 lists the properties of tungsten and molybdenum.

Aluminum. In recent years, because of its good electrical and mechanical properties, ready availability, and favorable cost, aluminum has gained importance as a conductor material. It has replaced copper in many applications, such as rectangular, tubular, and channel bus conductors. It has advantages over copper in density, mass, electrical conductivity, availability, and cost; it is lighter, and the same mass will conduct more current for the same voltage drop. Aluminum 1350 is preferred for contact materials because of its high conductivity (61.8 percent IACS), but it is low in strength and, for some designs, requires additional support. Where strength and resistance to joint relaxation are important, heat-treated 6101 is better suited, although there is some sacrifice in electrical conductivity (57 to 60 percent IACS for 6101). As a contact metal, aluminum is generally poor because it oxidizes readily. Where aluminum is used in contacting joints, it should be plated or clad with copper, silver, or tin. Aluminum should never be used for power applications where arcing is present.

Composite Materials. There are three major groups of composite contact materials made by powder metallurgy methods: refractory and carbide base, silver base, and copper base. Because manufacturing methods affect properties of materials with the same composition, the most common methods of producing composite electrical contact materials are discussed in this section.

Infiltration is used exclusively for making refractory metal and carbide-base composite contact materials. Silver or copper is infiltrated into the pores of a compacted powder preform. This method produces the most densified composites, generally 97 percent or more of theoretical density. After infiltration, the contact is sometimes

TABLE 5.7 Typical Properties of Tungsten and Molybdenum*

Tungsten	
Hardness	70 HRA, 385 HV
Modulus of elasticity	
At 20°C (68°F)	405 GPa (59 × 10^6 lb/in^2)
At 1000°C (1830°F)	325 GPa (47 × 10^6 lb/in^2)
Density	19.3 Mg/m^3 (0.697 lb/in^3)
Melting point	3410°C (6170°F)
Boiling point	5900°C (10,650°F)
Specific heat at 20°C (68°F)	140 J/kg (0.033 Btu/lb • °F)
Thermal conductivity at 20°C (68°F)	130 W/m • K (75 Btu/ft • h • °F)
Coefficient of linear thermal expansion at 20°C (68°F)	4.43 μm/m • °C (2.46 μin/in • °F)
Specific resistance at 20°C (68°F)	5.5 nΩ • m
Electrical conductivity at 20°C (68°F)	31% IACS
Molybdenum	
Hardness	58 HRA, 210 HV
Modulus of elasticity	
At 20°C (68°F)	325 GPa (47 × 10^6 lb/in^2)
At 1000°C (1830°F)	270 GPa (39 × 10^6 lb/in^2)
Density	10.22 Mg/m^3 (0.369 lb/in^3)
Melting point	2622°C (4750°F)
Boiling point	4800°C (8672°F)
Specific heat at 20°C (68°F)	270 J/kg (0.065 Btu/lb • °F)
Thermal conductivity at 20°C (68°F)	155 W/m • K (89 Btu/ft • h • °F)
Coefficient of linear thermal expansion at 20°C (68°F)	5.53 μm/m • °C (3.07 μin/in • °F)
Specific resistance at 20 °C (68°F)	5.2 nΩ • m
Electrical conductivity at 20°C (68°F)	33% IACS

*Some of the physical properties of tungsten and molybdenum vary considerably with cross-sectional area and grain structure.

Source: From Boyer and Gall.[1]

etched chemically or electrochemically so that only pure silver appears on the surface. The contact thus treated has better corrosion resistance and performs better in the early stages of use.

Press-sinter is used for small refractory-metal contacts (not exceeding about 25 mm). A high-density material can be obtained by pressing a blended powder of the desired composition into the required shape, then sintering it at the melting temperature of the low-melting-point component. In some cases an activating agent such as nickel, cobalt, or iron is added to improve the sintering effect on the refractory metal particles. However, the structure formed by this process is weaker than that made by the infiltration process.

Press-sinter-repress processing is used for all categories of contact materials, especially those in the silver-base category. Blended powders of the correct composition are compacted to the required shape and then sintered. Afterward the material is further densified by a second pressing. Sometimes the properties can be modified by a second sintering or annealing. However, it is difficult to obtain material with as high a density as is obtained with other processes.

Press-sinter-extrude adds an extrusion step to the press-sinter method to achieve a fully dense material in most cases. Blended powder of the final composition is pressed into an ingot and sintered. The ingot is then extruded into wires, slabs, or other desired shapes. The extruded material may be subsequently worked by rolling, swaging, or drawing. The press-sinter-extrude process is used mostly for silver-base composites.

Preoxidize-press-sinter-extrude is used exclusively for making silver–cadmium oxide contacts. Alloys are reduced to small particles in the shape of flakes, slugs, or shredded foil. These particles are oxidized and then consolidated with the press-sinter-extrude process. Mechanical properties are superior to those of the same material made by the press-sinter-repress method.

5.5.2 Electrical Resistance Alloys

Electrical resistance alloys include both the types used in instruments and control equipment to measure and regulate electrical characteristics and those used in furnaces and appliances to generate heat. In commercial terminology, electrical resistance alloys used for the control or regulation of electrical properties are called resistance alloys, and those used for the generation of heat are referred to as heating alloys. Electrical resistance materials used in applications where heat is converted into mechanical energy are termed thermostat metals. Nominal compositions and physical properties of metals and alloys used to make resistors for instruments and controls are listed in Table 5.8 (Boyer and Gall,[1] pp. 20.11–20.16).

Resistance Alloys. Applications of resistance alloys include wirewound resistance thermometers, ballast resistors, and reference resistors. Alloys used for precision resistors generally have resistivities ranging from 300 to 800 Ω • cmil/ft. To ensure freedom from thermoelectric effects, a resistor should have a small or negligible thermoelectric potential versus the material used for the connecting conductor. Ideally, a resistor should have a thermal coefficient of resistance equal to zero over the operating temperature range.

Resistance thermometers are commonly made of copper, nickel, or platinum. These devices are precision resistors whose resistance change with temperature is stable and reproducible over specified ranges of temperature.

Ballast resistors must be able to dissipate energy in such a way as to control current over a wide range of voltages. Typical materials used in ballast resistors are pure iron, pure nickel, and nickel-iron alloys such as 71Ni-29Fe (see Table 5.8).

Reference resistors must maintain resistances within narrow limits over long periods of time. Figure 5.13 shows the change in resistance with time for a 10-kΩ resistor made of a Ni-Cr-Al-Cu alloy. The principal sources of instability are:

1. Residual stresses. Figure 5.14 shows the effect of residual stresses on the stability of a manganese alloy. The top curve illustrates that a low-temperature stress-relieving treatment substantially eliminates these stresses.
2. Metallurgical changes. All resistance alloys are single-phase solid-solution alloys. Changes in resistance can result from internal changes in long-range order-disorder reactions in 71Ni-29Fe alloys, short-range order or clustering in quaternary nickel-chromium alloys, and even minor ordering in manganese alloys.
3. Corrosion or oxidation of the resistance element will decrease its effective cross section, resulting in a corresponding increase in resistance. If the corrosive attack

TABLE 5.8 Typical Properties of Electrical Resistance Alloys

Basic composition	Resistivity, * nΩ • m†	TCR, ppm/°C‡	Thermoelectric potential vs. Cu, μV/°C	Coefficient of thermal expansion,¶ μm/m • °C	Tensile strength		Density*	
					MPa	ksi	Mg/m^3	lb/in^3
Radio alloys								
98Cu-2Ni	50	+1350 (25–105°C)	−13 (25–105°C)	16.5	205–410	30–60	8.9	0.32
94Cu-6Ni	100	+550 (25–105°C)	−13 (25–105°C)	16.3	240–585	35–85	8.9	0.32
89Cu-11Ni	150	+430 (25–105°C)	−25 (25–105°C)	16.1	240–515	35–75	8.9	0.32
78Cu-22Ni	300	+160 (25–105°C)	−36 (0–75°C)	15.9	345–690	50–100	8.9	0.32
Manganins								
87Cu-13Mn	480	±15 (15–35°C)	+1 (0–50°C)	18.7	275–620	40–90	8.2	0.30
83Cu-13Mn-4Ni	480	±15 (15–35°C)	−1 (0–50°C)	18.7	275–620	49–90	8.4	0.31
85Cu-10Mn-4Ni§	380	±10 (20–45°C)	−1.5 (0–50°C)	18.7	345–690	50–100	8.4	0.31
Constantans								
57Cu-43Ni	490	±20 (25–105°C)	−43 (25–105°C)	14.9	410–930	60–135	8.9	0.32
55Cu-45Ni	500	±40 (20–1000°C)	−42 (0–75°C)	14.9	455–860	66–125	8.9	0.32
Nickel-chromium-aluminum alloys								
75Ni-20Cr-3Al-2(Cu, Fe, or Mn)	1330	±20 (−55–+105°C)	−0.1 (25–105°C)	12.6	690–1380	100–200	8.1	0.29
72Ni-20Cr-3Al-5Mn	1355	±20 (−55–+105°C)	−0.1 (25–105°C)	13	690–1380	100–200	7.1	0.26
Nickel-base alloys								
94Ni-3Mn-2Al-1Si	315	+2400 (20–100°C)		12.3	550–1035	80–150	8.5	0.31
80Ni-20Cr	1125	+85 (−55–+100°C)	+5 (0–100°C)	13	690–1380	100–200	8.4	0.31
78.5Ni-20Cr-1.5Si	1080	+85 (25–105°C)	+3.9 (25–105°C)	13.5	690–1380	100–200	8.3	0.30
76Ni-17Cr-4Si-3Mn	1330	±20 (−55–+150°C)	−1 (20–100°C)	15	900–1380	130–200	7.8	0.28
71Ni-29Fe	200	+4500 (25–105°C)	−40 (25–105°C)	15	480–1035	70–150	8.4	0.31
68.5Ni-30Cr-1.5Si	1180	+90 (25–105°C)	−1.2 (25–105°C)	12.2	825–1380	120–200	8.1	0.29
60Ni-16Cr-24Fe	1120	+150 (25–105°C)	+0.9 (25–105°C)	13.5	655–1200	95–175	8.4	0.30
35Ni-20Cr-45Fe	1015	+400 (25–105°C)	−1.1 (25–105°C)	15.6	550–1200	80–175	8.1	0.29

TABLE 5.8 Typical Properties of Electrical Resistance Alloys (*Continued*)

Basic composition	Resistivity, * nΩ • m†	TCR, ppm/°C‡	Thermoelectric potential vs. Cu, μV/°C	Coefficient of thermal expansion,¶ μm/m • °C	Tensile strength		Density*	
					MPa	ksi	Mg/m³	lb/in³
Iron-chromium-aluminum alloys								
73.5Fe-22Cr-4.5Al	1350	±50	−3.0 (0–100°C)	11	690–965	100–140	7.25	0.262
73Fe-22Cr-5Al	1390	±50	−2.8 (0–100°C)	11	690–965	100–140	7.15	0.258
72.5Fe-22Cr-5.5Al	1450	±50	−2.6 (0–100°C)	11	690–965	100–140	7.1	0.256
81Fe-15Cr-4Al	1250	±50	−1.2 (0–100°C)	11	620–900	90–130	7.43	0.268
Pure metals								
Aluminum (99.99+)	26.55	+4290*	−3.4 (0–50°C)	23.9*	50–110	7–16	2.70	0.098
Copper (99.99)	16.73	+4270 (0–50°C)	0	16.5*	115–130	17–19	8.96	0.324
Gold (99.99+)	23.50	+4000 (0–100°C)	+0.2 (0–100°C)	14.2*	130	19	19.32	0.698
Iron (99.94)	970	+5000*	+12.2 (0–100°C)	11.7*	180–220	26–32	7.87	0.284
Molybdenum (99.9)	52	+3300*	+6.9 (0–100°C)	4.9	690–2140	100–310	10.22	0.369
Nickel (99.8)	80	+6000 (20–35°C)	−22 (0–75°C)	15	345–760	50–110	8.90	0.322
Platinum (99.99+)	106	+3920 (0–100°C)	+7.6 (0–100°C)	8.9*	125	18	21.45	0.775
Silver (99.99)	16	+4100*	−0.2 (0–100°C)	19.7	125	18	10.49	0.379
Tantalum (99.96)	135	+3820 (0–100°C)	−4.3 (0–100°C)	6.5*	690–1240	100–180	16.6	0.600
Tungsten (99.9)	55	+4500*	+3.6 (0–100°C)	4.3*	1825–4050	265–590	19.25	0.695

*At 20°C (68°F).

†To convert to Ω • cmil/ft, multiply by 0.6015.

‡Temperature coefficient of resistance is $(R - R_0)/R_0(t - t_0)$, where R is resistance at t °C and R_0 is resistance at the reference temperature t_0 °C.

¶At 25–105°C.

§Shunt manganin.

Source: From Boyer and Gall.[1]

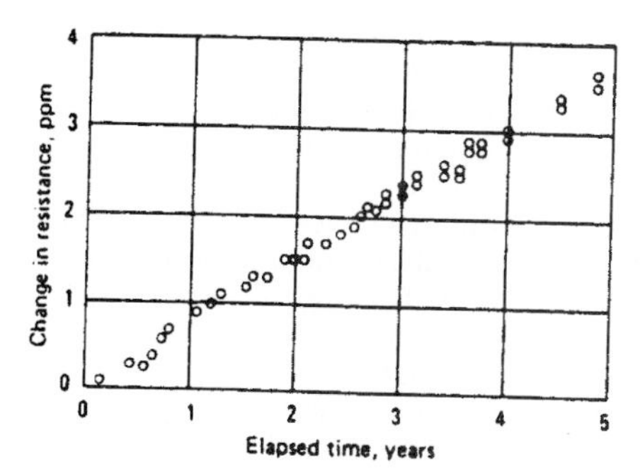

FIGURE 5.13 Change in resistance of 10-kΩ resistor with time.

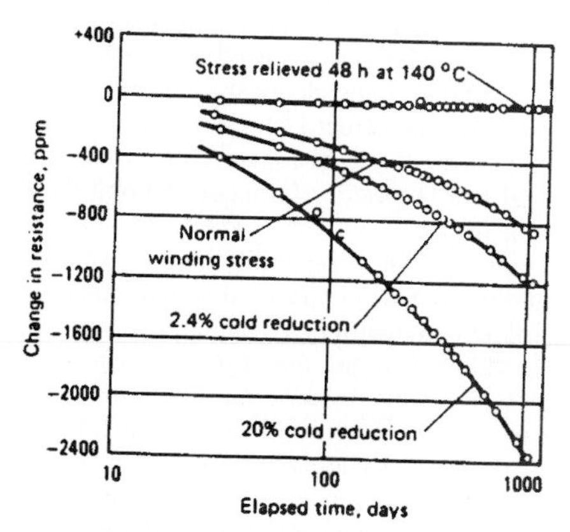

FIGURE 5.14 Change in resistance of manganin resistors upon aging at room temperature.

is selective, changes will occur in the temperature coefficient of resistance and the thermal electromotive force as well. One relatively common source of corrosive attack is flux residue at soldered or brazed joints. Another less obvious cause of instability is the presence of tin-containing solder. Intergranular stress corrosion may cause open circuits. In addition, the resistance value of a resistor may change if the hydrostatic pressure on the resistance element is changed; for manganese this change is about 0.155 $\mu\Omega/\Omega \cdot (\text{lb/in}^2)$.

Heating Alloys. Resistance heating alloys are used in various applications, from small household appliances to large industrial furnaces. In appliances, heating elements are designed for intermittent short-term service at about 100 to 2000°F. In industrial furnaces, elements often must operate continuously at temperatures as high as 2350°F for furnaces used in metal-treating industries, as high as 3100°F for kilns used for firing ceramics, and occasionally as high as 3600°F for special applications. The primary requirements of materials used for heating elements are high melting point, high electrical resistivity, reproducible temperature coefficient of resistance, good oxidation resistance in furnace environments, absence of volatile components, and resistance to contamination. Other desirable properties are good elevated-temperature creep strength, high emissivity, low thermal expansion and low modulus (both of which help minimize thermal fatigue), good resistance to thermal shock, and good strength and ductility at fabrication temperatures. Table 5.9 lists physical and mechanical properties and Table 5.10 presents recommended maximum operating temperatures for resistance heating materials. Of the four groups of materials listed in these tables, the first group (Ni-Cr and Ni-Cr-Fe alloys) serves by far the greatest number of applications. The ductile wrought alloys in the first group have properties that enable them to be used at both low and high temperatures in a wide variety of environments. The Fe-Cr-Al compositions (second group) are also ductile alloys. They play an important role in heaters for the higher temperature ranges, which are constructed to provide more effective mechanical support for the element. The pure metals that comprise the third group have much higher melting points. All of them except platinum are readily oxidized and are restricted to use in nonoxidizing environments. They are valuable for a limited range of applications, primarily for service above 2500°F. The cost of platinum prohibits its use except in small special furnaces. The fourth group, nonmetallic heating element materials, are used at still higher temperatures. Silicon carbide can be used in oxidizing atmospheres at temperatures up to 3000°F; two varieties of molybdenum disilicide are effective up to maximum temperatures of 3100 and 3270°F in air. Molybdenum disilicide heating elements are gaining increased acceptance for use in industrial and laboratory furnaces. Among the desirable properties of molybdenum disilicide heating elements are excellent oxidation resistance, long life, constant electrical resistance, self-healing ability, and resistance to thermal shock.

The effect of atmosphere on heating alloys at several heating element temperatures is shown in Table 5.11. The compositions of the atmospheres are given in Table 5.12. Element temperatures are always higher than furnace control temperatures; the difference depends on the power-density loading on the element surface. Thus when furnaces are operated near maximum element temperature in the more active atmospheres, the power-density loading should be lower and the element cross-sectional area should be higher. With the exception of molybdenum, tantalum, tungsten, and graphite, commonly used resistor materials have satisfactory life in air and in most other oxidizing atmospheres.

Sulfur, if present, will appear as hydrogen sulfide in reducing atmospheres and as sulfur dioxide in oxidizing atmospheres. Sulfur contamination usually comes from one or more of the following sources: high-sulfur fuel gas used to generate the protective atmosphere; residues of sulfur-base cutting oil on the metal being processed; high-sulfur refractories, clays or cements used for sealing carburizing boxes; and the metal being processed in the furnace. Sulfur is destructive to Ni-Cr and Ni-Cr-Fe heating elements, and the sulfur attack is greater the higher the nickel content.

Lead and zinc contamination of a furnace atmosphere may come from the work being processed. This is a common occurrence in sintering furnaces for processing powder metallurgy parts. In a reducing atmosphere, lead will vaporize from leaded bronze powders (such as those used to make sintered bronze bushings) and attack the heating elements, forming lead chromate. Metallic lead vapors are even more harmful than sulfur to nickel-chromium alloys and will cause severe damage to a heating element in a matter of hours if unfavorable conditions of concentration and temperature exist.

Thermostat Metals. A thermostat metal is a composite structure (usually in the form of sheet or strip) that consists of two or more materials bonded together, of which one may be a nonmetal. Because the materials bonded together to form the composite differ in thermal expansion, the curvature of the composite is altered by changes in tem-

TABLE 5.9 Typical Properties of Resistance Heating Materials

Basic composition	Resistivity,* nΩ • m[†]	Average change in resistance,[‡] % from 20°C to:				Thermal expansion, μm/m • °C from 20°C to:			Tensile strength		Density	
		260°C	540°C	815°C	1095°C	100°C	540°C	815°C	MPa	ksi	Mg/m^3	lb/in^3
Nickel-chromium and nickel-chromium-iron alloys												
78.5Ni-20Cr-1.5Si (80-20)	1080	+4.5	+7.0	+6.3	+7.6	13.5	15.1	17.6	690–1380	100–200	8.41	0.30
77.5Ni-20Cr-1.5Si-1Nb	1080	+4.6	+7.0	+6.4	+7.8	13.5	15.1	17.6	690–1380	100–200	8.41	0.30
68.5Ni-30Cr-1.5Si (70-30)	1180	+2.1	+4.8	+7.6	+9.8	12.2	—	—	825–1380	120–200	8.12	0.29
68Ni-20Cr-8.5Fe-2Si	1165	+3.9	+6.7	+6.0	+7.1	—	12.6	—	895–1240	130–180	8.33	0.30
60Ni-16Cr-22.5Fe-1.5Si	1120	+3.6	+6.5	+7.6	+10.2	13.5	15.1	17.6	655–1205	95–175	8.25	0.30
35Ni-30Cr-33.5Fe-1.5Si	1055	+7.95	+14.9	+18.0	+22.0	14.6	17.5	16.0	895–1380	130–200	7.90	0.29
35Ni-20Cr-43.5Fe-1.5Si	1015	+8.0	+15.4	+20.6	+23.5	15.7	15.7	—	550–1205	80–175	7.95	0.29
35Ni-20Cr-42.5Fe-1.5Si-1Nb	1015	+8.0	+15.4	+20.6	+23.5	15.7	15.7	—	550–1205	80–175	7.95	0.29
Iron-chromium-aluminum alloys												
83.5Fe-13Cr-3.25Al	1120	+7.0	+15.5	—	—	10.6	—	—	515–1035	75–150	7.30	0.26
81Fe-14.5Cr-4.25Al	1245	+3.0	+9.7	+16.5	—	10.8	11.5	12.2	550–1170	80–170	7.28	0.26
79.5Fe-15Cr-5.2Al	1370	+1.9	+5.5	+8.9	+9.6	11.3	12.6	—	550–895	80–130	7.12	0.26
73.5Fe-22Cr-4.5Al	1355	+0.3	+2.9	+4.3	+4.9	10.8	12.6	13.1	725–1205	105–175	7.15	0.26
72.5Fe-22Cr-5.5Al	1455	+0.2	+1.0	+2.8	+4.0	11.3	12.8	14.0	760–1205	110–175	7.10	0.26
Pure metals												
Molybdenum	52	+110	+238	+366	+508	4.8	5.8	—	690–2160	100–313	10.2	0.369
Platinum	105	+85	+175	+257	+305	9.0	9.7	10.1	345	50	21.5	0.775
Tantalum	125	+82	+169	+243	+317	6.5	6.6	—	345–1240	50–180	16.6	0.600
Tungsten	55	+91	+244	+396	+550	4.3	4.6	4.6	3380–6480	490–940	19.3	0.697

TABLE 5.9 Typical Properties of Resistance Heating Materials (*Continued*)

Basic composition	Resistivity,* nΩ • m†	Average change in resistance,‡ % from 20°C to:				Thermal expansion, μm/m • °C from 20°C to:			Tensile strength		Density	
		260°C	540°C	815°C	1095°C	100°C	540°C	815°C	MPa	ksi	Mg/m^3	lb/in^3
Nonmetallic heating-element materials												
Silicon carbide	995 to 1995	−33	−33	−28	−13	4.7	—	—	28	4	3.2	0.114
Molybdenum disilicide	370	+105	+222	+375	+523	9.2	—	—	185	27	6.2	0.212
$MoSi_2$ + 10% ceramic additives	270	+167	+370	+597	+853	13.1	14.2	14.8	—	—	5.6	0.202
Graphite	9100	−16	−18	−13	−8	1.3	—	—	1.8	0.26	2.3	0.057

*At 20°C (68°F).

† To convert to Ω • cmil/ft, multiply by 0.6015.

‡ Changes in resistance may vary somewhat, depending on cooling rate.

Source: From Boyer and Gall.[1]

TABLE 5.10 Recommended Maximum Furnace Operating Temperatures for Resistance Heating Materials

Basic composition	Approximate melting point		Maximum furnace operating temperature in air	
	°C	°F	°C	°F
Nickel-chromium and nickel-chromium-iron alloys				
78.5Ni-20Cr-1.5Si (80-20)	1400	2550	1150	2130
77.5Ni-20Cr-1.5Si-1Nb	1390	2540	—	—
68.5Ni-30Cr-1.5Si (70-30)	1380	2520	1200	2200
68Ni-20Cr-8.5Fe-2Si	1390	2540	1150	2100
60Ni-16Cr-22.5Fe-1.5Si	1350	2460	1000	1850
35Ni-30Cr-33.5Fe-1.5Si	1400	2550	—	—
35Ni-20Cr-43.5Fe-1.5Si	1380	2515	925	1700
35Ni-20Cr-42.5Fe-1.5Si-1Nb	1380	2515	—	—
Iron-chromium-aluminum alloys				
83.5Fe-13Cr-3.25Al	1510	2750	1050	1920
81Fe-14.5Cr-4.25Al	1510	2750	—	—
79.5Fe-15Cr-5.2Al	1510	2750	1260	2300
73.5Fe-22Cr-4.5Al	1510	2750	1280	2335
72.5Fe-22Cr-5.5Al	1510	2750	1375	2505
Pure metals				
Molybdenum	2605	4730	400*	750*
Platinum	1770	3216	1500	2750
Tantalum	2975	5390	500*	930*
Tungsten	3375	6116	300*	570*
Nonmetallic heating-element materials				
Silicon carbide	2410	4370	1600	2900
Molybdenum disilicide	2080	3775	1700–1900	3100–3270
$MoSi_2$ + 10% ceramic additives	1800	3270	—	—
Graphite	3650 to 3695†	6610 to 6690†	400‡	400‡

	Vacuum	Pure H_2	City gas
Mo	1650°C (3000°F)	1760°C (3200°F)	1700°C (3100°F)
Ta	2480°C (4500°F)	Not recommended	Not recommended
W	1650°C (3000°F)	2480°C (4500°F)	1700°C (3100°F)

*Recommended atmospheres for these metals are a vacuum of 10^{-4}–10^{-5} mm Hg, pure hydrogen, and partly combusted city gas dried to a dew point of +4°C (+40°F). In these atmospheres the recommended temperatures would be:

†Graphite volatilizes without melting at 3650–3695°C (6610–6690°F).

‡At approximately 400°C (750°F) (threshold oxidation temperature), graphite undergoes a weight loss of 1% in 24 h in air. Graphite elements can be operated at surface temperatures up to 2205°C (4000°F) in inert atmospheres.

Source: From Boyer and Gall.[1]

TABLE 5.11 Comparative Life and Maximum Operating Temperature of Heating-Element Materials in Various Furnace Atmospheres*,[†]

Element material	Oxidizing, air	Reducing, dry H_2 or type 501	Reducing, type 102 or 202	Reducing, type 301 or 402	Carburizing, type 307 or 309	Reducing or oxidizing, with sulfur	Reducing, with lead or zinc	Vacuum
Nickel-chromium and nickel-chromium-iron alloys								
80Ni-20Cr	Good to 1150°C	Good to 1175°C	Fair to 1150°C	Fair to 1000°C	Not recommended[‡]	Not recommended	Not recommended	Good to 1150°C
60Ni-16Cr-24Fe	Good to 1000°C	Good to 1000°C	Good to fair to 1000°C	Fair to poor to 925°C	Not recommended	Not recommended	Not recommended	—
35Ni-20Cr-45Fe	Good to 925°C	Good to 925°C	Good to fair to 925°C	Fair to poor to 870°C	Not recommended	Fair to 925°C	Fair to 925°C	
Iron-chromium-aluminum alloys								
Fe-23-Cr-4.5Al-1Co	Good to 1150°C	Fair to poor to 1150°C[¶]	Not recommended	Not recommended	Not recommended	Fair in oxidizing atmosphere	Not recommended	—
Fe-37-Cr-7.5Al	Good to 1320°C	Fair to poor to 1300°C[¶]	Not recommended	Not recommended	Not recommended	Fair in oxidizing atmosphere	Not recommended	—
Pure metals								
Molybdenum	Not recommended[§]	Good to 1650°C	Not recommended	Not recommended	Not recommended	Not recommended	Not recommended	Good to 1650°C
Platinum	Good to 1400°C	Not recommended	Not recommended	Not recommended	Not recommended	Not recommended	Not recommended	—
Tantalum	Not recommended	Not recommended	Not recommended	Not recommended	Not recommended	Not recommended	Not recommended	Good to 2500°C
Tungsten	Not recommended	Good to 2500°C[a]	Not recommended	Not recommended	Not recommended	Not recommended	Not recommended	Good to 1650°C

Nonmetallic heating element materials								
Silicon carbide	Good to 1600°C	Fair to poor to 1200°C	Fair to 1375°C	Fair to 1375°C	Not recommended	Good to 1375°C	Good to 1375°C	Not recommended
Graphite	Not recommended	Fair to 2500°C	Not recommended	Fair to 2500°C	Fair to poor to 2500°C	Fair to 2500°C in reducing atmosphere	Fair to 2500°C	—
Molybdenum disilicide	Good to 1800°C	—	—	—	—	—	—	—

*See Table 5.12 for atmosphere conditions.

†Inert atmosphere of argon or helium can be used with all materials. Nitrogen recommended only for the nickel-chromium group. Temperatures listed are element temperatures, not furnace temperatures.

‡Special 80Ni-20Cr elements with ceramic protective coatings designated for low voltage (8–16 V) can be used.

¶Must be oxidized first.

§Special molybdenum heating elements with $MoSi_2$ coating can be used in oxidizing atmospheres.

[a] Good with pure H_2 only.

Source: From Boyer and Gall.[1]

TABLE 5.12 Types and Compositions of Standard Furnace Atmospheres*

Type	Description	Composition, vol %					Typical dew point	
		N_2	CO	CO_2	H_2	CH_4	°C	°F
Reducing atmospheres								
102†	Exothermic unpurified	71.5	10.5	5.0	12.5	0.5	+27	+80
202	Exothermic purified	75.3	11.0	—	13.0	0.5	−40	−40
301	Endothermic	45.1	19.6	0.4	34.6	0.3	+10	+50
402	Charcoal	64.1	34.7	—	1.2	—	−29	−20
501	Dissociated ammonia	25	—	—	75	—	−51	−60
Carburizing atmospheres								
307	Endothermic + hydrocarbon	No standard composition					—	—
309	Endothermic + hydrocarbon + ammonia	No standard composition					—	—

*See Table 5.11 for comparative life of heating elements in these atmospheres.

†This atmosphere, refrigerated to obtain a dew point of +4°C (+40°F), is widely used.

Source: From Boyer and Gall.[1]

perature—this is the fundamental characteristic of any thermostat metal. For circuit breakers and similar devices there are thermostat metals that differ in electrical resistivity but are similar in other properties. Resistivity is varied by incorporating a layer of a low-resistivity metal between outer layers of two other metals that have high resistivities and that differ widely in expansion coefficient. In one series of commercial thermostat metals with resistivities ranging from 100 to 470 Ω • cmil/ft at 75°F, high-purity nickel is used for the intermediate layer. In another series with resistivities from 20 to 100 Ω • cmil/ft, high-conductivity copper alloys are employed for the intermediate layer. The use of a manganese-copper-nickel alloy having a resistivity of 1050 Ω • cmil/ft for one of the outer layers has extended the practical upper resistivity limit of thermostat metals to 975 Ω • cmil/ft at 75°F. Tolerances on resistivity at a standard temperature vary from +3 to +10 percent, depending on the type of thermostat metal and its resistivity. About 30 different alloys are used to make over 50 different thermostat metals. Most of these 30 alloys are nickel-iron, nickel-chromium-iron, chromium-iron, high-copper, and high-manganese alloys. Thermostat metals are available as strip or sheet in thicknesses ranging from 0.005 to 0.125 in and widths from 0.020 to 12 in. They are easily formed into the required shapes. Thermostat metals usually are selected on the basis of the temperature range in which they are required to operate. They are available for intermediate temperature intervals between −300 and +1000°F. Properties and typical bimetal combinations for several temperature ranges are given in Table 5.13.

5.5.3 Electromagnetic Shielding

The effectiveness of a metal to either absorb or reflect electromagnetic radiation depends primarily on its conductivity and permeability and, in turn, on how the permeability varies as a function of frequency, magnetic field strength, and induction (White[5]).

TABLE 5.13 Properties of Thermostat Metals Frequently Selected for Some Common Service Temperatures

Temperature range of maximum sensitivity		Composition		Resistivity at 24°C (75°F)		Flexivity*	
°C	°F	High-expanding side	Low-expanding side	nΩ • m	Ω • cmil/ft	μm/m • °C	μin/in • °F
−20 to +150	0 to 300	75Fe-22Ni-3Cr	64Fe-36Ni	780	470	26.3	14.6
−20 to +200	0 to 400	75Fe-22Ni-3Cr	Pure Ni	160	95	8.3	4.6
		72Mn-18Cu-10Ni	64Fe-36Ni	1120	675	38.5	21.4
120 to 290	250 to 550	67Ni-30Cu-1.4Fe-1Mn	60Fe-40Ni	565	340	16.6	9.2
150 to 450	300 to 850	66.5Fe-22Ni-8.5Cr	50Fe-50Ni	580	350	11.2	6.2

*At 40–150°C (100–300°F). See ASTM B-106 for standard test method for determining flexivity of thermostat metals.

Source: From Boyer and Gall.[1]

Table 5.14 lists the conductivities and permeabilities (relative to copper) of metals commonly used for shielding applications. The relative latent energy absorption loss of these metals is listed in column 4 as the square root of the product of conductivity and permeability. All the nonmagnetic metals with the exception of silver have absorption values relative to copper of less than 1.0, and they are relatively poor absorbers. On the other hand, magnetic materials have relatively high absorption values, and they are good absorbers of energy at low frequencies. However, at frequencies greater than a few hundred kilohertz, magnetic materials exhibit lower absorption losses than most nonmagnetic materials.

In the form of thin sheets (foils), where metal thickness is a small percentage of the skin depth at the frequency of interest, neither magnetic nor nonmagnetic materials are effective for shielding based on absorption loss alone.

The latent reflection loss of metals relative to copper is listed in column 5 of Table 5.14 as the square root of conductivity divided by permeability. Column 6 gives the latent reflection expressed in decibel form. These data illustrate that the nonmagnetic materials outperform nearly all magnetic materials for reflection-loss performance. For thin-sheet stock, the reflection loss is the only important term for overall shielding effectiveness. Consequently the better shielding materials are the nonmagnetic metals with higher conductivities.

For conductive coatings on glass or plastic substrates used for RF shields in which optical viewing is required, the more stable, higher conductive metals would be preferred, such as gold. The shielding effectiveness of gold films at frequencies between 10 Hz and 30 GHz is plotted in Figs. 5.15 and 5.16.

Besides permeability, conductivity, frequency, and thickness, a number of other factors must be known so that effective electromagnetic shielding can be achieved. They include eddy current losses, magnetic field strength, and distance to the metal barrier. Unless all of these factors are quantified, an equivalent permeability cannot be determined with accuracy.

The dependence of permeability on the magnetic environment is illustrated in Fig. 5.17 by the change in relative permeability as a function of applied magnetic field strength and induction. Figure 5.18 shows that the relative permeability of magnetic alloys decreases to 1 at frequencies greater than a few hundred kilohertz for an unde-

TABLE 5.14 Relative Conductivity and Permeability of Metals

(1) Metal	(2) Relative conductivity σ_r	(3) Relative permeability μ_r at ≤10 kHz	(4) Product $\sqrt{\sigma_r \mu_r}$ $A = k1\sqrt{\sigma_r \mu_r}$, dB	(5) Quotient $\sqrt{\sigma_r / \mu_r}$ $R = k\sqrt{\sigma_r / \mu_r}$	(6) Relative reflection R, dB
1. Silver	1.064	1	1.03	1.3	+0.3
2. Copper (solid)	1.00	1	1	1	0
3. Copper (flame spray)	0.10	1	0.32	0.32	−10.0
4. Gold	0.70	1	0.88	0.88	−1.1
5. Chromium	0.664	1	0.81	0.81	−1.8
6. Aluminum (soft)	0.63	1	0.78	0.78	−2.1
7. Aluminum (tempered)	0.40	1	0.63	0.63	−4.0
8. Aluminum (household foil, 1 mil)	0.53	1	0.73	0.73	−2.8
9. Aluminum (household foil, 1 mil)	0.61	1	0.78	0.78	−2.1
10. Aluminum (flame spray)	0.036	1	0.19	0.19	−14.4
11. Brass (91Cu-9Zn)	0.47	1	0.69	0.69	−3.3
12. Brass (66Cu-34Zn)	0.35	1	0.52	0.52	−5.7
13. Magnesium	0.38	1	0.61	0.61	−4.3
14. Zinc	0.305	1	0.57	0.57	−4.9
15. Tungsten	0.314	1	0.56	0.56	−5.0
16. Beryllium	0.33	1	0.53	0.53	−5.5
17. Cadmium	0.232	1	0.48	0.48	−6.3
18. Platinum	0.17	1	0.42	0.42	−7.6
19. Tin	0.151	1	0.39	0.39	−8.2
20. Tantalum	0.12	1	0.33	0.33	−9.6
21. Lead	0.079	1	0.28	0.28	−11.0
22. Monel (67Ni-30Cu-2Fe-1Mn)	0.041	1	0.20	0.20	−13.9
23. Manganese	0.039	1	0.20	0.20	−14.1

24.	Titanium	0.036	1	0.19	0.19	−14.4
25.	Mercury (liquid)	0.018	1	0.13	0.134	−17.4
26.	Nichrome (65Ni-12Cr-23Fe)	0.0012	1	0.035	0.035	−29.2
27.	Supermalloy	0.023	100000	53.7	0.0005	−65.4
28.	78 Permalloy	0.108	8000	29.4	0.0037	−48.7
29.	Purified iron	0.17	5000	29.2	0.0058	−44.7
30.	Conetic AA	0.031	20000	28.7	0.0011	−58.8
31.	4-79 Permalloy	0.0314	20000	25.1	0.0013	−58.0
32.	Mumetal	0.0289	20000	24.0	0.0012	−58.4
33.	Permendur (50Cu-1-2V,+Fe)	0.247	800	14.1	0.018	−35.1
34.	Hypernick	0.0345	4500	12.5	0.0028	−51.1
35.	45 Permalloy (1200 Anneal)	0.0384	4000	12.4	0.0031	−50.2
36.	45 Permalloy (1050 Anneal)	0.0384	2500	9.80	0.0039	−48.1
37.	Hot-rolled silicon steel	0.0384	1500	7.59	0.0051	−45.9
38.	Sinimax	0.0192	3000	7.59	0.0025	−51.9
39.	4% Silicon iron (grain oriented)	0.037	1500	7.45	0.0050	−46.1
40.	4% Silicon iron	0.029	500	3.81	0.0076	−42.4
41.	16 Alfenol	0.0113	4500	7.13	0.0016	−56.0
42.	Hiperco	0.069	650	6.70	0.010	−39.7
43.	Monimax	0.0216	2000	6.57	0.0033	−49.7
44.	50% nickel iron	0.0384	1000	6.20	0.0062	−44.2
45.	45-25 Perminvar	0.091	400	6.03	0.015	−36.4
46.	Commercial iron (0.2% impure)	0.17	200	5.83	0.29	−30.7
47.	Cold-rolled steel	0.17	180	5.53	0.031	−30.2
48.	Nickel	0.23	100	4.70	0.047	−26.6
49.	Stainless steel (1Cu-18Cr-8Ni-+Fe)	0.02	200	2.00	0.010	−40.0
50.	Rhometal (36Ni)	0.019	1000	4.36	0.0044	−47.2
51.	Netic 53-6	0.172	300	7.18	0.024	−32.4

Source: From White.[5]

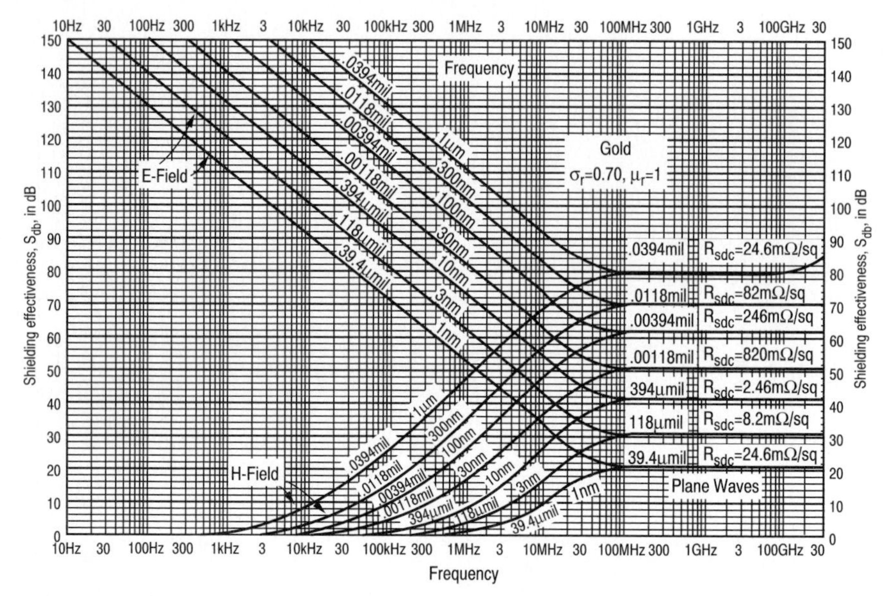

FIGURE 5.15 Shielding effectiveness of gold versus frequency for source-to-metal distance of 1 m.

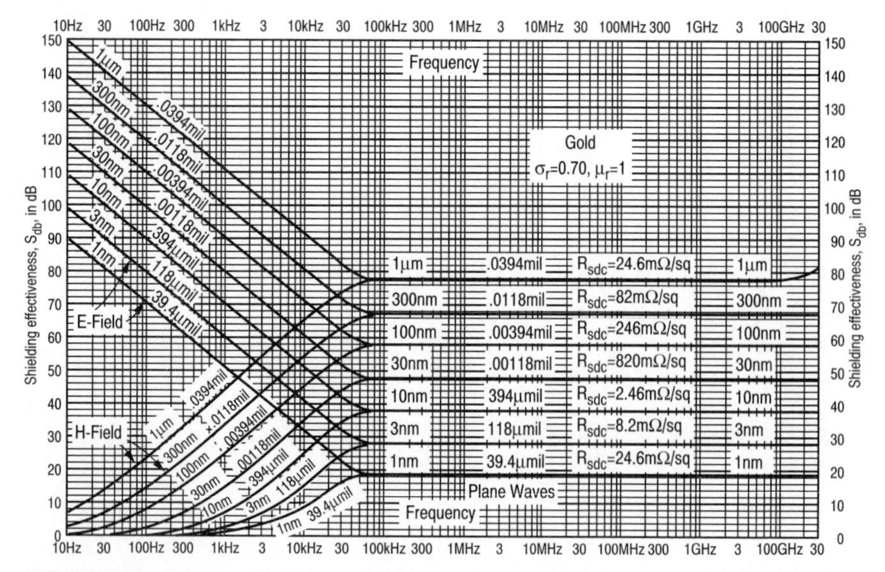

FIGURE 5.16 Shielding effectiveness of gold versus frequency for source-to-metal distance of 1 km.

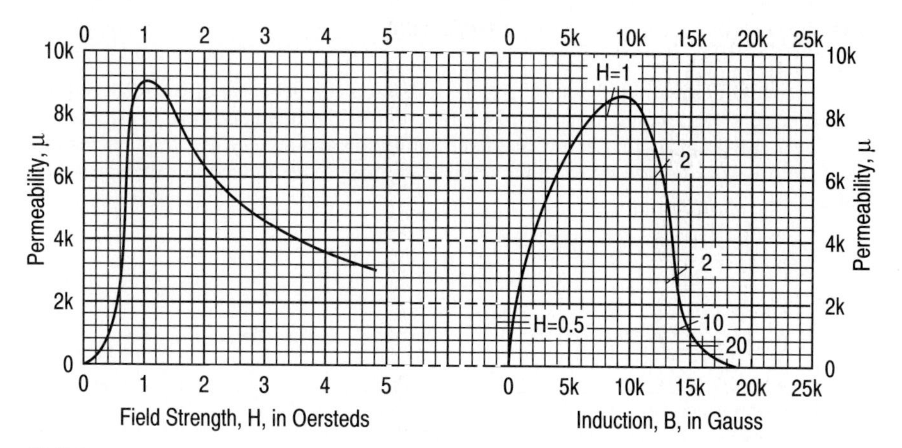

FIGURE 5.17 Permeability curves of iron, with permeability plotted against field strength H and induction I. (From Bozorth.[6]) (Courtesy of Bozorth, R. M., *Ferromagnetism,* D. Van Nostrand Co., Inc., New York.)

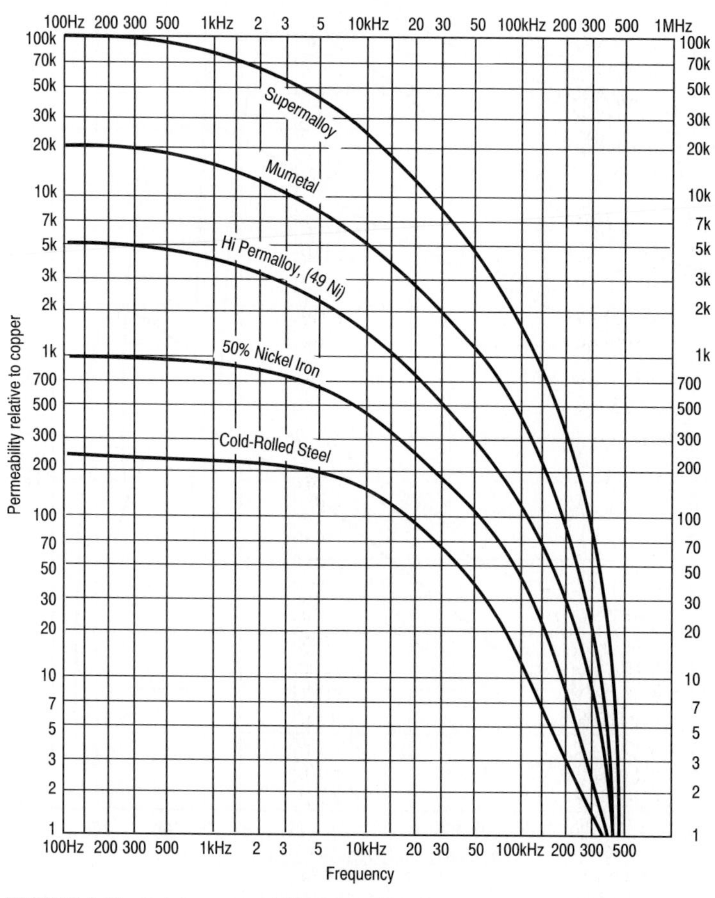

FIGURE 5.18 Relative permeability versus frequency.

fined magnetic field strength, and Fig. 5.19 illustrates the behavior of the relative permeability of several magnetic alloys as a function of the magnetic flux density at 60 Hz. These data suggest that below magnetic saturation and at very low frequencies the permeability can increase up to about an order of magnitude above the values listed in Table 5.14.

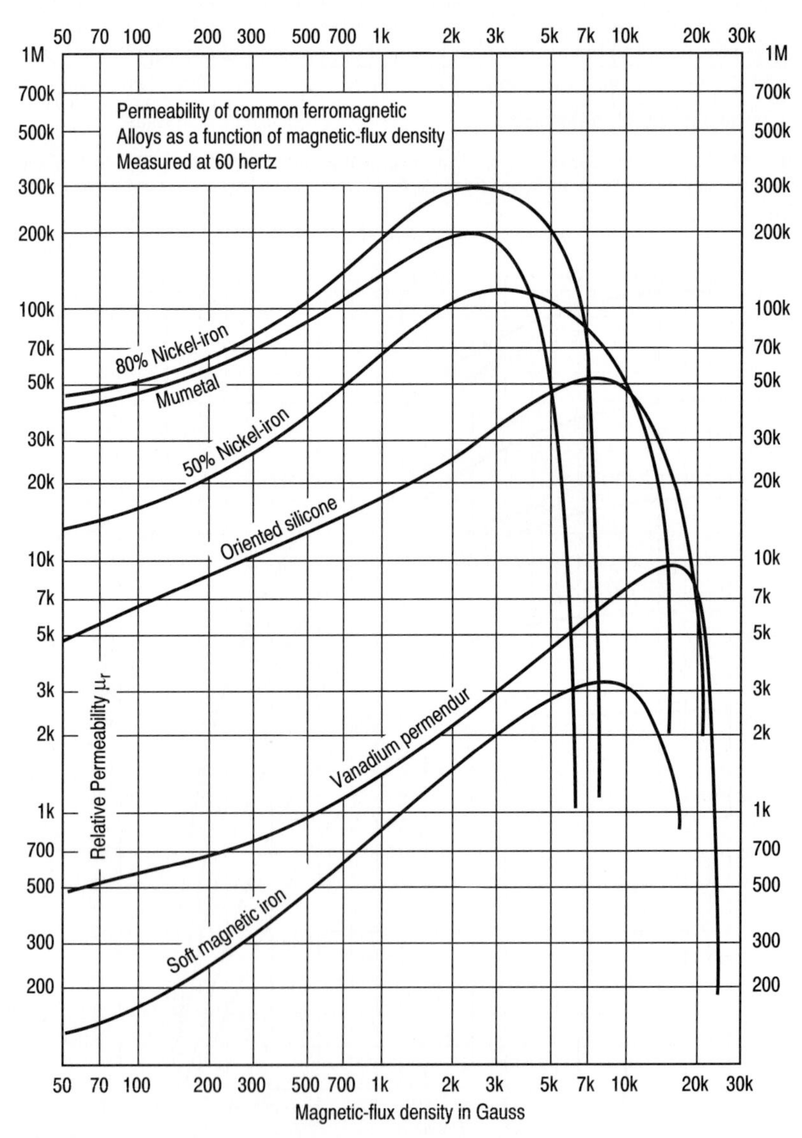

FIGURE 5.19 Relative permeability versus magnetic flux density.

5.5.4 Low-Expansion Alloys

Low-expansion alloys of iron and nickel and their various modifications are used in applications that include bimetal strip, glass-to-metal seals, thermostatic strip, integrated-circuit lead frames, and components for radios and other electronic devices (Boyer and Gall,[1] pp. 20.37–20.39).

Alloys of iron and nickel have many inconsistent properties depending on the relative proportions of the two elements. The coefficients of linear expansion range from a small negative value (−0.3 μin/in • °F) to a large positive value (11 μin/in • °F). Alloys that contain less than 36 percent nickel have much higher coefficients of expansion than alloys containing 36 percent nickel or more. Alloys that contain less than 36 percent nickel are of the so-called irreversible type, and are excluded from the present discussion. One alloy containing 36 percent nickel with small quantities of manganese, silicon, and carbon amounting to a total of less than 1 percent, has a coefficient of expansion so low that its length is almost invariable for ordinary changes in temperature. For this reason, the alloy was named Invar.* After the discovery of Invar, an intensive study was made of the thermal and elastic properties of similar alloys. Those with higher nickel contents were found to have higher coefficients of expansion. The alloy containing 39 percent nickel has a coefficient of expansion corresponding to that of low-expansion glasses. The 46 percent nickel alloy has a coefficient equivalent to that of platinum (5 μin/in • °F) and it is named Platinite. Dumet wire is an alloy containing 42 percent nickel that is coated with copper to prevent gasing at the seal. It is used for the "seal-in" wire in incandescent lamps and vacuum tubes. The 56 percent nickel alloy has a coefficient approaching that of ordinary steel (6.1 μin/in • °F). Nilvar† is identical to Invar (36 percent nickel). Elinvar, containing 36 percent nickel and 12 percent chromium, has a constant modulus of elasticity over a considerable range of temperature. It also has low thermal expansivity. There is an advantage in replacing some of the nickel with cobalt in the 36 percent nickel alloy. Substitution of 5 percent cobalt for 5 percent nickel provides an alloy with an expansion coefficient even lower than that of Invar. The 31Ni-5Co alloy is also less susceptible to variations in heat treatment.

Effect of Composition on Expansivity. The effect of variation in nickel content on linear expansivity is shown in Fig. 5.20. Minimum expansivity occurs at about 36 percent nickel. Small additions of other metals have considerable influence on the position of this minimum. The effects of additions of manganese, chromium, copper, and carbon are shown in Fig. 5.21. Minimum expansivity shifts toward higher nickel contents when manganese or chromium is added and toward lower nickel contents when copper or carbon is added. The value of the minimum expansivity for any of these ternary alloys is, in general, greater than that of a typical Invar. Addition of silicon, tungsten, and molybdenum produces effects similar to those produced by additions of manganese and chromium; the composition of minimum expansivity shifts toward higher nickel contents. Addition of carbon is said to produce instability in Invar. This instability is attributed to the changing solubility of carbon in the austenitic matrix during heat treatment.

Invar. Invar and related alloys have low coefficients of expansion only over a rather narrow range of temperature. At low temperatures, the coefficient of expansion is high

*Registered trademark of Carpenter Technology Corporation.
†Registered trademark of Driver-Harris Company.

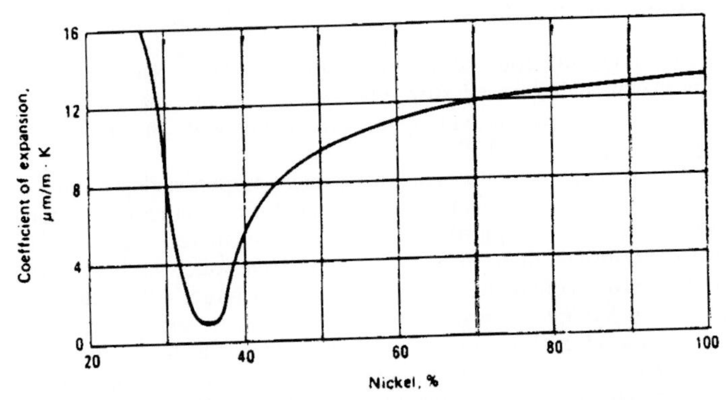

FIGURE 5.20 Coefficient of linear expansion at 20°C versus nickel content for iron-nickel alloys containing 0.4 percent manganese and 0.1 percent carbon.

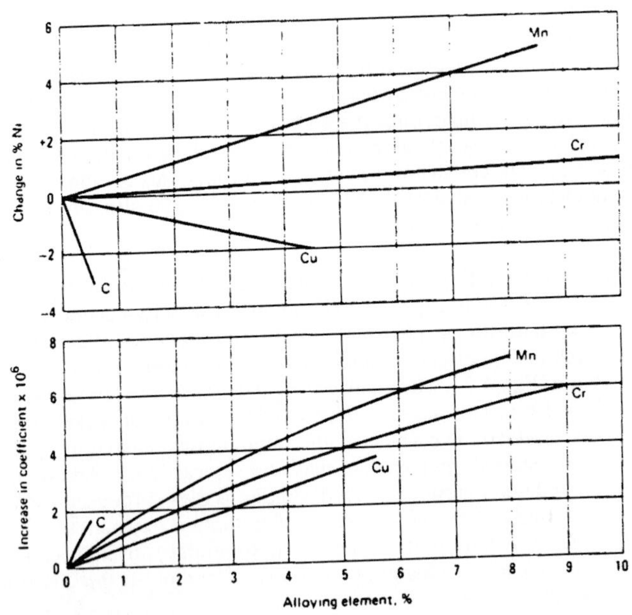

Top: Displacement of nickel content caused by additions of manganese, chromium, copper and carbon to alloy of minimum expansivity. Bottom: Change in value of minimum coefficient of expansion caused by additions of manganese, chromium, copper and carbon.

FIGURE 5.21 Effect of alloying elements on expansion characteristics of iron-nickel alloys.

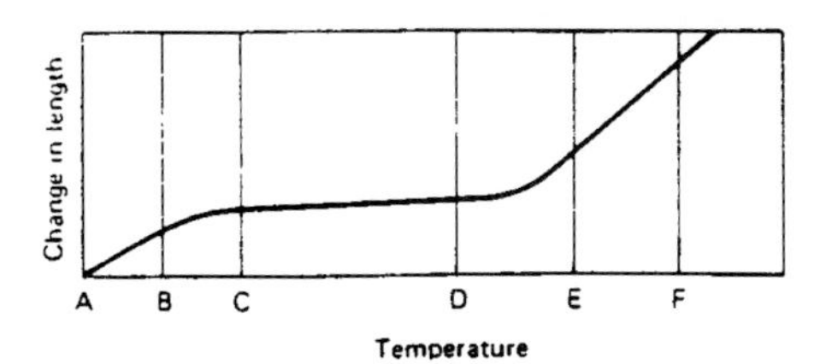

FIGURE 5.22 Change in length of typical Invar over different ranges of temperatures.

in the region from *A* to *B* as shown in Fig. 5.22. In the interval between *B* and *C*, the coefficient is lower, reaching a minimum in the region from *C* to *D*. With increasing temperature the coefficient begins again to increase in the range from *D* to *E*, and thereafter (from *E* to *F*) the expansion curve follows a trend similar to that of the nickel or iron of which the alloy is composed. The minimum expansivity prevails only in the range from *C* to *D*. Between *D* and *E* the coefficient is changing rapidly to a higher value. The temperature limits for a well-annealed 64 percent iron alloy are 324 and 520°F. These temperatures correspond to the initial and final losses of magnetism in the material. The slope of the curve between *C* and *D* is a measure of the coefficient of expansion over a limited range of temperature. Table 5.15 gives coefficients of linear expansion of iron-nickel alloys between 32 and 100°F. The expansion behavior of several iron-nickel alloys over wider ranges of temperature is represented by curves 1 to 5 in Fig. 5.23. For comparison, Fig. 5.23 also includes data for an ordinary carbon steel (curve 6). Heat treatment and cold work change the expansivity of Invar (or Nilvar) considerably. The expansivity is greatest in well-annealed material and least in quenched material. Cold drawing also decreases expansivity. The values for the coefficients shown in Table 5.16 are from experiments conducted on two heats of Invar. By

TABLE 5.15 Thermal Expansion of Nickel-Iron Alloys between 0 and 38°C

Nickel, %	Mean coefficient, μm/m • K
31.4	3.395 + 0.00885 *t*
34.6	1.373 + 0.00237 *t*
35.6	0.877 + 0.00127 *t*
37.3	3.457 − 0.00647 *t*
39.4	5.357 − 0.00448 *t*
43.6	7.992 − 0.00273 *t*
44.4	8.508 − 0.00251 *t*
48.7	9.901 − 0.00067 *t*
50.7	9.984 + 0.00243 *t*
53.2	10.045 + 0.00031 *t*

Source: Boyer and Gall.[1]

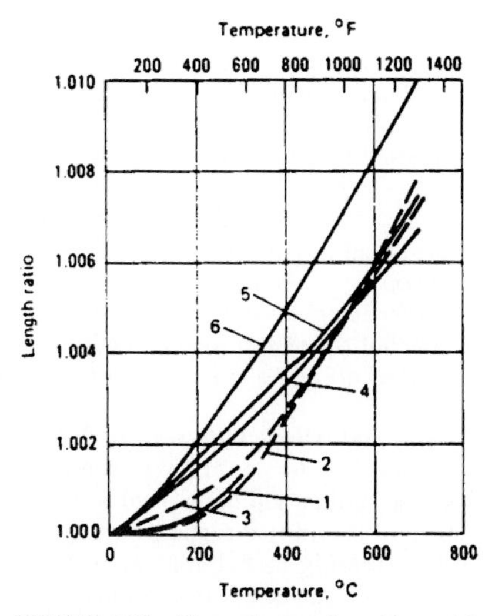

FIGURE 5.23 Thermal expansion of iron-nickel alloys.

TABLE 5.16 Effect of Metal Working on Expansivity

Material condition	Expansivity, μm/m • K*
Direct from hot mill	1.4
	1.4
Annealed and quenched	0.5
	0.8
Quenched and cold drawn†	0.14
	0.3

*Individual measurements for two heats of Invar.
†3.2–6.4 mm (0.125–0.250 in) diameter.
Source: From Boyer and Gall.[1]

cold working after quenching it is possible to produce material with a zero, or even a negative, coefficient of expansion. A negative coefficient may be increased to zero by careful annealing at a low temperature. However, these artificial methods of securing an exceptionally low coefficient produce instability in the material. With lapse of time and variation in temperature, exceptionally low coefficients usually revert to normal values. For special applications (geodetic tapes, for example) it is essential to stabilize the material by cooling it slowly from 212 to 68°F over a period of many months, followed by prolonged aging at room temperature. However, unless the material is to be used within the limits of normal atmospheric variation in temperature, such stabilization is of no value. Although these variations in heat-treating practice are important in special applications, they are of little significance for ordinary uses.

The magnetic properties of Invar are similar to those of other iron-nickel alloys. They are ferromagnetic at room temperature and become paramagnetic at higher temperatures. The points of inflection in the curves in Fig. 5.23 indicate the loss of magnetism. The loss of magnetism in a well-annealed sample of a true Invar begins at 324°F and ends at 520°F. In a quenched sample the loss begins at 400°F and ends at 520°F. Slow cooling through this range of temperature eliminates to a large extent the troublesome variability in the properties of materials of this class.

The electrical properties of Invar over the range of low expansivity are a resistivity between 750 and 850 megΩmeters at ordinary temperatures, a temperature coefficient of electrical resistivity of about 0.67 nΩ/Ω • °F, and a thermoelectric potential versus pure copper of about 5.6 mV/°F.

Other physical and mechanical property data of Invar are presented in Table 5.17 for the hot-rolled and forged conditions. The effects of temperature on the mechanical properties of forged 66Fe-34Ni are illustrated in Fig. 5.24.

Iron-Nickel Alloys Other Than Invar. Iron-nickel alloys that have higher nickel contents than Invar retain to some extent the expansion characteristics of Invar. Because further additions of nickel raise the temperature at which the inherent magnetism of the alloy disappears, the inflection temperature in the expansion curve rises with increasing nickel content. Although this increase in range is an advantage in some circumstances, it is accompanied by an increase in the coefficient of expansion. Table 5.18 and Fig. 5.25 present information on the coefficients of expansion of iron-nickel alloys at temperatures up to the inflection temperature. They also give data on alloys containing up to 68 percent nickel.

TABLE 5.17 Physical and Mechanical Properties of Invar

Solidus temperature	1425°C (2600°F)
Density	8.0 Mg/m^3 (0.29 lb/in^3)
Tensile strength	450–585 MPa (65–85 ksi)
Yield strength	275–415 MPa (40–60 ksi)
Elastic limit	140–205 MPa (20–30 ksi)
Elongation	30–45%
Reduction in area	55–70%
Scleroscope hardness	19
Brinell hardness	160
Modulus of elasticity	150 GPa (21.4 × 10^6 lb/in^2)
Thermoelastic coefficient	500 μm/m • K
Specific heat	515 J/kg • K (0.123 Btu/lb • °F) at 25–100°C (78–212°F)
Thermal conductivity	11 W/m • K (6.4 Btu/ft • h • °F) at 20–100°C (68 to 212 °F)
Thermoelectric potential (against copper)	9.8 μV/K at − 96°C (−140°F)

Source: From Boyer and Gall.[1]

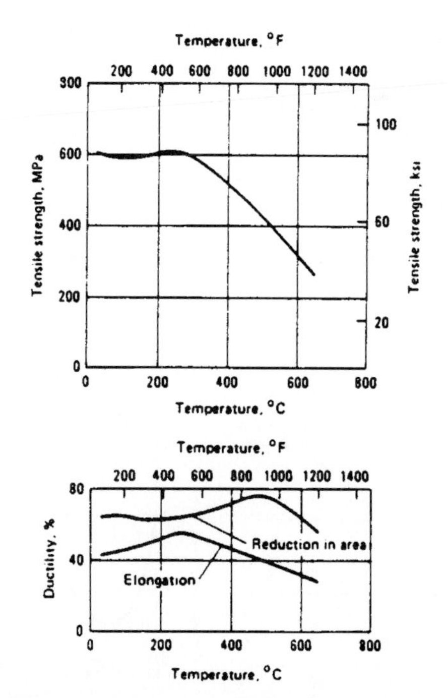

Alloy composition: 0.25 C, 0.55 Mn, 0.27 Si, 33.9 Ni, rem Fe. Heat treatment: annealed at 800 °C (1475 °F) and furnace cooled.

FIGURE 5.24 Mechanical properties of forged 34 percent nickel alloy.

TABLE 5.18 Expansion Characteristics of Iron-Nickel Alloys

Composition, %			Inflection temperature		Mean coefficient of expansion, from 20°C to inflection temperature, μm/m • K
Mn	Si	Ni	°C	°F	
0.11	0.02	30.14	155	310	9.2
0.15	0.33	35.65	215	420	1.54
0.12	0.07	38.70	340	645	2.50
0.24	0.03	41.88	375	705	4.85
—	—	42.31	380	715	5.07
—	—	43.01	410	770	5.71
—	—	45.16	425	795	7.25
0.35	—	45.22	425	795	6.75
0.24	0.11	46.00	465	870	7.61
—	—	47.37	465	870	8.04
0.09	0.03	48.10	497	925	8.79
0.75	0.00	49.90	500	930	8.84
—	—	50.00	515	960	9.18
0.25	0.20	50.05	527	980	9.46
0.01	0.18	51.70	545	1015	9.61
0.03	0.16	52.10	550	1020	10.28
0.35	0.04	52.25	550	1020	10.09
0.05	0.03	53.40	580	1075	10.63
0.12	0.07	55.20	590	1095	11.36
0.25	0.05	57.81	None		12.24
0.22	0.07	60.60	None		12.78
0.18	0.04	64.87	None		13.62
0.00	0.05	67.98	None		14.37

Source: From Boyer and Gall.[1]

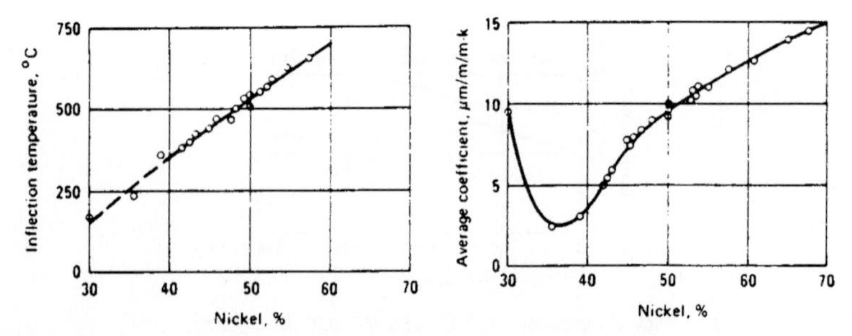

Left: Variation of inflection temperature. Right: Variation of average coefficient of expansion between room temperature and inflection temperature.

FIGURE 5.25 Effect of nickel content on expansion of iron-nickel alloys.

5.5.5 Magnetically Soft Materials

Magnetically soft materials are ferromagnetic and have little or no retentivity, that is, when removed from a magnetic field, they lose most, if not all, of the magnetism they exhibited while in the field. The most important characteristics of magnetically soft materials are (1) low hysteresis loss (easy domain movement during magnetization), (2) low eddy current loss from electric currents induced by flux changes, (3) high magnetic permeability, and sometimes constant permeability at low field strengths, (4) high magnetic saturation induction, and (5) minimum or definite change in permeability with temperature. Magnetically soft materials made in large quantities include high-purity iron, low-carbon steels, silicon steels, iron-nickel alloys, iron-cobalt alloys, and ferrites (Boyer and Gall,[1] pp. 20.1–20.5). For additional information on magnetics, consult Cullity.[7]

Impurities such as carbon, sulfur, nitrogen, and oxygen are especially deleterious to ferromagnetic properties because they distort the lattice of the crystal structure and even in small amounts may greatly interfere with easy movement of magnetic domains, which is the basis of the magnetic properties. A similar disturbance caused by carbon and nitrogen, known as "aging," occurs when low solubility at room temperature causes the excess solute to precipitate slowly as small particles within grains. In iron and iron-silicon alloys, the carbon content preferably should be less than 0.003 percent for best permeability, low hysteresis loss, and minimal aging. Figure 5.26 shows the relationship between carbon content and hysteresis loss for iron. Hysteresis loss is similarly related to sulfur and oxygen contents.

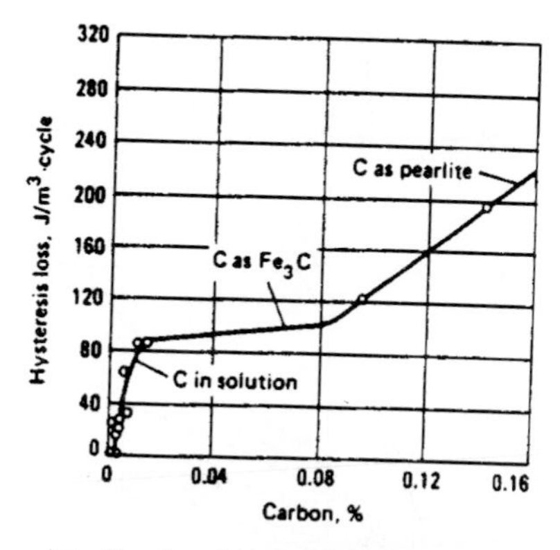

FIGURE 5.26 Relationship between carbon content and hysteresis loss for unalloyed iron.

The grain size for most applications should be as large as possible for nonoriented materials (Fig. 5.27). In oriented grades of silicon steel, optimum magnetic properties

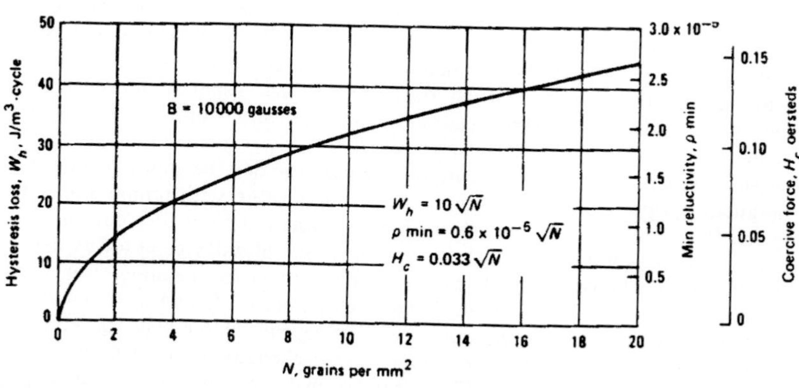

FIGURE 5.27 Effect of grain size on magnetic properties of pure iron and silicon iron.

are usually obtained with grain sizes of 80 to 400 mil, depending on the degree of crystal orientation. Increases in grain size above 400 mil are accompanied by significant increases in both domain-wall spacing and eddy current losses.

Grain orientation is a basic factor in the determination and control of magnetic properties. All ferromagnetic crystals are magnetically anisotropic, that is, they have different magnetic properties in different crystallographic directions. In nickel, the direction of easiest magnetization is the cube diagonal (111); in iron it is the cube edge (100).

Alloying elements increase the electrical resistivity of iron to give alloys that can be used in ac circuits. Pure iron is very soft magnetically, and it has a high saturation induction, but it requires special handling during manufacture due to its low mechanical strength. Pure iron is used extensively in dc applications and in some ac relays. However, its low electrical resistivity makes it unsuitable for use in ac circuits, which constitute most industrial applications of magnetic materials. Figure 5.28 shows the changes in resistivity that result from additions of silicon, aluminum, and other elements to iron.

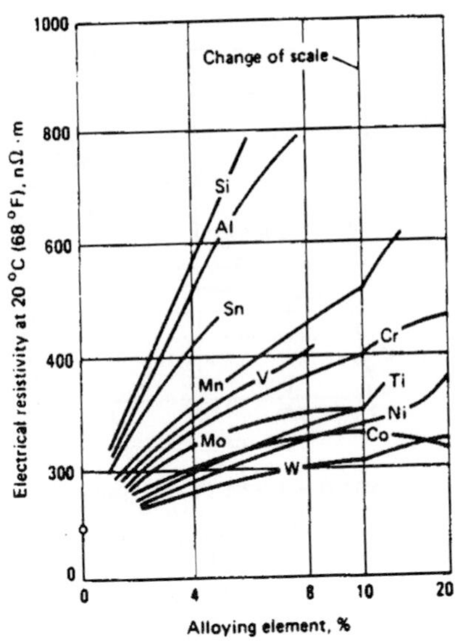

FIGURE 5.28 Effect of alloying elements on electrical resistivity of iron.

Grain growth in iron-silicon alloys is possible because the addition of silicon in sufficient amounts eliminates the allotropic transformation in iron. Consequently silicon-iron alloys can be annealed at high temperature to promote grain growth, thus facilitating the development of preferred grain orientation. However, room-temperature saturation induction is reduced by alloy additions other than cobalt (Fig. 5.29).

Stress affects power (core) loss and permeability. Plastically strained materials usually must be annealed unless the volume of strained material is only a small fraction of the total. In oriented silicon steels, a compressive stress of as little as 500 to 1000 lb/in^2 can increase core loss by 50 to 100 percent. Thus great care must be exercised in punching and assembling electrical steels to relieve stress before the unit is assembled, and to ensure that new stresses are not introduced during assembly. In some instances, however, small tensile stresses induced by coatings can improve the properties of grain-oriented silicon steel.

Constant permeability with changing temperature can only be approximated since the magnetic properties of all magnetic materials change with temperature. The change in flux density with temperature for iron tested at four different values of magnetizing force is plotted in Fig 5.30. The operation of a device at a flux density of 15,000 G would be only slightly affected by variations in operating temperature near ambient. There is a similar minimized temperature effect for all materials, except that the flux density for optimum operation depends on the material, given the proper flux density and temperature range. Large changes occur at temperatures approaching the Curie temperature (Fig. 5.30). For many reasons it is not always pos-

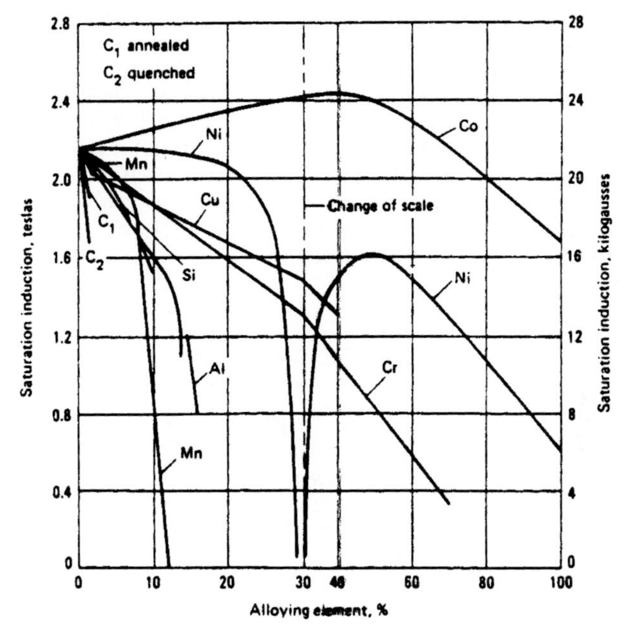

FIGURE 5.29 Effect of alloying elements on room-temperature saturation induction of iron.

sible to operate a material at the best flux density for temperature stability. One way of obtaining better temperature stability is to use magnetic materials in insulated powder form, such as pressed Permalloy powder cores, which have good temperature stability due to the presence of many built-in air gaps. Another method involves the use of iron-nickel alloys, such as Isoperm or Conpernik,* that have been drastically cold-rolled and then underannealed to produce a partly strained alloy less sensitive to temperature changes. In a third method, two alloy powders with opposite temperature coefficients are combined for use in the desired temperature range. Special Permalloy powder cores are combined with small amounts of iron-nickel-molybdenum powder having a low Curie temperature and a negative temperature coefficient near room temperature. The 30 percent nickel irons are of the low-Curie-temperature type.

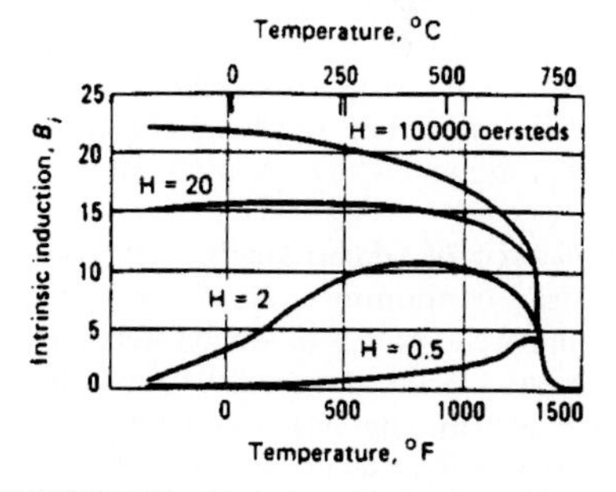

FIGURE 5.30 Variation of induction with temperature for iron at four different values of magnetizing force.

*Registered trademark of Westinghouse Electric Corporation.

High-Purity Irons. For experimental uses, 99.99 percent pure iron can be produced. Using high-temperature annealing for prolonged times, this material can achieve a maximum permeability of about 100,000 with a hysteresis loss of about 10^{-5} J/cm^3 per cycle at a flux density of 10,000 G (Fig. 5.31). The saturation induction of iron, based on a density of 0.2846 lb/in^3, is given as $4\pi I_s = 21{,}580 \pm 10$ G, where I_s is the intensity of magnetization or the magnetic moment per unit volume. The electrical resistivity is 59 Ω • cmil/ft at 68°F, and the temperature coefficient is 0.0036 per°F. Commercial irons with purities of 99.6 to 99.8 percent are available in a variety of shapes. These irons have a saturation induction of about 21,500 G, a specific gravity of 7.85, and a resistivity of 64 Ω • cmil/ft at 68°F.

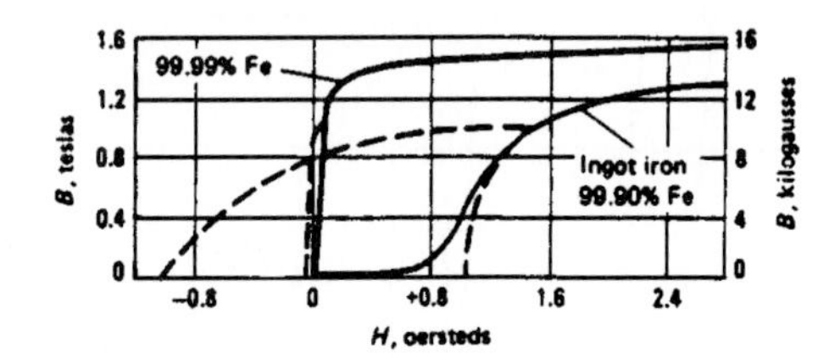

FIGURE 5.31 Partial hysteresis loops and *B–H* curves for two types of iron.

Low-Carbon Steels. For many applications that require less than superior magnetic properties, low-carbon steels (type 1010, for example) are used. Frequently higher than normal phosphorus and manganese contents are used to increase electrical resistivity. Such steels are not purchased to magnetic specifications. Although low-carbon steels exhibit higher power losses than silicon steels, they have better permeability at high flux density. This combination of magnetic properties coupled with low price makes low-carbon steels especially suitable for applications such as fractional-horsepower motors, which are used intermittently.

Nonoriented Silicon Steels. Except for saturation induction, the magnetic properties of iron containing a small amount of silicon are better than those of pure iron. Commercial grades of silicon steel contain 0.5 to 3.5 percent silicon. At levels above 4 percent silicon, the steel becomes brittle and difficult to process with cold-rolling methods. Silicon steel frequently is coated with organic or inorganic materials after annealing in order to reduce eddy currents in laminated stacks. The vast majority of finished nonoriented silicon steel is sold in either full-width coils of 34 to 48 in or slit-width coils, but some is sold as sheared sheets. Fully processed grades require only stress-relief annealing after fabrication or stamping. Semiprocessed grades must be decarburized by the customer to develop full magnetic properties.

Oriented Silicon Steels. High-flux-density permeability is achieved in silicon steels by crystallographic orientation. Like iron, silicon steels are more easily magnetized in the direction of the cube edge, (100). Rolling and heat treating are used to promote secondary recrystallization in the final anneal to produce a well-developed texture with the cube edge parallel to the rolling direction, {110} (001). Oriented steels contain from 2.9 to 3.2 percent silicon. Conventional grain-oriented silicon steel has grains of about 0.12 in in diameter. High-permeability silicon steel tends to have grains of about 0.31 in in diameter. Ideally the grain diameter should be less than 0.12

in to minimize excess eddy current effects from domain-wall motion. Special coatings provide electrical insulation and induced tensile stresses in the steel substrate. These induced stresses lower core loss and minimize noise in transformers. High-grade silicon electrical steel does not age significantly as received from the mill, because its carbon content has been reduced to about 0.003 percent or less. Table 5.19 lists typical applications of electrical steel.

Iron-Aluminum Alloys. Although aluminum and silicon have similar effects on electrical resistivity and some magnetic properties of iron, aluminum is seldom substituted for silicon because of the resulting difficulties in fabrication. Aluminum is used most commonly as small (<0.5 percent) additions to the better grades of nonoriented silicon steel to increase electrical resistivity and thereby reduce eddy currents. Tertiary alloys of iron, silicon, and aluminum have high resistivity and good permeability at low flux density. Increases in silicon and aluminum reduce saturation induction. At low flux densities the magnetic properties of these alloys can be made to approach those of some iron-nickel alloys.

TABLE 5.19 Silicon Contents, Densities, and Applications of Electrical Steel Sheet and Strip

AISI type	Nominal Si + Al content, %	Assumed density, Mg/m^3	Characteristics and applications
Lamination steel			
—	0	7.85	High magnetic saturation, properties may not be guaranteed; intermittent-duty small motors
Nonoriented electrical steels			
M47	1.05	7.80	Ductile, good stamping properties, good permeability at high inductions; small motors, ballasts, relays
M45	1.85	7.75	Good stamping properties, good permeability at moderate and high inductions, good core loss; small generators, high-efficiency continuous-duty rotating machines, ac and dc
M43	2.35	7.70	
M36	2.65	7.70	Good permeability at low and moderate inductions, low core loss; high-reactance cores, generators, stators of high-efficiency rotating machines
M27	2.80	7.70	
M22	3.20	7.65	Excellent permeability at low inductions, lowest core loss; small power transformers, high-efficiency rotating machines
M19	3.30	7.65	
M15	3.50	7.65	
Oriented electrical steels			
M6	3.15	7.65	Grain-oriented steel has highly directional magnetic properties with lowest core loss and highest permeability when flux path is parallel to rolling direction; heavier thicknesses used in power transformers, thinner gauges generally used in distribution transformers; energy savings improve with lower core loss
M5	3.15	7.65	
M4	3.15	7.65	
M3	3.15	7.65	
High-permeability oriented steel			
—	2.9–3.15	7.65	Low core loss at high operating inductions

Source: From Boyer and Gall.[1]

Iron-Nickel Alloys. Two broad classes of alloys have been developed in the iron-nickel system. The high-nickel alloys (about 79 percent nickel) have high initial permeability (Fig. 5.32) but a low saturation induction of about 9000 G, whereas the low-nickel alloys (about 50 percent nickel) are lower in initial permeability but higher in saturation, with an induction of about 16,000 G. The effects of nickel content in iron-nickel alloys on initial permeability and on saturation induction after annealing are illustrated in Figs. 5.32 and 5.33, respectively. The data plotted in Fig. 5.34 are from early laboratory studies and illustrate the effects of both composition and heat treatment on initial permeability. Values above 12,000 are now obtained commercially in 50 percent nickel alloys, and values above 60,000 are obtained in 79 percent nickel alloys containing 4 percent molybdenum. To obtain high initial permeability, both magnetocrystalline anisotropy and magnetostrictive anisotropy must be minimized.

Most applications of 50 percent nickel alloys are based on requirements for high

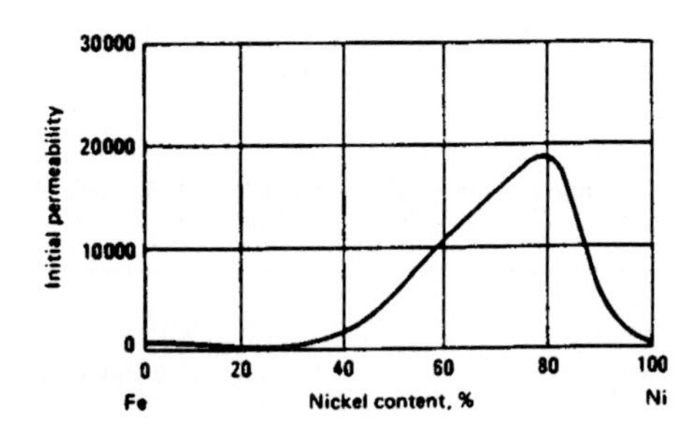

FIGURE 5.32 Initial permeability at 2 milli-tesla (20 G) for annealed iron-nickel alloys.

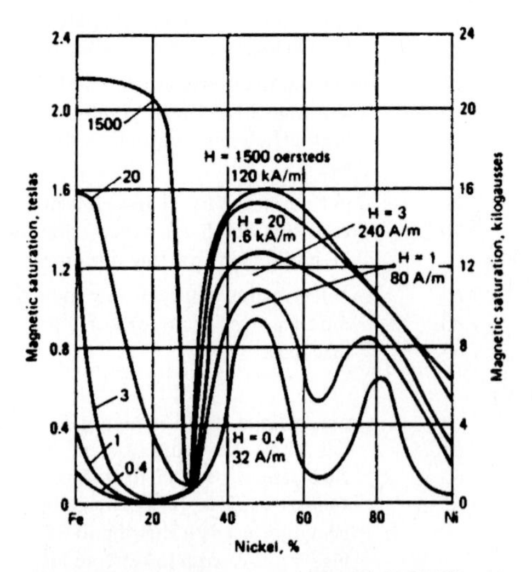

FIGURE 5.33 Magnetic saturation of iron-nickel alloys at various field strengths.

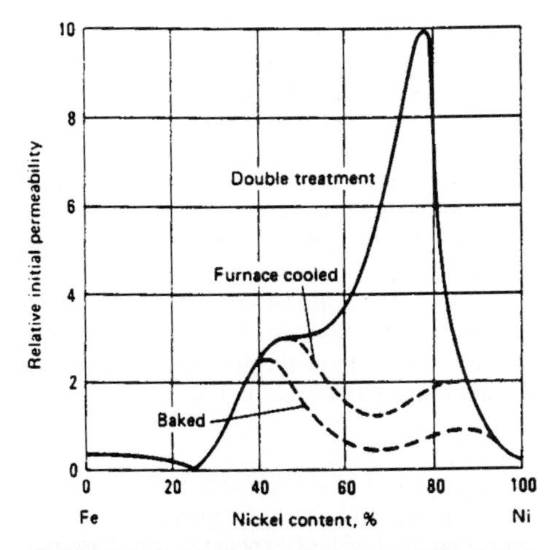

Treatments were as follows: furnace cooled—1 h at 900 to 950 °C (1650 to 1740 °F), cooled at 100 °C/h (180 °F/h); baked—furnace cooled plus 20 h at 450 °C (840 °F); double treatment—furnace cooled plus 1 h at 600 °C (1110 °F) and cooled at 1500 °C/min (2700 °F/min).

FIGURE 5.34 Relative initial permeability at 2 millitesla (20 G) for iron-nickel alloys given various heat treatments.

saturation induction. Nickel content is not critical near the middle of the iron-nickel series (50 percent nickel) and may be varied from 45 to 60 percent, but for highest saturation induction it should be held close to 50 percent (Fig. 5.33). Although the magnetocrystalline anisotropy factor κ equals zero, the value of the magnetostrictive constant (λ_{100}) in the <100> easy direction of magnetization is close to zero for these alloys. Therefore the initial permeability is still reasonable and in fact reaches a small maximum (Fig. 5.32).

Iron-Cobalt Alloys with High Magnetic Saturation. Pure iron has a saturation induction of 21,600 G. Higher values require cobalt additions of up to 65 percent. The highest values of about 24,200 G are obtained with cobalt contents of about 35 percent (Fig. 5.29). The use of alloys containing 25 to 50 percent cobalt is limited by low resistivity and high hysteresis loss, the high cost of cobalt, and the brittleness of alloys containing more than 30 percent cobalt. However, with small additions of vanadium and special treatment in processing, 50 percent cobalt alloys can be cold-rolled commercially to any gauge, and strip is ductile enough to be punched and sheared. The 27 percent cobalt alloy is more easily fabricated, and less subject to degradation by stresses, than the 50 percent cobalt alloy. Furthermore, proper annealing of the 27 percent cobalt alloy produces magnetic properties suitable for both dc and ac applications. Ac applications require low eddy current and hysteresis losses. Eddy current losses can be minimized by a proper combination of composition and thickness.

Amorphous Materials. Suitable alloys of iron prepared in amorphous, noncrystalline form have the attractive combination of high permeability and high volume resistivity. In the preferred method of production, the metal is rapidly quenched from the melt onto cooled rotating drums to form long ribbons about 1.6 mil thick. Attractive compositions contain 40 to 80 percent iron with various additions of carbon, cobalt, boron, silicon, nickel, and phosphorus. These materials are characterized by low hysteresis loss and low coercive force. However, the Curie temperature is limited to about 750°F, and magnetic saturation to about 16,000 G. Potential applications include substitution for conventional nickel-iron alloys in electronic devices. A major goal in distribution transformer design is to take advantage of the low core loss of amorphous material, which is less than one-half that of conventional grain-oriented, 0.010-in-thick silicon steel.

Compressed Powdered Iron. For applications having complicated magnetic circuits that would require considerable machining, the use of pressed-to-shape iron powder parts may save as much as 50 percent of the alternate cost. The densities of most iron-base powdered metal parts range from 0.224 to 0.253 lb/in^3. High densities and superior magnetic properties can be obtained by using higher molding pressures or higher sintering temperatures, or by making the parts from high-purity iron-base powders (with or without small additions of alloying elements such as silicon and phosphorus). The cost of such parts is higher, and unless magnetic flux density exceeds 10,000 G, performance is no better than that of less-expensive iron-base powdered metal magnet parts. This is especially true in magnetic circuits with short flux paths. Selection of a rolled or sintered iron for a specific application depends on die cost, required finish, tolerances, and number of parts.

Austenitic Stainless Steels. Austenitic stainless steels usually are not considered to be magnetic materials. However, increases in tensile strength due to cold working of these alloys are accompanied by increases in intrinsic permeability. This phenomenon is illustrated graphically in Fig. 5.35 for nine austenitic stainless steels.

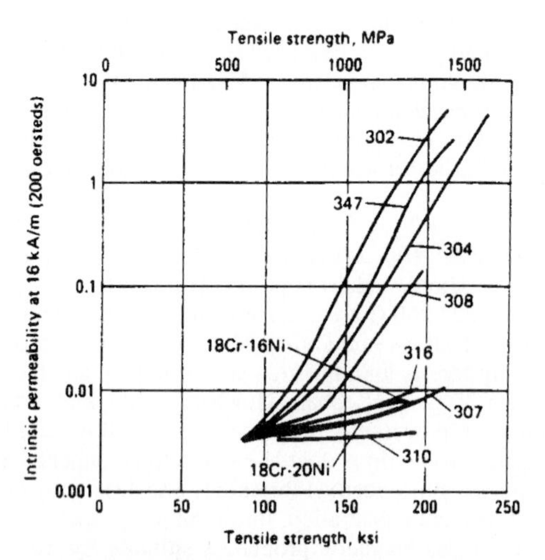

FIGURE 5.35 Correlation of increased tensile strength from cold working and permeability of cold-worked austenitic stainless steels.

Ferrites. Ferrites for high-frequency applications are ceramics with characteristic spinel-magnetic structures ($M \cdot Fe_2O_4$, where M is a metal) and usually comprise solid solutions of iron oxide and one or more oxides of other metals such as manganese, zinc, magnesium, copper, nickel, and cobalt. They are unique among magnetic materials in possessing outstanding magnetic properties at high frequencies, which result from very high resistivities ranging from about $10^8\ \Omega \cdot cm$ to as high as $10^{14}\ \Omega \cdot cm$. Hence at frequencies where eddy current losses for metals become excessive, ferrites make ideal soft magnetic materials. Because ferrites have inherently high corrosion resistance, parts made of these materials normally do not require protective finishing. Disadvantages of ferrites include low magnetic saturation, low Curie temperature, and relatively poor mechanical properties compared with those of metals. Ferrites are produced from powdered raw materials by mixing, calcining, ball milling, pressing to shape, and firing to the desired magnetic properties. The finished product is hard, brittle, and unmachinable, and thus close dimensional tolerances must be obtained by grinding. Many different types of ferrites are available for magnetic use. They can be classified into three general types: square-loop ferrites for computer memories, linear ferrites for transformers and for inductors in filters, and microwave ferrites for microwave devices. In recent years semiconductors have largely replaced square-loop ferrites for computer memory applications. Microstructure and composition have much stronger influences on the magnetic properties of ferrites than on those of metals. Hence properties of finished ferrite parts can vary drastically with the purity and structure of raw materials, with the nature of the binders used, and with the ceramic-processing technique employed. Linear ferrites comprise Mn-Zn and Ni-Zn ferrites. Mn-Zn ferrite is higher in saturation magnetization, but lower in resistivity, than Ni-Zn ferrite. Mn-Zn ferrite is preferred for frequencies up to about 2 MHz. For microwave applications Ni-Zn, Mg-Mn-Al, and Mg-Mn-Cu ferrites are used, as well as garnets of the type $M_{3+x}Fe_{5-x}O_{12}$ (where M = Y + Al or Y + Gd + Al).

5.5.6 Permanent Magnets

Permanent-magnet materials are developed to enhance high induction, high resistance to demagnetization, and maximum energy content (Boyer and Gall,[1] pp. 20.5–20.11). Magnetic induction is limited by composition; the highest saturation induction is found in binary iron-cobat alloys. Resistance to demagnetization is conditioned less by composition than by shape or crystal anisotropies and the mechanisms that subdivide materials into microscopic regions. Maximum energy content is the most important characteristic because permanent magnets are used primarily to produce a magnetic flux field (which is a form of potential energy). Maximum energy content and certain other characteristics of materials used for magnets are best described by a hysteresis loop. Hysteresis is measured by successively applying magnetizing and demagnetizing fields to a sample and observing the related magnetic induction.

Commercial Permanent-Magnet Materials. Table 5.20 lists most of the permanent-magnet materials available commercially in the United States, together with their nominal compositions. Magnetic properties are presented in Table 5.21, and physical and mechanical properties are summarized in Table 5.22. Generally the production of permanent-magnet materials is controlled to achieve magnetic characteristics, and other properties are allowed to vary according to the manufacturing process used. The selection of materials and the design of permanent magnets for particular applications comprise a well-defined engineering art; design assistance is available from most major producers. Permanent magnets are superior to electromagnets for many uses because they maintain their fields without an expenditure of electric power and with-

TABLE 5.20 Nominal Compositions, Curie Temperatures, and Magnetic Orientations of Selected Permanent-Magnet Materials

Designation	Nominal composition	Approximate Curie temperature °C	Approximate Curie temperature °F	Magnetic orientation*
3½% Cr steel	Fe-3.5Cr-1C	745	1370	No
6% W steel	Fe-6W-0.5Cr-0.7C	760	1400	No
17% Co steel	Fe-17Co-8.25W-2.5Cr-0.7C	—	—	No
36% Co steel	Fe-36Co-3.75W-5.75Cr-0.8C	890	1630	No
Cast Alnico 1	Fe-12Al-21Ni-5Co-3Cu	780	1440	No
Cast Alnico 2	Fe-10Al-19Ni-13Co-3cu	810	1490	No
Cast Alnico 3	Fe-12Al-25Ni-3Cu	760	1400	No
Cast Alnico 4	Fe-12Al-27Ni-5Co	800	1475	No
Cast Alnico 5	Fe-8.5Al-14.5Ni-24Co-3Cu	900	1650	Y,H
Cast Alnico 5DG	Fe-8.5Al-14.5Ni-24Co-3Cu	900	1650	Y,H,C
Cast Alnico 5-7	Fe-8.5Al-14.5Ni-24Co-3Cu	900	1650	Y,H,C
Cast Alnico 6	Fe-8Al-16Ni-24Co-3Cu-2Ti	860	1580	Y,H
Cast Alnico 7	Fe-8Al-18Ni-24Co-4Cu-5Ti	840	1540	Y,H
Cast Alnico 8	Fe-7Al-15Ni-35Co-4Cu-5Ti	860	1580	Y,H
Cast Alnico 9	Fe-7Al-15Ni-35Co-4Cu-5Ti	—	—	Y,H,C
Cast Alnico 12	Fe-6Al-18Ni-35Co-8Ti	—	—	No
Sintered Alnico 2	Fe-10Al-17Ni-12.5Co-6Cu	810	1490	No
Sintered Alnico 4	Fe-12Al-28Ni-5Co	800	1475	No
Sintered Alnico 5	Fe-8.5Al-14.5Ni-24Co-3Cu	900	1650	Y,H
Sintered Alnico 6	Fe-8Al-16Ni-24Co-3Cu-2Ti	860	1580	Y,H
Sintered Alnico 8	Fe-7Al-15Ni-35Co-4Cu-5Ti	860	1580	Y,H
Bonded ferrite A	$BaO\text{-}6Fe_2O_3$ + organics	450	840	No,P
Bonded ferrite B	$BaO\text{-}6Fe_2O_3$ + organics	450	840	No
Sintered ferrite 1	$BaO\text{-}6Fe_2O_3$	450	840	No,P
Sintered ferrite 2	$BaO\text{-}6Fe_2O_3$	450	840	Y,A
Sintered ferrite 3	$BaO\text{-}6Fe_2O_3$	450	840	Y,A
Sintered ferrite 4	$SrO\text{-}6Fe_2O_3$	460	860	Yes
Sintered ferrite 5	$SrO\text{-}6Fe_2O_3$	460	860	Yes
Lodex 30	9.9Fe-5.5Co-77.0Pb-8.6Sb	980	1800	Y,A
Lodex 31	16.0Fe-9.0Co-67.5Pb-7.5Sb	980	1800	Y,A
Lodex 32	19.2Fe-10.8Co-63.0Pb-7.0Sb	980	1800	Y,A
Lodex 33	21.9Fe-12.3Co-59.2Pb-6.6Sb	980	1800	Y,A
Lodex 36	9.9Fe-5.5Co-77Pb-8.6Sb	980	1800	No,E
Lodex 37	16Fe-9Co-67.5Pb-7.5Sb	980	1800	No,E
Lodex 38	19.2Fe-10.8Co-63Pb-7.0Sb	980	1800	No,E
Lodex 40	9.9Fe-5.5Co-77Pb-8.6Sb	980	1800	No,P
Lodex 41	16Fe-9Co-67.5Pb-7.5Sb	980	1800	No,P
Lodex 42	19.2Fe-10.8Co-63.0Pb-7.0Sb	980	1800	No,P
Lodex 43	21.9Fe-12.3Co-59.2Pb-6.6Sb	980	1800	No,P

TABLE 5.20 Nominal Compositions, Curie Temperatures, and Magnetic Orientations of Selected Permanent-Magnet Materials (*Continued*)

Designation	Nominal composition	Approximate Curie temperature °C	Approximate Curie temperature °F	Magnetic orientation*
P-6 alloy	45 Fe-45Co-6Ni-4V	—	—	No
Cunife	20Fe-20Ni-60Cu	410	770	Y,R
Cunico	29Co-21Ni-50Cu	860	1580	No
Vicalloy I	39Fe-51Co-10V	855	1570	No
Vicalloy II	35Fe-52Co-13V	855	1570	Y,R
Remalloy 1	17Mo-12Co-71Fe	900	1650	No
Remalloy 2	20Mo-12Co-68Fe	900	1650	No
Platinum cobalt	76.7Pt-23.3Co	480	900	No
Cobalt–rare earth 1	Co_5Sm	725	1340	Y,A
Cobalt–rare earth 2	Co_5Sm	725	1340	Y,A
Cobalt–rare earth 3	Co_5Sm	725	1340	Y,A
Cobalt–rare earth 4	$(Co,Cu,Fe)_7Sm$	—	—	Y,A

*Y—yes; H—orientation developed during heat treatment; C—columnar crystal developed; P or E—some orientation developed during pressing or extrusion; R—orientation developed by rolling or other mechanical working; A—orientation developed predominantly by magnetic alignment of powder prior to compacting but alignment influenced by pressing forces also.

Source: From Boyer and Gall.[1]

out the generation of heat. Tables 5.21 and 5.22 give nominal properties only. Even under the most carefully controlled manufacturing conditions, some variation from these nominal values must be expected and considered in practical applications.

Stabilization and Stability. For an important group of permanent-magnet applications the accuracy or performance of the device is drastically affected by very small changes (1 percent or less) in the strength of the magnet. These applications include braking magnets for watt-hour meters, magnetron magnets, special torque motor magnets, and most dc panel and switchboard instrument magnets. Operation of these devices requires extreme accuracy over a moderate range of conditions, or moderate accuracy over an extreme range of conditions. If the nature and magnitude of the conditions are known, it often is possible to predict the flux change. It also may be possible, by exposing the magnet to certain influences in advance, to render the magnet insensitive to subsequent changes in service. For many years, permanent magnets in instruments have exhibited long-term stability on the order of one part per thousand (0.1 percent). More recently, investigations in conjunction with inertial guidance systems for space vehicles have shown that long-term stability on the order of 1 to 10 ppm (0.0001 to 0.001 percent) can be achieved. This incredible stability of a magnetic field achieved with modern permanent magnets contrasts sharply with the stability of very early permanent magnets, in which both structural and magnetic changes caused a significant loss of magnetization with time.

Irreversible Changes. Losses in magnetization with time can be classified as either reversible or irreversible. Irreversible changes are defined as changes where the affected properties remain altered after the influence responsible for the change is

TABLE 5.21 Nominal Magnetic Properties of Selected Permanent-Magnet Materials

Designation	H_c, Oe	H_{cf}, Oe	B_r, G	B_{is}, G	$(BH)_{max}$, MG • Oe	B_d, G	H_d, Oe	Required magnetizing field, Oe	Permeance coefficient at $(BH)_{max}$	Average recoil permeability
$3^1/_2$% Cr steel	66	—	9500	—	0.29	—	—	—	—	—
6% W steel	74	—	9500	—	0.33	—	—	—	—	—
17% Co steel	170	—	9500	—	0.65	—	—	—	—	—
36% Co steel	240	—	9750	—	0.93	—	—	—	—	—
Cast Alnico 1	440	455	7100	10500	1.4	4500	305	2000	14	6.8
Cast Alnico 2	550	580	7250	10900	1.6	4500	350	2500	12	6.4
Cast Alnico 3	470	485	7000	10000	1.4	4300	2500	13	6.5	
Cast Alnico 4	730	770	5350	8600	1.3	3000	420	3500	8.0	4.1
Cast Alnico 5	620	625	12500	13500	5.25	10200	525	3000	18	4.3
Cast Alnico 5DG	650	655	12900	14000	6.1	10500	580	3500	17	4.0
Cast Alnico 5-7	730	735	13200	14000	7.4	11500	640	3500	17	3.8
Cast Alnico 6	750	—	10500	13000	3.7	7100	525	4000	13	5.3
Cast Alnico 7	1050	—	8570	9450	3.7	—	—	5000	8.2	
Cast Alnico 8	1600	1720	8300	10500	5.0	5060	950	8000	5.0	3.0
Cast Alnico 9	1450	—	10500	—	8.5	—	—	7000	7.0	—
Cast Alnico 12	950	—	6000	—	1.7	3150	540	5000	5.6	—
Sintered Alnico 2	525	545	6700	11000	1.5	4300	345	2500	12	6.4
Sintered Alnico 4	700	760	5200	—	1.2	3000	400	3500	—	7.5
Sintered Alnico 5	600	605	10400	12050	3.60	7850	465	3000	18	4.0
Sintered Alnico 6	760	790	8800	11500	2.75	5500	500	4000	12	4.5
Sintered Alnico 8	1550	1675	7600	9400	4.5	4600	1000	8000	5.0	2.1
Bonded ferrite A	1940	—	2140	—	1.0	1160	—	12000	1.3	1.1
Bonded ferrite B	1150	—	1400	—	0.4	—	—	8000	1.2	1.1
Sintered ferrite 1	1800	3450	2200	—	1.0	1100	900	10000	1.2	1.2

Sintered ferrite 2	2200	2300	3800	—	3.4	1850	1650	10000	1.1	1.1
Sintered ferrite 3	3000	3650	3200	—	2.5	1600	1600	10000	1.1	1.1
Sintered ferrite 4	2200	2300	4000	—	3.7	2150	1700	12000	1.2	1.05
Sintered ferrite 5	3150	3590	3550	—	3.0	1730	1730	15000	1.0	1.05
Lodex 30	1250	1470	4000	4400	1.6	2200	750	6000	3.4	1.5
Lodex 31	1140	1180	6300	7000	3.4	4400	770	6000	5.3	1.9
Lodex 32	940	960	7350	8300	3.5	5400	650	5000	8.2	2.6
Lodex 33	865	875	8000	9200	3.2	5850	545	5000	10.5	3.0
Lodex 36	1210	1380	3500	4400	1.5	1850	800	5000	2.0	2.0
Lodex 37	1000	1080	5450	7000	2.1	3150	670	5000	5.8	3.0
Lodex 38	850	890	6200	8300	2.2	3700	600	5000	7.0	3.5
Lodex 40	1100	1400	2700	4400	0.8	1400	600	5000	2.0	2.5
Lodex 41	990	1100	4350	7000	1.4	2400	600	5000	3.8	3.2
Lodex 42	845	920	5300	8300	1.4	2750	510	5000	7.6	3.5
Lodex 43	710	750	6000	9200	1.3	3300	400	5000	10	3.8
P-6 alloy	58	—	14000	19000	0.5	10500	48	300	220	23
Cunife	550	555	5400	5900	1.5	4000	325	2500	12	3.7
Cunico	680	750	3400	4500	0.8	1950	390	3000	5.0	3.2
Vicalloy I	240	242	8400	12900	0.9	5600	160	1000	—	—
Vicalloy II	415	420	9050	—	2.3	7000	325	2000	—	—
Remalloy 1	250	—	9700	14200	1.0	6100	155	1000	40	13
Remalloy 2	340	345	8550	—	1.2	5400	220	2000	—	—
Platinum cobalt	4450	5400	6450	—	9.2	3500	2700	20000	1.2	1.2
Cobalt–rare earth 1	9000	20000	9200	9800	21	—	—	30000	—	—
Cobalt–rare earth 2	8000	>25000	8600	—	18	4400	4100	30000	—	1.05
Cobalt–rare earth 3	6700	>15000	8000	—	15	4000	3700	30000	—	1.1
Cobalt–rare earth 4	5700	6500	9400	—	21	4600	4600	>15000	—	—

Source: Boyer and Gall.[1]

TABLE 5.22 Nominal Mechanical and Physical Properties of Selected Permanent-Magnet Materials

Designation	Density, Mg/m^3	Tensile strength		Transverse modulus of rupture		Hardness, HRC	Coefficient of linear, expansion, μm/m • K	Electrical resistivity, nΩ • m	Maximum service temperature	
		MPa	ksi	MPa	ksi				°C	°F
3$^1/_2$% Cr steel	7.77	—	—	—	—	60–65	12.6	290	—	—
6% W steel	8.12	—	—	—	—	60–65	14.5	300	—	—
17% Co steel	8.35	—	—	—	—	60–65	15.9	280	—	—
36% Co steel	8.18	—	—	—	—	60–65	17.2	270	—	—
Cast Alnico 1	6.9	28	4.1	96	14	45	12.6	750	540	1004
Cast Alnico 2*	7.1	21	3.1	52	7.5	45	12.4	650	540	1004
Cast Alnico 3	6.9	83	12	157	23	45	13.0	600	480	896
Cast Alnico 4	7.0	63	9.1	167	24	45	13.1	750	590	1094
Cast Alnico 5*,†	7.3	37	5.4	73	11	50	11.4	470	540	1004
Cast Alnico 5DG	7.3	36	5.2	62	9.0	50	11.4	470	—	—
Cast Alnico 5-7	7.3	34	4.9	55	8.0	50	11.4	470	540	1004
Cast Alnico 6(a)	7.4	157	23	314	46	50	11.4	500	540	1004
Cast Alnico 7	7.3	108	16	—	—	60	11.4	580	—	—
Cast Alnico 8	7.3	64	9.3	—	—	56	11.0	500	540	1004
Cast Alnico 9	7.3	48	6.9	55	8.0	56	11.0	—	—	—
Cast Alnico 12	7.4	275	40	343	50	58	11.0	620	480	896
Sintered Alnico 2	6.8	451	65	480	70	43	12.4	680	480	896
Sintered Alnico 4	6.9	412	60	588	85	—	13.1	680	590	1094
Sintered Alnico 5	7.0	343	50	392	57	44	11.3	500	540	1004
Sintered Alnico 6	6.9	382	55	755	110	44	11.3	530	540	1004
Sintered Alnico 8	7.0	—	—	382	55	43	—	—	—	—
Bonded ferrite A‡	3.7	4.4	0.63	—	—	—	94	~10^{13}	95	203

Sintered ferrite 1[¶]	4.8	49	7.1	—	—	—	10	~10^{13}	400	752
Sintered ferrite 2	5.0	—	—	—	—	—	10	~10^{13}	400	752
Sintered ferrite 3	4.5	—	—	—	—	—	18	~10^{13}	400	752
Sintered ferrite 4	4.8	—	—	—	—	—	—	10^{13}	400	752
Sintered ferrite 5	4.5	—	—	—	—	—	—	10^{13}	—	—
Lodex 30	10.1	—	—	31	4.5	—	18	1200	200	392
Lodex 31	9.6	6.9	1.0	31	4.5	—	18	1200	200	392
Lodex 32	9.3	6.9	1.0	31	4.5	—	18	1200	200	392
Lodex 33	9.2	—	—	31	4.5	—	18	1200	200	392
Lodex 36	10.2	—	—	108	16	—	18	1200	200	392
Lodex 37	9.7	—	—	108	16	—	18	1200	200	392
Lodex 38	9.6	—	—	108	16	—	18	1200	200	392
Lodex 40	10.2	—	—	27	3.9	—	18	1200	200	392
Lodex 41	10.1	6.9	1.0	27	3.9	—	18	1200	200	392
Lodex 42	9.8	6.9	1.0	27	3.9	—	18	1200	200	392
Lodex 43	9.4	—	—	27	3.9	—	18	2000	200	392
P-6 alloy	7.9	2160	313	1180	170	65	11	300	—	—
Cunife	8.6	686	99	—	—	95 HRB	12	180	350	662
Cunico	8.3	588	85	—	—	95 HRB	14	240	500	932
Vicalloy I	8.2	2060	299	—	—	62	7	630	450	842
Remalloy 1	8.2	882	128	—	—	60	9.3	450	500	932
Platinum cobalt	15.5	1370	199	1570	230	26	11	280	350	662
Cobalt–rare earth[§]	8.2	3430	498	13730	1990	50	511; 131	500	250	482

*Specific heat 460 J/kg • K (0.11 Btu/lb • °F).

†Thermal conductivity 25 W/m • K (170 Btu • in/ft^2 • h • °F) at room temperature.

‡Thermal conductivity 0.62 W/m • K (4.3 Btu • in/ft^2 • h • °F).

¶Thermal conductivity 5.5 W/m • K (38 Btu • in/ft^2 • h • °F).

§Specific heat 375 J/kg • K (0.09 Btu/lb • °F); thermal conductivity 15 W/m • K (104 Btu • in/ft^2 • h • °F).

Source: From Boyer and Gall.[1]

removed. For example, if a magnet loses field strength under the influence of elevated temperature, and if the field strength does not return to its original value when the magnet is cooled to room temperature, the change is considered irreversible.

Irreversible changes begin to occur at different temperatures for different alloys. These changes usually depend on both time and temperature, and thus short exposures above the recommended temperatures may be tolerated. These changes may take the form of growth of the precipitate phase, such as in Alnico, Cunife, and Cunico; precipitation of another phase, such as gamma precipitation in Alnico; an increase in the amount of an ordered phase, such as in PtCo; stress-relief effects, such as in quenched steels and Vicalloy; an increase in grain size, as in $BaO \cdot Fe_2O_3$; oxidation as occurs with metals; or reduction as occurs with oxides; radiation damage; cracking; or changes in dimensions. Irreversible metallurgical changes often can be counteracted, and original properties restored, by a suitably chosen thermal treatment. For example, if Alnico S is degraded by exposure to 1290°F, it may be solution-treated at 2370°F, cooled in a magnetic field, and aged at 1110°F to re-attain the optimum metallurgical structure. A nuclear environment is known to cause changes in metallurgical structure, and thus may cause changes in magnetic properties. Permanent-magnet materials tested were not affected by neutron irradiation at levels below about 1.9×10^{18} neutrons/in^2. Results of later work at levels up to 6×10^{20} neutrons/in^2 showed some degradation. The Alnicos are not affected by radiation up to 3.2×10^{21} neutrons/in^2 at neutron energies greater than 0.4 eV, and up to 1.3×10^{20} neutrons/in^2 for neutron energies greater than 2.9 MeV. Radiation effects were found to be independent of temperature, but high temperatures tended to counteract radiation effects. Changes in magnetic state may be caused by temperature effects, such as ambient temperature changes or statistical local temperature fluctuations within the material; mechanical effects, such as mechanical shock or acoustical noise; or magnetic field effects, such as external fields, circuit reluctance changes, or magnetic surface contacts. In all of these situations, the loss in magnetization may be restored by remagnetizing. Mechanical shock and vibration add energy to a permanent magnet, and decrease the magnetization in the same manner as thermal energy. The only difference is that the energy imparted thermally to the magnet is precisely kT, whereas the energy imparted mechanically usually is not known.

Reversible Changes. A loss in magnetization caused by a disturbing influence, such as temperature or an external magnetic field, is considered reversible if the original properties of the magnet return when the disturbing influence is removed.

Temperature affects the properties of a magnet in a manner that often can be predicted. The variation of saturation induction B_{is} with temperature can be calculated from theory, provided detailed knowledge of the crystallographic and magnetic structure of the magnetic phase is available. In many other instances such information is not yet available, but direct measurements of B_{is} versus T have been made. Changes in the intrinsic coercive force H_{ci} with temperature can be predicted from the changes with temperature of anisotropy and magnetization. This assumes knowledge of the physical origin of all anisotropies contributing to H_{ci}. For a case where uniaxial anisotropy predominates, as in $BaO \cdot 6Fe_2O_3$, quite good agreement is obtained between calculated and experimental results. When shape anisotropy is dominant, as it is for Lodex elongated particles with various cobalt contents, calculated and experimental results also are in good agreement, especially when the small crystal anisotropy contributions are considered. The case of Alnico is similar to that of Lodex, but crystal anisotropy is more in evidence. In addition, there is greater uncertainty as to the effect of the so-called nonmagnetic phase, especially at lower temperatures where the nonmagnetic phase may contribute appreciable magnetization. In the case of steels, the temperature dependence based on the inclusion mechanism is difficult to predict. Demagnetization curves may change in both shape and peak values with changes in temperature.

Time effects at constant temperature occur in ferromagnetic materials because the intensity of magnetization does not instantly attain its equilibrium value when the applied field is suddenly changed. This time dependence may be due to eddy current effects or to reversible or irreversible magnetic viscosity. In general, eddy current effects are important only for a very short time, normally less than a second after a change in the applied field. "Reversible" magnetic viscosity is due to ionic diffusion in the crystal lattice and thus has a time-temperature dependence characteristic of diffusion processes. Irreversible magnetic viscosity is important to the stability of permanent magnets. It is due to the influence of thermal fluctuations on magnetization or the domain process responsible for magnetization.

Effects of temperature variations on the magnetization are related to the dimensions of the magnet. Various permanent-magnet materials undergo changes in magnetization as the temperature is cycled above and below room temperature. For a long bar operating above $(BH)_{max}$ the change in magnetization is reversible. For a short bar operating below $(BH)_{max}$ the first cooling cycle results in a substantial loss in magnetization. After the initial low-temperature exposure, the changes in magnetization are reversible, but at a level below the initial magnetization. These data suggest that by proper choice of dimensions, a reversible coefficient of approximately zero could be achieved over a limited range of temperature.

5.5.7 Spring Materials

Spring materials are high-strength alloys which often exhibit the greatest strength in the alloy system (Wahl[8]). The energy storage capacity of a spring is proportional to the square of the maximum operating stress level divided by the modulus. Thus an ideal spring material has high strength, a high elastic limit, and a low modulus. Because springs are resilient structures, designed to undergo large deflections, spring materials must have an extensive elastic range. Other factors such as fatigue strength, cost, availability, formability, corrosion resistance, magnetic permeability, and electrical conductivity can be important. Table 5.23 lists typical physical properties of some commonly used spring materials, and Table 5.24 gives typical mechanical properties of spring temper alloy strip.

Steels for Cold-Wound Springs. There are a variety of applications for cold-wound spring wire. *Music wire* is high-quality carbon steel used widely for small helical springs, particularly when subjected to severe stress conditions. It is comparable to valve-spring wire in surface quality.

Oil-tempered spring wire is a good-quality high-carbon steel wire. Ordinary oil-tempered wire may have some surface defects and hence should be used only where long fatigue life is not needed. For applications requiring fatigue endurance, carbon-steel valve-spring-quality wire (ASTM A-230) should be used.

Hard-drawn spring wire is the cheapest spring wire, but is of lower quality and may contain surface defects such as hairline seams. It is ordinarily used in cases where the stresses are low or where only static loading is involved. This wire should not be used where fatigue loading is present. It is also considered better for plating than oil-tempered wire.

Chrome-vanadium steel wire is frequently specified for cases where a high-quality material is needed and temperatures are somewhat higher than normal. It may be obtained either in the oil-tempered or in the annealed condition to permit severe forming.

TABLE 5.23 Typical Properties of Common Spring Materials

	Young's modulus E†		Modulus of rigidity G†		Density†		Electrical conduc-tivity,	Sizes normally available‡				Typical surface qual-ity¶	Maximum service temperature§	
								Minimum		Maximum				
Common name*	MPa 10^3	10^6 lb/in^2	Mpa 10^3	10^6 lb/in^2	g/cm^3	lb/in^3	% IACS	mm	(in)	mm	(in)		°C	°F
Carbon steel wires														
Music	207	30	79.3	11.5	7.86	0.284	7	0.10	(0.004)	6.35	(0.250)	*a*	120	250
Hard drawn	207	30	79.3	11.5	7.86	0.284	7	0.13	(0.005)	16	(0.625)	*c*	150	250
Oil tempered	207	30	79.3	11.5	7.86	0.284	7	0.50	(0.020)	16	(0.625)	*c*	150	300
Valve spring	207	30	79.3	11.5	7.86	0.284	7	1.3	(0.050)	6.35	(0.250)	*a*	150	300
Alloy steel wires														
Chrome vanadium	207	30	79.3	11.5	7.86	0.284	7	0.50	(0.020)	11	(0.435)	*a,b*	220	425
Chrome silicon	207	30	79.3	11.5	7.86	0.284	5	0.50	(0.020)	9.5	(0.375)	*a,b*	245	475
Stainless steel wires														
Austenitic type 302	193	28	69.0	10.	7.92	0.286	2	0.13	(0.005)	9.5	(0.375)	*b*	260	500
Precipitation hardening 17-7 PH	203	29.5	75.8	11	7.81	0.282	2	0.08	(0.002)	12.5	(0.500)	*b*	315	600
NiCr A286	200	29	71.7	10.4	8.03	0.290	2	0.40	(0.016)	5	(0.200)	*b*	510	950
Copper-base alloy wires														
Phosphor bronze (A)	103	15	43.4	6.3	8.86	0.320	15	0.10	(0.004)	12.5	(0.500)	*b*	95	200
Silicon bronze (A)	103	15	38.6	5.6	8.53	0.308	7	0.10	(0.004)	12.5	(0.500)	*b*	95	200
Silicon bronze (B)	117	17	44.1	6.4	8.75	0.316	12	0.10	(0.004)	12.5	(0.500)	*b*	95	200
Beryllium copper	128	18.5	48.3	7.0	8.26	0.298	21	0.08	(0.003)	12.5	(0.500)	*b*	205	400
Spring brass, CA260	110	16	42.0	6.0	8.53	0.308	17	0.10	(0.004)	12.5	(0.500)	*b*	95	200
Nickel-base alloys														
Inconel alloy 600	214	31	75.8	11	8.43	0.304	1.5	0.10	(0.004)	12.5	(0.500)	*b*	320	700
Inconel alloy X750	214	31	79.3	11.5	8.25	0.298	1	0.10	(0.004)	12.5	(0.500)	*b*	595	1100
Ni-Span-C	186	27	62.9	9.7	8.14	0.294	1.6	0.10	(0.004)	12.5	(0.500)	*b*	95	200
Monel alloy 400	179	26	66.2	9.6	8.83	0.319	3.5	0.05	(0.002)	9.5	(0.375)	*b*	230	450
Monel alloy K500	179	26	66.2	9.6	8.46	0.306	3	0.05	(0.002)	9.5	(0.375)	*b*	260	500

Carbon steel strip														
AISI 1050	207	30	79.3	11.5	7.86	0.284	7	0.25	(0.010)	3	(0.125)	*b*	95	200
AISI 1065	207	30	79.3	11.5	7.86	0.284	7	0.08	(0.003)	3	(0.125)	*b*	95	200
AISI 1074, 1075	207	30	79.3	11.5	7.86	0.284	7	0.08	(0.003)	3	(0.125)	*b*	120	250
AISI 1095	207	30	79.3	11.5	7.86	0.284	7	0.08	(0.003)	3	(0.125)	*b*	120	250
Bartex	207	30	79.3	11.5	7.86	0.284	7	0.10	(0.004)	1	(0.040)	*a*	95	200
Stainless-steel strip														
Austenitic types 301, 302	193	28	69.0	10	7.92	0.286	2	0.08	(0.003)	1.5	(0.063)	*b*	315	600
Precipitation hardening 17-7 PH	203	29.5	75.8	11	7.81	0.282	2	0.08	(0.003)	2	(0.125)	*b*	370	700
Copper-base alloy strip														
Phosphor bronze (A)	103	15	43	6.3	8.86	0.320	15	0.08	(0.003)	5	(0.188)	*b*	95	200
Beryllium copper	128	18.5	48	7.0	8.26	0.298	21	0.08	(0.003)	9.5	(0.375)	*b*	205	400

*Inconel, Monel, and Ni-Span-C are registered trademarks of International Nickel Company, Inc. Bartex is a registered trademark of Theis of America, Inc. Music and hard drawn are commercial terms for patented and cold-drawn carbon steel spring wire.

†Elastic moduli, density, and electrical conductivity can vary with cold work, heat treatment, and operating stress. These variations are usually minor but should be considered if one or more of these properties are critical.

‡Diameters for wire; thicknesses for strip.

¶Typical surface quality ratings. (For most materials, special processes can be specified to upgrade typical values.)

a. Maximum defect depth 0–0.5% of d or t.

b. Maximum defect depth 1.0% of d or t.

c. Defect depth less than 3.5% of d or t.

§Maximum service temperatures are guidelines and may vary due to operating stress and allowable relaxation.

Source: From *Metals Handbook,* American Society of Metals, 1978.

TABLE 5.24 Typical Properties of Spring Temper Alloy Strip

Material	Tensile strength MPa	Tensile strength ksi	Rockwell hardness	Elongation,* %	Bend factor* (2r/t trans. bends)	Modulus of elasticity 10^4 MPa	Modulus of elasticity 10^6 lb/in^2	Poisson's ratio
Steel, spring temper	1700	246	C50	2	5	20.7	30	0.30
Stainless 301	1300	189	C40	8	3	19.3	28	0.31
Stainless 302	1300	189	C40	5	4	19.3	28	0.31
Monel 400	690	100	B95	2	5	17.9	26	0.32
Monel K500	1200	174	C34	40	5	17.9	26	0.29
Inconel 600	1040	151	C30	2	2	21.4	31	0.29
Inconel X-750	1050	152	C35	20	3	21.4	31	0.29
Copper-beryllium	1300	189	C40	2	5	12.8	18.5	0.33
Ni-Span-C	1400	203	C42	6	2	18.6	27	—
Brass CA 260	620	90	B90	3	3	11	16	0.33
Phosphor bronze	690	100	B90	3	2.5	10.3	15	0.20
17-7 PH RH950	1450	210	C44	6	Flat	20.3	29.5	0.34
17-7 PH condition C	1650	239	C46	1	2.5	20.3	29.5	0.34

*Before heat treatment.

Source: From *Metals Handbook,* American Society of Metals, 1978.

Chrome-silicon wire (SAE 9254) is somewhat more resistant to relaxation at temperatures around 250 to 450°F than is the chrome-vanadium material. Valve-spring-quality wire of this material generally has a higher fatigue life in the low-cycle range from 10^4 to about 10^5 cycles than other alloys.

Stainless-steel spring wire is used widely where resistance to corrosion or relaxation to creep at elevated temperatures is required. Type 316 has superior resistance to saltwater corrosion. Type 302 and 316 steels are magnetic in spring tempers, since they are hardened by cold working or drawing. Stainless steel, type 17-7 PH, is precipitation-hardenable. Tensile strengths comparable to those of music wire can be obtained, and this material may be used at temperatures up to 700°F.

Nonferrous Spring Materials. Nonferrous spring materials are available as different alloys.

Phosphor bronze has its greatest use for cases where good electrical conductivity is desired. It is also used where corrosion resistance is important.

Beryllium copper containing about 2 percent beryllium is used where high electrical conductivity combined with high strength is desired. Beryllium copper will not resist corrosion by sulfides present in some waters or by halides at slightly elevated temperature, but it will resist common acids, seawater, salts, and fresh water about as well as pure copper.

*Monel** is a copper-nickel alloy that has good resistance to corrosion, is slightly magnetic, and may be used up to temperatures of 400°F.

*Monel is a registered trademark of International Nickel Company.

"K" Monel is essentially the same as Monel, except with about 3 percent aluminum added to give a precipitation-hardened material that can give higher tensile strengths. It is nonmagnetic to −150°F and may be used at temperatures up to 700°F.

*Inconel** is hardened by cold working and is nonmagnetic to 40°F. It is resistant to heat and corrosion and may be used at temperatures up to 700°F.

Inconel "X"* is a precipitation-hardened alloy that may be used for temperatures up to 900°F. It is nonmagnetic down to −280°F. This precipitation-hardened alloy is frequently used for applications where it is desirable to maintain a constant modulus of elasticity in the range of 50 to 150°F.

REFERENCES

1. H. E. Boyer and T. L. Gall (eds.), *Metals Handbook,* desk ed., American Soc. for Metals, Metals Park, Ohio, 1985.
2. U. V. Evans, *The Corrosion and Oxidation of Metals,* Edward Arnold, London, 1967.
3. H. H. Uhlig, (ed.), *The Corrosion Handbook,* Wiley, New York, 1969.
4. L. B. Hunt, *Electrical Contacts,* Johnson-Matthey & Co., London, 1946.
5. D. J. White, *A Handbook on Electromagnetic Shielding Materials and Performance,* 2d ed., Interference Control Technologies, Don White Consultants Subsidiary, Gainesville, Va., 1980, chap. 2.
6. R. M. Bozorth, *Ferromagnetism,* Van Nostrand, New York, 1951.
7. B. D. Cullity, *Introduction to Magnetic Materials,* Addison-Wesley, Reading, Mass., 1972.
8. A. M. Wahl, *Mechanical Springs,* 2d ed., McGraw-Hill, New York, 1963, chap. 2.
9. *Design Handbook,* Associated Spring, Barnes Group Inc., Bristol, Conn., 1987.

*Inconel is a registered trademark of International Nickel Company.

CHAPTER 6
WIRE AND CABLE

Edward J. Croop

6.1 INTRODUCTION

The selection of suitable hookup and interconnecting wires or cables is an important step in the overall design of electronic packaging. Often these wires have been the last areas to be considered, and designs were based on historical or rule-of-thumb ratings.

Hookup wire can be thought of as connecting the components within electronic "black boxes" or chassis wiring, with insulation wall thicknesses below 20 mil, and usually rated at 250 to 600 V. Interconnecting wire usually indicates a cable or harness between black boxes or equipment, more rugged, and meeting specifications more rigorous than those for hookup wire.

Recent years have seen many improvements in electronic packaging technology, with a much higher density of components on printed-wiring boards, a vastly increased number of input and output leads on integrated-circuit devices, an increased speed of chips and devices (VHSIC), an increased number of layers of copper-clad wiring per board, and an entire new technology of surface-mounting devices replacing through-hole mounting and interconnecting, to name a few. As a result, the selection of wiring and cabling can be critical to the success of package and systems designs.

Because of the many advances in this technology, more and more engineers have recognized the importance of optimizing conductors, insulation, shielding, jacketing, and cabling. They are seeking those that offer the best combinations of size, weight, environmental protection, and ease of shop handling, together with the lowest possible cost, best availability, and least maintenance. If the intended environment is abnormal, proper wire and cable selection should be verified by adequate environmental testing. This should include extended life testing to determine a probable service life extrapolation. In addition, the designer must adequately evaluate the environmental, mechanical, and electrical stresses and consider the compatibility of the materials with all possible encapsulants, fuels, chemicals, explosives, as well as the new and varied types of associated hardware and fabrication techniques.

The purpose of this chapter is to provide the designer and user with sufficient background information and guideposts for the effective selection of hookup and interconnection wire and cable to meet the demands of specific applications.

6.2 WIRE CONDUCTORS

6.2.1 Materials

The following is a brief discussion of the electrical conductor materials used in the electronics and aerospace industries. Table 6.1 presents a summary of conductor materials and their metallic coatings.

Copper

Electrolytic Tough Pitch (ETP). Electrolytic tough pitch (ETP) copper makes up the majority of conductor strands used in industry. ETP copper is specified in composition, conductivity, and purity by ASTM B-3. The properties of individual strands such as tensile strength, elongation, and dimensions are covered by ASTM B-33 or QQ-W-343. ETP is more notch-sensitive than other forms, and care is required to prevent failure due to unusual nicks and severe abrasion of the copper during application.

Oxygen-Free High Conductivity. Oxygen-free high conductivity (OFHC) copper differs from ETP copper in its fabrication. Electrolytic slabs are melted into bars for extrusion or rolling in an inert atmosphere that excludes oxygen. OFHC copper, covered by ASTM B-170, has improved properties at temperatures above 1000°F and is resistant to hydrogen embrittlement. It has excellent flexibility, but is difficult to weld. It is sold at a premium price.

Continuous-Melt, Cast and Rolled. Continuous-melt cast and rolled (CMCR) copper is a more recent development in the copper industry and has many advantages over conventionally produced copper rods:

1. Fewer welds, resulting in improved reliability of the final drawn wire.
2. Reduced oxides and segregation of oxides which cause wire breakage and hot spots in equipment.
3. Considerably better impurity distribution. This directly affects, and is the significant factor in, uniform grain size.
4. Consistently superior physical and metallurgical properties unattainable in regular hot-rolled rods.

Copper Alloys. Pure copper has poor mechanical characteristics, especially in small wire sizes. Breakage will occur from tensile pull, flexing, and vibration. In response to progress in the miniaturization of equipment, various high-strength copper alloys have been developed.[1] Figure 6.1 compares flex endurance between high-strength copper alloys and copper. Since drawing and annealing techniques used by different suppliers play a significant part in the final physical and electrical properties of high-strength alloys, the recommended practice is to specify the performance requirements of the conductor instead of specifying a particular alloy. One exception to this practice would be the use of zirconium alloy for high-temperature applications.

Because of high annealing temperatures during processing, high-strength alloys are available only with high-temperature coatings such as silver or nickel. Tin-coated high-strength-alloy conductors are available but are temperature-limited due to the lower melting point of tin. A description of the more popular high-strength alloys follows.

Cadmium-copper alloy (cadmium-bronze). This is used in fine-wire applications for increased strength. ASTM B-105 covers hard-drawn temper. No specification is

TABLE 6.1 Bare and Coated Conductor Properties

Conductor material and coating*	Tensile strength, lb/in^2	Elongation, %	Flex. life	Conductivity, min. %	Continuous operation temperature, max., °C	Oxidation resistance	Galv. corrosion resistance	Solderability	Availability	Cost	Specifications
Bare copper (ann.)	34,000	10	Fair	100	150	Poor	Good	Fair	Good	Low	QQ-W-343 ASTM B-3
TC copper (ann.)	34,000	10	Fair	100	150	Good	Good	Good	Good	Low	QQ-W-343 ASTM B-33
SC copper	34,000	10	Fair	102	200	Good	Poor	Good	Good	Medium	ASTM B-298
NC copper	34,000	10	Fair	96	260	Good	Good	Poor	Fair	Medium	ASTM B-355
Aluminum	20,000–32,000	15	Poor	61	150	Poor	Good	Poor	Good	Low	MIL-W-7071 (insulated)
SC aluminum	20,000–32,000	15	Poor	63	150	Good	Poor	Good	Fair	Medium	
Silver	25,000	15	Fair	104	260	Fair	Good	Good	Poor	High	
SC copper-clad steel (HD)	110,000–125,000	1	Poor	30–40	200	Good	Medium	Good	Fair	Medium	QQ-W-345
SC copper-clad steel (ann.)	55,000–65,000	8	Good	30–40	200	Good	Medium	Good	Fair	Medium	
SC cadmium-copper (HD)	90,000	1	Poor	80	200	Good	Poor	Good	Fair	Medium	
SC cadmium-copper (ann.)	55,000	5	Good	80	200	Good	Poor	Good	Fair	Medium	
SC zirconium	55,000	5	Good	85	200	Good	Poor	Good	Fair	Medium	
SC chrome (ann.)	55,000	6	Good	85	200	Good	Poor	Good	Fair	Medium	ASTM B-268
SC cadmium-chrome	60,000	5	Good	84	200	Good	Poor	Good	Fair	Medium	

Source: Martin Marietta Corporation

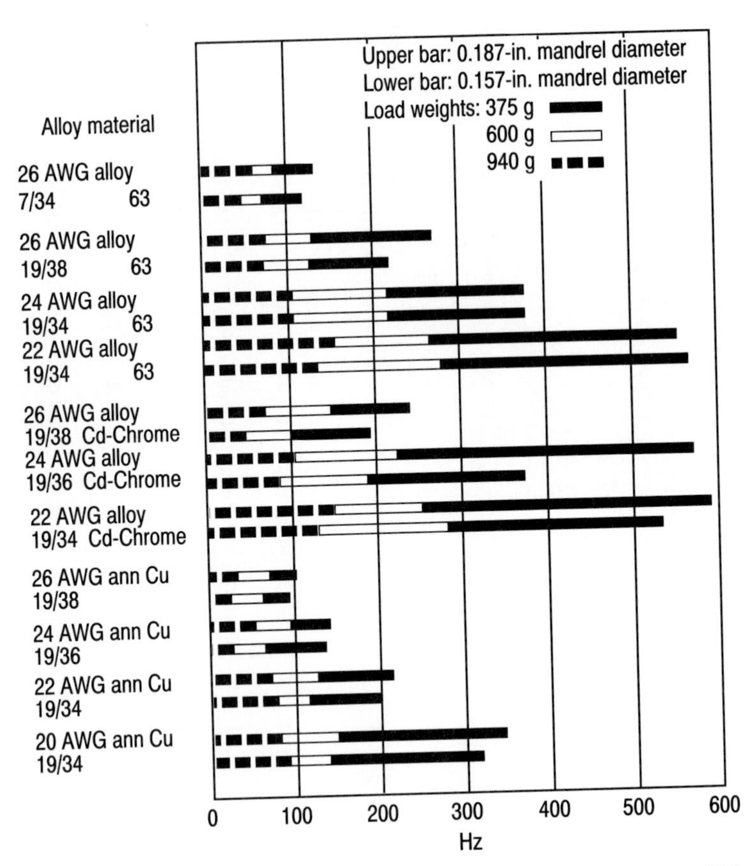

FIGURE 6.1 Flex endurance tests (concentric stranding), annealed copper versus high-strength alloys.[1]

available for temper applicable to high-strength-alloy conductors of higher elongation. This alloy is not recommended for temperature applications exceeding 400°F.

Cadmium-chromium-copper alloy. This alloy appears to present the most consistent and best physical properties, although the conductivity is slightly less than that of the copper-chrome alloy.

Zirconium-copper alloy. The defined physical and electrical characteristics of high-strength-alloy conductors are difficult to meet with zirconium alloy. However, zirconium-copper alloy shows best retention of physical and electrical properties at temperatures up to 400°C (752°F).

Chromium-copper alloy. This provides the best conductivity of the high-strength-alloy conductors. However, it is the most difficult alloy to process consistently. Chromium-copper alloy is a precipitation-hardened alloy. If improperly heat-treated, the chromium is not retained in solution, causing inconsistency in physical properties.

Copper-Covered Steel. This conductor is available in two consistencies: hard drawn for high strength (125,000 to 150,000 lb/in^2) and annealed (50,000 to 80,000 lb/in^2) for greater ductility; and in two conductivities: 40 and 30 percent. At high frequencies, either 30 or 40 percent copper-covered steel has a conductance almost equivalent to that of copper because of the skin effect. Therefore it is commonly used in coaxial cables. At low frequencies or with direct current, the conductance is 30 or 40 percent of an equivalent-size copper conductor. Specifications for copper-covered steel wire appear in QQ-W-345.

Aluminum. The major use of aluminum conductors has been in large-gauge power conductors and in some chip carrier interconnections. Aluminum offers a weight advantage of approximately 3:1, with a 62 percent lower conductance than an equivalent-size IACS copper conductor. Compared with copper, weight savings of approximately 2:1 can be attained for equivalent current-carrying capacity, thus still realizing a cost savings over copper. The major disadvantages of aluminum conductors versus copper are excessive creep, flow, and stress relaxation under pressure, which are time- and temperature-dependent. These adversely affect yield strength and contact pressure in mechanical joints. Shorter fatigue life, poor corrosion resistance, and poor solderability can also be expected. Aluminum is very difficult to join. Iron-aluminum alloys and copper-clad aluminum alleviate a few of these disadvantages somewhat. The military specification for insulated aluminum wire is MIL-W-5088.

Stainless Steel. Although very seldom used by itself as a conductor because of its low conductivity, stainless steel (in strands) has been used to reinforce stranded copper conductors. This reinforcement is normally used in small hookup wire (AWG 24 and smaller) and in ground-support cable (AWG 18 and smaller).

Silver. Recent experiences with corrosion problems of silver-plated copper have resulted in the use of pure silver and silver-alloy conductors. The physical properties of some silver-coated materials are covered in Table 6.1.

6.2.2 Metallic Coatings

Bare copper conductor is rarely used in the aerospace and military electronics industries since copper will oxidize on exposure to the atmosphere, especially if the temperature is elevated. Corrosion will degrade the conductivity and reduce solderability. Some insulating materials also tend to corrode bare copper, especially at elevated temperatures. The following is a list of the most widely used metallic coatings for protecting copper conductors.

Tin. For temperatures below 150°C the least expensive protective metallic coating is tin. In addition to protecting copper from oxidation, it facilitates soldering. Tin should not be used as a protective coating for service above 150°C. Protective tin coating can be applied to a copper conductor by one of two methods: dipping or electroplating.

Tin Dip. Tin dip is the process of passing strands through a molten tin bath. The major disadvantage of tin dip is poor control of coating thickness. Thickness can vary from 20 to 200 μin. Specifications for tin-coated wire are covered in ASTM B-33 and QQ-W-343.

Electroplated Tin. The electroplating process applies a tin coating of controlled thickness to the copper strand and is becoming increasingly important for automated stripping and termination equipment. There is no individual specification covering the application of tin coating by electroplating.

Several variations of the standard tin-coated copper conductor have been used as fabrication aids to eliminate strand twisting and pretinning. Some of the available variations are as follows:

Heavy tin. This is applied up to a minimum thickness of 100 μin to AWG 31 wire and smaller, and up to 150 μin to AWG 30 and larger. Applied with the aid of high-frequency induction heaters, tin is melted and flowed, tacking the strands together in the area where the conductor is to be stripped.

Prefused conductor. This utilizes heavy tin-coated strands that are melted and fused together for the entire length of the insulated conductor. Although fabrication costs may be reduced, conductor flexibility is impaired. This process defeats the purpose of stranded conductors for increased flex life.

Overcoated conductor. Here individually tinned copper strands receive an overcoating of tin over the conductor, bonding all strands together. Conductor flexibility is drastically reduced.

Top-coated conductor. This uses bare, untinned copper stranded wire that is given an overall coating of tin. The finished conductor is bonded along its entire length. It is less expensive than the overcoated conductor but has less corrosion resistance.

Silver. This coating is utilized for continuous service of conductors in excess of 150°C to a maximum of 200°C and is required for the application of certain high-temperature insulation materials to prevent their interaction with bare copper. Silver-coated conductors are widely used with fluorocarbon, polyimide, and silicone rubber insulation. Silver is very readily soldered. This at times is considered a disadvantage in that solder flows under the insulation and potentially reduces flex life at the conductor termination.

Silver-coated copper is susceptible to corrosion caused by the galvanic interaction between copper and silver. Microscopic breaks in the silver coating, caused by stranding or coating porosity, allow galvanic reaction to take place between bare copper and silver in the presence of moisture or other electrolytes. Considerable work has been under way to solve this problem. Protective coatings applied over the silver, increased silver thickness, and intermediate barrier platings are the subjects of continuing research. These approaches have been successful in reducing but not eliminating this problem. ASTM B-298 contains the specifications for silver-coated copper conductors.

Nickel. This is a good high-temperature protective coating, suitable for continuous use up to 300°C, and not susceptible to corrosion as is silver. The main disadvantage of nickel is its poor solderability. Even with activated fluxes and high soldering temperatures, good solder terminations are extremely difficult to achieve. Nickel coating can be applied to copper conductors by two methods: cladding and electroplating.

Cladding. A relatively thick nickel outer coating or tube is applied under heat and pressure over the copper billet before rolling and drawing into wire. The process is similar to the fabrication of copper-covered steel wire. Cladding offers superior protection at extremely high temperature applications (1000°F maximum, continuous). Cladding is not advantageous for lower temperatures because of the higher resistance of nickel.

Electroplating. Electroplating a thin nickel coating over copper is done by standard electroplating techniques, normally with 50-μin minimum thickness.

6.2.3 Construction

Conductor construction plays a significant role in the functioning and reliability of any selected conductor. Solid wire, while low in cost and weight, will not withstand much flexure without breaking because of metal work hardening and fatigue. Bunch stranding offers extreme flexibility, but does not provide a consistent circular cross section and thus is not recommended for thin-wall extruded wire insulation. It does offer low cost and fatigue resistance. Major conductor constructions are discussed in the following.

Solid. Solid conductor use is still limited for the aforementioned reasons. The major application of solid wire, with the exception of magnet wire, is for short jumper and bus wires not subjected to vibration or flexing. With recent developments in wire wrapping methods for termination, solid conductors are being used to a greater extent for hookup wires. Dimensional properties of solid conductors such as AWG, nominal diameter, weight (lb/1000 ft), length (ft/lb), and resistance as Ω/1000 ft, ft/Ω, and Ω/lb are available from all the wire suppliers.

Stranded. Despite increased flexibility, the construction characteristics of stranded conductors may limit the application of some insulating materials. Furthermore, in most electronics applications, stranding is kept to a minimum consistent with application requirements, that is, flex life. Even for primary power conductors, stranding is kept to a minimum. The need for a large number of strands is required only for specialty applications such as cables for welding or other high-power applications that involve repeated flexing. Table 6.2 contains detailed characteristics of stranded conductors that meet military specifications. The following is a brief description of various stranding constructions:

True Concentric. This construction presents the most consistent circular cross section of all available strandings. Alternate layers, applied in opposite directions with increasing lay, hold strands in place and prevent "strand popping" and high strands. Insulated wire manufacturers prefer this stranding for more uniform extruded thin-wall electrical insulation. Construction of concentric stranded conductors is illustrated in Fig. 6.2.[2]

Unidirectional Concentric. This construction differs from true concentric in that the lay of successive layers is applied over a core in the same direction; it does not alternate directions. Unidirectional lay is more flexible and has greater flex endurance than true concentric. A comparison of flex endurance between true concentric, unidirectional concentric, bunch, and unilay conductor constructions is shown in Fig. 6.3.

Bunch. Since bunch stranding consists in twisting a group of strands with the same lay length in the same direction, without regard to geometric arrangement, the strands are susceptible to movement, and circular cross section is not assured. Bunch-stranded conductor is not recommended if conductor circularity is important, as required in extruded thin-wall insulating materials.

Unilay. The advantages of unilay are a smaller diameter, flexibility approaching bunch construction, and superior flex endurance (Fig. 6.3).

Equilay. This is a variation of concentric construction that has reversed lay of layers but equal length of lay and a stiffer construction than unidirectional concentric lay.

Rope Lay. Rope lay is basically a large-gauge (AWG 10 and larger) conductor construction that consists of a central core stranded member with one or more layers of stranded members surrounding the core. The stranded members may be bunch-stranded (ASTM B-172) or concentric (ASTM B-173). Rope construction, shown in Fig. 6.2, provides a uniform circular cross section with good flexibility.

TABLE 6.2 Details of Stranded Conductors

Size designation, AWG	Nominal conductor area, cmil	Number of strands	Allowable no. of missing strands	Nominal diameter of individual strands, in	Maximum diameter of stranded conductor, in	Maximum resistance of finished wire, Ω/1000 ft at 20°C			
						Tin-coated copper	Silver-plated copper	Nickel-plated copper	Silver-plated high-strength copper alloy
30	112	7	0	0.0040	0.013	107.0	101.0	109.0	116.0
28	175	7	0	0.0050	0.016	67.6	62.9	68.3	72.2
26	304	19	0	0.0040	0.021	39.3	36.2	40.1	41.5
24	475	19	0	0.0050	0.026	24.9	23.2	25.1	26.6
22	754	19	0	0.0063	0.033	15.5	14.6	15.5	16.8
20	1,216	19	0	0.0080	0.041	9.70	9.05	9.79	10.4
18	1,900	19	0	0.0100	0.052	6.08	5.80	6.08	6.65
16	2,426	19	0	0.0113	0.060	4.76	4.54	4.76	5.23
14	3,831	19	0	0.0142	0.074	2.99	2.87	3.00	3.30
12	7,474	37	0	0.0142	0.102	1.58	1.48	1.59	1.70
10	9,2361	37	0	0.0159	0.118	1.27	1.20	1.27	1.38
8	16,983	133	0	0.0113	0.176	0.700	0.661	0.680	0. 760
6	26,818	133	0	0.0142	0.218	0.436	0.419	0.428	0. 483
4	42,615	133	0	0.0179	0.272	0.274	0.263	0.269	0. 302
2	66,500	665	2	0.0100	0.345	0.179	0.169	0.174	0. 194
0	104,500	1045	3	0.0100	0.432	0.114	0.105	0.109	0.123

Source: From Schuh.[1]

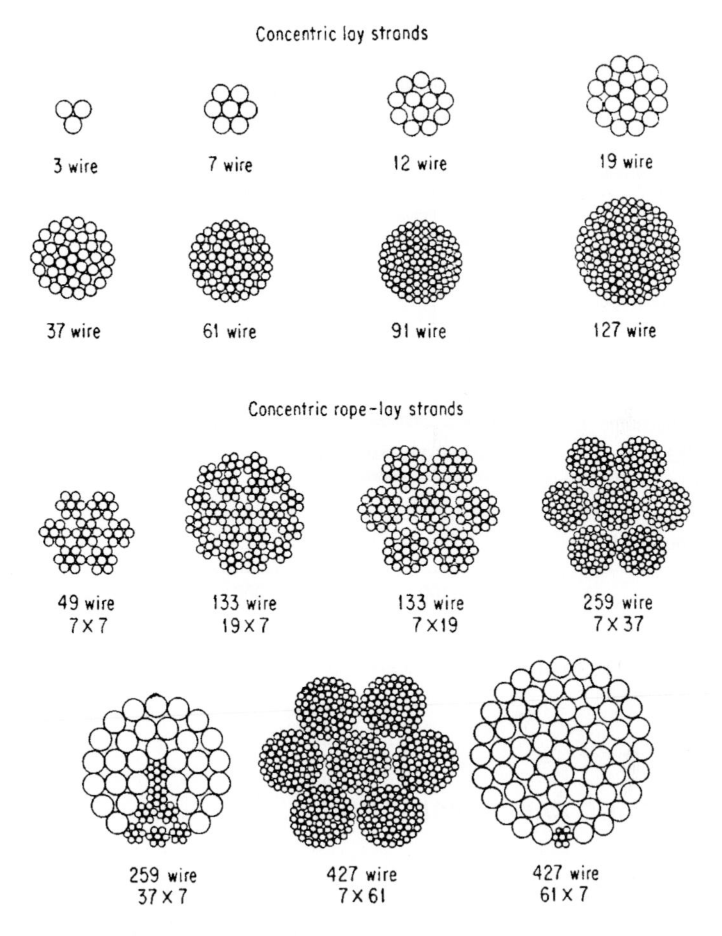

FIGURE 6.2 Stranded wire and cable construction.[2]

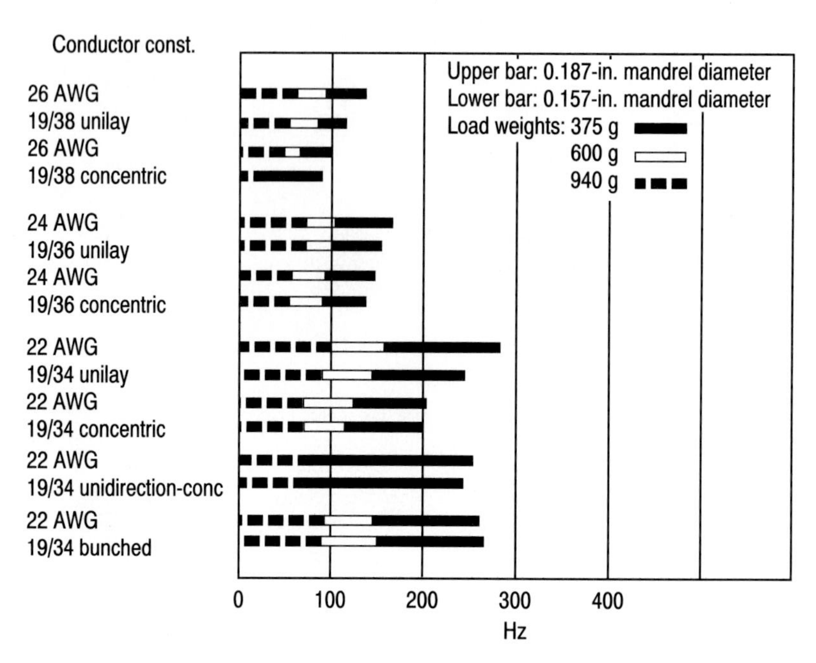

FIGURE 6.3 Flex endurance tests on annealed conductors (copper) in different stranding configurations.[1]

6.3 WIRE AND CABLE INSULATION

6.3.1 Materials

The major properties of insulating and jacketing materials used with hookup and interconnecting wire and cable can be divided into three main categories: mechanical, electrical, and chemical.

1. The mechanical and physical properties to consider are tensile strength, elongation, specific gravity, abrasion resistance, cut-through resistance, and mechanical temperature resistance (cold bend and deformation under heat).
2. The electrical properties are dielectric strength, dielectric constant, loss factor, and insulation resistance.
3. The chemical properties are fluid resistance, flammability, temperature resistance, and radiation resistance.

Polymers and plastics represent the largest class of wire-insulating materials (Table 6.3).[3] Thermoplastics have by far the greatest volume usage for hookup and interconnection wire insulation as well for jacketing and cabling. A more complete description of some of these polymers appears in Chap. 1.

Tables 6.4 to 6.6 list some major properties for the most commonly used wire-insulating and jacketing materials. The data are presented only as a guide; variations from

TABLE 6.3 Some Common Wire Insulating Materials

Butyl rubber (butyl)
Ethylene propylene diene monomer rubber (EPDM)
Fluorinated ethylene propylene (Teflon* FEP)
Fluorocarbon rubber (Viton*)
Monochlorotrifluoroethylene (Kel-F†)
Neoprene rubber (Neoprene*)
Polyalkene
Polyamide (nylon)
Polyethylene (PE)
Polyethylene terephthalate polyester (Mylar*)
Polyimide (Kapton*)
Polypropylene
Polysulfone
Polytetrafluoroethylene (Teflon* TFE)
Polyurethane elastomers (urethane)
Polyvinyl chloride (PVC)
Polyvinylidene fluoride (Kynar‡)
Silicone rubber (silicone)

*Registered trademarks of E. I du Pont de Nemours & Company, Inc.
†Registered trademark of 3M Company, Inc.
‡Registered trademark of Pennwalt, Inc.
Source: From Schuh.[3]

the given values result from varying wall thicknesses, processing methods, test methods, and so on. Figures 6.1 to 6.3 show typical physical design characteristics of hookup and interconnection wire conductors.

6.3.2 Construction and Application Methods

Reference has been made to extrusion, tape wrapping, and coating as methods of applying specific materials to a conductor. Certain insulating materials are available only in a tape or solution form and cannot be obtained in extruded form. Modern engineering thermoplastics such as PEEK and PES have outstanding thermal capabilities, but are difficult to apply to wire because of their high melt temperatures and their insolubility in all but the most powerful (and expensive) solvents. Following is a brief description of methods of application of conventional materials.

Extrusion. Many of the materials listed in Table 6.3 are thermoplastic; that is, they are heated to softness, formed into shape, then cooled to become solid again.

TABLE 6.4 Mechanical and Physical Properties of Insulating Materials

Insulation	Common designation	Tensile² strength, lb/in²	Elongation, %	Specific gravity	Abrasion resistance	Cut-through resistance	Temperature resistance (mechanical)
Polyvinyl chloride	PVC	2400	260	1.2–1.5	Poor	Poor	Fair
Polyethylene	PE	1400	300	0.92	Poor	Poor	Good
Polypropylene		6000	25	1.4	Good	Good	Poor
Cross-linked polyethylene	IMP	3000	120	1.2	Fair	Fair	Good
Polytetrafluoroethylene	TFE	3000	150	2.15	Fair	Fair	Excellent
Fluorinated ethylene propylene	FEP	3000	150	2.15	Poor	Poor	Excellent
Monochlorotrifluoroethylene	Kel-F	5000	120	2.13	Good	Good	Good
Polyvinylidene fluoride	Kynar	7100	300	1.76	Good	Good	Fair
Silicone rubber	Silicone	800–1800	100–800	1.15–1.38	Fair	Poor	Good
Polychloroprene rubber	Neoprene	150–4000	60–700	1.23	Good	Good	Fair
Butyl rubber	Butyl	700–1500	500–700	0.92	Fair	Fair	Fair-good
Fluorocarbon rubber	Viton	2400	350	1.4–1.95	Fair	Fair	Fair-good
Polyurethane	Urethane	5000–8000	100–600	1.24– 1.26	Good	Good	Fair-good
Polyamide	Nylon	4000–7000	300–600	1.10	Good	Good	Poor
Polyimide film	Kapton	18,000	707	1.42	Excellent	Excellent	Good
Polyester film	Mylar	13,000	185	1.39	Excellent	Excellent	Good
Polyalkene		2000–7000	200–300	1.76	Good	Good	Fair-good
Polysulfone		10,000	50–100	1.24	Good	Good	Good
Polyimide-coated TFE	TFE/ML	3000	150	2.2	Good	Good	Good
Polyimide-coated FEP	FEP/ML	3000	150	2.2	Good	Good	Good

Source: Martin Marietta Corporation.

TABLE 6.5 Electrical Properties of Insulating Materials

Insulation	Common designation	Dielectric strength, V/mil	Dielectric constant at 10^3 Hz	Loss factor at 10^3 Hz	Volume resistivity, $\Omega \cdot$ cm
Polyvinyl chloride	PVC	400	5–7	0.02	2×10^{14}
Polyethylene	PE	480	2.3	0.005	10^{16}
Polypropylene		750	2.54	0.006	10^{16}
Cross-linked polyethylene	IMP	700	2.3	0.005	10^{16}
Polytetrafluoroethylene	TFE	480	2.1	0.0003	10^{18}
Fluorinated ethylene propylene	FEP	500	2.1	0.0003	10^{18}
Monochlorotrifluoroethylene	Kel-F	431	2.45	0.025	2.5×10^{16}
Polyvinylidene fluoride	Kynar	1280 (8 mil)	7.7	0.02	2×10^{14}
Silicone rubber	Silicone	575–700	3–3.6	0.003	2×10^{15}
Polychloroprene rubber	Neoprene	113	9.0	0.030	10^{11}
Butyl rubber	Butyl	600	2.3	0.003	10^{17}
Fluorocarbon rubber	Viton	500	4.2	0.14	2×10^{13}
Polyurethane	Urethane	450–500	6.7–7.5	0.055	2×10^{11}
Polyamide	Nylon	385	4–10	0.02	4.5×10^{13}
Polyimide film	Kapton	5400 (2 mil)	3.5	0.003	10^{18}
Polyester film	Mylar	2600	3.1	0.005	6×10^{16}
Polyalkene		1870	3.5	0.028	6×10^{13}
Polysulfone		425	3.13	0.0011	5×10^{16}
Polyimide-coated TFE	TFE/ML	480	2.2	0.0003	10^{18}
Polyimide-coated FEP	FEP/ML	480	2.2	0.0003	10^{18}

Source: Martin Marietta Corporation.

Thermoplastics are shaped around a conductor by the extrusion process. This consists in forcing the plastic material under pressure and heat through an orifice concurrently with the conductor.

Tape Wrap. Tape wrapping is the application of insulation in the form of a thin film or tape. The tape is normally applied to the conductor with minimum overlap to provide wire flexibility without baring the conductor. Layers of tape can be built up to achieve the desired insulation wall thickness. Successive layers are wrapped in opposite directions, ensuring coverage.

Coating. Insulation by dip coating is limited almost exclusively to magnet wire, glass, other insulating fabrics, and braided sleeving. For small-diameter Teflon TFE and Teflon FEP applications, a polyimide overcoating of less than 0.001 in enhances the mechanical strength, abrasion protection, and solder resistance of the insulation significantly.

TABLE 6.6 Chemical Properties of Insulating Materials

Insulation	Common designation	Fluid resistance	Flammability	Radiation resistance, rads gamma exposure	Temperature resistance, °C	Comments
Polyvinyl chloride	PVC	Good	Slow to self-extinguishing	10^6–10^7	−55–105	
Polyethylene	PE	Excellent	Flammable	10^8	−65–8 0	
Polypropylene	—	Good	Self-extinguishing	10^8	−20–125	
Cross-linked polyethylene	IMP	Good	Self-extinguishing	10^8	−65–150	
Polytetrafluoroethylene	TFE	Excellent	Nonflammable	10^6	−80–260	
Fluorinated ethylene propylene	FEP	Excellent	Nonflammable	10^6	−80– 200	
Monochlorotrifluoroethylene	Kel-F	Good	Nonflammable	10^6	−80–200	Fluids tend to permeate at high temperature
Polyvinylidene fluoride	Kynar	Good	Self-extinguishing	10^8	−65–130	
Silicone rubber	Silicone	Poor	Flammable	10^8	−65–200	
Polychloroprene rubber	Neoprene	Good oil resistance	Self-extinguishing	10^7	−55–80	
Butyl rubber	Butyl	Good	Flammable	10^6	−55–85	Poor resistance to hydrocarbons
Fluorocarbon rubber	Viton	Excellent	Self-extinguishing	10^6–10^7	−40–200	Poor resistance to oxygenated alcohols
Polyurethane	Urethane	Good	Flammable	10^7–10^8	−55–85	
Polyamide	Nylon	Good	Self-extinguishing	10^7	−55–105	Soluble in alcohol
Polyimide film	Kapton	Excellent	Nonflammable	10^9	−80–260	
Polyester film	Mylar	Good	Flammable	10^8	−65–120	
Polyalkene	—	Good	Self-extinguishing	10^8	−65–135	
Polysulfone	Polysulfone	Fair	Self-extinguishing	—	−65–150	Soluble in chlorinated hydrocarbon
Polyimide-coated TFE	TFE/ML	Excellent	Nonflammable	10^6	−80–260	
Polyimide-coated FEP	FEP/ML	Excellent	Nonflammable	10^6	−80–260	

Source: Martin Marietta Corporation.

6.4 THERMAL STABILITY AND THERMAL RATINGS OF INSULATION

6.4.1 Thermal Rating Background

Around 1913 it became apparent[4] that operating temperatures, hot-spot temperatures, and environmental ambient temperatures had a degrading effect with time on the ability of organic materials to perform their designed electrical, mechanical, and chemical functions. Originally this was believed to be a linear effect, and a "10-degree rule" was later developed[5]—an increase of 10°C in integrated operating temperature would reduce life by 50 percent.

Modern practice treats thermal and other aging phenomena of organic materials as a chemical rate reaction, based on the Arrhenius rate equation

$$L = Ae^{B/T}$$

which plots as a straight line as

$$\ln L = \ln A + B/T$$

where L is the life (usually hours), T the absolute temperature (°K), and A and B are constants determined by the activation energy of the particular reaction.[6] Data are derived by plotting the change in a property (electrical, mechanical, chemical, weight loss, and so on) of the material on aging at various temperatures to an arbitrarily selected end point and then logarithmically plotting the times to reach the end point versus the reciprocal of the absolute temperatures of the thermal aging. This usually results in a straight line. Extrapolation of the line can be used to predict life expectancy at various integrated operating temperatures.

6.4.2 Methods of Determining Thermal Stability

ASTM D-2307 is the most widely used and perhaps one of the simplest methods (twisted pair) of determining thermal stability of insulated wire and cable. The entire test procedure is also described in military (federal) specification J-W-1177. IEEE Std. 1 defines the temperature index and IEEE Std. 98 explains its test procedure.

6.4.3 Thermal Classifications

Table 6.7 summarizes the evolution of insulation temperature classifications.

TABLE 6.7 Insulation Temperature Classifications

Thermal class (obsolete)	Limiting hot-spot temperature, °C	Old temperature ratings	New temperature ratings	Number range, °C	Temperature index
0	90	Class 90	90	90–104	90
A	105	Class 105	105	105–129	105
B	130	Class 130	130	130–154	130
F	155	Class 155	155	155–179	155
H	180	Class 180	180	180–199	180
>H, H+	Over 180	Class 200	200	200–219	200
		Class 220	220	200+	—

Source: From Croop.[7]

6.5 SHIELDING

6.5.1 Materials and Construction

Although there is no real substitute for braided copper as a general-purpose shield, certain applications may allow the use of a lighter, less bulky shield construction. The purpose of a shield is to (1) prevent external fields from adversely affecting signals transmitted over the center conductor, (2) prevent undesirable radiation of a signal into nearby or adjacent conductors, or (3) act as a second conductor in matched or tuned lines. Table 6.8 presents the shielding effectiveness at tested frequency ranges of the systems described. A list of various available shield types follows, with descriptions where necessary.

Braided round shield. Braiding consists in interweaving groups (carriers) of strands (ends) of metal over an insulated conductor. The braid angle at which the carriers are applied with reference to the axis of the core, the number of ends, the number of carriers, and the number of carrier crossovers per inch (picks) determine the percent coverage, which is a measure of the shielding gap (Fig. 6.4). For reference, a solid metal tube is equivalent to 100 percent coverage.

Flat braided shield

Metal tape

Solid shield. Solid or tubular shields may be applied by tube swaging or by forming interlocked, seam welded, or soldered tape around the dielectric, offering 100 percent shielding as well as a protective armor.

Served shield. A served or spiral shield consists of a number of metal strands wrapped flat as a ribbon over a dielectric in one direction. The weight and size of shielding are approximately one-half those of braided shields, as there is no strand crossover or overlap.

Foil shield

Conductive plastics. PVC and PE compounds have been formulated with conductive additives for the purpose of shielding. Effectiveness has been poor, normally limited to the low audio-frequency range. As with foil, termination is a problem; a drain wire can be used for terminating.

Conductive yarns. Conductive yarns, such as impregnated glass, provide weight reduction, but low conductivity and difficulty in termination make them suited for special applications only.

Metal shield and conductive yarn. An effective combination of metal strands and conductive yarns has been developed. It provides effective shielding when braided and reduces shield weight. Terminations can be accomplished by normal techniques.

6.5.2 Shield Jackets

It is the basic function of a shield jacket to insulate a shield from ground (structure), other shields, or conductors. Secondarily, a shield jacket may serve as a lubricant in multiconductor cables, enabling free movement during bending; as an insulator, to allow varying potentials between adjacent shields; and as a moisture and abrasion bar-

TABLE 6.8 Shield Effectivity, Volts Peak-to-Peak

Sample description	Frequency									
	30 Hz	100 Hz	300 Hz	1 kHz	3 kHz	10 kHz	30 kHz	100 kHz	300 kHz	1 MHz
Control unshielded	<0.600	<1.6	4.0	10.0	12.0	12.5	13.5	12.5	12.5	14.0
Braid, no. 36 AWG, tinned copper, 90% coverage	<0.001	<0.001	0.0025	0.005	0.006	0.00625	0.0075	0.0075	0.007	0.008
Aluminum tape, no. 22 AWG drain wire	<0.001	<0.001	0.001	0.002	0.002	0.002	0 .005	0.0085	0.012	0.014
Semiconductive PVC, no. 26 AWG drain wire	<0.001	<0.001	<0.00325	0.020	0.060	0. 120	0.240	0.450	0.540	1.85
Serve, no. 36 AWG tinned copper, 90% coverage, 4 ends reversed	<0.001	<0.001	0.001	0.002	0.0025	0.0025	0.004	0.006	0.0065	0.013
Braid, 8 carriers, no. 36 AWG tinned copper, 8 carriers, conductive glass yarn	<0.001	<0.001	<0.001	0.001	0.00125	0. 00125	0.002	0.003	0.005	0.012
Braid, no. 36 AWG flat ribbon silver-plated copper	0.00125	0.003	0.008	0.018	0.024	0.024	0.028	0.028	0.027	0.029
Serve, no. 36 AWG tinned copper	<0.001	0.003	0.008	0.018	0.026	0.026	0.029	0.029	0.030	0.031
Solid cadmium-plated copper	<0.001	<0.001	<0.001	<0.001	<0.001	<0.001	<0.001	<0.001	<0.001	<0.001

Source: Martin Marietta Corporation.[8]

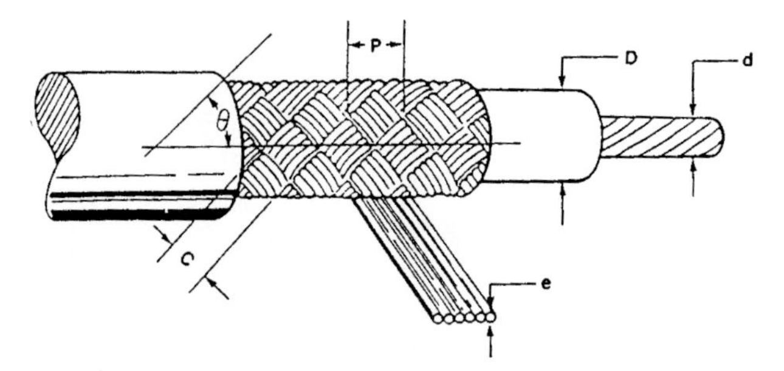

1 *Braid angle*

$$\theta = \tan^{-1}\left[\frac{2\pi(D + e)P}{C}\right], \quad \text{deg}$$

2 *Braid picks per inch*

$$P = \frac{C \tan \theta}{2\pi M}, \quad \text{picks/in}$$

3 *Braid shield weight*

$$w = \frac{nCl}{\cos \theta}, \quad \text{lb/Mft}$$

4 *Braid shield dc resistance*

$$R_{dc} = \frac{r_{dc}}{nC \cos \theta}, \quad \Omega/\text{Mft}$$

where D = diameter under shield, in
d = diameter of center conductor, in
C = number of carriers
e = diameter of end
P = pick (measured in picks per linear inch)
θ = braid angle, deg
w = weight of shield, lb/Mft
n = number of ends in one carrier
l = weight of one end, lb/Mft
M = D + buildup of braid on one shield wall, in
R_{dc} = dc resistance of braid shield, Ω/Mft
r_{dc} = dc resistance of one strand (end) of shield, Ω/Mft

FIGURE 6.4 Braid shield cable design equations. (*Courtesy of Carol Cable Company.*)

rier. A few of the materials listed in Table 6.3 are appropriately used as jacketing materials, because of either superior mechanical strength or compatibility with the chosen primary insulation system.

Extrusion, braiding, and taping techniques are used in forming shield jackets. Fused tapes and extruded materials are the only reliable methods for assuring a moisture-proof jacket. Consistent wall thickness is best maintained with tapes. However, reliable fusion across interstices of multiple shielded conductors is frequently a problem. Varnished fiber braids allow a large conductor or multiple conductors more flexibility than extruded or taped jackets, but offer limited moisture and humidity resistance. Figure 6.5 provides abrasion resistance data for shield jackets. In comparing abrasion resistance values, differences in wall thickness should be considered. Some available shield jacket materials follow:

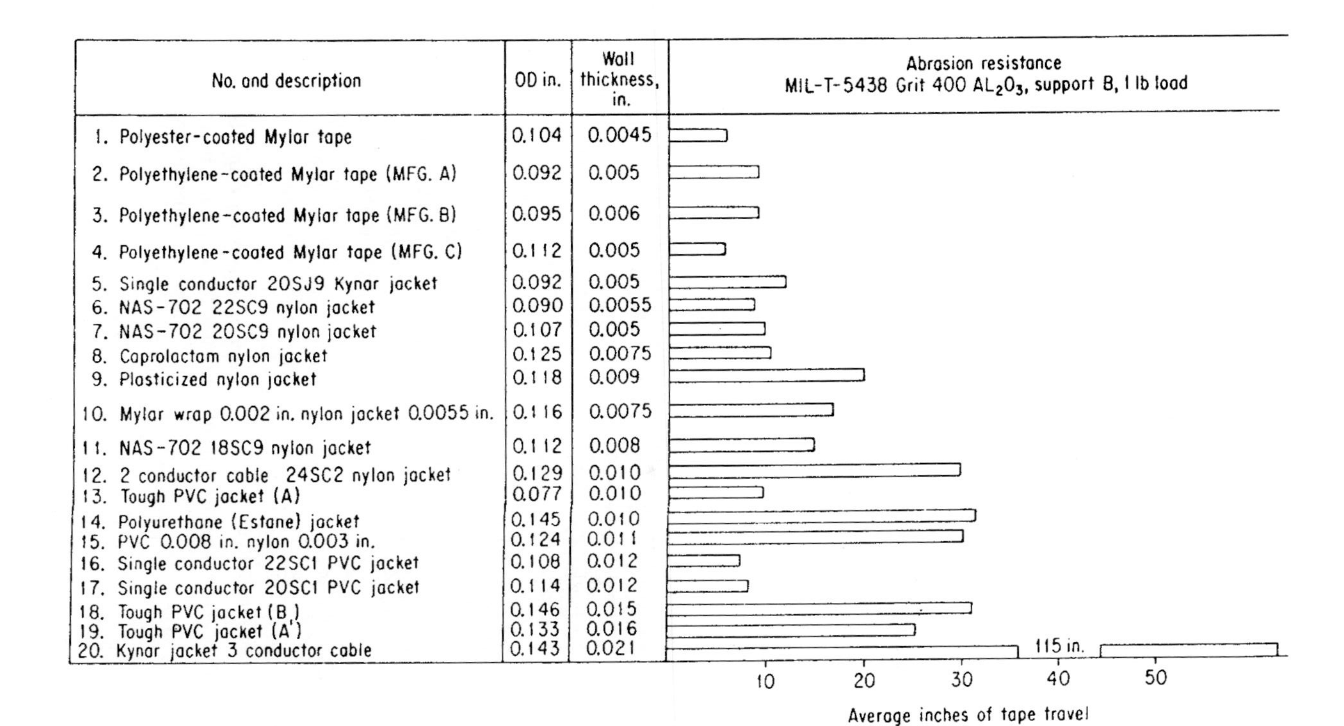

No. and description	OD in.	Wall thickness, in.
1. Polyester-coated Mylar tape	0.104	0.0045
2. Polyethylene-coated Mylar tape (MFG. A)	0.092	0.005
3. Polyethylene-coated Mylar tape (MFG. B)	0.095	0.006
4. Polyethylene-coated Mylar tape (MFG. C)	0.112	0.005
5. Single conductor 20SJ9 Kynar jacket	0.092	0.005
6. NAS-702 22SC9 nylon jacket	0.090	0.0055
7. NAS-702 20SC9 nylon jacket	0.107	0.005
8. Caprolactam nylon jacket	0.125	0.0075
9. Plasticized nylon jacket	0.118	0.009
10. Mylar wrap 0.002 in. nylon jacket 0.0055 in.	0.116	0.0075
11. NAS-702 18SC9 nylon jacket	0.112	0.008
12. 2 conductor cable 24SC2 nylon jacket	0.129	0.010
13. Tough PVC jacket (A)	0.077	0.010
14. Polyurethane (Estane) jacket	0.145	0.010
15. PVC 0.008 in. nylon 0.003 in.	0.124	0.011
16. Single conductor 22SC1 PVC jacket	0.108	0.012
17. Single conductor 20SC1 PVC jacket	0.114	0.012
18. Tough PVC jacket (B)	0.146	0.015
19. Tough PVC jacket (A')	0.133	0.016
20. Kynar jacket 3 conductor cable	0.143	0.021

FIGURE 6.5 Shield jacket abrasion resistance. (*Courtesy of Martin Marietta Corporation.*)

Dacron* braid
Fluorinated ethylene propylene (FEP)
Polyethylene-coated Mylar
Glass braid
Kynar
Polyamide (nylon)
Polyvinyl chloride (PVC)
Teflon (TFE)

6.6 DESIGN CONSIDERATIONS

6.6.1 Wire Gauge Selection

The following factors must be considered for proper wire gauge selection:

1. Voltage drop
2. Current-carrying capacity
3. Circuit-protector characteristics

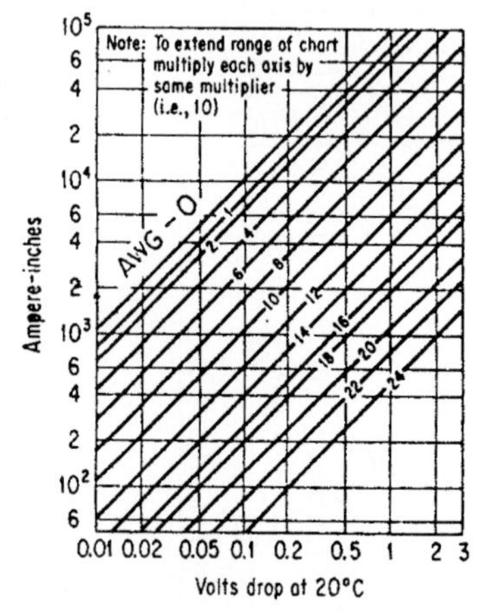

FIGURE 6.6 Direct-current voltage drop of copper wire.[1,9]

Voltage Drop. Voltage-drop calculations should be based on anticipated load current at nominal system voltage. Voltage drop through an aluminum structural ground return can be considered zero for all practical purposes. Normal voltage-drop limits do not apply to starting currents of equipment such as motors. The voltage at load during start-up should be considered to assure proper operation of the equipment. Figure 6.6 may be used to select the wire gauge. Ampere-inches is the product of wire length between terminations (inches) and wire current (amperes).

Current-Carrying Capacity. The following factors must be considered in determining the current-carrying capacity of a wire:

*Dacron is a registered trademark of E.I. du Pont de Nemours & Company, Inc.

1. Continuous-duty rating
2. Short-time rating
3. Effects of ambient temperature
4. Effects of wire grouping
5. Effects of altitude

Continuous-Duty Rating. This factor applies if a wire is to carry current for 1000 s or longer. The continuous current-carrying capacity of copper and aluminum wire in amperes for aerospace applications is shown in Table 6.9. The following criteria apply to the ratings in the table:

- Rated ambient temperatures:

  57.2°C (135°F) for 105°C rated insulated wire

  92°C (197.6°F) for 135°C rated insulated wire

  107°C (225°F) for 150°C rated insulated wire

  157°C (315°F) for 200°C rated insulated wire

- "Wire bundled" indicates 15 or more wires in a group.
- The sum of all currents in a bundle is not more than 20 percent of the theoretical capacity of the bundle, which is calculated by adding the bundle ratings of the individual wires.

Short-Time Current Ratings. These apply when a wire is to carry a current for less than 1000 s. The short-time rating is generally applicable to starting loads. Figure 6.7 shows the current ratings for various wire gauges in harness (bundles).

Temperature-Current Relationship. The following equation provides a means of rerating the current-carrying capacity of wire and cable at any anticipated ambient temperature:

$$I = I_r \sqrt{\frac{t_c - t}{t_c - t_r}}$$

where I = current rating at ambient temperature t
I_r = current rating at rated ambient temperature t_r (Table 6.9)
t = ambient temperature
t_r = rated ambient temperature (continuous-duty rating)
t_c = temperature rating of insulated wire or cable (continuous-duty rating)

Figures 6.8 and 6.9 present curves showing the effect of temperature on the current-carrying capacity of copper wire with 10- and 15-mil-thick insulation, respectively.

Effects of Wire Grouping. When wires are grouped (bundled, harnessed) together, their current ratings must be reduced because of restricted heat loss. Table 6.9 and Fig. 6.7 take into account reduced ratings based on grouping 15 or more wires.

Effects of Altitude. Air density decreases with increasing altitude. Since lower density reduces the dielectric properties of air, trapped air between conductors and insulators presents a problem. In addition, air is retained in voids that occur primarily in stranded conductors, although voids cannot be entirely eliminated from insulated solid-wire construction either. A direct effect of increased altitude on insulated wire is lower corona threshold voltage, resulting in lower peak operating voltage. Wire insulated with organic (polymers, plastics) materials should always be operated below the

TABLE 6.9 Maximum Current-Carrying Capacity, Amperes

Size, AWG	Copper (MIL-W-5088)		Aluminum (MIL-W-5088)		National Electrical Code	Underwriters Laboratory		American Insurance Association	500 cmil/A
	Single wire	Wire bundled	Single wire	Wire bundled		+60°C	+80°C		
30	—	—	—	—	—	0.2	0.4	—	0.20
28	—	—	—	—	—	0.4	0.6	—	0.32
26	—	—	—	—	—	0.6	1.0	—	0.51
24	—	—	—	—	—	1.0	1.6	—	0.81
22	9	5	—	—	—	1.6	2.5	—	1.28
20	11	7.5	—	—	—	2.5	4.0	3	2.04
18	16	10	—	—	6	4.0	6.0	5	3.24
16	22	13	—	—	10	6.0	10.0	7	5.16
14	32	17	—	—	20	10.0	16.0	15	8.22
12	41	23	—	—	30	16.0	26.0	20	13.05
10	55	33	—	—	35	—	—	25	20.8
8	73	46	58	36	50	—	—	35	33.0
6	101	60	86	51	70	—	—	50	52.6
4	135	80	108	64	90	—	—	70	83.4
2	181	100	149	82	125	—	—	90	132.8
1	211	125	177	105	150	—	—	100	167.5
0	245	150	204	125	200	—	—	125	212.0
00	283	175	237	146	225	—	—	150	266.0
000	328	200	—	—	275	—	—	175	336.0
0000	380	225	—	—	325	—	—	225	424.0

Source: Martin Marietta Corporation.

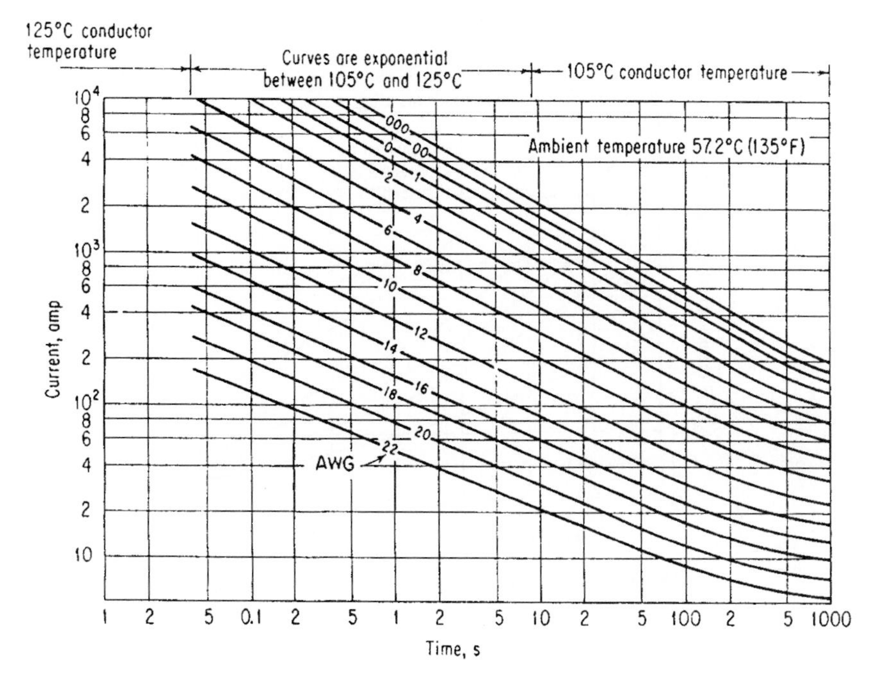

FIGURE 6.7 Short-time working curves for 105°C insulated copper wire in bundles.[1,9]

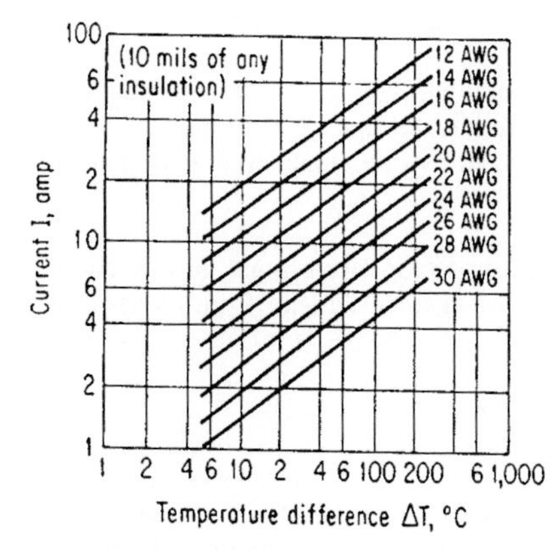

FIGURE 6.8 Effects of temperature on current-carrying capacity of copper conductors (10-mil insulation).[1,10]

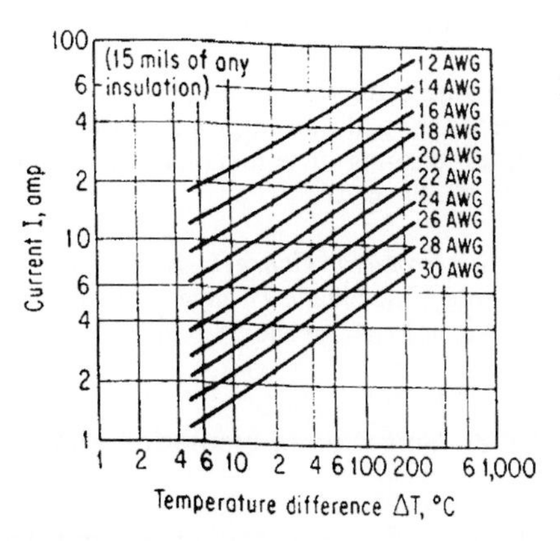

FIGURE 6.9 Effects of temperature on current-carrying capacity of copper conductors (15-mil insulation).[1,10]

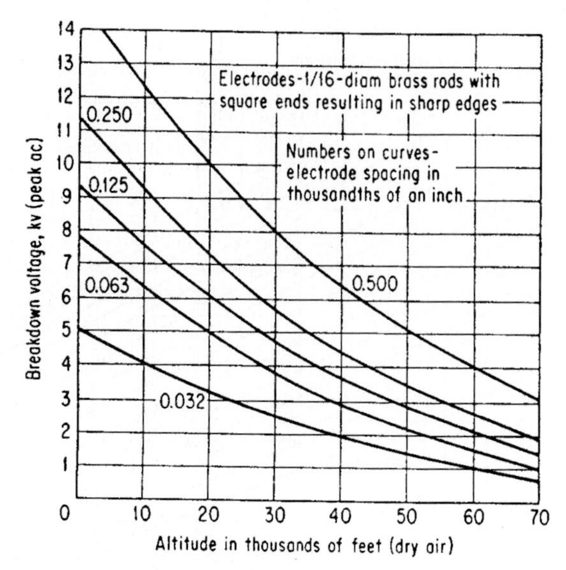

FIGURE 6.10 Average breakdown voltage versus altitude. (*Courtesy of Martin Marietta Corporation.*)

corona extinction voltage, as corona has a degrading effect on all organic materials. Operating below corona threshold but above corona extinction is risky, for a surge may start corona, which will continue until the voltage is lowered to the extinction level.

Figure 6.10 presents curves on insulation breakdown voltage of air as a function of altitude. In addition to proper derating of operating voltage on insulated wires and cables, derating of termination spacing in air must also be considered at higher altitudes (Table 6.20).

Circuit Protector Characteristics. The circuit protector rating must be low enough to protect the smallest-gauge wire connected to it against damage from overheating, smoke, or fire from short circuits.

6.6.2 Selection of Insulation Materials

Factors governing the selection of an optimum insulation material cover a wide range of properties and are less precise in nature than those for wire-gauge selection. In addition to meeting specific system electrical and environmental requirements, related fabrication techniques must be considered. The insulation must withstand mechanical abuse and the heat from soldering or application of associated heat-shrinkable devices. Furthermore the insulation must be compatible with encapsulants, potting compounds, conformal coatings, and adhesives. Table 6.10 is a checklist for the selection of insulation.

TABLE 6.10 Insulation Selection Checklist

Requirement	Considerations
Environment	
Temperature extremes:	
Continuous operating	Refer to Table 6.6.
Short-term operating	May require test that simulates specific application.
Fabrication temperature	Check for soldering-iron resistance in high-density packaging; cure temperatures of encapsulant; compatibility with shrinkable devices, if used.
Storage	Check for embrittlement, long-term storage, low-humidity conditions.
Altitude:	
Outgassing	Weight loss, smoke, condensation.
Corona	Maintain voltage below corona level, especially with insulations susceptible to erosion (Fig. 6.14).
Radiation	
Weather	Moisture resistance, aging, ultraviolet radiation.
Flame	
Fluids	
Electrical	
Capacitance	$C = \frac{7.36K}{\log_{10}(D/d)}$ where C = capacitance, pF/ft; K = dielectric constant (Table 6.5); D = insulated wire diameter, in; d = conductor diameter, in
Dielectric strength	Refer to Table 6.5.
Volume resistivity	Refer to Table 6.5.
Loss factor	Refer to Table 6.5.
Mechanical	
Installation and handling	Check for minimum bend radius, special tooling, clamping, stresses, chafing. Refer to Table 6.4 for abrasion, cut-through, and mechanical resistance.
Operation	Refer to Table 6.4.
Size	Refer to applicable specification for outside dimensions.
Weight	Refer to applicable specification for maximum weight. If not listed, use the following equation for insulation weight: $W = \frac{Dd^2}{2} KG$ lb/1000 ft where D = diameter over insulation, in; d = diameter over conductor, in; $K = 680$; G = specific gravity of insulation (Table 6.4)

Source: From Schuh.[1]

6.7 INTERCONNECTION AND HOOKUP WIRE

6.7.1 Definitions

For purposes of this discussion, the difference between interconnection and hookup wire is determined by the amount of mechanical stress applied to the wire. Interconnection wire is used to connect electric circuits between pieces of equipment and must withstand rough handling, abrasion when pulling through conduit, and potential accidental damage during installation that results from the slip impact of hand tools. Hookup wire is used to connect electric circuits within a unit of equipment or black box. This type of equipment may range from miniaturized airborne equipment to a massive ground installation. With today's emphasis on miniaturization, many applications specify the use of hookup wire where interconnection wire was used in the past.

6.7.2 Interconnection Wire

Interconnection wire is by nature bulky; it must withstand severe mechanical abuse. For its major application in the aircraft industry, it must have minimum weight and size to allow greater payload. Insulation suppliers and wire manufacturers are continuing to develop new interconnection wire insulations achieving the primary requisite of mechanical resistance consistent with minimum bulk and weight.

In addition to the development of higher-strength insulation materials, innovations such as the addition of reinforcing mineral fillers to existing resin systems have proved feasible (MS-17411, MS-17412, MIL-W-22759/7B and /8). The addition of filler materials enhances the abrasion resistance of the basic resin system without an increase in wire diameter and with minimum weight penalty. Table 6.11 lists the interconnection wires most widely used in industry. The tabulation is made on the basis of AWG 22 insulated conductor so that weight and size can be analyzed and compared for the different types.

6.7.3 Hookup Wire

Although the name hookup wire is likely to become a catchall for all wire constructions, it represents the area where by far the highest volume (footage) of military and commercial electronic wire is consumed. Under the demands of miniaturization, many aerospace and missile designs call for hookup wire rather than the stronger, heavier, and bulkier interconnection wire.

The hookup wire field is still the most dynamic in the wire and cable industry. Manufacturers must be cognizant of all newly developed resin (polymer) systems and fabrication techniques, and their applicability as a wire insulation. In addition, the manufacturer must be adaptable to the many and varied user requirements, ranging from a conventional back-panel application to the complicated environment of spaceflight. Working in a changeable and uncertain environment imposes substantial burden on both user and manufacturer. Many requirements are necessarily established without the ability to thoroughly analyze either the requirement or the capability of a selected insulation to withstand the required environment. The cost and availability of test equipment to simulate the many user environments is for all practical purposes prohibitive to a wire manufacturer, and in many cases manufacturers must depend on user evaluation to determine the suitability of their products.

Table 6.12 presents a summary of hookup wire types, including thin-wall insulations, which have been covered in specifications.

TABLE 6.11 Interconnection Wire Data, Copper Conductors

Basic specification	Class type or MS no.	Size range, AWG	Conductor coating	Primary insulation	Jacket	Voltage rating, V rms	Temperature rating, °C	Diameter rating, maximum	Weight rating, maximum
MIL-W-5086	MS-25190, Ty 1	22–12	Tin	PVC	Nylon	600	−55 to 105	1.0	1.0
	MS-25190, Ty 2	22–4/0	Tin	PVC	Glass nylon	600	−55 to 105	1.1	1.07
	MS-25190, Ty 3	22–4/0	Tin	PVC	Glass PVC nylon	600	−55 to 105	1.25	1.3
	MS-25190, Ty 4	22–16	Tin	PVC	Nylon	3000	−55 to 105	1.46	1.64
MIL-W-8777	MS-25471	22–2/0	Silver	Silicone rubber	Dacron braid	600	−55 to 200	1.25	1.32
	MS-27110	22–4	Silver	Silicone rubber	FEP	600	−55 to 200	1.25	1.61
MIL-W-22759	MS-17411	24–4	Silver	Reinforced TFE	—	600	−55 to 200	1.25	1.82
	MS-17412	24–4	Nickel	Reinforced TFE	—	600	−55 to 260	1.25	1.82
	/7	24–4	Silver	Reinforced TFE	—	600	−55 to 200	1.04	1.36
	/8	24–4	Nickel	Reinforced TFE	—	600	−55 to 260	1.04	1.36
	/2	22–2/0	Silver	TFE and glass tape	Glass braid FEP	600	−55 to 200	1.07	1.3
MIL-W-81044	/1	24–4	Silver	Cross-linked polyalkene	Cross-linked Kynar	600	−55 to 135	1.0	0.98
	/2	24–4	Tin	Cross-linked polyalkene	Cross-linked Kynar	600	−55 to 135	1.0	0.98
	/5	24–0	Silver	Cross-linked polyalkene	Cross-linked Kynar	600	−55 to 150	1.0	0.925
	/6	24–0	Tin	Cross-linked polyalkene	Cross-linked Kynar	600	−55 to 150	1.0	0.925
	/7	26–20	Silver-high-strength alloy	Cross-linked polyalkene	Cross-linked Kynar	600	−55 to 150	1.0	0.925
	/8	24–0	Silver	Cross-linked polyalkene	Cross-linked Kynar	600	−55 to 150	0.90	
	/9	24–0	Tin	Cross-linked polyalkene	Cross-linked Kynar	600	−55 to 150	0.90	
	/10	26–20	Silver-high-strength alloy	Cross-linked polyalkene	Cross-linked Kynar	600	−55 to 150	0.90	
MIL-W-25038	/1	22–4/0	Nickel clad	TFE and glass tape	Glass braid	600	−55 to 288	1.62	2.26
MIL-W-22759	Class 1	22–4/0	Silver	TFE and glass	Glass braid	600	−55 to 200	1.25	1.82
	Class 2	22–4/0	Nickel	TFE and glass	Glass braid	600	−55 to 260	1.25	1.82
MIL-W-81381	/3	26–2	Silver	Polyimide FEP film	FEP dispersion	600	−55 to 200	0.84	0.89
	/4	26–2	Nickel	Polyimide FEP film	FEP dispersion	600	−55 to 260	0.84	0.89

TABLE 6.11 Interconnection Wire Data, Copper Conductors (*Continued*)

Basic specification	Class type or MS no.	Duty rating*	Cost rating	Availability*	Mechanical properties	Electrical properties	Chemical resistance	Processing characteristics			
								Solder	Bond-ability	Strip-pability	Mark-ing
MIL-W-5086	MS-25190 Ty 1	M	1.0	RA	Fair	Fair	Good	Poor	Good	Good	Good
	MS-25190 TY 2	M	1.5	RA	Fair	Fair	Good	Poor	Good	Fair	Good
	MS-25190 Ty 3	H	1.7	RA	Good	Fair	Good	Poor	Good	Fair	Good
	MS-25190 Ty 4	M	1.3	RA	Fair	Fair	Good	Poor	Good	Good	Good
MIL-W-8777	MS-25471	H	9.6	LS	Good	Good	Fair	Fair	Good	Fair	Poor
	MS-27110	H	9.1	LS	Good	Good	Good	Fair	Poor	Good	Good
MIL-W-22759	MS-17411	H	10.0	RA	Excellent	Good	Excellent	Good	Poor	Good	Poor
	MS-17412	H	11.0	RA	Excellent	Good	Excellent	Good	Poor	Good	Poor
	/7	M	7.8	RA	Good	Good	Excellent	Good	Poor	Good	Poor
	/8	M	8.6	RA	Good	Good	Excellent	Good	Poor	Good	Poor
	/2	M	9.3	LS	Good	Good	Good	Good	Poor	Fair	Good
MIL-W-81044	/1	M	5.6	LS	Good	Good	Good	Fair	Good	Good	Good
	/2	M	4.8	LS	Good	Good	Good	Fair	Good	Good	Good
	/5	M	5.6	LS	Good	Good	Good	Fair	Good	Good	Good
	/6	M	4.8	LS	Good	Good	Good	Fair	Good	Good	Good
	/7	M	7.0	LS	Good	Good	Good	Fair	Good	Good	Good
	/8	L	5.4	LS	Good	Good	Good	Fair	Good	Good	Good
	/9	L	4.6	LS	Good	Good	Good	Fair	Good	Good	Good
	/10	L	6.8	LS	Good	Good	Good	Fair	Good	Good	Good
MIL-W-25038	/1	H	15.0	LS	Good	Fair	Good	Good	Poor	Fair	Poor
MIL-W-22759	Class 1	H	12.2	RA	Excellent	Good	Good	Good	Fair	Fair	Fair
	Class 2	H	12.3	RA	Excellent	Good	Good	Good	Fair	Fair	Fair
MIL-W-81381	/3	M	6.0	RA	Good	Good	Excellent	Good	Fair	Fair	Fair
	/4	M	6.7	RA	Good	Good	Excellent	Good	Fair	Fa ir	Fair

*L—light; M—medium; H—heavy; RA—readily available; LS—limited sources (fewer than four manufacturers).

Source: Martin Marietta Corporation.

TABLE 6.12 Hookup Wire Data, Copper Conductors

Basic specification	Class type or MS no.	Size range, AWG	Conductor coating	Primary insulation	Jacket material	Voltage rating, V rms	Temperature rating, °C	Diameter rating	Weight rating	Duty rating*
MIL-W-16878/1	Type B	32–14	Tin	PVC	—	600	−55 to 105	1.0	1.0	M
MIL-W-16878/1	Type B/N	32–14	Tin	PVC	Nylon	600	−55 to 105	1.15	1.15	H
MIL-W-16878/2	Type C	26–12	Tin	PVC	—	1000	−55 to 105	1.28	1.22	M
MIL-W-16878/3	Type D	24–1/0	Tin	PVC	—	3000	−55 to 105	1.81	1.88	M
MIL-W-16878/4A	Type E	32–10	Silver or nickel	TFE	—	600	−55 to 200 or 260	1.02	1.22	M
MIL-W-16878/5A	Type EE	32–8	Silver or nickel	TFE	—	1000	−55 to 200 or 260	1.21	1.48	M
MIL-W-16878/6A	Type ET	32–20	Silver or nickel	TFE	—	250	−55 to 200 or 260	0.86	0.97	L
MIL-W-16878/7	Type F	24–4/0	Tin, silver, or nickel	Silicone rubber	—	600	−55 to 200	1.21	1.1	M
MIL-W-16878/8	Type FF	24–4/0	Tin, silver, or nickel	Silicone rubber	—	1000	−55 to 200	1.83	1.75	M
MIL-W-16878/10A	Type J	24–4/0	Tin	Polyethylene	—	600	−55 to 75	1.13	0.98	M
MIL-W-16878/11	Type K	32–10	Silver	FEP	—	600	−55 to 200	1.02	1.22	M
MIL-W-16878/12	Type KK	32–8	Silver	FEP	—	1000	−55 to 200	1.21	1.48	M
MIL-W-16878/13	Type KT	32–20	Silver	FEP	—	250	−55 to 200	0.86	0.97	L
MIL-W-22759	MS-21985	28–12	Silver	TFE	—	600	−55 to 200	0.98	1.24	M
MIL-W-22759	MS-21986	28–12	Nickel	TFE	—	600	−55 to 260	0.98	1.24	M
MIL-W-22759	MS-18113	28–8	Silver	TFE	—	1000	−55 to 200	1.17	1.48	M
MIL-W-22759	MS-18114	28–8	Nickel	TFE	—	1000	−55 to 260	1.17	1.48	M
MIL-W-81381	/3	30–12	Silver	Cross-linked poly-alkene	Cross-linked Kynar	600	−55 to 135	0.92	0.99	H
MIL-W-81381	/4	30–12	Tin	Cross-linked poly-alkene	Cross-linked Kynar	600	−55 to 135	0.92	0.99	H
	/11	30–12	Silver	Cross-linked pol-alkene	Cross-linked Kynar	600	−55 to 150	0.92	0.99	H
	/12	30–12	Tin	Cross-linked poly-alkene	Cross-linked Kynar	600	−55 to 150	0.92	0.99	H
	/13	30–20	Silver	Cross-linked poly-alkene	Cross-linked Kynar	600	−55 to 150	0.92	0.99	H
MIL-W-81381	/1	26–10	Silver	Polyimide/FEP film	FEP dispersion	600	−55 to 200	0.95	1.0	H
MIL-W-81381	/2	26–10	Nickel	Polyimide/FEP film	TFE dispersion	600	−55 to 260	0.95	1.0	H

TABLE 6.12 Hookup Wire Data, Copper Conductors (*Continued*)

Basic specifications	Class type of MS no.	Cost rating	Availability[†]	Mechanical properties	Electrical properties	Chemical properties	Solder iron resistance	Processing characteristics		
								Solderability	Strippability	Marking
MIL-W-16878/1	Type B	1.0	RA	Poor	Fair	Fair	Poor	Good	Good	Good
MIL-W-16878/1	Type B/N	1.2	RA	Good	Fair	Good	Poor	Good	Good	Good
MIL-W-16878/2	Type C	1.1	RA	Fair	Fair	Fair	Poor	Good	Good	Good
MIL-W-16878/3	Type D	1.5	RA	Fair	Fair	Fair	Poor	Good	Good	Good
MIL-W-16878/4A	Type E	6.4	RA	Poor	Excellent	Excellent	Excellent	Poor	Fair	Poor
MIL-W-16878/5A	Type EE	8.8	RA	Fair	Excellent	Excellent	Excellent	Poor	Fair	Poor
MIL-W-16878/6A	Type ET	6.4	RA	Fair	Excellent	Excellent	Excellent	Poor	Fair	Poor
MIL-W-16878/7	Type F	6.4	RA	Fair	Good	Poor	Fair	Fair	Good	Fair
MIL-W-16878/8	Type FF	9.0	LS	Fair	Good	Poor	Fair	Fair	Good	Fair
MIL-W-16878/10A	Type J	1.1	RA	Poor	Excellent	Good	Poor	Poor	Good	Good
MIL-W-16878/11	Type K	5.5	RA	Poor	Excellent	Excellent	Poor	Poor	Good	Fair
MIL-W-16878/12	Type KK	7.8	RA	Poor	Excellent	Excellent	Poor	Poor	Good	Fair
MIL-W-16878/13	Type KT	5.3	RA	Poor	Excellent	Excellent	Poor	Poor	Good	Fair
MIL-W-22759	MS-21985	6.2	RA	Fair	Excellent	Excellent	Excellent	Poor	Fair	Poor
MIL-W-22759	MS-21986	6.2	RA	Fair	Excellent	Excellent	Excellent	Poor	Fair	Poor
MIL-W-22759	MS-18113	8.9	RA	Fair	Excellent	Excellent	Excellent	Poor	Fair	Poor
MIL-W-22759	MS-18114	9.2	RA	Fair	Excellent	Excellent	Excellent	Poor	Fair	Poor
MIL-W-81381	/3	5.1	LS	Good	Good	Good	Fair	Good	Good	Good
MIL-W-81381	/4	4.2	LS	Good	Good	Good	Fair	Good	Good	Good
	/11	5.1	LS	Good	Good	Good	Fair	Good	Good	Good
	/12	4.2	LS	Good	Good	Good	Fair	Good	Good	Good
	/13	5.6	LS	Good	Good	Good	Fair	Good	Good	Good
MIL-W-81381	/1	8.3	RA	Good[†]	Good	Excellent	Good	Fair	Fair	Fair
MIL-W-81381	/2	9.0	RA	Good[†]	Good	Excellent	Good	Fair	Fair	Fair

*Light—light; M—medium; H—heavy.

†RA—readily available; LS—limited sources (fewer than four manufacturers).

Source: Martin Marietta Corporation.

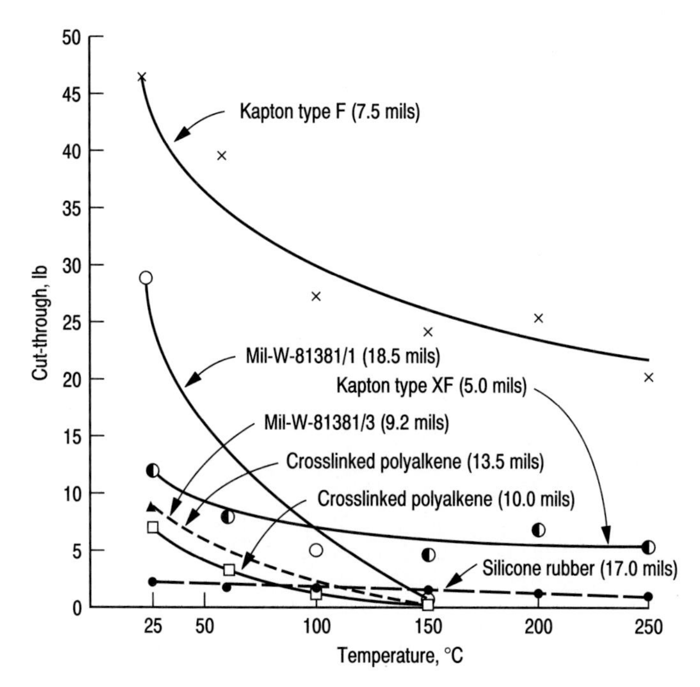

FIGURE 6.11 Cut-through resistance (dynamic).[1,11] Radius 0.005 in; edge 90° bevel; penetration rate 0.05 in/min.

Thin-Wall Insulations. Certain applications may require the use of ultrathin-walled insulation. Kapton type F (FN) film, used in the construction of MIL-W-81381/1–4 wire, is composed of 1-mil polyimide film coated on one or both sides with 0.5-mil Teflon FEP. The thin-walled wire construction using this Kapton film results in a nominal insulation thickness of 7.5 mil. The use of Kapton type XF (FN) film further reduces the thickness to 5.0 mil. Figure 6.11 compares the cut-through resistance of these insulation systems.

Another Kapton thin-film variation has become available: Kapton 120FN616 utilizing 0.1 mil of Teflon FEP coating on both sides of 1-mil polyimide film, used in a 5-mil nominal wall construction. This configuration offers greater cut-through and abrasion resistance than type XF (FN) film because of the increased thickness of the stronger polyimide film.

Kapton 120FN616 film also offers the potential of an even thinner wall construction (3-mil nominal wall thickness) with the single-layer wrap construction. Use of the Teflon FEP coating on both sides of the polyimide film makes a single-layer construction with a good seal between overlaps feasible. Thus the potential disadvantage of poor strippability caused by placing the Teflon FEP next to the conductor is greatly reduced because of the small amount of FEP used for sealing.

Automated Termination. Table 6.13 shows the properties and characteristics of a specialized hookup wire used with automated termination techniques, such as wire

TABLE 6.13 Automated-Termination Wire Data (no. 30 AWG Conductor, Nominal 0.005-in Wall Thickness)

Insulation material	Teflon TFE	TFE/ML	FEP/ML	Vinylidene fluoride	Polyethylene-coated Mylar	Polysulfone	Polyalkene + Kynar	Kapton
Conductor coating	Silver or nickel	Silver or nickel	Silver	Tin	Tin	Tin	Tin or silver	Silver or nickel
Temperature rating, °C	200 or 260	200 or 260	200	135	125	125	135	200 or 260
Cut-through resistance	Poor	Fair	Fair	Good	Good	Excellent	Good	Excellent
Abrasion resistance	Poor	Fair	Fair	Good	Good	Good	Good	Excellent
Dielectric constant	2.1	2.1	2.1	7.7	2.8	3.2	3.4	3.2
Dielectric strength	Good	Good	Good	Good	Good	Good	Good	Good
Flexibility (stiffness)	Good	Good	Good	Fair	Fair	Fair	Fair	Fair
Chemical resistance	Excellent	Excellent	Excellent	Fair	Good	Poor	Fair	Excellent
Cost	Medium	Medium to high	Medium	Low	Low	Low	Low to medium	High
Availability*	RA	LS	LS	RA	LS	LS	LS	DR
Long lengths	Poor	Poor	Fair	Good	Fair	Good	Good	Fair

*RA—readily available; LS—limited sources (fewer than three manufacturers); DR—development required.

Source: Martin Marietta Corporation.

wrap and Termi-Point* (Fig. 6.19). These termination techniques require a special set of criteria. The major conductor size is AWG 24 with a nominal 10-mil wall of insulation. AWG 30 conductor with nominal 5-mil wall of insulation is gaining increasing acceptance.

There are several important considerations for automated termination wire:

1. *Stiffness.* It is undesirable to have wire that "takes a set" and "pops up" on wiring panels.
2. *Cut-through.* To achieve a satisfactory wire wrap, termination pins have very sharp edges.
3. *Long lengths.* Long uninterrupted wire lengths are desirable for increased efficiency of operation.
4. *Strippability.* For machine stripping of insulated wire, wall thickness and concentricity must be controlled to close tolerances.

6.7.4 Outer-Space Applications

Light hookup wire configurations lend themselves to outer-space usage because of the prime importance of weight which directly affects the useful load that can be carried by the space vehicle. The requirements for an outer-space wire are directly related to the performance of the insulation, since insulation is the component most susceptible to the environmental extremes of space. Proper selection of insulated wire requires that the environment be well defined for a specific vehicle, mission, and trajectory.

Environment. The environment discussed here will be limited to those factors peculiar to spaceflight. The ability of an insulated wire to meet previously discussed criteria such as abrasion, rough handling, vibration, and shock are considerations common to all applications and not limited to space usage. The principal space environmental conditions that may have damaging effects in wire insulations are temperature, pressure, and radiation. Figure 6.12 gives a composite picture of the space environment. The extent of environmental extremes to which wiring may be subjected will depend on the specific design of the vehicle and the location of the wire within the vehicle.

Design Considerations

Temperature Environment. If insulated wire is to be used in nonenvironmental controlled areas, Teflon TFE, Teflon FEP, and Kapton insulations offer the broadest range of temperature resistance. These insulations will withstand 180° bending at −184°C without insulation fracture, and maximum continuous temperatures of 260°C for Teflon TFE and 200°C for Kapton and Teflon FEP. Cross-linked polyolefin insulation can be utilized at 184°C; however, it will not take such severe bending and its temperature rating is only 135°C.

Pressure Environment. The effect of the extremely high vacuum of space on wire insulations manifests itself primarily as an initial weight loss due to loss of water and absorbed gases. Thereafter a weight-loss rate that does not rapidly approach zero is a serious indication of possible long-term continuous weight loss and slow volatilization

*Termi-Point (a wire-post termination using metal clips) is a registered trademark of Amp, Inc.

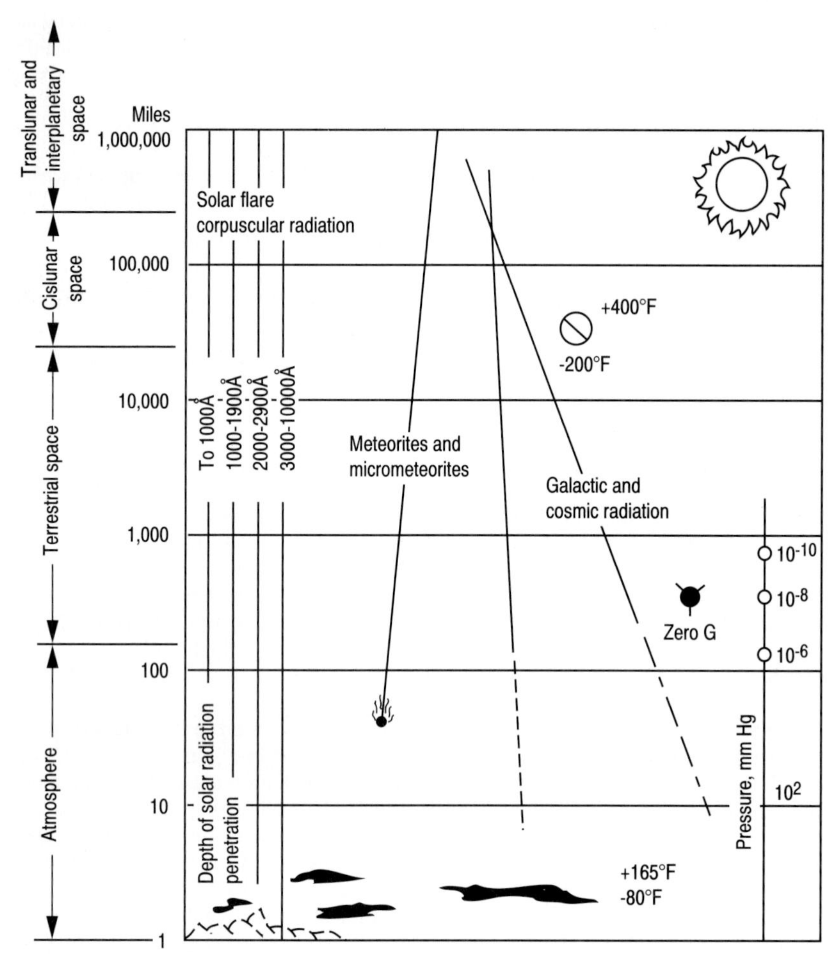

FIGURE 6.12 Space environment.[1,12]

of organic materials. This may eventually degrade both the physical and the electrical properties of the wire insulation or seriously affect other equipment in the vehicle. The secondary effect of insulation volatilization of condensable materials is extremely serious for any vehicle with optical systems that may be fogged by these condensables in the spacecraft atmosphere. Table 6.14 presents a comparison of percent weight loss for various insulation materials in ultrahigh vacuum at varying temperatures and times. For comparative purposes, this table has been compiled from the results published in Jolley and Reed[13] and Lanza and Halperin,[14] and additional data on Kapton were obtained from du Pont. Novathene is a specially formulated radiation cross-linked polyalkene system designed for space applications. It exhibits somewhat higher weight loss than Teflon or Kapton.

Radiation Environment. All space vehicles are subjected to some form or radiation: solar, Van Allen belt radiation, and radiation from a possible nuclear power source. The effects of solar radiation seem to be negligible within the confines of a

TABLE 6.14 A Comparison of Gross Weight Loss in Ultrahigh Vacuum (at 10^{-6} mm Hg)

Sample	Temperature, °F	Time in vacuum, h	Weight loss, %*
Irradiated modified polyolefin	77	240	0.053
Polytetrafluoroethylene	77	240	0.006
Novathene	78	96	0.09
Irradiated modified polyolefin	122	138	0.30
Polytetrafluoroethylene	212	100	0.012
Novathene	212	200	0.66
Novathene	232	42	0.78
Polytetrafluoroethylene	250	120	0.018
Novathene	250	200	0.99
Irradiated modified polyolefin	300	240	1.30
Kapton	392	1/2	0.036
Kapton	392	30	0.036

*All data based upon insulation weight only.

Source: From Schuh.[1]

spacecraft, and degradation of insulation material is a secondary thermal effect from absorption of electromagnetic energy.

The radiation resistance of materials varies widely. Figure 6.13 presents a generally broad spectrum of the effects of radiation on humans, electronic components, organics, and inorganics. Of all the organic materials used as hookup and interconnection wire insulation, Teflon TFE shows the lowest resistance to radiation. The threshold dose of Teflon TFE resin in air is approximately 7×10^4 rad. At 1×10^6 rad the tensile strength is about 50 percent of original and the elongation less than 5 percent of original. In the absence of oxygen, Teflon TFE shows less degradation of tensile strength and elongation.

Teflon FEP reacts somewhat differently than Teflon TFE. Cross-linking can occur, and when it is irradiated in the absence of oxygen with doses greater than 2.6×10^6 rad, some initial improvement of the physical properties is noted, namely, yield stress and deformation resistance. In terms of elongation, which is an important characteristic for a flexible wire insulation, the radiation tolerance of Teflon FEP is approximately 10 times that of Teflon TFE.[16]

The effects of radiation on the electrical properties of Teflon TFE and Teflon FEP can be summarized as follows:

1. *Volume resistivity.* After the threshold dose rate is reached, resistivity decreases rapidly until a dose of 5.5×10^5 rad, where equilibrium sets in. The equilibrium value is greater than 1×10^{12} $\Omega \cdot$ cm for a 20-mil specimen.[16]

2. *Dielectric strength.* The dielectric strength of 3-, 5-, and 11-mil specimens showed no change at a dose of 5.7×10^7 rad.[16]

3. *Dielectric constant and dissipation factor.* The dielectric constant and the dissipation factor of FEP are unaffected when irradiated to a dose of 8×10^6 rad by x rays in the absence of oxygen, at measured frequencies of 100 Hz to 100 kHz.[13] The foregoing data indicate that Teflon TFE and Teflon FEP insulation may be used satisfactorily in limited-radiation environments and that the limiting level of usage may be

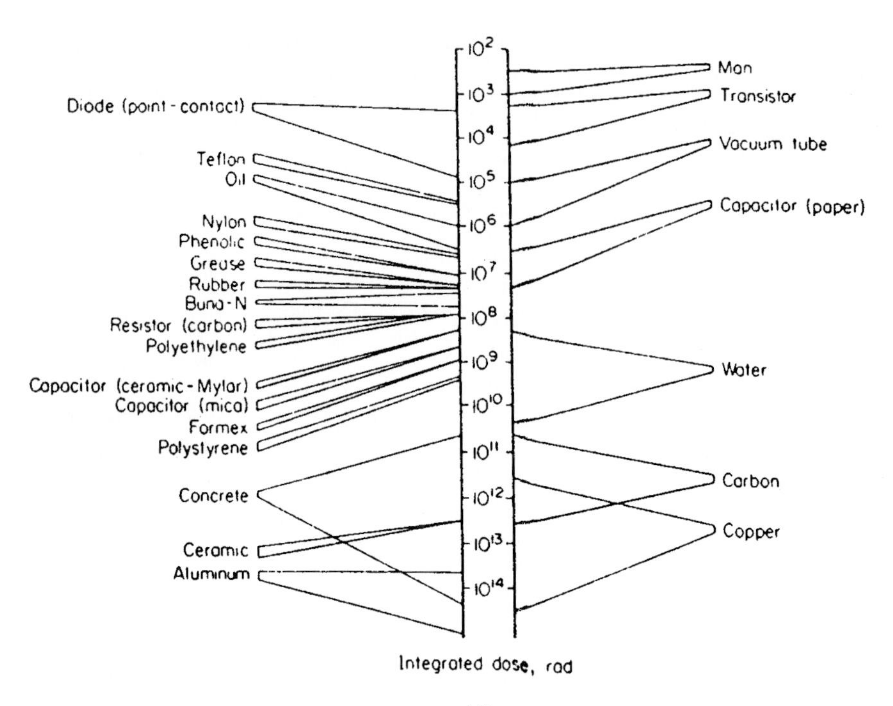

FIGURE 6.13 Functional radiation-dose thresholds.[1,15]

increased if the material is kept in an oxygen-free environment, as far as these electrical properties are concerned.

6.8 COAXIAL CABLES

6.8.1 Design Considerations

Coaxial cable consists of a center conductor, an insulation, a shield, and usually an outer jacket. It is essentially a shielded and jacketed insulated wire. The term coaxial not only implies construction, but also connotes usage at radio frequencies.

Background. The purpose of a coaxial cable is to transmit radio-frequency (RF) energy from one point to another with minimum loss (attenuation). Loss of RF energy in a coaxial cable can occur (1) in the conductor, which is a power loss due to heating caused by currents passing through a finite resistance; and (2) in the dielectric, caused by the use of materials with high power factor (dielectric losses). High-frequency transmission invokes a phenomenon called skin effect, where currents travel on the outer surface (skin) of a conductor and partly through the adjacent insulation material. Hence loss in the insulation itself becomes more significant.

Electrical. In addition to loss, other important electrical characteristics of coaxial cables are velocity of propagation, impedance, capacitance, and corona extinction point. These are now discussed in detail.

Velocity of Propagation. Velocity is an inverse function of the insulation dielectric constant, where $V = 1/K$, and is expressed in percentage of the speed of light.

Impedance. The three common impedance values for coaxial cables are 50, 75, and 90 Ω. Impedance can be determined by the following formula:

$$Z_0 = \frac{138}{\sqrt{K}} \log_{10}\left(\frac{D}{d}\right)$$

where Z_0 = characteristic impedance, Ω
D = diameter over insulation, in
d = diameter over conductor, in
K = dielectric constant of insulation

Capacitance. It is usually desirable to have minimum capacitance for minimum coupling and crosstalk. Capacitance, like impedance, is a logarithmic function of dimensions and is also dependent on the dielectric constant. The equation for calculating capacitance is

$$C = \frac{7.36K}{\log_{10}(D/d)}$$

where C = capacitance, pF/ft
D = diameter over insulation, in
d = diameter over conductor, in
K = dielectric constant of insulation

Corona Extinction Point. This determines the maximum voltage at which a coaxial cable may be operated. The corona extinction point of a cable is determined experimentally by gradually raising the voltage on a sample of cable until corona is detected (Fig. 6.14) and then lowering the voltage until no further ionization is present. If the cable is operated consistently below this level, corona will not occur within the cable. Corona can cause noise at higher frequencies and eventual degradation of organic insulating materials.

Mechanical. The dependence of significant electrical properties, such as attenuation, capacitance, and impedance, on the relative sizes of conductor and insulation has ramifications with respect to the mechanical strength of coaxial cables. Attenuation can be reduced by increasing conductor size, which in turn forces an increase in insulation wall thickness if capacitance and impedance are to be maintained.

An additional means of maintaining low capacitance with increased conductor size is by foamed or air dielectrics. The introduction of air into a solid insulation material such as polyethylene or Teflon FEP can reduce the dielectric constant to as low as 1.4.

Important electrical properties are dependent on the dimensions of conductor and insulation. Any flow or movement of the conductor can affect the electrical properties seriously. For installation, the minimum allowable bend radius should be at least 10 times the cable diameter in order to minimize stresses and preclude any cable deformation.

Environmental. Coaxial cables are normally fabricated with low-loss low-dielectric-constant insulation materials, as covered in Table 6.5. Teflon TFE, Teflon FEP, PE, and irradiated PE, including foamed versions, are the most common dielectric materials. For coaxial cable application, one additional requirement is imposed on shield or jacket materials: they must be noncontaminating. Shield jackets must withstand the

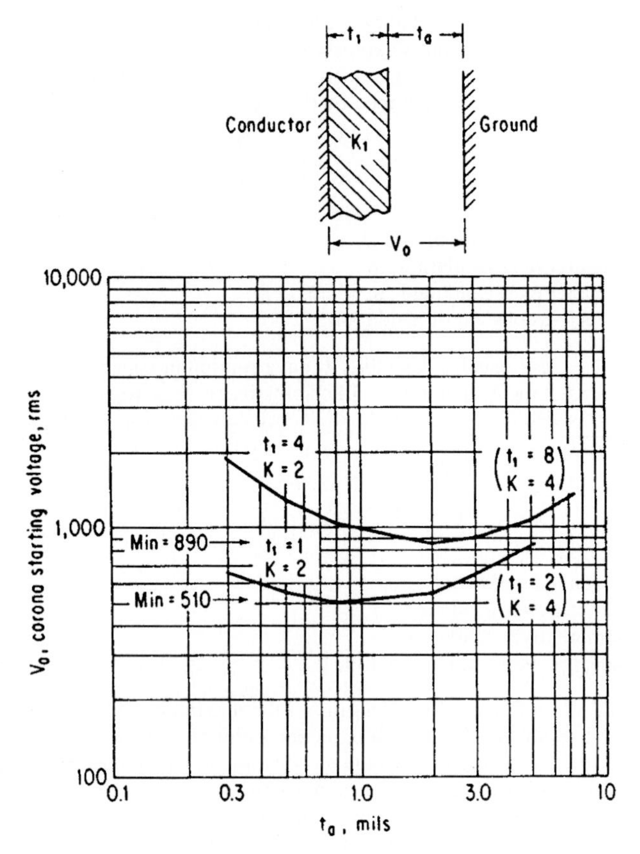

FIGURE 6.14 Corona starting voltage at sea level and 25°C.[1,17,18]

required environment without allowing any contamination of the dielectric which might affect its loss characteristics. Contamination is usually associated with PVC compounds, which contain plasticizers that can migrate into the dielectric material. In coaxial cable applications, moisture resistance of the cable jacket is important. If water, which has a relatively high dielectric constant, penetrates the core, cable performance can be affected seriously.

6.8.2 Cable Selection

A guide to military coaxial cable selection is MIL-HDBK-216. Specific cable types are documented in MIL-C-17, which covers requirements for approximately 150 different cable configurations. Power ratings of MIL-C-17 cable types are covered in Table 6.15. Table 6.16 presents nominal attenuation figures for MIL-C-17 cables at specific frequencies ranging from 1 to 10,000 MHz. MIL C-23806 and MIL-C-22931 are specifications covering semiflexible cables with foamed dielectric and air-spaced dielectric.

TABLE 6.15 Maximum Input Power Ratings* of Coaxial Cable at Different Frequencies, Watts

	Frequency, MHz									
RG/U cable	1.0	10	50	100	200	400	1000	3000	5000	10,000
5, 5A, 5B, 6, 6A, 212	4,000	1,500	800	550	360	250	150	65	50	25
7	4,100	1,550	810	540	370	250	140	70	50	30
8, 8A, 10, 10A, 213, 215	11,000	3,500	1,500	975	685	450	230	115	70	
9, 9A, 9B, 214	9,000	2,700	1,120	780	550	360	200	100	65	40
11, 11A, 12, 12A, 13, 13A, 216	8,000	2,500	1,000	690	490	340	200	100	60	
14, 14A, 74, 74A, 217, 224	20,000	6,000	2,400	1,600	1,000	680	380	170	110	40
17, 17A, 18, 18A, 177, 218, 219	50,000	14,000	5,400	3,600	2,300	1,400	780	360	230	
19, 19A, 20, 20A, 220, 221	110,000	28,000	10,500	6,800	4,200	2,600	1,300	620	410	
21, 21A, 222	1,000	340	160	115	83	60	35	15		
22, 22B, 111, 111A	7,000	1,700	650	430	290	190	110	50		
29	3,500	1,150	510	340	230	150	95	50	35	
34, 34A. 34B	19,000	7,200	2,700	1,650	1,100	700	390	140	8 0	
35, 35A, 35B, 164	40,000	13,500	5,500	3,800	2,500	1,650	925	370	210	
54, 54A	4,400	1,580	675	450	310	210	120	60	40	
55, 55A, 55B, 223	5,600	1,700	2,700	480	320	215	120	60	40	
57, 57A, 130, 131	10,000	3,000	1,250	830	570	370	205	95		
58, 58B	3,500	1,000	450	300	200	135	80	40	20	
58A, 58C	3,200	1,000	425	290	190	105	60	25	20	
59, 59A, 59B	3,900	1,200	540	380	270	185	110	50	30	
62, 62A, 71, 71A, 71B	4,500	1,400	630	440	320	230	140	65	40	15
62B	3,800	1,350	600	410	285	195	110	50	31	15
63, 63B, 79, 79B	8,200	3,000	1,300	1,000	685	455	270	130	75	35
87A, 116, 165, 166, 226, 227	42,000	15,000	6,250	4,300	3,000	2,050	1,200	620	480	250
94	62,000	15,500	5,900	4,300	2,900	1,900	1,400	650	480	200
94A, 226	64,000	18,000	9,600	6,800	4,600	3,300	1,750	775	540	250

TABLE 6.15 Maximum Input Power Ratings* of Coaxial Cable at Different Frequencies, Watts (*Continued*)

	Frequency, MHz									
RG/U cable	1.0	10	50	100	200	400	1000	3000	5000	10,000
108, 108A	1,300	360	145	100	70	45	30	15	5	
114, 114A	5,300	1,350	475	345	230	150	85	40	25	15
115, 115A, 235	33,000	9,900	4,200	2,900	2,000	1,380	830	600	450	170
117, 118, 211, 228	200,000	66,000	25,000	19,000	12,800	8,500	4,800	2,200	1,400	490
119, 120	100,000	31,000	13,000	9,000	6,100	4,100	2,400	1 ,100	770	250
122	1,000	240	100	65	45	30	15	10	5	
125	8,500	2,300	910	620	435	285	165	75	45	
140, 141, 141A	19,000	6,300	2,700	1,700	1,200	830	450	220	140	65
142, 142A, 142B	19,000	5,700	2,600	1,800	1,300	900	530	265	175	100
143, 143A	26,000	8,700	3,750	2,600	1,800	1,250	750	390	275	160
144	51,000	17,000	7,500	5,400	3,700	2,500	1,400	700	440	20
149, 150	7,100	1,900	740	485	315	200	105	45	25	
161, 174	1000	350	160	110	80	60	35	15	10	
178, 178A, 196	1,300	640	330	240	180	120	75	40		
179, 179A 187	3,000	1,400	750	480	420	320	190	100	73	
180, 180A, 195	4,500	2,000	1,100	800	570	400	240	130	90	50
188, 188A	1,500	770	480	400	325	275	150	80	55	
209	180,000	55,000	22,000	15,000	8,500	6,000	3,400	1,600	1,000	310
281	150,000	47,000	19,000	13,500	8,800	6,000	3,300	1,650	1,150	625

*Power-rating conditions: ambient temperature 104°F; center-conductor temperature 175°F with polyethylene dielectric, 400°F with Teflon dielectric. Altitude: sea level.

Source: Amphenol Corporation.[19]

TABLE 6.16 Attenuation Ratings of Coaxial Cables at Different Frequencies, dB/100 ft

	Frequency, MHz									
RG/U cable	1.0	10	50	100	200	400	1000	3000	5000	10,000
5, 5A, 5B, 6, 6A, 212	0.26	0.83	1.9	2.7	4.1	5.9	9.6	23.0	32.0	56.0
7	0.18	0.64	1.6	2.4	3.5	5.2	9.0	18.0	25.0	43. 0
8, 8A, 10, 10A, 213, 215	0.15	0.55	1.3	1.9	2.7	4.1	8.0	16.0	27.0	>100.0
9, 9A, 9B, 214	0.21	0.66	1.5	2.3	3.3	5.0	8.8	18.0	27.0	45.0
11, 11A, 12, 12A, 13, 13A, 216	0.19	0.66	1.6	2.3	3.3	4.8	7.8	16.5	26.5	>100.0
14, 14A, 74, 74A, 217, 224	0.12	0.41	1.0	1.4	2.0	3.1	5.5	12.4	19.0	50.0
17, 17A, 18, 18A, 177, 218, 219	0.06	0.24	0.62	0.95	1.5	2.4	4.4	9.5	15.3	>100.0
19, 19A, 20, 20A, 220, 221	0.04	0.17	0.45	0.69	1.12	1.85	3.6	7.7	11.5	>100.0
21, 21A, 222	1.5	4.4	9.3	13.0	18.0	26.0	43.0	85.0	>100.0	>100.0
22, 22B, 111, 111A	0.24	0.80	2.0	3.0	4.5	6.8	12.0	25.0	>100.0	>100.0
29	0.32	1.20	2.95	4.4	6.5	9.6	16.2	30.0	44.0	>100.0
34, 34A, 34B	0.08	0.32	0.85	1.4	2.1	3.3	5.8	16.0	28.0	>100.0
35, 35A, 35B, 164	0.06	0.24	0.58	0.85	1.27	1.95	3.5	8.6	15.5	>100.0
54, 54A	0.33	0.92	2.15	3.2	4.7	6.8	13.0	25.0	37.0	>100.0
55, 55A, 55B, 223	0.30	1.2	3.2	4.8	7.0	10.0	16.5	30.5	46.0	>100.0
57, 57A, 130, 131	0.18	0.65	1.6	2.4	3.5	5.4	9.8	21.0	>100 .0	>100.0
58, 58B	0.33	1.25	3.15	4.6	6.9	10.5	17.5	37.5	60.0	>100.0
58A, 58C	0.44	1.4	3.3	4.9	7.4	12.0	24.0	54.0	83.0	>100.0
59, 59A, 59B	0.33	1.1	2.4	3.4	4.9	7.0	12.0	26.5	42.0	>100.0
62, 62A, 71, 71A, 71B	0.25	0.85	1.9	2.7	3.8	5.3	8.7	18.5	30.0	83.0
62B	0.31	0.90	2.0	2.9	4.2	6.2	11.0	24.0	38.0	92.0
63, 63B, 79, 79B	0.19	0.52	1.1	1.5	2.3	3.4	5.8	12.0	20.5	44.0
87A, 116, 165, 166, 225, 227	0.18	0.60	1.4	2.1	3.0	4.5	7.6	15.0	21.5	36.5
94	0.15	0.60	1.6	2.2	3.3	5.0	7.0	16.0	25.0	60 .0
94A, 226	0.15	0.55	1.2	1.7	2.5	3.5	6.6	15.0	23.0	50.0

TABLE 6.16 Attenuation Ratings of Coaxial Cables at Different Frequencies, dB/100 ft (*Continued*)

	Frequency, MHz									
RG/U cable	1.0	10	50	100	200	400	1000	3000	5000	10,000
108, 108A	0.70	2.3	5.2	7.5	11.0	16.0	26.0	54.0	86.0	>100.0
114, 114A	0.95	1.3	2.1	2.9	4.4	6.7	11.6	26.0	40.0	65.0
115, 115A, 235	0.17	0.60	1.4	2.0	2.9	4.2	7.0	13.0	20.0	33.0
117, 118, 211, 228	0.09	0.24	0.6	0.9	1.35	2.0	3.5	7.5	12.0	37.0
119, 120	0.12	0.43	1.0	1.5	2.2	3.3	5.5	12.0	17.5	54.0
122	0.40	1.7	4.5	7.0	11.0	16.5	29.0	57.0	87.0	>100.0
125	0.17	0.50	1.1	1.6	2.3	3.5	6.0	13.5	23.0	>100.0
140, 141, 141A	0.30	0.90	2.1	3.3	4.7	6.9	13.0	26.0	40.0	90.0
142, 142A, 142B	0.34	1.1	2.7	3.9	5.6	8.0	13.5	27.0	39.0	70.0
143, 143A	0.25	0.85	1.9	2.8	4.0	5.8	9.5	18.0	25.5	52.0
144	0.19	0.60	1.3	1.8	2.6	3.9	7.0	14.0	22.0	50.0
149, 150	0.24	0.88	2.3	3.5	5.4	8.5	16.0	38.0	65.0	>100.0
161, 174	2.3	3.9	6.6	8.9	12.0	17.5	30.0	64.0	99.0	>100.0
178, 178A, 196	2.6	5.6	10.5	14.0	19.0	28.0	46.0	85.0	> 100.0	>100.0
179, 179A, 187	3.0	5.3	8.5	10.0	12.5	16.0	24.0	44.0	64.0	>100.0
180, 180A, 195	2.4	3.3	4.6	5.7	7.6	10.8	17.0	35.0	50.0	88.0
188, 188A	3.1	6.0	9.6	11.4	14.2	16.7	31.0	60.0	82.0	>100.0
209	0.06	0.27	0.68	1.0	1.6	2.5	4.4	9.5	15.0	48.0
281	0.09	0.32	0.78	1.1	1.7	2.5	4.5	9.0	13.0	24.0

Source: Amphenol Corporation.[19]

6.9 MULTICONDUCTOR CABLES

Multiconductor cables fall into four general categories: airborne, ground electronics, ground support, and miscellaneous.[20]

6.9.1 Airborne Cables

The primary design considerations for airborne multiconductor cables are size and weight. Airborne cables are often fabricated by a user who selects the appropriate interconnection or hookup wires, lays the insulated wires in a bundle or harness, then laces, spot ties, and applies insulating tubing over the wiring assembly (see Fig 6.23). Several variations of this type of harness construction have been utilized in an effort to reduce size and weight and increase mechanical protection.

Specifications MIL-C-7078 and MIL-C-27500 cover multiconductor cables utilizing interconnection and hookup wires in a round configuration. These specifications include single shielded and up to seven multiconductor cables with or without an overall shield and with or without an overall jacket. All conductors must be of the same gauge; no individually shielded conductors are permitted. Various shield and jacket options are available offering compatibility with the chosen primary wire. In the interest of minimum weight and size, fillers are not used. Table 6.17 presents recommended MIL-C-27500 options, including construction details and a mechanical usage rating.

6.9.2 Ground-Electronics Cables

The term ground-electronics multiconductor cabling encompasses rack and panel interconnection, equipment cabling installed in conduit (as used with fixed computer and data-processing installations), or cabling placed beneath flooring that is not subjected to extreme mechanical abuse or environment.

Military specification coverage for multiconductor cables in this area of usage is limited; MIL-C-7078, MIL-C-27500, and MIL-C-27072 are frequently used.

The most commonly used heavy-duty construction for this category of cabling contains PVC insulated primary conductors jacketed with nylon. These are then cabled and jacketed with a PVC sheath.

6.9.3 Ground-Support Cables

Ground-support multiconductor cables for tactical systems should receive early attention from the designer. These cables are tailored to system needs, require considerable lead time for delivery, and can amount to a considerable system cost if no effort at standardization is made. A major manufacturing cost in the production of this type of multiconductor cable is cabling machine setup. Many cable manufacturers require a minimum order of 500 to 1000 ft of cable for a given configuration. After setup, additional cable footage can be produced more economically. Some ground-support multiconductor cable applications follow:

Permanent Installation. Cables that are buried or placed in conduits, open ducts, troughs, or tunnels are considered permanent cables. These cables are not handled,

TABLE 6.17 Multiconductor Cable Options, Recommended by MIL-C-27500

Basic primary wire specifications	Specification symbol	Size range, AWG	Shielded		Jacketed		Shielded-jacketed		Voltage rating, V rms	Temperature rating, °C
			Shield style	Jacket style	Shield style	Jacket style	Shield style	Jacket style		
MIL-W-5086:										
MS-25190, Ty 1	A	22–12	T	O	U	1 or 3	T	1, 2, or 3	600	−55 to 105
MS-25190, Ty 2	B	22–4/0	T	O	U	3	T	1 or 3	600	−55 to 105
MS-25190, Ty 3	C	22–4/0	T	O	U	3	T	1 or 3	600	−55 to 105
MS-25190, Ty 4	P	22–16	T	O	U	3	T	1 or 3	3000	−55 to 105
MIL-W-22759/1	D	22–4/0	S	O	U	7	S	6 or 7	600	−55 to 200
MIL-W-22759/2	E	22–4/0	N	O	U	7	N	6 or 7	600	−55 to 200
MIL-C-8777:										
MS-25471	H	22–2/0	S	O	U	4	S	4	600	−55 to 150
MS-27110	F	22–4	S	O	U	5	S	5	600	−55 to 200
MIL-W-22759:										
/5	V	24–4	S	O	U	6 or 7	S	6 or 7	600	−55 to 200
MS-17412	W	24–4	N	O	U	6 or 7	N	6 or 7	600	−55 to 260
/7	S	24–4	S	O	U	6 or 7	S	6 or 7	600	−55 to 200
/8	T	24–4	N	O	U	6 or 7	N	6 or 7	600	−55 to 260
MS-18113	LA	28–8	S	O	U	6	S	6	1000	−55 to 200
/10	LB	28–8	N	O	U	6	N	6	1000	−55 to 200
MS-21985	R	28–12	S	O	U	6	S	6	600	−55 to 200
MS-21986	L	28–12	N	O	U	6	N	6	600	−55 to 260
/2	N	22–2/0	S	O	U	6 or 7	S	6 or 7	600	−55 to 200
MS-W-25038/1	J	22–4/0	F	O	U	7	F	7	600	−55 to 750
MIL-W-81044/6	M	24–4	S	O	U	4 or 5	S	4 or 5	600	−55 to 135
	MA	24–4	T	O	U	4 or 5	T	4 or 5	600	−55 to 135
	MB	30–12	S	O	U	4 or 5	S	4 or 5	600	−55 to 135
	MC	30–12	T	O	U	4 or 5	T	4 or 5	600	−55 to 135

TABLE 6.17 Multiconductor Cable Options, Recommended by MIL-C-27500 (*Continued*)

Basic primary wire specifications	Specification symbol	Mechanical duty rating	Conductor	Construction details		
				Primary insulation	Shield	Jacket
MIL-W-5086:						
MS-25190, Ty 1	A	Medium	Tinned copper	PVC/nylon	Tinned copper	(1) PVC
MS-25190, Ty 2	B	Medium; fire resistant	Tinned copper	PVC/glass, nylon	Tinned copper	(2)Extruded nylon
MS-25190, Ty 3	C	Heavy	Tinned copper	PVC/glass PVC/nylon	Tinned copper	(3) Nylon braid
MS-25190, Ty 4	P	Medium	Tinned copper	PVC/nylon	Tinned copper	
MIL-W-22759/1	D	Heavy	Silver-coated copper	Teflon TFE	Silver-coated copper	(6) Taped Teflon TFE
MIL-W-22759/2	E	Heavy	Nickel-coated copper	Tapes and glass braid	Nickel-coated copper	(7) Glass braid
MIL-C-8777:						
MS-25471	H	Heavy	Silver-coated copper	Silicone rubber	Silver-coated copper	(4) Dacron braid*
MS-27110	F	Medium	—	—	—	(5) Extruded Teflon FEP
MIL-W-22759:						
/5	V	Heavy	Silver-copper	Mineral-filled	Silver-copper	(6) Taped Teflon TFE
MS-17412	W	Heavy	Nickel-copper	Teflon TFE	Nickel-copper	(7) Glass braid
/7	S	Medium	Silver-copper	—	Silver-copper	
/8	T	Medium	Nickel-copper	—	Nickel-copper	
MS-18113	LA	Light	Silver-copper	Extruded Teflon TFE	Silver-copper	(6) Taped Teflon TFE
/10	LB	Light	Nickel-copper	—	Nickel-copper	
MS-21985	R	Light	Silver-copper	—	Silver-copper	
MS-21986	L	Light	Nickel-copper	—	Nickel-copper	
/2	N	Medium	Silver-copper	TFE-glass-FEP	Silver-copper	(6) Taped Teflon TFE (7) Glass braid
MIL-W-25038/1	J	Heavy; fire resistant	Nickel-clad copper	TFE tapes and glass braid	Stainless steel	(7) Glass braid
MIL-W-81044/6	M	Medium	Silver-copper	Polyalkene and Kynar (cross-linked)	Silver-copper	(4) Dacron braid*
	MA	Medium	Tinned copper	—	Tinned copper	(5)Extruded Teflon FEP
	MB	Light	Silver-copper	—	Silver-copper	
	MC	Light	Tinned copper	—	Tinned copper	

*Jacket compatible with primary insulation system not available to date. Kynar, polyethylene-coated Mylar, and cross-linked Kynar are proposed additions to specification.

Source: Martin Marietta Corporation.

flexed, reeled, or dereeled except at the time of installation. Either neoprene or polyethylene cable sheaths are preferred.

Portable Installation. MIL-C-13777 is the basis for the design of heavy-duty portable cable. Thin-wall cable insulation is appropriate for applications of a lighter-duty nature, especially where weight is a critical factor.

6.9.4 Flat Flexible Cable

There are two basic types of flat flexible cables, often referred to as tape cable and flexible printed wiring. The major difference between the two types of cable is their manner of construction and general application. Tape cable is used primarily to interconnect individual electronic units; flexible printed wiring is used to provide interconnection within a unit.

Flat-Conductor Cable

Construction. Flat conductor cable is constructed by the encapsulation of flat rectangular conductors between layers of dielectric film. Figure 6.15 depicts a manufacturing technique for laminating flat-conductor cable. Conductor ribbons are positioned parallel to each other with a uniform spacing.

Materials. Conductors are usually rolled copper, annealed in accordance with QQ-C-576. Other materials and protective coatings may be obtained for conductors to meet unusual application requirements. For selecting the proper conductor size to satisfy current and voltage requirements, refer to Table 6.18 and Figs. 6.16 and 6.17. Table 6.19 gives mechanical, electrical, and environmental data for several flat flexible cable insulating materials.

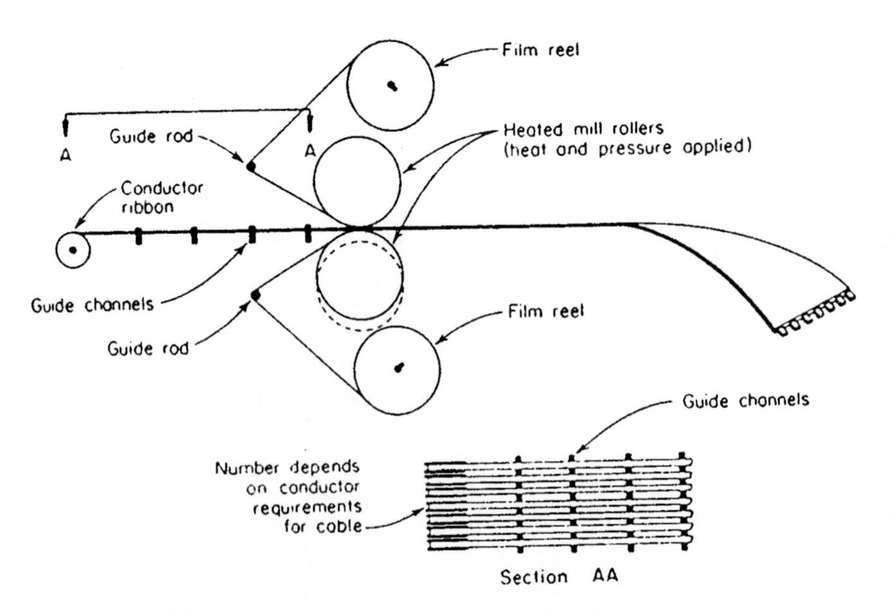

FIGURE 6.15 Flat-conductor laminating process.[1,21]

TABLE 6.18 Copper Conductor Characteristics, Flat Conductor

Flat conductor dimensions		Cross section		Nearest AWG wire size based on equivalent:		Resistance at 20°C, mΩ/ft	Current for 30°C rise, A
Thickness, in	Width, in	mil²	cmil	Cross section	Current rating		
0.0027	0.030	81	102	30	28	100	3.4
	0.045	122	154	28	27	67	3.8
	0.060	162	204	27	25	50	5.1
	0.075	202	254	26	24	40	5.8
	0.090	243	306	25	23	34	6.5
	0.125	338	425	24	22	24	8.2
	0.155	418	527	23	21	19.5	9.2
	0.185	500	630	22	20	16.2	10.7
	0.250	675	850	21	18	12	13.5
0.004	0.030	120	151	28	27	67	4.0
	0.045	180	227	26	25	45	5.2
	0.060	240	302	25	24	34	6.0
	0.075	300	378	24	23	27	7.0
	0.090	360	454	23	22	22.5	7.8
	0.125	500	630	22	20	16.2	10.0
	0.155	620	780	21	19	13	11.8
	0.185	740	930	20	18	11	13.5
	0.250	1000	1260	19	17	8	17.0
0.0055	0.045	248	312	25	24	33	6.0
	0.060	330	415	24	23	25	7.2
	0.075	412	520	23	22	20	8.2
	0.090	495	624	22	21	16.5	9.5
	0.125	687	865	21	20	12	12.2
	0.155	852	1075	20	19	9.5	14.8
	0.185	1020	1285	19	17	8	17
	0.250	1375	1730	18	16	6	21
0.008	0.045	360	454	23	22	23	7.8
	0.060	480	605	22	21	17	9.8
	0.075	600	755	21	20	13.5	11.5
	0.090	720	905	20	19	11.2	13.2
	0.125	1000	1260	19	17	8	17
	0.155	1240	1560	18	16	6.5	20
	0.185	1480	1860	17	14	5.5	23
	0.250	2000	2520	16	13	4.1	26

Source: From *Flexprint Circuit Design Handbook.*[22]

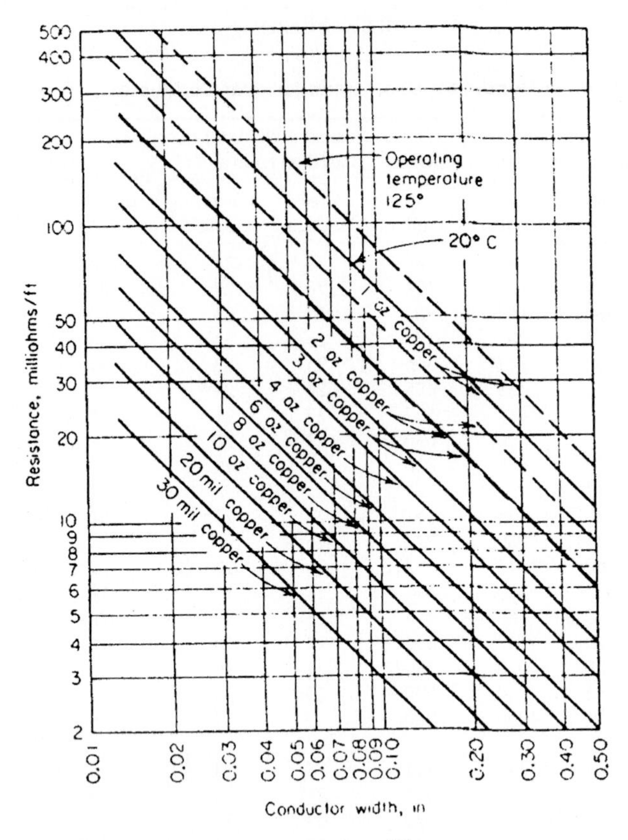

FIGURE 6.16 Etched-conductor resistance.[1,23]

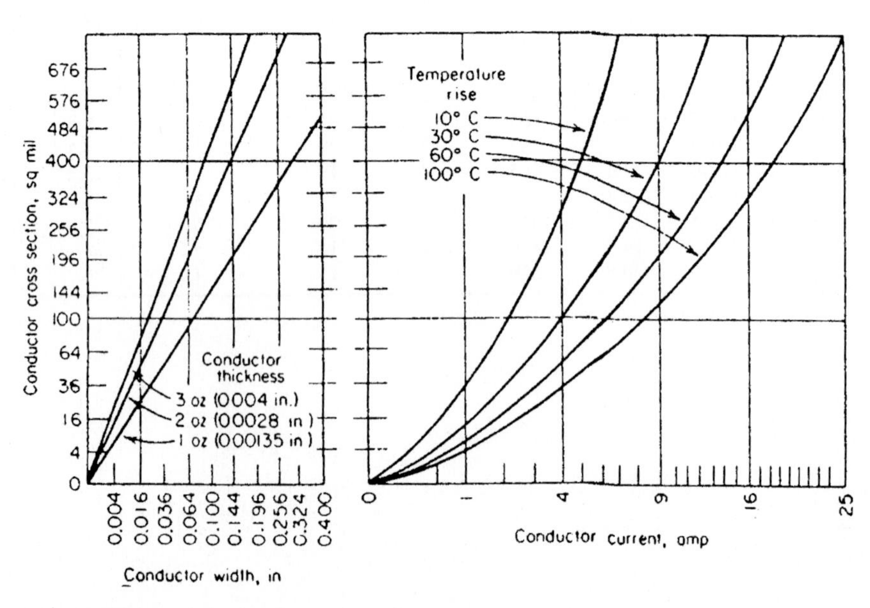

FIGURE 6.17 Etched-conductor current capacity.[1,23]

TABLE 6.19 Flat Flexible Cable Insulating Materials

Property	TFE fluoro-carbon	TFE glass cloth	FEP fluoro-carbon	FEP glass cloth	Poly-imide	Poly-chloro-trifluoro-ethylene	Poly-vinyl fluoride	Poly-propylene	Poly-ester	Poly-vinyl chloride	Poly-ethylene
Specific gravity	2.15	2.2	2.15	2.2	1.42	2.10	1.38	0.905	1 .395	1.25	0.93
Square inches of 1-mil film per lb	12,800	13,000	12,900	13,000	19,450	12,000	20,000	31,000	21,500	22,000	30,100
Service temperature, °C											
Minimum	−70	−70	−225	−70	−250	−70	−70	−55	−60	−40	−20
Maximum	250	250	200	250	+250	150	105	125	150	85	60
Flammability	Nil	Nil	Nil	Nil	Nil	Nil	Yes	Yes	Yes	Slight	Yes
Appearance	Translucent	Tan	Clearbluish	Tan	Amber	Clear	Clear	Clear	Clear	Translucent	Clear
Thermal expansion, in/in, °F × 10^6	70	Low*	50	Low*	11	45	28	61	15		
Bondability with adhesives	Good†	Good†	Good†	Good†	Good†	Good†	Good	Poor	Good	Good	Poor
Bondability to itself	Good	Poor	Good	Good	Poor	Good	Good	Good	Poor	Good	Good
Tensile strength at 77°F, lb/in^2	3000	20,000	3000	20,000	20,000	4500	8000	5700	20,000	3000	2000
Modulus of elasticity, lb/in^2	80,000	3	70,000	3	430,000	200,000	280,000	170,000	550,000	—	50,000
Volume resistivity, Ω • cm	2×10^{16}	1×10^{16}	1×10^{17}	1×10^{16}	1×10^{15}	1×10^{18}	3×10^{13}	1×10^{16}	1×10^{18}	1×10^{10}	1×10^{16}
Dielectric constant at 10^2–10^8 Hz	2.2	2.5/5³	2.1	2.5/5³	3.5	2.5	7.0	2. 0	2.8–3.7	3–4	2.2
Dissipation factor at 10^2–10^8 Hz	0.0002	0.0007/0.001*	0.0002	0.0001/0.001*	0.002/0.014	0.015	0.009–0.041	0.0002/0.0003	0.002–0.016	0.14	0.0006
Dielectric strength at 5-mil thickness, V/mil	800	650/1600	3000	650/1600	3500	2000	2000	750 ‡	3500	800	1500
Chemical resistance	Excellent	Excellent	Excellent	Excellent	Excellent	Excellent	Good	Excellent	Excellent	Good	Excellent
Water absorption, %	0	0.10/68	0	0.18/30	3	0	15	0.01	0.5	0.10	0
Sunlight resistance	Excellent	Excellent	Excellent	Excellent	Excellent	Excellent	Excellent	Low	Fair	Fair	Low

* Depends on percent glass cloth.

† Must be treated.

‡ At 0.125-in thickness.

Source: *Flexprint Circuit Design Handbook.*[22]

Design. The following important criteria should be considered by the designer and used to formulate optimum cable construction, materials selection, terminations, installation, fabrication, and handling:

1. Electrical requirements
  a. Current-carrying capacity
  b. Voltage drop
  c. Overload rating
  d. Impedance
  e. Capacitance
  f. Shielding requirements
  g. Derating due to stacking of cables

2. Mechanical requirements
  a. Conductor width
  b. Conductor thickness
  c. Conductor spacing
  d. Insulation tear resistance
  e. Insulation puncture or cut-through resistance
  f. Insulation and conductor flexure resistance

3. Environmental requirements
  a. Temperature extremes
  b. Flame resistance
  c. Vibration and shock requirements
  d. Temperature cycling
  e. Humidity requirements
  f. Altitude requirements
  g. Materials compatibility
  h. Fungus resistance
  i. Radiation resistance
  j. Aging resistance

4. Termination requirements
  a. Spacing compatible with connector
  b. Type of termination, such as eyelet, rivet, welded, soldered, brazed, or crimped
  c. Termination type compatible with material

5. Fabrication requirements
  a. Insulation strippability technique, such as mechanical abrasion, chemical, thermal, or piercing
  b. Conductor support

6. Insulation requirements
  a. Cable support
  b. Cable routing; avoid interference with high-heat sources, sharp edges, and components

Flexible Printed Wiring

Construction. The majority of flexible printed-wiring cables are manufactured by standard printed-circuit etching techniques.

Materials. As was true for flat-conductor cable, copper is the most widely used printed-wiring conductor material. Two basic types of copper are available: rolled and electrodeposited copper foil.

Design. The design considerations applicable to printed flexible cable are analogs of those previously established for flat-conductor cable. Since the fabrication techniques involved are essentially identical to those used for conventional rigid printed wiring, the design standards apply for conductor spacing, line width, and so on, as specified in MIL-STD-275. Figures 6.16 and 6.17 present design criteria for selecting the optimum conductor size.

Ribbon Cable. Ribbon cable consists of conventional wires (usually stranded, for flexibility) in single or multiple parallel layers encapsulated and bonded within a ribbon of flexible insulation (usually PVC, PE, or Teflon TFE). It offers particularly good geometric and flexibility advantages in many interconnection and hookup applications.

6.9.5 Design Considerations

Design of the cabling installation is an integral part of the mechanical design of any equipment or system requiring electrical interconnection. Design planning for electrical installation should be concurrent with the layout of the mechanical design. Quality installation design requires that each of the following major considerations be thoroughly evaluated and a positive design approach determined.

Environment. Vibration, acceleration, and shock are dynamic environments that are controlling from a design viewpoint. Acceleration places a load on cables, supports, brackets, connectors, and mounting points for black boxes. Wherever possible, these items must be so designed as to be in compression against a structural member. Connectors should be so oriented that the possibility of inadvertent disconnection is minimized. Consideration shall also be given to deceleration forces. Vibration sets up varying stresses in cables and supports, proportional to the mass being supported. Shock, including acceleration and deceleration, also contributes to an overstressed condition.

Other environmental conditions include high dynamic pressure, elevated temperatures, and lowered atmospheric pressure. High temperature has the most severe effect on network installations. Insulating against high temperature and selection of materials capable of withstanding high temperature are the basic approaches for controlling problems caused by temperature extremes. Low atmospheric pressure is insignificant except in outer-space applications. There the effects can be serious: outgasing and deterioration of plastics and insulation with time.

Several different phenomena are generated by a nuclear burst: radiation, heat pulse, shock wave, and electromagnetic pulses (EMPs). The initial radiation may be of such type and intensity over a sufficient time interval that it may dangerously degrade the quality of materials, including metals. The heat pulse and shock wave generate conditions similar to those already described. The EMP can produce voltages and currents through the metallic structure. Thus installation hardware can become electrically energized with electric stresses far higher than those encountered normally.

Ground environmental factors include high and low temperatures, humidity, and dynamic parameters caused by handling and transportation. The levels of these ground environmental factors are usually far less severe than those of flight environment. But despite the lower level, the duration of these stresses is far in excess of normal flight time. The accumulated stress or degradation under these conditions may be quite appreciable.

Routing and Grouping. Interconnecting cables and networks should be designed and installed to minimize the adverse effects of electromagnetic interference and to control crosstalk between circuits. To eliminate the adverse effects, special grouping, separation, and shielding practices should be followed, for which the following general guidelines are recommended:

1. Dc supply lines. Use twisted pair; separate from ac power and control lines.
2. Ac power lines. Use twisted lines; separate from susceptible lines; shield ac circuits in which switching transients occur, and ground the shield at both ends.
3. Low-level signals. Use shielded twisted pair, and ground the shield at one end.
4. High-level signals. Use shielded twisted pair, and ground the shield at both ends.
5. Provide adequate filtering to prevent conducted noise problems.
6. Follow a single-point ground concept where possible. Analyze flow of parasitic chassis currents, and design the ground conductor for worst case.
7. Plan the separation of signal and power circuits with maximum distance between runs.
8. Keep wire and cable length to a minimum.
9. Locate high-heat-generating wires on the outside.
10. Plan the cable routing in coordination with the structural design effort. Plan cable runs and tie down points in the early design phase for incorporation into the structure. Plan for minimum length; attempt to optimize cable installation; compromise only in the solution of installation and maintenance problems.
11. Hold bend radius of coaxial and multiconductor cables to at least 10 times the cable outside diameter.

6.10 MAGNET WIRE

This field of film-insulated wire is enormous and impacts the insulation materials and practices of electronic packaging in many ways. Magnet wire is used in printed-wiring-board jumpers, RF coils, relays, transformers, and inductors, to name a few. Some descriptions of conductors and insulation must be restated in this section in a form applicable to magnet wire.

6.10.1 Conductors

Materials and Construction

Copper. The most common conductor is bare round solid annealed copper wire in accordance with ANSI C7.1. Square and rectangular copper wire is available as described in ANSI C7.9. Copper strip can be obtained in accordance with federal specification QQ-C-576. Rounded edges should be specified to preclude any roughness or sharp projections.

Aluminum. Although copper is the most widely used conductor material, there is an increasing application of aluminum magnet-wire conductors. Unfortunately in most applications, direct substitution of copper is impossible without significant change in design because aluminum conductors have lower conductivity (62 percent IACS) and increased brittleness and cold flow under pressure than copper.

Anodized (chemically surface-oxidized) aluminum conductors present a unique approach to both mechanical protection and specialized electrical insulation. Figure 6.18 shows the dielectric strength of various anodized aluminum surface thicknesses. Aluminum oxide is inorganic and possesses many desirable electrical insulation properties, such as resistance to radiation, to aging at high temperatures (melting point 3600°F), and to chemical attack. The film is very brittle and somewhat porous, but organic sealing treatments for protection against moisture are available. Anodized aluminum is a very specialized material. It cannot be a direct substitute for conventional wire enamel, tape, or served insulation.

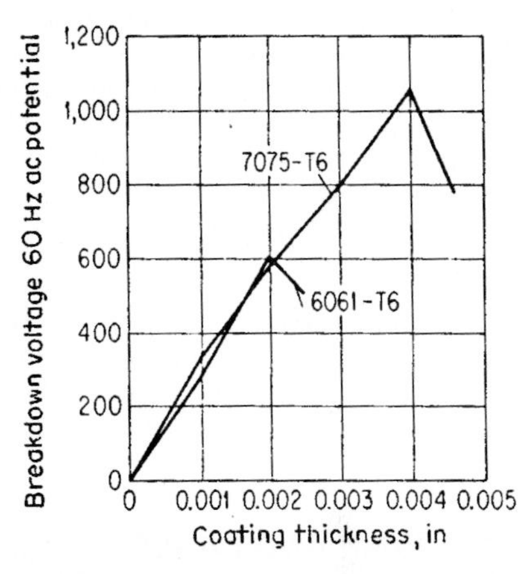

FIGURE 6.18 Dielectric strength of anodized aluminum. (*Courtesy of Martin Marietta Corporation.*)

Conductors for High-Temperature Applications. The usable temperature range of copper conductors (bare and with protective coatings) is evaluated on the basis of oxidation, melting point, grain growth, and solid-state diffusion (Table 6.1).

6.10.2 Insulation Materials

Enameled Film Insulation. A list of magnet-wire insulating enamel materials follows:

Acrylic
Ceramic
Ceramic with overcoat
Epoxy
Oleoresinous (plain enamel, black enamel)
Polyamide (nylon)
Polyamide-polyimide (amide-imide)
Polyester
Polyimide (PYRE-ML*)
Polytetrafluoroethylene (Teflon TFE)
Polyurethane
Polyvinyl formal (Formvar†)

*PYRE-ML, Dacron, and Nomex are registered trademarks of E. I. du Pont de Nemours & Company, Inc.
†Formvar (a polyvinyl formal polymer wire enamel) is a registered trademark of Monsanto Inc.

PYRE-ML is a more recent, advanced, outstanding magnet-wire insulation, the only unvarnished wire enamel which has a 220°C thermal classification. It is more chemically resistant and is compatible with practically all varnishes and encapsulating compounds. As an overcoating, polyimide provides an extremely tough, abrasion-resistant film which exhibits high resistance to nuclear radiation, excellent thermal resistance, and good windability. NEMA MW-16 covers polyimide-coated round wire. Federal specification J-W-1177/15B, class 220, type M is also applicable to polyimide-coated round wire.

Textile and Composite Insulation. Glass fibers, Dacron, nylon, acrylic, cotton, Kraft paper, and Nomex* are common fibrous magnet-wire insulations. Silk and rayon are seldom used today. Asbestos is no longer used, except for very limited and OSHA-specified strictly controlled applications because of its severe health hazards.

6.11 WIRE AND CABLE TERMINATIONS

6.11.1 Terminating Hardware

Wire and cable terminations are of major importance to design reliability. Careful selection of proper terminating hardware and the reduction of the number of terminations to a minimum should be primary design goals.[24]

The following conditions must be evaluated in the selection and use of terminations: (1) termination life, (2) connection density, (3) compatibility, (4) environment, (5) preparations, (6) mass production, (7) process control, (8) inspectability, (9) current, voltage, and resistance limits, (10) maintenance tools, (11) repairability, including time and skill requirements, (12) contractual constraints.

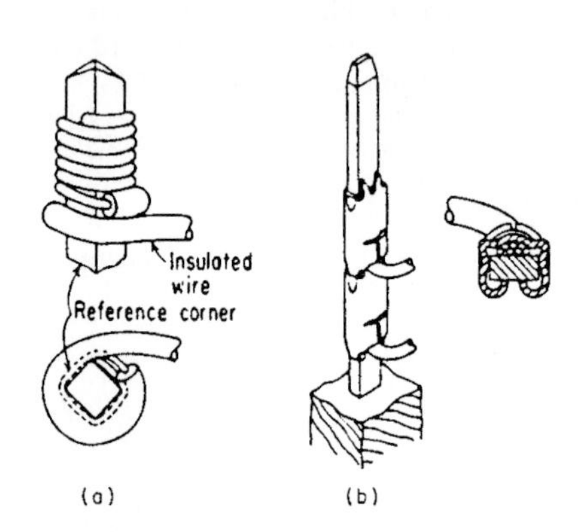

FIGURE 6.19 Wire-post termination systems.[17] (a) Gardner-Denver's wire wrap. (b) Amp's Termi-Point.

A number of wire attachment methods are used in hookup and interconnecting wire terminations such as (1) crimping, (2) soldering, (3) clamping, (4) welding, (5) wire wrapping,[25] and (6) friction (Fig. 6.19).

Terminating devices normally used in electrical installation are (1) studs, (2) lugs (crimp), (3) terminal posts (solder or wire wrap), (4) connectors (with solder or crimp-type contact terminal), (5) splices (crimp or solder), (6) compression screw lugs, (7) screw terminal (usually limited to use on barrier strips), (8) ferrules, (9) taper pins, and (10) pads and eyelets.

*Nomex is a registered trademark of E. I. du Pont de Nemours & Company, Inc.

Terminals

Terminal Lugs. Terminal lugs are designed to establish electrical connection between a wire and a connection point such as a stud.

Terminal Posts. Terminal posts are used on terminal boards in assembly-type wiring and on many components, such as electric connectors (solder type), relays, transformers, lamp holders, and switches.

Design Guides for Terminals. Some rules of terminal and wire termination design follow:

1. *Do*
- ***a.*** Use special prebused connector terminals where required.
- ***b.*** Apply supplementary insulation sleeving over axial terminations where continuous insulation is not provided between adjacent terminations.
- ***c.*** Ensure that electrical spacings between terminals conform to Table 6.20.

2. *Don't*
- ***a.*** Use solder cap adapters to accommodate additional connectors or larger gauges in connector terminals.
- ***b.*** Connect more than three leads to one terminal.
- ***c.*** Twist multiple wires or leads to effect terminations.
- ***d.*** Terminate more than one wire in a connector terminal.

Terminal Boards. These boards are used for junctions or terminations of wire or cable assemblies as an aid to installation and maintenance.

Stud Terminal Board. The stud terminal board is generally a threaded post with the axial portion of its body firmly anchored into a mounting panel. It requires the use of tools for the attachment of wire lugs.

Barrier Terminal Board. The barrier terminal board is molded of thermoset insulating material. It has integral raised barriers between pairs of screw terminals. Its features are:

1. Longer leakage path than the stud type between adjacent terminals
2. More connections in a given length of board
3. Limited current-carrying capacity
4. Poor adaptability to applications with high levels of dynamic stress

Taper Pin Terminal Block. This terminal block is composed of molded insulating material containing metal inserts designed to hold taper pins.

Specifications. Terminal boards should be installed in accordance with MIL-E-7080 for aircraft and MIL-E-25366 for missiles unless other specific requirements are established.

Splices. Permanent splices, available for both shielded and unshielded cables, should be used only when absolutely required. Conductor splices in interconnecting wiring should be grouped and located in designated areas selected for ready access. Where leads from electrical equipment are spliced into a cable assembly, the splice area should be located as near to the equipment as practical. Nonpermanent splices should be avoided; however, certain special applications may require their use.

Shield Wiring Terminations. There are two basic shield terminations: terminated to a shield common and floating.

TABLE 6.20 Allowable Voltage Between Terminals,* Volts

Minimum air space, in	Creepage distance, in	At sea level			At 50,000 ft			At 70,000 ft		
		Flashover, V rms	Working dc	Working ac	Flashover, V rms	Working dc	Working ac	Flashover, V rms	Working dc	Working ac
†	3/64	800	280	200	300	100	75	200	70	50
1/32†	1/16	1400	490	350	500	190	125	375	125	90
3/64	5/64	2000	700	500	700	210	175	500	175	125
1/16	7/64	2500	840	600	900	315	225	600	210	150
5/64	1/8	3000	1050	750	1050	360	260	675	230	165
3/32	5/32	3600	1260	900	1200	420	300	750	260	185
1/8	3/16	4500	1550	1100	1400	490	350	900	310	225
3/16	1/4	6100	2000	1500	1800	630	450	1100	375	275
1/4	5/16	7300	2500	1800	2000	700	500	1300	455	325
5/16	3/8	8500	2900	2100	2300	810	575	1420	500	355

*The allowable voltage is determined by the actual creepage distance or the minimum air space, whichever provides a lower rating. At 70,000 ft visible corona has been recorded by voltages as low as 350 V rms. Consequently, at these elevations corona may be the limiting factor rather than flashover.

†Continuous insulation should be provided between electrical connections of 1/32 in or less.

Source: Martin Marietta Corporation.

Shield termination merits careful consideration from the design phase through production. A judicious grouping of wires and careful examination of the need for shielding will alleviate termination problems. Some generalized design recommendations follow:

1. Minimize the use of shielded wiring.
2. Avoid shielding of leads less than 4 in long.
3. Provide lead segregation instead of shielding where this is practical.
4. Become familiar with all facets of shield termination techniques (see subsequent paragraphs) to ensure that the techniques fully satisfy the design environment.

Some of the preceding is, of course, not applicable to ac, pulsed, or RF leads and cables with significant EMI radiation potential. The following shield terminations are in general use:

1. Direct shield termination (pigtail)
2. Ferrule termination (crimp attachment)
3. Solder sleeve termination (solder attachment)

Direct Shield Termination. An established practice is to form the shield braid into a pigtail, as illustrated in Fig. 6.20. No special tooling is required, but the pigtail must be the braid at the breakout point or the shield braid strands. No external pressure should be applied to the breakout point by clamp, tie, or flexure. Supplementary insulation (sleeving) over the breakout of the braid and the braid itself is required.

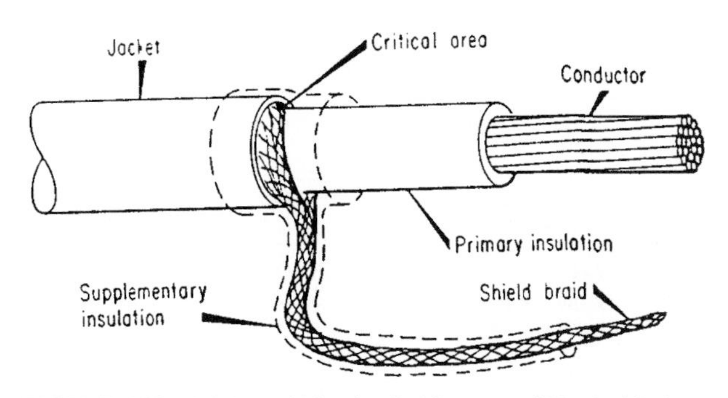

FIGURE 6.20 Shield braid in pigtail. (*Courtesy of Martin Marietta Corporation.*)

Ferrule Termination. Shield termination ferrules are available in two basic types: two-piece preinsulated and single-piece uninsulated. A typical application of a two-piece insulated ferrule is shown in Fig. 6.21.

Solder Sleeve Termination. Before this termination is selected, compatibility between the solder melting temperature, the shrink temperature of the insulating sleeve, and the temperature resistance of the primary wire insulation as well as that of the jacket over the shield must be determined. The grouping of shield conductors for

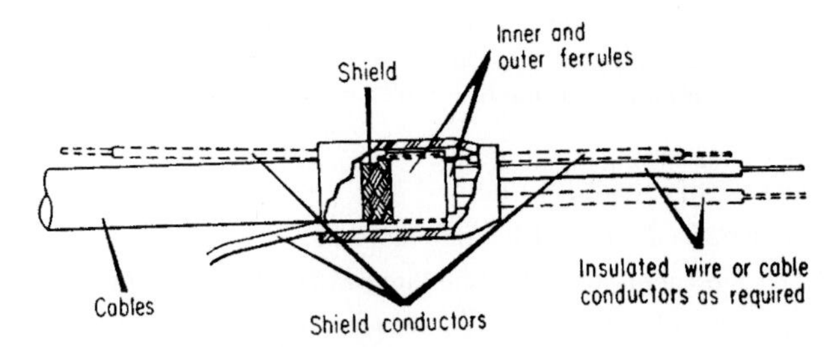

FIGURE 6.21 Insulated ferrule shield termination. (*Courtesy of Martin Marietta Corporation.*)

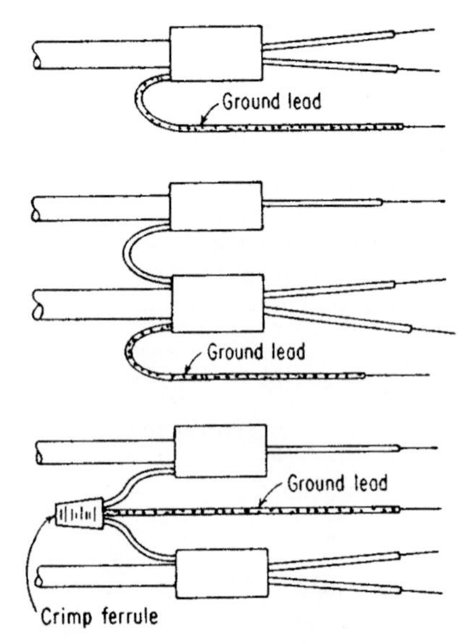

FIGURE 6.22 Typical shield-termination grouping practice. (*Courtesy of Martin Marietta Corporation.*)

solder sleeve applications is shown in Fig. 6.22. Solder sleeves permit the use of center strip terminations to minimize the bulk of shield terminations at connector back shells and to allow continuation of the shields closer to the point of termination for the shielded conductor.

RF Cable (Coaxial) Termination. Terminations for RF cables may be selected from MIL-HDBK-216. Straight RF connectors of the TNC (threaded coupling) type are desirable. The right-angle type of RF connectors and adapters should be avoided because of the inherent mechanical weakness of many designs which use brazed metal housings.

Additional interconnections and terminations are described in Godwin.[17]

6.11.2 Identification

Identification of wiring and cabling includes marking or coding individual wire leads, harness, cables, and termination devices. Wire identification facilitates design control and traceability (such as wiring diagram to hardware), manufacturing efficiency, and maintenance (trouble shooting). Identifying markings on harness and cable assemblies usually provide usage information, interconnection instructions, and part numbers. In addition, the marking may also include serial number, source, assembly date, lot number, and so on. Marking is also useful for inventory control and supply stock records.

Wire Marking. MIL-STD-681 is applicable to various wire marking methods. Color, stripes, bands, and numbers are acceptable. Numbering can be a relatively simple sequential matter, beginning with number 1 and progressing consecutively to the highest number required for an assembly.

Certain military requirements specify a coded marking for individual wires, which includes (1) unit number, (2) equipment identity or circuit function, (3) wire number, (4) wire segment letter, (5) wire gauge, and (6) ground, phase, or thermocouple letter. Hot-impression stamping or color banding of the required wire identification is a practical user production marking method.

Harness Marking. Wire harness assembly marking should be as simple as possible to convey the information required. A simple method is to use a short length of close-fitting insulating sleeving over the harness trunk, adjacent to each termination. Identification can be applied by hot-impression stamping of the thermoplastic sleeve.

Cable Marking. Sheathed cables are identified in a manner similar to wiring harness assemblies. The most significant difference between the two involves the materials used in the actual marking. A cable marking method for production cables uses a reflective label with pressure-sensitive adhesive backing.

6.11.3 Associated Hardware and Materials

Insulation Sleeving and Tubing. Insulation sleeving and tubing serve multiple purposes in electrical assembly and harness fabrication. They are used for insulation, protection from chafing or abrasion, jacketing, strain relief, thermal or chemical protection, and identification. Extruded tubings are made from all the plastic and rubber materials listed in Table 6.3.

Braided Insulating Sleeving. Braided sleeving is made from basic uncoated yarns, lightly treated yarns, or yarns heavily coated with various insulating varnishes or resins. Although practically any yarn can be used for braiding, the material most frequently used in the United States is fiberglass, with some polyethylene terephthalate (Dacron), acrylic, and nylon yarn also available.

Shrinkable Tubing. Shrinkable tubing is based on the elastic memory of plastics. Under specific thermal and mechanical conditions, molecules of certain polymers may be overexpanded and then fixed in place in a strained condition. When heated, the material tends to return to its original shape and size as strain is relieved. Solvents may also be used with some materials to relieve strain to shrink tubing.

Materials. Heat-shrinkable tubing is available in many of the thermoplastics and elastomers listed in Table 6.3. The properties of these heat-shrinkable materials are comparable to those of the same basic conventional materials. Heat-shrinkable tubing is available in the following materials:

Butyl rubber
Fluorinated ethylene propylene (Teflon FEP)
Polychloroprene (Neoprene)
Polyolefin (irradiated) (PE, etc.)
Polytetrafluoroethylene (Teflon TFE)
Polyvinyl chloride (PVC)
Polyvinylidene fluoride (Kynar)
Silicone rubber

Table 6.21 presents typical properties of heat-shrinkable tubings.

Specifications. Military specification MIL-I-23053 covers heat-shrinkable polyvinyl chloride, polyolefin, and Teflon TFE tubing materials.

Wire and Cable Mounting and Spacing Hardware. Good installation design practice requires adequate space not only for wiring and cabling (Fig. 6.23) but also for the supporting hardware (clamps, sleeving, grommets, guides, etc.; Fig. 6.24). Space is also needed for manipulating tools during initial installation as well as during maintenance and replacement.[25]

Hardware. Cable mounting with the MS type of cable clamp is a proven method. Many variations of this clamp are available from specialty suppliers. The principal advantages of MS clamps are low cost, light weight, high strength, ready adaptability, and ease of installation and servicing. Clamps can be installed on any structure or skin of adequate strength that can be drilled. If the structure or skin cannot be drilled, bonding is recommended. One technique is to bond a cable-supporting device or pad to the supporting area, and then strap, tie, or clamp the cable to it. An alternative technique is to bond the entire cable to the supporting area for a very secure installation. If bonding is the only possible means of attachment, MS nylon, reinforced nylon, or Kynar harness straps, mounting plates, and a compatible bonding material can be used. Specification MIL-S-23190 covers adjustable plastic cable straps for military use.

Support Spacing. The spacing of clamps and other cable support tie-down devices (Fig. 6.24) can be determined from experience and developmental mock-ups. Applicable electrical system specifications generally establish bundle tie-down spacing by stating a maximum distance between supports (MIL-W-8160 maximum spacing is 24 in). For adequate design of electric cable installations in missiles and space vehicles, the spacing must be resolved analytically and tested for verification. The spacing will be determined by the dynamic environment in which the cabling must perform reliably. Complete design coordination must exist between the structures, dynamics, and electrical installations in order to meet system requirements. Dynamic tests on development hardware are recommended early in a program for verifying the installation.

Drastic changes in cable stiffness or section size caused by the ending or branching of wires may lead to points of dynamic weakness. Firm support is recommended on both sides immediately adjacent to these points, regardless of the spacing of other tie downs or clamps.

TABLE 6.21 Typical Heat-Shrinkable Tubing Properties

Properties	Irradiated polyolefin												
	Flexible opaque	Flexible clear	Semirigid opaque	Semirigid clear	Dual wall	Irradiated PVF_2	Flexible PVC	Flexible irradiated PVC	Semirigid irradiated PVC	Neo-prene rubber	Sili-cone rubber	Butyl rubber	PTFE
Tensile strength, lb/in^2	2500	2500	3000	3000	2000	7000	3000	3000	5000	1900	900	1600	4500
Ultimate elongation, %	400	400	400	400	400	300	300	300	250	220	300	350	250
Brittleness temperature, °C	−60	−85	−60	−90	—	−73	−20	−20	−20	−40	−75	—	−90
Hardness	98A	90A	—	—	—	—	85A	—	—	85A	70A	80A	—
Specific gravity	1.3	0.93	1.3	0.95	0.94	1.76	1.4	1.35	1.4	1.4	1.2	1.2	2.2
Water absorption, %	0.05	0.01	0.05	0.01	0.1	0.1	—	0.6	0.6	0.5	0.5	0.1	0.01
Dielectric strength, V/mil	1300	1300	1300	1300	1100	1500	750	750	900	300	300	130	1200
Volume resistance, Ω • cm	10^{15}	10^{17}	10^{15}	10^{17}	10^{16}	—	10^{12}	10^{12}	$>10^{13}$	10^{11}	10^{15}	10^{12}	10^{18}
Dielectric constant	2.7	2.3	2.7	2.4	2.4	—	5.4	—	—	—	3.3	—	2.1
Power factor	0.003	0.0003	0.003	0.0003	0.0005	—	0.12	—	—	—	—	—	0.0002
Fungus resistance	Inert	Inert	Inert	Inert	Inert	Inert	Inert	Inert	Inert	Inert	Inert	Inert	Inert
Fuel and oil resistance	Excellent	Excellent	Excellent	Excellent	Excellent	Excellent	—	Excellent	Excellent	Good	Fair	Fair	Excellent
Hydraulic fluid resistance	Excellent	Excellent	Excellent	Excellent	Excellent	Excellent	—	Excellent	Excellent	Fair	Poor	Good	Excellent
Solvent resistance	Good	Good	Good	Good	Good	Excellent	—	Excellent	Excellent	Fair	Fair	Fair	Excellent
Acid and alkali resistance	Excellent	Excellent	Excellent	Excellent	Excellent	Excellent	—	Excellent	Excellent	Good	Good	Good	Excellent
Flammability	Self-extin-guishing	Burns slowly	Self-extin-guishing	Burns slowly	—	Nonburning extin-guishing	Self-extin-guishing	Self-extin-guishing	Self-extin-guishing	Self-extin-guishing	Self-extin-guishing	Burns slowly	Non-burning

Source: From Schuh.[1]

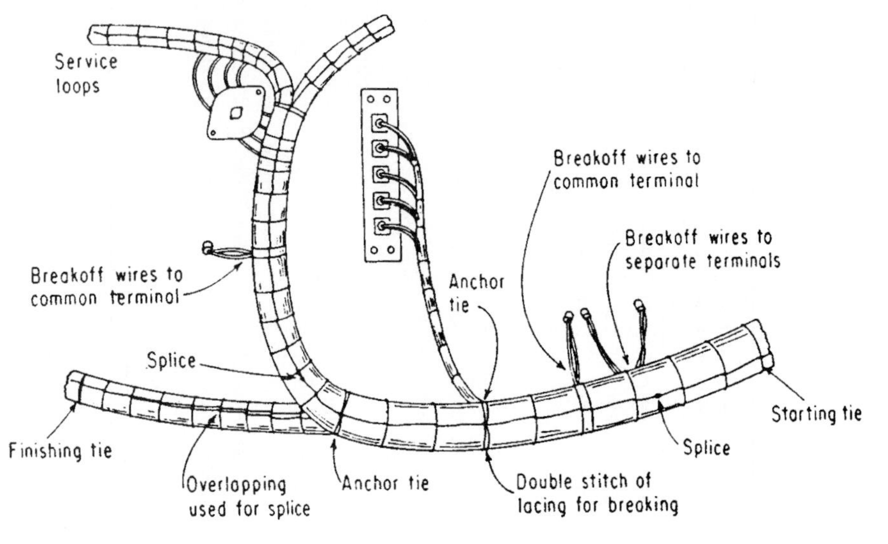

FIGURE 6.23 Cable harness lacing details.[17]

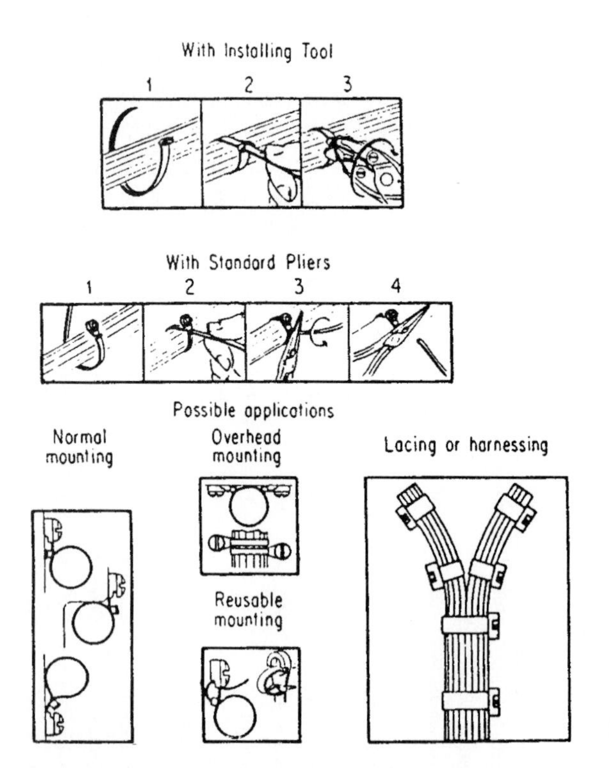

FIGURE 6.24 Cable straps for clamping and harnessing.[17] Installation: 1—slip strap around wire bundle, rib side inside; 2—thread tip through eye, draw up snug; 3,4—apply tool, cinch tight, twist 120°, squeeze to cut off excess.

The spacing of supporting devices on a high-acceleration missile system can determined by the formula

$$F = \Sigma LANG$$

where F = design load of attachment device, lb
L = unsupported length, in
A = unit weight per length for each wire size, lb/in
l = number of each wire size in bundle
G = maximum dynamic environmental load, g's (gravitational)

A sample calculation will illustrate a design example. Given the following harness parameters, find the unsupported length:

10 unshielded wires size AWG 20

10 shielded and jacketed cables size AWG 20

10 twisted, shielded, jacketed pairs size AWG 26

$$G \text{ (load)} = 150\text{g}$$

$$F = 50 \text{ lb}$$

$$A_{20u} = 4.02 \times 10^{-4} \text{ lb/in}$$

$$A_{20} = 6.36 \times 10^{-4} \text{ lb/in}$$

$$A_{26w3} = 4.98 \times 10^{-4} \text{ lb/in}$$

F is indicated as 50 lb. However, a safety factor of 2:1 changes this value to

$$F = \frac{50}{2} = 25 \text{ lb}$$

From this,

$$L = \frac{25}{150} \times \frac{1}{10(4.02 \times 10^{-4}) + 10(6.36 \times 10^{-4}) + 10(4.98 \times 10^{-4})}$$

$$= \frac{1}{6} \times \frac{1}{1.536 \times 10^{-2}}$$

$$= 10.8 \text{ in}$$

The sample harness must be clamped or attached every 10.8 in to satisfy the given conditions.

These details indicate a technique evolved under a specific set of requirements and are presented as a guide only.

Support and Clamping of Cables to Connectors. The cable clamp associated with a multipin connector is used primarily to support the wires or the cable terminating at the connector and also to relieve strain from the terminations. Soft telescoping bushings (in accordance with federal specification AN 3420) are available for cables smaller than the cable clamp opening. The bushings permit the cable to be centered and anchored securely without excessive padding. There should be adequate clamping pressure without bottoming the two halves of the clamp. Clamp screw thread engagement should be equal to two-thirds to one-half times the major nominal screw diameter.

Major differences in size between cable and connector can be corrected with step-up or step-down telescoping extension sleeves instead of bushings.

6.12 SPECIFICATION SOURCES

The following are sources of specifications and standards pertaining to many of the materials discussed in this chapter:

American National Standards Institute (ANSI), 11 West 42 St., New York, NY 10036.

American Society for Testing and Materials (ASTM), 1916 Race St., Philadelphia, PA 19103.

Canadian Standards Association (CSA), 178 Rexdale Blvd., Rexdale, Ontario, M9W 1RC.

Institute of Electrical and Electronics Engineers (IEEE), 345 East 47 St., New York, NY 10017.

Insulated Cable Engineers Association (ICEA), P.O. Box P, South Yarmouth, MA 02644.

Military and Federal Specifications, DODSS Subscription System and Manual Index, U.S. Government Printing Office, Washington, DC 20401, (202)783-3238.

American Institute of Aeronautics and Astronautics (AIAA), 1250 Eye St. NW, Suite 1100, Washington, DC 20005, (202) 371-8400.

National Electrical Manufacturers Association (NEMA), 2101 L St. NW, Washington, DC 20037.

Underwriters Laboratories (UL), 207 East Ohio St., Chicago, IL 60611.

Since specifications are continually subject to change, it is advisable to consult the latest revisions in meeting technical and contractual requirements. A recent major change involved superseding the entire MIL-W-583C wire specifications with J-W-1177.

6.13 REFERENCES

1. A. G. Schuh, "Wires and Cables," in C. A. Harper (ed.), *Handbook of Materials and Processes for Electronics,* McGraw-Hill, New York, 1970, chap. 4.
2. *The Rome Cable Manual of Technical Information,* Rome Cable Corp., Rome, Ga., 1967.
3. A. G. Schuh, "Wires and Cables," in C. A. Harper (ed.), *Handbook of Electronic Packaging,* McGraw-Hill, New York, 1969, chap. 2.
4. C. P. Steinmetz and B. G. Lamme, *Trans. AIEE,* vol.32, pp. 79–89, 1913.
5. V. M. Montsinger, "Loading Transformers by Temperature," *Trans. AIEE,* vol. 49, pp. 776–792, 1930.
6. T. W. Dakin, "Electrical Insulation Deterioration Treated as a Chemical Rate Phenomenon," *Trans. AIEE,* vol. 67, pt. 1, pp. 113–118, 1948.

7. E. J. Croop, "Wiring and Cabling for Electronic Packaging," in C. A. Harper (ed.), *Electronic Packaging and Interconnection Handbook,* McGraw-Hill, New York, 1991, chap. 4.
8. "Extra Flexible Tactical Cable Report," No. 3, Martin Marietta Corp., Bethesda, Md., Dec. 1964.
9. *Electrical Design,* The Martin Company, Bethesda, Md., 1958.
10. J. C. Reed, "Save Space by Hookup Wire Insulated with Teflon," *J. Teflon,* 1964.
11. L. L. Lewis, "Ultra Thin-Wall Wire Insulation from Kapton Polyimide Film, Type XF," presented at the Naval Air Systems Command Symp., Oct. 1966.
12. S. Schwartz and D. L. Wells, "Processing of Plastics in Space," *J. Soc. Plastics Eng.,* Aug. 1962.
13. E. E. Jolley and J. C. Reed, "The Effects of Space Environments on Insulation of Teflon TFE and FEP Resins," presented at the Signal Corps Symp., Nov. 1962.
14. V. L. Lanza and R. M. Halperin, "The Design and Development of Wire Insulators for Use in the Environment of Outer Space," presented at the Signal Corps Symp., Dec. 1963.
15. W. J. Prise, "When the Gamma Heat Is on Insulators," *Electron. Des.,* May 23, 1968.
16. J. C. Reed and J. T. Walbert, "Teflon Fluorocarbon Resins in Space Environments," presented at the Signal Corps Symp., Nov. 1962.
17. E. F. Godwin, "Hookup Wires, Multiconductor Cables, and Associated Terminating Devices," in C. A. Harper (ed.), *Handbook of Wiring, Cabling, and Interconnecting for Electronics,* McGraw-Hill, New York, 1972, chap. 3.
18. N. J. Cotter and J. R. Perkins, "Life vs. Voltage Performance of Flat Conductor Cables and Light Weight Round Wire Systems," E. I. du Pont de Nemours & Co., Wilmington, Del., 1968.
19. *Cable Products Catalog,* ACD-5, Amphenol Corp., Oak Brook, Ill., 1970.
20. R. A. Bellino, "How to Select a Multiconductor Cable," American Enka Corp., Brand-Rex Div., Willimantic, Conn., 1970.
21. K. C. Byram, "Flat Conductor Cabling and Connectors," in *Flexible Flat Cable Handbook,* Inst. of Printed Circuits, Chicago, Ill., 1965.
22. *Flexprint Circuit Design Handbook,* Bull. FT-169, Sanders Assoc., Manchester, N.H., 1965.
23. *Flexible Flat Cable Handbook,* Inst. of Printed Circuits, Chicago, Ill., 1965.
24. "Reliable Integrated Wire Termination Devices," Final Rep. ECOM-0394, Contract DAAB07-69-C-0394, U.S. Army Electronics Command, Fort Monmouth, N.J., 1969.
25. A. Fox and J. H. Wisher, "Superior Hook-Up Wires for Miniaturized Solderless Wrapped Connections," *J. Inst. Metals,* vol. 100, pp. 30–32, 1972.

FURTHER READING

Beitel, J. J., et al.: "Hydrogen Chloride Transport and Decay in a Large Apparatus, I. Decomposition of Polyvinyl Chloride Wire Insulation in a Plenum by Current Overload," *J. Fire Sci.,* vol. 4, no. 1, pp. 15–27, 1986.

Cabey, M. A.: "New Silicone Rubber Cable Insulation Promises Circuit Integrity in Flaming Environment," NTIS, Springfield, Va., 1983, 8 pp.

Felsch, C.: "A New Way of Insulating Power Cables with Crosslinked Polyethylene," *Wire World Int.,* vol. 22, p. 294, 1980.

Fischer, T. M.: "Impact of Polyethylene Curing Methods on Wire and Cable Performance," *IEEE Elec. Insul. Mag.,* vol.5, no. 1, pp. 29–32, 1989.

Frasure, J. W., J. H. Snow, and D. A. Voltz: "Wire and Cable Update," *IEEE Trans. Ind. Appl., vol. IA-22, pp. 178–194, 1986.*

Gage, C. A., G. Carrillo, E. D. Newell, W. D. Brown, and P. Phelan (eds.): "Wire Harness Automation," in *Proc. Int SAMPE Symp. and Exhibition*, vol. 33, pp. 787–795, 1988.

Gross, B., J. E. West, H. Von Seggern, and D. A. Berkley: "Time-Dependent Radiation-Induced Conductivity in Electron Irradiated Teflon Foils," *J. Appl. Phys.,* vol. 51, pp. 4875–4881, 1980.

Hamer, P. S., and B. M. Wood: "Are Cable Shields Being Damaged During Ground Faults?," *IEEE Trans. Ind. Appl.,* vol. IA-22, pp. 1149–1155, 1986.

Ishibashi, M., T. Yamamoto, and S. Mogi: "Evaluation of Crosslinked Material for Insulated Electronic Appliance Hook-Up Wires," in *Proc. Int. Wire and Cable Symp.,* vol. 27 (1978) Am. Chem. Soc., 1989, pp. 213–219.

Meyer, F. K., and H. Linhart: "New Data on Long Term Stabilization of Polyethylene for Telecommunication Wire Insulation," in *Proc. Int. Wire and Cable Symp.,* (Cherry Hill, N.J.) NTIS, Springfield, Va., 1983, 10 pp.

Payette, L. J.: "Solderable Wire Enamel," *Electrical Installation,* vol. 6, no. 5, pp. 8–12, September 1990.

Samborsky, A. M.: "Current Rating for Bundled Wires—A Step by Step Procedure," U.S. Naval Electronic Laboratory, San Diego, Calif., 1967.

Sheppard, A. T., and R. G. Webber: "Polytetrafluoroethylene Insulated Cable for High Temperature Oxygen Aerospace Applications," in *Proc. Int. Wire and Cable Symp.,* (Cherry Hill, N.J.), NTIS, Springfield, Va., 1983, 10 pp.

Veda, K., "Progress of Magnet Wire Technology in Japan," *Electrical Insulation,* vol. 5, no. 3, pp. 18–26, January 1989.

Wilkens, W. D., "Telephone Cable—Overview and Dielectric Challenges," *Electrical Installation,* vol. 6, no. 2, pp. 23–28, March 1990.

Wolf, C. J., D. L. Fanter, and R. S. Soloman: "Environmental Degradation of Aromatic Polyimide Insulated Electrical Wire," *IEEE Trans. Elec. Insul.,* vol. EI-19, pp. 265–272, 1984.

Zeller, A. F., et al.: "Insulation on Potted Superconducting Coils," *IEEE Trans. Magn.,* vol. 25, pp. 1536–1537,1989

CHAPTER 7
METALS JOINING OF ELECTRONIC CIRCUITRY

James F. Maguire

7.1 INTRODUCTION

This chapter addresses the material and process characteristics used in metal joining during the fabrication of electronic devices. Metals joining methods used in electronics are summarized in Fig. 7.1. Discussions include wire bonding, soldering, and welding methods. During the writing of this chapter, special emphasis was given to those characteristics related to process control and reliability. Additionally, a very brief primer on experimental methodologies is also included. The technical terms used in this chapter are defined in IPC-T-50, Terms and Definitions for Interconnecting and Packaging Electronic Circuits.[1]

7.2 WIRE BONDING

After the silicon chip has been bonded to a base or substrate, the next step is to provide electric connectors with wire leads. The techniques used to attach gold or aluminum wires between the contact areas on the silicon chip and package leads or lands are thermocompression bonding, ultrasonic bonding, and resistance welding.

The lead-bonding processes depend on obtaining intimate contact between the two metallic surfaces so that the unsaturated atomic bonds on one surface interact with those of the second metallic surface and give a good bond. Atoms at the surface of any metal have atomic bonds that are not saturated by adjacent metal atoms of the same type. This is the definition of a surface or interface. If these unsaturated bonds become saturated with atoms of the same metal, then the surface or interface ceases to exist. If the bonds become saturated with metal atoms of a different type, then an intermetallic interface exists, and a true metallurgical bond is obtained. This type of bonding is, however, rare in normal practice. Even carefully ground metal surfaces have irregularities with peak-to-valley distances that average about 500 Å. More often the irregularities are two orders of magnitude greater than this. Thus, even if these metal surfaces were perfectly clean, contact between touching surfaces would be obtained on only a small fraction of the interface area.

Wire Bonding
- Thermocompression Bonding
  - Wedge Thermocompression Bonding
  - Ball or Nailhead Bonding
  - Stitch or Scissors Bonding
  - Bird Beak Bonding
- Ultrasonic Bonding
- Thermosonic Bonding

Soldering
- Solder Materials
  - Solder Alloys
  - Solder Fluxes
  - Solder Pastes
  - Solder Wire & Preforms
- Component and PWB Baking
- Soldering Processes
  - Hand Soldering
  - Mass Soldering
    - Wave Soldering
    - Vapor Phase Soldering
    - IR Soldering
    - Hot Gas Soldering
    - Resistance/Pulse Soldering
    - Laser Soldering
    - Rework/Repair
- Post Solder Cleaning
  - Solvent Mobility Factors
  - Solubility Factors
  - Solubilisation
  - Solvent Cleaning
  - Aqueuos Cleaning
  - Semi-Aqueous Cleaning
  - Ultrasonic Cleaning

Welding
- Resistance Welding
- Ultrasonic Welding

Other Mechanical Joining Methods
- Discrete Wiring Methods
- Compliant Pin

FIGURE 7.1 Summary of metal joining processes discussed.

In preventing intimate atomic contact between two metal surfaces, chemical contamination is just as important as surface irregularities. The surface atomic bonds attract and hold a variety of chemical species. Metal oxides are the most common form of this surface contamination, but many other types also exist. The nature of the oxide layers varies from one metal to another, but they are found even on gold. For a given metal in an oxidizing environment, the oxidation rate increases with temperature. Other forms of chemical contamination may either be substituted for surface oxides or form on top of oxides.

The contact areas on silicon devices, in general, have been of two types—direct contacts and expanded contacts. In the first of these, metallization is applied directly to the silicon, and the wire bond is formed directly on top of this metallization. The expanded contact is made with a metal film contact to the silicon and the wire bond located on the chip periphery. The metal film on top of the silicon dioxide passivation layer is extended to chip edge where a bonding pad is formed. The bonding pads are located on the edge of the chip in order to avoid contact between bare wires and metallization patterns. Only expanded contacts are used in integrated circuits. All contact points are located on the periphery of the chip, and a metallization pattern provides intraconnections between these and the active device structures. These contact structures are indicated schematically in Fig. 7.2. The oxide window providing a region for contact to the silicon typically is several tenths of a mil wide, while the diameter of the wire used in bonding is on the order of 1 mil (25.4 μm). The dimensions in the figure are not to scale.

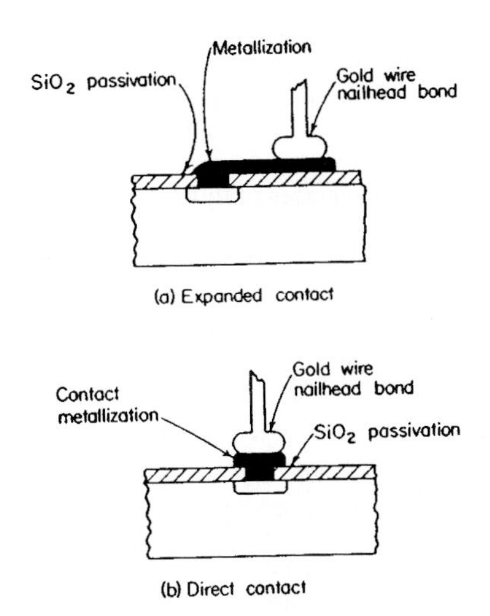

FIGURE 7.2 Expanded and direct contacts.

For best results, a bond wire deformation of 1.4 to 1.8 wire diameters for Al wire and 1.4 to 3 wire diameters for Au wire.[6]

7.2.1 Thermocompression Bonding

Thermocompression bonding is a process in which the metals being joined are brought into intimate contact by pressure, using a shaped, smooth bonding tool at a temperature below that required for interface melting. In the region to which pressure is applied, plastic deformation and diffusion occur at the interface during a controlled time, temperature, and pressure cycle. Plastic deformation at the interface is necessary not only to increase the contact area but also to destroy any interface films and bring the metal surfaces into intimate contact.

The amount of deformation that occurs in the wire being bonded to the silicon is dependent on the force applied through the bonding tool as well as on the bonding temperature. Consistency of the lead material and of the bonding configuration is a prerequisite for reproducible bonds. In most commercial equipments it is possible to set the bonding pressure over a considerable range.

The techniques for accomplishing thermocompression bonds are classified as follows:

1. Wedge bonding
2. Nailhead, or ball bonding
3. Stitch or scissors bonding
4. Bird-beak bonding

Generally speaking, thermocompression bonding is most commonly used on gold wire. Typically the gold wire is alloyed with a small amount of beryllium and copper to control grain growth during the formation of the bond with the substrate pad. The resistivity of a 25-μm gold wire is ~0.46 Ω/cm. Typical pull strengths for a good thermocompression gold wire bond is 8 to 10 g for a 25-μm gold wire. The advantages and disadvantages of this technique are summarized below:

Advantages	Disadvantages
The method is *omnidirectional* (i.e., after the first bond, the bonding head may move in any direction for the second bond).	Requires high working temperature (~300°C).
No unique fixturing is required.	Process is very sensitive to surface contamination on the substrate.
Easiest process to develop and control.	Requires large pad area, 1-mil wire requires 4 × 4 mil pad. Requires relatively long bond times (time during which force is applied to wire).

Wedge Thermocompression Bonding. *Wedge thermocompression bonds,* as the name implies, are made with a wedge or chisel-shaped tool. The end of this tool is rounded with a radius of one to four times that of the wire being bonded and is made of sapphire or similar hard material. It is used to apply pressure to the lead wire located on the bonding pad which has been heated to the bonding temperature. In the wedge-bonding procedure, the mounted silicon chip is positioned under a microscope. The wire is brought into the proper position on the silicon ship and held there by a glass capillary. After the wire is positioned, the wedge is brought into position and lowered to the wire. Thus, two precise positioning operations are required. Bonding is accomplished by an automatically controlled time sequence. In various equipments different methods are provided for precisely coaligning the bonding pad, wire, and wedge.

During the bonding operation, a gas curtain, either forming gas (a dilute mixture of hydrogen in nitrogen, 1 part hydrogen to 9 parts nitrogen) or dry nitrogen, surrounds the chip. This prevents oxidation of the aluminum or the gold from interfering with the bonding operation. Because the lead wire is very small and ductile, a special wire feed mechanism is required. Usually a spool of carefully specified wire is mounted in an enclosure which protects it from contamination. The wire from the spool is fed through the glass capillary for bonding. The bonding sequence is illustrated in Fig. 7.3.

Difficulties with wedge bonding may be traced to a number of sources. The most important of these are imprecise temperature control, poor wire, inadequately mounted silicon chips, or a poorly finished bonding tool. High temperatures are provided by means of a heat column which holds the package and sometimes by a heated wedge assembly. Uneven airflows and poor thermal contact between the package and the heat column are some of the difficulties that may be experienced in maintaining temperature control. The ductility of the wire is very important. Any bending or cold working of the wire is sufficient to cause poor bonds. The wire is annealed on the spools which fit the bonding apparatus, and if these spools are handled roughly or dropped, it may be sufficient reason to require that the wire be reannealed. Since it is very small, the wire is very weak. If it becomes exposed to a humid atmosphere or collects dust particles, it becomes difficult to use the wire without breaking. A common difficulty is that the wire sticks in the glass capillary through which it is fed to the work area. The wedge

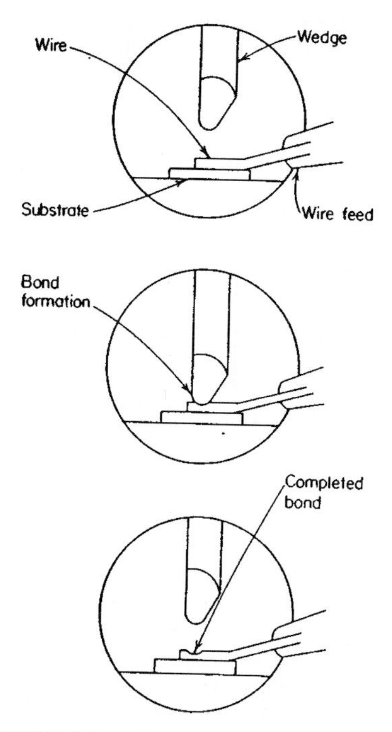

FIGURE 7.3 Thermocompression wedge bonding.

which is used for bonding must be polished to prevent sticking of the lead wire. At the bonding temperatures, the wire can adhere to a tool with sufficient force to break the bond when the tool is raised. A 10- to 20-nm finish on the tool surface is required.

Since in wedge bonding the cross-sectional area of the wire is reduced by the bonding operation, the wire is weakened. The shape of the wedge tool is an important factor in determining bond strength. The desired pressure is that which deforms the lead to one-half its original cross section. Excessive pressure may damage the silicon die or weaken the lead wire. An indication of the conditions required is shown in Fig. 7.4, which was obtained for wedge bonds on films on a glass substrate.

Ball or Nailhead Bonding. *Ball or nailhead bonding* is a technique for thermocompression bonding in which a small ball is formed on the end of the wire and deformed under pressure against the pad area on the silicon chip, giving a bond. It is used only with gold wire; the lead wire is perpendicular to the silicon chip as it leaves the bond area.

Ball bonding requires that the small wire be fed through a quartz or tungsten carbide thick-walled capillary tube. The capillary tube, with one end tapered to a few mils in diameter on the outside, is mounted in a suitable mechanical fixture so that it can be moved both vertically and horizontally. The horizontal positioning must be accomplished by means of precision manipulators while being observed through a microscope. Positioning accuracies on the order of 20 millionths of an inch are required.

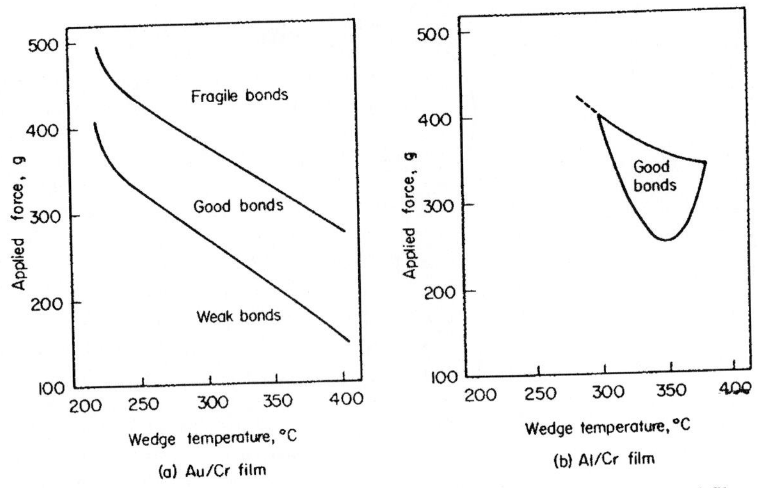

FIGURE 7.4 Pressure for wedge bonds (4-mil gold wire, 10-mil wedge), to metal films on glass.

Before bonding, a small spherical ball is formed on the end of the gold wire by a hydrogen flame. The silicon chip to which the wire will be bonded is positioned on the work stage below the bonding capillary. The capillary is positioned over the bonding pad of the silicon chip and lowered until the ball is brought into contact with the capillary tip. The ball is then brought into contact with the bonding pad where a predetermined amount of force is applied. This deforms the ball and establishes intimate contact between the gold ball and the bonding pad. These operations are illustrated in Fig. 7.5. Wire lead attachment to the other terminal may be accomplished by bonding the wire with the edge of the capillary providing bonding pressure. The capillary is then

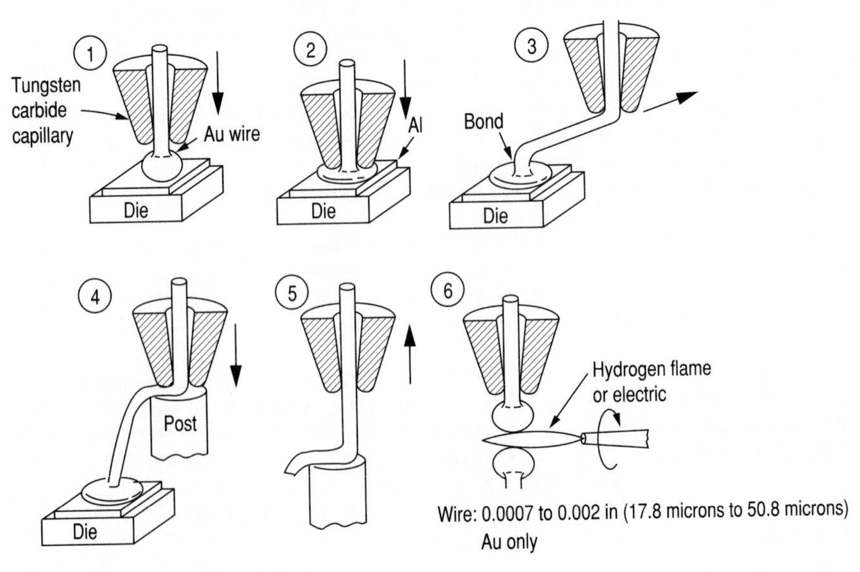

FIGURE 7.5 Ball bonding operation.

raised and the hydrogen flame used to cut it off while forming another ball for the next bonding operation.

Ball bonding is used because of its relatively high speeds, economy, and strength. The number of steps in this bonding operation are fewer than in alternative methods, and the strength of the bond obtained is superior to that of other methods. Aluminum wire cannot be used because of its inability to form a ball when severed with a flame. However, gold wire is an excellent electric conductor, is more ductile than aluminum, and is chemically inert. For ball bonding, hard gold wire may be used since the balling process determines the ductility of the gold to be deformed.

Among the disadvantages of ball bonding is the fact that a relatively large bonding pad is required. For a given device structure, the largest wire size practical should be used to give a higher bonding yield and greater mechanical strength. Two-mil gold wire typically requires a 5- by 5-mil (0.127-mm by 0.127-mm) aluminum pad; a $1^1/_2$-mil (0.09 mm) gold wire requires a 4- by 4-mil (100-μm by 100-μm) bonding pad. Other bonding techniques allow smaller bonding-pad sizes. The gold-wire-aluminum system is used because of the excellent properties of aluminum as a contact material for silicon and of the gold as a bonding wire (its ductility, electric conductivity, and corrosion resistance). However, when gold and aluminum are placed in intimate contact and the combination is heated to modest temperatures in excess of 200°C, detrimental changes take place: gold-aluminum intermetallic compounds are formed. This intermetallic compound formation is further complicated by the presence of silicon and oxygen. A purplish material begins to appear at the junction between the two metals.

Stitch or Scissors Bonding. *Stitch or scissors bonding* combines some of the advantages of both wedge and ball thermocompression bonding. The wire is fed through the bonding capillary, the bonding area is smaller than for ball bonds, and no hydrogen flame is required. The procedure is shown in Fig. 7.6. The wire is fed through the cap-

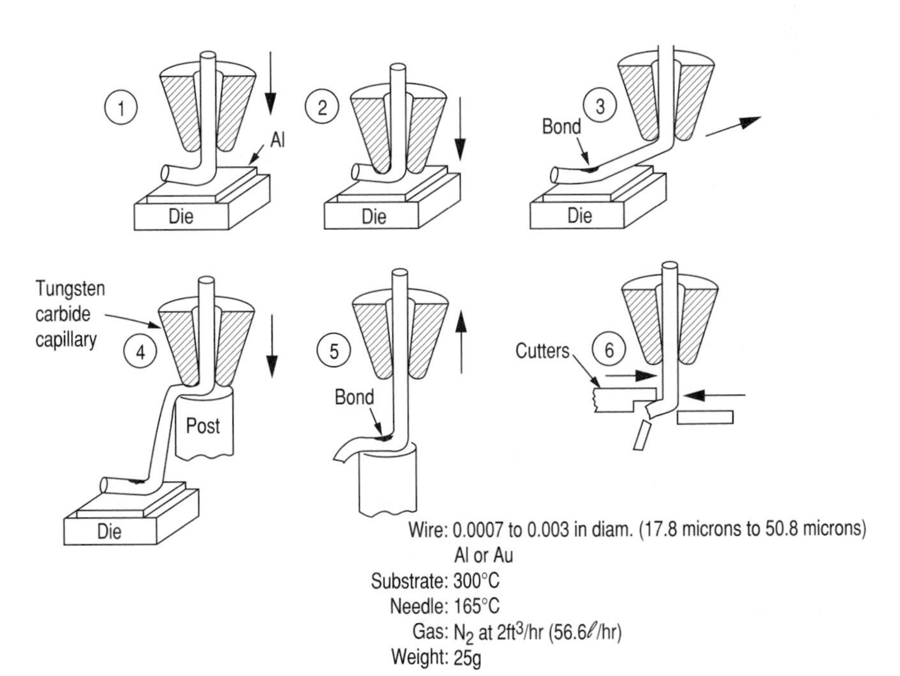

FIGURE 7.6 Stitch bonding.

illary and bent at a right angle by cutting with the scissors. As the capillary is lowered to the bonding pad, the bent wire is pulled against the capillary edge which performs the bonding operation much as a wedge. Either glass or tungsten carbide capillaries can be used. Capillary heating is possible. Either gold or aluminum wires can be bonded at a high rate.

Bird-Beak Bonding. The *bird-beak bond* is a form of thermocompression bond that is made with a split bonding tool. It feeds, bonds, and cuts the lead wire without requiring subsidiary equipment or leaving a pigtail. It is rapid, has a high yield, and requires small bond areas. A 1-mil (25.4-μm) wire gives a bond area of 1.3 by 1.0 (33 μm by 25.4 μm) mils. In the bonding tool, the lead wire is fed between a jeweled bonding tool and a jeweled holding tool. The underside of the bonding jewel is shaped so that it forms a rib, about 1 mil (25.4 μm) in diameter, on the wire as the bond is made. The rib adds mechanical strength and current-carrying capacity to the bond, while the flattened underside of the wire makes for excellent adhesion and electric contact to the underlying thin films. The bonding sequence is shown in Fig. 7.7. Once the bond is made, the vacuum clamp holds the wire. The wire breaks cleanly at the edge of the bond when the tool is moved horizontally away from the bond. This avoids short circuits due to tails or loose ends on the bonded wire.

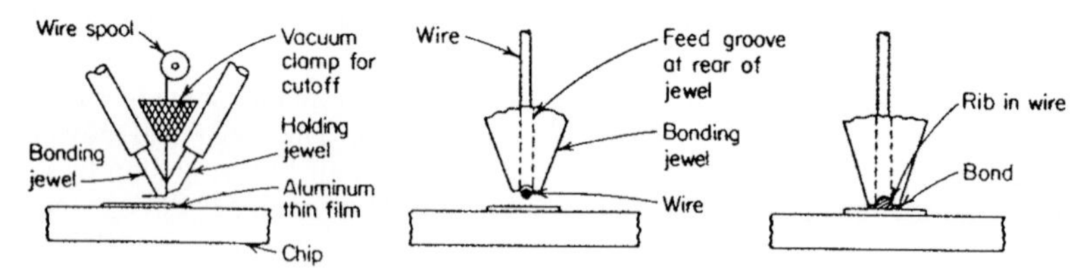

FIGURE 7.7 Bird-beak bonding.

Bonding pressures vary from 25 to 125 g and bonding times from 1 to 6 s, depending on the thickness, purity, and physical conditions of the aluminum film on the chip. Either gold or aluminum wire may be used. These are three methods by which heat is transmitted to the bond. Larsen[8] has described these:

Method 1—Heat Column. With heat-column thermocompression bonding equipment, the microcircuit or workpiece to be bonded is placed on an electrically heated column where it is allowed to "soak up" enough heat to permit the lead to be bonded when a preset force is applied. Usually, the wire is fed to the workpiece through a glass capillary tip. The same tip also is used to apply the pressure required for making the bond.

Method 2—Heated Tip. In the heated-tip method of thermocompression bonding, the heat for making the bond is applied directly through the capillary tip itself. The heat at the tip may be either continuous or intermittent. A continuously heated tip is, in a sense, like a heat column which is held at a continuous temperature and transfers its heat to the materials being joined. The intermittently or pulse-heated tip, on the other hand, is heated only during that time in which the bond actually is being made. Heat is produced by passing an electric current either directly through the tip or through a heating element surrounding the tip. In the former case, the tip usually is made of high-resistance materials, and heat is built up rapidly during the current pulse. The conduction of heat from the tip to the materials being joined is sufficient to make the thermocompression bond.

Method 3—Combination. The third method used in making thermocompression bonded joints is a combination of the two types just described. That is, the bonding equipment has both a heated tip and a heat column. This approach allows lower temperatures to be used at both the bonding tip and heat column. This condition helps to minimize the oxidation problem. Also, the lower temperatures make it possible to bond some delicate types of components that otherwise could not be joined. Reducing oxidation makes it possible to extend the useful life of tips before replacement or repair is needed.

A summary of the advantages and disadvantages of these methods is shown in Table 7.1.

TABLE 7.1 Summary of Advantages and Disadvantages for Bird-Beak Bonding Methods

Method	Advantages	Disadvantages
Heated column	Minimal oxidation Good thermal control	Sensitive to thermal conductivity of substrates
Heated tip	Lower tip oxidation Can be successfully used on substrates not compatible with heated column	Not compatible with thicker metal pads Rapid tip usage due to corrosion and oxidation.
Combination	Increased tip life compared to heated tip Able to bond to variety of materials	More complex setup and parameter selection Equipment more costly

7.2.2 Ultrasonic Bonding

Ultrasonic bonding involves the same mechanism as thermocompression bonding except that the source of energy is mechanical rather than thermal. Ultrasonic bonding is very successful with aluminum because aluminum is soft and deformation occurs at relatively low pressures. Aluminum is coated with a refractory oxide which readily breaks down under ultrasonic stressing to provide an abrasive flux.

In the ultrasonic bonding device, an elastic vibration is created by the rapid expansion and contraction of a magnetostrictive transducer driven by a source of high-frequency alternating current. These high-frequency stress waves travel through a coupler or mechanical transformer to the welding tip. The welding tip serves to transfer the vibrations to the materials to be welded. A portion of this coupler is tapered to provide a mechanical impedance match which increases the amplitude of vibration. The bonding tip undergoes excursions in a direction parallel to the interfaces of the weld, inducing a shear mode of vibration into the materials. A simplified schematic representation of a lateral-drive ultrasonic welder is shown in Fig. 7.8.

The materials to be bonded are clamped between the welding tip and the lower work stage called an *anvil,* so that the workpieces experience the stresses resulting from the clamping force and the superimposed vibration of the welding tip.

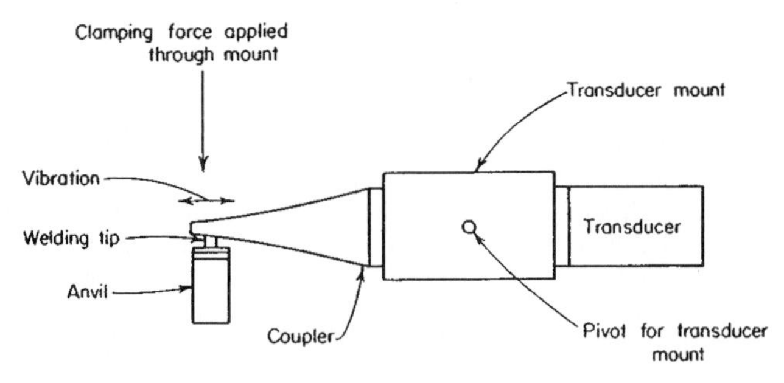

FIGURE 7.8 Ultrasonic bonding head.

Since the ultrasonic welding process requires the conversion of electric energy into acoustical vibratory energy by means of a transducer element, this energy must be transmitted efficiently to the interface through a mechanical transformer and a welding tip. The energy delivered to the weld interface depends on the characteristics of this coupling system. The optimum value of rigidity and energy transmission is achieved by using a Fourier mechanical transformer. The Fourier-shaped horn permits the use of lower welding power and pressure in bonding operations.

There are some supplementary phenomena which effect the welding process. The first is a temperature rise that results from the elastic hysteresis of the highly stressed portion of the weld zone during the welding sequence. The temperature rise has been measured by fine-wire thermocouple and fusible insert techniques and considered in the light of extensive electromicrography. These investigations revealed that the maximum transient temperature was between 30 and 50 percent of the absolute melting point of the metal in a similar metal joint. This temperature rise is highly localized in the weld metal and promotes plastic deformations which are associated with the process.

A second supplementary phenomenon derives from the fact that metallic crystal structures can be temporarily plasticized by high-frequency mechanical vibrations. This phenomenon has been observed in ultrasonically accelerated processes and in the nature and extent of the internal deformations in ultrasonic welds themselves. This plasticization occurs as a result of the acoustical excitation per se and is independent of temperature and the thermal plasticization previously described. However, it does accomplish the same purpose, that is, to facilitate the plastic deformation of the interface.

Ultrasonic bonding is most commonly used on aluminum wire. The resistivity of a 25-μm aluminum wire is ~0.58 Ω/cm. Typical pull strengths for a good thermocompression gold wire bond is 3 to 5 g for a 25-μm gold wire. The advantages and disadvantages of this technique are summarized below:

Advantages	Disadvantages
Essentially a room temperature process.	Unidirectional bonding method (i.e., orientation of the wire on the pad in first bond determines the direction wire can be run toward next bond).
Relatively small pad sizes (1-mil wire needs ~1.5 × 3 mil pad).	Component and/or package wall must allow for bonding tool and wire clamp tooling.
Least sensitive process to surface contamination.	

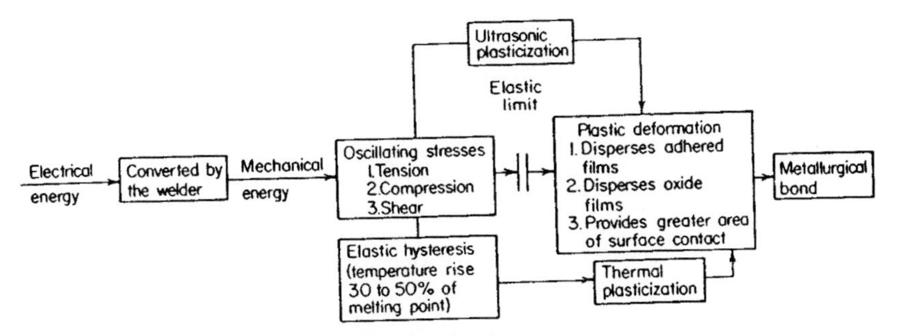

FIGURE 7.9 Block diagram of the ultrasonic bonding process.

Figure 7.9 is a block diagram of the ultrasonic welding process. There are four parameters that may be varied to produce a specified weld: the welding tip, clamping force, power, and weld time.

The geometry of the welding tip is determined by the diameter of the wire. For an acceptable weld, the working surface of the tip should have a semicylindrical groove with a radius of curvature equal to the radius of the wire to be welded. This groove aids in the positioning of the wire and is necessary for the formation of a nugget weld. In addition, the length of this groove must be twice the diameter of the wire. The tip must be positioned so that the long axis of the groove is parallel to the long axis of the coupler assembly. It is also very important to ensure that the working surface of the tip is parallel to the plane of the workpiece.

The clamping force used is of prime importance. This clamping force affords intimate contact among the tip, wire, and contact area of the device, thus providing good coupling. There is an optimum clamping force for each welding job.

The amplitude of tip vibration may be varied by means of power adjustment. There is a threshold for the minimum amount of stress required, and a power input lower than this minimum cannot produce an acceptable weld. It has been noted that there is also a maximum power input, which, if exceeded, will distort the materials being joined.

The time may also be varied and the two parameters, time and power, determine the energy (watt-seconds) applied to the weld. The physical characteristics (hardness, thickness) of the materials being welded determine the appropriate time setting.

Because ultrasonic welding can create bonds between a very wide variety of dissimilar materials, it is an extremely flexible tool when applied to the problem of welding interconnections to microminiature circuit elements. Table 7.2 summarizes some of the combinations of films and wires that have been successfully bonded.

Void-free junctions are produced by ultrasonic bonding with relatively few foreign-material inclusions, making it a desirable means of creating high-quality, low-resistance electric junctions. Welding dissimilar metals at low temperature eliminates or greatly decreases the formation of intermetallic compounds and allows bonds to be made in the immediate vicinity of temperature-sensitive materials without adverse effects. A further advantage of ultrasonic welding is that it requires no preheating, thus eliminating temperature rises that can damage the assemblies or create problems of thermal distortion or warping.

Surface cleaning is not highly critical in preparing most materials for ultrasonic bonding. The vibratory displacements occurring during the welding operation disrupt normal oxide layers and other surface films at the interfaces.

TABLE 7.2 Electric Conductors That Have Been Ultrasonically Bonded to Metallized Surfaces[18]

Conductor material	Lead material	Lead diameter, in (μm)
	On silicon substrate:	
Aluminum	Aluminum wire	0.0005 (12.7)
		0.001 (25.4)
		0.003 (76.2)
Aluminum	Gold wire	0.001 (25.4)
		0.002 (50.8)
	On ceramic substrate:	
Silver	Aluminum wire	0.010 (25.4)
		0.004 (101.6)
	On glass or glazed alumina substrates:	
Aluminum	Aluminum wire	0.002 (50.8)
		0.004 (101.6)
Aluminum	Gold wire	0.003 (76.2)
Nickel	Aluminum wire	0.002 (50.8)
		0.004 (101.6)
Nickel	Gold wire	0.002 (50.8)
		0.003 (76.2)
Copper	Aluminum wire	0.002 (50.8)
		0.004 (101.6)
Gold	Aluminum wire	0.002 (50.8)
		0.004 (101.6)
Gold	Gold wire	0.003 (76.2)
Tantalum	Aluminum wire	0.002 (50.8)
		0.004 (101.6)
Chromel*	Aluminum wire	0.002 (50.8)
		0.005 (127)
Chromel*	Gold wire	0.003 (76.2)
Nichrome†	Aluminum wire	0.002 (50.8)

*Trademark of Hoskins manufacturing Co., Detroit, Mich.

†Trademark of Driver-Harris Co., Harrison, N.J.

In general, results show that the pull strength and thus the bond efficiency is greater for gold films than for aluminum films.

7.2.3 Thermosonic Bonding

Thermosonic welding is a process that combines the best of both the thermocompression and ultrasonic wire bond methodologies. Essentially the use of ultrasonic excitation allows for a gold wire bond to be formed at lower temperatures and with less sensitivity to contamination than in the "pure" thermocompression case. A summary of the advantages and disadvantages of this system is shown below:

Advantages	Disadvantages
Uses lower temperatures (120 to 150°C) and faster bond times than thermocompression bonding (can be used on organic substrates).	Requires large pad area: 1-mil wire requires 4 × 4 mil pad.
Omnidirectional wire bonds.	Process requires control of both thermal and acoustic energy.
Less sensitive to surface contamination.	

7.3 SOLDERING

7.3.1 Solder Materials

Solder Alloys

Eutectic and Near-Eutectic Solders. Tin-lead solders used for electronics soldering are typically eutectic or near-eutectic alloys. The eutectic composition as shown in Fig. 7.10 shows the tin-lead eutectic at 63 percent tin by weight and 37 percent lead by weight (see point *B* of Fig. 7.10), and is referred to as Sn63Pb37.[8] In this diagram the area above the line *A-B-C* is always liquid (this line is referred to as the *liquidus*), and the area below the line *A-D-E-C* is always solid (this line is the *solidus*). The area between these two lines is a mixture of solid and liquid phases, often referred to as the *pasty region.* The selection of the near-eutectic condition is to promote rapid solidifi-

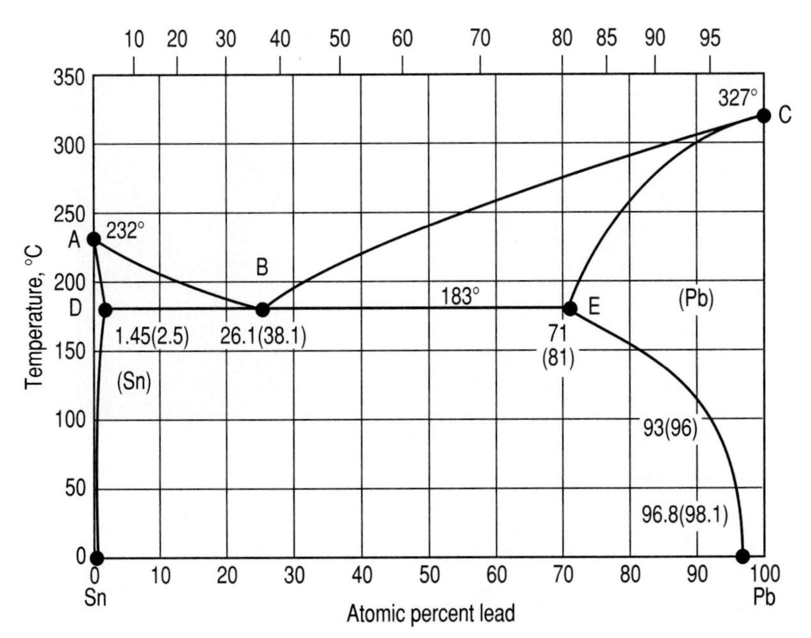

FIGURE 7.10 Phase diagram of tin-lead solder. (*From: Handbook of Materials and Processes for Electronics*, 1st ed., Fig. 13.51.)

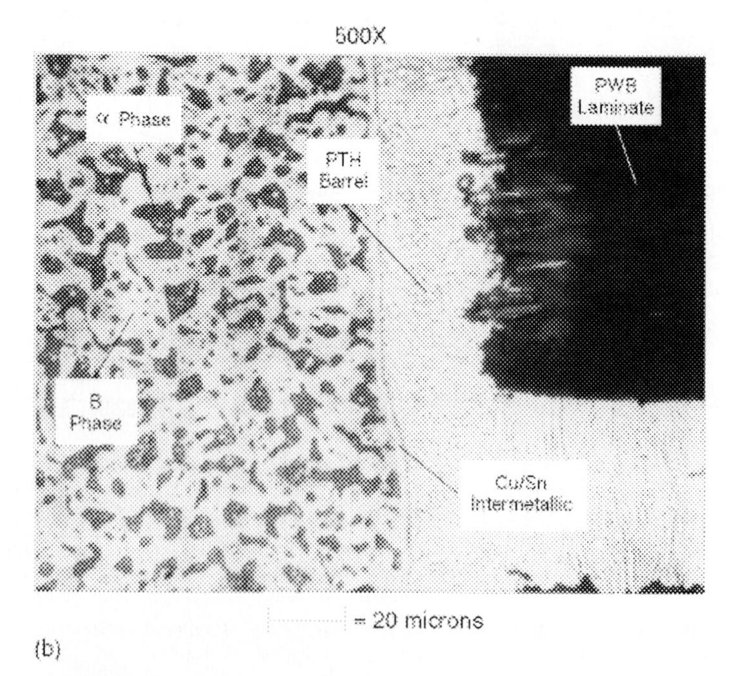

FIGURE 7.11 Metallograph of solder microstructure.

cation of the solder connection which minimizes grain growth during solidification, yielding a "shinier" solder joint.

Another common alloy used in electronics soldering is 60 percent tin and 40 percent lead, referred to as Sn60Pb40. This alloy is often preferred in commercial applications due to its lower cost and similar properties to Sn63Pb37. The tradeoff between Sn60Pb40 and Sn63Pb37 is the longer solidification time (due to the fact that the solder solidification process passes through the pasty region in Fig. 7.10). This can result in duller-appearing (although no less reliable) solder joints,[9] and an increased occurrence of disturbed solder connections (solder connections which move during solidification, resulting in a rough surface and dewetting).

The metallurgy of solder is a mixture of tin-rich crystals referred to as the *beta* (ß) *phase* with a composition of ~97.5 percent tin and 2.5 percent lead, and lead-rich crystals referred to as the *alpha* (α) *phase* with a composition of ~19 percent tin and 81 percent lead. The as-cooled appearance of the eutectic solder microstructure is seen in Fig. 7.11 as a two-phase structure containing a dispersion of darker platelets, which are the lead-rich α phase, mixed in with larger crystals of the tin-rich ß phase. [*Note:* When viewing solder using a scanning electron microscope (SEM), the solder matrix appears like a photographic negative of the optical appearance due to the relationship between apparent "brightness" in an SEM view and atomic weight.]

Solder morphology can be affected by a number of diffusion reactions which can occur while the solder is molten and after solidification. These reactions change the appearance of the solder microstructure and can be summarized as:

Leaching: The dissolution of substrate metallization(s) into the molten solder during soldering.

Coarsening: A solid-state diffusion reaction which causes the individual α and ß phases to coalesce in order to reduce total free energy of the system by minimizing the interfacial area of the separate phases. Visually the solder joints will appear to "dull." This process can be driven by either temperature or applied stress-plastic deformation.

Aging: The growth of intermetallic layers between the bulk solder and the substrate or component which is driven primarily by temperature.

The properties of these solder alloys are sensitive to temperature, strain rate, cyclic loading frequency, and grain size. A summary of these effects is given in Table 7.3.

A listing of common physical properties solder is given in Table 7.3.

The following environmental factors have a marked impact on the physical properties of eutectic tin-lead solder.

Factor	Effect
Temperature	Increasing temperature lowers strength (see Table 7.4) and changes solder deformation mechanism (see Fig. 7.12) and rate of stress relaxation (see Fig. 7.13).
Strain rate	Tensile and shear strength decrease with increasing strain rate (see Fig. 7.14).
Cycling loading rate	Fatigue life decreases with decreasing loading rate (see Fig. 7.15).

TABLE 7.3 Listing of Common Physical Property Values for Near-Eutectic Tin-Lead Solder

Physical property (alloy)	Value
Surface tension (Sn60Pb40)	0.41 J/m^2 @ 250°C in air
Elasticity (Sn63Pb37)	32×10^3 N/mm^2 @ 25°C
Viscosity (Sn63Pb37 and Sn60Pb40)	2 cP ($N \cdot s/m^2$)
Density: (Sn63Pb37)	8400 kg/m^3 at 20°C
(Sn63Pb37)	8000 kg/m^3 at 250°C
(Sn60Pb40)	8500 kg/m^3 at 20°C
Coefficient of thermal expansion (Sn63Pb37)	24.5×10^{-6}/K
Heat of fusion (Sn60Pb40)	4600 J/kg
Specific heat (Sn60Pb40)	176 J/(kg-K)
Electrical resistivity (Sn60Pb40)	0.17 $\mu\Omega \cdot m$ @ 25°C 0.32 $\mu\Omega \cdot m$ @ 100°C
Thermal conductivity (Sn60Pb40)	51 $J/(m \cdot s \cdot K)$ @ 25°C 49 $J/(m \cdot s \cdot K)$ @ 100°C
Thermal diffusivity (Sn63Pb37)	$3.4 * 10^{-5}$ m^2/s

Source: R. J. Klein Wassink, *Soldering in Electronics,* 2d, ed., Electrochemical Publications, 1989, pp. 162–167.

TABLE 7.4 Effect of Temperature on Lap Shear Strength psi (Pa) *[Testing Done at 0.020 in/min (0.5 mm/min)]*

	Temperature, °C		
Solder alloy	−55°C	25°C	100°C
Sn63Pb37	10,070 (69.4)	6060 (41.8)	3510 (24.2)
Sn40Pb60	10,320 (71.2)	6090 (42)	3260 (22.5)

Source: "Latent Defect Life Model and Data," *Air Force Wright Laboratories Report No. AFWALTR-86-3012,* App. A-6, March 1986.

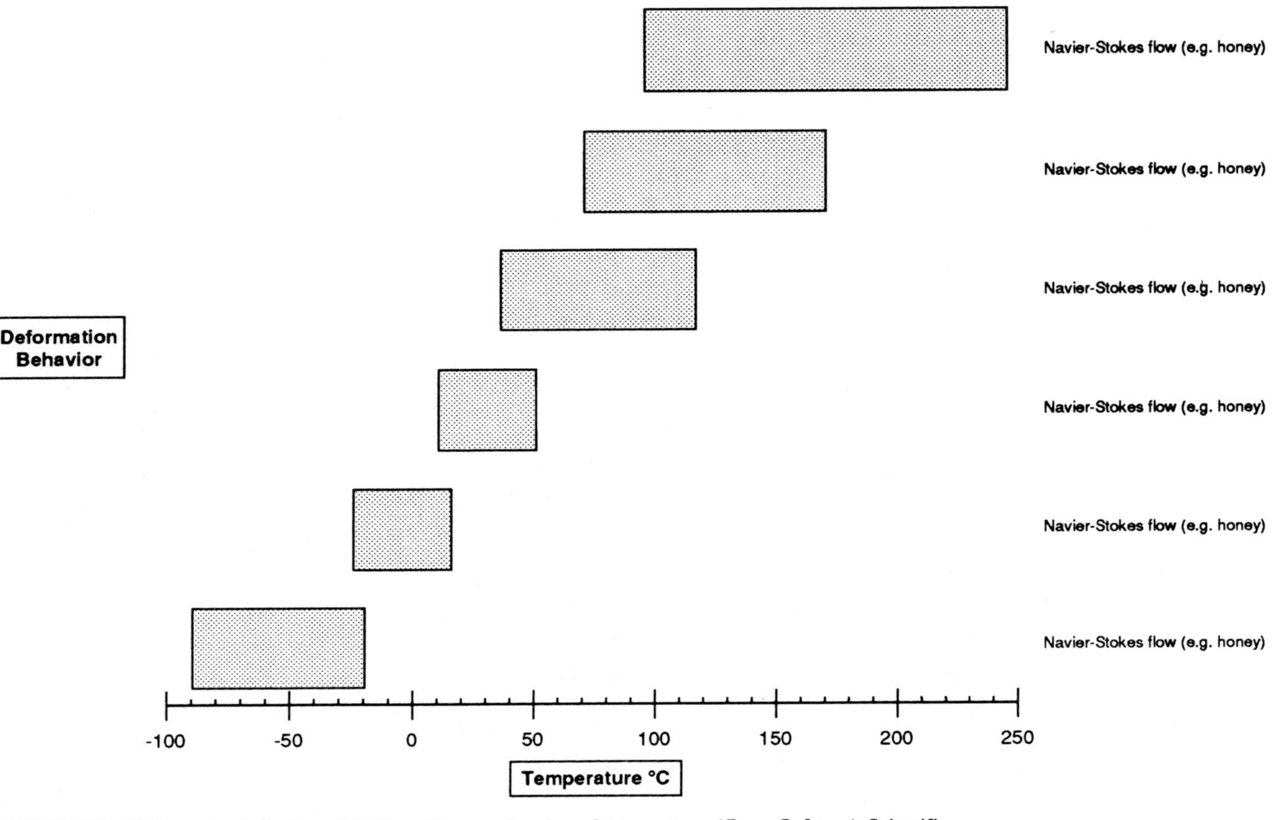

FIGURE 7.12 Deformation behavior of Sn/Pb solder as a function of temperature. (*From C. Lea,* A Scientific Guide to Surface Mount Technology, *Electrochemical Publications, 1988, p. 386.*)

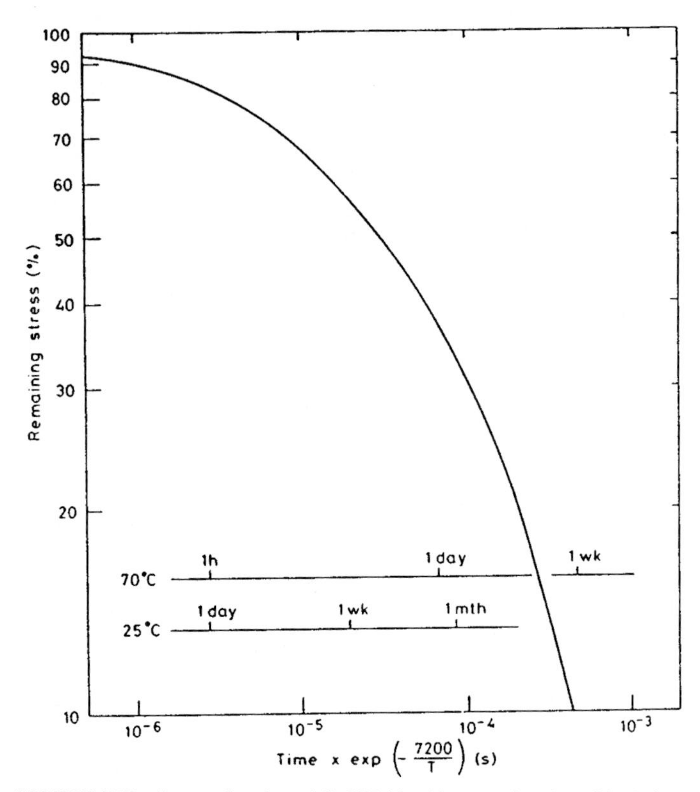

FIGURE 7.13 Stress relaxation of Sn60Pb40 solder as a function of both time and temperature. (*From C. Lea,* A Scientific Guide to Surface Mount Technology, *Electrochemical Publications, 1988, p. 406.*)

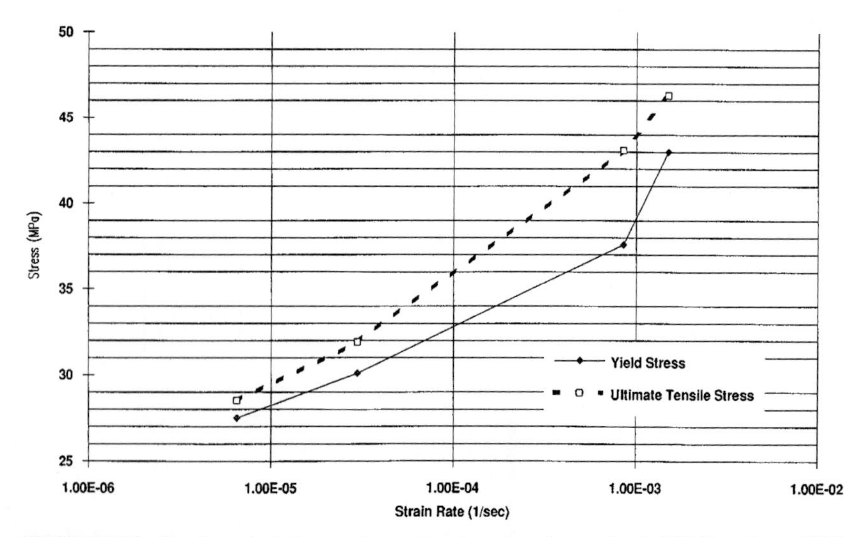

FIGURE 7.14 Tensile and yield strength as a function of strain rate for Sn60Pb40 solder at 25°C. (*From S. K. Kang and R. A. Pollack,* Solder Technology for Surface Mount Components, *5th International Electronics Packaging Conference, October 1985.*)

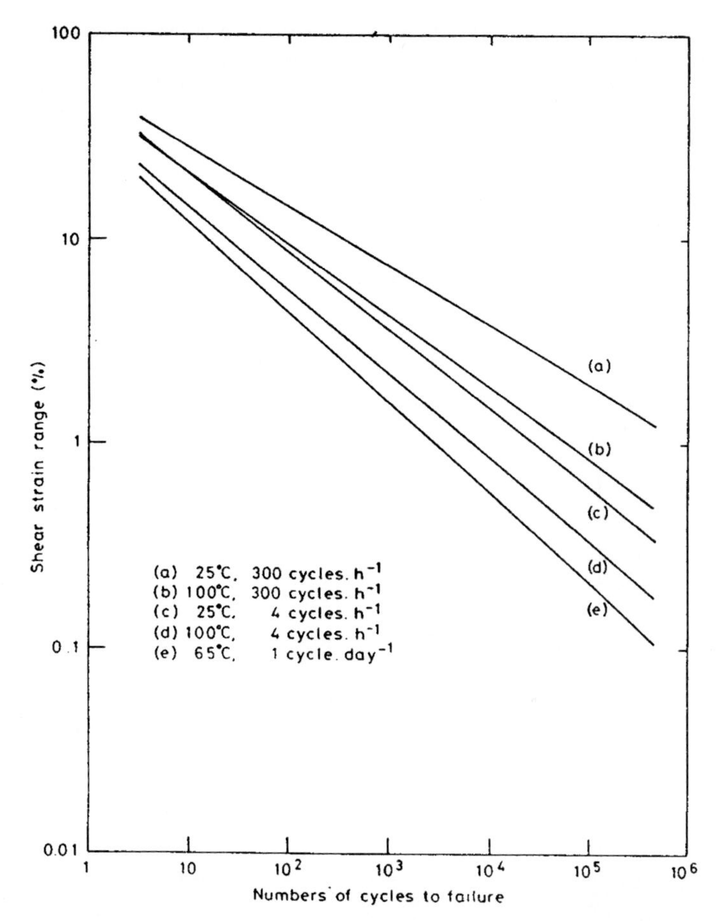

FIGURE 7.15 Fatigue data for Sn63Pb37 solder as a function of cyclic frequency, temperature, and shear strain. (*From C. Lea,* A Scientific Guide to Surface Mount Technology, *Electrochemical Publications 1988, p. 412.*)

The morphology of solder is a laminar structure and has no "grains." In this discussion, the *grain size* is actually a measure of the mean free path between lead-rich islands[10] defined by

$$\overline{L} = \frac{V\alpha}{N} \tag{7.1}$$

where $\overline{L}$ = mean intercept length
$V\alpha$ = volume fraction of the lead-rich α phase
N = number of interceptions of lead-rich phase per unit length of a random test line

Grain size, or, more properly, the size and distribution of the α and β phases in Sn-Pb solders, is initially a function of cooling rate with finer dispersions resulting from faster cooling rates. As grain size increases, both yield strength and tensile strength increase while total elongation decreases, which can result in a decrease in fatigue life as shown in Fig. 7.16.

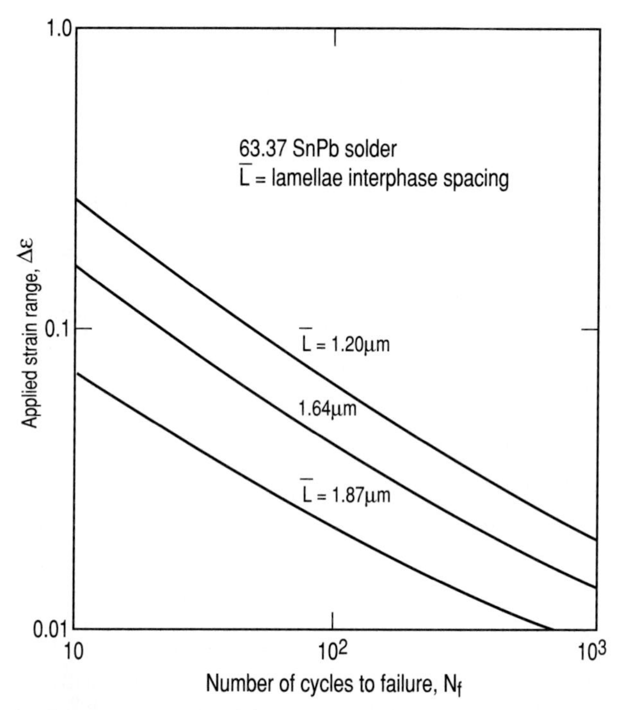

FIGURE 7.16 Fatigue life of Sn63Pb37 solder as a function of microstructure. (*From C. Lea,* A Scientific Guide to Surface Mount Technology, *Electrochemical Publications, 1988, p. 426.*)

It should be noted, however, that these effects are generally not stable over time. Harris et al. demonstrated that Sn60Pb40 solder morphology coarsens even during room temperature storage with grain sizes doubling in as little as 48 h at room temperature. It is also noted that this coarsening occurs more rapidly with finer (e.g., quenched) microstructures than with normally cooled (and coarser) microstructures.[11]

A summary of the recommended levels of solder impurities for Sn60Pb40 and Sn63Pb37 solders and their effect[12,13,14] are shown in Table 7.5. As an aid in determining the rate at which some of these materials may be introduced into the molten solder, the dissolution rates for a variety of materials are shown in Fig. 7.17.

TABLE 7.5 Recommended Levels of Solder Impurities for Sn60Pb40 and Sn63Pb37 Solders, Weight %

Material	Maximum allowable new solder,* %	Maximum allowable during use, %	Effect on solder joint if maximum allowable level is exceeded during usage
Silver (Ag)	0.10	0.10	Solder surface finish becomes gritty.
Aluminum (Al)	0.005	0.005**	Solder becomes sluggish, frosty, and porous; oxidation rate increases.
Arsenic (As)	0.03	0.03	Grainy, pitted appearance; edge dewetting and reduction in area of spread.
Gold (Au)	0.08	0.20**	Solder becomes sluggish and grainy; above 4% solder becomes brittle.
Bismuth (Bi)	0.10	0.50	Dulling of surface; may reduce ability of solder to wet brass and steel.
Cadmium (Cd)	0.005	0.005**	Oxidation rate increases; above 0.15% area of spread decreased by 25%.
Copper (Cu)	0.08	0.20**	Solder begins "sticking" to insulation materials; above 0.3% viscosity increases and solder becomes gritty.
Iron (Fe)†	0.020	0.020	Forms $FeSn_2$ which is insolderable.
Nickel (Ni)†	0.010	0.010	Blisters; formation of hard insoluble compounds.
Phosphorus (P)‡	0.010	0.010	Increased probability of dewetting.
Sulphur (S)‡	0.005	0.005	Solder becomes gritty due to formation of SnS and PbS particles.
Antimony (Sb)‡	0.50¶	0.05	Increased dewetting.
Zinc (Zn)	0.005	0.005**	Finish becomes dull; dewetting; oxidation rate increases.

*Data from QQ-S-571E Amendment 6, "Solder, Electronic (96 to 485°C)," for Class 3 (critical) use solders.

**The total of copper, gold, cadmium, zinc, and aluminum shall not exceed 0.4 percent.

†The solubility of these materials in solder at temperatures <260°C is very low, and problems arise only as a result of prolonged exposure.

‡These materials are used in the production of solder and are not normally added during use.

¶This value will be lowered to a maximum of 0.12 percent within 1 year of the publication of QQ-S-571F based on recent testing.

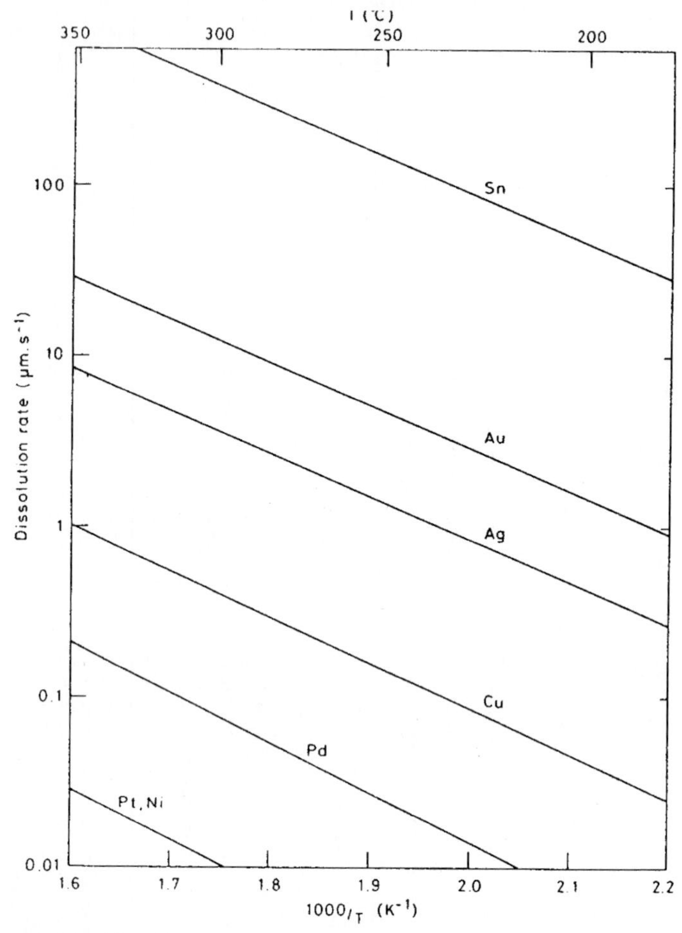

FIGURE 7.17 Dissolution rates of various metals in Sn60Pb40 solder as a function of temperature. (*From C. Lea,* A Scientific Guide to Surface Mount Technology, *Electrochemical Publications 1988, pg. 167.*)

The rates of growth of the intermetallic compounds which form between solder and various substrates can be estimated using the following equation:

$$z_0^2 = D_0 \cdot t \cdot \exp\left(\frac{-Q}{RT}\right) \tag{7.2}$$

where z_0 = intermetallic thickness after exposure at time t and temperature T
D_0 = diffusion coefficient, m^2/s
t = time of exposure, s
Q = activation energy, J/mol
R = molar gas constant, 8.314 J/(mol °K)
T = temperature, °K

It should be noted that this formula is only an approximation and the following caveats apply:

- The initial growth rate is faster during the initial exposure, and, therefore, these estimates should not be used to estimate growth over short aging times (yielding thicknesses ≤1.0 μm).
- Intermetallic layers tend to vary in thickness over the contact area, and these equations estimate average thickness only.
- These estimates were developed from test data on Sn60Pb40 and Sn63Pb37 solders and should not be applied to other solder alloys.

A summary of the diffusion constants D_0, and activation energies Q as well as the intermetallics formed are given in Table 7.6. The intermetallic compounds are listed in the order found when moving from the substrate toward the bulk of the solder.

TABLE 7.6 Intermetallic Compounds and Diffusion Constants for Near-Eutectic SnPb Solders

System	Intermetallic compounds	Diffusion coefficient, m^2/s	Activation energy, J/mol
Cu-Sn	Cu_6Sn_5, Cu_3Sn	1×10^6	80,000
Ni-Sn	Ni_3Sn_2, Ni_3Sn_4, Ni_3Sn_7	2×10^7	68,000
Au-Sn	AuSn, $AuSn_2$, $AuSn_4$	3×10^4	73,000
Fe-Sn	FeSn, $FeSn_2$	2×10^9	62,000
Ag-Sn	Ag_3Sn	8×10^9	64,000

Source: C. Lea, *A Scientific Guide to Surface Mount Technology,* Electrochemical Publications, 1988, pp. 333–336.

The specific intermetallic compound formed often depends on the temperature used. For example, the copper intermetallic Cu_6Sn_5 (η) forms at all temperatures and is relatively coarse grained while Cu_3Sn (ϵ) forms at temperatures >60°C at the interface of the Cu_6Sn_5 and the bulk solder.[16]

Low-Temperature Solders. *Low-melting-point solders* (those having a melting point below 183°C, which is the melting point for Sn63Pb37) are used for a number of applications, especially the soldering of heat-sensitive components. Typically these alloys are of tin and lead, tin, lead, and bismuth, tin and indium, or tin, lead, and indium systems. Some tin-lead-cadmium alloys also fulfill the requirement for low melting point but are not as commonly used due to the higher cost and the toxicity of cadmium.

Tin-bismuth solders are noted for their low melting point combined with higher strength, comparable to tin-lead solder. Some of the physical properties of these solders are listed in Table 7.7 along with Sn63Pb37 for comparison. A phase diagram for the tin-lead-bismuth system is shown in Fig. 7.18.

In use, the soldering of tin-lead-bismuth solders suffers from two constraints:

- The fluxes normally used for soldering near-eutectic tin-lead solders are often not fully activated at the melting point of these alloys, often causing poor wetting.[17]
- Severe oxidation occurs both in the liquid and solid states. This is important in thermal safety fuses where the oxide may prevent the fuse from "opening" even after the melting point is reached.[18]

TABLE 7.7 Properties of Some Common Tin-Lead-Bismuth Solders

Alloy composition				Melting point, °C		Shear strength, Pa	
Sn	Pb	Bi	Alloy name	Liquidus	Solidus	20°C	100°C
42	—	58	Sn42Bi48*	138	138	50	19.5
16	32	52	Sn16Pb32Bi52*	96	96	—	—
34	42	24	Bi24	146	100	34.3	17.5
43	43	14	Sn43Pb43Bi14*	163	144	—	—
34	20	46	Sn34Pb20Bi46*	100	100	—	—
60	40	—	Sn60Pb40*	188	183	33.6	21.6

*Alloy listed in QQ-S-571E, "Solder, Electronic (96 to 485°C)."

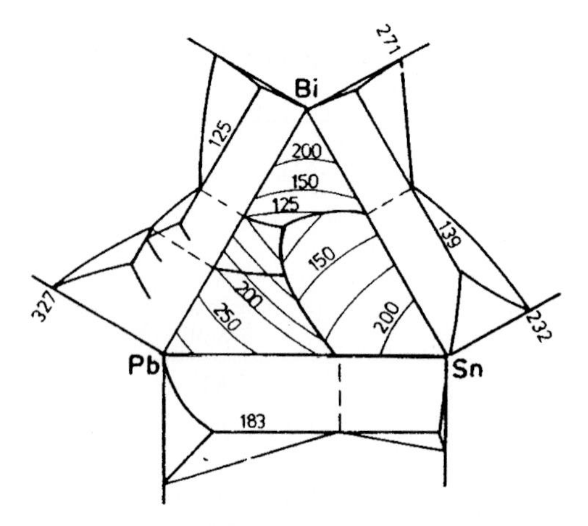

FIGURE 7.18 Phase diagram for tin-lead-bismuth solder. (*From R. J. Klein Wassink,* Soldering in Electronics, *2d ed., Electrochemical Publications, 1989, p. 197.*)

The resistance of the tin-lead-bismuth solders to thermal and mechanical fatigue failure is still a matter of much debate. It has been established that the shear strain at failure is very sensitive to strain rate (being reasonably good at small strain rates[19]), but actual testing of components and test specimens in thermal cycle,[20] and isothermal fatigue[21] have indicated that tin-lead-bismuth solders are more susceptible to fatigue failure than Sn60Pb40 solders.

The tin-indium solders are also low-temperature solders. They differ from the tin-lead-bismuth solders in that they typically have lower strength and higher cost but offer advantages such as:

- Excellent wetting properties.[22]
- Higher ductility (and thus better fatigue resistance) than the tin-lead-bismuth systems and in some cases superior to Sn60Pb40.[23]
- In the cases in which dissolution of gold into the solder is a problem, lead-indium solders are often used since the dissolution of gold into both lead and indium is low.[24]

It has been noted that the indium in these solders may be selectively dissolved in liquid rosin flux (colophony),[25] and this effect should be considered when developing soldering processes for these materials. The composition and melting points of some common lead-indium solders are given in Table 7.8 along with Sn60Pb40 for comparison.

TABLE 7.8 Composition, Alloy Name, and Melting Point of Tin-Lead-Indium Solders

Composition				Alloy	Melting point, °C	
Sn	Pb	In	Ag	name	Liquidus	Solidus
48	—	52	—	Sn48In52*	118	118
—	15	80	5	In80Pb15Ag5*	150	149
—	—	99	—	In99*	157	157
70	18	12	—	Sn70Pb18In12*	163	153
37	37	25	—	In25	138	138
60	40	—	—	Sn60Pb40*	188	183

High-Temperature Solders. A number of tin-lead or tin-lead-silver solders are used when high melting points are needed (e.g., either for step soldering or component assembly). These solders typically have poorer wetting, a duller surface finish, and lower mechanical strength than the eutectic tin-lead solders. A summary of the composition, melting points, and some mechanical properties are listed in Table 7.9.

TABLE 7.9 Properties of Some Common Tin-Lead-Silver Solders

Alloy composition				Melting point, °C		Shear strength, Pa	
Sn	Pb	Ag	Alloy name	Liquidus	Solidus	20°C	100°C
10	90	—	Sn10Pb90	234	183	34.3	13.7
5	95	—	Sn05Pb95*	312	308	—	—
10	88	2	Sn10Pb88Ag02*	290	268	—	—
5	93.5	1.5	Sn05Pb94Ag02*	301	296	23.8	15.7
60	40	—	Sn60Pb40*	188	183	33.6	21.6

*This alloy listed in QQ-S-571E, "Solder, Electronic (96 to 485°C).

Source: C. Lea, *A Scientific Guide to Surface Mount Technology,* Electrochemical Publications, 1988, p. 171.

Solder Fluxes

Purpose of Fluxing. A material will flow freely over a surface only if, in doing so, the total free energy of the system is reduced. In the case of soldering, the free energy of a clean surface is higher than a dirty one and therefore will be more likely to promote solder flow. The major purposes of a flux are:

- *Chemical:* To remove oxides and contaminants from the surface to be soldered and protect this surface (by covering it) from reoxidation

- *Thermal:* To assist in the transfer of heat from the heat source to the item being soldered (especially critical for hand and pulse soldering)
- *Physical:* To allow for transport of the oxides and other reaction products away from the area being soldered.

Fluxes normally consist of an active agent (to reduce or break up fluxes) dissolved in a liquid or paste.

In use, fluxes have two general criteria:

- *Flux activity:* The ability of the flux to remove oxides and other contaminants and promote soldering
- *Flux corrosivity:* The impact of the flux residues on the long-term reliability of the assembly or device being soldered.

Typically these two criteria oppose one another because active fluxes tend to be highly corrosive and fluxes that are not corrosive over the long term are also generally not very active in the short term. One method of overcoming this problem is, obviously, to remove the flux residue using some cleaning method, but there is always some potential for residue remaining (due to operator error or inadequate processing), and therefore extremely active fluxes are rarely used in electronics.

Flux activity can be measured in a number of ways (see below), the important issue when using these tests to compare fluxes is to ensure that all the samples used have the same initial solderability and that the level(s) of solderability used are representative of production hardware:

- Solder spread test (See MIL-F-14256. "Flux, Soldering, Liquid, Paste Flux, Solder Paste And Solder Paste Flux (For General Electronic/Electrical Use), General Specification For").
- Wetting Time - Globule Test (See ANSI/J-STD-002, "Solderability Tests for Component Leads, Terminations, Lugs, Terminals and Wires").
- Wetting Time - Wetting Balance Test (ANSI/J-STD-002, "Solderability Tests for Component Leads, Terminations, Lugs, Terminals and Wires").

Rosin Fluxes. Typically rosin fluxes are a combination of colophony combined with aliphatic alcohols. *Colophony* is distilled from pine tree sap and is a combination of several isomeric resin acids with the major component being abietic acid $C_{19}H_{29}COOH$ which has a melting point of 172°C.

Rosin fluxes are subject to degradation during use including:

- Oxidation which can reduce the solubility of the flux in some solvents and decrease the fluxes activity.
- Polymerization which will increase the melting point of the colophony and may contribute to the formation of white flux residues.
- Water absorption, which while it does not directly affect the flux activity, can lead to dilution of the flux. This is caused by the water increasing the specific gravity of the flux which is used to monitor solids content (a measure of the amount of actual rosin in the flux). As water is absorbed, the specific gravity increases, the operator adds thinner (typically isopropyl alcohol) to correct, and unintentionally dilutes the flux activity. Levels of water >5 percent can have a discernible effect on the activity of the flux.[26]

Synthetic resins have also been produced in an effort to allow for a more controllable source. (The characteristics of colophony often are related to the origin and harvest year of the pine tree used.) These resin materials are thought to be more consistent and offer less splattering and fuming during soldering.

The advantages of the colophony (rosin) fluxes are that they wet the surface to be soldered well at soldering temperatures and are comparatively nonreactive at room temperatures (although commercially available nonactivated rosin fluxes have been found to lead to low-electrical resistance failures during high humidity exposures[27,28]).

Organic acids or salts may be added to increase the activity of the flux. These materials (sometimes incorrectly referred to as *activators*) may include chloride, bromide, monocarboxylic acid, dicarboxylic acid, and hydroxyl substituted polybasic acids (see Table 7.10).

Often the designations R (for pure rosin), RMA (rosin, mildly activated), and RA (rosin, activated) are used to differentiate between levels.[29] These designations offer only a very coarse approximation of the activity of the flux, and it is not uncommon that, in practice, the apparent rankings are reversed, e.g., RMA fluxes proving more active than RA fluxes. Often, these rankings are based on factors other than corrosivity, e.g., the extract resistivity of the flux (the lower the resistivity, the more active the flux), and not on corrosivity. An alternative ranking system is being used in MIL-F-14256D and is proposed in IPC-SF-818 based on corrosivity with rankings of low, medium, and high. Each flux should be separately evaluated in the actual formulation to be used to determine its acceptability in a given soldering and cleaning process.

The designation SA is used for *synthetically activated* (or "*super-activated*") fluxes which contain organic derivatives of sulfur- and phosphorus-bearing acids. These fluxes have an extremely high activity, and flux removal is mandatory for most usage applications.

Water-Soluble Fluxes. Water-soluble fluxes are those designed to be removable using a nonsolvent cleaning process (typically water or water and detergent). These fluxes normally contain:

- A reactive chemical for removing oxides, i.e., "activator" (see Table 7.11)
- A wetting agent to promote flow of the flux over the surfaces to be soldered
- A solvent (typically isopropyl alcohol) to distribute these materials evenly across the surface to be soldered
- A material (typically a polyglycol or glycol) to protect the surface from oxidation during soldering and to keep the activator in contact with the surface to be soldered

In general, the fluxes using inorganic salts are not used for soldering electronics due to their excessive corrosivity but are sometimes used for nonelectrical soldering (e.g., lightening ground straps). The fluxes with organic salts generally also contain polyglycol or glycol vehicles. Some of these vehicles (especially polyethylene glycol and polyvinyl alcohol) have been found to degrade insulation resistance.[30] This is due to the fact that these materials can be absorbed into the substrate and are hygroscopic (absorbing water from the air at relatively low humidity levels). These residues can cause a substantial loss (three to six orders of magnitude) in the insulation resistance of the substrate at humidity levels greater than ~75 percent RH.[31]

No-Clean Fluxes. These are generally fluxes with moderate activity and a very low corrosivity. Typically these fluxes use weak organic acids for activators and do not use colophony, although synthetic resins may be used. The intent is that the product is soldered and the residue is left in place.

TABLE 7.10 Some Acids and Salts Used As Activators in Fluxes

Chemical nomenclature		Molecular	Melting	Boiling	Acid
Monobasic acids	Monocarboxylic acids	weight	point, °C	point, °C	value
Formic acid	Methanoic acid	46	8	101	
Acetic acid	Ethanoic acid	60	17	118.5	
Propionic acid	Propanoic acid	74	−21	141	
Butyric acid	Butanoic acid	88	−6	164	
Valeric acid	Pentanoic acid	102	−34	186	
Caproic acid	Hexanoic acid	116	−4	205	
Enanthic acid	Heptanoic acid	130	−10	233	
Caprylic acid	Octanoic acid	144	16	239	
Pelargonic acid	Nonanoic acid	158	12	253	
Capric acid	Decanoic acid	172	31	286	326
Lauric acid	Dodecanoic acid	200	44	300	280
Myristic acid	Tetradecanoic acid	228	54	300	246
Palmitic acid	Hexadecanoic acid	256	63	300	219
Stearic acid	Octadecanoic acid	284	70	300	197
Arachic acid	Eicosanoic acid	312	76	300	179
Behenic acid	Docosanoic acid	340	80	300	165
Dibasic acids	Dicarboxylic acids				
Oxalic acid	Ethanedioic acid	90	189d	157s	
Malonic acid	Propanedioic acid	104	135	140d	1079
Succinic acid	Butanedioic acid	118	182	235s	951
Glutaric acid	Pentanedioic acid	132	97.5	300	850
Adipic acid	Hexanedioic acid	146	153	300	768
Pimelic acid	Heptanedioic acid	160	106	300	701
Suberic acid	Octanedioic acid	174	140	300	645
Azelaic acid	Nonanedioic acid	188	106	300	597
Sebacic acid	Decanedioic acid	202	134	300	555
Fumaric acid	Trans-butenedioic acid	116	286	>300	967
Maleic acid	Cis-butenedioic acid	116	130		967
Tataric acid	2,3 dihydroxybutanedioic acid	150	173	d	748
Levulinic acid	4-oxo-pentanoic acid	116	37	245d	483
Lactic acid	2-hydroxy-propanoic acid	90	25	d	623
Acrylic acid	Propanoic acid	72	12	141	778
Benzoic acid	Benzenecarboxylic acid	122	122	s	460
Salicylic acid	2-hydroxy benzoic acid	138	159	211s	406
Anisic acid	4-methoxy benzoic acid	152	185	275	369
Citric acid	2-hydroxy propane tricarboxylic acid	192	153	d	
Organic salts					
	Dimethylammonium chloride	81.6	170		
	Diethylammonium chloride	109.6	223.5		

Source: R. J. Klein Wassink, *Soldering in Electronics,* Electrochemical Publications, 1989, p. 262.

TABLE 7.11 Common Activators Used for Water-Soluble Fluxes

Type		Activators used
Inorganic	Salts	Zinc chloride Ammonium chloride, zinc chloride
	Acids	Hydrochloric acid Orthophosphoric acid
Organic	Salts	Halide containing salts, e.g.: Aniline hydrochloride Glutamic acid hydrochloride Dimethylammonium chloride
	Acids	Lactic acid Glutamic acid Amino acids
	Amines	Urea Triethanolamine Amino ethyl ethanol amine

Source: R. J. Klein Wassink, *Soldering in Electronics,* Electrochemical Publications, 1989, p. 238.

Generally the fluxes fall into two categories with respect to the residue formed after soldering:

- The fluxes that contain 15 to 35 percent solids (typically synthetic resins) which essentially encapsulate the soldered surfaces
- Low-solids fluxes that contain 2 to 8 percent solids which leave little or no discernible residue

For the fluxes to be useful, it is necessary that these materials be evaluated for corrosivity and insulation resistance of the residue when exposed to life-cycle-type environments. It has been noted that in many cases "no-clean" fluxes are not compatible with cleaning attempts or even other no-clean fluxes.[32] The impact of materials used in subsequent processes (e.g., fluxes, solvents, or conformal coatings) needs to be evaluated in combination with the flux and soldering process under evaluation.

Solder Pastes. Solder paste is a mixture of solder particles with flux and materials used to control viscosity and flow/print characteristics.

Solder pastes are nonnewtonian fluids (composed of more than phase) and are also *thixotropic* (a material which viscosity decreases asymptotically over time under constant shear). The materials used to control viscosity and the thixotropic nature of the solder paste are generally proprietary.

Solder Powder. The mixture of the flux and thixotropic control materials are referred to as the *flux binder.* Occasionally solder powders are mixed such that the "bulk" composition is a common solder (e.g., Sn63Pb37) but is in fact a mix of particles of other compositions. An example of this is the solder powder sometimes used for leaded SMD soldering which is Sn63Pb37 when analyzed in bulk but is in fact a mix of pure tin and Sn10Pb90 powders. This mixture is used to slow the reflow of the joint during vapor phase soldering, thus reducing the number of open solder connections sometimes produced.[33] Although the melting points of both materials are above the typical vapor phase reflow temperature of 215°C, solder reflow is accomplished initially by solid-state diffusion between the individual particles creating a meltable alloy at the interface, and in this manner complete reflow is achieved.

The particles are generally formed by atomizing liquid solder in an inert atmosphere. The shape distribution of the resulting powder is determined during the atomization phase as sieving the material through a mesh doesn't necessarily eliminate elongated particles. This distribution is a function of the skill and processes used by the powder manufacturer, details of which are generally proprietary. Particle size is also impacted by the powder production process.

The particle shape of the solder powder affects a number of paste characteristics including:

- Oxide formation on the surface of the powder. The larger the surface-to-volume ratio, the larger the amount of oxides formed on the solder paste. These oxides will inhibit reflow, increase the formation of solder balls, and subsequently reduce the overall volume of solder in the point compared to the amount of solder printed (the difference being the amount of solder loss through solder ball formation).
- Large variations in particle size and shape will increase both the wear on the dispensing equipment (e.g., stencil, screen, or syringe) as well as increase the probability of small openings becoming blocked by solder particle "logjams."

A certain degree of nonuniformity of solder shape has been shown to increase the ability of the solder paste to resist solder paste "slump" during heating. In modern solder pastes, however, slump is controlled by the "binder." The uniformity of shape (spherical) is generally taken as one of the measures of the overall quality of the solder powder.[34]

Solder particle sizes are a result of the solder powder process combined with post-process sorting of the powder using mesh screens to sort the powder by size. Typically, solder powder sizes are categorized by a mesh value (which is nominally the number of lines per inch of wire in the mesh) preceded by a plus or minus sign which indicates whether the powder in question passed through the screen (a minus sign) or is retained on top of the screen (a plus sign). A summary of common mesh sizes used for solder powders is shown in Table 7.12.

TABLE 7.12 Mesh Size versus Particle Size for Solder Powders Used in Solder Paste

Mesh size	Particle sizes, μm
−325	Less than 45
−270	Less than 55
−200	Less than 75
−200, +325	Between 45 and 75

Source: C. Lea, *A Scientific Guide to Surface Mount Technology*, Electrochemical Publications, 1988, p. 156.

The choice of appropriate solder paste size is a tradeoff between the ease with which the solder paste can be printed, stenciled, or dispensed versus the amount of solder balling. Solder balling increases with decreasing particle size due to the increased ratio (surface area/volume) of the powders and the fact that the smaller particles are more apt to be washed away from the main deposition of solder paste. These isolated solder particles are unable to coalesce into the main solder joint and are subsequently washed away during post-solder cleaning. Fines may be reduced by requiring both an upper and lower bound for particle size (for example, −200, +325).[35]

Oxides are formed by the exposure of the solder powder to air either during the powder fabrication or the addition of the binder materials. Typically solder powders contain some oxides, but new manufacturing techniques have made it possible to fabricate the solder paste (from powder through mixing with the binder) with an oxide content of <0.03 percent by weight. When the oxide content increases to above 0.15 percent by weight, solder balling may result, especially with less active (e.g., RMA) fluxes. Solder oxides are less dense than solder (typically ~6500 kg/m^3) and will float to the surface of the molten solder.

Metals Content. Typically the metals content of solder pastes is expressed as a percentage by weight. For most electronics applications, the solder paste viscosity used is 80 to 95 percent by weight. The relationship between volume and weight percents is shown in Fig. 7.19. Usually, as the solder paste metals content decreases, the viscosity of the paste becomes more dependent on thixotropic control materials. These materials typically lose effectivity with increasing temperature, and therefore pastes with lower metals contents have an increased tendency toward slump. Typical metal contents as a function of paste application method is shown in Table 7.13.

Paste Viscosity. The viscosity of the solder paste is highly dependent upon:

- Temperature.
- Applied shear stress.
- The "work history" of the solder paste. This is a result of the fact that, as a solder paste is exposed to shear stresses, the viscosity drops and does not recover immediately.
- The shear velocity at which the viscosity is measured.

A demonstration of these effects is shown in Table 7.14.

As a result, all the above items must be tightly controlled in order to get repeatable results between various test sites. Typically two types of viscosity measurement equipment are used for solder pastes, a rotating concentric cylinder method and a rotating helipath with a paddle.

As viscosity changes with velocity, a measure of viscosity at two or more points in the paste's operating range (as determined, for example, from squeegee print speed) is the most informative. This information can be used to develop a *viscosity slope index* (VSI), which can be used to characterize solder pastes for printability. It has been suggested that a VSI of 3 to 6 are acceptable with VSIs as high as 8 being available and providing superior printability.[36]

Typical viscosities of solder pastes (as a function of metal content and application method) is given in Table 7.13.

Determination of Correct Paste Volume. The use of solder paste allows for a unique control of the volume of solder deposited for a given solder joint. In this fashion the volume of solder deposited can be tailored to reduce defects and compensate for other restraints (e.g., need for common thickness of a stencil or pad to pad spacings).

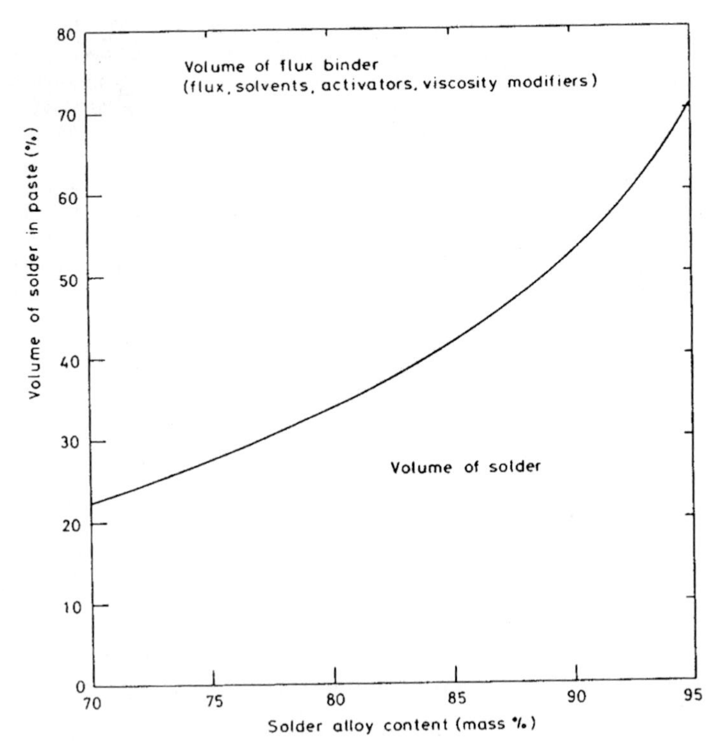

FIGURE 7.19 Typical relationship between volume percent and weight percent. (*From C. Lea,* A Scientific Guide to Surface Mount Technology, *Electrochemical Publications, 1988, Fig. 6.7.*)

TABLE 7.13 Recommended Viscosities of Solder Pastes
(@25°C; 1 Pa • s = 1000 cP)

Application method	Metal content, % by weight	Viscosity, Pa • s
Stencil print	90	600–1000
Screen print	85–90	400–700
Syringe dispenser	80–85	200–450
Pin transfer	75	50–250

TABLE 7.14 Effects of Parameters on Viscosity
(Shear rate is 1.5/s unless otherwise noted.)

Parameter	Change	Viscosity change, Pa • s
Particle size	1–45 μm to 40–80 μm	From 250 to 390
Metal content	86% to 90% (mass)	From 280 to 400
Temperature	From 18°C to 28°C	From 320 to 220
Shear rate	From 0.1/s to 50/s	From 2000 to 50

Source: R. J. Klein Wassink, *Soldering in Electronics,* Electrochemical Publications, 1989, p. 552.

Models for determining the volume of solder paste required for a specific solder joint vary from simple geometric models[37] to more complex functions which take into account such features as surface tension, solder density, gravity, and wetting angle.[38] In general, the geometric models perform well enough for most purposes and are readily adaptable to spreadsheet type "what-if" models.

Items that the model will need to take into consideration are:

- Lead shape (including lead-to-lead coplanarity and potential spacing of lead from pad)
- Proposed solder fillet shape
- Solder volume required for acceptable fillet
- Solder already present in the form of pretinned pads and leads
- Variability in lead dimensions, pad dimensions

It should be stressed that this model is only an approximation based on nominal conditions; allowances should be made for normal and expected process variations.

Application Methods. Solder pastes can be applied using a variety of applications techniques, depending on the process and manufacturing constraints.

Solder Wire and Solder Preforms. Typically solder wire is identified by flux type, flux form, and weight percent flux in the wire. The identification scheme used in QQ-S-571, for example, is shown below.

Flux Type

S = solid metal (no flux)

R = rosin flux (no activators)

RMA = rosin, mildly activated

RA = activated rosin or resin flux

Core (flux) condition and amount

P = plastic form

D = dry powder

Flux Percentage, by weight

Flux percentage, designator	Flux percentage, by weight		
	Nominal	Minimum	Maximum
1	1.1	0.8	1.5
2	2.2	1.6	2.6
3	3.3	2.7	3.9
4	4.5	4.0	5.0
6 (not for R, RMA, or RA)	6.0	5.1	7.0

Solder preforms are generally either solder wire or solder sheets formed to a specific shape (typically a toroid or washer) for use in reflow soldering (e.g., oven, vapor phase, or IR) of PTH devices. Preforms may contain flux but are typically either flux free or coated with a mild rosin flux (both for soldering and to prevent oxidation on the preform surface).

Component and PWB Baking

Often during fabrication and storage, both components and printed wiring boards will absorb water. If left in the device, this water will vaporize at soldering temperatures and can lead to PWB delamination, soldering voids (especially in PTHs), and device cracking.

For printed wiring boards, the bakeout is to remove water accumulated during the fabrication process as well as water absorbed during storage. Recommended baking times and temperatures[39] are given below. Longer bakeout times and higher temperatures are not recommended as they can degrade PWB and component solderability.

Bake temperature, °C	Time, h
120	3.5–7
100	8–16
80	18–48

Water reabsorption begins immediately upon removal of the printed wiring board from the oven and is linearly related to relative humidity. For a storage environment of 20°C and 30 percent RH a maximum interval of 2 to 3 days is recommended with the interval decreasing with increasing humidity.[40]

Plastic encapsulated devices, especially ICs, also have a tendency to absorb water from the air which is violently released during soldering. Typically 1000 ppm of absorbed moisture is considered a maximum content, beyond which device failure due to body cracking may result. Bakeouts similar to those used for printed wiring boards have been successful in eliminating these defects. After baking, the parts again begin to absorb water, and recommended maximum storage times after bakeout as a function of relative humidity at 25°C[41] are given below based on the time to achieve 800 ppm of water:

Relative humidity, %	Maximum recommended storage time, days
36	20
40	11
50	7

Pretinning of components is often done as a method of:

- Adding solder volume
- Protecting and enhancing the solderability
- Removing gold (to prevent gold embrittlement)

Tinning is essentially the immersion of the to-be-soldered portion of the lead into molten solder. Factors to be controlled include those listed for solderability testing (flux, solder temperature, dwell time, solder alloy).

More active fluxes are often used for tinning than would be used in assembly due to the fact that the loose parts may often be cleaned more thoroughly than would be possible for the assembly. The solder used for tinning should be essentially the same alloy used for soldering in order to prevent significant variations from the intended solder alloy mixture due to mixing of the tinning and soldering alloys, especially when using Sn/Pb/Bi solders.[42]

Typically, for gold removal to be effective, either two immersions in a static solder pot or a single immersion in a flowing solder pot is required.

Preheating of surface mount devices prior to tinning is recommended to avoid cracking of the device. As an example, chip capacitors should be preheated to within 80°C of the solder temperature directly prior to tinning and care should be taken that any tools used are thermally nonconductive to prevent "heat sinking" the part and cooling it when it is picked up to be immersed in the solder.

In handling devices, care should be taken to avoid damage to the device. In the case of fine pitch devices, this damage is most often lead deformations resulting in solder bridging or open solder connections after reflow. In the case of chip devices, damage often occurs as cracks on the chip body. Examples are shown in Fig. 7.20. It has been estimated, for instance, that as much as 75 percent of the damage and/or cracking in chip capacitors is caused by mishandling.[43]

7.3.3 Soldering Processes

Hand Soldering. Hand soldering is the least expensive and least controllable process used to fabricate solder joints. This process may be used either as the primary (original) soldering process or as a repair process.

Solder irons used for electronic assembly are electrically heated using one of two methods:

- *Resistance heating*

  A heating element connected to a fixed voltage with a resulting fixed equilibrium temperature. Heat is transferred by conduction to the tip which in turn achieves a temperature which is a function of the power input, solder iron tip mass, and thermal loss.

The tip temperature is typically monitored via a thermocouple and controlled by regulating the power input to the solder iron; generally switching is done when the ac voltage crosses zero voltage to reduce electrical noise.

- Skin effect heating where a high-frequency ac current is applied to a ferromagnetic material with a suitable Curie temperature. At temperature below the Curie point, the current travels almost exclusively on the outside edge of the tip, and this high current density will cause the tip to heat. As the temperature of the tip reaches the Curie temperature, the depth of the tip conducting current dramatically increases, the current density drops, and the heating drops off. These solder irons are of a fixed temperature design and are often referred to as *self-regulating*.

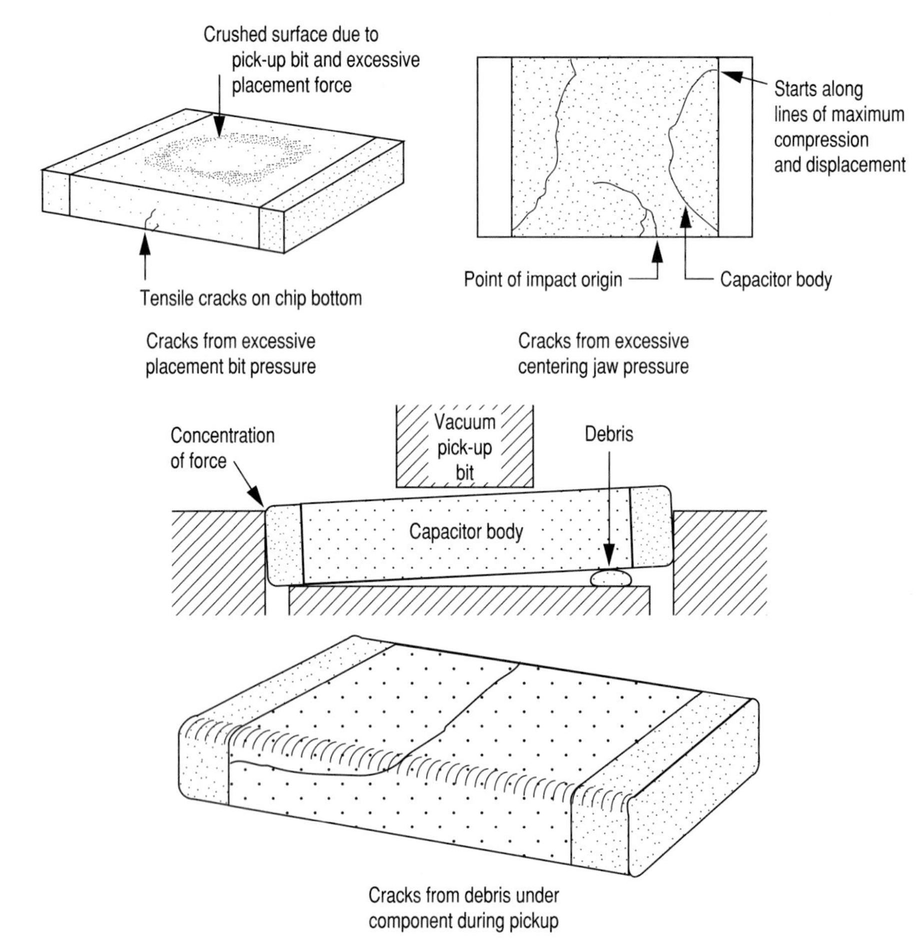

FIGURE 7.20 Placement equipment damage to chip capacitors. (*From J. Maxwell, "Cracks: The Hidden Defect,"* Circuits Manufacturing, *November 1988, pp. 28–33.*)

With the exception of skin effect soldering irons, solder iron types are typically copper potentially alloyed with small amounts (<0.15 percent) of zirconium, chromium, or vanadium. The tips are then plated with nickel, iron, or both to control the dissolution of the copper into the solder. Solder iron tips do inevitably age and must be replaced.

Regardless of the soldering iron type, successful hand soldering depends on the following conditions:

- Good thermal contact between the solder iron tip and the item to be soldered. This includes ensuring that the tip is clean and free of oxides and creating a thermal "bridge" of molten solder from the tinned tip of the solder iron to item to be soldered.
- Allow the flux (typically from cored solder wire or liquid flux) to flow over the area to be soldered in advance of the molten solder. The flux will both increase the ability of the surface to accept solder and act as a thermal transfer medium to aid in heating.
- Apply adequate solder to form an acceptable solder connection.
- Maintain thermal contact until good solder spread is obtained (but do not stay too long).
- Remove solder iron, but do not disturb parts until solder solidification is complete.

Typically, skin effect solder irons are more efficient in delivering heat and can perform soldering at lower tip temperatures than resistance soldering irons. Temperatures for resistance soldering irons are 650°F (343°C) to 800°F (426°C), and for skin effect solder irons typically 550°F (287°C) and 650°F (343°C). The increased effectivity of the skin effect soldering irons is due to their more efficient heating and heat delivery systems. It has been claimed that this increased efficiency can result in a decrease in PWB damage (e.g., delamination) caused by hand soldering. Studies have shown[44] that thermal damage is primarily the result of excessive tip pressure and excessive temperature and excessive dwell time (with tip pressure playing the predominant role). The selection of the tip size, temperature, and heater type should be done to reduce the time (and pressure due to operator fatigue) on the solder joint.

Mass Soldering

Wave Soldering. Wave soldering systems are commonly used for the mass soldering of plated-through hole (PTH), and surface mount devices (typically chip devices and leaded devices). When bottom side surface mount devices are soldered, they typically are placed in an adhesive and cured in place prior to soldering. These adhesives are specially formulated to have high "drop" heights (to bridge from the PWB surface to the part) as well as good bonding to component surfaces at temperatures to +260°C (although their after-solder characteristics are generally limited to being nonionic, with post-solder application strength being redundant to the solder connection and generally unspecified).

The basic components of a wave soldering system are shown in Fig. 7.21. The system automatically performs the soldering process as described previously for hand soldering, that is, flux application, heating of the area to be soldered, application of molten solder, and solidification. As all of these components act together to ensure proper soldering, a solder "schedule" is often developed recording the optimum process parameters selected for flux, preheat, conveyor speed, solder temperature, and (on some systems) the solder wave configuration used.

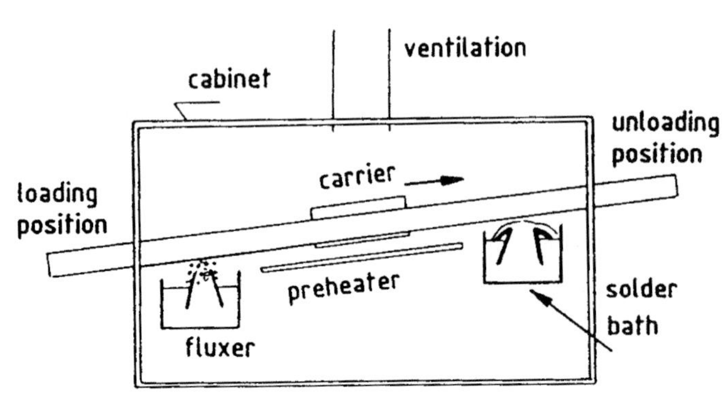

FIGURE 7.21 Components of a wave soldering system. (*From R. J. Klein Wassink,* Soldering in Electronics, *2d ed.,* Electrochemical Publications, 1989, p. 523, Fig. 9.42.)

Flux is generally applied by:

- *Foaming fluxing* where air is bubbled through the flux creating a foam head the PWA (printed wiring assemblies) then passes through
- *Wave fluxing* where a wave of liquid flux is pumped up into a wave over which the PWA passes
- *Spray fluxing*

Not all fluxes are compatible with all application methods, and, in fact, generally the fluxes are specially modified with a specific application method in mind. Typically wave fluxes add the most flux to the PWA followed by foam and then spray. The volume of flux applied, however, does not relate directly to its fluxing power, and again it is best to apply the flux in the manner for which it was designed.

Generally following flux application is some type of air knife or brush used to remove excess flux from the PWA. If not adequately removed, this material can affect preheating, solder joint quality (too much flux can inhibit soldering), and cleanability (what is placed on the PWA typically has to be removed later on). When an air knife is used at this point, it is typically not heated (due to the flammability of the flux solvents) and is often set to hit the PWA surface at an angle ~45 to 60° in order to both "squeegee" the excess flux away and drive some material up the PTH to aid in solder reflow on the topside connections.

The PWAs are heated using either calrod or quartz heating sources. These heating elements are typically mounted below the PWA (heating from the bottom side) although topside preheaters are available. Preheating acts to evaporate solvents from the flux (to prevent outgassing in the solder wave), thermally activates the flux (allowing its maximum activity prior to soldering), and reduces the thermal shock during soldering experienced by both the PWA and its components. When soldering surface mount devices, especially chip capacitors, adequate preheating is essential to reduce the thermal shock on the PWA and prevent thermal-shock-induced cracks in the capacitors.

The cracking of components, especially multilayer ceramic chip capacitors, is to be avoided at all costs as these cracks often do *not* appear during either external visual inspection (at up to 50X) or electrical testing but only show up as "shorted" capacitors after exposure to humidity and/or electrical bias during use.[45]

The "solder schedule" often records the preheater settings (preheater temperature and conveyor speed) versus a thermal profile developed by placing a thermocouple on the surface of the PWA during soldering. In order to maximize repeatability both within and between PWA designs, the thermocouple is generally placed on the surface of the laminate (as opposed to the circuit trace or part lead). An example of a wave solder preheat profile showing PWB laminate temperature and the temperature of a device (dual inline package on topside of PWA) is shown in Fig. 7.22.

A preheat schedule needs to address both ramping rate (excessive ramping can cause delamination of PWBs or cracking of parts) as well as the ultimate temperature. Typical thermal profiles for a PWA during wave solder result in a preheat achieved of from 65 to 121°C (150 to 250°F) as measured on the laminate on the topside of the PWA immediately before the PWA contacts the solder wave and utilizing a heat-up rate of 2 to 4°C/s (3.6 to 7.2°F/s).[46]

Insufficient preheating of the flux may not fully activate the flux and may leave the viscosity of the flux low (due to inadequate solvent removal), which in turn can create dewetting and solder icicles on the PWA.

The printed wiring assembly is then passed over the molten solder. Typically Sn60Pb40 and Sn63Pb37 solders are used with soldering temperatures of 260°C • 5°C (500°F • 10°F) although wave solder temperature setpoints may be varied from 232°C (450°F) to 287°C (550°F) with higher temperatures being used for more massive PWAs and lower temperatures for thermally sensitive devices (with Sn/Pb/Bi solders being used for cases of extreme thermal sensitivity). Dwell time of a single point on a PWA on the surface of the solder pot is generally limited to 3 to 6 s to prevent part

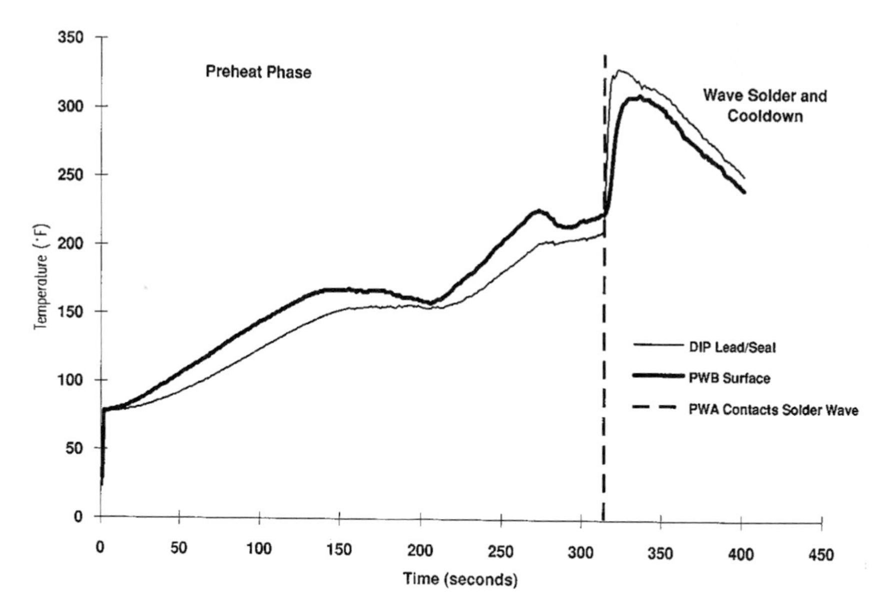

FIGURE 7.22 Wave solder preheat profile showing PWB laminate temperature and the temperature of a device (dual inline package on topside of PWA).

and PWB damage due to overheating. Dwell time is a function of conveyor speed, wave configuration, and immersion depth (the latter two defining the contact length of the PWA on the wave).

Molten solder is forced (generally by a mechanical impeller) from a sump (solder pot) through a channel and a series of baffles up through a nozzle where it forms a standing wave of molten solder. The molten solder then falls away on either side of the wave back into the solder pot. The configuration of the solder wave varies considerably, but essentially its purpose is to present a constantly renewed (oxide-free) source of solder to the bottom surface of the printed wiring board. In fact, a thin, static oxide film does form on the wave but is broken and skimmed away by the advancing PWA during soldering.[47]

A product of wave soldering is the formation of dross on the static portion of the wave solder pot. This *dross* is actually cells of solder metal encapsulated by tin oxide[48] (generally the actual amount of oxide in the dross is 1 to 5 percent with the remainder being solder). Additionally, a fine, black powder will often form near the impeller shaft and this too is SnO. It should be noted that this formation of tin oxides gradually diminishes the tin available in the solder and that the tin level needs to be frequently monitored and the tin replaced.

Shown below are two formulas for determining the amount of tin to be added to a solder pot to adjust its tin content. Two cases are shown depending on the resulting mass in the solder pot after the addition:

Case 1. Solder pot mass constant (i.e., solder is removed and replaced by new solder or tin):

$$\text{mass}_{\text{tin-add}} = \frac{\text{mass}_{\text{pot}}(\text{Sn\%}_{\text{desired}} - \text{Sn\%}_{\text{current}})}{(\text{Sn\%}_{\text{addition}} - \text{Sn\%}_{\text{current}})} \tag{7.3}$$

Case 2. Solder pot mass increases with added solder or tin:

$$M\ \text{mass}_{\text{tin-add}} = \frac{\text{mass}_{\text{pot}}(\text{Sn\%}_{\text{desired}} - \text{Sn\%}_{\text{current}})}{(\text{Sn\%}_{\text{addition}} - \text{Sn\%}_{\text{desired}})} \tag{7.4}$$

where $\text{mass}_{\text{tin-add}}$ = mass of tin to be added to the solder pot
mass_{pot} = mass of solder pot (prior to addition)
$\text{Sn\%}_{\text{desired}}$ = tin content desired in the solder pot, percent by weight
$\text{Sn\%}_{\text{current}}$ = tin content currently in solder pot, percent by weight
$\text{Sn\%}_{\text{addition}}$ = Tin content in solder or tin bar used for addition (e.g., 100 for pure tin bar)

A number of wave configurations are currently in use, and, in order to evaluate them, it is necessary to understand that the solder wave in contact with a PWA can be divided into three regions (see Fig. 7.23):

- The wave entrance or wetting region where initial contact with the solder is made. The final evaporation and dispersion of the flux and the initial wetting (especially to bottom side SMDs) is done here, and thus solder wetting and "skips" are generally related to this region. Often low-frequency vibrations are set up in this region (or a separate oscillating wave is used) to both "scrub" and to aid in the fluxing action and also to achieve a degree of initial flux distribution and solder wetting on bottom side SMDs to reduce shadowing or skipping of these connections. Skipping of bottom side SMD connections is aggravated by both increasing part size versus spacing between parts and the placement of the to-be-soldered lead too close to the part body.

- The middle, or heat transfer, region where the bulk of the heating and solder wetting and spreading occurs.
- The last region, the exit, or "break-away," region where the PWA exits the wave. This region is where the final formation of the solder joints (and solder defects) occur. Up until this point all the solder has been fluid and subject to change, and no "joints" per se have existed.

The rate at which the solder falls away from the PWA is a function of the inclination of the conveyor (typically tilted at 5 to 10° from the horizontal) and the wave shape. The rate of breakaway in the exit region impacts the formation of solder bridges and icicles. A rate that is too slow can lead to solder ball and dross formation and retention on the PWB and associated bridging; a change that is too abrupt can leave the solder "stranded." The rate of breakaway is controlled by conveyor speed and conveyor angle. Hot-air knives directly after the solder wave are also sometimes used to control or reduce solder bridges. (Cold or room-temperature air should not be used due to possible thermal shock damage to components, especially chip capacitors.)

Recently *controlled atmospheres* (typically N_2) have been used to exclude oxygen from the solder pot area. These systems are sometimes referred to as *inert,* or *nitrogen blanket, systems.*

By excluding the oxygen from the preheating and soldering areas of the system (reducing the oxygen content in these areas to 5 to 10 ppm[49]), adequate solder quality can be achieved using far less flux (and less active fluxes). Typically a weak flux (for example, 1 percent carboxylic acid) is used as a flux. Flux application is generally done by foaming, although spray fluxing is also being used. Some testing has indicated that N_2 systems are not compatible with normal (25 to 35 percent solids) rosin fluxes.[50] Formic acid is sometimes added to the N_2 (0.1 to 1 percent by volume)[51] to further reduce solder defects, but this material must be vented to keep the concentration in the work area to <5 ppm.

The major benefits of this system are the greatly reduced residues (due to the reduction in the need for flux), making post-solder cleaning relatively simple (and in some cases unnecessary). The solder joints resulting from such a system tend to have a much more uniform surface finish. (While not related to reliability, this shiny finish is generally more acceptable to the customer.) A more tangible benefit than shiny solder joints (although with a related cause) is that N_2 solder pots generate little if any dross.

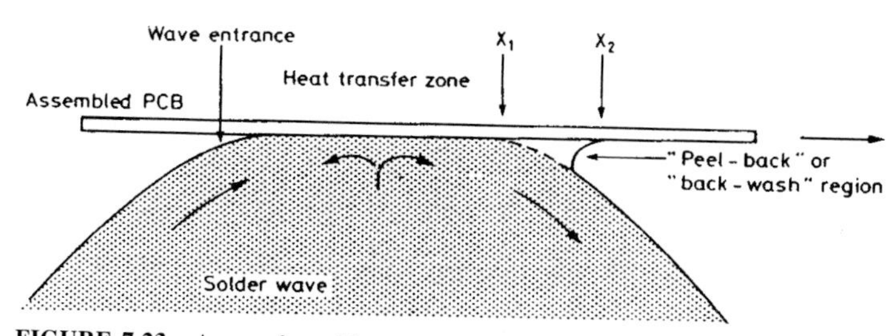

FIGURE 7.23 Areas of a solder wave relative to contact with a printed wiring board. (*From C. Lea,* A Scientific Guide to Surface Mount Technology,

One drawback of these systems is the appearance of small solder balls and bridges on the surface of soldermasked PWAs (rarely on nonsoldermasked PWAs). The reasons for this are not clear but seem to be related to the soldermask, its curing cycle, and the flux(es) used.

According to the critical parameters, process control in wave soldering should be concentrated on the following areas:

- Quality (e.g., solderability) of the incoming product
- Control and maintenance of the solder flux
- Evaluation and control of preheat to eliminate volatiles from the flux, achieve optimum flux activation, and minimize thermal shock to PWAs and parts
- Optimization of wave configuration to control dwell time in the wave and critical "wave exit" conditions (Even in a "fixed" wave, variations in pot height and conveyor angle can effect soldering quality.)
- Turbulence of the wave used for bottom side SMD soldering

These variables interact during the soldering process, and best results are often a compromise between competing effects as shown in Fig. 7.24.

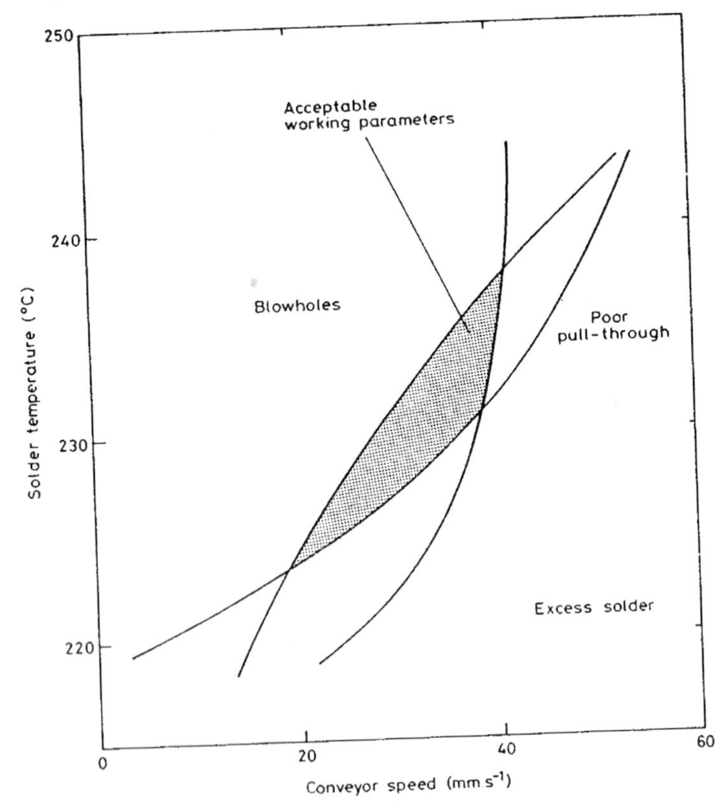

FIGURE 7.24 Interaction of wave solder parameters on solder defect rates. (*From C. Lea,* A Scientific Guide to Surface Mount Technology, *Electrochemical Publications, 1988, Fig. 5.26, p. 150.*)

Vapor Phase Soldering. Vapor phase soldering uses the condensation of vapor from a boiling fluid to heat the PWA and components up to solder reflow temperatures. In principle, the PWA is lowered into the vapor over the boiling fluid using either an elevator or conveyor. It dwells in this area until reflow is achieved and then is removed. Vapor phase soldering is used for both surface mount and plated-through hole assembly. In fact, this method was originally developed for the soldering of backplane PWAs using solder preforms.[52]

The vapor used is generated by boiling a *primary,* or *reflow,* fluid which is generally an organic compound in which the carbon-bound hydrogen atoms are replaced with fluorine atoms. These materials are colorless, odorless, nonflammable, chemically inert, nontoxic, and expensive ($500 to $1000 per gallon or $130 to $265 per liter).

Due to the expense of the reflow fluid, a variety of methods are used to keep the vapor contained in the system (vapor loss constitutes one of the largest operating expenses). These methods include cooling coils, long enclosed tunnels (inline systems), automated covers, and sacrificial vapor blankets (batch systems). The *sacrificial vapor blanket system* (or *dual-phase system*) uses a secondary vapor blanket of R113 (e.g., Freon TF and SF-2[53]) which is usually added over the boiling fluid and contained with a secondary set of cooling coils. The product generally dwells in this zone on the way out of the system to allow for the primary vapor to drain back into the reflow sump, preventing or reducing dragout losses. Due to the restrictions being placed on CFC materials such as Freon TF, non-CFC substitutes are available (for example, SF-2 from 3M), but most new systems rely on increased cooling coils and a system of tunnels and/or covers to control fluid loss. Some physical properties of common reflow fluids are given in Table 7.15.

TABLE 7.15 Physical Properties of Vapor Phase Reflow Fluids

Property	R113	LS 230, (Galden)	FC 5312, (3M)
Boiling point, °C	47.6	230 × 5	215
Molecular weight	187	~650	821
Pour point, °C	—	−80	−13
Density @ 25°C, g/cm³	—	1.82	1.93
Density of saturated vapor @ boiling point, mg/cm³	7.38	19.5	20.3
Viscosity of liquid @ 25°C, cP	0.7	8	30
Surface tension of liquid @ 25°C, 1000 * (N/m)			
Specific heat of liquid @ 25°C, J/(g • K)	0.95	1.00	1.05
Thermal conductivity @ 25°C, 1000 * [W/(m • K)]	74	70	70
Electrical resistivity, Ω • cm	—	2×10^{15}	1×10^{15}
Heat of Vaporization @ boiling point, J/g	—	63	67
Heat transfer coefficient, horizontal plate @ 200°C, W/(m² • K)	—	280	~300

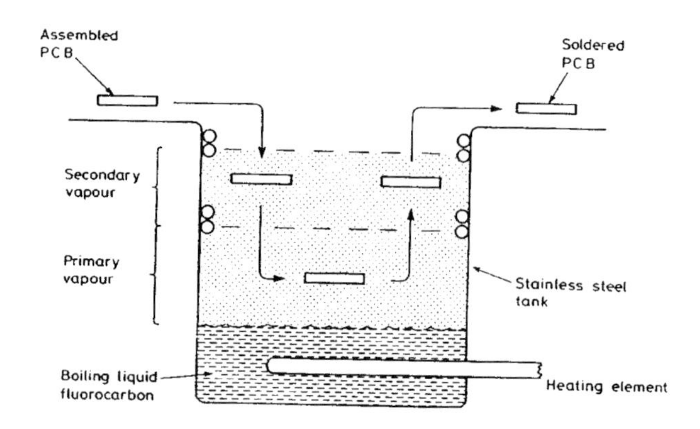

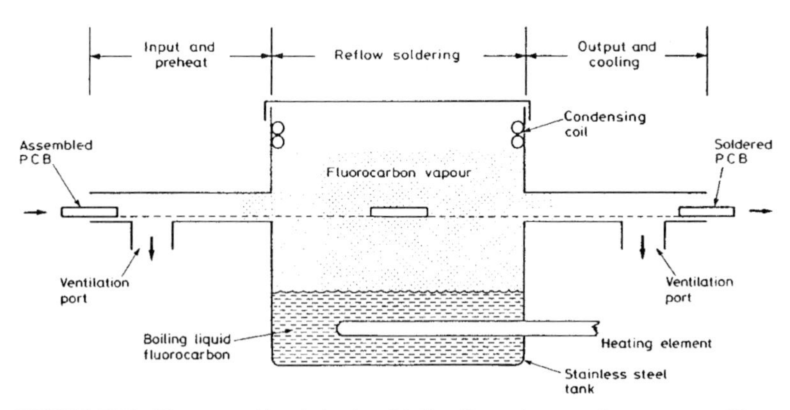

FIGURE 7.25 Diagrams of batch (*top*) and inline (*bottom*) vapor phase systems. (*From C. Lea,* A Scientific Guide to Surface Mount Technology, *Electrochemical Publications, 1988, Figs. 8.10 and 8.12, pp. 250–251.*)

Diagrams of basic vapor phase systems are shown in Fig. 7.25 with a dual-vapor system on the top and an inline single-vapor system on the bottom. The advantages of this method of soldering are summarized below:

Advantages	Disadvantages
Simple to control, and easy to adapt to new designs; solders both SMT and PTH connections.	Process more likely to induce solder wicking.
Absolute control of maximum temperature regardless of design or thermal load.	More expensive to purchase and maintain than other mass reflow systems.
Rapid heatup rate with uniform application of heat for all sides.	Increased intermetallic thicknesses and a dull finish.
Small, unbonded devices are "self-centering."	More likely to induce "drawbridging" of chip devices.

The heatup rate for a solid body in a vapor phase system can be modeled using the following equation:[54]

$$T_S = (T_F - T_0) \bullet (1 - e^{(-t/t_0)}) + T_0 \qquad \text{when } t_0 = \frac{(\delta \bullet c \bullet V)}{(h \bullet A)} \tag{7.5}$$

where T_S = temperature of solid at time t, °C
T_F = boiling/vapor point of reflow fluid, °C
t = time, s
T_0 = starting temperature, °C
c = specific heat, J/(kg • K)
h = heat transfer coefficient, W/(m^2 • K)
t_0 = characteristic time where T_S = 63 percent of T_F, s
δ = density of solid, kg/m^3
V = volume, m^3
A = area of body, m^2

A listing of some thermal diffusivity data for materials used in electronics soldering is shown in Table 7.16.

TABLE 7.16 Thermal Data for Electronic Materials

Material	Specific heat, J/kg • K	Density, kg/m^3
Copper	390	8920
Solder	180	8155
Alumina	1000	4020
Glass	630	2250
Epoxy	1500	1350

It should be noted that, when soldering is involved, the heatup profile predicted by this equation will also have a delay at approximately the melting point whose length is proportional to the difference between the melting point of the solder and the boiling point of the fluid. This delay is predicted by the equation:

$$t_L = \frac{(\lambda_S \bullet \delta \bullet V)}{[h \bullet A(T_F - T_S)]} \tag{7.6}$$

where t_L = time for solder to change phase from solid to liquid, s
λ_s = heat of fusion of solder (see Table 7.3), J/kg
δ = density of solid, kg/m^3
h = heat transfer coefficient, W/(m^2 • K)
T_S = temperature of solid at melting point of solder, °C
V = volume, m^3
A = area of body, m^2
T_F = boiling/vapor point of reflow fluid, °C

Typically problems in vapor phase soldering take three forms:

- Dissolution of termination materials (especially on chip devices and gold leaded parts), leading to poor solderability and lowered solder joint strength
- Open solder connections caused by the solder's not staying in the pad area but flowing up the lead (also known as *solder wicking*)
- Lifting of one end of chip devices creating an open circuit (also known as *drawbridging* and *tombstoning*)

Typical solutions to these problems are shown in Table 7.17.

Recent developments in vapor phase system design have begun to include inline IR preheaters. These preheaters both bake out solder paste and raise the initial temperature of the PWA which in turn lowers the heating rate. It has been reported that the use of this hybrid IR/vapor phase system combines the best of both IR and vapor phase as it greatly reduces the incidence of drawbridging and open solder joints.[55]

By its very nature the vapor phase process has few critical process parameters. The heatup rate is primarily controlled by the design of the part being soldered (size, material, mass, and so on). In setting up the parameters for a new assembly, the operator controls only the speed at which the hardware is introduced to the vapor and the time it dwells during reflow and cool-down.

The dwell time to achieve reflow should be kept as short as possible while achieving complete reflow (usually monitored by either visually monitoring the progress of

TABLE 7.17 Problems and Solutions in Vapor Phase Soldering

Problem	Probable cause	Typical solutions
Poor solderability on leads	Excessive dissolution of termination material into solder.	Reduce dwell time after achieving reflow. Use parts with barrier layer (typically Ni) between termination material and solder joint.
"Crusting"/lack of solder paste reflow	Dissolution of flux by condensing reflow fluid.	Lower mass of work piece (e.g., remove or reduce tooling). Change to a reflow fluid with lower rosin solubility.
Drawbridging or tipping up of chip devices	Solder reflows at one end of device before the other end and wetting force tips part.	Revise pad size to reduce pad extension and thus tipping force. Use solder paste which slows rate of reflow, giving other side of chip time to "catch up."*
Open circuits on leaded SMD devices	Leads not in contact with paste or pad having lower effective mass and heat up faster than substrate; solder flows preferentially to hotter area.	Use solder paste which slows rate of reflow, giving other side of chip time to "catch up."

*R. N. McLellan and W. H. Schroen, "TI Solves Wicking Problem," *Circuits Manufacturing,* September 1987, pp. 78–85.

reflow or thermocoupling the PWA). The dwell time in the system after reflow (out of the reflow vapor zone) is critical to ensure that the assembly is not inadvertently shaken prior to solder solidification.

Most of the fluids also break down to some degree under thermal loading (use), releasing *perfluoroisobutylene* (PFIB), a pulmonary irritant with an accumulated lethal concentration of 0.5 ppm over 6 h.[56] Vapor phase systems incorporate venting systems to control PFIB release, but the rate of PFIB generation can be increased by overheating. The heating elements in the vapor phase system should be run at an "idle" system when product is not being actually soldered to reduce PFIB generation. Rosin contamination can build up on heater surfaces, causing localized "hot spots", and should be monitored and controlled. Contaminated fluid can usually be reprocessed by the fluid vendor or locally by distillation and/or filtration to control rosin accumulation.

Fluid loss is also critical both as an operating cost and as a safety hazard (reflow fluids are typically slippery, and condensation on surfaces like floors around the system can be dangerous.) Fluid loss is generally caused by dragout (due to trapping of the fluid in cavities of the part being soldered or excessive speed during removal of the part) or excessive venting.

IR Soldering. *Infrared* (IR) *soldering* relies on the absorption of infrared radiation into the substrate, components, solder, and flux to heat the assembly to soldering temperatures. IR radiation is sometimes broken down into classes by wavelength as shown below, although for practical purposes IR wavelengths greater than 100 μm are not used.

Class	Wavelength, μm
Near infrared	0.72–1.5
Middle infrared	1.5–5.6
Far infrared	5.6–1000

A summary of the advantages and disadvantages are shown below:

Advantages	Disadvantages
Low equipment maintenance costs	Peak temperatures not limited
Less likely to create drawbridging and open connections than vapor phase	More likely to cause discoloration of PWB
Can be run with a variety of processing atmospheres including inert gas	Can be used to solder only SMD devices (no PTH)

The principle of operation of IR soldering systems is that at a surface at a temperature T will emit radiation and the heat flux generated by this radiation is given by the Stefan-Boltzmann law:

$$q = \alpha \cdot T^4 \tag{7.7}$$

where q = heat flux density, W/m^2
α = constant (for "black body"), 5.7×10^{-8} W/(m^2 • °K^4)
T = temperature of source, °K

The energy transfer between this source and another body (assuming both are black bodies or perfect absorbers of IR) is given by

$$\dot{Q} = F_{1,2} \bullet A_1 \bullet \sigma \bullet (T_1^4 \bullet T_2^4) \tag{7.8}$$

where $\dot{Q}$ = thermal energy, W
$F_{1,2}$ = view factor, unitless
A_1 = area absorbing the IR energy
σ = constant (for black body), 5.7×10^{-8} W/(m^2•°K^4)
T_1 = temperature of source 1 (emitter), °K
T_4 = temperature of source 2 (recipient), °K

View factors for a number of potential arrangements are shown in Fig. 7.26. The absorption factor α varies as a function of surface texture (rougher surfaces absorb more energy), material, and wavelength of IR radiation. The effect of material and IR wavelength on IR absorption are shown in Fig. 7.27 for common electronics materials. It should be noted that Fig. 7.27 shows the general shape only and omits localized "peaks" caused by the excitation of various molecular groups.

Infrared sources for SMT soldering and their characteristics are shown in Table 7.18. The first three sources rely on filaments wrapped in a tube directly irradiating the item to be heated. The last type (area source secondary emitter) uses a filament buried in a thermally conductive material and backed (on the side away from items to be soldered) with a refractory material to provide a more uniform emitter.

A schematic of an IR reflow oven is shown in Fig. 7.28 with a corresponding typical thermal profile shown in Fig. 7.29. It should be noted that the heating rate of a PWA may be directly controlled in an IR oven, although excessive heating rates may lead to substrate damage, especially on glass-epoxy printed wiring boards as shown in Fig. 7.30.

The addition of inert atmospheres for soldering using IR are also being introduced. The benefits derived are:

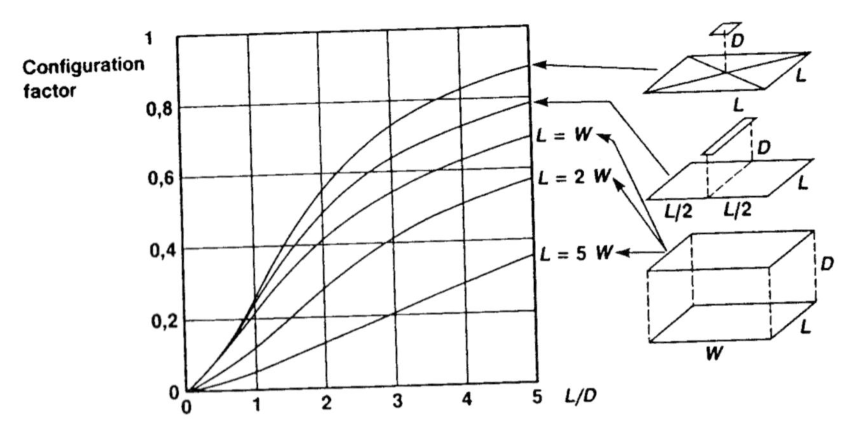

FIGURE 7.26 View factors as a function of configuration. (*From R. J. Klein Wassink,* Soldering in Electronics, *2d ed., Electrochemical Publications, 1989, Fig. 3.13, p. 102.*)

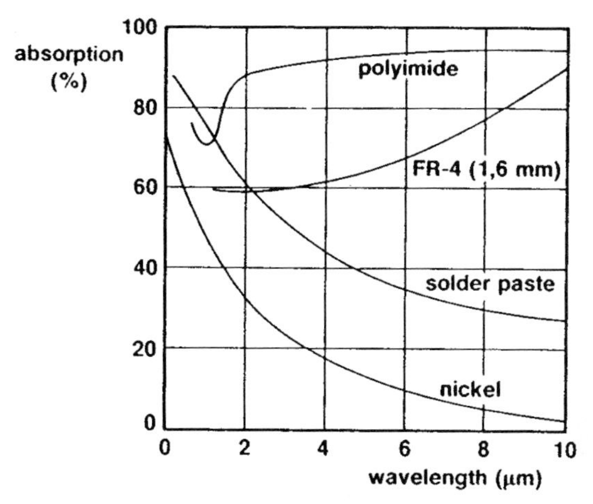

FIGURE 7.27 Absorption coefficient for soldering materials as a function of wavelength of IR radiation. (*From R. J. Klein Wassink,* Soldering in Electronics, *2d ed., Electrochemical Publications, 1989, Fig. 3.15, p. 104.*)

TABLE 7.18 Characteristics of Infrared Sources for SMT Soldering

Emitter type	IR class	Wattage, W/cm	Suitability
Focused tungsten tube filament lamps	Near IR	300	Not suitable; problems with shadowing and sensitivity to color as well as overheating
Diffuse array of tungsten filaments	Near IR	50–100	May be used, but heating rate sensitive to color of components
Diffuse array of nichrome tubes filament lamps	Near to middle IR	15–50	Better source; little sensitivity to color and less sensitive to shadowing
Area source secondary emitter	Middle to far IR	1–4	Best type; no shadowing or color sensitivity

Source: C. Lea, *A Scientific Guide to Surface Mount Technology,* Electrochemical Publications, 1988, p. 211.

- There is lower oxidation of the devices, and thus the less active fluxes can be used. In addition, the area of flux spread is reduced, which decreases the area needing to be cleaned and reduces the impact of solderballing.
- As the combustibility of the fluxes are reduced in a nitrogen atmosphere, higher processing temperatures (up to 300°C with <5 ppm O_2) may be realized.
- The discoloration of the epoxy surfaces is reduced.

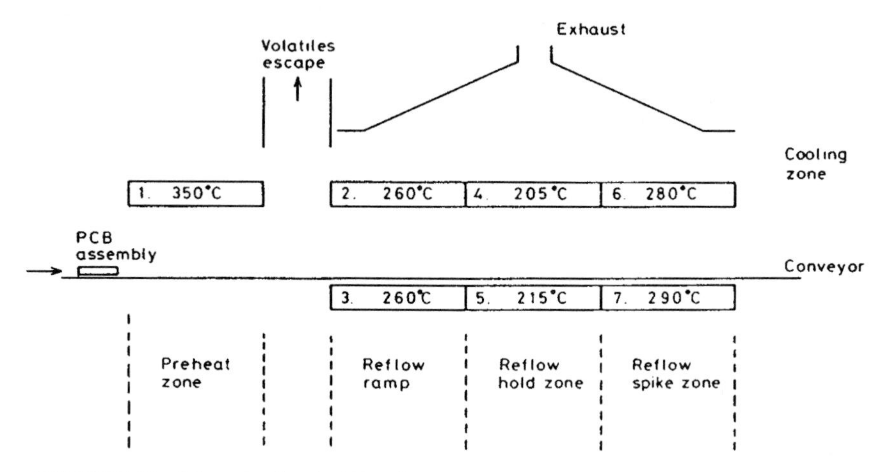

FIGURE 7.28 Schematic of typical IR reflow oven. (*From C. Lea,* A Scientific Guide to Surface Mount Technology, *Electrochemical Publications, 1988, Figs. 7.32, p. 232.*)

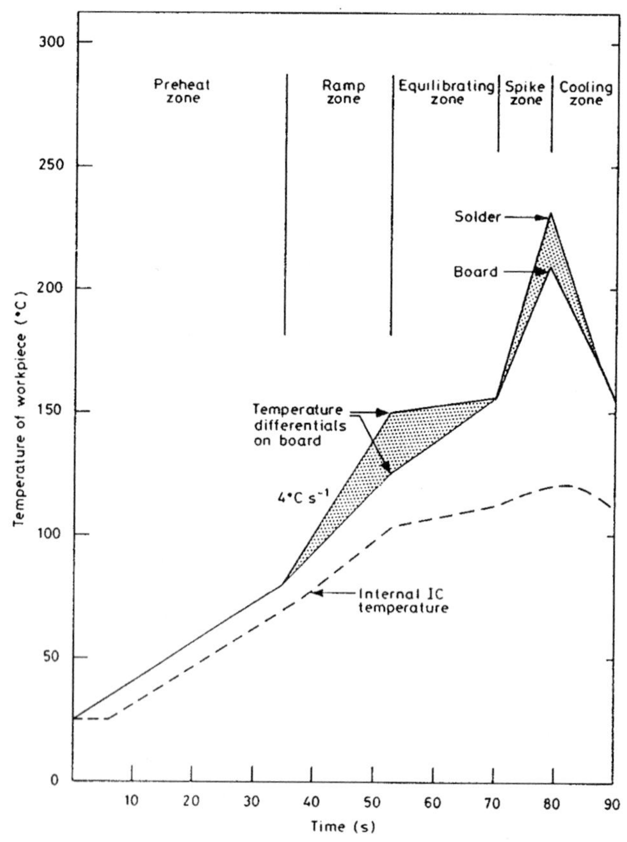

FIGURE 7.29 Typical temperature profile for PWA in an IR reflow oven. (*From C. Lea,* A Scientific Guide to Surface Mount Technology, *Electrochemical Publications, 1988, Fig. 7.31, p. 231.*)

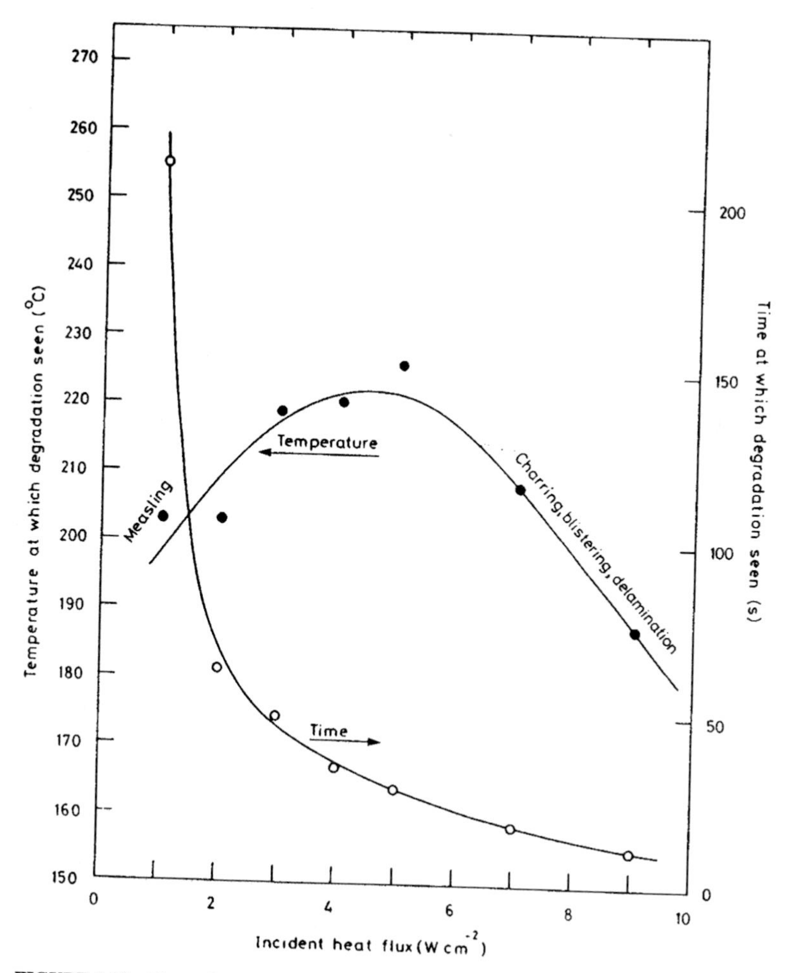

FIGURE 7.30 Thermal damage to printed wiring board as a function of heat flux. (*From C. Lea,* A Scientific Guide to Surface Mount Technology, *Electrochemical Publications, 1988, Fig. 7.23, p. 222.*)

In the establishment of the solder schedule, the main issues (critical parameters) in process control are to create a schedule with an effective tradeoff between rapid heating rates to reduce oxidation and affect reflow on all device types and the need to protect substrates and parts.

Once a solder schedule or thermal profile has been established for an oven, the main concerns are simply to ensure that the air (or gas) flow rate and heater efficiencies remain constant to ensure repeatable processing. If gas mixtures are used, they too would be monitored.

Hot Gas Soldering. Soldering by hot gas can be done in a number of atmospheres (typically air or nitrogen) with the hot air applied either locally for rework or limited assembly, or generally (as in a convection oven). Hot air soldering can be used for either plated-through hole or surface mount soldering.

An advantage of hot gas soldering over IR soldering is that hot gas soldering can approach vapor phase soldering heating rates while avoiding the problems associated with variations in the heating rates of different parts of the assembly due to variations in IR absorption by different materials. A typical heating profile is shown in Fig. 7.31.

Process control typically consists of creating solder schedules (relating setpoints within the oven zones) and monitoring the repeatability of the oven control zones. Repeatability is measured both over time and across the chain in the direction perpendicular to the direction of travel.

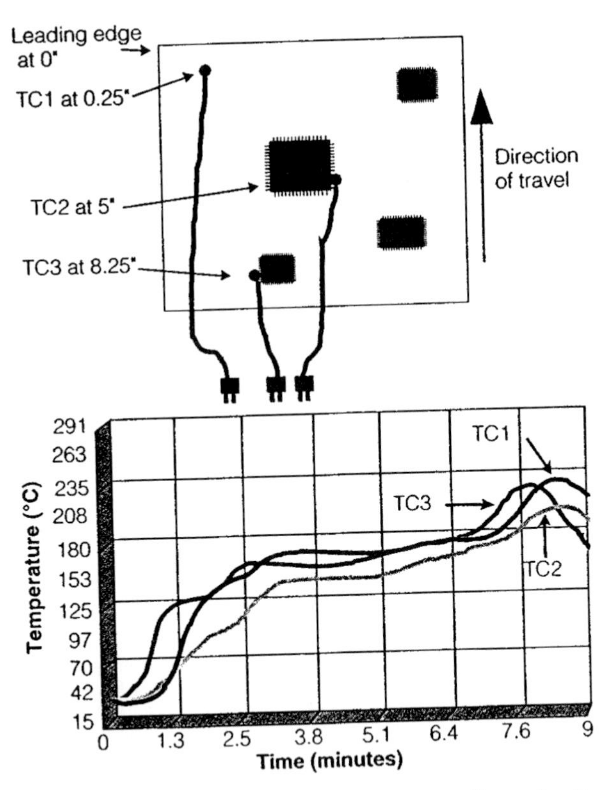

FIGURE 7.31 Typical convection (hot air) oven profile as a function of location of thermocouples on the PWA and location in the oven. (*From P. C. Kazmierowicz,* Thermal Profiling Reflow Solder, *Circuits Assembly, August 1992, pp. 59–62.*)

Resistance/"Hot Bar"/Pulse Soldering. In this process an electric current is passed either through the device leads or through a "heater bar," and the resistance of the leads (generally of the material or the contact resistance) is used to produce the heat required for soldering. This method can be applied to either individual joints (called *single-point soldering*) or multiple joints along one or multiple sides of a leaded component. A thermocouple is generally attached to the heater bar (or thermode) near where the leads will be soldered and a "feedback" control system is used to control temperature.

Resistance soldering is used almost exclusively for the soldering of surface mount leaded devices, typically with the leads extending away from the device body (e.g., the "gull wing"). In general, there are two types of resistance soldering systems:

- Single-point soldering systems designed to solder one lead at a time. These use either the heater bar (see Fig. 7.32, left) or a "parallel gap" soldering system (see Fig. 7.32, right) where the current passes through the device leads.

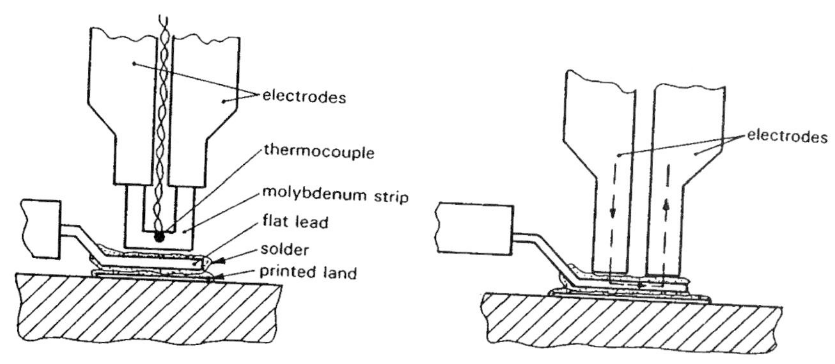

FIGURE 7.32 Single-point soldering systems using a heater bar (*left*) or a parallel gap (*right*) heating method. (*From R. J. Klein Wassink,* Soldering in Electronics, *2d ed., Electrochemical Publications, 1989, Fig. 10.39, p. 588.*)

- Multiple-lead soldering systems similar in method to the resistance bar system except that the bar extends across a number of leads on one side of a part. (Some systems allow for soldering all four sides using typically four independently operated bars.)

Solder is supplied by either pretinning the device or pad (a reflowed solder thickness of 15 to 30 μm is generally required for leaded surface mount devices[57]) or is added by using solder paste or preforms. Typically already reflowed materials (solder from tinning or preforms) is preferred over paste due to the lower outgassing and lower volume required.

The critical parameters for process control in resistance solder schedules typically have three stages: heatup, solder reflow, and cool-down. The optimum schedule is one in which the time and force on the pads is minimized. A summary of the characteristics of these three stages is shown below:

- The heatup stage is when the heat and pressure are initially applied to the lead and solder. Heatup rates can run up to 300°C/s (as measured on the heater bar) for FR-4 material. Due to the thermal mass of the device being soldered, the actual lead or laminate temperature may lag the heater bar by 35 to 80°C. Typically the heatup stage should last 1 to 2 s (with 1 s being preferable). The degree of lag is a function of:[58]

Heater bar-to-lead contact area (Higher areas yield lower thermal lags.)

Heater bar-to-lead contact force (Higher force will reduce thermal lag, but too much force will cause delamination of the PWB and lead misalignment.) Typical hold-down forces are in the range of 4.2 N/mm^2 (600 lb/in^2).[59]

The relative location of the heater bar control thermocouple to the leads being reflowed (Smaller distances result in lower lags.)

- Reflow or time-at-temperature is the stage when actual solder flow and solder wetting occur and generally runs 4 to 6 s. A good measure of adequate reflow time is to look for the *reflow line,* or demarcation between the solder reflowed onto the lead and the solder of the pad which remains unreflowed. When this line is consistently well away from the leads, the time at temperature is sufficient.[60]
- Cool-down is the stage when the solder solidifies. The thermode should be removed from the joint *prior* to solder solidification to prevent excessive residual stresses from causing solder joint failures.

Hold-down force is also critical to the quality of the solder joint as noted above. For best results it is recommended that initially a low pressure be used and that this be increased after solder reflow to a maximum and then tapered off rapidly during cool-down.

Using the thermode to "hold down" leads badly out of alignment will result in residual stress in the solder joint which in turn can lead to creep rupture failures in the solder connection. By removing the thermode well before solder solidification, these stresses are allowed to dissipate while the solder is still molten. An alternative approach is to ensure that all leads are in contact with the pad prior to application of the thermode to ensure that the residual force is minimized. (Generally a residual force of less than 200 psi in either tension or shear is recommended.[61])

Laser Soldering. *Laser* (light amplification by stimulated emission of radiation) *soldering* is used for the soldering of surface mount solder joints of both the leaded and leadless varieties. The main advantage to laser soldering is its rapid heating rate (lasers typically produce about 100 kW/m^2) without any applied thermal mass (e.g., a solder tip). This leads to very rapid heating and cooling combined with the ability to only reflow small, selected areas while leaving the balance of the assembly at essentially ambient temperatures.

Typically the laser is directed either onto the lead (Fig. 7.33, no. 1) or onto the pad near the lead/metallization (Fig. 7.33, nos. 2 and 3). Solder is supplied either by pretinning the lead/metallization or pad or by the use of solder paste. When solder paste is used, care must be taken to avoid solder balling (due to rapid solvent evaporation during soldering), and solder balling.

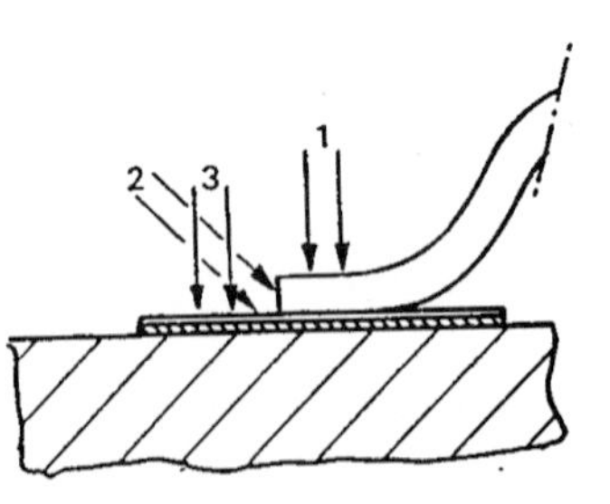

FIGURE 7.33 Directions of laser impingement onto leaded surface mount device. (*From R. J. Klein Wassink,* Soldering in Electronics, *2d ed., Electrochemical Publications, 1989, p. 583.*)

Two types of laser are used for soldering, CO_2 and Neodymium-Yttrium-Aluminum-Garnet (Nd:YAG) laser. While all CO_2 lasers are continuous pumped systems, Nd:YAG lasers are subdivided into two types, continuous pumped and pulse pumped. In general, continuous pumped systems are used for soldering.[62]

TABLE 7.19 Comparison of Laser Soldering Systems

Nd:YAG	CO_2
Radiation in the near infrared; wavelength 1.06 μm.	Radiation in the middle infrared, wavelength 10.6 μm.
Transmitted by glass and plastics:	Absorbed by glass and plastics:
Opaque safety screens must be used.	Glass safety screens allow visual access.
Beam need not be blanked off between joints; beam unlikely to damage PCB.	Laser will burn PCB; beam should be blanked off during movement between joints.
Cheap and good materials for optical systems; hence coincident viewing and multiple beaming.	More expensive and difficult materials required for optics (zinc selenide and germanium). Coincident viewing capability rare. Multiple beaming not possible at present.
Absorbed by solder (and most metals): Good thermal coupling to solder and pad. Fluxless soldering possible.	Reflected by solder: Thermal coupling relies on flux in the paste which absorbs well.
Minimum focused spot size: 10–20 μm.	Minimum focused spot size: 50–100 μm.
Power available: 1 kW: 20 W adequate for surface mounting. Soldering of wires and pins using preforms not usually possible.	Power available: 200 kW: No restrictions on electronic assembly solder joints.
Capital cost: Higher.	Capital cost: Lower.
Running/maintenance costs: Slightly higher.	Running/maintenance costs: Slightly lower.

Source: S. B. Dunkerton, "A Comparison of Laser Types for Reflow Soldering," *Soldering and Surface Mount Technology,* No. 10, February 1992, pp. 9–13.

The operating wavelengths are the main differences between the two lasers. The CO_2 laser operates at 10.6 μm while the Nd:YAG operates at a wavelength of 1.06 μm. The difference in wavelengths was important as it affected the amount of laser energy absorbed by the materials.

At a wavelength of 10.6 μm (CO_2), ~74 percent of the laser energy is reflected by solder and only ~2 percent by polyimide whereas at 1.05 μm only ~26 percent is reflected by solder and 21 percent by polyimide. This difference has not been shown to be that important in soldering as the absorption of laser energy by rosin at 10.6 μm is ~95 percent, and after reflow the reflectivity of solder drops dramatically.[63] In practice both systems are used. An overall comparison of the two systems is shown in Table 7.19.

Critical process parameters for laser soldering are:

- Beam power (measured in watts)
- Beam duration (in milliseconds)
- Beam direction (for example, 30°, 45°, 90°)
- Beam target (i.e., does the beam hit the lead or the pad near the lead)
- Solder application method (paste or reflowed solder plus flux)

While lasers have been successfully used for soldering leadless and gull wing leaded devices, there are reports of problems on J-Lead devices. Problems are related to PWB damage caused by excessive beam duration times and power settings needed to achieve solder reflow on lead under parts where most of the joint is "hidden" from the laser.[64]

Rework and Repair. Rework and repair are similar operations in that both are generally "unplanned" operations used to at least allow for a partial recovery of the value of the circuit card.

The primary difference between rework and repair is that *rework* involves returning the assembly to full drawing compliance while *repair* restores (at least partially) the form, fit, and function of the assembly but does not return it to full compliance with the drawing. An example would be that the resoldering of a defective solder connection is considered rework while the rebonding of a lifted pad is considered repair in that, after the operation, it is still obvious a nonstandard operation has taken place.

Soldering operations are generally considered rework in that, when properly accomplished, they leave no evidence that the joint has been resoldered. Solder rework can be accomplished by a number of solder methods depending on the device. Rework and repair of soldered connections fall into two categories:

- Those involving just reflow of the solder joint (i.e., resoldering) to correct marginal poor wetting (add solder and touch up solder bridges)
- Those involving removal and replacement of all the solder in the connection (i.e., desoldering)

During desoldering, the temperatures and dwell times used often exceed those used during initial soldering due to the increased thermal mass present. In order to prevent damage to the parts, two parameters must be controlled:

- The ultimate temperature and rate of temperature increase must be controlled to prevent thermal stress to the die (cracking or diffusion reactions), reflow of the die bond material, or charring of the device or substrate. This temperature is generally measured as close as possible to the active portion of the device, e.g., at the lead-body interface. Typically a temperature of 215°C for "normal" parts and 150°C for "thermally sensitive" parts is used. Additionally, heating in populated areas adjacent to the part should be kept below 150°C to prevent accidental reflow of solder joints.[65]
- The rate of increase of temperature during rework should be kept below 4°C/s during heating and to less than 10°C/s during cooling. This may require preheating of sensitive devices (e.g., chip capacitors and large ceramic chip carriers) prior to soldering. This is necessary to prevent delamination of the part or PWB.

A variety of tools are available for manual rework of solder connections. Typically normal solder irons and standard tips are used for resoldering or "touch-up" of both surface mount and plated-through hole connections.

Removal of solder form plated-through hole connections is generally accomplished using either a solder iron and *braid* (usually a fluxed, fine copper braid that is used to wick up the molten solder) or a *solder sucker* which removes the solder with a vacuum. The suction device can either be a manually activated device using a spring-driven "syringe" to gather the solder, or it may be incorporated into an open cylindrical tip that fits over the lead and seats on the solder pad. After solder removal the device is removed and another inserted in the hole and the leads resoldered.

Removal of surface mount devices with manual tools is typically done using special solder iron tips, "heated" tweezers, or small-area hot air wands. On multileaded surface mount devices and chip devices, these tips are often formed as a U to allow for heating of both sides of the part. This design both allows for reflow of all terminations at once (helping in the removal of the part) and minimizes damage to the device (by evenly heating from both sides). Similar arrangements are made for four-sided *quad packages* using a solder iron with a heated collet that heats on all four sides of the device at once.

Semi-automated or automated hot air systems are also used for desoldering of surface mount devices. These devices use heated air or nitrogen gas and may incorporate pick-and-place systems with thermocouple controlled hot gas heaters and a computer interface to record and "replay" successful solder schedules.

One of the advantages of this type of system is that it can be used in areas where direct contact with the soldered terminations is either not possible (due to design density) or not advisable (as is the case with fine pitch leaded devices) with an increase in process control and repeatability over the manual methods. As the heated gas contacts the PWB surface adjacent to the device and can cause heating of these areas, a variety of shrouds are used to limit the affected area. The parameters generally controlled for hot air soldering are:

- Gas flow rate
- Heater power
- Substrate temperature (Often the substrate is preheated and continues to be heated during rework to both reduce thermal energy required for rework and to reduce heating of adjacent areas by reducing the thermal gradient.)
- Process time (At least during initial attempts, this is judged visually by the operator as the completion of solder reflow at which time the part may be removed.)

Replacement of the device is typically done using the same equipment by simply reversing the removal procedure.

7.3.4 Post-Solder Cleaning

The first question to be answered when addressing cleaning is whether or not cleaning is actually required for a specific product. This decision should be based on:

- *The corrosivity of the flux residues:* This is a function of the corrosivity of the fluxes and soldering process used. For example a no-clean process with inert atmosphere uses a very corrosive flux but in very small amounts and may not require cleaning where hand soldering with RA flux generally would require cleaning.
- *The environment in which the hardware is to be used*
- *The required reliability of the hardware*

Typical effects of excessive flux residues are corrosion, leakage currents and/or shorts (generally associated with humidity exposure), coating debonding, poor appearance (e.g., white residues), and insulating contact surfaces (by covering up contacts or test points with an insulative layer of contaminate). A general guideline for cleaning requirements is shown in Table 7.20 for high-reliability designs based on a number of sources and the author's experiences. This table is meant as a general guideline and is

TABLE 7.20 Cleaning Requirements As a Function of Environment

Temp, °C	Relative humidity, %	Condensing humidity	Voltage, V	Conformal coating	Clean*	Cleaning test**
<65	<65	Yes	<100	Yes	Yes	3–4
<65	<65	Yes	>100	Yes	Yes	3–4
<65	<65	No	<100	No	No	1–2
<65	<65	No	>100	Yes	Yes	2–3
<65	65–98	Yes	<100	Yes	Yes	3–4
<65	65–98	Yes	>100	Yes	Yes	4
<65	65–98	No	<100	Yes	Yes	2–3
<65	65–98	No	>100	Yes	Yes	2–3
>65	<65	Yes	<100	Yes	Yes	3–4
>65	<65	Yes	>100	Yes	Yes	4
>65	<65	No	<100	No	No	1–2
>65	<65	No	>100	Yes	Yes	2–3
>65	65–98	Yes	<100	Yes	Yes	4
>65	65–98	Yes	>100	Yes	Yes	4
>65	65–98	No	<100	Yes	Yes	3–4
>65	65–98	No	>100	Yes	Yes	3–4

*If flux is rosin (without activator) or an equivalent no-clean process, cleaning may not be required.

**Test key is: 1 = no test, 2 = visual only (e.g., to prevent insulating test points), 3 = visual + ionic contamination (extractive test), and 4 = visual + ionic contamination + moisture insulation resistance (pass/fail criteria based on electrical design parameters such as impedance). (*Moisture insulation resistance* is sometimes referred to as *surface insulation resistance,* but, in fact, more than just surface contaminants are evaluated and moisture insulation resistance is a more accurate description.)

not meant to be rigidly applied. Process effects such as poor removal techniques or workmanship can easily overcome the environmental effects, causing the test and coating requirements to shift toward more protective levels with careful workmanship.

There are three fundamental steps in the cleaning process:

- Penetration of the cleaning agent into the area to be cleaned
- Dissolution of the contaminant(s)
- Removal of the dissolved contaminants and cleaning solution from the cleaned area

Thus it can be seen that the ability of a solvent to remove a specific residue is a function of both mobility (or penetration capability for getting under components) and solubility.

Solvent Mobility Factors. Factors affecting mobility are listed in Table 7.21. The relative merits of these parameters is a complex issue and arguments can be made for the importance of each. When discussing how to model solvent penetration under components, both the surface tension and capillary penetration have their proponents.

In the case of surface tension, the most desirable case is when the solvent wets easily to the surface, thus wetting easily under parts. For spontaneous wetting to occur,

TABLE 7.21 Factors Affecting Solvent Mobility

Factor	Common symbol	Effect on mobility of increase in numerical value of factor
Component spacing	d	Increase
Component size	L	Decrease
Solvent viscosity	η	Decrease
Impact pressure	P	Increase
Solvent contact angle	θ	Decrease
Surface tension	σ	Increase
Interfacial tension	γ	Decrease

Source: M. E. Hayes, Ph.D., "Physiochemical Aspects of Electronics Assembly Cleaning and Their Implications for Halogen-Free Solvent Selection," *Third International SAMPE Conference,* June 1989, pp. 998–1010.

the critical surface free energy γ_c must be less than the surface tension of the solvent (between the liquid and the vapor), γ_{LV}. Some common values of γ_c for materials used in electronic assembly are shown in Table 7.22 with surface tensions for some common solvents shown in Table 7.23.

Capillary penetration, on the other hand, is the penetration of a solvent into an enclosed area (e.g., under a component) and is similar to capillary rise in a tube. The driving force for this penetration is the pressure differential across the meniscus of the fluid as it wets across the entrance of the capillary , and the penetration of the solvent is proportional to surface tension of the fluid. The rate at which the fluid flows into the capillary is a function of both the surface tension and the viscosity of the fluid where the flow rate is proportional to ($\gamma_{LV} \cdot \cos \theta/\eta$).[66] A tabulation of some capillary flow parameters for common solvents is shown in Table 7.24.

TABLE 7.22 Critical Surface Free Energy for Electronic Materials

Material	γ_c, mN/m
Polytetrafluoroethylene (PTFE)	18.5
Polypropylene	29
Polyethylene	31
Polystyrene	30–36
Polyvinylchloride (PVC)	39
Polycarbonate	42
Polyamide resin	52
Urea formaldehyde resin	61
Glass (fused silica)	~300
Ceramics	~1000

Source: C. Lea, *A Scientific Guide to Surface Mount Technology,* Electrochemical Publications, 1988, p. 446.

TABLE 7.23 Typical Surface Tensions of Common Solvents

Solvent	γ_{LV} @ 25°C
1,1,2 trichlorotrifluoroethane (R113)	17.7
Ethyl alcohol	22.3
1,1,1 trichloroethane	25.1
Water	72

Source: C. Lea, *A Scientific Guide to Surface Mount Technology,* Electrochemical Publications, 1988, p. 446.

TABLE 7.24 Flow Parameters for Common Solvents

Solvent	Temp., °C	θ°	γ_{LV}, mN/m	η, mPa • s	$\frac{\gamma_{LV}, \cos\theta}{\eta}$, m/s
Fluorocarbon-alcohol blend	25	~0	17.4	0.66	26.4
Chlorocarbon-alcohol blend	25	~0	25.2	0.80	31.5
Water	25	60	72.0	0.89	40.4
Fluorocarbon-alcohol blend	40	~0	14.7	0.52	28.3
Chlorocarbon-alcohol blend	73	~0	17.0	0.42	40.5
Water	70	45	64.4	0.40	113.8

Source: C. Lea, *A Scientific Guide to Surface Mount Technology,* Electrochemical Publications, 1988, p. 447.

Spray pressure is often used to increase the penetration under components above that achieved by capillary penetration. In this case the flow rate is not determined by surface tension and viscosity but rather by the applied pressure differential across the gap under the component. The velocity of the fluid flow is given by Eq. (7.9) while the drag force applied to a contaminant (e.g., a solder ball) is given by Eq. (7.10).

$$v = \frac{\Delta p \cdot h^2}{12 \cdot l \cdot \eta} \tag{7.9}$$

where v = fluid velocity
Δp = applied pressure differential
h = height of the gap
l = length of the gap
η = viscosity of the fluid

$$F_{\text{drag}} = k \cdot \vartheta \cdot A \cdot v^2 \tag{7.10}$$

where F_{drag} = drag force
k = drag coefficient, dimensionless, determined experimentally
A = projected area of the contaminant

Solubility Factors. A contaminant's solubility in a solvent can be generally predicted based on the solubility parameter theory. This theory assumes that the solubility of a material in another material S_t is composed of three partial solubility parameters: S_D which is the nonionic (dispersion) term, S_P which is the polar (ionic) term, and S_H which is the hydrogen bonding term. These values are tabulated for various pure solvents.[67] The solubility parameter for mixtures of pure solvents can be determined by summing the volume fraction of the pure solvent multiplied by the respective solubility parameter of the pure solvent. Solubility theory would predict that, as the difference between solubility parameters between the solvent and the contaminant decrease [as calculated using Eq. (7.11)], the solubility increases:

$$\Delta S = \sqrt{(S_D^i - S_D^j)^2 + (S_P^i - S_P^j)^2) + (S_H^i - S_H^j)^2} \tag{7.11}$$

where the superscript i refers to the solvent and j to the solute (contaminant). Plots of the saturation solubility of abietic acid (a common flux activator) as a function of ΔS are shown in Fig. 7.34.

An alternative measure sometimes used for comparing solvent effectivity is a kauri-butanol value, but this measure is based on the solubility of Kauri gum rosin (which is a hard, fossilized rosin) in the solvent and was originally conceived for hydrocarbon solvents. Its applicability to rosin fluxes has been questioned.[68]

Solubilization. Another approach to dissolving contaminants is solubilization where a normally insoluble material is rendered soluble.

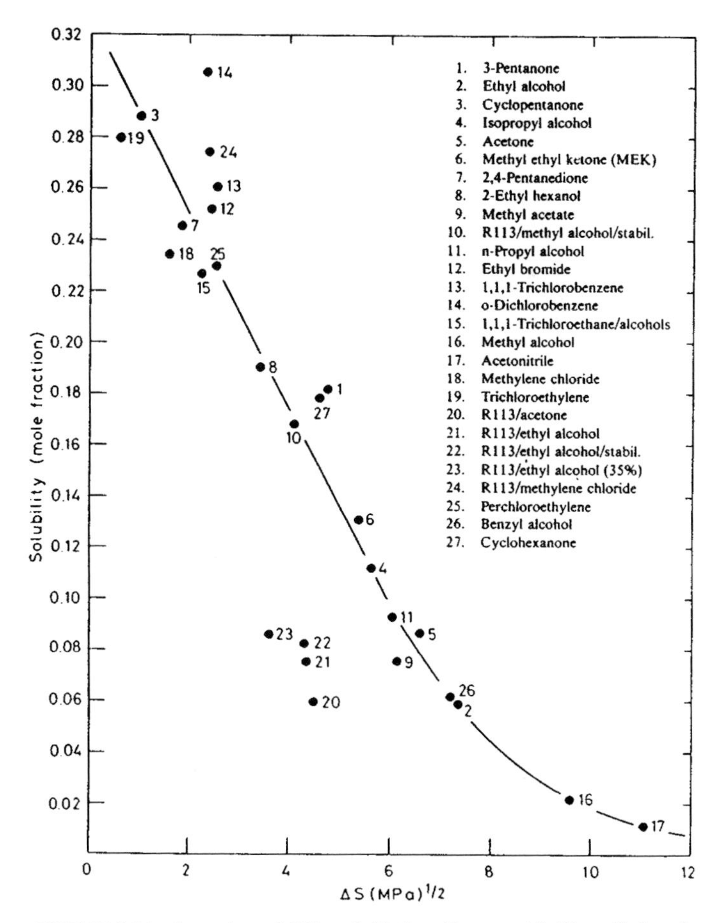

FIGURE 7.34 Saturation solubility of abietic acid versus ΔS. (*From C. Lea,* A Scientific Guide to Surface Mount Technology, *Electrochemical Publications, 1988, Fig. 12.4, p. 443.*)

Saponification is one method of solubilization where a saponifier is used to convert the rosin (or similar contaminate) to a soap which is then more readily dissolved in water. A typical saponifier process would be shown chemically as:

$$\underset{\text{rosin}}{C_{19}H_{29}COOH} + \underset{\text{monoethanolamine}}{H_2NCH_2CH_2OH} = \underset{\text{rosin-soap}}{C_{19}H_{29}COOCH_2CH_2NH_2} + \underset{\text{water}}{H_2O} \quad (7.12)$$

This process involves the use of long chain molecules which are strongly hydrophobic on one end and strongly hydrophilic on the other. At some critical value of concentration (depending on the saponifier), these molecules bond together to form micelles which have the property of reducing the surface tension of the water (see Fig. 7.35). As these micelles come in contact with a contaminated surface, the micelles unwind (to increase the amount of contact of the hydrophilic ends with water), and, in doing so, form around the contaminate, breaking it up (see Fig. 7.36). This reaction

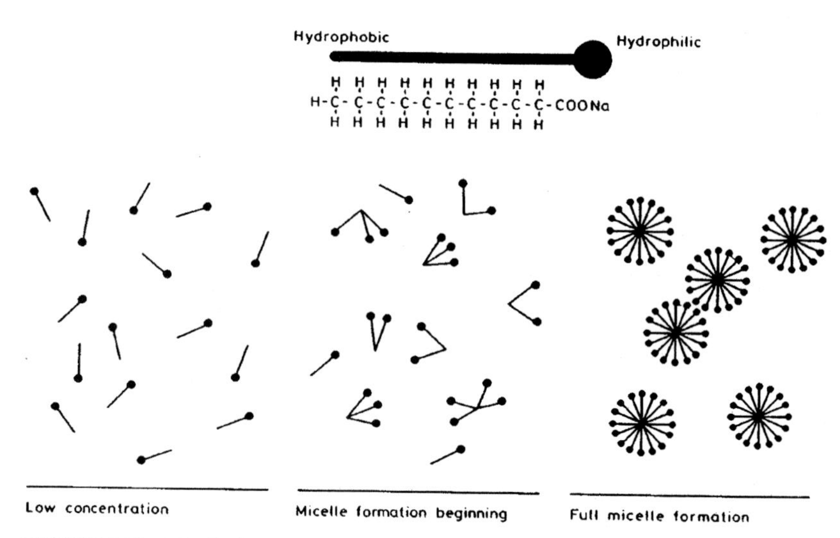

FIGURE 7.35 Micelle formation during saponification process. (*From C. Lea,* A Scientific Guide to Surface Mount Technology, *Electrochemical Publications, 1988, Fig. 12.2, p. 441.*)

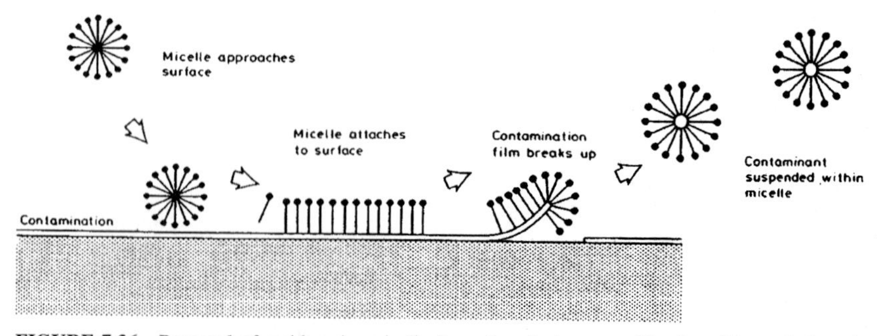

FIGURE 7.36 Removal of residues by micelle formation during saponification. (*From C. Lea,* A Scientific Guide to Surface Mount Technology, *Electrochemical Publications, 1988, Fig. 12.3, p. 441.*)

depends on the adsorption bond between the contaminate and the assembly being weaker than both the ionic attraction between the water and the soap molecules and the attraction between the soap molecules and the contaminate.

The use of a material to slightly acidify the washing solution to reduce otherwise insoluble contaminates is demonstrated by the use of a weak flux solution being used to dissolve some forms of "white residue." In this process a weak, activated flux (typically RMA) is applied to the contaminated surface, allowed to stand briefly (typically 1 to 5 min), and then the assembly is cleaned again using a typical rosin cleaning process.

Additional Cleaning Process Parameters. Aside from the theoretical factors cited in the previous sections, a number of processing variables can also strongly influence the effectivity of any given cleaning process. These include:

- Flux type used (both activator, base flux (rosin or nonrosin), carrier, form (paste, liquid, or wire)
- Volume and spread of flux used
- Soldering process used (typically temperature/time profile and atmosphere in which reflow occurs such as oxygen or nitrogen)
- Time delay from end of soldering to cleaning
- In a manual or operator-dependent process, the aggressiveness of the operator in utilizing the solvent and mechanical application of the solvent (e.g., brushing, spraying, or rinsing)

The proper selection of these parameters can easily make a relatively weak solvent perform admirably while inattention to these details can render even the most powerful solvents ineffective.

An additional concern in the selection of solvents is the compatibility of the solvent/cleaning process with the devices being cleaned. Among the items that should be considered are:

- Compatibility of the solvent with the items being cleaned (Many solvents attack various plastics and elastomers.)
- Compatibility with unsealed devices or switches which may be either damaged or "overcleaned" (e.g., sealing material or lubricants dissolved)
- Compatibility with the production flow (e.g., inline versus batch systems and cleaning process time)
- Need for follow on processes (rinses, drying steps, and so on)
- Operator safety considerations such as solvent toxicity and ability to control or contain exposure to the cleaning materials
- Environmental compatibility of the process and its waste stream such as:

  Ozone depletion potential (ODP)[69] and photochemical reactivity (influencing smog formation) which are typically associated with solvents

  pH, water temperature, heavy-metal concentration, *biological oxygen demand* (BOD), *chemical oxygen demand* (COD), "grease, oil, and fat," and *total toxic organics* (TTO) which are typically associated with aqueous and semiaqueous systems

Due to the influence of parameters cited above, it is difficult to quantify in any absolute sense the effectivity of any given cleaning process without complete knowledge and consideration of the preceding process(es). For this reason the discussion which follows on specific cleaners will be limited to their physical properties and general processing guidelines, and no specific recommendations will be made.

Solvent Cleaning. A number of solvents are available for the cleaning of electronic assemblies. These solvents are typically used in one of the following ways:

- *Cold cleaners* used in a tank or batch process (often accompanied by spray or brushing action) at room temperature
- *Heated cleaners* designed to be used at above room temperature but well below their flash point or boiling point
- *Vapor cleaners* which rely primarily on the condensation of vapors from the boiling cleaning agent to remove contaminants (e.g., vapor degreasers)

While in accordance with the Montreal Protocol, the use of CFC and other ozone-depleting materials is being curtailed and production will cease in the 1995–2000 timeframe, these solvents are included as reference materials due to their extensive past usage:

Cold Cleaning Solvents. The cold cleaning solvents currently in use are generally based on either alcohols or terpenes. Terpenes are naturally occurring solvents found in many plants and are generally regarded as derivatives of isoprene (2-methyl-1,3-butadiene). Terpenes have a similar molecular structure to rosin flux constituents (see Fig. 7.37), thus explaining their ability to readily dissolve rosin flux residues. A summary of characteristics of a variety of solvents used for cold cleaning of electronics is shown in Table 7.25.

When using cold cleaning solvents, it is important to both monitor the contaminate level in the cleaning tank and to ensure a clean final rinse. Monitoring of the solvent for contamination may be accomplished by checking specific gravity, boiling point, color, or turbidity. Often filtration or a multiple-tank system with a counterflow replenishment (for example, three tanks used with last, cleanest tank being poured into middle, middle into first, and first tank discarded or recycled) is used to control contamination levels.

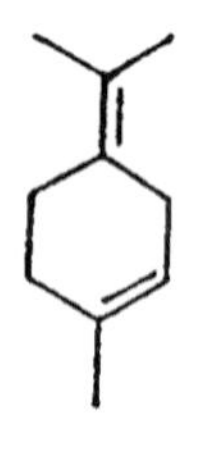

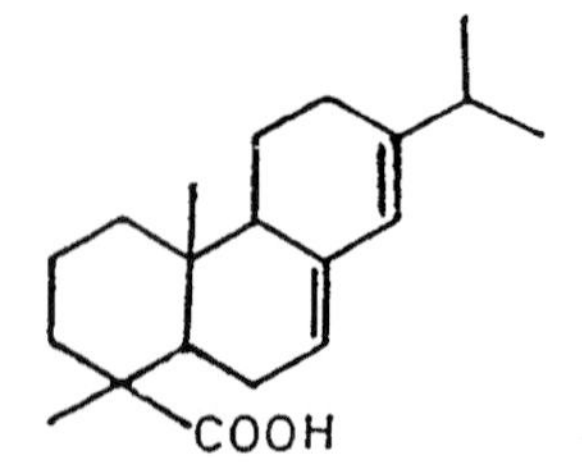

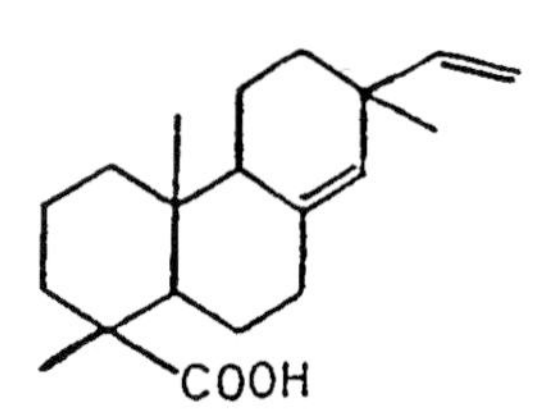

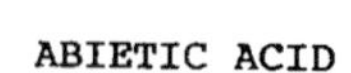

PIMARIC ACID

FIGURE 7.37 Molecular structure of a typical terpene (terpinolene) and components of rosin flux (abietic acid and pimaric acid). (*From M. E. Hayes, Ph.D., "Cleaning SMT Assemblies Without Halogenated Solvents,"* Surface Mount Technology, *December 1988, pp. 37–40.*)

TABLE 7.25 Physical and Usage Characteristics of Cold Cleaning Solvents

Trade name	Vendor	Type	Specific gravity	Viscosity @25°C, cP	Surface tension @25°C, dyn/cm	Flash point, closed cup, °C	Incompatible with:
EC-7M	Alpha metals	Terpene	0.84 @25°C	0.8	—	47	Polystyrene, polycarbonate, RTV silicone
FluxOff II	Chemtronics	Alcohol blend	0.788 @20°C	—	—	21	Polystyrene, polycarbonate
Methanol	—	Alcohol	0.791 @20°C	0.59	22.6	12	—
Ethanol	—	Alcohol	0.789 @20°C	1.2	22.8	14	—
Isopropyl alcohol (2-Propanol)	—	Alcohol	0.786 @20°C	2.32	21.7	13	—

Heated Solvent Cleaners. Heated solvent cleaners can be used to increase the effectivity of the solvent, but extreme care must be used to ensure that the solvents (many of which have low flash points) are not ignited during use. Typically, in heated solvent systems, the solvent is either mixed with a nonflammable material (for example, IPA and water) or is used in nonoxygen atmosphere (generally a vapor blanket of a perfluorocarbon material). As an example, the flammability of an alcohol-perfluorocarbon mix is controlled by carefully monitoring their mixing ratio.

Currently systems are being manufactured employing a perfluorocarbon-alcohol mix,[70] and a mix of aliphatic esters and perfluorocarbons.[71] Advantages of these systems are good flux removal properties combined with low solvent losses and the elimination of a water rinse step after use.

Vapor Cleaners (Degreasers). Vapor degreasers have been used for some time in electronics cleaning and have offered one of the most effective methods of cleaning of electronics. Typically the solvents used for this purpose have been chlorofluorocarbons and chlorinated solvents (e.g., trichloroethane). Unfortunately, the solvents used for degreasing have been found to adversely affect the ozone layer and are being phased out with the production of CFCs being halted in 1995 and of other chlorinated solvents (e.g., trichloroethane and HCFCs) about the year 2000.

Typically degreasers of either an inline or batch configuration were used, and these systems often employed immersion in the vapor and liquid phases of the solvent for cleaning purposes with a spray assist. A listing of physical properties for some degreasing solvents used for electronics cleaning (including their ozone depletion potential) is shown in Table 7.26.

A major distinction between the chlorinated solvents such as 1,1,1 trichloroethane and trichloroethylene and the CFC/HCFC solvents is in the area of material compatibility. As shown in Table 7.27, the CFC/HCFC materials are generally more compatible with the materials used in electronics fabrication than are the chlorinated solvents.

Aqueous Cleaning. Fully aqueous cleaners are used for removal of ether with water-soluble fluxes or with rosin fluxes using saponifiers. Aqueous systems can be run as either batch or inline systems. Aqueous cleaning has been effective in cleaning both plated-through hole and surface mount designs.

TABLE 7.26 Physical and Usage Characteristics of Vapor Degreaser Solvents

Material	Specific gravity @25°C	Surface tension @25°C, dyn/cm	Boiling point, °C	ODP	Last year of production
CFC-113	1.565	18.7	47.6	~0.8	1995
CFC-113/ethanol blend	1.50	18.2	43.6	~0.8	1995
CFC-113/methanol blend	1.48	—	39.7	~0.8	1995
Trichloroethylene	1.322	32	87.2	—	2000
1,1,1 trichloroethane	1.456	25.9	74.1	0.1	2000
HCFC 141b/alcohol blend	1.213	18.5	29.4	~0.14	2000

TABLE 7.27 Compatibility of Vapor Degreaser Solvents and Common Electronic Materials

Material	Incompatible with:
Trichloroethylene and 1,1,1 trichloroethane	Polyoxymethylene polyacetal (Delrin, Hostaform, etc.); polymethylmethacrylate (Plexiglas, Perspex, etc.); acrylonitrilebutadeinestyrol copolymer (Vestodur, Terluran, etc.); polycarbonate (Makrolon, Lexan, etc.); polyethylene (Polythene, Alkathene, etc.); polypropylene (Hostalen PP, Daplen, etc.); polystyrol (Styrom, Lustrex, etc.); polyphenylene oxide; styrolacrylnitrile copolymer (Luran, Vestoran, etc.); polyvinylchloride (Trovidur, Hostatit, etc.); natural rubber; polyurethane rubbers (Adiprene, Vulkallan, etc.); butyl rubbers; ethylenepropylene rubbers (Dutral, Nordel, etc.); polychloroprene (Neoprene, Bayprene, etc.); nitrile rubbers (Buna N, Perbunan, etc.); epichlorohydrin rubbers (Herclor, Hydrin, etc.); silicone rubbers; chlorosulphonopolyethylene (Hypalan).
CFC 113-ethanol azeotrope and CFC 113-methanol azeotrope	Polymethylmethacrylate (Plexiglas, Perspex, etc.)l; polystyrol (Styrom, Lustrex, etc.).
HCFC 141b-alcohol mixture	Acrylonitrilebutadeinestyrol copolymer (Vestodur, Terluran, etc.); acrylic; styrene.

The effectivity of the process is a function of the compatibility of the flux or contaminate, the saponifier (type and concentration), and the PWA design. Typical processes run at 48 to 80°C and saponifier concentrations of 2 to 10 percent by volume. Two final factors often overlooked by users of aqueous (and semiaqueous) system involving the rinse section and its impact on cleanliness:

- Inadequate rinsing can leave saponifier, solvent, and/or flux residues on the circuit card, which can in turn lead to field failures through excessive leakage current and corrosion.

- Inadequate removal of rinse water (e.g., lack of air knives) prior to drying can lead to failure from both.
- Coating delamination (e.g., mealing) at the edges of water spots (One theory relates this problem to the concentration of soluble surface contaminates as the water drop evaporates.)
- Corrosion caused by dissolution of copper, exposed on the sides of many circuit traces, after exposure to high-purity deionized water; typically >10 MΩ; which is in itself a corrosive material.[72]

Semiaqueous Cleaning. *Semiaqueous processing* refers to a two-step cleaning process consisting of a washing section (generally a combination of terpenes or hydrocarbons and water) and a rinsing section (typically DI water). Usually the initial washing section contains a 10 to 50 percent concentration of solvent with the balance being water.

As many of the solvents used for semiaqueous cleaning are highly flammable, the cleaning systems often use a nitrogen atmosphere and/or extensive temperature controls with fire suppression systems to control fire risk. A summary of physical properties of some semiaqueous solvents is shown in Table 7.28.

Ultrasonic Cleaning. The effectivity of cleaning with either solvents or aqueous solutions can be significantly enhanced by the introduction of ultrasonic energy into the cleaning medium. The effect of this ultrasonic energy is to introduce a mechanical oscillation which is so fast that the fluid cannot flow with it. This oscillation first causes a reduction of pressure in the liquid, which can, if the pressure falls below the vapor pressure of the liquid, spontaneously generate cavities or bubbles in the solution. The second half of the mechanical cycle causes the pressure to rise again, imploding the bubbles and causing a local temperature rise and mechanical shock waves which are extremely efficient at dislodging contaminates. Ultrasonic cleaning has been found to be of great benefit, especially in the removal of contaminates under surface mount components. Recommended exposure settings are:

- Maximum frequency of 40 kHz
- Maximum exposure time of 1.5 min (total)
- Maximum power of 10 W/liter
- Boards should be racked and placed such that they cannot be touched, which can change characteristic resonances

TABLE 7.28 Physical Characteristics of Semiaqueous Cleaner Solvents

Material	Vendor	Type	Flash point (closed cup)	Vapor pressure, mm Hg	Viscosity, cP	Surface tension, dyn/cm	Specific gravity @ 25°C	Odor
EC-7	Alpha metals	Terpene	47°C	1.5	0.8	—	0.84	Oranges
Axarel 38	DuPont	Hydrocarbon blend	71°C	0.2	1.4	28	0.85	Mild to low
KCD-9577	DuPont	Hydrocarbon blend	99°C	<0.1	2.8	27	0.84	Low
Ionox LC	Kyzen	Low vapor pressure alcohol	None to boiling	17	—	19–21	1.032	Mild to low

The effect of ultrasonic cleaning on long-term component reliability is currently a question seeing a great deal of interest. Testing has shown that many typical parts can survive a properly configured ultrasonic cleaning system without immediate effect (see Fig. 7.38). It should be stressed, however, that this testing is valid only for the part types tested [typically TO-18 cans, dual inline packages (plastic and ceramic), and SOIC/SOTs].[73] Testing has shown that ultrasonic cleaning can impact the reliability of quartz crystal devices.[74] Additionally, it has been noted that the much longer wire bond lengths used in large hybrid devices can result in failures in wire bonds.[75] As the impact of ultrasonics can be affected by the configuration of the cleaner, stiffness (thickness and anneal) of the wire, cleaning media, and wire bond lengths, it is felt that the only definitive answer on the impact of component reliability is a direct test of the component to be used as mounted on the assembly cleaned in the cleaner (or a similarly configured system).

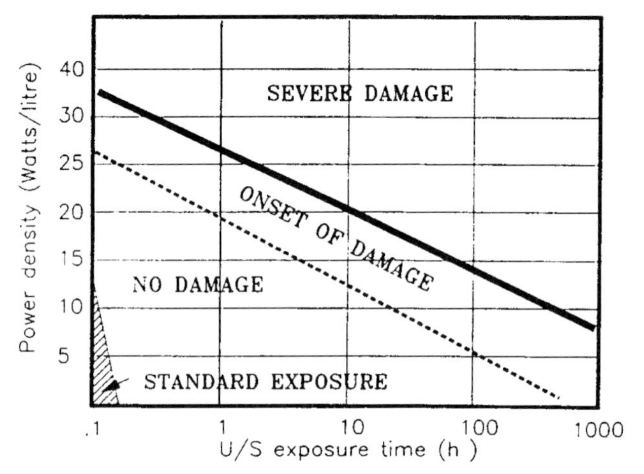

FIGURE 7.38 Effect of *ultrasonic* (U/S) exposure time and power density on wire bond damage in TO-18 cans, dual inline packages (plastic and ceramic), and SOIC/SOTs (as noted visually immediately after exposure). (*From B. P. Richards, P. Burton, and P. K. Footner, "The Effects of Ultrasonic Cleaning on Device Degradation—An Update,"* Circuit World, *Vol. 17, No. 4, 1991, pp. 29–34.*)

7.4 WELDING

7.4.1 Resistance Welding

In resistance welding, an electrode is used to mechanically hold two or more metals together, and then heat is generated by passing a high current through the work pieces causing the material to bond. The reason that the bonding occurs is that at a microscopic level the materials are not smooth and in fact only make contact at selected points (see Fig. 7.39). At this contact point, the pressure applied to the metal pieces is

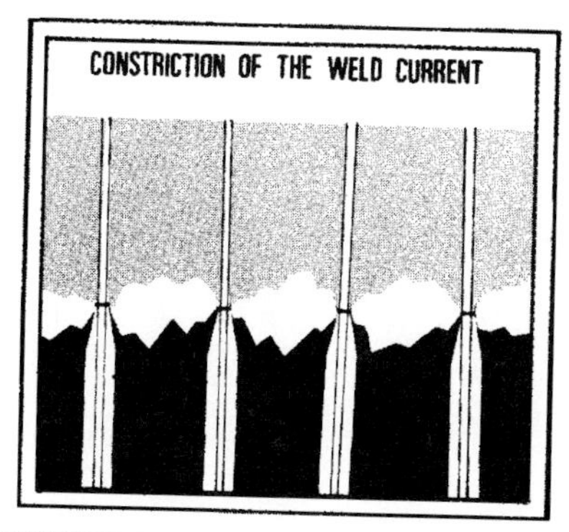

FIGURE 7.39 Exaggerated cross section of two metal pieces (gray and black) being welded showing the small contact areas and restricted current flow. (*From "Resistance Welding Systems,"* Unitek Technical Bulletin.)

magnified by a small contact area, and any surface oxides rupture allowing for metal-to-metal contact. The weld current is concentrated by being forced to go through these restricted areas, which increases current density, which in turn generates enough heat to initiate reflow of the metals. As the contact and current are maintained, the area of reflow increases.

There are four basic types of weld:

- *Soldered or brazed bond* where resistance heating is used to melt an intermediary material (e.g., solder), which in turn forms an alloy with both pieces. Brazing is generally differentiated from soldering by the temperature at which the alloy melts; brazing is generally considered to be done with materials that melt above 400°C.
- *Forged weld* where the surfaces of the work piece are bonded together without any apparent or large-scale melting. This is done by using a very short weld time and is advantageous when bonding dissimilar metals with radically different structures.
- *Diffusion welding* which refers to a process in which the two metals are heated to a plastic state and solid-state diffusion creates a bond but no melting actually occurs. This type of bonding is used with dissimilar metals which have similar structures.
- *Fusion welds* refer to welds that involve complete melting of the two metals being joined. During subsequent cooling, a "nugget" is formed which contains all alloys of the two materials. This type of weld is typical of long-pulse welding and is used to bond similar metals.

Two types of welding systems are used: direct energy and stored energy systems. *Direct energy systems* use alternating current to supply energy to the weld, and *stored energy systems* supply energy by discharging a capacitor. The two methods are compared in Fig. 7.40. Both systems have their advantages as shown in Table 7.29. A number of different types of weld may be formed using these systems as shown in Fig. 7.41.

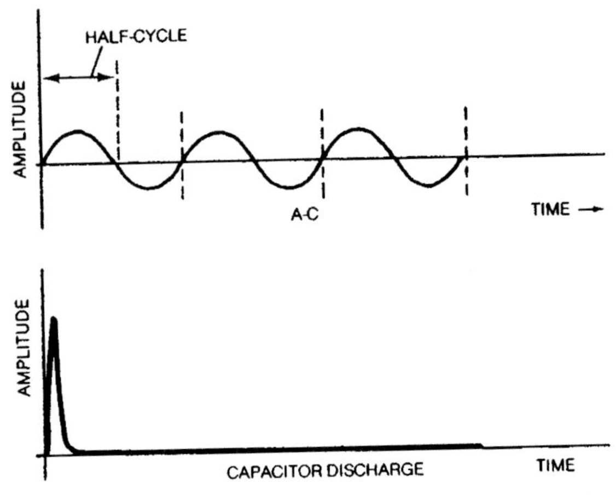

FIGURE 7.40 Comparison for energy versus time for the direct energy system (*top*) and stored energy system (*bottom*). (*From "Resistance Welding Systems,"* Unitek Technical Bulletin.)

TABLE 7.29 Comparison of Resistance Welding Systems

Stored energy systems	Direct energy (AC) systems
Independence from line voltage fluctuations	Provide large amounts of energy economically
Lower power line requirements	Increased welding rates (up to 5 welds per minute)
Can consistently weld small and dissimilar materials	Supply energy over long periods for control in multilead soldering or brazing

Source: "Resistance Welding Systems," *Unitek Technical Bulletin.*

Ultrasonic Welding

Ultrasonic welding is also being investigated for use in attaching solder tinned components to solder plated circuit boards.[76] Initial testing has shown promising results on chip devices, and the process offers the advantages of low temperature and no flux–no clean while retaining the reworkability of a standard solder connection.

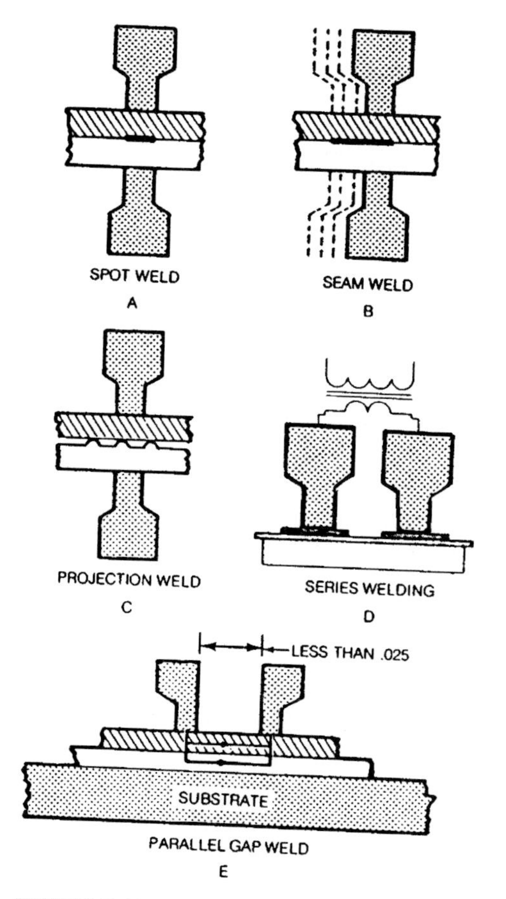

FIGURE 7.41 Weld types made using resistance welding systems. (*From "Resistance Welding Systems,"* Unitek Technical Bulletin.)

7.5 OTHER MECHANICAL JOINING METHODS

7.5.1 Discrete Wiring Methods

Discrete wiring is a term used to describe a solderless connection formed by one of three methods: wire wrap, clip termination, and insulation displacement (see Fig. 7.42). The advantages of all these systems is in the ease of modification and rework that is available. As no solder or welded connection is formed, the old wire may be mechanically removed and replaced with a new one. Integrity of the connections is checked using pull tests and a check of the integrity of the gas tight seal by suspending the samples of an *aqua regia* (a 1-to-1 mixture of concentrated hydrochloric and

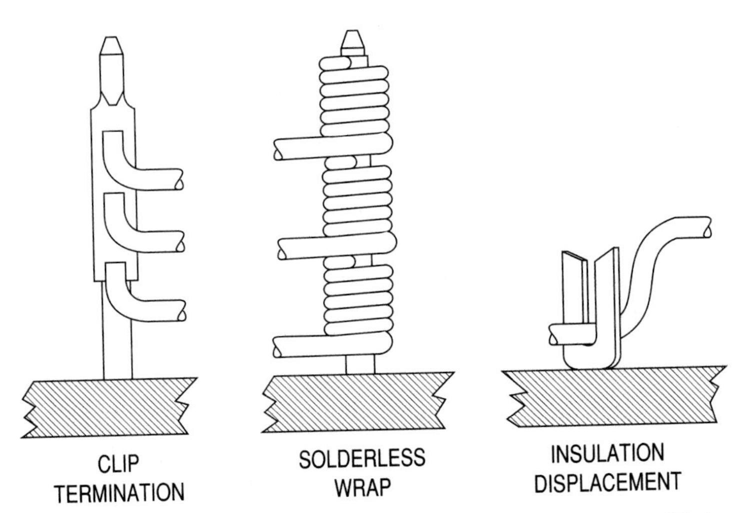

FIGURE 7.42 Three types of discrete wiring technology. [*From IPC-TR-474,* An Overview of Discrete Wiring Techniques, *Institute for Interconnecting and Packaging Electronic Circuits* (*IPC*).]

nitric acid). After a period of time (typically 30 min), the sample is removed and dried and the contact resistance checked (resistance for a typical connection is on the order of $10^{-4}\ \Omega$).

Wire Wrap. In wire wrap the connection wraps a solid wire conductor tightly around a square post. As the tool pulls the wire tight, the corners of the post "bite" into the wire, forming an electrically conductive, gas-tight interface. (The gas-tight feature is useful in preventing oxidation in this area which could lead to intermittent opens or high-resistance paths in use.) Both wrap tightness and number of wraps are related to connection integrity. The posts used on 0.025-in centers with 26- to 30-gauge, tin-plated copper wire used. For further information see MIL-STD-1130, Connection, Electrical, Solderless Wrapped.

Clip Termination. In clip terminations a tool is used to slide a clip containing a wire onto a post with a tapered end. As the tool pushes the clip onto the post, the clip is spread and the edges of the clip bite into the wire, again forming a gas-tight fit. The clip is generally a phosphor-bronze alloy, and the posts are generally set up on 0.100-in centers. This method can be used with either solid or stranded wire with gauges from 22 to 32. For additional information, see MIL-STD-1664, Connection, Electrical Clip Termination.

Insulation Displacement. With this technique an insulated wire is pressed into a clip, typically mounted in the end of a post which in turn is placed in a PTH on the board. The advantage of this method is that the post used is usually of a lower profile than the posts required for the other methods (however, it is wider.) The wire used is generally 24 to 30 gauge with PVC or FEP insulation.

7.5.2 Compliant Pin

Compliant pin technology refers to a specially designed pin used for interconnecting to plated-through holes on thick printed wiring boards where soldering is either difficult or impossible to achieve. The printed wiring boards used with compliant pins are designed to special criteria such as those used in MIL-STD-2166, Connections, Electrical, Compliant Pin. The pins are typically of a C or S configuration relying on a temporary compression of a wide portion of the pin during insertion which expands against the whole wall, holding the pin in place.

Testing of compliant pins is done to verify performance in humid, mechanical shock, and thermal shock environments with retention force (typically 5 lb for a standard pin and 7.5 lb for a pin containing a wire wrap post), and electrical resistance (2 to 7 mΩ) being used to evaluate the connection integrity.[77] Testing in thermal shock, salt spray, and extended humidity has also been accomplished with encouraging results.[78]

REFERENCES

1. This publication is available through the Institute for Interconnecting and Packaging Electronic Circuits (IPC), 7380 North Lincoln Avenue, Lincolnwood, IL 60646.
2. R. L. Beadles, "Interconnections and Encapsulation," AD 654–630, Vol. 14 of *Integrated Silicone Device Technology,* ASD-TDR-63-316, Research Triangle Institute, May 1967.
3. G. G. Harman, "Metallurgical Failure Modes of Wire Bonds," 12th Annual Reliability Physics Symposium, April 1974, pp. 131–141.
4. H. K. Dicken and D. B. Kret, "Assembling Integrated Circuits," *Electronics Engineering,* October 1966.
5. R. B. Larson, "Microjoining Processes for Electronic Packaging," *Assembly Engineering,* November 1966. Copyright 1966. Reprinted by permission of Hitchcock Publishing Co., Wheaton, IL.
6. E. M. Ruggiero, "Aluminum Bonding Is Key To 40 watt Microcircuits," *Electronics,* August 23, 1965.
7. McKaig, Peterson, and DePrisco, "Ultrasonic Welding in Electronic Devices," Part 6, *IRE International Convention Record,* 1962.
8. For the naming convention used for solder alloys, see QQ-S-571, "Solder, Electronic, 96 to 485°C," Interim Amendment 6, dated December 10, 1990.
9. R. J. Klein Wassink, *Soldering in Electronics,* 2d ed., Electrochemical Publications, 1989, p. 649.
10. H. J. Rack and J. K. Maurin, "Mechanical Properties of Cast Tin-Lead Solders," *Journal of Testing and Evaluation,* Vol. 2, No. 5, September 1974, pp. 351–353.
11. P. G. Harris, K. S. Chaggar, and M. A. Whitmore, "The Effect of Aging on the Microstructure of 60:40 Tin-lead Solders," *Soldering and Surface Mount Technology,* No. 7, February 1991.
12. M. L. Ackroyd, C. A. Mackay, and C. J. Thwaites, "Effect of Certain Impurity Elements On the Wetting Properties of 60% Tin–40% Lead Solders," *Metals Technology,* Vol. 2, 1975, pp. 73–85.
13. R. J. Klein Wassink, *Soldering in Electronics,* 2d ed., Electrochemical Publications, 1989, pp. 167–174.
14. MIL-STD-2000A, Standard Requirements for Soldered Electrical and Electronic Assemblies, p. 24.

15. J. Fischer, et al., "Reliability Assessment of Antimony Removal from Tin-Lead Solders," *16th Annual Electronics Manufacturing Seminar, Naval Weapons Center, China Lake CA,* NWC Publication 7163, pp. 219–235.
16. C. Lea, *A Scientific Guide to Surface Mount Technology,* Electrochemical Publications, 1988, p. 334.
17. L. E. Felton, D. L. Millard, C. H. Raeder, V. A. Tanzi, D. B. Knorr, and C. Havesy, "Pb Free Solders and Advanced Joining," *Electronic Manufacturing Program Report,* Rensselaer Polytechnic Institute, May 1992, pp. 16–18.
18. R. J. Klein Wassink, *Soldering in Electronics,* 2d ed., Electrochemical Publications, 1989, p. 197.
19. L. E. Felton, D. L. Millard, C. H. Raeder, V. A. Tanzi, D. B. Knorr, and C. Havesy, "Pb Free Solders and Advanced Joining," *Electronic Manufacturing Program Report,* Rensselaer Polytechnic Institute, May 1992, p. 16.
20. J. L. Marshall, J. Calderon, and J. Sees, "Scanning Electron Microscopy Characterizations of Solder Joint Failure—Standard and Non-Standard Solders," NEPCON West 1990, pp. 1361–1366.
21. Z. Mei and J. W. Morris, "Characterization of Eutectic Sn-Bi Solder Joints," *Journal of Electronics Materials,* Vol. 12, No. 6, June 1992, pp. 599–607.
22. R. J. Klein Wassink, *Soldering in Electronics,* 2d ed., Electrochemical Publications, 1989, p. 195.
23. Z. Mei and J. W. Morris, "Characterization of Eutectic Sn-Bi Solder Joints," *Journal of Electronic Materials,* vol. 21, No. 6, pp. 599–607.
24. C. Lea, *A Scientific Guide to Surface Mount Technology,* Electrochemical Publications, 1988, p. 334.
25. R. J. Klein Wassink, *Soldering in Electronics,* 2d ed., Electrochemical Publications, 1989, p. 195.
26. C. Lea, *A Scientific Guide to Surface Mount Technology,* Electrochemical Publications, 1988, p. 130.
27. S. Zayic, "R vs. RMA Cleaning for Non-PWAs," *Boeing DS&G Electronics M&P Report EM/P 390,* December 14, 1990.
28. S. Zayic, "RMA Cleaning for PWAs and Wire," *Boeing DS&G Electronics M&P Report EM/P 390,* April 18, 1990.
29. See MIL-STD-14256, Flux, Soldering, Liquid (Rosin Base).
30. F. M. Zado, "Effects of Non-Ionic Water Soluble Flux Residues," *Western Electric Engineer,* Vol. 1, No. 1, 1983.
31. F. M. Zado, "Effects of Non-Ionic Water Soluble Flux Residues," *Western Electric Engineer,* Vol. 1, No 1, 1983.
32. J. Savi, "No-Clean Flux Incompatibilities," *Circuits Assembly,* June 1992, pp. 48–50.
33. R. N. McLellan and W. H. Schroen, "TI Solves Wicking Problem," *Circuits Manufacturing,* September 1987, pp. 78–85.
34. C. Lea, *A Scientific Guide to Surface Mount Technology,* Electrochemical Publications, 1988, p. 155.
35. Dr. R. P. Anjard, *Solder Paste, Basic and Intermediate Primer,* Johnson Matthey.
36. Dr. R. P. Anjard, *Solder Paste, Basic and Intermediate Primer,* Johnson Matthey.
37. M. J. Remmler, A. Modl, and F. Hermann, "Calculation of Solder Joint Volumes for TAB and MCR Components," *Soldering and Surface Mount Technology,* No. 11, May 1992, pp. 22–25.
38. S. Jopek, "Solder Volume Determination for Fine Pitch Packages," *Printed Circuit Assembly,* March 1989, pp. 15–20.

39. C. Lea, F. H. Howie, and M. P. Seah, "Blowholing in PTH Solder Fillets—The Scientific Framework Leading to Recommendations for Its Elimination," *Circuits World,* Vol. 13, No. 3, 1987.

40. C. Lea, F. H. Howie, and M. P. Seah, "Blowholing in PTH Solder Fillets—The Scientific Framework Leading to Recommendations for Its Elimination," *Circuits World,* Vol. 13, No. 3, 1987.

41. D. Morency, "PLCC Packaging Cracking, Is Pre-Assembly Baking Really the Answer?," *Surface Mount Technology,* March 1991, pp. 62–64.

42. J. H. Lau, *Solder Joint Reliability—Theory and Applications,* Van Nostrand Reinhold, 1991, p. 183.

43. J. Maxwell, "Cracks: The Hidden Defect," *Circuits Manufacturing,* Vol. 28, No. 11, November 1988, pp. 28–33.

44. P. Prasad, *EM/P 153, Contributing Factors to Thermal Damage in Printed Wiring Assemblies during Hand Soldering,* Boeing Technical Report, dated March 9, 1981.

45. V. Meeldijk, "Latent Defects Result in Multilayer Ceramic Capacitor Failures," *Electronic Packaging and Production,* January 1984, pp. 261–265.

46. J. Maxwell, "Surface Mount Soldering Techniques and Thermal Shock in Multilayer Ceramic Capacitors," *AVX Technical Bulletin,* AVX Corporation.

47. C. Lea, *A Scientific Guide to Surface Mount Technology,* Electrochemical Publications, 1988, p. 133.

48. R. J. Klein Wassink, *Soldering in Electronics,* 2d ed., Electrochemical Publications, 1989, p. 176.

49. D. A. Elliott, "Q&A: Nitrogen Wave Soldering," *Circuits Assembly,* October 1991, pp. 55–59.

50. D. A. Elliott, "Q&A: Nitrogen Wave Soldering," *Circuits Assembly,* October 1991, pp. 55–59.

51. H. J. Hartmann, "Nitrogen Atmosphere Soldering," *Circuits Assembly,* January 1991, pp. 60–66.

52. A. W. Wright, R. L. Mahajan, and G. M. Wenger, "Thermal and Soldering Characteristics of Condensation Heating Fluids," Proceedings of NEPCON West, p62, 1985.

53. Product of 3M.

54. C. Lea, *A Scientific Guide to Surface Mount Technology,* Electrochemical Publications, 1988, Figs. 5.6 and 5.7, pp. 235–254.

55. N. E. Plapp, "Vapor Phase Soldering with IR Preheat—The Key to Trouble Free SMT Production and Reliable Processes," *Nepcon West Proceedings,* February 1990.

56. C. Lea, *A Scientific Guide to Surface Mount Technology,* Electrochemical Publications, 1988, p. 261.

57. C. Lea, *A Scientific Guide to Surface Mount Technology,* Electrochemical Publications, 1988, p. 281.

58. L. Roberts, "Developments in Hot Bar Reflow," *Circuits Assembly,* August 1991, pp. 56–64.

59. L. Roberts, "Developments in Hot Bar Reflow," *Circuits Assembly,* August 1991, pp. 56–64.

60. L. Roberts, "Developments in Hot Bar Reflow," *Circuits Assembly,* August 1991, pp. 56–64.

61. R. N. Wild, "Page Flat Pack Solder Joint Failures," *IBM Report #77TPA0021,* April 1977.

62. C. Lea, "Laser Soldering: Production and Microstructural Benefits for SMT," *Soldering and Surface Mount Technology,* No. 2, June 1989, pp. 13–21.

63. C. Lea, *A Scientific Guide to Surface Mount Technology,* Electrochemical Publications, 1988, p. 293.

64. S. B. Dunkerton, "A Comparison of Laser Types for Reflow Soldering," *Soldering and Surface Mount Technology,* No. 10, February 1992, pp. 9–13.

65. L. Abbagnaro, "Rework and Repair Revisited," *Circuits Assembly,* May 1992, pp. 26–35.

66. C. Lea, *A Scientific Guide to Surface Mount Technology,* Electrochemical Publications, 1988, p. 447.

67. A. F. M. Barton, *Handbook of Solubility Parameters and Other Cohesion Parameters,* CRC Press, Boca Raton, FL, 1983.

68. C. Lea, *A Scientific Guide to Surface Mount Technology,* Electrochemical Publications, 1988, p. 441.

69. ODP is a relative scale weighing the effectivity of the solvent in attacking ozone in the upper atmosphere and its "life" in the atmosphere when released with CFC-11 with the ranking of 1.0.

70. D. S. L. Slinn and B. H. Baxter, *Alcohol Cleaning Under A Non-Flammable Vapor Blanket,* NEPCON West, 1990, pp. 1810–1819.

71. M. E. Hayes, Ph.D., "A New Process Alternative for Replacing CFC Solvents Cleaning," *Proceedings of the International CFC and Halon Alternatives Conference, September 29–October 1, 1992,* ARF Conference Services.

72. M. Kozicki, S. Hoenig, and P. Robinson, "Cleanrooms—Facilities and Practices," Van Nostrand Reinhold, New York, NY, p. 180, 1991.

73. B. P. Richards, P. Burton, and P. K. Footner, "The Effects of Ultrasonic Cleaning on Device Degradation—An Update," *Circuit World,* Vol. 17, No. 4, 1991, pp. 29–34.

74. B. P. Richards, P. Burton, and P. K. Footner, "The Effects of Ultrasonic Cleaning on Device Degradation—Quartz Crystal Devices," *Circuit World,* Vol. 18, No. 4, 1992, pp. 47–54.

75. B. P. Richards, P. Burton, and P. K. Footner, "Does Ultrasonic Cleaning of PCBs Cause Component Problems: An Appraisal," *IPC Technical Review,* June 1990, pp. 15–27.

76. E. Goold, "The Feasibility of Ultrasonically Bonding Surface Mount Components to Printed Circuit Boards," *Circuit World,* Vol. 15, No. 3, 1989, pp. 33–39.

77. MIL-STD-2166, Connections, Electrical, Compliant Pin, December 5, 1984.

78. F. J. Dance, "Compliant Press-Fit Pins for Solderless Backplane Assembly," *Connection Technology,* May 1988, pp. 26–30.

CHAPTER 8
THIN AND THICK FILMS

J. J. Licari

8.1 INTRODUCTION

This chapter addresses thin and thick films as applied to microelectronic devices and circuits and focuses on materials and processes. Characteristics, comparisons, design guidelines, materials selection criteria, and deposition processes are discussed. However, the chapter stops short of a full treatment of hybrid microcircuits, multichip modules, and integrated circuits, such as assembly, sealing, and testing—subjects that are fully treated in other books and articles.[1–4]

Thin and thick films, functioning as electrical conductor traces for interconnecting electronic devices, as passive components (resistors, capacitors, and inductors), and as insulation and passivation coatings, have served as the foundation for all microelectronic circuits for the past 40 years. With continued refinements and advancements in multilayering and fine-line patterning, thin and thick films continue to meet emerging requirements for higher frequency and speed performance and for further miniaturization.

Thin and thick films are defined not only in terms of their thicknesses but also by the process used to deposit them. Thin films range from low angstrom thicknesses ($1\ \text{Å} = 1 \times 10^{-8}$ cm) to several micrometers ($1\ \mu\text{m} = 10{,}000\ \text{Å}$). Aluminum metallization used in manufacturing *integrated-circuit* (IC) devices, for example, may range from 5000 to 10,000 Å. Metal, metal alloy, and inorganic thin films used in electronics are generally deposited by vapor deposition under high vacuum or by sputtering. Several variations of each process exist: for evaporation, e beam, induction heating, and resistance heating; for sputtering, DC, RF, and reactive sputtering. *Chemical vapor deposition* (CVD) is also widely used. In recent years several enhancements to CVD, primarily plasma-assisted CVD, have resulted in lowering deposition temperatures, thus rendering the processes more compatible with temperature-sensitive devices and circuits. A thorough treatment of thin-film deposition methods may be found in Bunshah's book.[5]

Thin organic-polymer films are also used in fabricating electronics but are generally applied by spin coating. Both temporary thin films, such as photoresists used in the processing of devices and hybrid circuits, and permanent films, such as polyimides used as interlayer dielectrics for multichip modules (see Chap. 11), are applied by spinning at 1000 to 5000 rev/min. However, some older methods—spraying and flow coating—are being used for high-production, low-cost commercial applications.

Thick films differ from thin films not only in being thicker (0.5 mils up to several mils) but also in the method used for their deposition (screen-printing and firing of pastes). Thick-film pastes are compositions containing not only the functional material but also glass constituents, binders, organic vehicles, adhesion promoters, and other ingredients. This heterogeneous composition coupled with the limitations of screen-printing does not provide the precision resistors and fine-line definition obtained from thin-film circuits. However, where such precision is not a requirement, which is the case for most commercial and even military uses, thick films have the overriding advantage of being less costly from a material, process, and equipment standpoint. Besides low cost, another valuable characteristic of thick films is the capability of producing multilayer structures, so what is lost in precision fine lines is often gained by the ability to produce five to seven layers of circuitry. Furthermore, with thick films a wide range of resistor values is possible by separately screen-printing and firing resistor pastes having different sheet resistances. Thick films are primarily used in fabricating hybrid and microwave circuits. Table 8.1 compares thin-film properties with those of screen-printed and fired thick films.

8.2 SUBSTRATE MATERIALS FOR THIN AND THICK FILMS

All major classifications of solid materials—metals, ceramics, metal-ceramic composites, glasses, and plastics—are being used as basic substrates upon which interconnect circuits are formed and active and passive devices are interconnected.

Substrates for thin- and thick-film circuits serve four main functions:

Mechanical support for interconnecting and assembling devices

Electrical insulation medium on which conductor traces and passive devices are deposited and patterned

Thermal medium for conducting and transferring heat from the devices

Dielectric layer to control impedance, crosstalk, and signal propagation for high-speed circuits.

Though for many years the main functions of hybrid microcircuit substrates have been mechanical support and electrical insulation, the thermal function has now become a key driver for *multichip modules* (MCMs) (see Chap. 11), *high-density interconnect substrates* (HDMIs), and 3-D packaging. In these cases closely packed devices generate more heat per unit area, and reliability depends on drawing the heat away from devices so that they can operate at low junction temperatures. Some very high-density IC devices may dissipate over 15 W/cm^2. Furthermore, the substrate now plays an active role in controlling impedance and matching impedance to devices for high-speed performance.

8.2.1 Engineering Considerations

The substrate can either improve or adversely affect the performance and reliability of the circuit. The major engineering considerations for substrate selection are listed in Table 8.2. They fall into four categories: electrical, thermal, chemical, and physical.

TABLE 8.1 Comparison of Thin Films with Thick Films

	Thin films	Thick films
Process	Indirect (subtractive)	Direct, screen-print, and fire pastes
	Evaporate, then photoetch	No etchants required
	Problem with safe disposal and handling of chemicals, etchants, developer	
Film thickness	50 to 24,000 Å	24,000 (0.1 mils) to 240,000 Å (1 mil)
Cost	Relatively high cost—batch process	Low cost—continuous conveyorized belt process
	Precious metals must be recovered from etching solutions to reduce costs	No recovery of precious or noble metals required
	Initial equipment investment high	Initial equipment investment moderate to low
	Class 1000 or better clean rooms required Class 100 or better laminar flow stations	Class 10,000 clean rooms
	Expensive, long lead time masks	Short design cycle; low cost masks and quick turnaround iterations
	Requires higher-cost, smooth-surface, ceramic substrates	Uses low-cost, rougher-surface, ceramic substrates
Characteristics	Mature, single-layer process (emerging MCM-D multilayer process)	Mature, multilayer process (to seven conductor layers)
	Sheet resistances and resistance values limited to low values: 100–300 Ω/square	Wide range or resistor values by using several pastes of different sheet resistances to over 1 MΩ/square
	Precision resistors TCRs 0 ± 50 ppm/°C	TCRs 100 to 500 ppm/°C
	Resistor tolerances ±0.1%	Resistor tolerances ±2%
	Power handling ~15 W/in^2	Power handling ~50 W/in^2
	Line definition 1 mil Micrometer and submicrometer for ICs	Line definition 5–10 mils
Reliability	Wire bondability and/or strengths affected by contaminants in plating baths	Wire bondability and/or strengths affected by composition of the paste and impurities in paste compositions
	Potential intermetallic problem in bonding Au wire to thin-film Al	Potential intermetallic problem in bonding Al wire to thick-film gold
	Resistors susceptible to moisture and chemical corrosion—require hermetic sealing for best reliability	More rugged resistors can withstand harsher environments

TABLE 8.2 Substrate Engineering Requirements

Electrical	High electrical insulation resistance
	Low/uniform dielectric constant
Thermal	Thermal stability
	High thermal conductivity
	Matched expansion coefficients
Physical	Low porosity
	Smoothness
	Camber
	Strength
Chemical	Chemical inertness
	Resistor-conductor compatibility

Electrical. From an electrical standpoint, substrates must be insulating. Generally, ceramic substrates used for thin- or thick-film circuits have an initially high insulation resistance of 10^{12} Ω-cm or greater. This high insulation resistance, however, must be retained at the elevated temperatures that the circuit might be exposed to (generally 80 to 200°C for hybrid circuits) and under humid exposures, especially if the circuit is not hermetically sealed. Fortunately, alumina and most ceramic substrates can meet these electrical requirements. Plastic substrates, on the other hand, must be chosen with care. Most plastics have very high initial insulation resistances, but, depending on their molecular structure (type of polymer), composition ingredients, and degree of cure, insulation resistance may drop significantly under humid, high-temperature conditions.[6] As a general rule, both ceramics and plastics have negative *temperature coefficients of electrical resistivity* (TCR).

Dielectric constants and dissipation factors are electrical parameters that have not heretofore been critical, but again, with the emergence of high-speed, high-performance multichip modules, these parameters are assuming greater importance. Reducing capacitance and controlling impedance are critical to the performance of high-speed (>100-MHz) circuits. The interconnect substrate is now more an integral portion of the electrical functioning of the circuit than previously where it served merely as a mechanical support. Since capacitance is directly proportional to the dielectric constant, substrates and dielectric insulating layers having low dielectric constants (<4) are required. The dielectric constant of high-purity alumina ceramic is 9 to 10. As the glass content of ceramics increases, the dielectric constant decreases (see Fig. 8.1). Very low k (4 to 5) alumina ceramics have been produced but mostly at the expense of increased porosity, lower thermal conductivity, and reduced strength. Dupont has developed a low-k (4 to 5), *low-temperature cofired ceramic* (LTCC) tape whose CTE is also low (4.5 ppm/°C), making it compatible for circuits having large-area bare silicon devices.

The best low-k materials are high-purity polymers such as polyimides, fluorocarbons, silicones, and some epoxies. Dielectric constants in the 2 to 3 range are possible with these materials, but, as with low-k ceramics, polymers are also poor thermal conductors. Unfortunately, a material combining high thermal conductivity, high electrical resistivity, low dielectric constant, and ease of processing does not exist. Ultimately, the electrical design engineer must select a compromise.

Thermal. Though the thermal conductivities of alumina ceramics (see Chap. 3) are relatively low, they are satisfactory for most hybrid microcircuits being built today. For power hybrid circuits, the thermal conductivity of the alumina substrate can be

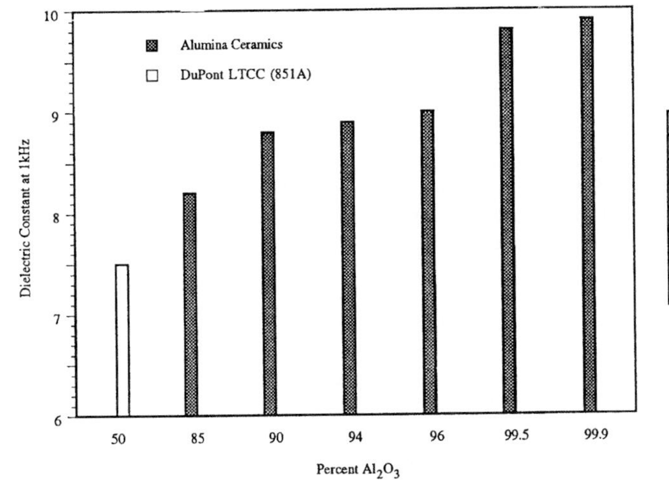

% Alumina	K at 1 kHz
~50	~7.5
85	8.2
90	8.8
94	8.9
96	9.0
99.5	9.8
99.9	9.9

FIGURE 8.1 Percent aluminum oxide in alumina ceramics and in DuPont low-temperature cofired ceramic as a function of dielectric constant.

assisted by using metal, beryllia, or diamond heat spreaders beneath the heat-generating devices. Beryllia substrates are widely used for both thin- and thick-film power circuits, but beryllia has always presented the problem of high cost primarily because of the safety requirements in processing. Industry is now looking forward to aluminum nitride as an emerging substrate material that has almost the high thermal conductivity of beryllia without the toxicity issue.

Besides thermal conductivity, thermal expansion is also drawing increased attention, especially as silicon IC devices approach 1 in^2. A close CTE match of the device material with the substrate material is required to avoid fracturing the die or the interconnections (bumped flip chip interconnects, especially) during temperature cycling. This is another reason for the interest in AlN since its CTE of approximately 4.4 ppm/°C closely matches that of silicon (3 ppm/°C). Of course, silicon substrates are ideal in providing an exact CTE match with silicon chips and in having a fairly high thermal conductivity. Gallium arsenide devices, whose CTE is approximately 6, present a unique substrate situation, especially if they are to be assembled along with silicon devices on the same substrate.

Glass-ceramic and cordierite-glass materials (Chap. 3) have been formulated and patented by IBM where both the CTE and dielectric constants have been tailored to meet specific circuit requirements for mainframe computers.

Thermal stability is still another thermal requirement, though one that is easily met. Substrates must not decompose, outgass, change in dimensions, or fracture during any of the thermal processing steps or subsequent screen testing. Ceramics can withstand the highest firing temperatures that are required for thick-film processing (850 to 1000°C). Thin-film processing temperatures are far lower—the highest that might be encountered during extended sputtering of metal films might be 200°C. Thus ceramics are widely used for both thick and thin films, but lower-temperature stable substrates such as plastics could be used for thin-film circuits. Plastics may also be used for the low-cost polymer thick films (PTF) since they are processed (cured) at temperatures of 125 to 175°C instead of being fired.

Chemical. Chemical inertness of the substrate is important for thin-film processing because of the number and nature of chemical solutions that are used in etching metal conductors and resistors, developing and removing photoresists, and in electroplating. For example, the hydrofluoric acid etching of tantalum nitride resistors on an alumina substrate having a high glass content will certainly also etch the glassy constituent of the substrate. A porous substrate entraps and retains chemical solutions, subsequently causing electrical leakage currents, metal migration, or corrosion. Even a high-grade alumina, as used in thin-film circuit processing, can absorb and retain ionic contaminants from etching or plating solutions, if not thoroughly cleaned with deionized water and the rinses monitored by electrical resistivity measurements.

Physical. The key physical attribute of substrates is surface smoothness. Surface smoothness is essential in fabricating thin-film precision circuits. Smoothness is reported as the average deviation from an arbitrary centerline that traverses the hills and valleys of a surface profile (measured using a profilometer). One measure in reporting smoothness (or conversely roughness) is microinches *centerline average* (CLA). In fabricating precision fine-line conductors (25 μm or less), a surface smoothness of 6 μin or less is required. For precision, reproducible resistors must have smooth, highly polished surfaces. If one attempts to use a rough-surface ceramic, for example, one having a CLA of 20 μin (5000 Å), to deposit thin-film nichrome or tantalum nitride resistors that are only 200 Å thick, it is apparent that the thin film would have to conform to the many hills and valleys, some spanning 10,000 Å. The length of the resistor deposited on a rough surface will, therefore, vary from area to area, and since resistance is directly proportional to length, both the length and resistance values would be unpredictable and variable (Fig. 8.2):

$$R = \frac{rl}{A}$$

where R = resistance
l = length
r = resistivity
A = area

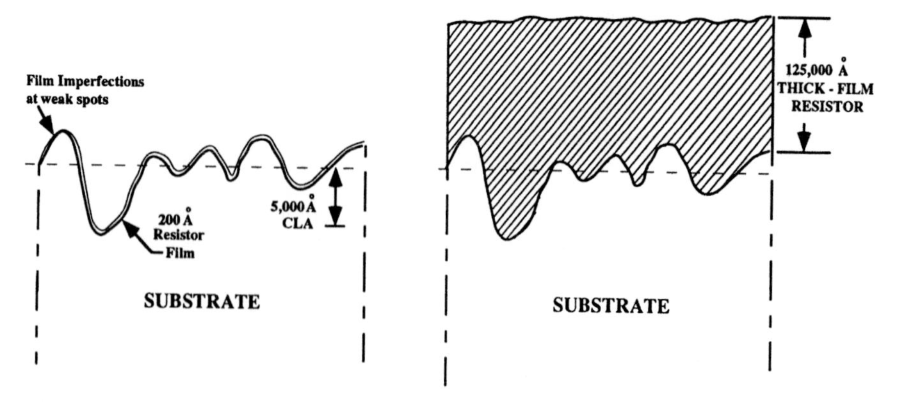

FIGURE 8.2 Resistor values depend on surface smoothness.

Furthermore, weak spots and discontinuities often occur in thin films at the peaks, resulting in opens and in poor-quality resistors. With thick-film resistors, the opposite is true. A rougher surface is actually desirable to achieve some mechanical interlocking and thus enhance adhesion of the thick film to the substrate. Assuming the above scenario for a surface having a 20-μin CLA, a thick-film resistor, approximately 0.5 mils thick (125,000 Å), would completely fill in and encapsulate the surface. The length of the resistor and resistor value would therefore be reproducible.

Camber is yet another substrate characteristic that can affect the processing of thin and thick films. *Camber,* defined as the warpage or bowing of a substrate or the total deviation from perfect flatness, is important especially for large-area substrates. Camber is reported in inches per inch deviation from flat (Fig. 8.3). Substrates having a high camber are difficult to attach inside packages. They require thicker adhesive preform which, in turn, reduces thermal transfer. Substrate camber and waviness (Fig. 8.4) also affect thick-film screen-printing, resulting in poor print resolution. Typically, ceramic substrates have a camber of 0.001 to 0.003 in/in when measured across the diagonal. Both camber and waviness may be reduced by lapping and polishing. Test procedures and further definitions for camber, waviness, and surface roughness may be found in ASTM F865-84 and ASTM F109-73.

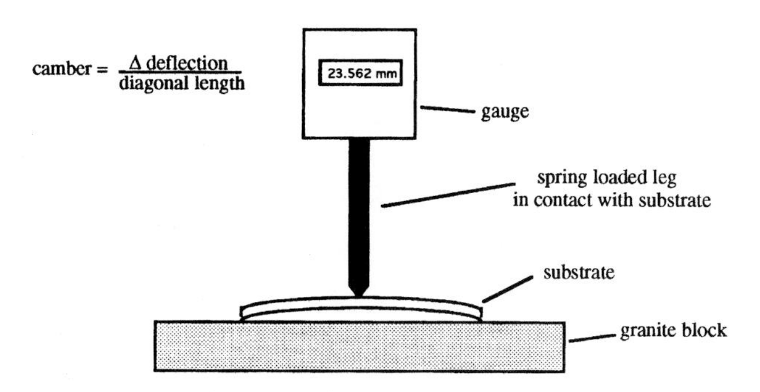

FIGURE 8.3 Measurement setup for substrate camber.

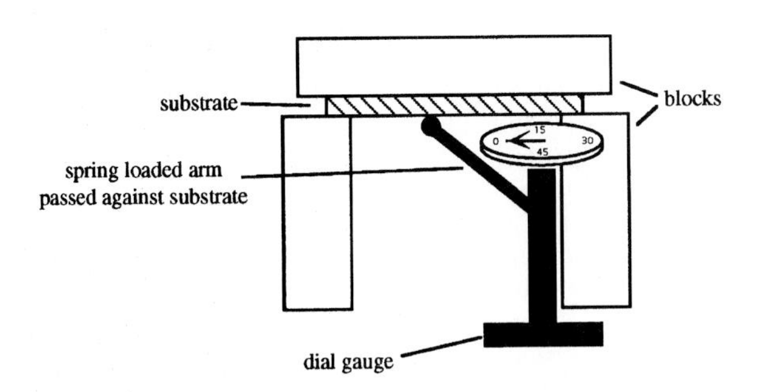

FIGURE 8.4 Measurement setup for substrate waviness.

Finally, high mechanical strength is required for ceramic substrates. Mechanical strength becomes critical as substrates increase in size, complexity, and shape. High-stress areas at corners or nicks at edges can be the foci for fracturing. Generally alumina, beryllia, and aluminum nitride ceramics are quite tough, but, as more glass is used in their compositions, flexural strength decreases.

8.2.2 Alumina

Ceramics, particularly alumina, are the most widely used substrates for both thin- and thick-film circuits. However, depending on whether the circuit is thin or thick film, different grades of alumina must be used. Chemical composition and surface smoothness are two key factors that impact performance and cost. Thin films require a substrate of high alumina content (99 + percent) and a very smooth surface finish (<6 μin CLA); otherwise, the key benefit of fine-line definition and precision derived from thin films cannot be achieved. Table 8.3 gives typical characteristics of alumina substrates used for thin-film circuits. On the other hand, thick films actually benefit from a rougher surface (20 to 50 μin CLA) and from a lower alumina content (96 percent), both of which ensure better adhesion of thick-film pastes and reduce overall costs. Table 8.4 gives properties of alumina substrates used for thick-film circuits.

The balance of the composition of alumina ceramics consists of glassy materials. The glass content of alumina substrates can vary from very high (approximately 50 percent), as in the case of low-temperature cofired ceramics, to less than 1 percent for alumina used for thin-film circuits. Because of the extremely poor thermal conductivity of glass, the low thermal conductivity of high-glass-content ceramics must be taken into account in designing high-density, high-power circuits. Figure 8.5 shows thermal conductivity as a function of alumina content. The dielectric constant of alumina ceramics is also affected by the amount of glass, varying inversely as the glass content increases (Fig. 8.1).

8.2.3 Beryllia

Next to alumina, beryllia (BeO) is the most widely used substrate for hybrid microcircuits, but, because of its higher cost, it is used only for specific high-power circuit applications. Beryllia belongs to a limited class of materials that combine high thermal conductivity with high electrical insulation resistance. Others in this class include aluminum nitride, boron nitride, and diamond. Beryllia is used as a substrate material for both thin and thick film circuits, as a heat spreader beneath power devices, and as a package material. Its key attribute—high thermal conductivity (approximately 240 W/m K)—approaches that of aluminum metal yet it has the electrical resistivity of the best of the plastics. High-purity beryllia ceramic (99.5 percent) has a thermal conductivity approximately 1200 times that of a typical epoxy plastic, 200 times that of most glasses, and 6 times better than alumina ceramic. Figure 8.6 compares the thermal conductivity of beryllia with other commonly used electronic materials. The electrical properties of beryllia are given in Figs. 8.7 and 8.8 and in Tables 8.5 and 8.6. Other physical properties are given in Table 8.7. Electrical, thermal, and physical properties improve as the beryllium oxide content of the ceramic increases, as shown in Table 8.8. The coefficient of thermal expansion of beryllia closely matches that of GaAs (approximately 6).

TABLE 8.3 Characteristics of Alumina Substrates for Thin-Film Circuits

Characteristic	Unit	Test method	ADS-995	ADS-996	Substrates® 996
Alumina content	Weight, %	ASTM D2442	99.5	99.6	99.6
Color	—	—	White	White	White
Nominal density	g/cm³	ASTM C373	3.88	3.88	3.87
Density range	g/cm³	ASTM C373	3.86–3.90	3.86–3.90	3.85–3.89
Hardness—Rockwell	R45N	ASTM E18	87	87	87
Surface finish (Working Surface—"A" Side)	Microinches, centerline avg.	Profilometer—0.030 in cutoff, 0.0004 in dia. stylus	6	3	2
Average grain size	Micrometers	Intercept method	<2.2	<1.2	<1.0
Water absorption	%	ASTM C373	NIL	NIL	NIL
Gas permeability	—	*	NIL	NIL	NIL
Flexural strength	KPSI	ASTM F394	83	86	90
Elastic modulus	MPSI	ASTM C623	54	54	54
Poisson's ratio	—	ASTM C623	0.20	0.20	0.20
Coefficient of linear thermal expansion	ppm/°C	ASTM C372			
25–300°C			7.0	7.0	7.0
25–600°C			7.5	7.5	7.5
25–800°C			8.0	8.0	8.0
25–1000°C			8.3	8.3	8.3
Thermal conductivity	W/m-°K	Various			
20°C			33.5	34.7	35.0
100°C			25.5	26.6	26.9
400°C			—	—	—
Dielectric strength (60 cycles AC avg. RMS)	V/mil	ASTM D149			
0.025 in thick			600	600	600
0.050 in thick			450	450	450
Dielectric constant (relative permittivity)	—	ASTM D150			
1 KHz			9.9 (±1%)	9.9 (±1%)	9.9 (±1%)
1 MHz			9.9 (±1%)	9.9 (±1%)	9.9 (±1%)
Dissipation factor (loss tangent)	—	ASTM D150			
1 KHz			0.0003	0.0003	0.0001
1 MHz			0.0001	0.0001	0.0001
Loss index (loss factor)	—	ASTM D150			
1 KHz			0.003	0.003	0.001
1 MHz			0.001	0.001	0.001
Volume resistivity	Ω-cm	ASTM D257			
25°C			$>10^{+14}$	$>10^{+14}$	$>10^{+14}$
100°C			$>10^{+14}$	$>10^{+14}$	$>10^{+14}$
300°C			$>10^{+12}$	$>10^{+12}$	$>10^{+13}$
500°C			$>10^{+9}$	$>10^{+9}$	$>10^{+10}$
700°C			$>10^{+8}$	$>10^{+8}$	$>10^{+9}$

*Helium leak through a plate 25.4 mm diameter by 0.25 mm thick measured at 3×10^{-7} torr vacuum versus approximately 1 atm of helium pressure for 15 s at room temperature.

Source: Courtesy Coors Ceramic.

TABLE 8.4 Characteristics of Alumina Substrates for Thick-Film Circuits

Property		Unit	Test method	ADOS-90R	ADS-96R
Alumina content		Weight, %	ASTM D2442	91	96
Color		—	—	Dark brown	White
Bulk density		g/cm^3	ASTM C373	3.72 min	3.75 min
Hardness—Rockwell		R45N	ASTM E18	78	82
Surface finish—CLA (as fired)		μin	Profilometer; Cutoff: 0.030 in; Stylus diameter: 0.0004 in	≤45	≤35
Grain size (avg)		μm	Intercept method	5–7	5–7
Water absorption		%	ASTM C373	None	None
Flexural strength		kpsi	ASTM F394	53	58
Modulus of elasticity		Mpsi	ASTM C623	45	44
Poisson's ratio		—	ASTM C623	0.24	0.21
Coefficient of linear thermal expansion	25–200°C	ppm/°C	ASTM C372	6.4	6.3
	25–500°C			7.3	7.1
	25–800°C			8.0	7.6
	25–1000°C			8.4	8.0
Thermal conductivity	20°C	W/m-K	Various	13	26
	100°C			12	20
	400°C			8	12
Dielectric strength (60 cycle ac avg RMS)	0.025 in thick	V/mil	ASTM D116	540	600
Dielectric constant (relative permittivity) @ 25°C	1 KHz		ASTM D150	11.8	9.5
	1 MHz			10.3	9.5
Dissipation factor (loss tangent) @ 25°C	1 KHz		ASTM D150	0.1	0.0010
	1 MHz			0.005	0.0004
Loss index (loss factor) @ 25°C	1 KHz		ASTM D150	1.2	0.009
	1 MHz		ASTM D2520	0.05	0.004
Volume resistivity	25°C	Ω-cm	ASTM D1829	$>10^{14}$	$>10^{14}$
	300°C			4×10^8	5.0×10^{10}
	500°C			—	—
	700°C			7×10^6	4.0×10^7

Source: Courtesy Coors Ceramics.

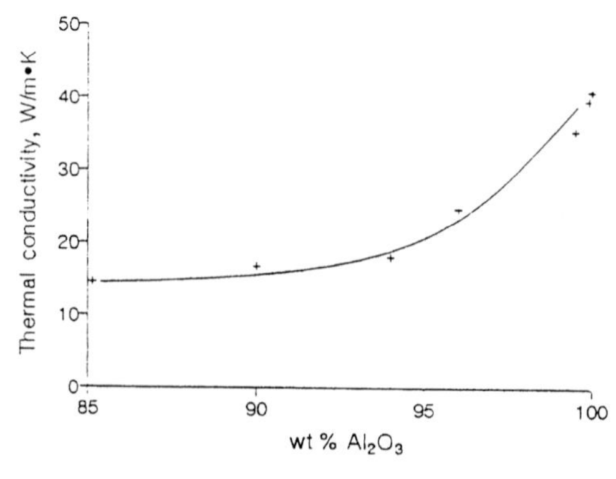

FIGURE 8.5 Effect of weight percent alumina on thermal conductivity.

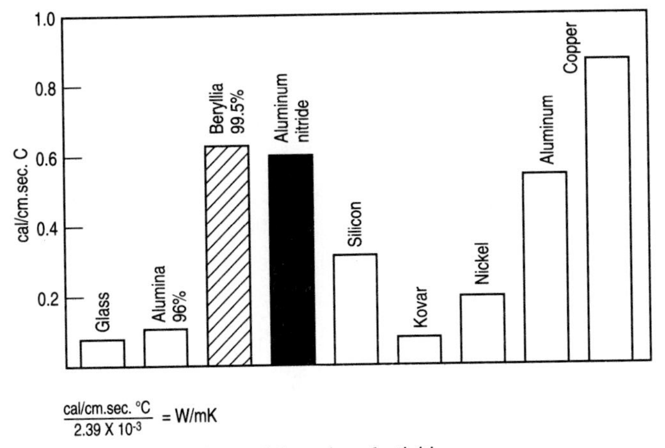

FIGURE 8.6 Comparisons of thermal conductivities.

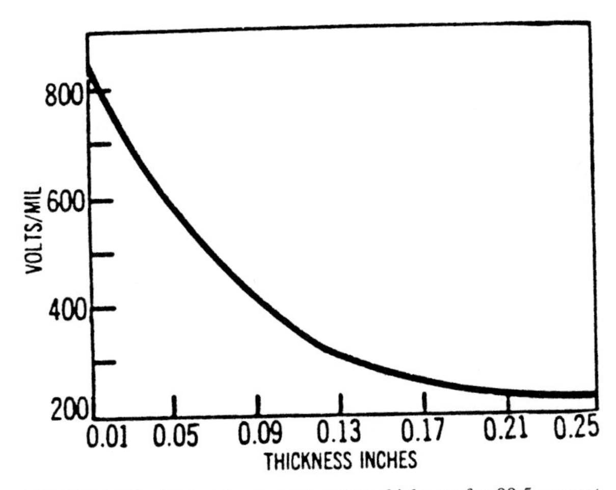

FIGURE 8.7 Dielectric strength versus thickness for 99.5 percent beryllia. (*Courtesy of National Beryllia, Division of General Ceramics.*)

Beryllia would be more widely used were it not for its inherent toxicity which results in added costs due to the safety measures required in processing it. Toxicity arises from inhaling beryllia dust, particles, or fumes. However, once the beryllia has been processed as a ceramic, there is little danger unless it is subsequently machined, drilled, or otherwise processed in such a way as to produce particles or fumes. Adequate exhaust and safety precautions must be provided where resistors deposited on beryllia are being laser trimmed.

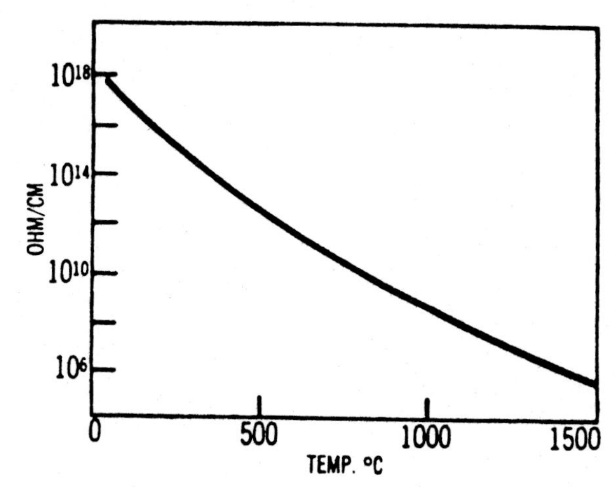

FIGURE 8.8 Electrical resistivity versus temperature for 99.5 percent beryllia. (*Courtesy of National Beryllia, Division of General Ceramics.*)

TABLE 8.5 Loss Tangent versus Temperature and Frequency for Beryllia[7] (Berlox K-150*)

Temperature, °C	Loss tangent			
	1 kHz	1 MHz	1 GHz	10 GHz
25	0.0002	0.0002	0.0003	0.0003
300	0.0003	0.0003	0.0005	0.0005
500	0.0005	0.0005	0.0008	0.0008

*Tradename for 99.5 percent beryllia of National Beryllia, Division of General Ceramics.
Source: Courtesy National Beryllia, Division of General Ceramics.

TABLE 8.6 Dielectric Constant versus Temperature and Frequency for Beryllia[7] (Berlox K-150*)

Temperature, °C	Dielectric constant			
	1 kHz	1 MHz	1 GHz	10 GHz
25	6.7	6.7	6.7	6.6
300	6.8	6.8	6.8	6.7
500	7.0	7.0	6.9	6.9

*Tradename for 99.5 percent beryllia of National Beryllia, Division of General Ceramics.
Source: Courtesy National Beryllia, Division of General Ceramics.

TABLE 8.7 Properties of 99.5 Percent Beryllia[7]

Specific heat, cal/°C-g	0.25
Dielectric constant, 1 GHz	6.7
Dielectric loss, 1 GHz	0.0003
Compressive strength, psi	225,000
Flexural strength, psi	30,000
Tensile strength, psi	20,000
Young's modulus, psi	50,000,000
Density, g/cc minimum	2.85
Permeability	Impervious to gases and liquids
Vapor pressure at 1500°C, atm	10^{-12}
Maximum use temperature	1800°C
Radiation hardness	Excellent

TABLE 8.8 Average Physical Property Values as a Function of Purity for Beryllia Ceramics[7]

	95% BeO	98% BeO	99.5% BeO	99.9% BeO
Flexural strength, psi	20,000	25,000	35,000	36,000
Electric resistivity, Ω-cm	10^{14}	10^{15}	10^{17}	10^{18}
Young's modulus, psi	44×10^6	45×10^6	50×10^6	55×10^6
Thermal conductivity, cal/cm-s-°C	0.35	0.52	0.62	0.66
Dissipation factor at 1 GHz at 25°C	0.0005	0.0005	0.0003	0.0002

8.2.4 Aluminum Nitride

The key benefit of using aluminum nitride (AlN) as a substrate or package material is its high thermal conductivity (170 to 200 W/m K) coupled with its excellent electrical insulating and mechanical properties. In this respect AlN is similar to beryllia but without the toxicity concern. A further benefit is AlN's low expansion coefficient (approximately 4.4 ppm/°C) closely matching that of silicon (3 ppm/°C). As silicon integrated circuits become more complex and larger and contain higher numbers of I/Os (some over 500), CTE compatibility becomes essential to minimize or avoid stresses that can fracture the die or the interconnect bonds. Thus AlN is being investigated intensively for multichip modules and three-dimensional packaging.

Considerable progress has been made during the past 5 years in producing and processing AlN powder into ceramic that has reproducible thermal, mechanical, and electrical properties so that today AlN substrates are readily available from many suppliers, both Japanese and American. Among the suppliers of AlN powder are Tokuyama Soda and Dow Chemical. Ceramic suppliers include Kyocera, Toshiba, Tokuyama Soda, NTK, General Ceramics, Coors Ceramics, Carborundum, and Sumitomo.

Both pressureless and hot-press processes are used to sinter aluminum nitride "green tape." Sintering is performed at 1800 to 1900°C in a reducing (hydrogen or forming gas) atmosphere to avoid any oxide formation, a small amount of which significantly degrades the thermal conductivity. The pressureless process has the advantage of lower cost, but, as with alumina cofired tape, *x-y* shrinkage must be taken into account to control the final dimensional tolerances. In the hot-press process the tape is constrained in the *x-y* plane during pressurization and sintering; hence virtually no *x-y* shrinkage occurs. Shrinkage occurs totally in the *z* direction. Hot pressing has successfully been used to produce large 4 × 4 in AlN cofired substrates and package bases with 612 I/Os on 25-mil pitch.[8,9]

Plots of thermal conductivity, thermal expansion, dielectric constant, and dissipation factor for pressureless sintered AlN as a function of temperature are given in Figs. 8.9 through 8.12. Properties of hot-pressed AlN ceramic are given in Table 8.9.

8.2.5 Silicon

Silicon has been the basic substrate material for the fabrication of almost all semiconductor devices and integrated circuits since their invention. This is due to the ease of doping silicon with elements from periodic table groups III or V to form *p* or *n* junctions. Semiconductor device is a mature technology and constitutes the basis for the electronics revolution that has occurred in the past 30 years. Attempts at extending the single-chip technology to wafer scale integration of many circuit functions has not yet materialized because of the low yields and inability to rework at the monolithic level.

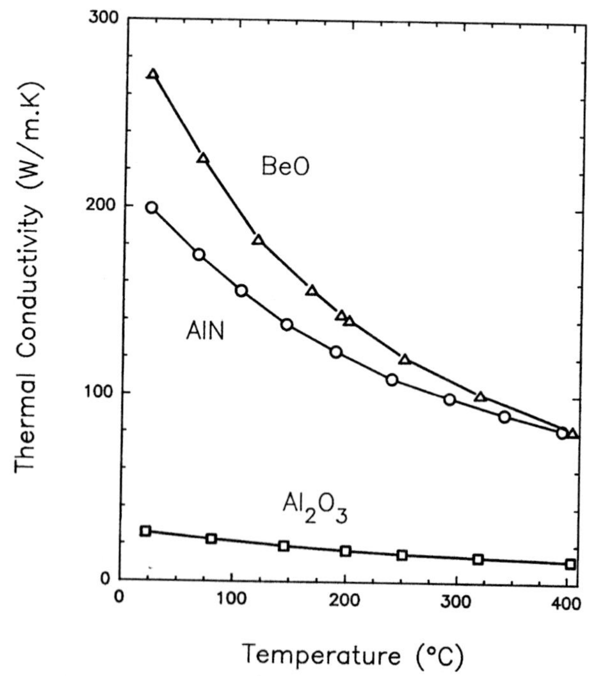

FIGURE 8.9 Thermal conductivity of AlN beryllia and alumina versus temperature. (*Courtesy of The Carborundum Company.*)

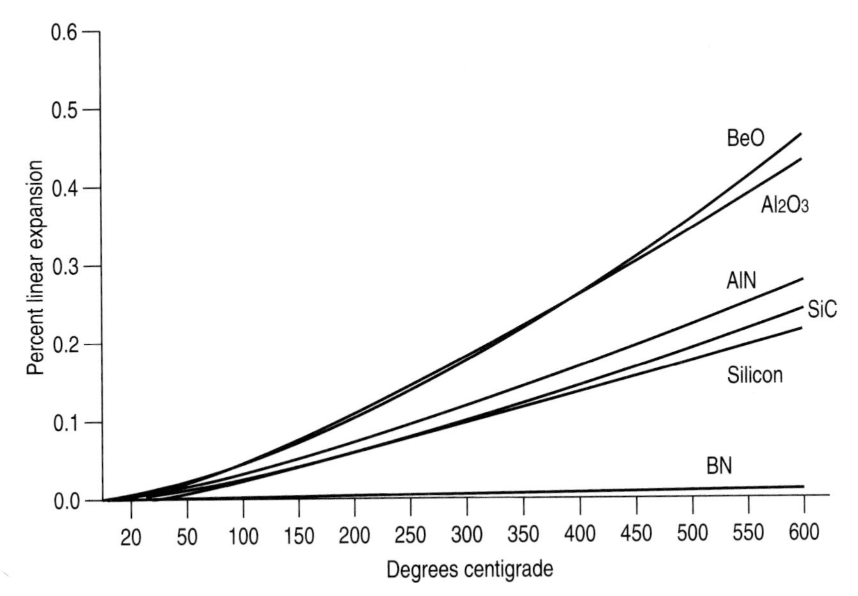

FIGURE 8.10 Thermal expansion of AlN and other ceramics versus temperature.

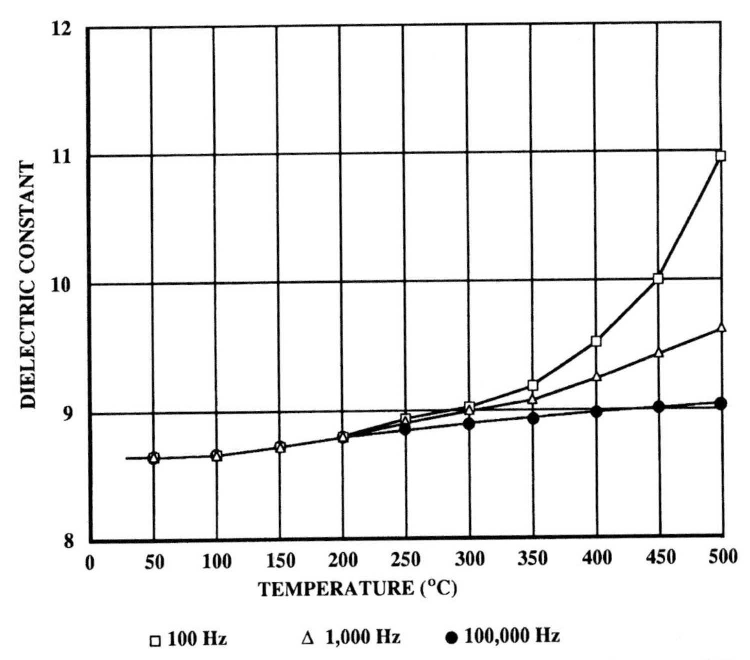

FIGURE 8.11 Dielectric constant of AlN versus temperature. (*Courtesy of The Carborundum Company.*)

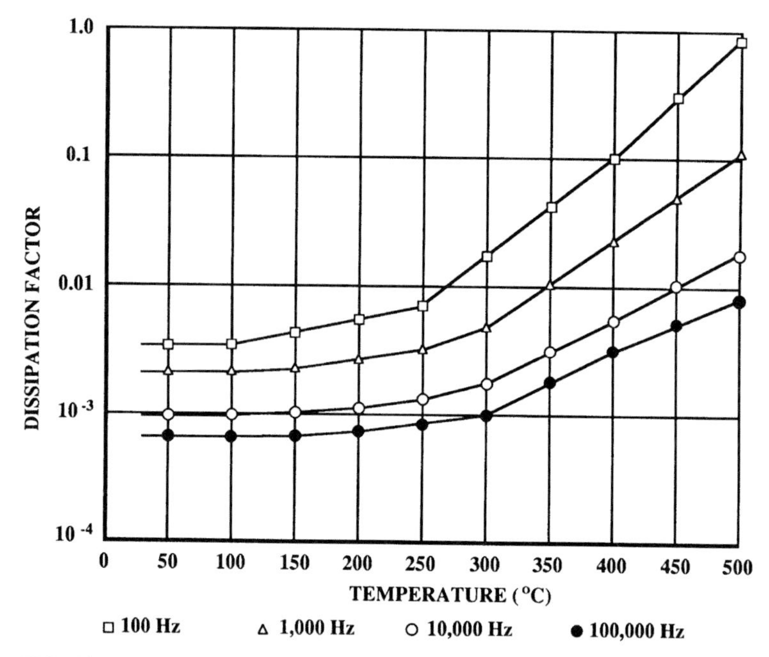

FIGURE 8.12 Dissipation factor of AlN versus temperature. (*Courtesy of The Carborundum Company.*)

However, as a compromise, high-density multichip interconnect (HDMI) substrates have been developed that start with silicon wafers as the substrate and use thin-film multilayering processes to form a high-density, high-performance interconnect structure. As in hybrid microcircuits, uncased bare die are then interconnected to these substrates. Besides silicon, other materials are also being used as the basic substrate—for example, alumina, alumina-glass, and aluminum nitride. The key advantage of silicon, besides its low cost and commercial availability, is its capability of monolithically integrating passive components such as decoupling capacitors and resistors and to some extent active devices. SEMI-grade silicon wafers (MI STD 8-85) up to 6 inches in diameter, 25 mils thick, are commercially available. These monocrystalline wafers have a smooth polished surface and are flat to 60 μm over a 6-in span. The thermal conductivity of silicon is variously reported between 84 and 135 W/m K, depending on its purity. Silicon is easily metallized with aluminum and with other metals but may require an adhesion-promoting layer.

8.2.6 Diamond

Diamond is often referred to as the ideal material for the most advanced high performance microcircuits (also see Chaps. 3 and 4). It certainly is unique among materials in combining the highest thermal conductivity (over 2000 W/m K for single-crystal, natural diamond Type IIa) with high electrical resistivity ($>10^{13}$ Ω-cm), low dielectric constant (5.6), high thermal and radiation resistance, and excellent passivation proper-

TABLE 8.9 Properties of Hot-Pressed Aluminum Nitride*

Property	Value
Mechanical properties (room temperature):	
Young's modulus, GPa/psi	339/48,400,000
Shear modulus, GPa/psi	137/19,600,000
Poisson's ratio	0.24
Modulus of rupture, MPa/psi	280/40,000
Vicker's hardness, kg/mm^2	120 (100 g loaded)
Thermal properties:	
Thermal expansion, ppm/°C	3.2 (RT to 100°C)
	3.7 (RT to 200°C)
	4.3 (RT to 400°C)
	4.7 (RT to 600°C)
Specific heat, J/(g)(K)	0.74
Thermal conductivity, W/m K	170–190
Physical properties:	
Density, g/cm^3	3.25–3.26
Microstructure, grain size, μm	5–10
Surface roughness† R_a, μin	30
Surface flatness,‡ in/in	<0.0005
Electrical properties	
Dielectric constant @ 25°C	8.7 @ 1 KHz
	8.5 @ 1 MHz
	8.5 @ 10 MHz
	8.3 @ 9.3 GHz
Dielectric loss	0.0002 @ 1 KHz
	0.0001 @ 1 MHz
	0.0001 @ 10 MHz
	0.0012 @ 9.3 GHz
ac breakdown, V/mil	330 (94 mil thick)

* AlN powder manufactured by Dow Chemical Company and processed by W. R. Grace & Co.
† After grinding with resin-bonded diamond wheel.
‡ After grinding.
Source: Courtesy W. R. Grace & Co.

ties. Synthetic diamond thin films and even thick substrates (up to 40 mils thick) are being produced by several commercial firms.[10] Plasma-assisted chemical vapor deposition (CVD), microwave plasma CVD, dc arc jet, hot filament, and other low-pressure methods have been developed for converting methane and other hydrocarbon gases into polycrystalline diamond.[11–14]

Theoretically, CVD diamond should be inexpensive since the starting materials, methane and hydrogen, are relatively inexpensive. However, because of the very slow deposition rates, long times required to deposit even several micrometers of film, and the high consumption of energy, the cost of the final product remains high. Cost, however, is a function of the quality of the diamond. To date, the highest optical and electronic grade diamond can be produced only at slow rates (less than 1 μm/h). At higher rates (1 to 10 μm/h), lesser-quality, darker-color diamond is produced. This lower-quality diamond still has a fairly high thermal conductivity and thus can be used for heat spreaders and multichip module substrates.

It is expected that, as very high density multichip modules are developed, diamond wafers will be essential in removing heat generated, for example, from three-dimensional stacked circuits and from high-power multichip modules.[15] Diamond wafers 4

TABLE 8.10 Properties of CVD Diamond Compared with Other Electronic Packaging Materials

	CVD Diamond	BeO	AlN	Alumina, 96%	Copper
Approximate thermal conductivity, W/m K @ 100°C	1300–1500	200–250	170–200	18–20	400
Thermal diffusivity, cm^2/s	7.4	0.67	0.65	0.05	1.2
Coefficient of thermal, expansion (CTE), ppm/°C	0.8–2.0	6.4	4.5	8	18.8
Thermal shock, relative to BeO	926	1	2.1	0.07	—
Dielectric constant	5.2	6.7	8.8	8.9	—
Electrical resistivity, Ω-cm	10^{12}–10^{14}	10^{14}	10^{13}	10^{13}	1.6×10^{-6}
Dielectric strength, V/mil	8750	850	1275	850	—
Density, g/cm^3	3.5	3.0	3.3	3.7	8.92
Young's modulus, M psi	145	45	39	52	—

Source: Courtesy Crystallume Corp.

inches in diameter and 16 mils thick have already been produced by Norton. These are of high quality, translucent to white, and have a thermal conductivity of approximately 1400 W/m K. Properties of CVD diamond compared to other materials are listed in Table 8.10. There is considerable activity in optimizing adhesion, fabricating larger substrates (6-in diameter), increasing the rates of deposition, and reducing the deposition temperatures to extend diamond's usefulness in electronic applications.[16]

8.2.7 Silicon Carbide–Reinforced Aluminum

Silicon carbide–reinforced aluminum belongs to a general class of materials known as *metal matrix composites.* This unique material, commercially available under the tradename of Lanxide (Lanxide Electronic Components), combines four desirable properties as a substrate or package material:

Very lightweight—slightly higher than aluminum

High thermal conductivity—similar to aluminum nitride or beryllia

CTE closely matching those of GaAs and alumina (Figs. 8.13 and 8.14 and Table 8.11)

High stiffness

Lanxide may be used as a substrate material for high-power circuits or devices, as a heat spreader, or as a SEM-E frame material.

Lanxide is being produced in net shapes, thus obviating one of its early drawbacks—its extreme hardness and difficulty in machining and cutting to final shape. Lanxide is produced by a pressureless infiltration process. According to this process, silicon carbide particles are packed into a graphite mold, topped with an aluminum

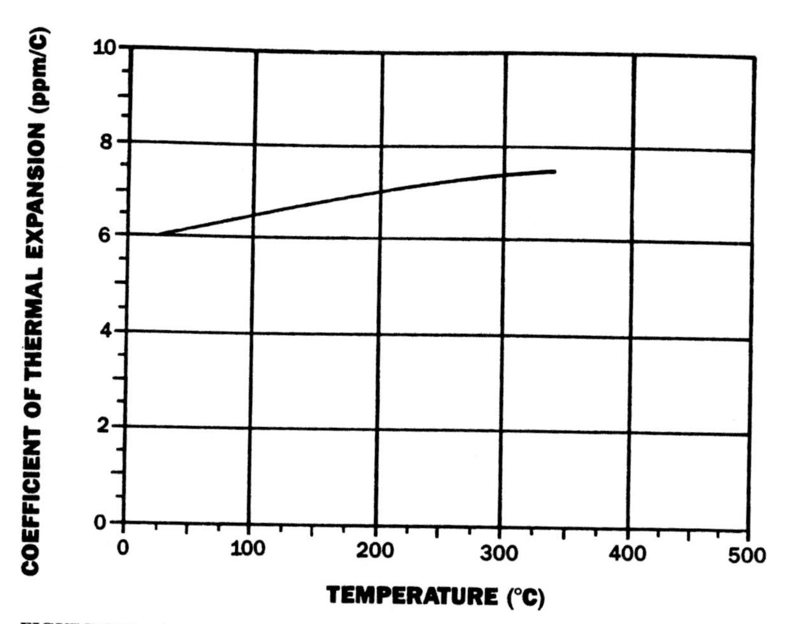

FIGURE 8.13 Average coefficient of thermal expansion of MCX-622, silicon carbide reinforced aluminum. (*Courtesy of Lanxide Electronic Components.*)

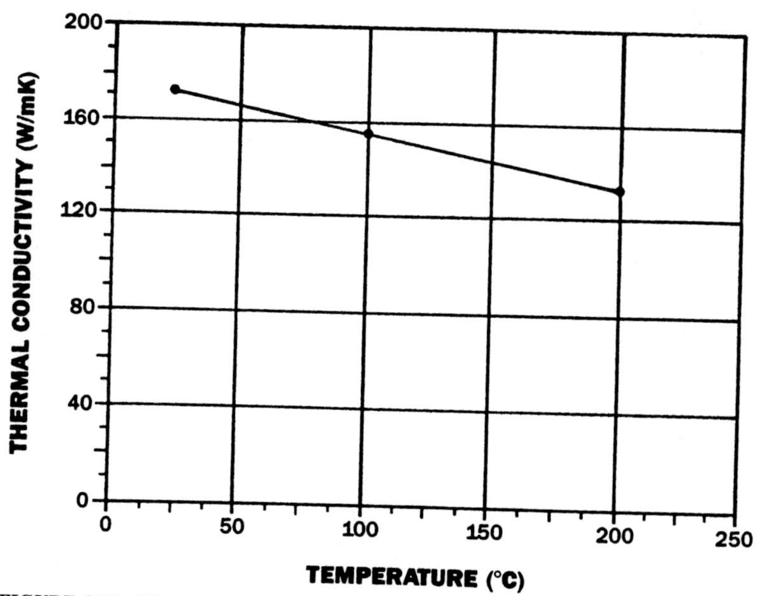

FIGURE 8.14 Thermal conductivity versus temperature of MCX-622, silicon carbide reinforced aluminum.

TABLE 8.11 LEC Silicon Carbide–Reinforced Aluminum Material Properties

	MCX-622	MCX-693
Coefficient of thermal expansion	6.2×10^{-6}/°C	6.9×10^{-6}/°C
Density	3.0 g/cm^3	3.0 g/cm^3
Thermal conductivity	170 W/m K	180 W/m K
Young's modulus	265 GPa	235 GPa
Flexural strength	300 MPa	300 MPa
Fracture toughness	10 MPa-m$^{1/2}$	9 MPa-m$^{1/2}$
Poisson's ratio	0.22	0.24
% SiC loading	70	65

Note: MCX-622 and MCX-693 are trademarks of Lanxide Electronic Components, L.P.
Source: Courtesy Lanxide Electronic Components.

alloy ingot, placed in a controlled atmosphere furnace, and heated to the melt temperature of aluminum. The molten aluminum wets and infiltrates the SiC powder forming the composite.[17,18] Ceramic feedthroughs can be incorporated by positioning them in the filler prior to heating since they too are wetted by the aluminum, thus forming reliable hermetic joints. When removed from the graphite mold, the package is a finished part. Tolerances are ±0.003 in up to a 2-in span and surface finish is 64 μin.

A coefficient of thermal expansion (CTE) of 6.2 ppm/°C is reported for a 70 percent volume of SiC, closely matching that for GaAs devices. However, CTEs ranging from 6 to 14 can be engineered depending on the volume percent of SiC (see Figure 8.13).[19] Thermal conductivity is also quite high and close to that for either aluminum, aluminum nitride, or beryllia (Fig. 8.14).

8.2.8 Other Substrate Materials

Besides the substrate materials already discussed, other materials can and are being used in selected applications. Highly polished quartz (single-crystal silicon dioxide) with photodelineated thin-film aluminum conductor traces is used for surface acoustical wave devices. Sapphire (single-crystal aluminum oxide) is used as a wafer to fabricate radiation-resistant ICs—for example, *silicon on sapphire* (SOS) devices.

A series of low-temperature cofired ceramics (LTCCs) offer a low-cost, in-house, fast-turnaround capability to fabricate multilayer substrates. LTCC green tape was developed by DuPont[20,21] and is now commercially available from several sources. The tape consists of alumina and large amounts of glass and other binders. A circuit layer is processed from the tape by drilling vias, filling them with gold or silver conductive paste, and screen-printing conductor lines. The individual circuit layers (separate processed tapes) are then inspected, aligned, stacked, and laminated at 70°C and 3000 psi. The stacks are then placed in an oven to burn out most of the organic binders, then fired in an air furnace at 850 to 900°C. Multilayer interconnect substrates with as many as 40 conductor layers have been produced.

The key advantages of the LTCC process over the high-temperature cofired ceramic process is the ability to fire high electrical conductivity metals such as gold and silver in air because of the low firing temperatures required for LTCC. Screen-printable thick-film gold or silver pastes compatible with the ceramic are employed.[22] A general comparison of physical and electrical properties of LTCC with those for conventional thick-film circuits and with high-temperature cofired ceramics is given in Table 8.12. Electrical and physical characteristics for a specific cofired ceramic tape are given in Table 8.13.

Specifically engineered glass-ceramic compositions, known as *cordierites,* have been formulated by IBM for their mainframe computer electronic modules. Multilayer substrates having 63 layers of copper conductors and CTEs matching that of silicon chips have been introduced.[23,24] It should be noted that all these high-glass-content ceramics display very poor thermal conductivities and that thermal vias, metal inserts, and other heat-sinking techniques are required for some circuits.

Finally, metals such as aluminum, copper, and copper alloys can be used and can serve both as high thermal conductivity bases for multilayer interconnects and as ground planes.

TABLE 8.12 Properties of Low-Temperature Cofired Ceramic Compared to Thick-Film and High-Temperature Cofired Ceramic

	LTCC	Thick film on 96% alumina	High-temperature cofired package, 92% alumina
	Physical properties		
CTE at 300°C, ppm/°C	7.9	6.4	6.0
Density, g/cm³	2.9	3.7	3.6
Camber, mils/in	1–4*	1–2	1–4*
Surface smoothness, μin	8.7	14.5	20.0
Thermal conductivity, W/m K	3–5†	20‡	14–18
Flexural strength, kpsi	22	40‡	46
Thickness/layer after firing, mils	3.5–10.0	0.5–1.0§	5–20
	Dimensional tolerances		
Length and width	±0.2%	NA	±1.0%
Thickness	±0.5%	NA	±5.0%
	Electrical properties		
Insulation resistance, Ω at 100 V dc	$>10^{12}$	$>10^{12}$	$>10^{12}$
Breakdown field, V/mil	>1,000	>1,000	>700
Dielectric constant, 1 MHz	7.1	9.3	8.9
Dissipation factor, %	0.3	0.3	0.03

*Function of firing setter and part design.
†With thermal vias, 16–20.
‡Property of substrate.
§Property of printed dielectric.

TABLE 8.13 Properties for Low-Temperature Cofired Tape Dielectric, Dupont 845

FIRED PROPERTIES*

Electrical Properties		Physical Properties	
Dielectric Constant (K)	≤4.8 (5 kHz to 5 GHz)	Thermal Expansion (25°C to 300°C)	4.5 ppm/°C
Dissipation Factor (DF)	≤0.3% (5 kHz to 5 GHz)	Density	2.4 g/cm³ (>95% theoretical density)
Insulation Resistance (at 100 VDC)	>10^{12} ohms	Camber†	Conforms to setter
Breakdown Voltage	>500 volts/25 μm (1 mil)	Surface Smoothness	20 to 25 μcm (8 to 10 microinches)
		Thermal Conductivity	2.0 W/m °K
		Flexural Strength	240 MPa (35 kpsi)

Dielectric Constant vs. Frequency

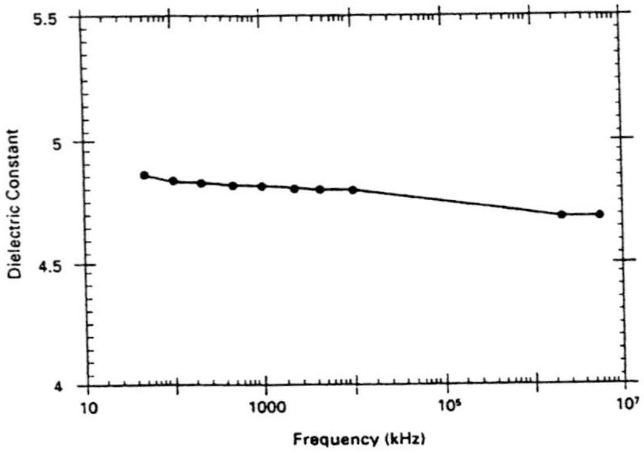

Loss Tangent vs. Frequency

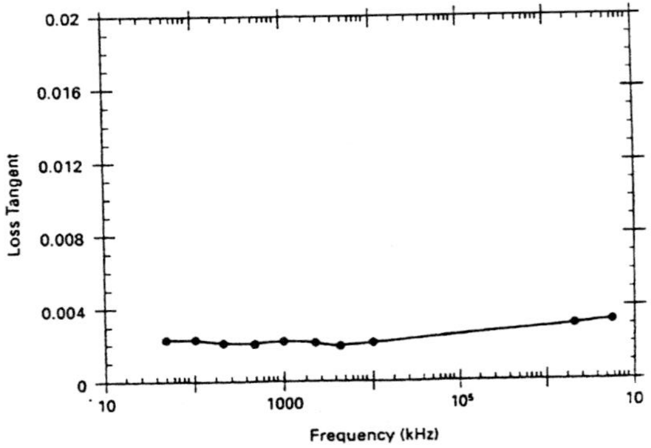

* Typical properties are based on laboratory tests using recommended processing procedures.

+ Camber of the co-fired laminate is determined by the setter the substrate is fired on.

Source: Courtesy of DuPont.

8.3 THIN FILMS

8.3.1 Applications of Thin Films in Electronics

Thin-film conductors, resistors, and dielectrics are essential in manufacturing electronic devices, integrated circuits, high-density hybrid microcircuits, and, more recently, multichip modules. Single-layer, thin-film hybrid circuits have been produced for over 30 years and are used for the most precise and reliable circuit functions. For example, over 10,000 hybrid circuits of both thin- and thick-film types are used in each space shuttle. (Fig. 8.15).

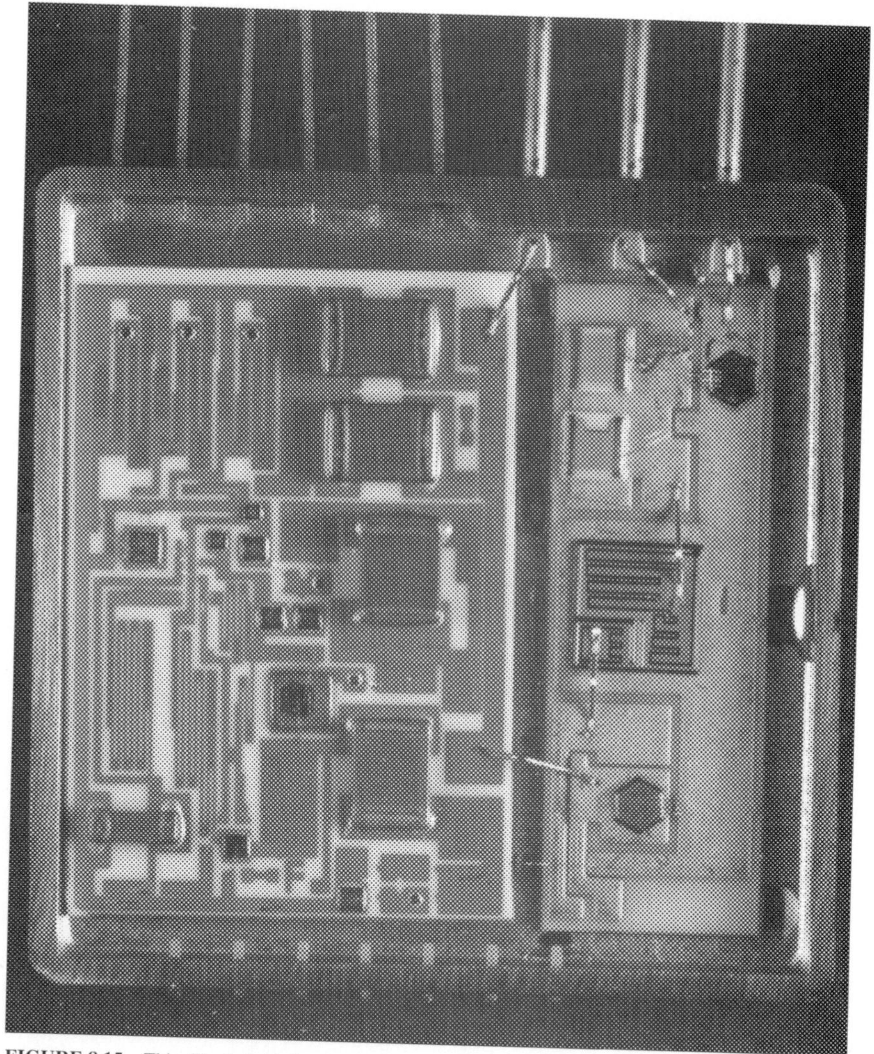

FIGURE 8.15 Thin-film hybrid circuit; 5-A switch used in space shuttle. (*Courtesy of Rockwell Intl.*)

Photolithographic and electron-beam lithography patterning of thin films provide the finest lines and spacings and highest circuit densities of any current technology. Dimensions in the submicrometer range are being achieved routinely for *very large scale integrated circuits* (VLSICs). Thin photopatterned aluminum is used almost exclusively to interconnect active junctions in integrated circuits. Aluminum is also used as the conductor metallization for multichip module interconnect substrates (MCM-Ds) to form ground planes, voltage power planes, and signal layers (Fig. 8.16). A 2 × 4 in multichip module composed of eight layers of thin aluminum isolated by eight layers of thin-film polyimide is shown in Fig. 8.17. This high-density analog/digital processor circuit has 25-μm lines and 75-μm spacings and over 3000 wire-bonding sites. Copper is also used for conductor traces in multichip modules, and in some circuits it is required for its electrical conductivity and current-carrying capacity, which is greater than that of aluminum. For an equivalent thickness, copper has approximately twice the electrical conductivity of aluminum (Table 8.14). However, both copper and aluminum are susceptible to corrosion, so generally the top metal layer of a multilayer

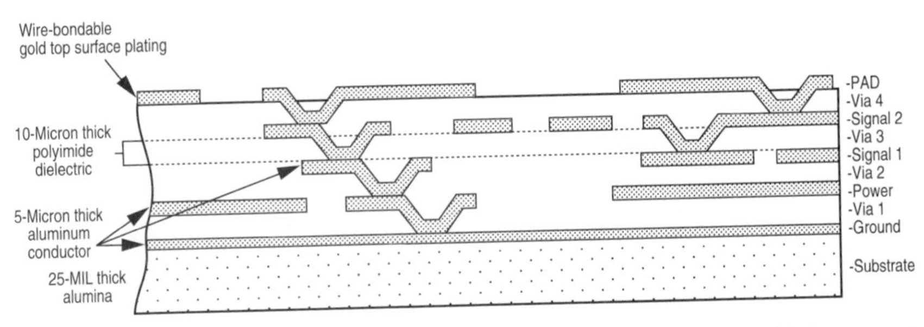

FIGURE 8.16 Cross-section schematic of five-conductor-layer high-density multichip interconnect substrate.

FIGURE 8.17 High-density multichip module. Thin-film polyimide-aluminum multilayer substrate. (*Courtesy of Hughes Microelectronics Division.*)

substrate is transitioned to gold or solder for environmental protection but also to provide a compatible metallization for die attachment and interconnection.

Refractory metals such as titanium, tungsten, and molybdenum are used as barriers separating aluminum from gold, thus preventing the well-known Al-Au intermetallic formation at elevated temperatures.[25] Refractory metals also serve as adhesion layers. Other excellent adhesion and barrier layers are chromium, nickel, and nichrome (nickel-chromium). These metals are widely used as adhesion layers on cured polyimide prior to depositing copper and as barrier layers encapsulating the copper to avoid copper diffusion into the polyimide.[25,26] Refractory metals, however, are generally not used as the primary conductor metallization for thin-film circuits because of their relatively poor electrical conductivities (Table 8.14).

Thin metal-alloy films such as nickel chromium and inorganic compounds such as tantalum nitride are extensively used as precision resistors. Other inorganic compounds are used as capacitors and dielectric insulators including silicon dioxide, silicon oxide, silicon nitride, tantalum pentoxide, and titanium dioxide. Some widely used thin-film metals, alloys, and inorganic compounds, their uses, and properties are listed in Table 8.14.

Polymer thin films are being used in microelectronic circuits more extensively today than some 20 years ago, largely because of advancements that have been made by resin manufacturers. Early formulations developed for the commercial market were not reliable when applied to microelectronic circuits and devices. These early polymers contained ionic contaminants and low-molecular-weight species which resulted in corrosion and outgassing failures especially of thin-film metallization. Most of the polymer coatings now sold to the microelectronic market are highly purified, meeting rigid government specifications for total halide ions, sodium and potassium ions, total extractable ions, amount of outgassing, and nature of outgassing products. Polymers have been synthesized whose molecular structures are highly dense and contain thermally stable backbone structures such as aromatic benzene rings or nitrogen heterocyclic rings. Some polymers such as polyimides are thermally stable up to 500°C and are also radiation resistant. Other features of these custom-designed, high-purity polymers include low CTEs, low stress, improved adhesion to substrates, and excellent electrical properties including low dielectric constants, low dissipation factors, and high electrical breakdown voltages over a wide temperature and frequency range. Among thin-film generic polymer types widely used today are polyimides, parylenes, silicones, epoxies, and fluorocarbons.

8.3.2 Deposition Methods: Metals, Alloys, and Inorganic Compounds

Thin-film deposition processes have been extensively covered in numerous articles and books;[1,5,28] hence only a cursory treatment focusing on applications to electronic circuits will be presented.

Generally all deposition methods for metals, alloys, and inorganic compounds involve vapor phase reactions whether they are physical, such as the simple high-temperature vaporization of metal, or chemical, such as the reactions of two gases at elevated temperatures or in a plasma.

Vacuum Evaporation. *Vacuum evaporation,* also called *vapor deposition,* is the oldest process for depositing thin metal films. Most metals can be vaporized at reasonably low temperatures (less than 1000°C) under high vacuum (10^{-5} to 10^{-6} torr). Gold, silver, copper, aluminum, nickel, and chromium are easily deposited in this manner. The metal as powder, wire segments, or foil is contained in a crucible or boat and, under vacuum, heated by passing a current through a high-resistance wire or filament

TABLE 8.14 Thin-Film Metals, Alloys, and Inorganic Compounds—Their Uses and Properties*

Thin film	Application	Electrical resistivity, μΩ-cm	Dielectric constant	CTE, ppm/°C	Thermal conductivity, W/m K	Chemical/physical properties
Aluminum	Widely used conductor metallization for ICs, hybrid circuits, and multichip modules	2.6–2.8		23–23.9	240	Easily forms oxide layer, protecting it from further oxidation
						Easily etched with mineral acids such as HCl and H_3PO_4
						Susceptible to chloride ion corrosion in presence of moisture
						Good adhesion to polymer dielectrics
Aluminum oxide, 99%	Dielectric film, capacitor film, protective film for Al		9.9	6.2–6.6	37	Protects Al from further oxidation
Chromium	Adhesion layer for gold and beneath resistors; thin-film resistor	13.0		6.3	66	Etches in ammonium cerium nitrate and inHCl/H_3PO_4
Chromium oxide	Resistors					
Copper	Conductor traces; thermal conductor and heat spreader	1.7		16.5–17.6	393	Corrodible in presence of chloride ions and moisture; diffuses into some polyimides; oxidizes in air at high temperatures; soluble in hot H_2SO_4 and HNO_3; good solderability
Diamond (synthetic film)	High thermal conductivity coatings; wear and protective coatings	10^{10}–10^{14} (Ω-cm)	5.7	0.8–2	1000–1500	Low-pressure CVD deposition from methane or other hydrocarbons and hydrogen

Gold	Conductor traces; wire bonding pads; solder attach pads; low-current electrical contacts; electroplate for corrosion protection of copper and other metals	2.2–2.4	14.2	297	Chemically inert; noble metal; good corrosion resistance Poor adhesion to polymer dielectrics
Indium	Contacts for infrared detector arrays; cryogenic applications; low-temperature solder for device attachment	8.4–9.0	33		Low melting solder wets most surfaces including ceramics and glass; does not embrittle at low temperatures; does not degrade near melt temperature
Indium tin oxide (ITO)	Transparent electrical contacts for liquid crystal displays				
Molybdenum	Adhesion and barrier layer	5.2–5.7	4.9–5.4	146	
Nickel	Conductor traces; adhesion and barrier layer; electroplating on Kovar for lid sealing; plating for corrosion and wear resistance; termination metallization for chip capacitors	6.9	13.3	92	Etchable in mineral acids
Nickel/chromium (nichrome)	Thin-film precision resistors	100 (Ω-cm)			Easily corrodes in presence of chloride ions and other ionic salts and moisture
Palladium	Adhesion and barrier layer	9.9–10.8	11–11.8	70	Chemically inert; noble metal; reduces solder leaching of gold; reduces metal migration of silver Can be deposited by electroless or electrolytic plating Forms ductile solid solutions with other metals

TABLE 8.14 Thin-Film Metals, Alloys, and Inorganic Compounds—Their Uses and Properties* (*Continued*)

Thin film	Application	Electrical resistivity, μΩ-cm	Dielectric constant	CTE, ppm/°C	Thermal conductivity, W/m K	Chemical/physical properties
Platinum	Adhesion and barrier layer	9.8–10.6		89	71	Chemically inert; noble metal; reduces solder leaching of gold; reduces metal migration of silver
Silver	Conductor traces; capacitor terminals; low-resistance contact pads used in relays	1.5–1.6		19.7	418	Tarnishes with sulfur compounds; migrates in presence of moisture, ionic contaminants, and voltage
Silicon dioxide	Passivation layer for ICs and resistors; capacitor dielectric; interlayer dielectric for multichip modules	10^{-16} (Ω-cm)	4	0.3–0.5	2.1	Etchable in HF
Silicon nitride	Passivation layer for ICs	$>10^{14}$ (Ω-cm)	7	2.3	30	Very low sodium and potassium ion mobilities
Tantalum	Capacitor material		14	6.5		
Tantalum nitride	Thin-film precision resistors	100(Ω-cm)				Etchable in HF
Tantalum pentoxide	High-value capacitors		20–27			Etchable in HF
Tin		11.5				Liquefies at 232°C; forms tin "whiskers," causing shorts under some conditions
Tin/lead (60/40)	Most widely used solder	14.9–16.0		24.5	51	Most widely used solder
	Solder coating for device attachment; plating for printed circuit boards; capacitor terminal material; solder thick-film coating					Scavenges gold and silver Liquefies at 183°C Good tarnish resistance

Titanium	Excellent adhesion layer under Pt or Au	41	62	8.5–8.9	168	Refractory metal; resists corrosion
Titanium dioxide	High-value capacitors		~100			
Titanium/tungsten	Adhesion layer for gold; barrier between TaN resistors and gold; barrier preventing copper diffusion into polyimide	65–75				
Tungsten	Adhesion and barrier layer	5.5–5.6		4.0–4.5	167–200	Refractory metal; oxidizes in air at high temperatures; prevents interdiffusion between gold and aluminum

*Ranges represent the spread of values reported from several sources.

(resistance heating). Vaporization may also be effected by induction heating or by exposure to an electron beam. The simplest process consists in resistance heating the evaporant in the form of a wire or foil (Fig. 8.18). It is preferable that the material not melt before it vaporizes, that is, that it *sublimes* (transitions from the solid-state directly to the vapor state). This avoids interaction of the melt with the filament or crucible. For example, silicon as a melt rapidly alloys with and destroys a refractory filament. The container, of course, must consist of a material that itself will not melt or vaporize at the vaporization temperature of the charge, nor must it alloy with or contaminate the charge. High-temperature stable refractory metals such as tungsten or molybdenum, and ceramics such as alumina, zirconia, boron nitride, or graphite are often used as container materials. Vaporization temperatures and corresponding vacuum pressures for some of the more popularly used metals are given in Table 8.15. Estimates of

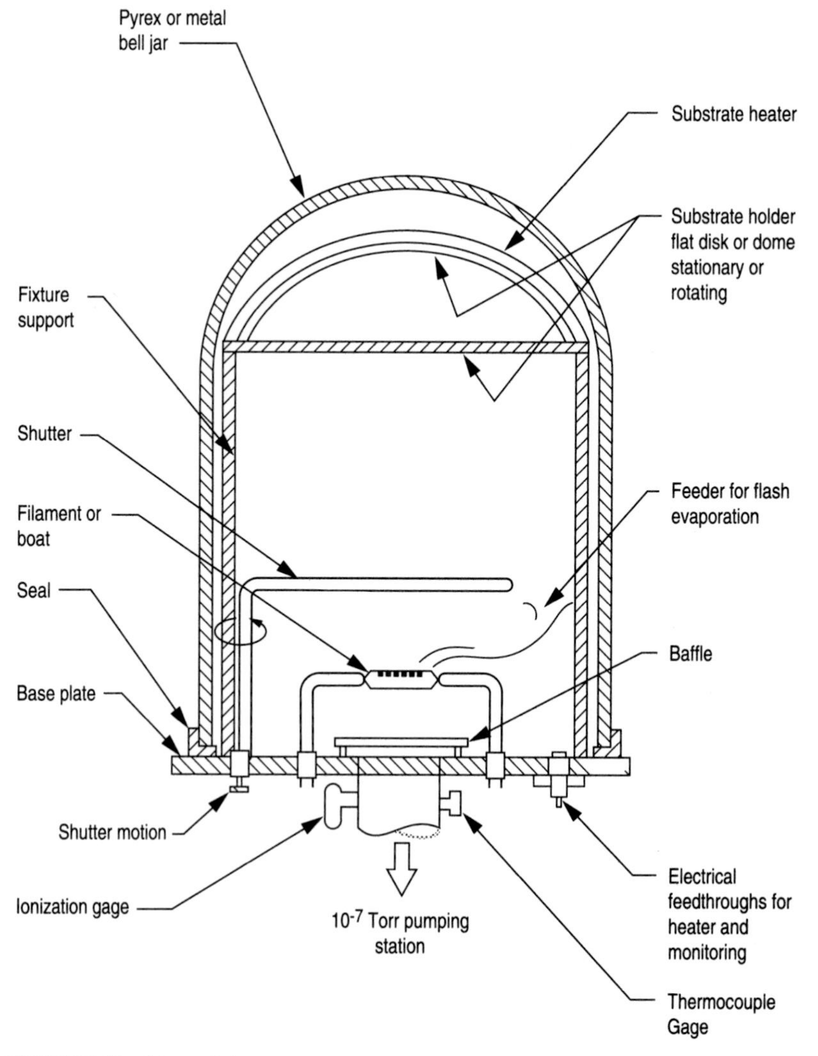

FIGURE 8.18 Schematic for vacuum evaporator.

TABLE 8.15 Vapor Pressure and Temperature Data for Some Metals Used in Electronic Circuits*

Metal	Melting temp, °C	Pressure, torr											
		10^{-8}	10^{-7}	10^{-6}	10^{-5}	10^{-4}	10^{-3}	10^{-2}	10^{-1}	1	10	10^2	10^3
Aluminum	659	685	742	812	887	972	1082	1217	1367	1557	1777	2097	2527
Cadmium	321	74	95	119	146	177	217	265	320	392	489	612	787
Chromium	1900	837	902	977	1062	1157	1267	1397	1552	1737	1967	2277	2727
Copper	1084	722	787	852	937	1027	1132	1257	1417	1617	1867	2187	2647
Gold	1063	807	877	947	1032	1132	1252	1397	1567	1767	2047	2407	2857
Indium	156	488	539	597	664	742	837	947	1082	1247	1467	1757	2157
Lead	328	342	383	429	485	547	625	715	832	977	1162	1427	1797
Molybdenum	2620	1592	1702	1822	1957	2117	2307	2527	2787	3117	3517	4027	4747
Nickel	1450	927	997	1072	1157	1262	1382	1527	1697	1907	2157	2497	2957
Palladium	1550	842	912	992	1082	1192	1317	1462	1647	1877	2177	2567	3107
Platinum	1770	1290	1382	1492	1612	1747	1907	2097	2317	2587	2917	3337	3897
Rhodium	1966	1277	1367	1472	1582	1707	1857	2037	2247	2507	2837	3247	3797
Ruthenium	2347	1507	1607	1717	1847	1987	2147	2347	2587	2857	3207	3627	4177
Silicon	1410	992	1067	1147	1237	1337	1472	1632	1817	2057	2347	2717	3217
Silver	961	574	626	685	752	832	922	1027	1162	1332	1542	1827	2217
Tantalum	3000	1957	2097	2237	2407	2587	2807	3057	3357	3707	4127	4657	5307
Tin	232	682	747	807	897	997	1107	1247	1412	1612	1867	2227	2687
Titanium	1700	1062	1137	1227	1327	1442	1577	1737	1937	2177	2487	2857	3367
Tungsten	3380	2117	2247	2407	2567	2757	2977	3227	3537	3907	4357	4927	5627
Zirconium	1850	1482	1582	1702	1837	1987	2177	2397	2657	2977	3377	3897	4557

*Data abstracted and reconstructed from R. E. Honig, *RCA Rev*, 23, 567 (1962).

source temperatures are based on the assumption that vapor pressures of 10^{-2} torr must be achieved to produce reasonable evaporation rates. Practically, these temperatures range from 600 to 2000°C depending on the metal. However, the lowest temperature possible, commensurate with a reasonable evaporation rate, should be used to avoid interaction of the container with the charge such as alloying, contamination, or vaporization. Refractory metals such as tungsten, molybdenum, and tantalum—due to their very high melting points (3380, 2610, and 3000°C, respectively), and low vapor pressures—are difficult to evaporate by normal resistance heating. They are best deposited by electron beam evaporation or by sputtering.

A key advantage of vacuum evaporation is the very high deposition rates that can be achieved with some metals—in some cases hundreds of thousands of angstroms per minute. On the other hand, because vacuum evaporation is a line-of-sight process, complex and irregularly shaped items cannot be uniformly or completely coated.

Flash Evaporation. In the direct evaporation of metal alloys and metal-dielectric mixtures, the composition of the deposit is invariably different from that of the starting material and is also difficult to control and reproduce. This is because the constituents of the mixture have different vapor pressures and the one with the higher pressure will evaporate sooner and at a higher rate. Fractionation or disproportionation occurs during evaporation. Thus starting with a 80/20 mixture of nickel chromium (nichrome) results in a final deposit that may be 60 percent chromium and 40 percent nickel or even higher in chromium since chromium has a higher vaporization pressure than nickel (Table 8.15). The final ratio and hence sheet resistance of nichrome is therefore hard to control. However, if the alloy mixture in small quantities is dropped into an empty container that is preheated to a temperature higher than the vaporization temperature of the less volatile component, the deposited film's composition will closely resemble that of the starting material.[1] This variation of vacuum evaporation is called *flash evaporation* and is widely used in depositing nichrome and chromium silicon monoxide resistors (Fig. 8.19). Although some fractionation occurs during the evaporation of each small increment, it is so small and does not appreciably change the ratio of components of the bulk film. Nichrome may be deposited from mixed powders by feeding the powder to a vibrating chute onto a filament heated at 1800°C. Deposition rates vary from 1 to 10 Å/s. Several variations of feeder mechanisms are used including moving belt, spool and guide, vibrating chute, rotating tube, and worm drive.[28]

Sputtering. Sputtering is rapidly supplanting vacuum evaporation for metal deposition because it is a more controllable process, provides more adherent films, and enables better coverage of complex-shaped objects. Sputtering is also a much more versatile process, allowing not only high-vapor-pressure metals to be deposited but also refractory metals, alloys, inorganic compounds, and ceramics. Reactive sputtering (a variation of sputtering) also permits the chemical formation of compounds during deposition. Comprehensive and theoretical treatments of sputtering may be found in several books.[5,29,30]

Sputtering, like vacuum evaporation, is a physical vapor deposition process but, unlike vacuum evaporation, is electronically not thermally induced. In its simplest form, dc sputtering, the material to be deposited is attached to a cathode, referred to as the *target*. The object to be coated is placed on a holder that is rendered anodic or grounded. After evacuation of the chamber to remove oxygen, moisture, and other volatiles, argon or other inert gas is introduced, and a bias voltage of 1000 to 2000 V dc is applied across the electrodes. Exposure of the argon to this potential causes it to ionize to positively charged argon ions (Ar^+), simultaneously emitting an electron.

Process

Boat resistance heated to temp higher than vaporization temp of each constituent powder vibrated down an incline into boat.

Boat consists of W or other refractory.

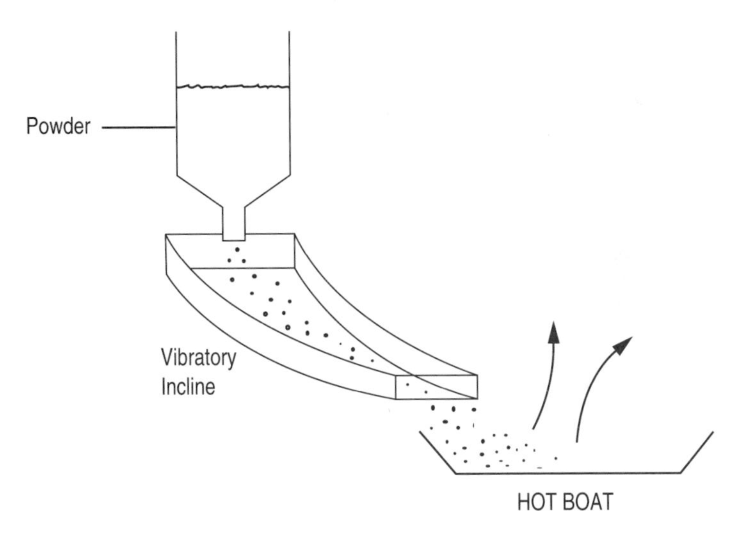

Advantages

No alloying with or contamination from boat material.

Little disproportionation of mixture.

FIGURE 8.19 Flash evaporation.

The argon ions are attracted by and accelerated to the cathode target bombarding it with great kinetic energy which sloughs off (sputters) particles of the target material and generates secondary electrons. These energetic electrons in turn interact with more argon atoms, creating more ions; thus a sustaining plasma of ions, electrons, atoms, and photons is formed. The sputtered particles traverse the plasma and deposit onto the substrate, again with high kinetic energy, resulting in a highly dense adherent film. Besides the adhesion that is imparted due to the inherent nature of sputtering, surfaces may be sputter-etched by reversing polarity on the substrate. Thus in the same pumpdown surfaces may be atomically cleaned and roughened prior to sputtering—enhancing adhesion even further. A four-stage sputtering system in which the first stage is used to sputter-etch-clean the surface, followed by the sequential sputtering of aluminum, titanium-tungsten, and gold is shown in Fig. 8.20

In rf (radio frequency) sputtering, an rf instead of direct current is applied to the cathode (Fig. 8.21). The advantage here is that materials other than electrically conductive metals can be deposited. With conductive materials the positive charge that builds up on the cathode is quickly dissipated. However, with insulative materials such as inorganic compounds or ceramics, the positive charge builds up on the cathode and prevents further sputtering from positively charged ions—quickly quenching the plasma. Alternating the current from positive to negative cleanses the positively charged layer during the positive cycle and allows sputtering during the negative cycle.

FIGURE 8.20 Four-stage sputtering system.

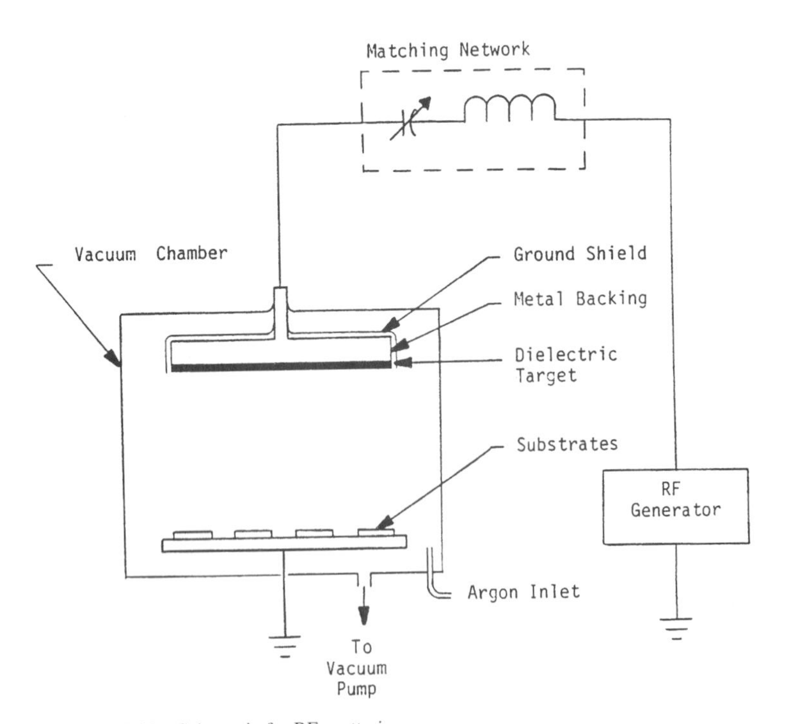

FIGURE 8.21 Schematic for RF sputtering.

Finally, reactive sputtering allows new chemical compounds to be formed in the plasma and deposited. This occurs by simply introducing a reactive gas during the sputtering of a metal. For example, introducing some nitrogen along with argon during the sputtering of tantalum, produces tantalum nitride (TaN) films—a process widely used for the thin-film deposition of TaN resistors. Similarly oxides are produced by introducing oxygen. Other possibilities are given in Table 8.16.

TABLE 8.16 Films Produced by Reactive Sputtering: Sputtering Process in which a Reactive Gas Is Introduced Alone or along with the Ar Such That the Sputtered Particles React to Form a New Compound

Target	Reactive gas	Film
Ta	N_2	Ta_3N_2
Ta	O_2	TaO
Al	N_2	AlN
Al	O_2	Al_2O_3
Si	N_2	Si_3N_4
Si	O_2	SiO_2
Ti	O_2	TiO_2

Metallo-organic Deposition (MOD). Liquid solutions of metallo-organic compounds may be spin-coated, sprayed, or screen-printed onto a substrate, dried, then decomposed to the free metal. Decomposition is effected by heating the entire coated surface to the decomposition temperature of the metallo-organic compound, generally no higher than 850°C. Decomposition may also be selective by programming a laser beam so that conductor traces are formed by direct writing. In both cases the decomposition products are carbon dioxide, water, and some volatile low-molecular-weight organics. Silver, nickel, gold, and indium-tin oxide have been deposited from their respective salts or complexes with organic compounds. The metal films are extremely thin (several thousand angstroms) and hence must be subsequently thickened by electrolytic or electroless plating to provide sufficient electrical conductance.

A typical metallo-organic compound used in depositing gold is shown in Fig. 8.22. Silver can be deposited from silver neodecanoate.[31,32] Metallo-organic solutions are commercially available from *Electro Science Laboratories* (ESL) and Engelhard Corp. Resistors and dielectrics have also been deposited by MOD.

8.3.3 Deposition Methods: Organic Polymers

The main methods for depositing thin polymer coatings include spraying, dipping, flow coating, spin coating, vapor deposition, and screen-printing. Automatic spray coating and flow coating are popular for high-volume, low-cost production. Spin coating, screen-printing, and vapor deposition methods are extensively used in semiconductor, integrated circuit, and hybrid microcircuit processing to apply photoresists, permanent insulation, interlayer dielectrics, environmental protective and passivation coatings, and particle immobilizing coatings.

Spin Coating. The key advantage of *spin coating* is the deposition of ultrathin, uniform, and pinhole-free coatings. Thicknesses can be accurately controlled by control-

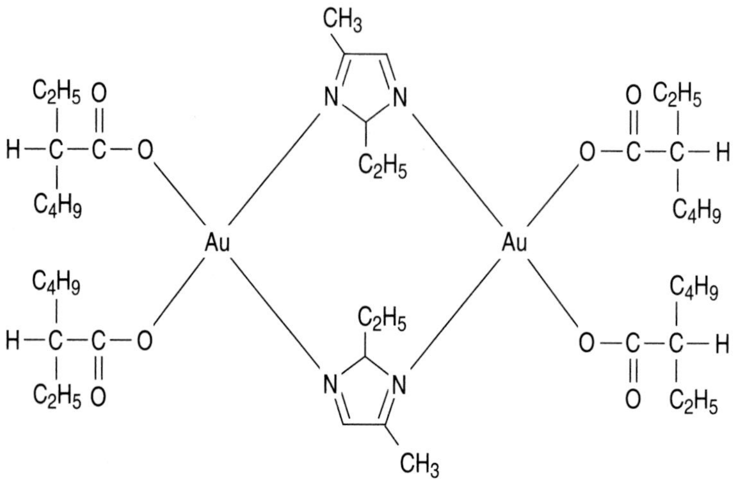

FIGURE 8.22 Example of a gold metallo-organic compound.

ling the centrifugal speed (revolutions per minute) that the part is subjected to and the viscosity of the coating solution. By diluting the coating (reducing the resin content and increasing the solvent content) and by increasing the spin speed, angstrom-thin uniform coatings can be deposited. Figure 8.23 shows the relationship of thickness to solids content (viscosity) and to spin speed for an addition-type polyimide. Spin coating is suitable only for flat surfaces and generally limited to round wafers up to 8 in in diameter. Loss of coating material, often as high as 90 percent, is not uncommon because the centrifugal force of spinning forces the liquid outward beyond the edges of the wafer.

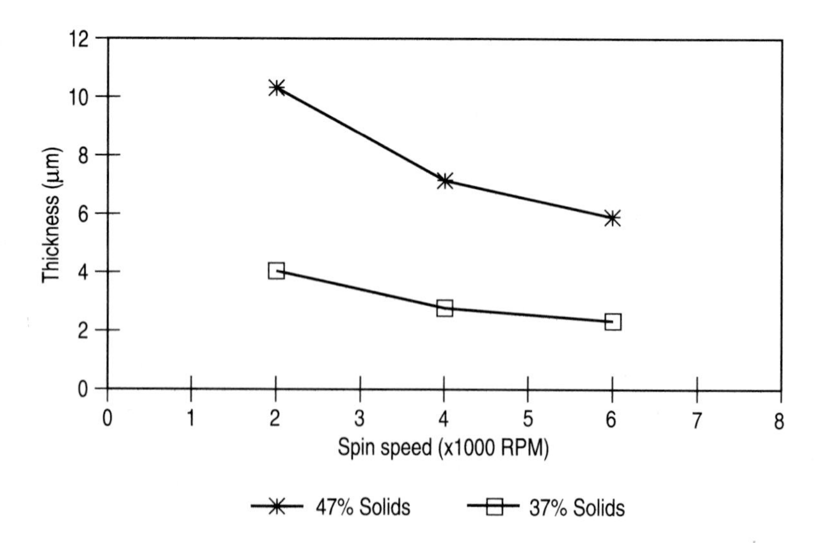

FIGURE 8.23 Thickness versus spin speed for Thermid EL-5512 polyimide. (*Courtesy of National Starch and Chemical Corp.*)

Flow Coating. In *flow coating,* also called *curtain coating,* the coating solution is pumped through a narrow horizontal slit so that it flows downward in a continuous stream giving the appearance of a curtain—thus the nomer "curtain coating." The parts to be coated are moved on a conveyor belt beneath the flow and can then be conveyorized through a curing oven. One variation of flow coating is *extrusion coating* in which a pressurized liquid resin is programmed to flow over the whole surface or onto selected portions of a surface. A commercially available extrusion coater is the FAS-COAT (Figs. 8.24 and 8.25). As opposed to spin coating, extrusion coating provides several benefits among which are:

Very little material is wasted (5 to 10 percent compared with 80 to 95 percent for spin coating).

Substrate sizes and shapes are not limited; large square panels can be coated.

No edge beading or corner effects occur.

Film thicknesses up to 100 μm in a single pass can be obtained.

Throughput is rapid and high.

Extrusion coating has been applied to dispense high-viscosity polyimides as interlayer dielectrics for multichip modules.[33]

Screen-Printing. In screen-printing, the resin coating is squeegeed through a stainless-steel mesh. Depending on the mesh size and the viscosity of the coating, ultrathin (5 to 10 μm) coatings can be deposited. However, to achieve pinhole-free coatings, several superimposed layers may be necessary. There are many epoxy and polyimide screen-printable coatings on the market, some of which are ultrapure designed for high-reliability electronic applications such as passivation of active devices and humidity protection of closely spaced conductor circuit traces.

Vapor Deposition. Vapor deposition is a highly desirable method because of the ability to completely cover assembled electronic circuits—not only top surfaces but also penetrating beneath devices and completely coating wire bonds. Unfortunately, organic monomers and polymers decompose before they vaporize. The best mechanism for vapor-depositing polymer coatings is to synthesize the polymer in the vapor phase by generating free radicals or other reactive species from low-molecular-weight monomers. An outstanding example of this approach is *poly paraxylylene,* commercially known as *Parylene.** The *xylene dimer,* a white solid, is first sublimed, then thermally decomposed into gaseous diradicals which then rapidly recombine on a surface at almost room temperature (Fig. 8.26).

8.3.4 Thin-Film Conductors

The thin-film conductors most widely used in microelectronic device and circuit manufacture are aluminum, gold, copper, and silver. Among the metals, these four have the highest electrical conductivities. They are easily deposited by sputtering or by vacuum evaporation and, with the exception of aluminum, may also be deposited by electroplating. Less conductive metals such as nickel and chromium are primarily used as barrier layers preventing interdiffusion of two other metals, for example, nickel to prevent diffusion between gold and aluminum. Tungsten, molybdenum, and mixtures of the two are used as inner conductor traces in a cofired ceramic substrate because only these refractory metals can withstand the high firing temperatures required to sinter

*Parylene is a tradename of Union Carbide.

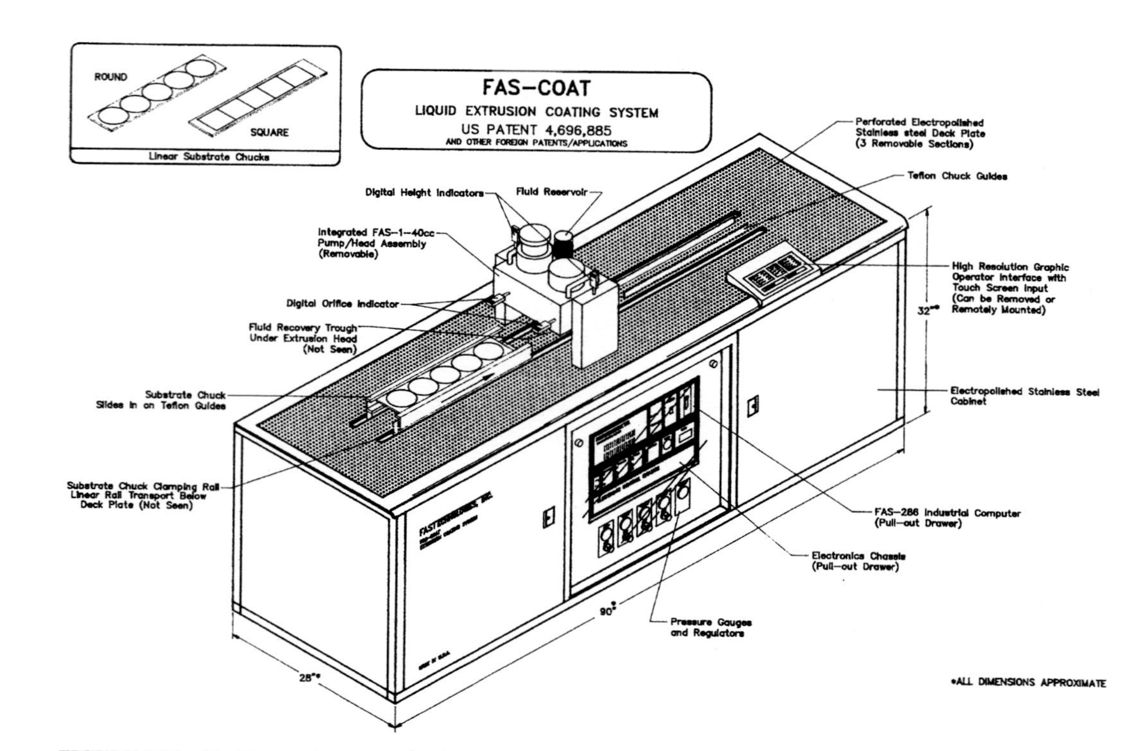

FIGURE 8.24 Liquid extrusion coater sketch. (*Courtesy of FAS Technologies Inc.*)

FIGURE 8.25 Extrusion coating system. (*Courtesy of FAS Technologies Inc.*)

the ceramic compositions. Refractory metals are also not as conductive as gold, and therefore the buried lines must be designed wider and thicker to provide the same current-carrying capacity as narrower gold lines. The outer (top) tungsten metallization is nickel plated, followed by gold plating to further augment electrical conductance and to provide a compatible metal for the attachment of devices. Properties of some metal conductors used in circuit fabrication are given in Table 8.14.

8.3.5 Thin-Film Resistors

Though there are many thin-film resistor materials available, only three are in wide use—nichrome, tantalum nitride, and chromium silicon oxide cermets. All use gold terminations and a barrier metallization. Barrier metals are required to prevent resistor constituents—for example, chromium in nichrome resistors—from diffusing into gold. Diffusion not only can change the resistor value but also interferes with wire bonding

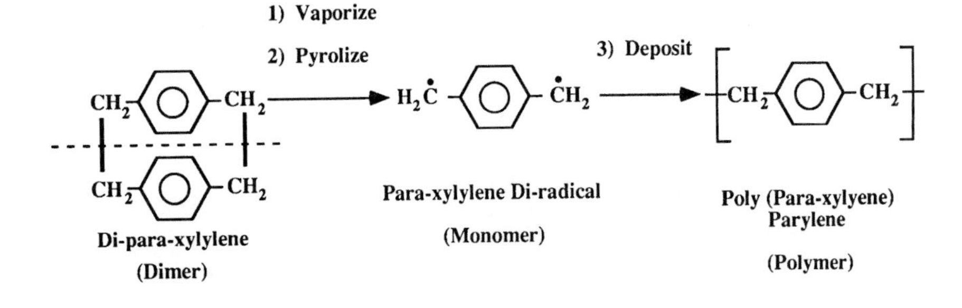

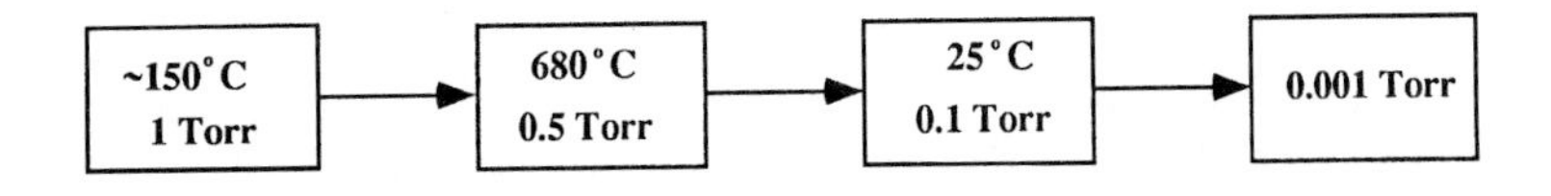

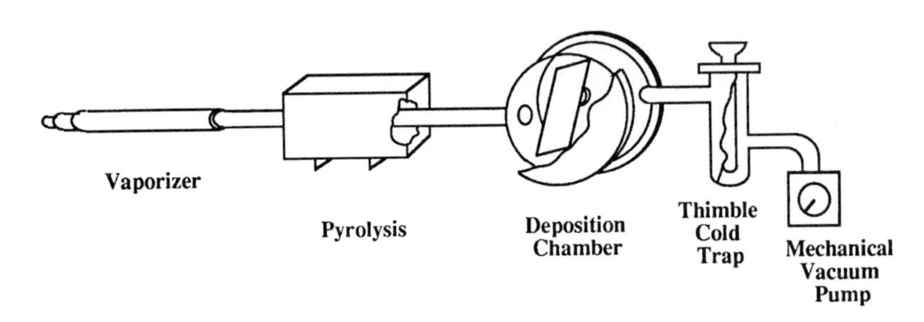

FIGURE 8.26 Parylene vapor deposition process.

or soldering to the top conductor pads. Nickel is generally used as the barrier metal for nichrome resistors. Deposited in thicknesses of 400 to 1000 Å, it effectively prevents interdiffusion. Nichrome also functions as an adhesion layer, a so-called tie layer, to enhance adhesion of gold to the ceramic substrate since gold is difficult to adhere by itself. In the tantalum nitride process, titanium is the tie layer and palladium serves as the barrier diffusion layer.

Both nichrome and tantalum nitride offer similar resistor capabilities: sheet resistances of approximately 100 to 300 Ω/square and low-temperature coefficients of resistances (TCRs) 0 ±100 ppm/°C. Cermet resistors, however, provide an extended capability of sheet resistances of one thousand to several thousand ohms per square, thus offering the design engineer the freedom to lay out resistors of very high values

(100,000 Ω to the megaohm range), albeit at the expense of higher TCRs. Chromium silicon oxide compositions are also more difficult to deposit, and small changes in the oxide percentage will make large changes in values.

Nichrome. Regardless of the resistor material used, the fabrication steps are similar (Fig. 8.27). First, a sandwich structure of resistor film, barrier film, and top conductor is sequentially deposited by vacuum evaporation or preferably sputtering. It is important that all three films be deposited in the same closed-system operation to avoid air oxidation or contamination of the layers. Then, through precision photolithography involving a series of photoresist application, exposure, developing, and etching steps, an intricate pattern of conductors (1- to 5-mil lines and spacings) and resistors (5 to 40 mils wide) can be formed.

Typical properties of nichrome resistors are given in Table 8.17. Thin-film resistor properties and performance are highly dependent on the surface characteristics of the substrate—the smoother the surface finish, the more reproducible and stable the resistance values. However, other factors affect resistor stability, chief among which are annealing conditions, stabilization baking, and trimming process. Annealing is extremely important in fabricating precision resistors. Nichrome resistors are generally annealed at 300 to 350°C in air for several hours whereupon the chromium constituent oxidizes to some extent, creating a passivating layer that slows further oxidation and stabilizes resistor values. Sheet resistances will increase approximately 10 to 20 percent during annealing. Long-term stability measured as resistance drift after aging 1000 h at 125 or 150°C, depends largely on the annealing schedule used; the higher temperature and longer anneal times provide greater stabilities, lower TCRs, and better resistance tracking (Table 8.18). Sputtered nichrome resistors annealed 4 to 5 h at 350°C drifted less than 100 ppm after 1000 hours aging at 150°C and exhibited TCRs of only 0 ± 3 ppm/°C.

Though nichrome resistors are among the most precise, thermally stable resistors, they are also among the most susceptible to chemical and electrolytic corrosion. Chloride ions transferred from fingerprints combined with moisture will quickly corrode and etch nichrome films. Under such conditions, nichrome resistors on integrated circuits when passivated with silicon dioxide have been reported to disappear. This was attributed to acid etching, in which the phosphorus contained in the silicon dioxide layer reacted with moisture to form phosphorous acid. Some organic coatings also react with nichrome, producing resistance changes. Therefore, any coating, whether inorganic or organic, should first be tested to ensure compatibility with nichrome or any other thin-film resistor. Generally, in military and space applications thin-film resistors are left uncoated because the hybrids or multichip modules containing them are hermetically sealed in metal or ceramic packages in a dry nitrogen ambient.

Nichrome resistors are further limited in their low sheet resistances—about 100 to 300 Ω/square. Higher sheet resistances are possible by depositing ultrathin layers of less than 100 Å since sheet resistance is inversely proportional to thickness. However, the deposition of such extremely thin films is not practical because discontinuous, imperfect films are formed, giving rise to open circuits and unstable values. High-value resistors using low sheet resistance materials such as nichrome and tantalum nitride can be designed only by using large numbers of squares having narrow widths and serpentining the resistor lines to conserve space. Even so, designing a 100,000-Ω resistor with a 200 Ω/square material takes up considerable space. Fortunately, for many thin-film circuits, the number of high-valued resistors needed is small and can be satisfied by using chip resistors.

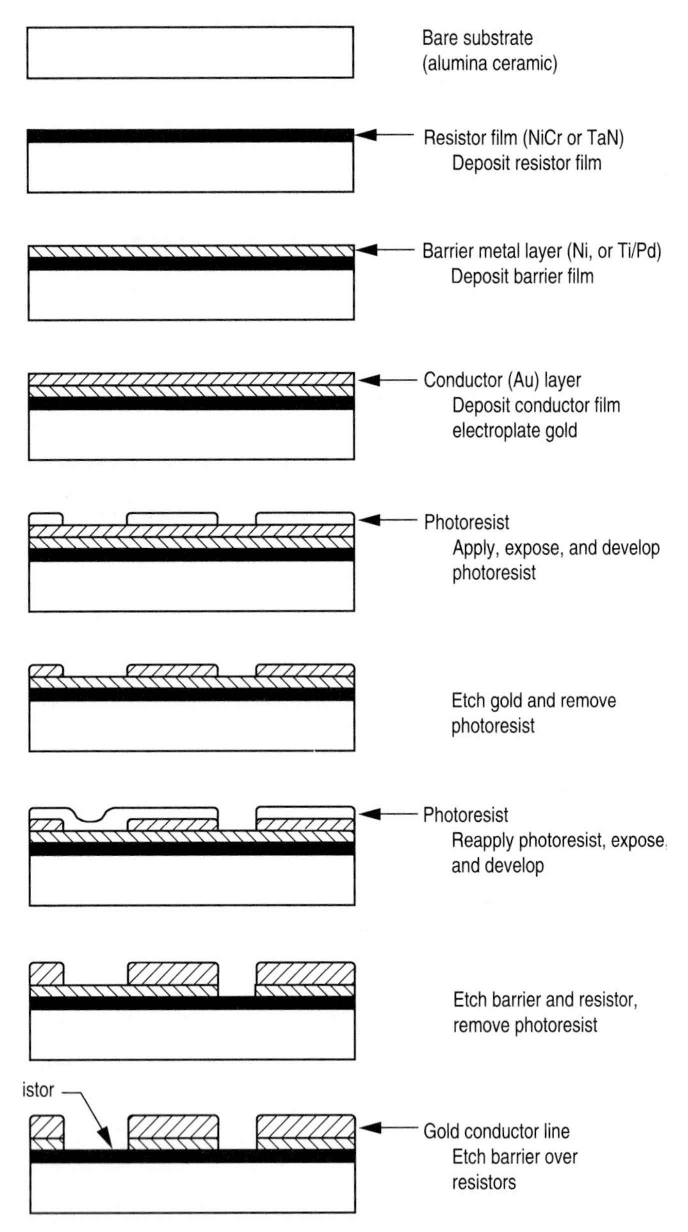

FIGURE 8.27 Process sequence for depositing and photodelineating thin-film conductor-resistor networks.

TABLE 8.17 Characteristics of Nichrome Resistors

Sheet resistance	25–300 Ω/square, 100–200 Ω/square (typical)
Sheet resistance tolerance	±10% of nominal value
TCR	0 ± 50 ppm/°C, 0 ± 25 ppm/°C (with special anneal)
TCR tracking*	2 ppm
Resistance drift	<2000 ppm after 1000 h at 150°C <1000 ppm with special anneal <200 ppm, sputtered films with 350°C anneal
Ratio tracking	5 ppm
Resistor tolerance after anneal and laser trim	±0.1%
Noise (100 Hz to 1 MHz)	−35 dB (maximum)

* −55 to +125°C.

TABLE 8.18 Nickel-Chromium Resistor Stability and TCR Data

Deposition method	Sheet resistance of NiCr as deposited, Ω/square	Annealing conditions (air) °C	Annealing conditions (air) min	Resistance change after aging, ppm: 50 h, N_2 150°C	168 h, N_2 125°C	500 h, N_2 125°C	Final TCR
Flash evaporated without Ni barrier	165	225	120	2000	500	700	−20
Flash evaporated with 400-Å Ni barrier	165	225	120	1700	350	630	−16
Flash evaporated with 800-Å Ni barrier	165	225	120	1400	280	400	−12
Flash evaporated with 800-Å Ni barrier	165	300	120	1310	260	300	−10
Flash evaporated with 800-Å Ni barrier	165	350	120	460	70	110	+34
Sputtered with 800-Å Ni barrier	125	350	235*	71	41	—	+3
Sputtered with 800-Å Ni barrier	105	350	315†	96	−96	+80	−3

* In four steps.
† In six steps.
Source: Courtesy Rockwell International.

Tantalum Nitride. The processes for fabricating tantalum nitride resistor/gold conductor circuits patterns are similar to those for nichrome resistors. In both cases photolithography involving selective etching of multilayered metal structures is used. Several process differences, however, exist: (1) Tantalum nitride is deposited by reactive sputtering of tantalum in a nitrogen atmosphere instead of by direct sputtering or flash evaporation as with nichrome. (2) Titanium and palladium are used as the tie layer and barrier layer, respectively, between the tantalum nitride and gold whereas nickel is used as the barrier layer between nichrome and gold. (3) The titanium and

TABLE 8.19 Characteristics of Tantalum Nitride Resistors

Sheet resistance	20–150 Ω/square, 100 Ω/square (typical)
Sheet resistance tolerance	±10% of nominal value
TCR	−75 ± 50 ppm/°C (typical), 0 ± 25 ppm/°C, with vacuum anneal
TCR tracking*	<2 ppm
Resistance drift (1000 h at 150°C in air)	<1000 ppm
Ratio tracking	5 ppm
Resistor tolerance after anneal and laser trim	±0.10% standard, ±0.03% bridge trim
Noise (100 Hz to 1 MHz)	<−40 dB

*−55° to 125°C.

tantalum nitride are etched with hydrofluoric acid–nitric acid solution or may be sputter-etched for very fine line definition.

The partial pressure of nitrogen gas introduced during reactive sputtering affects both the sheet resistivity and TCR of the deposited resistors since tantalum nitride undergoes several crystallographic changes as the concentration of nitrogen is increased. Besides this batch process for depositing tantalum nitride resistors on alumina substrates, tantalum nitride resistors may also be obtained in chip form. Films are sputtered onto highly polished silicon, quartz, or sapphire wafers, then batch processed with aluminum-nickel barrier terminations and sputter etched to yield fine line resistors having high resistance values— up to 12 MΩ.[34]

Tantalum nitride resistors are similar to nichrome resistors in their electrical properties, including sheet resistance range, TCRs, resistor tracking, and long-term resistance drift after elevated temperature aging. Some electrical characteristics are provided in Table 8.19. Tantalum nitride resistors are more rugged and more chemically and thermally resistant than nichrome resistors. A passivating layer of tantalum pentoxide is formed during annealing in air or oxygen or by introducing controlled amounts of oxygen during the reactive sputtering process. Annealing is performed at 450°C or higher. This *inherent* oxide is quite stable to moisture and other hostile environments.

Cermet. *Cermet,* a term coined from the words "ceramic" and "metal," is a composition of metal oxides and metals. They may be thick-film pastes or evaporated thin films and are useful in forming high-valued resistors in a small area. The most widely used thin-film cermet resistors consist of silicon monoxide and chromium (SiO-Cr) and are used primarily because of their high sheet resistances. By varying the ratio of silicon oxide to chromium, a wide range of sheet resistances is possible, from several hundred to tens of thousands of ohms per square. However, it is difficult to control the stability and reproducibility of the very high resistance values. The most practical range is 1000 to 3000 Ω/square.

8.3.6 Thin-Film Capacitors

Fabrication of thin-film capacitors requires three deposition steps—the lower electrode plate, the dielectric film, and the upper electrode plate—resulting in a sandwich structure. The mathematical expression for capacitance is

$$C = \frac{0.225K\,(N-1)\,A}{t}$$

where C = capacitance, pF
K = dielectric constant
N = number of plates
A = area, in^2
t = thickness of dielectric, in

In comparing the properties of different dielectrics, it is convenient to use capacitance density (capacitance per unit area). Hence for a given material, the film thickness alone determines capacitance density.

In choosing capacitor materials for thin-film circuits and devices, the dielectric constant of the material need not be as high as for its thick-film counterpart because the key to high capacitance in thin films is producing very thin pinhole-free films in large areas. However, films should be thicker than 1000 Å to avoid defects and pinholes that will cause dielectric breakdown. The dielectric must also withstand temperatures of at least 125 to 150°C, which is the range for burn-in and even higher temperatures if solder attachment or other high-temperature assembly steps are used. Thus, evaporated or sputtered inorganic films or oxidized metal films are preferred over polymeric films. In using evaporated films, the following criteria for the dielectric material should be considered:

1. Must not decompose during the evaporation-deposition process
2. Should not be hygroscopic
3. Should be integral pinhole-free films of controlled thickness
4. Must adhere well to substrates and metals
5. Must be stress free
6. Must be stable over the temperature range that the circuit will be exposed to during fabrication and assembly and during subsequent operation

Silicon oxide, silicon dioxide, silicon nitride, silicon oxynitrides, tantalum pentoxide, aluminum oxide, titanium dioxide, and a number of titanates meet these requirements, depending on their deposition conditions. A list of other dielectrics may be found in Holland[35] and Maissel and Glang.[28]

Silicon oxides are among the more popular capacitor dielectrics. Silicon monoxide's dielectric constant is approximately 6 but can vary depending on its deposition conditions. Its dielectric breakdown voltage is also high, approximately 2×10^6 V/cm, but also can vary depending on thickness and deposition conditions. Silicon dioxide has a lower k (4) and is more useful as a passivation coating than as a capacitor. A key advantage of silicon oxide as well as oxides of aluminum, titanium, and tantalum is their ability to be produced by several processes. They may be grown from the base metal by thermal oxidation or electrochemical oxidation or deposited by evaporation, sputtering, reactive sputtering, chemical vapor deposition, and plasma-enhanced chemical vapor deposition. The dielectric properties of some thin-film dielectrics useful as capacitors are given in Tables 8.20 and 8.21.

Thin-film tantalum capacitors are also widely used in electronic circuits. A further advancement in the development of tantalum capacitors is obtained by the thermal or electrolytic oxidation of sputtered tantalum films. An example of the duplex configuration is shown in Fig. 8.28. While SiO is fundamentally limited by the incidence of pin-

TABLE 8.20 Properties of Various Dielectric Films[36]

Material	Method of deposition	Dielectric constant, ϵ	tan δ	Breakdown stress, MV/cm	Capacitance, μF/cm^2
Silicon dioxide (SiO_2)	Reactive sputtering	4	0.001	3.0	0.015
Magnesium fluoride (MgF_2)	Evaporation	5	0.016	1.0	0.01
Silicon monoxide (SiOx)	Evaporation	5–7	0.010	1.2	0.01
Aluminum oxide (Al_2O_3)	Plasma oxidation	8	0.005	2.0	0.10
Aluminum oxide (Al_2O_3)	Anodic oxidation	8	0.005	4.0	0.20
Tantalum oxide (Ta_2O_5)	Reactive sputtering	20	0.003	1.0	0.10
Tantalum oxide (Ta_2O_5)	Anodic oxidation	27	0.005	3.0	0.15
Titanium oxide (TiOx)	Anodic oxidation	30–40	0.030	1.0	0.30
Lead titanate ($PbTiO_3$)	Reactive sputtering	80	0.040	0.6	0.20

TABLE 8.21 Characteristics of Thin-Film Capacitor Materials[37]

Material	How formed	Dielectric constant at 25°C	Breakdown voltage above 1000 Å, kv/cm	Dissipation factor, (Hz at 25°C)
SiO	Evaporated from powder with Al electrodes	6.0	1.6	$0.025(10^3)$
Ta_2O_5	Anodize, sputtered, or evaporated, Ta with Ta, Au, or Al electrodes			$0.008(10^3)$
SiO_2	Thermal oxidation or plasma enhanced CVD			$0.7(10^6)$
Al_2O_3	Anodize; evaporated Al with evaporated counterelectrode		8(9)	$0.5(10^6)$
				$0.01(10^3)$
$BaTiO_3$	Evaporated	Approx 1000		
ZnS	Evaporated	8.2	0.2	
MgF_2	Evaporated	6.5	2	
TiO_2	Oxidized evaporated Ti; sputtered			$0.004(10^3)$

hole defects, by depositing SiO on Ta_20_5 to form a duplex dielectric structure, low-value capacitors (0.01 μF/cm^2) and large-area distributed RC networks can be realized and incorporated into tantalum integrated circuitry. Duplex capacitors are commonly produced on glass substrates by anodization of sputtered tantalum electrodes and vacuum sublimation of bulk silicon monoxide. These capacitors perform reliably at 50 V and 85°C. Capacitors with gold counterelectrodes, as seen in Fig. 8.28, and the thicker films of silicon oxide exhibited a lower failure rate at the higher life-test voltages and temperatures.

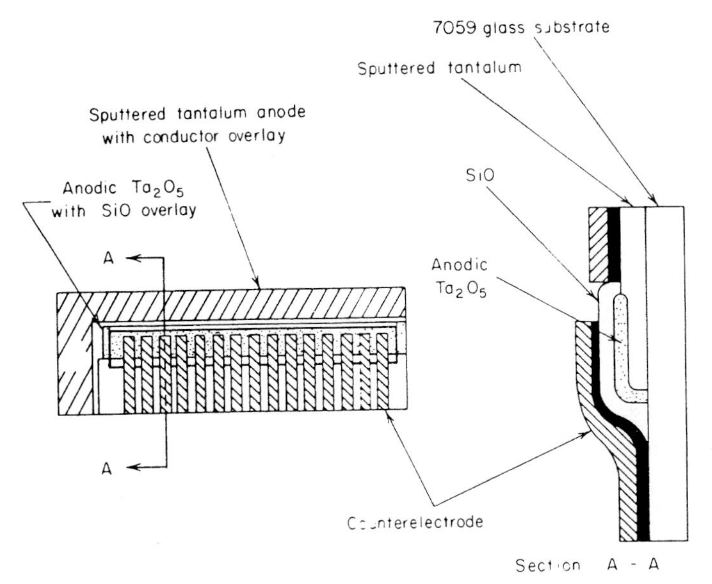

FIGURE 8.28 Sketch of tantalum oxide–silicon oxide duplex dielectric thin-film capacitor.

Thin-film capacitors are of several types and may be designed in different configurations. The sketch shown in Fig. 8.29 illustrates a widely used construction where the capacitor is formed on a silicon or alumina ceramic substrate. *Metal oxide semiconductor* (MOS) capacitors are formed by utilizing a highly doped silicon substrate as the bottom electrode, thermally growing an oxide, then vapor depositing or sputtering the top electrode metallization. Photolithographic processes are used to delineate precise geometries. The emitter diffusion in semiconductor processing may be used to provide a low sheet resistance under the dielectric. Thermally grown silicon dioxide or a thin film of spun-on glass may be used as the dielectric. The upper electrode is usually the same metal as the interconnection metallization and simultaneously deposited.

A diffused junction capacitor differs from the MOS capacitor in being a reverse biased *p-n* junction, usually the base-collector junction. Several limitations of diffused capacitors are

1. Low capacitance per unit area
2. High parasitic capacitance
3. High voltage dependency
4. High series resistance

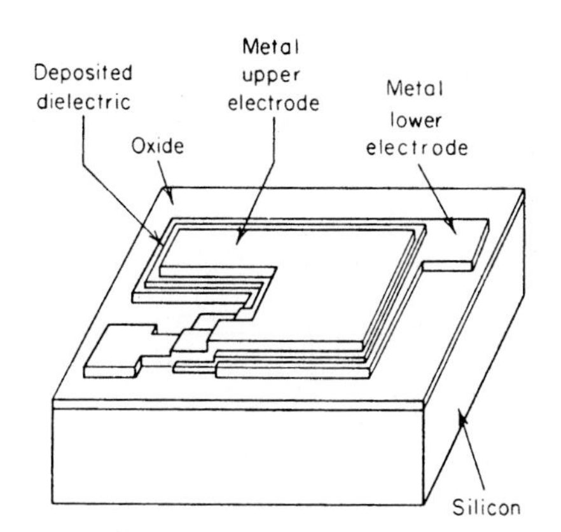

FIGURE 8.29 Typical configuration for thin-film capacitor.

For example: For a 0.5-Ω-cm *n*-type silicon:

Volts	Capacitance, pF/mils^2
−1	0.2
−6	0.085
−10	0.06

Thus, to obtain large capacitance values, large areas are required: 100 pF with 6 V requires 1200 mil^2 of area. A cross section of a *p-n* junction capacitor compared with that for an MOS capacitor is shown in Fig. 8.30. The *p-n* junction is essentially a diode which functions as a capacitor as long as it is reverse-biased. Metal contacts to the electrodes of the capacitive elements continue over the passivating oxide to other parts of the circuit.

The processes just described for the fabrication of thin-film capacitors have been used largely for producing chip capacitors. Hybrid circuit design engineers have preferred to use chip capacitors instead of batch fabricating them as an integral part of the interconnect substrate since in most circuits only a few capacitors are needed. However, in multichip modules that require large numbers of decoupling capacitors (in some cases over 20 percent of the surface must be devoted to capacitors), integration of these capacitors within the substrate has cost, reliability, and density advantages. Batch fabricating the capacitors avoids the handling and attachment of a large number of chips and integrates the capacitors within the substrate multilayer structure. Where silicon substrates are used, the MOS capacitor construction can be used effectively.

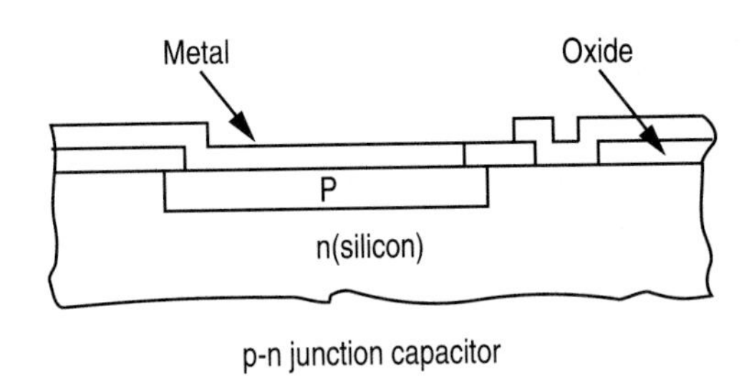

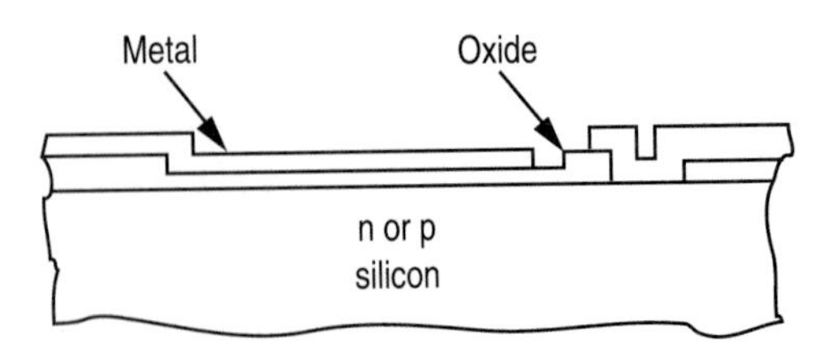

FIGURE 8.30 Comparison sketches for *p-n* junction and MOS capacitors.

8.3.7 Thin-Film Polymer Dielectrics

Thin-film organic polymers have played a key role in the development of modern-day electronics. They are used as electrical insulation, protective coatings for printed circuit assemblies, hybrid microcircuits, and chip devices, capacitors, particle immobilizers, multilayer dielectrics for ICs and interconnect substrates, and flexible cable and wire insulation.

Recently, polymer thin films having low dielectric constants (5 or less) are being used to fabricate high-performance, high-speed digital and analog circuits generally referred to as *high-density multichip interconnect* (HDMI) or *high-density interconnect* (HDI) substrates. These interconnect substrates are the basis for one type of high-speed multichip module (MCM-D). In one widely used process, multilayer thin-film circuits are formed on a substrate base such as alumina or silicon by depositing a thin film of organic polymer, etching vias in the polymer, metallizing, photopatterning the metal, then repeating the sequence (Fig. 8.16). Film thicknesses ranging from 2 to 20 μm may be used depending on the polymer's dielectric constant and the desired capacitance and impedance properties of the circuit. The most widely used high-performance dielectrics are polyimides because they combine excellent electrical properties with high thermal stability and good mechanical behavior. Also polyimides have a history of over 30 years of experience and use in military, space, and commercial electronics and extensive technical support from manufacturers and suppliers such as DuPont. In spite of this, considerable current activity exists in evaluating other polymers that may provide the same engineering attributes as polyimides yet be less costly and require fewer process steps, resulting in faster turnaround. Polyimides and other polymer types are discussed and compared in the next section.

Polyimide Dielectric Films. Polyimides have been used for decades as high-performance dielectrics for electronic applications. Single- and multilayer laminates for printed circuit boards and flexible cables for interconnections, wire insulation, conformal protective coatings, particle getters, and thin IC dielectrics are just some of its many uses.

Three basic types of polyimides differing in properties and processing conditions are available (see Chap. 1). The type most commonly used cures by a condensation mechanism; that is, the applied resin, an amide-acid polymer, eliminates water on heating, forming the cyclized imide structure along the polymer chain—a process called *imidization.* One widely used condensation-type polyimide is referred to by chemists as *PMDA-ODA* because of its synthesis from pyromellitic dianhydride (PMDA) and oxydianiline (ODA).[38] Polyamide-acid resins must be heated as high as 425°C to effect complete cure. If only partly imidized (cured) water can continue to be evolved, and this will affect the reliability of the circuit by loss of metallization adhesion, blistering, and degradation of electrical properties. Three-stage step curing culminating in the 425°C cure is effective in completing the cure. Incompletely cured polyimides may also react with copper metallization, causing corrosion and diffusion. Copper has been shown to migrate even in fully cured polyimides, hence the need for a barrier layer of titanium-tungsten, nickel, or chromium to separate the copper from the polyimide. The barrier layer also serves as an adhesion layer.

A widely used condensation type polyimide is DuPont's 2525 whose structure is a variation of the PMDA-ODA molecule (Fig. 8.31). DuPont has also synthesized and formulated a polyimide having a very compact molecular structure (PI-2611D) which is popular because of its low coefficient of thermal expansion, approximately 3 to 5 ppm/°C, thus closely matching that for silicon and obviating thermal mechanical stresses in large silicon wafer substrates and ICs (Fig. 8.32).

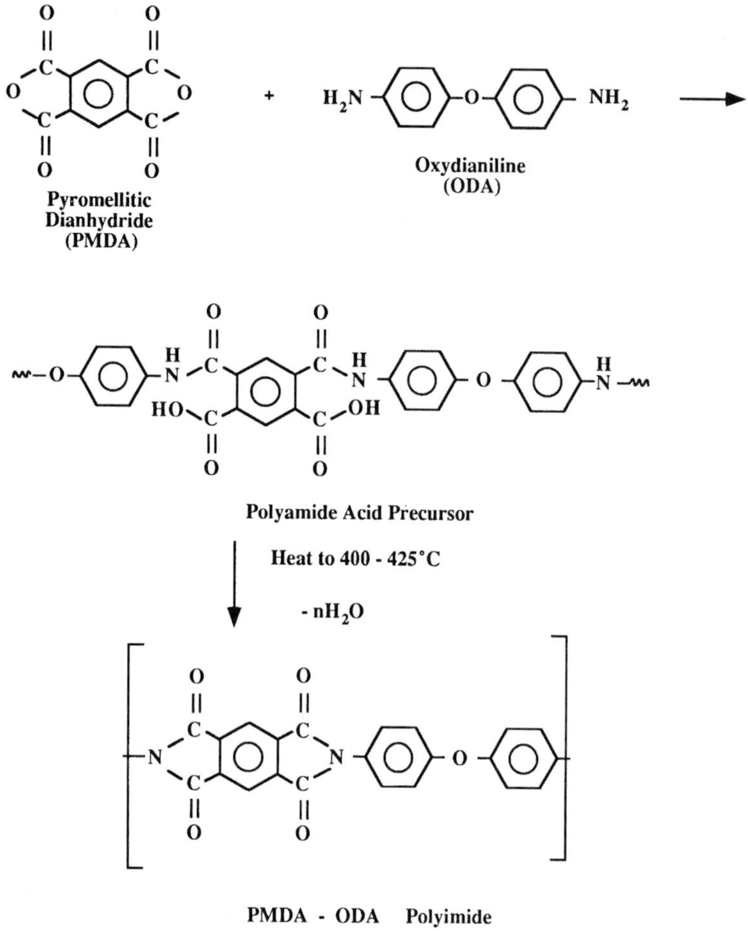

Pyromellitic Dianhydride (PMDA)

Oxydianiline (ODA)

Polyamide Acid Precursor

Heat to 400 - 425°C

- nH_2O

PMDA - ODA Polyimide

FIGURE 8.31 Reactions and molecular structures for condensation-type polyimides.

An important feature of condensation-cured polyimides is the ease of forming vias (used for *z*-direction interconnections) either by dry plasma etching the cured polyimide or by wet etching the amide-acid precursor, then completing the cure. This process is possible because the amide-acid structure is readily soluble in alkaline solutions such as ammonium hydroxide, potassium hydroxide, or tetramethyl ammonium hydroxide. (Fig. 8.33). The process consists of depositing photoresist over the prepolymer, exposing and patterning the photoresist to form the via pattern, then etching the exposed polymer. In fact, if a positive photoresist is used, the same aqueous alkaline solution used to develop the resist simultaneously etches the vias. After removing the photoresist, the prepolymer is imidized at 400°C.[39,40] Properties for several condensation-cured polyimides are given in Table 8.22.

A second polyimide type cures by an addition mechanism. *Oligomers* (low-molecular-weight prepolymers) in which the imidization reaction has already been complet-

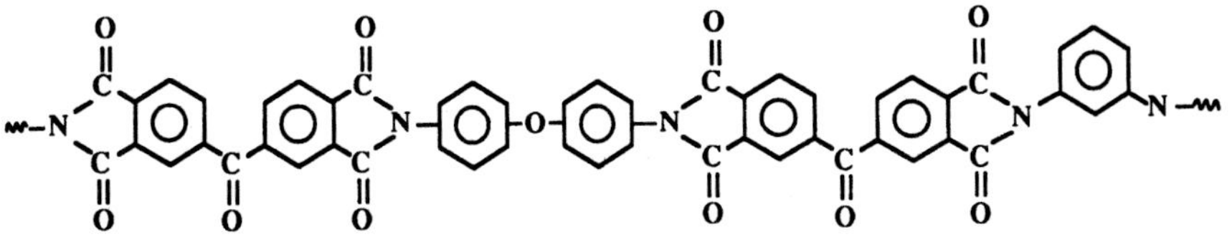

Benzophenone Tetracarboxylic Dianhydride / Oxydianiline / m-phenylene Diamine

BTDA / ODA / MPD duPont 2525

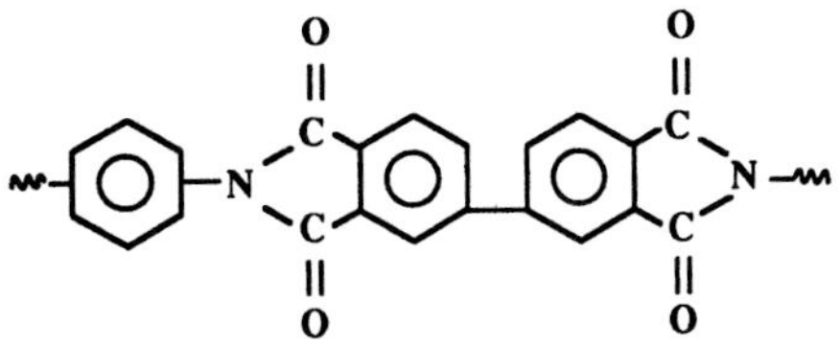

Biphenyl dianhydride p - phenylene diamine

BPDA / PPD duPont 2611D

FIGURE 8.32 Condensation curing polyimides: DuPont 2525 and 2611D.

ed are used. These oligomers contain acetylenic groups (-C ≡ CH), which, on heating, join together "head to tail" to form long-chain, high-molecular-weight polymers without eliminating water or any other compound (Fig. 8.34). Acetylene terminated polyimides were first synthesized by investigators at Hughes Aircraft[41,42] and later licensed and marketed by National Starch and Chemical Corp. under the tradename Thermid. It is reported that these polyimides, unlike the condensation-cured types, are compatible with copper without the need for a barrier metal. Properties of addition-cured polyimides may be found in Table 8.23.

Photodefinable Polyimides. Photodefinable polyimides behave like photoresists in that they can be polymerized (hardened) on exposure to ultraviolet light through a mask. Their use provides both cost and quick turnaround advantages over the conventional polyimides by eliminating approximately 75 percent of the process steps required to form vias (Table 8.24, Fig. 8.35).

The first photosensitive polyimides were developed at Siemens in the late 1970s,[43] but they have only recently been modified and reformulated to meet the requirements for inner-layer dielectrics for multichip modules.[44,45] The major types of photodefinable polyimides are *negative acting,* that is, they crosslink and harden on exposure to UV light. In these compositions the basic polyimide molecular structure has been altered to incorporate photoreactive unsaturated ester groups (acrylates) and sensitizers. Once the coating has been imaged, the photoactive ester groups are burned out during the final curing (imidization) of the polyimide. This results in a high shrinkage (as high as 60 percent being reported), resulting in stresses, poor via geometries, and degradation of mechanical properties. A second photocuring mechanism involves

Amide-Acid Precursor + $(CH_3)_4\overset{+}{N}\,\overset{-}{O}H$ (Tetramethyl Ammonium Hydroxide) → Aqueous Soluble Salt

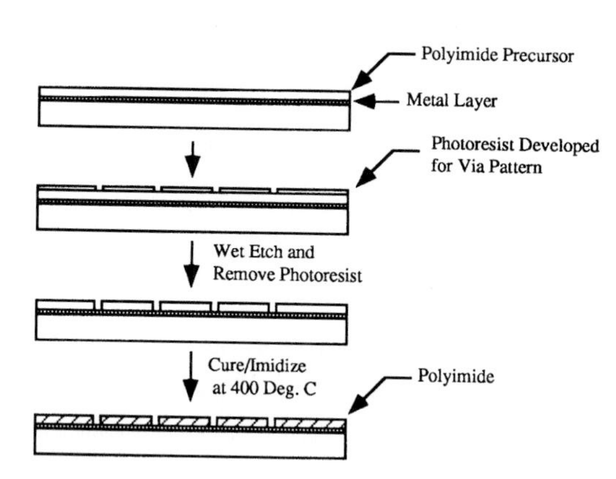

FIGURE 8.33 Chemistry and mechanism for wet etching vias in polyimide.

TABLE 8.22 Comparison of Several Polyimide Dielectrics

Property	PI-2611d	WE-1111	PI-2525
Solids, %	12.5–14.5	19.0–21.0	24.0–26.0
Viscosity, P	110–135	90–110	50–70
Tensile modulus, GPa	7.0	4.4	2.9
Tensile strength, MPa	580	300	150
Break elongation, %	60	55	25
CTE, ppm/°C	5.0	19	50
TGA, °C for 5% loss	610	580	550
Tg, °C	350	385	>310
K, 1 MHz, 0% RH	2.9	2.8	3.5
Water uptake, %, 50% RH	0.8	0.9	3.0
Self-priming	No	Yes	No
Wet etch process latitude	None	Broad	Narrow

Source: Courtesy DuPont Electronics. (The information given herein is based on data believed to be reliable, but the authors, manufacturers of products, and publisher make no warranties express or implied as to its accuracy and assume no liability arising out of its use by others. This publication is not to be taken as a license to operate under, or recommendation to infringe any patents.)

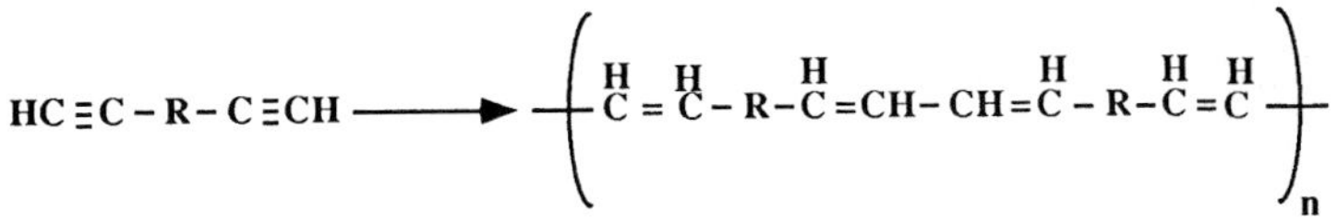

$$HC \equiv C-R-C \equiv CH \longrightarrow \left(\overset{H}{C} = \overset{H}{C} - R - \overset{H}{C} = CH - CH = \overset{H}{C} - R - \overset{H}{C} = \overset{H}{C} \right)_n$$

GENERIC ADDITION CURE MECHANISM

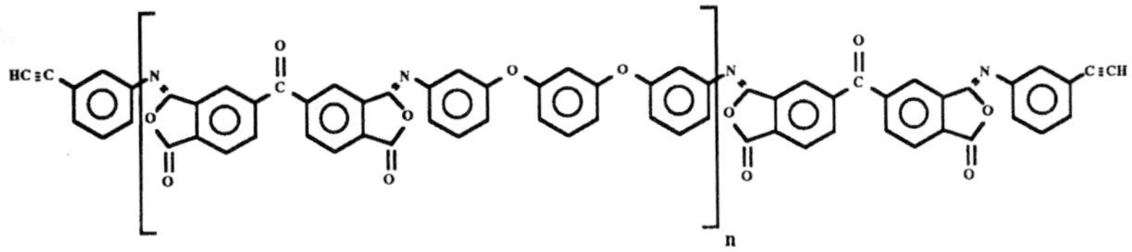

STRUCTURE OF CONVENTIONAL THERMID POLYIMIDE

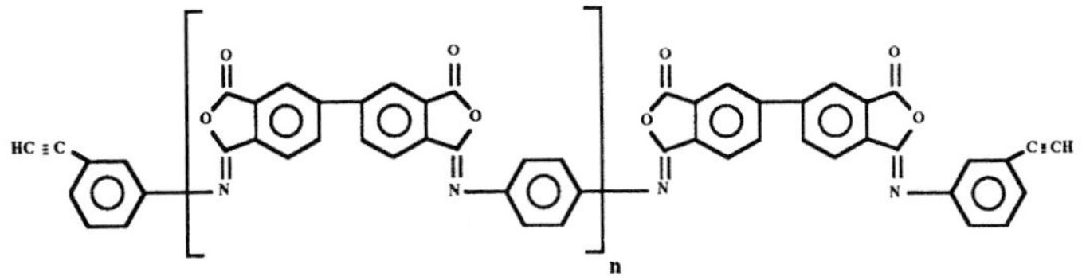

DEVELOPMENTAL LOW-STRESS THERMID 144B

FIGURE 8.34 Reactions and molecular structures for addition-type polyimides.

ionic salt formations which are also photoactive. These coatings have reduced shrinkage and have much improved mechanical properties, though at the expense of poorer resolution for thicker layers (20 μm). A further drawback to all negative-acting polyimides is the need to use organic solvents to develop the via patterns. In spite of these limitations, many firms (Boeing, NTT, NEC, Toshiba, and Mitsubishi) are successfully using photosensitive polyimides for their high-density multichip interconnect substrates.[46,47]

Positive-acting polyimides are also being developed by the major suppliers. When available, these can be patterned using environmentally friendly aqueous solutions instead of volatile organic solvents. A comparison of the properties of a nonphotosensitive polyimide (DuPont PI2611) with a photodefinable one (DuPont PI2741) is given in Table 8.25.

TABLE 8.23 Properties of Addition Cured Polyimides

Typical properties: THERMID EL-5000 series	Viscosity/ % solids	Filtration, μm	Suggested uses/properties
EL-5010	1500 cps/35% @ 25°C	0.5	General-purpose formulation. Suggested for passivation coating and as an interlayer dielectric; K of 3.2 @ 10 KHz. Excellent planarization. Will adhere direct to microcircuitry oxides and nitrides; also excellent intercoat adhesion.
			Recommended for spin coating; at as-received viscosity, will deposit 15 μm and can be varied to 1 μm with dilution.
EL-5501	1000 cps/53% @ 25°C	0.5	Fluorinated polyimide formulation. Suggested for passivation coating and as an interlayer dielectric. Forms a rigid, low interfacial stress film having a high Tg and modulus with superior solvent resistance. K of 2.8 @ 10 KHz and low moisture absorption. Excellent adhesion and planarization.
EL-5512	1100 cps/47% @ 25°C	0.5	Fluorinated polyimide formulation. Suggested for same applications as EL-5501. Forms solvent-resistant film with good tensile properties. K of 2.8 @ 10 KHz and low moisture absorption. Excellent planarization and adhesion.
			Both EL-5501 and EL-5512 are recommended for spin coating; at as-received viscosity, will deposit 15 μm and can be varied to 1 μm with dilution.
EL-5330	980 cps/24% @ 25°C	0.2	Suggested as alpha particle barrier and for heavy deposition protective applications. K of 3.2 @ 10 KHz. Excellent adhesion.
			Recommended for syringe application on DRAM IC chips.
Suggested cure schedule* for all THERMID EL formulations (in nitrogen or air)			Soft bake @ 180°C for 30 min. Bake @ 300°C for 60 min. Postcure 400°C for 15 min.

*Cure schedule—i.e., total time at temperature—is for the THERMID material. Heat sink effects due to the substrate may require adjustment of total schedule.

Source: Courtesy National Starch & Chemical Co.

Benzocyclobutene Dielectric Films. *Benzocyclobutene* (BCB) *resins* are a promising series of prepolymers that may be used as inner-layer dielectrics for high-speed, multichip module interconnect substrates, coatings for flat panel liquid-crystal displays, inner-layer dielectrics for gallium arsenide devices, and coatings to replace nickel alloy in magnetic disks. Polymers of Cyclotene* 3022, a divinyl siloxane bis-benzocyclobutene resin, are reported to have excellent planarization (>90 percent in a single coating), low moisture absorption (<0.2 percent), and excellent electrical properties

*Tradename of Dow Chemical Co.

TABLE 8.24 Low-Cost–Fast Turnaround Via Formation

In-contact hard mask process	Photoformed vias
1. Metallize cured dielectric. 2. Apply photoresist. 3. Expose. 4. Develop. 5. Etch via pattern in metal. 6. Remove photoresist. 7. Laser ablate or plasma etch. 8. Etch away metal.	1. Expose photosensitive dielectric. 2. Develop (form vias).
For a five-conductor-layer HDMI	
32 steps	8 steps

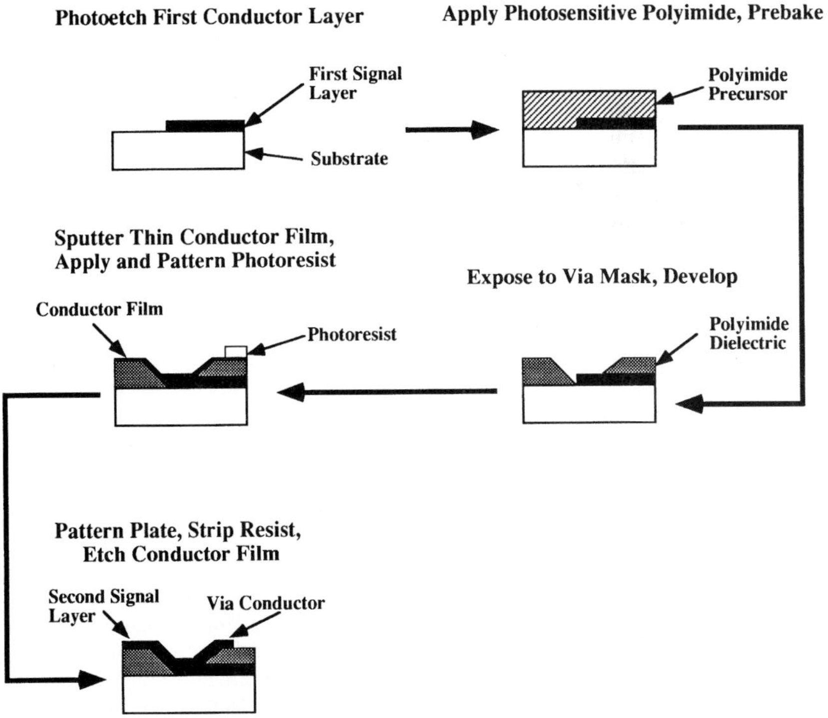

FIGURE 8.35 Process steps for photoforming vias.

TABLE 8.25 Cured Film* Properties of DuPont's Photodefinable and Nonphotodefinable Polyimides

	Nonphotodefined, PI-2611	Photodefined, PI-2741
Thermal properties		
Glass transition temperature, °C	350	365
Decomposition temperature, °C	620	620
% weight loss, 500°C in air, 2 h	1.0	1.5
Thermal expansion coeff., ppm/°C	5	10
Residual; stress, MPa	10	20
Mechanical properties		
Young's modulus, GPa	6.6	6.1
Tensile strength, MPa	600	330
Elongation to break, %	60	50
Electrical properties		
Dielectric constant, @ 1kHz:0% RH	2.9	3.0
Water absorption, w+% at 50% RH	0.8	1.2

*Cured 400°C in nitrogen.

Source: A. E. Nader et al., *ICMCM Proceedings,* 1992.

even into the GHz region (Table 8.26). For thin-film electronic applications, BCB resins are B staged (partially polymerized), producing low-molecular-weight prepolymers that are soluble in several organic solvents. Solutions of various solids content (up to 62 percent) are possible allowing the viscosity to be adjusted to control thickness and planarization. Resins are generally applied by spin coating from mesitylene solutions.

Curing temperatures for BCB are relatively low (200 to 300°C) compared with polyimides (approximately 400°C), an important feature in its use as a coating for heat-sensitive parts such as flat panel displays and assembled microcircuits. A further advantage is that BCB resins cure by an addition mechanism—thus they do not emit water or other volatiles as do most of the polyimides that cure by condensation. However, curing should be performed in a nitrogen atmosphere since there is a tendency for oxidation.[48] The mechanism and kinetics of curing are somewhat complex, involving the thermal opening of both the vinyl bonds and cyclobutane rings to form free radicals (Fig. 8.36). These free radicals then couple in several ways to form long-chain, high-molecular-weight polymers.[49] The curing mechanism is somewhat analogous to that for Parylene.†

Poly Paraxylylenes (Parylenes). *Poly paraxylylenes* coatings, known commercially as *Parylenes* (tradename of Union Carbide) are unique in being able to be vapor deposited, providing complete coverage even around and beneath intricate and closely spaced components. It is an almost ideal coating for the protection of electronic circuits because of its high purity, thorough coverage, being pinhole free even in thicknesses of a few hundred angstroms, excellent protection from moisture, and excellent electrical insulation properties. The deposition temperatures are almost at room temperature. Though most organic coatings cannot be vaporized by resistance heating because of their low decomposition temperatures, Parylene deposits by an entirely different mechanism—the poly-

†Tradename of Union Carbide.

TABLE 8.26 Typical Properties of CYCLOTENE* 3022 Advanced Electronics Resin**

Property	CYCLOTENE 3022 resin
Dielectric constant, 1 MHz	2.6
Dissipation factor, 1 MHz	0.0005
Breakdown voltage, V/cm	3×10^6
Volume resistivity, Ω-cm	1×10^{19}
Coefficient of thermal expansion, ppm °C	51
Glass transition temperature, °C	>350
Modulus, GPa	2.0
Elongation, %	8
Water absorption, % (24-h boil)	0.23
Planarization, %	>90
Stress, MPa	38
Thermal stability, °C	350
Refractive index	1.56
Poisson's ratio	0.34
Tensile strength, MPa	85

*Trademark of the Dow Chemical Company.
**Typical properties only—not to be construed as sales specifications.
Source: Dow Chemical Company.

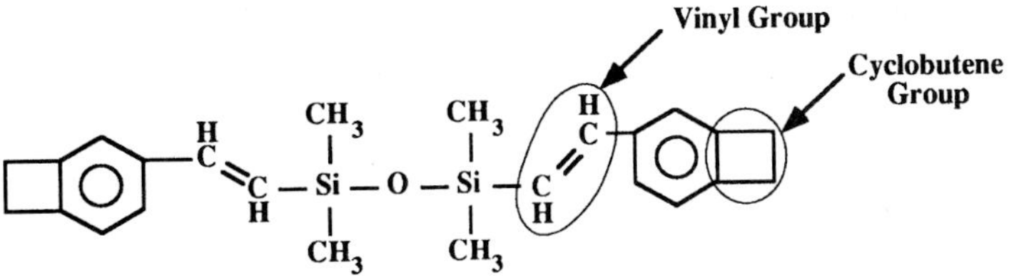

FIGURE 8.36 Divinyl siloxane of bis-benzocyclobutene monomer.

mer forms in situ on the surface by the combining of gaseous diradical monomers (Fig. 8.26). The basic polymer, Parylene N, is a pure hydrocarbon (N = no added groups), while Parylene C is the monochlorinated version, and Parylene D is the dichlorinated version. Parylene C has been the popular type because of its faster rate of deposition, though the N version has somewhat better electrical and moisture protective properties.

Parylenes were first evaluated as passivation coatings for semiconductor devices in 1968 by Licari and Lee through a NASA-funded program.[50] Metal oxide semiconductor devices coated with the Parylenes N, C, and D were tested electrically before and after coating and after extended thermal and environmental exposures. No significant changes were noted in MOS *field effect transistor* (FET) device parameters such as I_{DSS}, BV_{SDS}, V_{GST}, and BV_{DSS}.[51] Subsequently, Parylenes were found to be effective in conformally coating printed wiring boards and immobilizing loose particles in hybrid microcircuits.[6] Electrical, thermal, and physical properties of Parylene are given in Tables 8.27, 8.28, and 8.29.

TABLE 8.27 Electrical Properties of Parylenes Compared to Other Coating Types

Properties*	Method or conditions	Parylene C	Parylene N	Epoxies†	Silicones†	Urethanes†
Dielectric strength						
Short-time, V/mil						
1-mil films	ASTM D 149-64	5600	7000			
Corrected to $^1/_8$ in	ASTM D 149-64	590	700	400–500	550	450–500
Step-by-step						
1-mil films	ASTM D 149-64	4700	6000			
Corrected to $^1/_8$ in	ASTM D 149-64	550	550	380	550	450–500
Volume resistivity (23°C, 50% humidity, Ω-cm	ASTM D 257-61 (1-in² mercury electrodes)	8.8×10^{16}	1.4×10^{17}	10^{12}–10^{17}	2×10^{15}	2×10^{11}–10^{15}
Surface resistivity (23°C, 50% humidity), Ω	ASTM D 257-61 (1-in² mercury electrodes)	10^{14}	10^{13}			
Dielectric constant						
60 Hz	ASTM D 150-65T	3.15	2.65	3.5–5.0	2.75–3.05	4–7.5
10^3 Hz	(1-in² mercury	3.10	2.65	3.5–4.5		4–7.5
10^6 Hz	electrodes)	2.95	2.65	3.3–4.0	2.6–2.7	6.5–7.1
Dissipation factor						
60 Hz	ASTM D 150-65T	0.020	0.0002	0.002–0.01	0.007–0.001	0.015–0.017
10^3 Hz	(1-in² mercury	0.019	0.0002	0.002–0.02		0.05–0.06
10^6 Hz	electrodes)	0.013	0.0006	0.03–0.05	0.001–0.002	

*Properties measured on Parylene films 0.001 in thick.

†Properties and methods as reported in *Modern Plastics Encyclopedia,* issue for 1968, Vol. 45, No. 1A, McGraw-Hill, New York, 1967.

Source: By permission of Union Carbide Corporation.

TABLE 8.28 Physical and Thermal Properties of Parylenes*[6]

Property	Parylene N	Parylene C
Coefficient of friction	0.25	0.29
Abrasion index, mg loss per 100 cycles	2.6	4.4
Water absorption (2-mil films), %/24 h	0.01	0.06
Thermal conductivity, cal/(s)(cm)(°C)	3×10^{-4}	
Coefficient of thermal expansion, ppm/°C	69	35
Nitrogen gas permeability*	15	0.6
Oxygen gas permeability	55	5
Carbon dioxide gas permeability	420	14
Moisture-vapor transmission, g-mil/(100 in^2)(24 h)	15	0.6

*This and other gas permeability values are in cm^3 mils/(100 in^2)(24 h)(atm).

TABLE 8.29 Thermal-Stability Data for Parylenes[6]

	Temperature, °C	
Material	In air	In inert environment
Parylene N		
Long-term*	60	220
Short-term†	95	270
Melting temperature‡	400 (Decomposes)	—
Parylene C		
Long-term	80	230
Short-term	115	265
Melting temperature	280–300	280

*About 10 years
†About 1000 h.
‡Criterion for failure was 50 percent loss of tensile strength.

8.4 THICK FILMS

8.4.1 Applications of Thick Films in Electronics

Thick-film materials have been used widely in electronics for over 30 years, and thick-film processes are now very mature. In fact, because of their capability to produce multilayer, high-density interconnect substrates at a low cost, thick-film circuits have supplanted many single-layer, thin-film circuits except where high-precision resistors and very fine line conductors are involved. The thick-film process that is almost universally used for hybrid and microwave circuits consists of screen-printing, drying, and firing or curing conductor, resistor, and dielectric pastes (also called *inks*). Thick-film pastes may be cermets (ceramic-metal compositions) screen-printed onto ceramic or metal substrates and fired at 850 to 1000°C, or *polymer thick films* (PTFs) which contain a polymer resin which are screen-printed on either ceramic or plastic substrates and cured at temperatures of 125 to 400°C. Though PTF pastes may be applied to almost any substrate material, they are generally screen-printed onto plastic

FIGURE 8.37 Single-layer, thick-film ceramic printed circuit assemblies with solder-attached prepackaged components. (*Courtesy of Hughes Aircraft Microelectronics Division.*)

laminate boards and used for very low cost consumer applications. Numerous paste suppliers carry a repertoire of pastes formulated for a wide range of applications.

Some of the uses of thick films in electronics are: conductor lines, resistors, interlayer dielectrics, crossover dielectrics, overglazes and underglazes, capacitors, heaters, sensors, thermistors, potentiometers, via fill, terminations for resistors, inductors, and capacitors. Commercial thick-film circuits are being produced in large volumes at low cost for telecommunications, automobiles, solar cells, thermal printheads, displays, and other consumer products. High-reliability thick-film circuits are used in military, space, and medical products. Configurations range from small single-layer ceramic circuit cards onto which are solder-attached, prepackaged components (Fig. 8.37) to large high-density multilayer interconnect substrates containing bare chip devices (hybrid circuits) (Fig. 8.38) and more recently to very high density, fine-line multichip modules (Fig. 8.39).

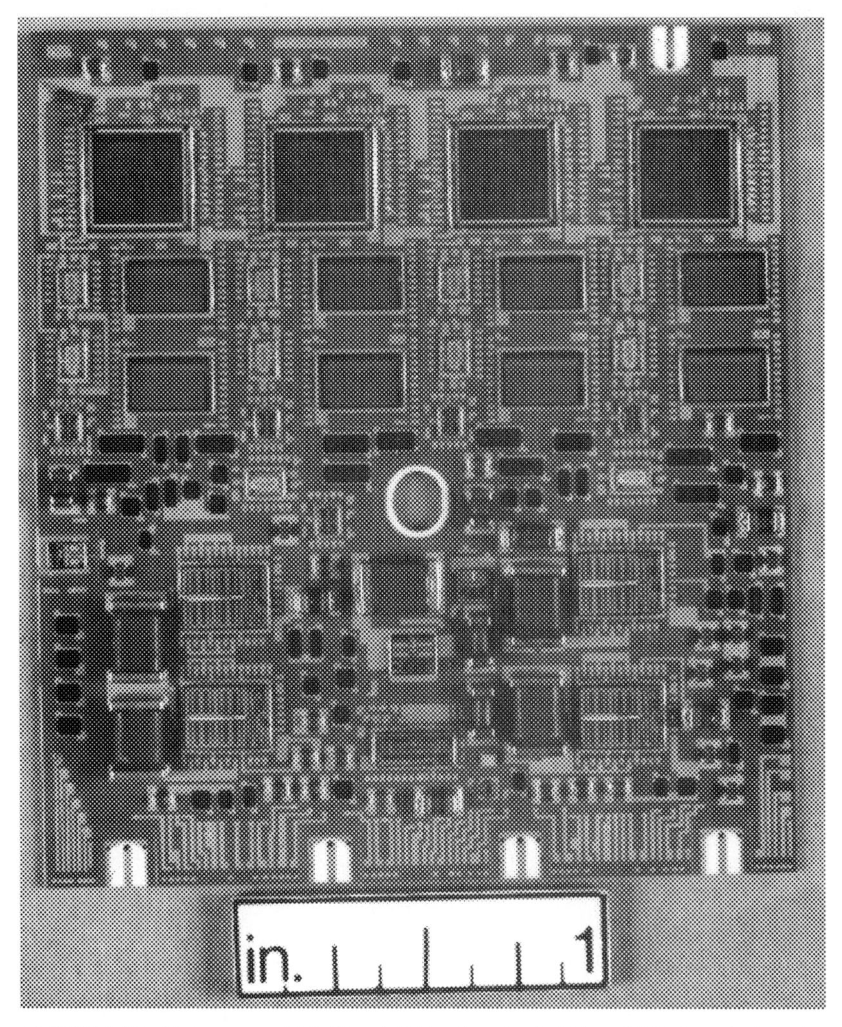

FIGURE 8.38 Sixteen-layer thick-film digital/analog circuit. Drives 4 RF channels in active array T/R module. (*Courtesy of Hughes Aircraft Microelectronics Division.*)

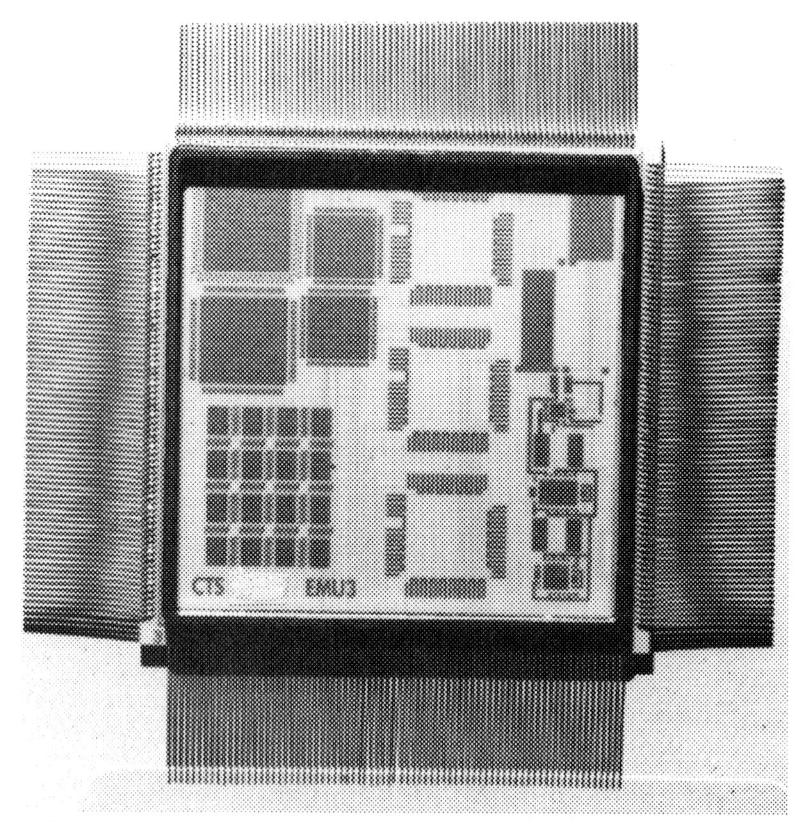

FIGURE 8.39 High-density multichip module using fine-line thick-film processes. (*Courtesy of CTS Corp.*)

8.4.2 Thick-Film Processing Methods

Sequential Screen-Printing and Firing. Almost all paste manufacturers have standardized the conditions for processing thick-film pastes. The almost universal method consists of:

1. Screen-printing the pastes (conductors, resistors, dielectrics) through a 200- to 325-mesh stainless-steel screen onto a prefired ceramic (generally alumina) substrate (Fig. 8.40).[52]
2. Allowing the print to level 5 to 10 min at room temperature.
3. Drying 10 to 20 min at 125 to 150°C to remove organic solvents.
4. Firing in a conveyorized furnace to fuse particles together and effect adhesion to the substrate (Fig. 8.41). The firing cycle is generally 60 min with a 10-min dwell at a peak temperature of 850 to 950°C.

Screen-printing is an art as old as the ancient Egyptians, but, of course, the equipment and materials have been developed and advanced to the current state where high-

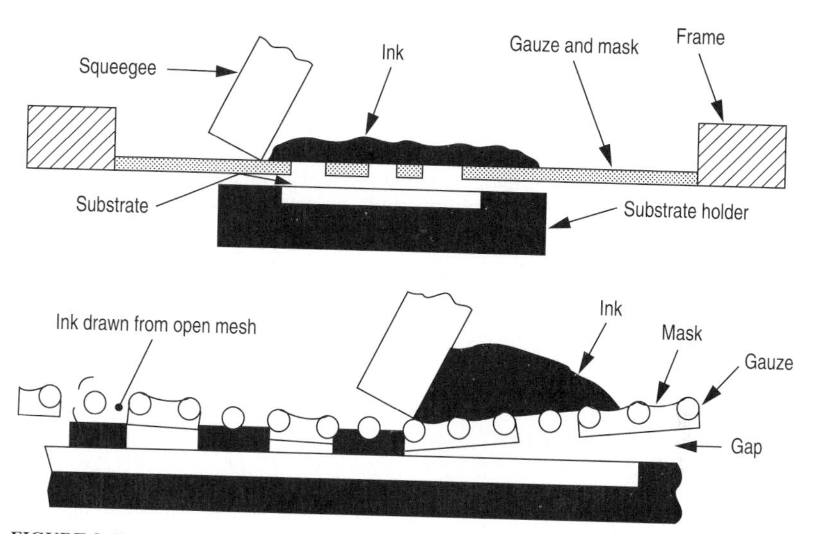

FIGURE 8.40 Thick-film screen-printing.

FIGURE 8.41 Thick-film firing furnaces.

ly reproducible and reliable microelectronic circuits are produced. Computer-controlled screen printers and firing furnaces are now available which take most of the human element and "black magic" that previously existed out of thick-film processing. The key features of thick-film processing include:

- Fast turnaround (screens and changes in design and artwork can be processed in a matter of several hours on site).
- The process is additive, thus avoiding environmental concerns related to the handling and disposing of etching solutions and the loss of precious metals.
- The equipment is mechanized and conveyorized, requiring minimum labor.
- The process is capable of multilayering to provide high-density circuits.
- Resistors covering a wide range of resistance values can be batch fabricated at low cost.

Other advantages and limitations of thick-film processing compared with thin-film processing were given in Table 8.1 and may also be found in several references.[1,53]

Multilayering. A key advantage of thick-film processing is its capability of fabricating multilayer high-density interconnect substrates and also double-sided circuit boards. The process is sequential; each conductor, dielectric, and resistor layer is screen-printed, dried, and fired separately. A flow diagram for multilayer substrate fabrication is shown in Fig. 8.42. Dielectrics are deposited as double layers to assure the absence of pinholes. Each dielectric layer may be dried and fired separately, or two layers may be screen-printed and dried separately, then cofired to eliminate one process step for each interlayer dielectric. Small apertures (typically 10 mils diameter) are formed in the dielectric during screen-printing. These vias are filled with metal paste simultaneously with the screen-printing of the next conductor layer and serve as z-direction interconnects. However, separate via-fill screen-printing steps may be used for multilayers having more than three conductor layers in order to maintain surface planarity and improve yields. Larger apertures are also left in the dielectric so that resistors can subsequently be deposited directly onto the base substrate. Resistors deposited and fired directly onto the as-fired alumina substrate are more stable and have better heat dissipation than those fired onto the top dielectric. Interconnect substrates having three to seven conductor layers and meeting most circuit requirements are easily fabricated. Beyond seven layers, yields begin to drop largely because of the increasing nonplanarity of the surfaces. Cofired ceramic tape processes (see the following section) should then be considered.

Direct Writing. Besides screen-printing, thick-film pastes may be dispensed onto a substrate by direct writing, thus avoiding the need for hard tooling (masks) and providing fast-turnaround fabrication of prototypes and of small to intermediate quantities of circuits.[54,55,56] Direct pattern writing systems employ a fine nozzle through which the paste flow is controlled from a computer database and programmed to flow in selected patterns. Thus a direct link exists between *computer-aided-design* (CAD) and hybrid circuit production. Separate cartridges are used for each paste composition (conductors, resistors of various sheet resistances, dielectrics). Flat surfaces are not required as in screen-printing since direct writing conforms and adapts to the surface topography. An engineer may also make design changes and quickly produce a new prototype, even in one day. A further advantage arises in the dispensing of resistor pastes having several sheet resistances. In screen-printing separate masks and separate screen-printing, drying, and firing steps are required for each sheet resistance. Direct

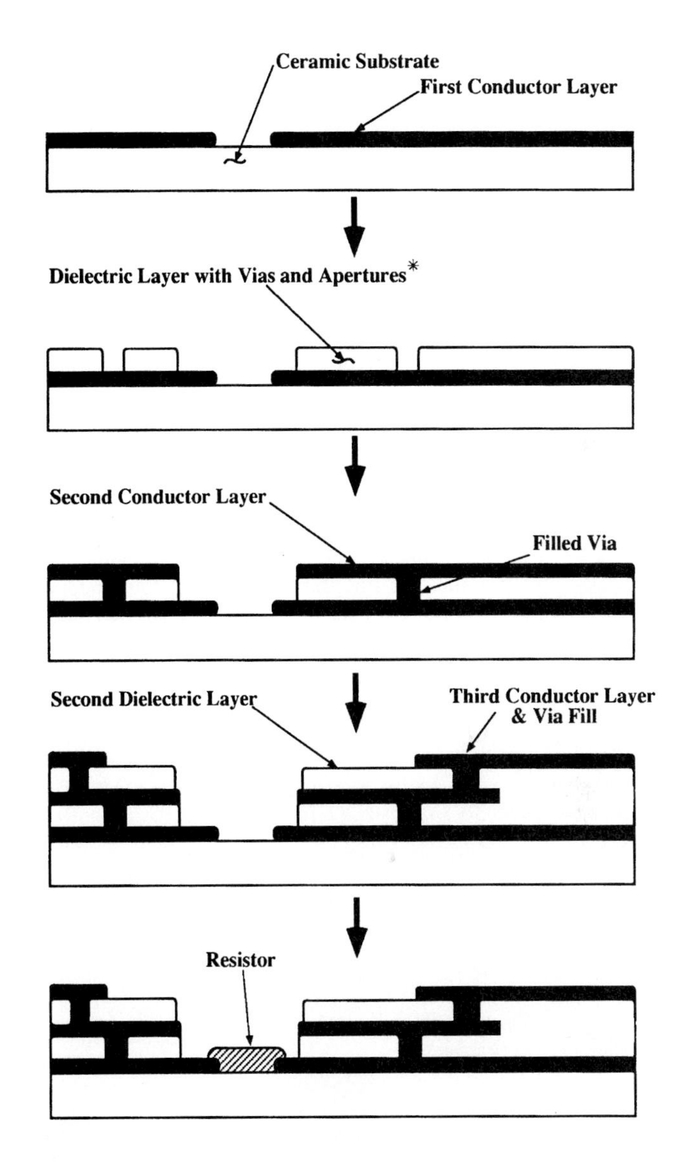

FIGURE 8.42 Thick-film multilayer fabrication steps.

pattern writing, however, can dispense all sheet resistance pastes by replacing the cartridges, thus avoiding separate drying and firing steps between cartridge changes.

Precision line widths of 4 mils for gold conductors and 6 mils for silver conductors have been reported.[57] With certain pastes even finer lines of 3 mils are achievable. Tighter variability of as-deposited resistors is also achieved, minimizing the extent of resistor trimming. Direct writing also has the benefit of being able to program the thickness as well as the width of the film on the same substrate. For high-frequency microwave and analog circuits, direct pattern writing provides precisely controlled line widths, spacings, and thicknesses and even curvilinear lines. Two limitations of direct writing are its high initial equipment cost and the fact that it is slow for high-volume production. It is especially adapted to producing, testing, and reiterating prototypes during the initial phase of a program and to producing small to moderate quantities. Direct writing should be considered complementary to screen-printing as a program evolves into high production. Standard equipment capable of printing up to 6 × 6 in panels is commercially available (Fig. 8.43 and 8.44). A general comparison of screen-printing with direct writing is given in Table 8.30.

Fine-Line, Thick-Film Processes. Though generally the pitch of conductor lines and spacings and the size of vias that can be produced by screen-printing is limited, several approaches to producing thick-film circuits in dimensions approaching those of thin films are available, and others are being explored especially for high-density multichip modules.

By using finer-mesh, stainless-steel screens (400 mesh), controlling the screen emulsion thickness, and using paste formulations based on finer, more controlled metal particles, line widths and spacings as low as 2 mils are reported.[58] Fritless gold pastes, especially, can provide fine-line definition. This, combined with their high electrical conductivity, makes fritless gold pastes suitable for microstrip transmission lines and low-loss microwave circuits.

A second approach—etchable thick films, in which even finer lines can be produced—involves a combination of thick- and thin-film processing. Specially formulated fine-particle thick-film gold pastes are first screen-printed over the entire surface,

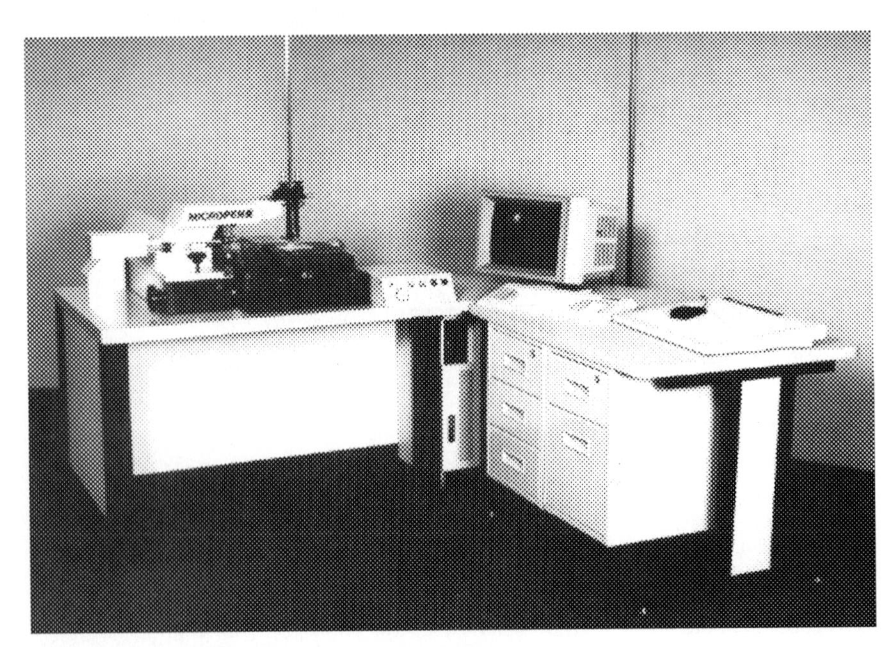

FIGURE 8.43 Direct write machine. (*Courtesy of MicroPen Inc.*)

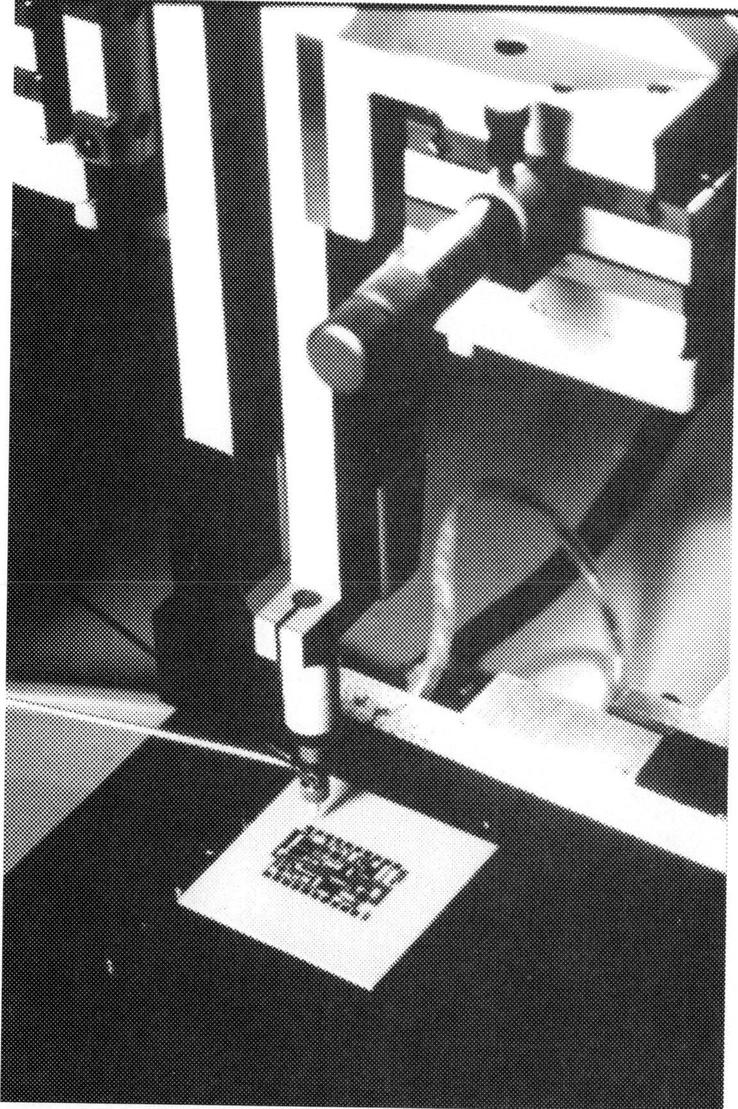

FIGURE 8.44 Close-up of direct writing of thick-film circuits. (*Courtesy of MicroPen Inc.*)

dried and fired. Photoresist is then applied, exposed to ultraviolet light through the desired pattern, and developed. The exposed thick film is then etched using an aqueous solution of potassium iodide (250 g) and iodine (60 g) in 2000 mL of water. Using this photoetch process, line widths and spacings of 15 μm (0.6 mil) are reported.[59]

Still another photoetch process avoids the need for separate photoresist application and removal steps by rendering the paste composition itself photosensitive. Photopatternable pastes tradenamed Fodel were first introduced by DuPont in 1972.[60]

TABLE 8.30 Comparison of Screen-Printing and Direct Writing of Thick Films

Screen-printing	Direct pattern writing
Low initial capital equipment investment.	High initial capital equipment investment.
No CAD system required.	CAD system required for operation.
Separate screens needed for each ink used.	No screens needed.
Multiple drying and firing processes required (one between each print).	One drying/firing process after printing.
Operator setup and cleanup required for each screen.	Change ink cartridge only setup needed; no cleanup.
Printing capability: 6-mil pitch (3-mil lines, 3-mil spacing).	Printing capability: 4-mil line width (100 μm) with gold paste; 3-mil line width (75 μm) with silver or silver palladium pastes.
Pretrimmed resistor value ±20%.	Pretrimmed resistor value ±10%.
Time-consuming prototype production turnaround.	Fast prototype turnaround; lower associated engineering costs
High production capacity.	Limited production capacity.
Limited to nonpopulated surfaces.	Can apply patterns on populated surfaces.

Those compositions contained negative-acting photosensitive materials, allowing the dried conductor layer to be exposed to a fine-line photomask, developed by dissolving away the unexposed portions and firing. Several drawbacks hindered the implementation of this early technology.

The polymerization of the exposed paste was oxygen sensitive and had to be performed in vacuum. Secondly, 80 to 90 percent of the gold paste was lost or had to be recovered after developing because the pastes were negative-acting types; and thirdly, organic solvents were needed to develop the pattern. Recently, improved versions of photopatternable pastes that obviate these problems have been introduced by DuPont. Both conductors and dielectrics based on positive-acting photolithography and developed using aqueous alkaline solutions of sodium carbonate are available. Vias of 3 mils diameter are reported for the dielectric and line widths of 25 μm and spacings of 50 μm for the compatible gold paste when fired to a thickness of 7 to 9 μm.[61]

Another option to fine-line circuitry utilizes metallo-organic inks. Gold compositions are first screen-printed over the entire substrate surface, then fired to decompose and remove the organic portion of the metallo-organic complex, leaving a thin metal film. Photoresist is then applied, exposed through a mask, and developed. The conductor lines are then pattern plated, the photoresist removed, and the original thin metal film etched away (Fig. 8.45). Metallo-organic films may also be decomposed in selected areas by exposure to a programmed laser beam. This is a computerized direct writing approach, but, because of the very thin metal, the film produced would require subsequent plating.

High-Temperature Cofired Ceramic. *High-temperature cofired ceramic* (HTCC) is a thick-film technology differing significantly from the sequential screen-print and fire process already described. HTCC has been used for many years to fabricate both multilayer ceramic interconnect substrates and integral lead high-density hermetic packages.

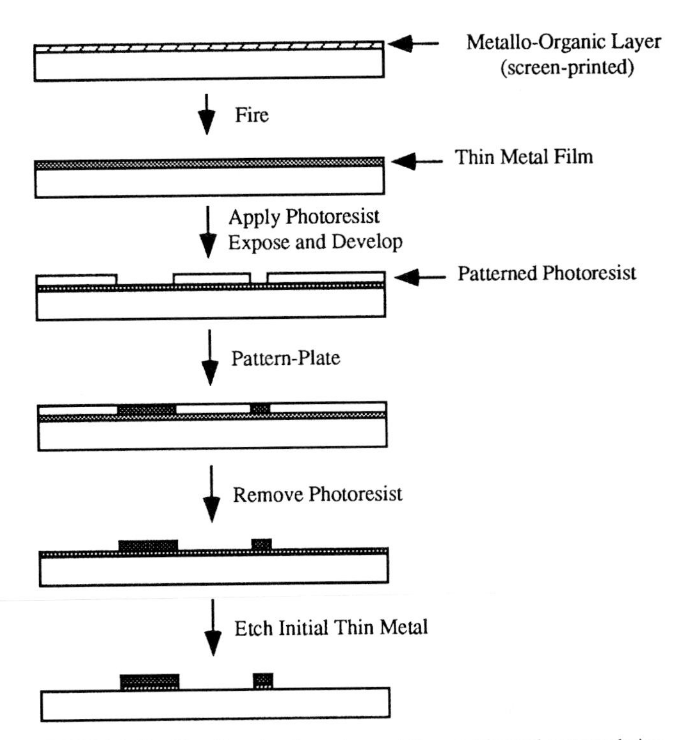

FIGURE 8.45 Fine-line circuitry using metallo-organics and pattern plating.

In this the process "green" tape (unfired) is first prepared by casting and drying a slurry of alumina powder, glass frit, organic binders, and solvents to form a rubberlike, easy-to-handle ceramic tape. The green tape is continuously cast and dried onto a Mylar film on a conveyor belt. The tape is then cut into smaller sections, vias and registration holes are punched or precision drilled, and a thick-film conductor paste consisting of a refractory metal is screen-printed and dried to form individual circuit layers. The layers are then stacked and aligned and laminated, analogous to epoxy multilayer circuit board fabrication. After cutting to shape, the laminates are prefired at 800°C to burn off the organic binders, then finally fired at approximately 1500°C to sinter the mass. Considerable shrinkage occurs during firing, but the shrinkage is predictable and can be taken into account in designing the final dimensions.

Gold, silver, or other high electrically conductive pastes cannot be used in HTCC because of the very high temperatures required for sintering. Therefore, refractory metals such as tungsten, molybdenum, or mixtures are used, but even so they must be fired in a reducing atmosphere of hydrogen or forming gas to prevent oxidation. Because refractory metallizations are poor in electrical conductivity, wire bondability, and solderability, the top layer is plated with nickel and gold. Figure 8.46 depicts a general HTCC process flow.

The HTCC process is applicable and cost effective for high-volume production of ceramic packages and ceramic interconnect substrates after the design is frozen and the relatively high cost of hard tooling can be amortized.

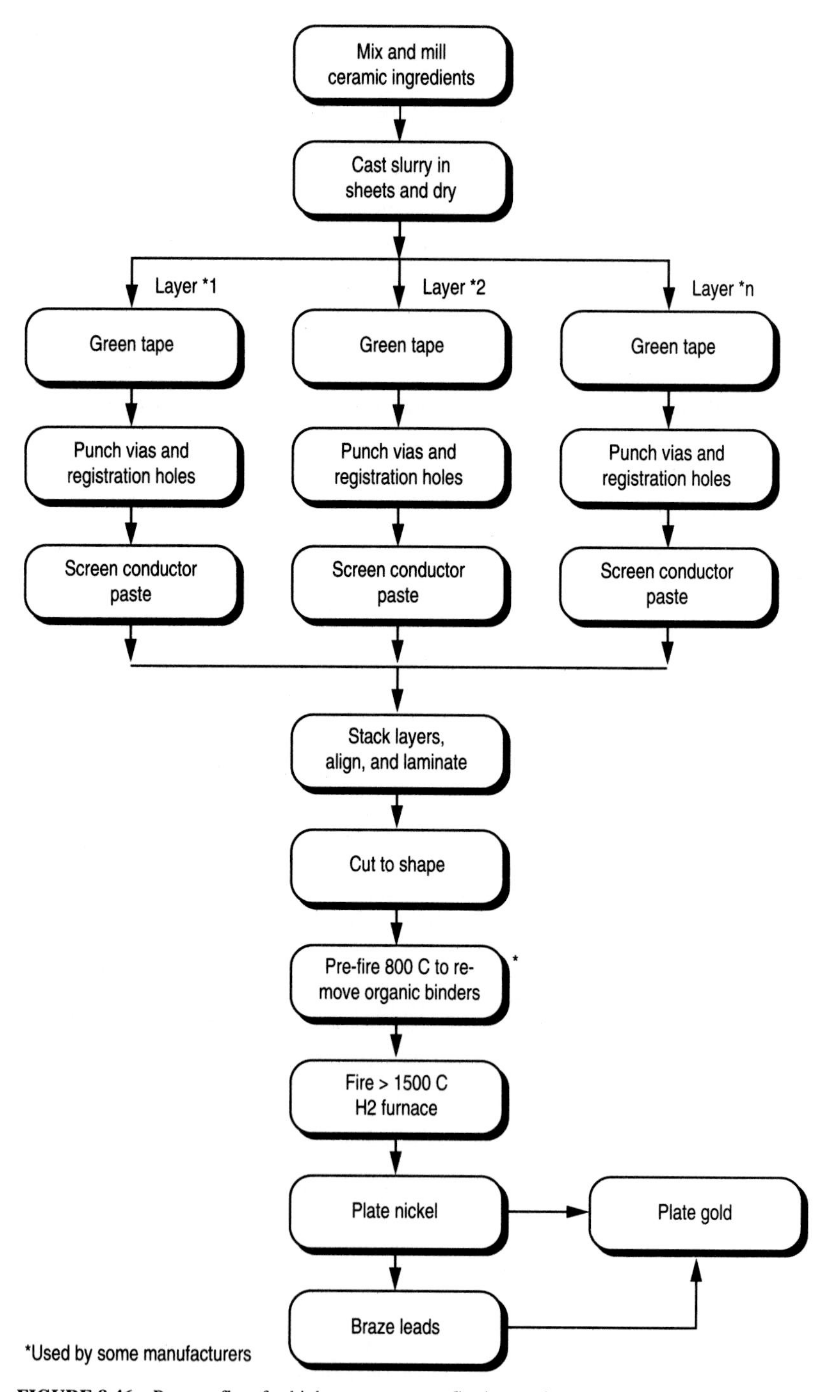

FIGURE 8.46 Process flow for high-temperature cofired ceramic.

TABLE 8.31 Characteristics of Conventional Thick Film and LTCC

Multilayer thick film	LTCC
Requires base to support layers.	Monolithic structure from laminated, individually processed layers.
Requires serial process.	Parallel process.
5 print and fires per layer, 2 dielectrics, 2 via fills, 1 conductor.	Single print and dry per layer.
75 print and fires for 15 layers.	15 print and dry for 15 layers—one firing.
Defects covered over by next layer. Requires 5 inspections.	Each complete layer inspected independently.
Buried defect irreparable.	Defective layer can be replaced before lamination.
Multiple firing of layers results in warped substrate, increased resistor variability.	Single firing of monolithic structure results in flat package.
Impedance control of buried transmission lines difficult due to variable dielectric thickness.	Constant layer thickness ensures good impedance control.
Cavities for devices very inefficient due to 5-mil setback per layer.	Vertical wall cavities easily achieved.
Thick-film dielectric loss.	LTCC dielectric lower loss.
Cost increases substantially with increased layer count.	Cost increases linearly with increased layer count.

Low-Temperature Cofired Ceramic. *Low-temperature cofired ceramic* (LTCC), a relatively new thick-film process technology, was reported in 1982[62] and commercialized by DuPont in 1985.[63] LTCC has many engineering and manufacturing advantages over both sequential thick-film processing (Table 8.31) and HTCC (Table 8.32). Like HTCC, the process starts with a green unfired ceramic tape into which vias, registration holes, and apertures are formed by drilling, punching, or laser ablation (Fig. 8.47). However, the LTCC tape differs significantly in composition, properties, and processing from that used in HTCC. LTCC tape is fired at much lower temperatures (850 to 950°C) than HTCC (1500°C) and can be fired in conventional air furnaces, already available and used by many firms in processing conventional thick films. Thus the processing of cofired multilayer substrates, integral-lead packages, and substrate-package combinations can be performed by the microcircuit manufacturers, resulting in faster turnaround, lower tooling costs, and greater versatility in designs. The ceramic tape is commercially available from several manufacturers—DuPont, Electro-Science Laboratories, EMCA-REMEX, and Ferro, in particular. It is sold in sheets or rolls supported on a polyester Mylar film.

From a performance standpoint, LTCC offers significant advantages. Unlike HTCC, where only refractory metal pastes can be used, in LTCC, high electrical conductivity pastes based on gold, silver, and silver alloys are used. Top-layer, inner-layer, via-fill conductor pastes compatible with the cofired tape as well as resistor pastes are commercially available from the same suppliers as the tape. LTCC tapes have lower dielectric constants than conventional dielectric pastes and HTCC, some as low as 4.8, and the thicknesses of the inner layers can be controlled and varied to control the imped-

TABLE 8.32 Comparison of Cofired Ceramic Technologies

	High fire	Low fire
Electrical characteristics:		
Conductor resistance	Tungsten, 15 mΩ/square	Gold, 5 mΩ/square Silver, 3 mΩ/square
Dielectric constant	9.7	4.0–8.0 (typical)
Thick-film resistors	No	Yes
Dissipation factor, %	<0.03	0.03–0.20
Dielectric strength	550 V/mil	>1000 V/mil
Volume resistivity	$>10^{14}\Omega$-cm	$>10^{14}$ Ω-cm
Conductor width	5 mils and less	5 mils and less
Via size	4 mils diameter	4 mils diameter
Thermal/mechanical characteristics:		
Flexural strength	60 kpsi	20–30 kpsi
Thermal conductivity, W/m K	~20	~3–5
Bulk density, g/cm^3	3.7	2.9
Coefficient of thermal expansion	6.5 ppm/°C	5–8 ppm/°C

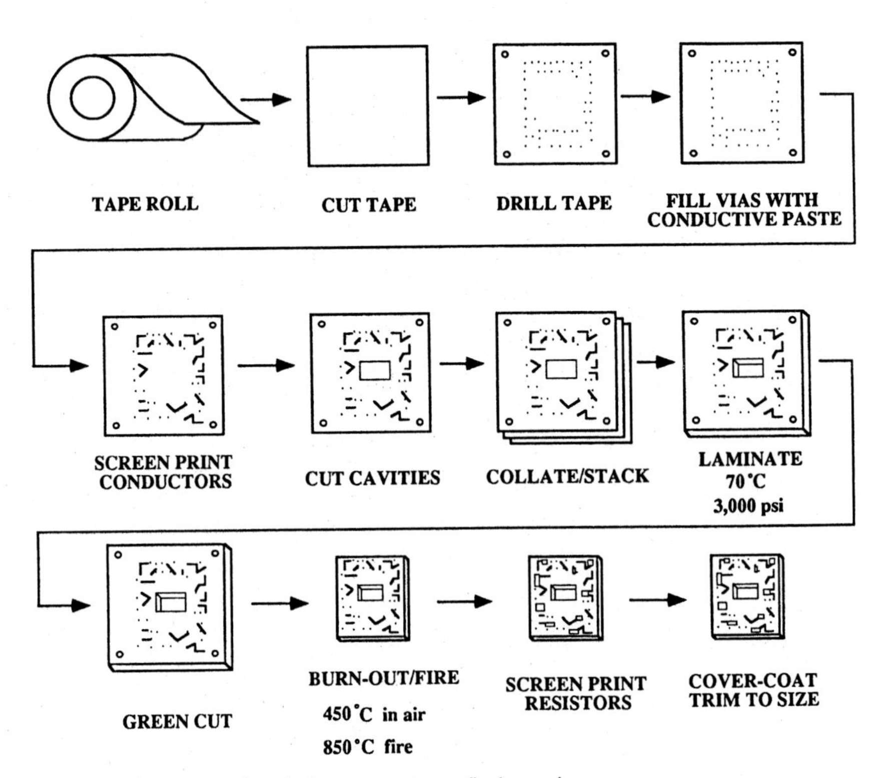

FIGURE 8.47 Process flow for low-temperature cofired ceramic.

ance of the interconnect structure. Cofired tape also provides greater isolation between conductor layers than sequentially screen-printed dielectric pastes. Dielectric strengths (breakdown voltage) are high (>1000 V/mil), and reliability is assured because of the controlled thickness of the tape and absence of pinholes. This electrical integrity is essential for high-current, high-voltage circuits. For these reasons there is increasing interest in LTCC to design and produce high-performance, high-speed, and high-density multichip modules (Table 8.33).

The design versatility of LTCC is clearly shown in Fig. 8.48. First, integral lead, hermetic cavity package/substrate combinations can be produced (Fig. 8.49). This significantly reduces volume, weight, and cost since a separate metal or ceramic package is not required. Secondly, cavities can be formed in the ceramic simultaneously with

TABLE 8.33 DuPont LTCC Tape Comparison

Tape	851	951	845
Available thickness, mil	4.5, 6.5, 10	4.5, 6.5, 10	5.0
Shrinkage, *X, Y, Z*	12 ± 0.2% 17 ± 0.5%	13% 15%	15 ± 0.2% 15 ± 0.5%
Tensile strength, MPa	1.4	1.7	
Young's modulus, MPa	123	152	103
Flexural strength, MPa	250	320	240
Density, g/cm^3	2.89	3.1	2.4
CTE, ppm/°C	7.0	5.7	4.5
Thermal conductivity, W/m K	2.4	3.0	2.0
Peak firing temperature, °C	850	875	920
Breakdown voltage, V/mil	>1000	>1000	>500
Dielectric constant, @ 1 MHz	7.3	7.9	4.8
Dissipation factor, % @ 1 MHz	0.3	0.15	0.3

Source: Courtesy DuPont.

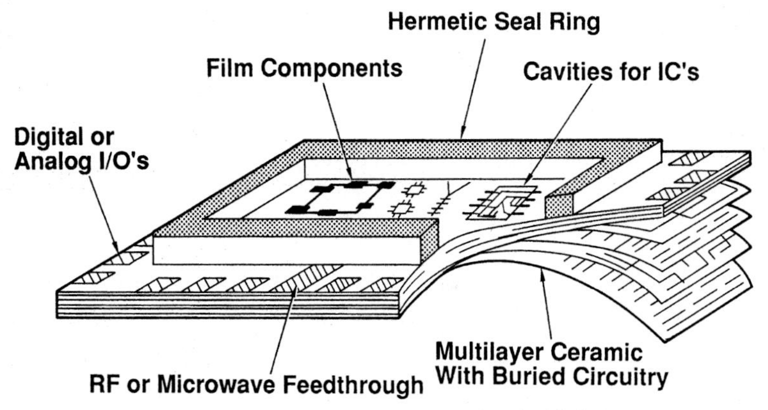

FIGURE 8.48 Low-temperature cofired ceramic integral package/substrate structure.

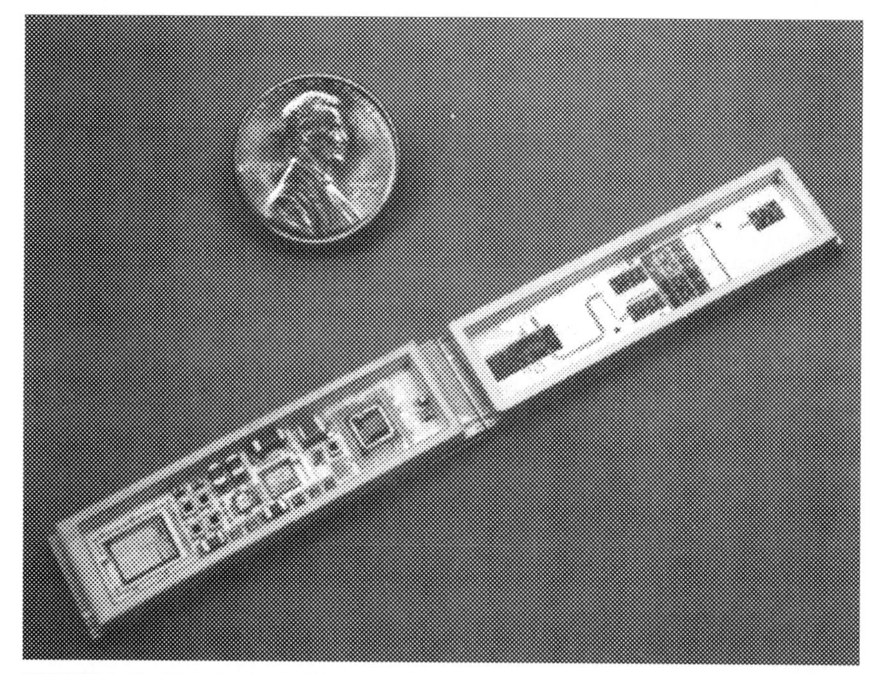

FIGURE 8.49 High-density LTCC package/substrate (transmit/receive module for active phased array radar). (*Courtesy of Hughes Aircraft Microelectronics Division.*)

its fabrication so that ICs or other components can be inserted and recessed, allowing for a low-height package. Metal heat sinks can also be bonded to the bottom of the package and the "hot" devices mounted through the apertures directly to the heat sink. Furthermore, thermal plugs and thermal vias can also be produced in the ceramic to draw heat away from devices. Finally, passive components can be integrated into the cofired structure—not possible with HTCC. Compatible resistor and capacitor pastes are available and can be embedded within the cofired structure or post-screen-printed on the top layer. LTCC tapes having very high dielectric constants (high k) can also be cofired with low-k tapes. Though, in view of the above, LTCC may seem an ideal packaging material, it has several limitations that a design engineer must consider:

1. After firing, the tape shrinks. The amount of shrinkage can be accurately measured, controlled, and compensated for in the design of the artwork and masks.
2. The high glass content (over 50 percent) results in a very low thermal conductivity (2 to 3 W/m K), but this can be obviated by constructing thermal vias or plugs through the tape or, if flip-chip bonded devices are used, by removing the heat from the backs of the devices.
3. The high glass content has also been a concern in embrittling the structure; however, there are several reports on the reliability of LTCC under high vibration and mechanical shock conditions.[22,64]
4. LTCC, like other thick-film processes requiring screen-printing and where shrinkage after firing occurs, is limited in the widths and spacings of conductor lines and diameter of vias that can be produced.

TABLE 8.34 Comparison of PTF with Cermet Thick Films and with PWBs

Cermet thick films	Polymer thick films	Printed wiring boards
Screen-print and fire conductor pastes (additive process).	Screen-print and cure conductor pastes (additive process).	Photoetch conductors from copper-clad plastic laminates (subtractive process).
Screen-print and fire resistor pastes.	Screen-print and cure resistor pastes.	Solder attach or wire bond chip resistors.
Screen-print and fire dielectric.	Screen-print and cure dielectric insulation.	Laminate prepregs.
Form vias during dielectric processing and fill with metal paste.	Form vias during dielectric processing, then fill with metal paste.	Drill through-holes and electroplate.
Substrates limited to ceramics or metals; must withstand firing temperatures.	Wide variety of substrates can be used, but generally plastics or ceramics.	Substrates limited to reinforced plastic laminates.

Polymer Thick Films. *Polymer thick-film* (PTF) pastes are screen-printed and dried similar to cermet thick films, but they differ in the permanent binder used in the paste formulation. PTF pastes contain a polymer resin binder instead of a glass frit or inorganic binder as in cermets. Polymer resins being used include epoxies, polyimides, phenolics, acrylics, and modified silicones. Because of the plastic binder, the screen-printed patterns are cured instead of fired or sintered. Two key advantages of PTFs are therefore their much lower processing temperatures and lower cost. Processing temperatures range from 100 to 400°C depending on the cure temperature of the polymer resin used. Generally, curing at 125 to 175°C for 1 to 2 h in an air convection oven is adequate. However, as with polymer adhesives, curing times may be reduced by increasing the temperature or by infrared heating. Besides screen-printing, PTF pastes may be applied by direct writing, and in both cases the fabrication of multilayer interconnect substrates is possible. The PTF process may be considered an extension of the printed circuit board process and a technology that is intermediate between PWBs and cermet thick films shown by the comparison given in Table 8.34.[65,66]

Conductor, resistor, and dielectric polymer thick-film pastes are commercially available from most cermet thick-film paste suppliers. Conductor pastes are generally based on silver or silver alloys and screen-printed through a 200-mesh stainless-steel screen. Conductor lines and spacings of 7 mils or less are reported.[67] PTF dielectrics may be purchased in a wide range of viscosities and colors. Dielectric constants are fairly low (4 to 6 at 1 KHz), and, when screen-printed through a 200-mesh screen, provide vias that are typically 15 mils in diameter. Properties for some representative PTF conductors, resistors, and dielectrics, are given, respectively, in Tables 8.35, 8.36, and 8.37.

Polymer thick films are primarily used to produce high-volume, low-cost printed circuit assemblies for commercial products such as personal computers, television, radio, cellular phones, membrane switches for keyboards, musical instruments, potentiometers, and flexible circuits. There are reports of their use in military and other high-reliability electronics, but for these applications caution should be taken because of the inherent thermal and moisture limitations of the polymeric portion of the film. Resistor TCRs, tracking, and tolerances are not as well controlled nor as stable as for cermet resistors. Solderability and solder leach resistance can also be problems, though reported to be resolved by using special PTF conductor pastes or plating the conductors.

TABLE 8.35 Characteristics of Polymer Thick-Film Silver Conductor Pastes

Designation	P2422	P2607
Application	Crossovers with P7130 dielectric; termination for P3900 resistors on PC boards	Flexible conductor for polyester or polyimide films used for membrane switches, keyboard, and flexible circuits
Sheet resistance	0.035 Ω/square (mil)	≤0.023 Ω/square (mil)
Viscosity	50 kcps ± 20 kcps @ 25°C Brookfield HBT viscometer, no. 27 spindle at 10 rev/min	40 kcps ± 10 kcps @ 25°C Brookfield RVT viscometer, no. 5 spindle at 5 rev/min
Solids	70% ± 2%	72% ± 1.5%
Cure schedule	165°C/30 min	100–120°C/5–10 min
Cure atmosphere	Circulating air convection oven or belt furnace	Circulating air convection oven or belt furnace
Line definition	15 mils (stainless-steel mesh) 10 mils (polyester mesh)	≤10 mils, both stainless-steel and polyester mesh
Adhesion; pull strength	>5 lb on 100 × 100 mil pad	Crosshatch = 5B per ASTM, D 3359, method B

Source: Courtesy EMCA-REMEX.

TABLE 8.36 Characteristics of P3900 Series Polymer Thick-Film Resistor Pastes

Application: P3900 series materials are designed to replace discrete carbon composition and/or metal film resistors for PCB applications. The P3900 series requires termination to polymer silver conductor P2422, on rigid boards, and can be used in most applications where resistor tolerances are ±20 percent.

Designation	Resistivity on FR-4	TCR, +25°C to +125°C
P3911*	10 Ω	+600 ppm
P3912	100 Ω	±600 ppm
P3913	1,000 Ω	±600 ppm
P3914	10,000 Ω	±600 ppm
P3915	100,000 Ω	±700 ppm

Viscosity: 35 ± 15 kcps at 25°C, using a Brookfield HBT viscometer with 13R small-sample adaptor chamber and no. 27 spindle, at 10 rev/min.

Cure schedule: 210–220°C for 15–20 min in a circulating air convection oven or belt furnace.

Wet thickness: 37 ± 2 μm.

Cured thickness: 20 ± 2 μm.

Power rating: 5 W/in^2.

Drift after 5 s in 235°C solder: ΔR < 3%, 1,000, 10,000, 100,000 Ω
ΔR < 5%, 10 and 100 Ω

Substrates: FR-4. Resistance and TCR will vary when used on other substrates.

*P3911 contains silver in the formulation.

Source: Courtesy EMCA-REMEX.

TABLE 8.37 Characteristics of P7130 Series Polymer Thick-Film Dielectric Pastes

Application: P7130 series is an ideal material for multilayer work. It is compatible with all rigid substrates with Tg above 110°C and EMCA-REMEX P2422, P2134, and P2136 polymer conductors.

Color: Blue: P7134
Red: P7135

Viscosity: 55 ± 15 kcps at 25°C, using a Brookfield HBT viscometer with 13R small-sample adaptor chamber and no. 27 spindle, at 10 rev/min.

Voltage breakdown: Greater than 750 V/mil.

Dielectric constant: 4–7 at 1 kHz.

Printing:

Screen type	Emulsion	Maximum wire/thread dia.	Maximum mesh opening
Stainless steel	0.001 in	0.0016 in	0.0034 in

Cure schedule: 165°C for 35 min in a circulating air convection oven or belt furnace.

Wet thickness: 32 ± 4 μm.

Cured thickness: 20 ± 3 μm (two layers are recommended).

Coverage: 225 cm^2/g at 30 μm wet thickness.

Shelf life: 4 months from date of shipment when refrigerated at 5°C. Allow unopened jars to stabilize to room temperature after removal from the refrigerator to prevent condensation.

Thinner: Emflow 102.

Source: Courtesy EMCA-REMEX.

8.4.3 Thick-Film Paste Materials

Thick-film pastes, also referred to as *inks,* consist of four basic ingredients (Fig. 8.50):

1. A functional material that provides the fired thick film its electrical function—for example, a metal in a conductor paste, a metal and/or a metal oxide in a resistor composition, or an oxide or glass in a dielectric paste. The functional material is sintered during the last stage of firing, typically at temperatures of 850 to 1000°C for 10 minutes (Fig. 8.51).
2. A solvent such as terpineol. The solvent provides viscosity control during screen-printing but is easily volatilized during the initial drying phase at approximately 125°C for 10 to 20 min.
3. A temporary binder such as ethyl cellulose. The binder is an organic polymeric material that provides thixotropic properties to the paste, allowing it to flow through the screen mesh under pressure of the squeegee but to cease flowing on removal of the squeegee. Temporary binders are easily oxidized and burned out during the early stages of firing (200 to 500°C), provided that sufficient air flow is fanned through that section of the furnace.
4. A permanent binder such as glass frit. The glass frit melts at 500 to 600°C and, on cooling, binds the functional material particles together and to the substrate. The binder remains part of the fused thick-film mass after sintering. Oxides of copper and cadmium are also used in molecularly bonded conductor pastes. In polymer thick films, the permanent binder is a polymer resin such as epoxy or polyimide.

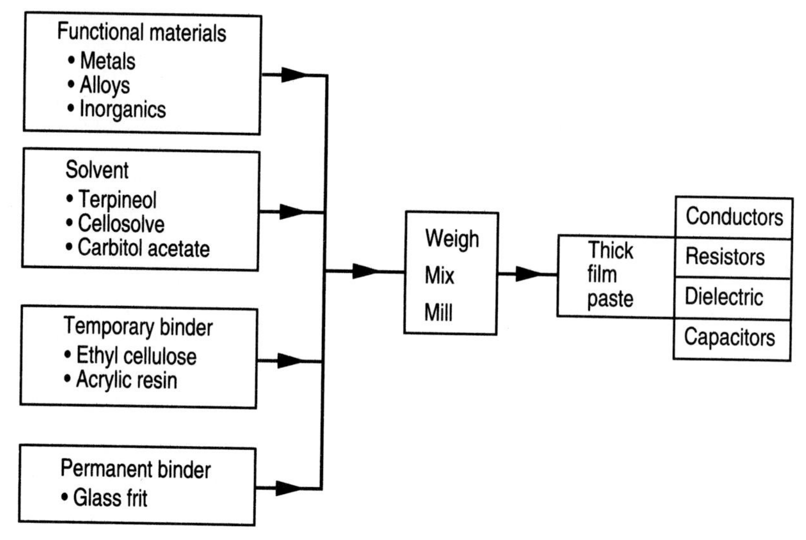

FIGURE 8.50 Thick-film paste compositions.

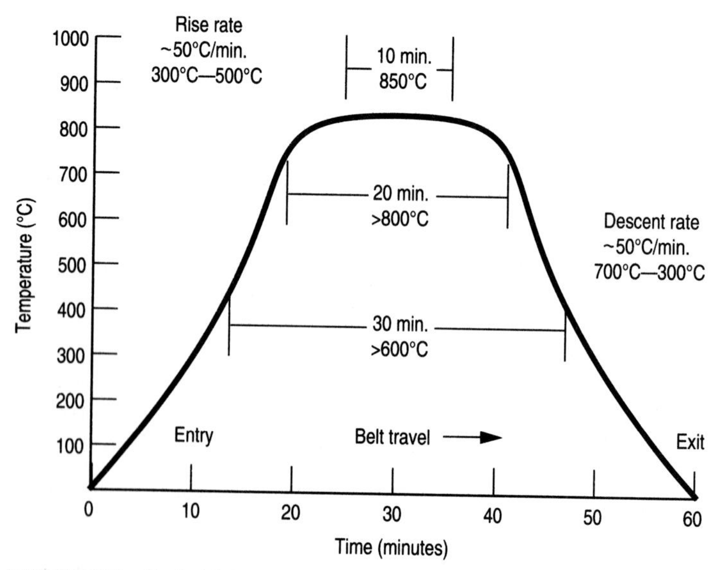

FIGURE 8.51 Typical firing temperature profile for thick films.

Paste manufacturers often add other proprietary ingredients such as wetting agents and flow control agents to improve application and performance properties.

Thicknesses of fired thick films are seldom less than 0.5 mils nor greater than 2 mils. For cermet pastes there is approximately a 50 percent reduction in thickness from the dry state to the fired stage due to burn-out and removal of the organic binders during firing.

Thick-Film Conductors. Thick-film conductor pastes are classified as *fritted, fritless,* or *mixed bonded* depending on their composition and mechanisms used for adhesion to alumina substrates. Fritted pastes contain reprecipitated glass (frit) along with the functional metal or alloy, organic binders, and solvent. The glass frit allows the composition to adhere by a physical mechanical mechanism in which the glassy phase melts during firing at the same time that some of the glass in the alumina ceramic substrate also melts, fusing the two materials. Unfortunately, the early fritted compositions could not produce very fine lines (less than 5 mils), and the glassy heterogeneous surface often obstructed wire bondability and solderability.

Fritless pastes next appeared on the market to solve these problems. These formulations contain no glass frit. They adhere by a *molecular-chemical bonding mechanism*—also called *reactive* or *chemical bonding.* To effect this bonding mechanism, small amounts of copper oxide or cadmium oxides (about 1 percent) are added to the gold paste. When fired at 930 to 1000°C, the oxides chemically react and interlock with the alumina of the substrate forming a spinel compound (copper aluminate, for example) which bridges and locks the fired metal to the substrate. Fine-line pitch, improved wire bondability, and more reliable aluminum-to-gold wire bonds have been achieved. However, to assure optimum adhesion, the firing temperature and time must be rigidly controlled because of the narrow window in which chemical bonding occurs.[68] Since reactive bonding does not depend on the glass content or surface smoothness of the substrate, excellent adhesion is attained on 99 percent alumina.

Finally, to meet most requirements, mixed bonded pastes were introduced as a compromise. In mixed bonded pastes, both mechanical and chemical adhesion mechanisms were employed by including some fritted glass and some copper oxide. These pastes can be fired at the more standard 850°C conditions and combine the desirable attributes of both the fritless and fritted pastes. Properties of some fritted and mixed bonded conductor pastes are given in Table 8.38. Properties of some fritless conductors are given in Table 8.39. For high yields, design guidelines specify 10 mils minimum for line widths and spacings; however, 5-mil dimensions are routinely being produced, and, with special pastes, screens, and printing controls, even 2- to 3-mil lines and spacings have been reported. Conductor pastes may be based on noble metals such as gold or alloys of gold with palladium or platinum, precious metals such as silver or alloys of silver with palladium or platinum, nonnoble metals such as copper or nickel, or refractory metals such as tungsten or molybdenum.

Gold and Gold Alloy Conductors. High-reliability (military, space, medical) applications still require the use of gold or gold alloy conductors because of their chemical inertness, stability, and long history of wire bond reliability. Gold alloys with Pt or Pd are useful for top-layer solderable pads since they do not readily leach as does pure gold. Because adding Pt or Pd to gold degrades gold's excellent electrical conductivity, these alloys are used only for the top layer and may even be screen-printed selectively just on the pads that will be solder coated.

Silver and Silver Alloy Conductors. For consumer products the preferred conductor materials are low-cost silver or silver alloys with palladium or platinum. Because

TABLE 8.38 Properties of Some Fritted and Mixed Bonded Gold Conductors

Typical properties	8831	8831-A	8831-B	8835	8835-A	8835-1A	8835-1B	8835-1BH
Type of metal	Au	Au	Au	Au	Alloy Au	Alloy Au	Au	Au
Bonding system	Fritted	Fritted	Fritted	Fritted	Fritted	Mixed bonded	Mixed bonded	Mixed bonded
Peak firing temperature, °C:								
On glass				550–650			550–650	
On alumina	850–1000	850–1000	850–1000	850–1000	850–1000	850–1000	850–1000	850–1000
Solderability:								
Excellent with ESL pastes nos.	SP-3601 (80Au/20Sn); SP-3501 (251n/75Pb) or SP-3502 (501n/50Pb)							
Viscosity,* kcps @ 25–26°C	325 ± 25	325 ± 25	300 ± 25	350 ± 25	350 ± 25	350 ± 25	300 ± 25	300 ± 25
Fired thickness, μm:	10–15	8–13	8–13	10–15	8–13	8–13	10–15	10–15
Resistivity, mΩ/square: for a 8–15 μm-thick, 0.25-mm-wide, line	2.5–4	2–3	1.8–2.8	2.5–4	5–7	5–7	2–3	2–3
Adhesion, 90° peel, kg/2.5 × 2.5 mm pad; 96% alumina substrate; 80Au/20Sn solder								
Initial peel strength	4–7	3–6	2–3	3–5	3–6	3–6	4–7	2–3
Aged peel strength, 48 h, @ 150°C	3–5	2–4	1–2	3–5	2–5	2–5	4–6	1–2
Bondability:								
Thermosonic, 25 μm, Au wire/125°C bonding temperature:								
Initial, g:	5–8	7–10	9–11	5–8	6–9	5–8	6–9	7–10

Ultrasonic, 25 μm Al wire:								
Initial, g:	8–10	10–12	11–12	7–9	8–11	7–11	8–10	8–12
Aged, g (24 h @ 300°C)	—	—	—	—	5–6	4–6	—	—
Fine-line definition, μm: line and space widths	100–125	100–125	100–125	100–125	100–125	100–125	75–100	75–100
Coverage, cm^2/g	70 to 90							
Thinner	ESL 401							
Compatibility:								
Resistors	Compatible with ESL 2700, 2900, 2900M, 3800, and 3900 series.							
Multilayer dielectrics	Compatible with ESL 4608, 4611, 4612, 4901, 4901-H, 4903, separately fired.							

*RVT viscometer, 10 rev/min, no. 7 special spindle with $^3/_4$-in immersion.

Source: Courtesy Electro-Science Laboratories, Inc.

TABLE 8.39 Properties of Fritless Gold Conductors

Typical properties Gold Micro-Lok Pastes	8880	8880-H	8881	8882 8883
Peak firing temperature, °C:				
Optimum temperature	980	980	850	850
Useful range	930–1030	930–1030	850–1000	850–1000
Solderability excellent with:	ESL SP-3601, 80 Au/20 Sn; or ESL SP-3501, 25 In/75 Pb; or ESL SP-3502, 50 In/50 Pb			
Substrate type *Note:* Pressed or molded thick substrates may give different results than slip-cast (tape process) substrates.	96 to 99.5% Alumina, 99.5% Beryllia			
Adhesion, 90° peel, kg per 2.5 × 2.5 mm pad, (96% Al_2O_3) 80 Au/20 Sn Solder				
Initial peel strength	5–6	5–6	5–6	5–6
Age 1000 h at 150°C	4–5	4–5	4–5	4–5
Wire bondability, thermocompression:				
Au Wire, 50 μm	30–35 g	30–35 g	30–35 g	30–35 g
Ultrasonic, 25 μm, Al wire	8–12 g	8–12 g	8–12 g	8–11 g
Al bond retention: 1000 h (150°C)	4–6 g	5–7 g	4–5 g	5–7 g
Resistivity, Ω/square for 12–15 μm film	0.002–0.004	0.0015–0.003	0.002–0.004	0.004–0.006
Thickness fired, μm	10–15	12–17	10–15	10–15
Fine-line definition (minimum line and space widths, μm)	100–125	75–100	100–125	100–125
Compatibility:				
Resistors	Compatible with 2900, 3900, and 3800 series.			
Multilayer	Compatible with 4608, 4901, 4901-H, 4903; separately fired.			
Dielectrics	Dielectrics and conductors should be separately fired.			
Thick Film Capacitor dielectrics	These golds will give about 10–20% lower capacitance values than fritted palladium silvers when used as capacitor electrodes.			

Source: Courtesy Electro-Science Laboratories, Inc.

of silver's propensity to migrate under moist, biased conditions, it is generally used only for inner (encapsulated) layers of a multilayer circuit. The outer layers are printed with alloys of silver and palladium or platinum. These noble metal additions suppress silver migration and improve solderability and solder leach resistance.[69] Silver migration is further suppressed by coating the conductors with high-purity organic coatings such as Parylene, silicones, polyimides, or with overglazes, by efficient surface cleaning to assure the absence of ionic contaminants, and by preventing the condensation of moisture.[1,70] Typical properties of silver and silver alloy conductors are given in Tables 8.40 and 8.41.

TABLE 8.40 Typical Properties[a] for Silver-Palladium-Platinum Fired Conductors, DuPont 4093

Line resolution	μm	150–200	
	mil	6–8	
Fired thickness	μm	15–17	
	mil	0.6–0.7	
Soldering:			
Initial acceptance[b]			
62Sn/36Pb/2Ag		Excellent	
63Sn/37Pb		Excellent	
Resistance to leaching[c]			
62Sn/36Pb/2Ag	cycles	15	
63Sn/37Pb	cycles	11	
Adhesion:[d]			
1. Shipping specifications			
Initial	N	≥20	
	lb	≥4.5	
Aged 48 h at 150°C	N	≥17.7	
	lb	≥4.0	
		1 firing[e]	5 firings[e]
2. Typical data			
Initial	N	28 (24–35)	26 (24–29)
	lb	6.4 (5.5–8.0)	5.9 (5.5–6.6)
Aged 100 h at 150°C	N	22 (17–27)	24 (21–29)
	lb	5.0 (4.2–6.2)	5.5 (4.8–6.6)
Resistivity	mΩ/square	42	
Resistance to silver migration:			
Failure time in water drop test[f]	min	>20	
Failure time in 85% RH/85°C[g]	h	>300	

[a]Typical conductor properties are based on laboratory tests using recommended processing procedures: printing—200-mesh stainless-steel screen; firing—60-min cycle to peak temperature of 850°C for 10 min.

[b]Excellent characterized as complete wetting with smooth solder film after 5-s dip at 220°C using mildly activated flux (Alpha 611).

[c]Cycle consists of dip in mildly activated flux (Alpha 611), 10-s dip in solder at 220°C, and washing off flux residue.

[d]90° wire peel test on 2 × 2 mm pads soldered with 62Sn/36Pb/2Ag solder at 220°C and mildly activated flux (Alpha 611). See Wire Peel Adhesion Test (Bulletin A-74672 for details).

[e]Firing cycle: 60-min total firing time with 10 min at 850°C.

[f]Space between conductor lines: 250 μm (10 mils); bias voltage: 5 V.

[g]Space between conductor lines: 250 μm (10 mils); bias voltage: 60 V.

Source: Courtesy DuPont.

Copper Conductors. Screen-printed and fired copper pastes are low-cost alternatives to gold. However, unlike gold, copper pastes cannot be fired in air since they will quickly oxidize and lose their electrical conductivity. They must be fired in an inert atmosphere, such as nitrogen, under very controlled conditions. The oxygen content of the furnace must be accurately controlled in the low parts per million range.[71] Hence the cost of continuously flowing nitrogen through the furnace and the strict monitoring of oxygen somewhat offsets the savings from the use of copper and must be taken into account in determining the net savings. An early deterrent to the use of copper in multilayer circuits was the need for dielectric and resistor pastes that could

TABLE 8.41 Properties of Silver-Palladium–Fired Conductor, DuPont 9476

Fired thickness	11–15 μm (0.4–0.6 mil)
Line resolution	150–200 μm (6–8 mil)
Resistivity	32–40 mΩ/square
Solderability:	
Initial acceptance	Complete wetting with 10/90 Sn/Pb solder at 325°C using mildly activated flux.
Resistance to leaching	No observable deterioration after thirty 10-s immersion cycles in 10/90 Sn/Pb solder at 325°C.
Adhesion:	
Initial	>17 N (3.8 lb)
Aged	>15.6 N (3.5 lb)
Compatibility:	
Resistors	Makes smooth, bubble-free overlaps with DuPont Birox* and Certi-fired* Resistor Compositions.
Dielectrics	Makes smooth, contiguous crossovers with DuPont dielectric compositions.

*Typical fired conductor properties are based on laboratory tests of 9476 using recommended printing and firing procedures. Prints were made with 200-mesh screens on 96 percent Al_2O_3 substrates.

Source: Courtesy DuPont.

also be fired in nitrogen and be compatible with the copper. The organic binder of dielectric pastes could not be completely burned out without sufficient oxygen, and thus films were porous and contaminated with carbonaceous material. These early formulations required three and even four layers of dielectric to achieve interlayer integral insulation. Subsequent improvements by DuPont resulted in compatible dielectric and resistor pastes. Copper thick-film pastes are available as fritted, fritless, and mixed bonded compositions. Properties of several fired copper conductors are given in Tables 8.42 and 8.43.

Thick-Film Resistors. Thick-film resistor pastes are commercially available in a wide range of sheet resistances in decade values from 10 Ω/square to 10 MΩ/square. Intermediate values can be obtained by blending pastes according to the manufacturer's directions. Resistors are screen-printed, dried, and fired under conditions similar to conductors, typically screen-printed through a 200- or 325-mesh stainless-steel screen, dried at 125°C for 10 to 15 min and fired at a peak temperature of 850°C for 10 to 12 min. After burning out the organic binders and sintering, the thickness of the dried film is reduced by approximately 50 percent (Table 8.44).

Almost all high-reliability resistor pastes are now based on compounds of ruthenium—for example, ruthenium oxide or bismuth ruthenate. As-fired tolerances are ±10 percent, but somewhat higher for the low and very high valued resistors (Table 8.45). After laser trimming, however, tolerances of ±1 percent are achievable.

Unlike thin-film resistors, annealing of thick-film resistors is normally not required to stabilize them. Also, though not normally needed, thick-film resistors may be protected by screen-printing a glassy overglaze and firing at 500 to 520°C or by coating with an organic polymer. TCRs are typically ±100 ppm/°C except again for the very low and high valued resistors where TCRs may be as high as ±200 ppm/°C. In all, thick-film resistors are more rugged and more thermally and chemically stable than thin-film resistors though not as precise and low in TCRs and tolerances. A comparison of some of the resistor properties of thick films with thin films is given in Table 8.46.

TABLE 8.42 Typical Properties of Copper Thick-Film Pastes

ESL designation	2310	2311	2312
Rheology	Highly viscous materials with pseudoplastic behavior		
Type	Fritted, copper	Micro-Lok®, fritless, copper	Mixed-bonded, copper
Thinner	ESL 401	ESL 401	ESL 401
Drying, in air	125°C, 10–15 min	125°C, 10–15 min	125°C, 10–15 min
Peak firing temperature in nitrogen atmosphere, °C:			
Firing range	850–1050	900–1050	900–1050
Optimum firing	900–950	925–950	925–950
Fired thickness, μm	12–18	12–18	12–18
Resistivity, milliohms per square for 15-μm thickness	1.5–2.5	1.0–1.5	1.0–2.0
Printability resolution (line width and spaces), μm	175	125	125
Solderability, in 5–10 s mildly activated rosin flux:			
62/36/2 Sn/Pb/Ag 220°C	Good	Excellent	Excellent
63/37 Sn/Pb 250°C	Good	Excellent	Excellent
10/90 Sn/Pb 330°C	Good	Excellent	Excellent
Adhesion to 96% Al_2O_3, 90° peel, Sn/Pb/Ag solder: kg per 2.5 × 2.5 mm pad	4–5	5–7	5–7
Bondability:			
Ultrasonic, Hugle model 1300 25 μm Al—1% Si wire, after copper cleaned with ESL 641 cleaner, g	5–8	5–8.5	6–9
Thermocompression, K & S Model 421: 350°C heat column temperature, 50 μm Au wire, g	20–25	25–30	25–30

Source: Courtesy Electro-Science Laboratories, Inc.

Resistor characteristics are dependent on the use of compatible conductor terminations and dielectrics. Hence, the use of a system of compatible resistor, conductor, and dielectric pastes from the same manufacturer is recommended. For example, DuPont has several series of resistor pastes (tradenamed Birox) that are formulated for specific conductor and dielectric pastes and can be selected based on cost and performance (TCRs, voltage overload, power dissipation). The Birox 1900 series (pastes range from 10 Ω/square to 1 MΩ/square) are designed for Pd-Ag conductor multilayer circuits (Table 8.47) while Birox 1400 series are designed for compatibility with gold and gold-platinum conductors and used mainly for high-performance aerospace, telecommunications, and space and medical electronics.

The assembly and packaging of hybrid microcircuits using thick-film resistors are discussed in detail in the *Hybrid Microcircuit Technology Handbook.*[1]

Thick-Film Dielectrics. Dielectric pastes are of two types: those having a low dielectric constant (K = 7 to 15), used primarily as electrical insulation especially as interlayer dielectrics and crossovers for multilayer circuits and those having a high dielectric constant (20 to 1500) used as capacitors.

TABLE 8.43 Properties of Fired Copper Conductor 9922[a]

Line Resolution: 125-μm (5-mil) line-space pattern	125-μm (5-mil) lines and spaces over Alumina and Dielectric 4575		
Fired thickness	15–18 μm (0.6–0.7 mil)		
Resistivity	1.6–1.9 mΩ/square at recommended fired thickness		
Solder acceptance	Very good, using 60 Sn/40 Pb solder at 240°C and Alpha 611 flux		
Adhesion:[b]	Initial	Aged[c]	Multiple fired[d]
On Alumina	25–29 N (5.6–6.5 lb)	22–27 N (5.0–6.0 lb)	19–22 N (4.3–5.0 lb)
Over dielectric 4575	20–25 N (4.5–5.6 lb)	16–20 N (3.6–4.5 lb)	16–20 N (3.6–4.5 lb)
Wire pull strengths:	Initial	Aged[c]	
Ultrasonic aluminum[e]	70–80 mN (7–8 g)	45–55 mN (4.5–5.5 g)	
Thermosonic gold[f]	80–90 mN (8–9 g)	65–75 mN (6.5–7.5 g)	

[a]Typical conductor properties are based on laboratory tests using recommended processing procedures.
[b]DuPont Test Bulletin A-74672.
[c]Aged 2000 h at 150°C in air.
[d]Refired 4 times at 900°C.
[e]25 μm (1 mil) Al—1% Si wire.
[f]25-μm (1-mil) Au wire; matte-finish bonding tool required.
Source: Courtesy DuPont.

TABLE 8.44 General Characteristics of ESL 3980 Series Resistors

Print thickness: Dried	25 ± 2.5 μm.
Fired	10–12 μm.
Coefficient of variation (resistors with same geometry)	Usually <4%.
Protection:	(Not normally needed.)
Overglaze	ESL 4775 (fire at 500–520°C).
Organic overcoat	ESL 240-SB (Cure at 200–250°C for best solvent resistance.)
Terminations	Pd/Ag, Pt/Au, Au
Viscosity: (Brookfield Model RVT, 26 ± 1°C, 10 rev/min, 7 spindle, 3/4-in immersion)	225 ± 25 kcps
Thinner	ESL 401; use sparingly.
Shelf life, refrigerated	12 months.
Stabilization:	(Not normally needed.)
Heat	200°C, 100 h or 450°C, 2 h. (May affect some oxidizable conductors.)
Voltage	Overload at 400 V/mm of resistor length for 5 s. This may be repeated several times.
Substrates	Alumina, beryllia, and over some dielectric pastes (e.g., ESL 4903 and 4905).

Source: Courtesy Electro-Science Laboratories, Inc.

TABLE 8.45 Typical Fired Resistor Properties

	3981	3982	3983	3984	3985	3986	3987
Resistivity, Ω/square	10	100	1000	10,000	100,000	1 megohm	10 MΩ
Fired tolerances, %*	±15	±10	±10	±10	±10	±15	±25
Average TCR, ppm/°C:							
+25 to +125°C	0 to +100	0 to +100	0 to +50	0 to +50	0 to +50	−50 to +50	−50 to +50
−55 to +25°C	−50 to +50	−50 to +50	−50 to 0	−50 to 0	−50 to 0	−100 to −50	−150 to −50
Average VCR [ppm/V(cm) 2 × 2 mm]	—	—	—	−10	−20	−40	−60
High-voltage capacitor discharge pulsing, 10 kV/cm, 20 pulses, % Δ R	—	—	—	<0.2	<0.2	<0.2	<0.2
Wattage rating,† 150°C max hot-spot temperature	200 W/in^2 (0.31 W/mm^2)						
Current noise (Quan-Tech, dB, 2 × 2 mm)	−30	−20	−15	−10	−5	−3	+5
Solder dip, % Δ R, 225°C, 10 s:							
Unoverglazed	±0.5	±0.3	±0.1	±0.1	±0.1	±0.1	±0.1
Overglazed	±0.1	±0.1	±0.05	±0.05	±0.05	±0.05	±0.05
Laser trimming stability‡	Less than 0.2% Δ R after YAG laser trimming						

* Based on 1.25 × 1.25 mm square resistors terminated with 9669, Pd/Ag, printed and fired according to our standard processing.

† Wattage rating may be exceeded as long as the hottest portion (hot spot) of the resistors does not exceed 150°C. If 150°C is reached, the resistors should be derated to a lower power level. Heat sinking the substrates permits much higher power loadings to be used.

‡ When faster-than-normal firing cycles are used with extremely rapid cooling from peak temperatures (such as with radiant energy furnaces), strains may be introduced into the glasses of the resistors, due to inadequate annealing. In such cases, laser trimming or other extreme thermal shocks may cause larger changes than those shown above.

Source: Courtesy Electro-Science Laboratories, Inc.

TABLE 8.46 Typical Thick- and Thin-Film Resistor Characteristics

Criterion	Thick film	Thin film
Initial resistor tolerances, %	±10–20 as fired	±5 as deposited
Trimming tolerance, %	±0.5	±0.1
Power handling capabilities, W/in^2	~50	~15
Resistor temperature coefficients, ppm/°C	±100	0 ± 50
Resistance, Ω/square	$1 - 10^6$	0.1 – 1,000
Line-width capabilities, mils	5 ± 1	0.2 ± 0.02
10,000-h drift, %	±1	±0.1

Low-k Dielectrics. Low-k dielectric pastes consist of oxides (glasses and ceramics), organic solvent vehicles, and organic binders. The functional materials are complex mixtures of oxides of, for example, aluminum, silicon, barium, magnesium, titanium, and bismuth. Devitrifying (crystallizable) glasses are used so that, during subsequent firings, as is necessary for multilayering, the previously fired dielectrics will not soften or flow. Dielectrics isolating conductor lines are applied in two and sometimes three separately processed layers to assure a lack of pinholes and the integrity of the film. The after-fired thicknesses may be 2 mils or even greater. Colorants (blue, green, yellow) are often added to the dielectric paste so that it can easily be inspected for coverage and pinholes and so that the registration of the dielectric layers with previously screen-printed layers can be facilitated. In some cases the color remains after firing, while in other formulations it burns out and changes to off-white or amber. Pastes must be formulated so that their expansion coefficients closely match that of the substrate. Shrinkage stresses must be minimized; otherwise bowing of the substrate occurs. The properties of some commercially available dielectric pastes are given in Tables 8.48 and 8.49.

High-k Dielectrics. High-k dielectric pastes having dielectric constants from 25 to 1500 are available from the major paste suppliers and are used to produce capacitor chips and capacitors that are batch fabricated on hybrid circuit substrates. The functional constituents of the pastes are inorganic compounds having high dielectric constants such as titanates, niobates, titanium oxides, and ferroelectric materials. Capacitors are produced by screen-printing and firing processes similar to those used for conductors and interlayer dielectrics, except that the dielectric is sandwiched between bottom and top metallization acting as the electrodes.

The dielectric constant of the dielectric and dimensions of the structure determine its capacitance value. Capacitance is directly proportional to the dielectric constant and area and inversely proportional to thickness. Generally, very high k materials require sintering temperatures of 1300 to 1400°C in order to produce nonporous films. Such high temperatures, of course, are incompatible with normal thick-film hybrid circuit processing. Furthermore, the high-temperature-fired films are marginally adherent to alumina. Fortunately, dielectric pastes having moderately high dielectric constants (25 to 1500) that can be processed at temperatures of 850 to 1050°C have been developed. Nevertheless, most capacitors require an overglaze to encapsulate any porosity and prevent failures under humidity-bias testing. Adding glass to the compositions reduces porosity, lowers the firing temperatures to below 1000°C, and improves adhesion to the substrate, but at the expense of drastic reductions in dielectric constant. The dielectric constants are reduced several orders of magnitude to the point where

TABLE 8.47 Typical Fired Resistor Properties for Birox 1900 Series

Birox 1900-series*	1911	1921	1931	1933	1935	1939	1949	1959
Resistivity, Ω/square	10 ± 20%	100 ± 20%	1 k ± 20%	3 k ± 20%	3 k ± 20%	10 k ± 20%	100 k ± 20%	1 M ± 20%
TCR†, ppm/°C	0 ± 100	0 ± 100	0 ± 100	0 ± 100	0 ± 100	0 ± 100	0 ± 100	0 ± 100
Short-term overload voltage,‡ V/mm	45	22	62	77	90	145	245	310
Standard working voltage,§ V/mm	1.8	9	25	30	36	58	98	124
Maximum rated power dissipation,¶ mW/mm^2	320	810	625	300	430	340	96	15
Blendable series	A	A	A	A	B	B	B	B

*Typical resistor properties based on laboratory test using recommended processing conditions; termination—DuPont Pd/Ag Conductor Composition 6134 prefired over DuPont Dielectric 5704 at 850°C; substrate—96 percent alumina; printing—200-mesh stainless steel screen (8–12 μm emulsion thickness) to a dried thickness of 25 ± 3 μm; firing—30-min cycle to peak temperature of 850°C for 10 min.

† Shipping specifications. Resistor geometry: 1.5 × 1.5 mm. Temperature coefficient of resistance: −55–+25°C and +25–+125°C.

‡ Short-term overload voltage; tested under MIL specification of MIL-R-83401D, Paragraph 3.15.

§ Standard working voltage: 0.4 × short-term overload voltage.

¶ Maximum rated power dissipation: $\frac{(\text{Standard working voltage})^2}{\text{Resistance}}$

Source: Courtesy DuPont.

TABLE 8.48 Typical Properties of Dielectric Pastes

Typical properties	4608	4608C	4608FB	4608CFB	4608CFB-M2	4611	4612	4612C
Dielectric constant, KHz	8–15	8–15	8–15	8–15	8–15	7–13	7–13	7–14
Dissipation factor, 1 kHz	<0.4%	<0.4%	<0.6%	<0.6%	<0.6%	<0.3%	<0.6%	<0.6%
Color	Ivory to yellow*	Ivory to yellow	Blue (dried) ivory to yellow	Blue (dried) ivory to yellow	Blue (dried) ivory to yellow	Ivory to yellow	Green	Green
Viscosity: (26°C; Brookfield RVT: no. 7 spindle $^3/_4$-in depth)	250,000± 25,000 cps	325,000± 25,000 cps	325,000± 25,000 cps	450,000± 50,000 cps	450,000± 50,000 cps	250,000± 25,000 cps	250,000± 25,000 cps	400,000± 35,000 cps
Fine-line definition (minimum printable via sizes)	Fair (10–15 mil) (250–375 μm)	Good (7–12 mil) (175–300 μm)	Fair to Good (10–15 mil) (250–375 μm)	Excellent (5–10 mil) (125–250 μm)	Excellent (5–10 mil) (125–250 μm)	Fair (10–15 mil) (250–375 μm)	Fair to Good (10–15 mil) (250–375 μm)	Very Good (7–12 mil) (175–300 μm)
Leveling	Excellent	Fair	Good	Good	Excellent	Excellent	Excellent	Very Good
Firing range, °C:								
For crossovers	850–1000	850–1000	850–1000	850–1000	850–1000	825–1000	800–1000	800–1000
For multilayers (10–15 min at peak temperature)	930–1000	930–1000	930–1000	930–1000	930–1000	900–1000	850–1000	850–1000

*Permanent green or yellow versions of 4608 are available on special large-quantity orders.

Source: Courtesy Electro-Science Laboratories, Inc.

TABLE 8.49 Properties of Multilayer Dielectric C8301

Application: The C8301 multilayer dielectric was designed to closely match expansion characteristics of alumina substrates. Optimum results are achieved in the manufacture of sophisticated multilayer hybrid circuits due to the unique low-warp characteristics and excellent electrical properties of this material. The C8301 dielectric is compatible with all the Gold MULTISYST™ conductors, resistors, and overglazes.

Fired film properties
Camber: <1 mil/in
Dielectric constant, K: 6–10 @ 1 kHz
Dissipation factor DF: <0.2%
Voltage breakdown: >600 V/mil
Insulation resistance: >5 × 10^{10} Ω @ 100 V dc
Formulation properties
Solids: 75 ± 2%
Viscosity: 340 ± 50 kcps
Coverage: ≈90 cm^2/g at 40 μm wet print thickness
Shelf life: 6 months at 25°C

Source: Courtesy EMCA-REMEX.

they may not be useful as capacitors. The porosity problem can be alleviated by sealing the fired capacitor with a screen printed and fired overglaze. Several layers (sometimes up to 4) or combinations of encapsulants and overglazes are required to assure complete passivation. Even so, shifts in capacitance occur during the firing of the overglazes and must be measured and taken into account in the electrical functioning of the circuit. The greatest success in fabricating capacitors compatible with thick-film circuits has been achieved with pastes having low to moderate dielectric constants (Tables 8.50 and 8.51). Even here it has been shown that capacitance is quite sensitive to firing temperatures and electrode metallizations (Fig 8.52).[72]

TABLE 8.50 Properties of Capacitor Dielectrics

Material	4113	4114	4115
Dielectric constant (Pd/Ag electrodes)	90–120	40–60	20–30
Capacitance (pF/cm^2, 1 kHz)	10,000	5000	2500
Insulation resistance, Ω	10^{11}	10^{11}	10^{11}
Dissipation factor, % (overglazed)	0.2–0.4	0.1–0.2	0.02–0.06
Breakdown voltage V/25 μm (for >50 μm thickness)	500	600	700
TCC (+25–+125°C) 980°C firing	300 ppm/°C (+100 ppm/°C avail.)		
Electrode	ESL 9635-C; Pd/Ag; 980°		
Viscosity, Brookfield	250 + 25 kcps		
Substrate	96% alumina		
Thinner	ESL 401 or 404		

Source: Courtesy Electro-Science Laboratories, Inc.

TABLE 8.51 Characteristics of High-k Dielectrics

Electrode	7231D		7233D		7236D	
	K	DF, %	*K*	DF, %	*K*	DF, %
2660D Ag	1200–1700	1.0–2.0	3000–4500	1.8–2.8	5000–7000	1.2–2.2
2533D Pt/Ag	1000–1500	1.0–2.0	3000–4500	1.5–3.0	5000–7000	1.2–2.5
2039 Pd/Ag	1000–1500	1.0–2.0	3000–4500	1.5–3.0	5000–7000	1.2–2.5
2036D Pd/Ag	800–1200	1.0–2.0	1800–2500	1.5–3.0	2500–3500	1.0–2.0
2033D Pd/Ag	600–1000	1.0–2.0	1800–2500	1.5–3.0	2500–3500	1.0–2.0
3246 Au	800–1200	1.0–2.0	1800–3000	1.5–3.0	3000–4000	1.0–2.0

Notes: 1. The data was obtained after a 24-h stabilization period.
2. In all cases IR ranged from 0.5 to 100 GΩ, depending on capacitor size and dielectric thickness.
3. Breakdown voltage ranged from 400–700 V dc.
4. The high-k dielectrics are critically balanced for optimum performance. Using conductors or overglazes other than those listed above can severely degrade them.
5. 2660D and 2039 are newer versions of 2633 and C3302, respectively. They incorporate powders with better morphology for improved conductor performance. Capacitor results are identical between the two versions.

Source: Courtesy EMCA-REMEX.

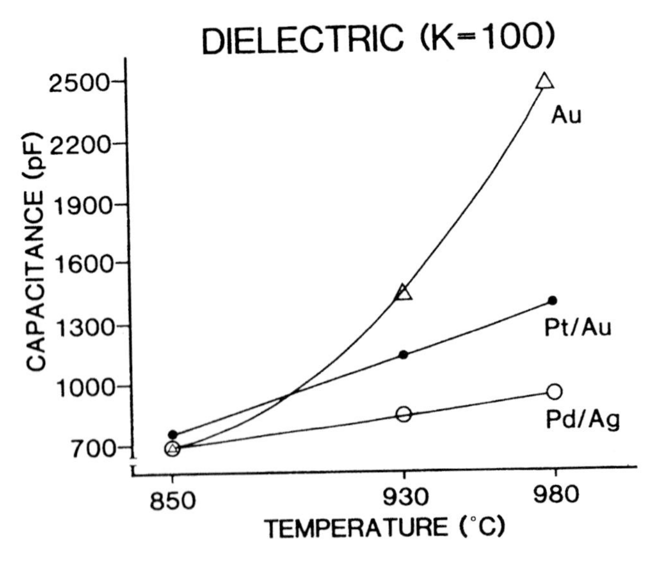

FIGURE 8.52 Capacitance for 0.25-cm^2 area for various electrodes and firing temperatures. (*Courtesy of Electro Science Laboratories.*)

On the whole, though screen printing and firing of thick-film capacitors is widely used for producing chip capacitors, capacitor arrays, and RC networks, batch fabricating capacitors during the processing of thick-film circuits has not materialized largely because of the sensitivity of capacitance values to the processing conditions, the controls that are needed, and the extra steps involved. Almost all manufacturers of hybrid circuits find it simpler and less costly to purchase chip capacitors and to assemble them with conductive silver-filled epoxies—especially if only a few capacitors are required for each hybrid.

8.4.4 Design Criteria for Thick Films

Operating rules for layout criteria must be established that are compatible with the equipment and processes in place at a given installation. Where a line-width minimum of 0.003 in may be satisfactory for one facility, a minimum line width of 0.010 in may be the least acceptable for another facility. This logic extends to resistors, capacitors, and even the insulation, since all process criteria are "layout-dependent" if reasonable yields are to be expected. The question of yield is most important to the success of any operation, and it is especially important in the field of thick-film hybrid circuits. One major factor affecting the process yield can cause many repercussions in the cost and the delivery of the end item. For these reasons layout criteria should be carefully established with extreme consideration for each and every process step. Some typical guidelines for layout criteria are listed in Table 8.52.

It is important to know the available area that a given substrate or package allows for the necessary active or passive circuitry. Obviously, the area available for resistors, capacitors, and add-on parts is completely dependent on the total substrate area that can be used. Input-output pad areas must be subtracted from this total area available.

The following guidelines have been prepared to assist the circuit designer in making

TABLE 8.52 Typical Guidelines for Layout of Thick Films

Criterion	Design	Minimum
Conductor width, in	0.010–0.020	0.005
Conductor spacing, in	0.010 or larger	0.005
Resistor length, in	0.040	0.020
Resistor width, in	0.040	0.010
Resistor spacings, in		0.025
Resistor dissipation, W/in^2		25–50
	(Dependent on heat-sinking methods)	
Solder land, in	0.050 × 0.050	
	(Dependent on component lead size)	
Wire-bond land, in		0.010 × 0.010
		(0.020 × 0.020 preferred)
Die-bond land, in		0.005 on each side, larger than the chip
Edge of substrate clearance (for resistors), in		0.020
Resistor-conductor overlap, in		0.010

the transition from the breadboard to the hybrid thick film. These guidelines set forth the proper design of conductor patterns, cermet resistors, and semiconductor mounting pads, and outline the proper techniques for attaching die to the substrate, and for wire bonding where necessary. These are practical guides and are not meant to describe current state-of-the-art techniques but rather to establish a set of rules which, if followed, will allow the resultant layout to be manufactured with good yield at a reasonable cost.

Thick-Film Resistor Value Calculations. the basic equation for resistance is

$$R = \frac{pL}{A} \tag{8.1}$$

where R = resistance, Ω
p = bulk resistivity
L = length of resistor
A = area of resistor cross section

Expanding gives

$$R = \frac{pL}{A} = \frac{pL}{tW}$$

where t = thickness of resistor
W = width of resistor

Also, the resistance is inversely proportional to the thickness, or

$$p_s = \frac{p}{t} \qquad \text{or} \qquad t = \frac{p}{p_s}$$

where p_s = effective sheet resistivity, Ω/square.
Substitution provides

$$R\,\Omega = p_s \frac{L}{W}$$

$$= p_s \times \text{number of squares } N$$

and

$$N \text{ squares} = \frac{R}{p_s} = \frac{L}{W}$$

Example. If an ink with sheet resistivity p_s = 12.5 kΩ/square were screened with dimensions of L = 0.090 and W = 0.030 in, the resistance would be

$$R = 12.5 \text{ k}\Omega \times \frac{0.090}{0.030}$$

$$= 12.5 \text{ k}\Omega \times 3 \text{ squares or } 37.5 \text{ k}\Omega$$

Example. If an ink with sheet resistivity p_s = 12.5 kΩ/square is to be used to screen a resistor of 37.5 kΩ, the number of squares will be

$$N = \frac{R}{p_s} = \frac{37.5}{12.5} = 3 \text{ squares}$$

If the maximum resistor width can be 0.050 in, then the length would be

$$L = W \times N = 0.050 \times 3 = 0.150 \text{ in}$$

As a general practice, resistor geometry should be made as large as the available substrate will allow.

Allowance for Trimming. Thick-film resistors can be screened and fired to an accuracy of approximately 10 to 20 percent. Subsequently, if tighter tolerances are required, it is necessary to trim the resistor, usually by laser cutting into the width. It is standard practice to trim all resistors, and for this reason the area of the resistor should be chosen such that its value is approximately 85 percent of the desired circuit value. The resistor will then be brought to the correct value by trimming.

Example. If the final circuit value of a resistor is to be 25 kΩ, the screened and untrimmed resistor should be

$$R = 25K \times 85 \text{ percent} = 21.3 \text{ k}\Omega$$

If it is assumed that the sheet resistivity of the ink with which the resistor is to be printed is 12.5 kΩ/square, the number of squares N should be

$$N = \frac{R}{p_s} = \frac{21.3}{12.5} = 1.7 \text{ squares}$$

If the resistor is 0.050 in wide, the length will be

$$L = W \times \text{N} = 0.050 \times 1.7 = 0.085 \text{ in long}$$

The final value will be attained by trimming a portion of the resistance material away from the substrate, as shown in Fig. 8.53. This has the effect of increasing the number of squares, and thus the resistance.

Resistor Aspect Ratio. The aspect ratio of a film resistor is the ratio of its length to its width, or L/W. This is also equal to the number of squares N in the resistor. For the condition where the length of the resistor is greater than the width, the maximum aspect ratio should not exceed 10:1, and for good design, it should be 5:1 or less.

For resistors where the width is greater than the length, the aspect ratio should be 1:2 or greater, and never more than 1:3.

Number of Resistor Inks per Substrate. The number of resistor inks per substrate should be kept to a minimum and should never exceed three. Each resistor ink requires separate artwork, its own screen, and a separate screening operation. For this reason, inks should be selected so that their sheet resistivities p_s cover the largest range of required values within the proper range of aspect ratios.

Example. Resistors in a circuit include:

$$R_1 = 100 \text{ k}\Omega$$

$$R_2 = 50 \text{ k}\Omega$$

$$R_3 = 10 \text{ k}\Omega$$

$$R_4 = 500 \text{ }\Omega$$

FIGURE 8.53 Thick-film resistors showing laser trim cuts.

The ink combinations could be:

Combination 1: 37.5 kΩ/square and 375 Ω/square

Combination 2: 125 kΩ/square, 37.5 kΩ/square, and 375 Ω/square

Combination 3: 125 kΩ/square, 1.25 kΩ/square, and 375 Ω/square

The best combination would be the first since it has one less screening operation.

On many occasions several possible ink combinations will appear equally advantageous. Under such a condition, the total area required for each combination should be calculated, after which a more meaningful decision can be made, based on considerations of substrate size and available resistor area.

Power Dissipation

Resistors. A safe value for heat dissipation in thick-film resistors is 35 W/in^2 of resistor at 125°C. The power to be dissipated by a resistor in a circuit may be calculated from

$$P = I^2R = EI = \frac{E^2}{R}$$

The area of a resistor required to dissipate its generated heat may be calculated from

$$A = \frac{P}{P_r}$$

where A = resistor area, in^2
P = power dissipated by resistor, W
P_r = rated resistor power, W/in^2

or if the area is known, the power that a resistor can dissipate can be readily calculated from

$$P = A \times P_r$$

Example. Area of resistor is 0.060 by 0.030 in, and rated power is 35 W/in^2. The maximum power that the resistor can safely dissipate is

$$P = A \times P_r = (0.060 \times 0.030) \times 35 = 0.063 \text{ W}$$

If the voltage drop across the resistor is 12 V, and the current is 4 mA, the power to be dissipated is

$$P = EI = (12)(0.04) = 0.048 \text{ W}$$

The resistor can dissipate the heat since its capacity is 0.063 W.

Substrates. Similar calculations must be made for the completed substrate. A good design factor is 5 W/in^2. The total heat dissipated by each component must be calculated, totaled, and compared to the total dissipation capability for that particular size substrate.

REFERENCES

1. J. J. Licari and L. R. Enlow, *Hybrid Microcircuit Technology Handbook,* Noyes, Park Ridge, NJ, 1988.
2. R. R. Tummala (ed.), *Microelectronics Packaging Handbook,* Van Nostrand Reinhold, 1989.
3. R. Jones, *Hybrid Circuit Design and Manufacture,* Marcel Dekker, 1982.
4. G. E. Messner, I. Turlik, J. W. Balde, and P. E. Garrou, *Thin Film Multichip Modules,* International Society for Hybrid Microelectronics, 1992.
5. R. F. Bunshah (ed.), *Deposition Technologies for Films and Coatings,* Noyes, Park Ridge, NJ, 1982.
6. J. J. Licari and L. Hughes, *Handbook of Polymer Coatings for Electronics,* Noyes, Park Ridge, NJ, 1990.
7. P. L. Fleischner, "Beryllia Ceramics in Microelectronic Applications," *Solid State Technology,* Vol. 20, No. 1, 1977.
8. Naval Command Control and Ocean Surveillance Center, Contract N66001-88-C-0181, "VLSI Packaging Technology," 1988–92.
9. E. Y. Luh, J. H. Enloe, A. Kovacs, and R. Lucernoni, "Metallization of Aluminum Nitride Packages," *IEPS Journal,* Vol. 13, No. 2, 1991.
10. F. Borchelt and G. Lu, "Use of CVD Diamond Substrates in Electronic Applications," 6th International SAMPE Electronics Conference, Baltimore, 1992.
11. K. V. Ravi, "Plasma Enhanced Chemical Vapor Deposition of Diamond Films-Technology and Commercial Applications, *American Vacuum Society, Proceedings,* 1988.
12. J. A. Herb, M. G. Peter, and T. J. Fileds, "Diamond Films for Thermal Management Applications," *Proceedings of the ASM International 3rd Electronic Materials and Processes Congress,* San Francisco, 1990.

13. M. I. Landstrass and K. V. Ravi, "The Resistivity of CVD Diamond Films," *Applied Physics Letters,* 1989.

14. J. C. Angus, "History and Current Status of Diamond Growth at Metastable Conditions," *Proceedings of the 1st International Symposium on Diamond and Diamond-Like Films,* Vols. 89-12, Electrochemical Society, 1989.

15. R. C. Eden, "Applicability of Diamond Substrates to Multichip Modules," *ISHM Proceedings,* 1991.

16. ISHM, *Abstracts,* Diamond and Diamond-Like Film Workshop, Breckenridge, CO, March 1991.

17. M. K. Aghajanian, "A New Infiltration Process for the Fabrication of Metal Matrix Composites," *SAMPE Quarterly,* vol. 20, No. 4, 1989.

18. D. R. White, U. S. Patent 4828008, "Metal Matrix Composites," May 9, 1989.

19. D. White, S. Keck, I. Smith, and A. Silzars, "New High Ground in Hybrid Packaging," *Hybrid Circuit Technology,* December 1990.

20. A. L. Eustice, S. J. Horowitz, J. J. Steward, A. R. Travis, and H. T. Sawhill, "Low Temperature Co-Fireable Ceramics: A New Approach for Electronic Packaging," *36th Electronic Components Conference,* Seattle, 1986.

21. J. I. Steinberg, S. J. Horowitz, and R. J. Bacher, "Low Temperature Co-Fired Tape Dielectric Material Systems for Multilayer Interconnections," *Solid State Technology,* January 1986.

22. D. Schroeder, "The Use of Low Temperature Cofired Ceramic for MCM Fabrication," *ICMCM Proceedings,* 1992.

23. R. R. Tummala, H. R. Potts, and S. Ahmed, "Packaging Technology for IBM's Latest Mainframe Computers," *41st Electronic Components & Technology Conference,* 1991.

24. R. R. Tummala and J. Knickerbocker, "Advanced Cofire Multichip Technology at IBM" *IEPS Proceedings,* San Diego, September 1991.

25. G. Harman, "*Reliability and Yield Problems of Wire Bonding in Microelectronics,*" *ISHM,* 1989.

26. F. K. Legoues, B. D. Silverman, and P. S. Ho, *Journal of Vacuum Science Technology,* A6, 2200, 1988.

27. F. Faupel, D. Gupta, B. D. Silverman, and P. S. Ho, "Direct Measurements of Cu Diffusion Into a Polyimide Below the Glass Transition Temperature," *Applied Physics Letters,* Vol. 55, No. 4, July 1989.

28. L. Maissel and R. Glang (eds.), *Handbook of Thin Film Technology,* McGraw-Hill, New York, 1970.

29. J. J. Licari and L. R. Enlow, *Hybrid Microcircuit Technology Handbook,* Noyes, Park Ridge, NJ, 1988.

30. K. Wasa and S. Hayakawa, *Handbook of Sputter Deposition Technology,* Noyes, Park Ridge, NJ, 1991.

31. R. W. Vest, *Metallo-Organic Materials for Ink Jet Printing,* Final report, Navy Contract N00163-81-C-0350, 1982.

32. W. Havey, "Formation of Conductive Patterns on Hybrid Circuits and PWBs Using Alternative Techniques," *Hybrid Microcircuit Technology,* August 1985.

33. T. Snodgrass and G. Blackwell, "Advanced Dispensing and Coating Technologies for Polyimide Films," *ICMCM Proceedings,* 1992.

34. N. Puri and T. Yaser, "Tantalum Nitride Chip Resistors for High Reliability Hybrid Microcircuits," *Proceedings ISHM,* September 1978.

35. L. Holland, "Vacuum Deposition of Thin Films," *Vacuum,* Vol. 1, 1951.

36. P. Lloyd, "Review of Thin Film Techniques for Microelectronics," *Microelectronics and Reliability,* Vol. 6, 1967.

37. E. L. Kern and L. A. Teichthesen, "Passive and Process Materials for Semiconductor Device Fabrication," *Solid State Technology,* October 1966.
38. J. D. Craig, "Polyimide Coatings," *Electronic Materials Handbook,* Vol. 1, *Packaging,* ASM, 1989.
39. H. J. Neuhaus, "A High Resolution Anisotropic Wet Patterning Process Technology for MCM Production," *ICMCM Proceedings,* 1992.
40. J. Summers, et al., "Wet Etching Polyimides for Multichip Module Applications," *ICMCM Proceedings,* 1992.
41. N. Bilow, A. Landis, et al., "Acetylene-Substituted Polyimide Adhesives," *SAMPE Journal,* January/February 1982.
42. N. Bilow, *Acetylene-Substituted Polyimides as Potential High Temperature Coatings,* ACS Symposium on Organic Coatings, 177th ACS National Meeting, Honolulu, HA, 1979.
43. R. Rubner, *Siemens Forsch-u.Entwicki-Ber.,* 5235, 1976.
44. A. E. Nader, et al., "Photodefinable Polyimides Designed for Use As Multilayer Dielectrics For Multichip Modules," *Proceedings ICMCM,* 1992.
45. J. Pfeifer and O. Rohde, *Proceedings 2nd International Conference on Polyimides,* 130, 1985.
46. K. K. Chakravorty and J. M. Cech, "Photosensitive Polyimide As a Dielectric in High Density Thin Film Copper Polyimide Interconnect Structures," *Journal of the Electrochemical Society,* 137 (3), 1990.
47. T. Ohsaki, T. Yasuda, S. Yamaguchi, and T. Kon, *Proceedings of the IEEE International Electronics Manufacturers Technology,* 1987.
48. M. G. Dibbs, P. H. Townsend, et al., "Cure Management of Benzocyclobutene Dielectrics for Electronic Applications," *Proceedings of the 6th International SAMPE Conference,* Baltimore, MD, 1992.
49. T. M. Stokich, W. M. Lee, and R. A. Peters, "Real Time FT-IR Studies of the Reaction Kinetics for Polymerization of Divinyl Siloxane Bis-Benzocyclobutene Monomer," *Proceedings of the Materials Research Society,* 1991.
50. S. M. Lee, *Polymeric Films for Semiconductor Passivation,* Final report, NASA Contract NAS12-2011, 1969.
51. J. J. Licari, S. M. Lee, and I. Litant, "Reliability of Parylene Films," *Proceedings of the Metals Society Technical Conference, Defects Electronic Materials and Devices,* Boston, 1970.
52. R. Heimsch, "Screen Printing in Thick Film Hybrid Production," *Hybrid Circuit Technology,* July 1991.
53. C. Harper (ed.), *Handbook of Thick Film Hybrid Microelectronics,* McGraw-Hill, New York, 1982.
54. D. F. Zarnow, "An Introduction to the Navy Manufacturing Technology Program for Computerized Thick-Film Printing," *IEEE Trans, Components, Hybrids, and Manufacturing Technology,* September 1988.
55. W. J. Havey, "The Formation of Conductive Patterns on Hybrid Circuits and PWBs Using Alternative Techniques," *Hybrid Circuit Technology,* August 1985.
56. M. Shankin, "Write It, Don't Screen It," *Proceedings ISHM,* 1978.
57. R. Tryggestad and G. Freeman," Direct Writing vs. Screen Printing: Complementary Technologies," February 1989.
58. Johnson Matthey, *Thick Film Products Bulletin,* JM1301-Fine-Line Gold Conductor.
59. Johnson Matthey, *Thick Film Products Bulletin,* JM1202-Ultra Fine Line Etchable Gold Conductor.
60. D. H. Scheiber and R. M. Rosenberg, "Circuitry from Photoprintable Paste—A New Technology," *Proceedings ISHM,* 1972.

61. DuPont Experimental Data Sheets, Photo Patterning System—6050D Dielectric Paste and 1056-2 Gold Conductor.
62. W. Vitriol and J. I. Steinberg, "Development of a Low Temperature Cofired Multilayer Ceramic Technology," *Proceedings ISHM,* 1982.
63. J. I. Steinberg, S. J. Horowitz, and R. J. Bacher, "Low Temperature Cofired Tape Dielectric Material Systems for Multilayer Interconnections," *Solid State Technology,* January 1986.
64. R. Brown and P. Polinski, "Manufacturing of Low-Temperature Cofired Ceramic Modules for Advanced Radar Applications," *Proceedings GOMAC,* Las Vegas, 1992.
65. F. W. Martin, "Polymer Thick Film Extends Options for Hybrid and PCB Fabrication," *Circuits Manufacturing,* Vol. 17, No. 5, 1977.
66. F. W. Martin, "Low Firing Polymer Thick Film Enables Screen Printing of Resistors and Conductors on PC Boards," *Insulation/Circuits,* Vol. 23, No. 2, 1975.
67. *EMCA-REMEX Bulletin 513, Solderable Polymer Silvers.*
68. S. J. Stein, C. Huang, and L. Cang, "Fine Wire and Ribbon Attachments to Thick Film Conductors," International Microelectronics Conference, Tokyo, May 1982.
69. S. J. Stein, C. Huang, and P. Bless, "Low-Cost, High Reliability Multilayers Made With Silver Containing Thick Films." *Proceedings 4th International Microelectronics Conference,* ISHM-Japan pp. 287[en dash]296, Kobe, Japan, May 1986.
70. D. E. Riemer, "Material Selection and Design Guidelines for Migration-Resistant Thick Film Circuits with Silver-Bearing Conductors," *Proceedings of the 31st Electronic Components Conference,* 1981.
71. E. A. Hayduk and D. R. Taschler, "Atmosphere Control and Copper Thick Film Firing Process," *Hybrid Circuit Technology,* January 1986.
72. S. J. Stein, C. Huang, and P. Bless, "New Thick Film Dielectrics," International Microelectronics Conference, Tokyo, May 1984.

CHAPTER 9

ELECTRODEPOSITION: MATERIALS AND PROCESSES

Donald Baudrand

9.1 INTRODUCTION

Plating and surface finishing are a crucial part of the manufacture of many electronic devices.

Plating is any process which deposits a relatively thin metal coating onto a substrate which enhances the performance and/or decorative appeal of the substrate.

Not all metals can be electrodeposited; thus the selection of metals is limited to practical deposition methods of a select group of metals.

In electronic devices, metals commonly plated are gold, copper, nickel, silver, tin, various solder alloys, rhodium, palladium and palladium nickel alloys, and to a lesser extent platinum, indium, ruthenium, tin nickel alloy, and some rare earth metals. Some of these metals and alloys can be deposited autocatalytically without external electrical current. Common are electroless nickel-phosphorus, electroless nickel boron, electroless copper, and, to a lesser extent, electroless palladium.

Plating is used to improve conductivity; to make a surface solderable, weldable, die bondable, or brazeable; to improve corrosion resistance or wear characteristics; to provide lower contact resistance; to make electrical connections as in through-hole plating of printed wiring boards and via hole connections; and to provide conductive paths, etch resists, diffusion barriers, and so on (see Figs. 9.1 and 9.2).

Without plating, many devices would not be able to perform to expectations or specifications, and others could not be produced at all.

9.1.1 Electronic Plating Applications

A wide variety of components require plating. They vary from parts so small that assembly is done under a microscope to components so large that they require heavy-duty handling equipment. Examples are PC boards (PWB); connectors made from steel, copper alloys, aluminum, and Kovar*; transistor bases and leads; relay switches;

*Trade name of Westinghouse Electric Corp.

FIGURE 9.1 Plated electronic products. (*Reprinted with permission from the* AESF Illustrated Lecture, Printed Circuit/Wiring Board Manufacture and Plating for Electronic Applications.)

FIGURE 9.2 Electroplating process equipment. (*Reprinted with permission from the* AESF Illustrated Lecture, Printed Circuit/Wiring Board Manufacture and Plating for Electronic Applications.)

FIGURE 9.3 Continuous strip plating. (*Reprinted with permission from the* AESF Illustrated Lecture, Printed Circuit/Wiring Board Manufacture and Plating for Electronic Applications.)

chassis; housings; tubing; wire; pin contacts; lead frames; semiconductors; hybrid circuits on substrates such as metallized ceramic, ceramic, and polyimide; and numerous hardware devices.

Plated deposit requirements differ for each type of device to be plated. Lead frames are often nickel and gold plated. Gold is often spot plated only where required for its functional properties. Lead frames are often plated on automatic reel-to-reel machines with all the processes required for preparation and plating on one machine. Silver, palladium, tin, tin lead, nickel boron, and nickel phosphorus electroless deposits are used for these devices. Aluminum circular connectors are electroless nickel phosphorus plated, and some are also cadmium plated and chromated over the electroless nickel to provide additional corrosion protection. Nickel first, then rhodium are often plated on plug-in connectors, as are nickel and gold. Figure 9.3 shows a continuous strip plating line.

9.2 PLATING PRINTED CIRCUIT (PRINTED WIRING) BOARDS

A printed circuit consists of a pattern of conductors bonded onto a nonconductive base, onto which electronic components may be attached. The most frequent conductor used is copper. Kovar,* Inar,† aluminum, chrome, nichrome, nickel, and brass have been used to a much lesser extent.

† Trade name of Soc. Anon. de Commentry-Fourchambault et Decazivillle.

9.2.1 Substrate Materials

The nonconducting substrate of a printed wiring board is usually phenolic-paper, epoxy-glass, fabric, epoxy-paper, polyimide-glass fabric, PTFE-glass fabric, and other moldable plastics. These are described in Chapters 1 and 2. Selection of which substrate to use depends on the requirements of the specific application and impacts on the electrodisposition methods.

9.2.2 Copper Foil

In most cases, copper foil is bonded to the substrate by direct bonding to the plastic. Copper foil is available in $^1/_2$ oz/ft^2, 1 oz/ft^2, and 2 oz/ft^2 to provide various conductor thicknesses (see Chap. 2).

9.2.3 Types of Printed Wiring Boards

1. Print and etch, one-sided or two-sided
2. Plated-through hole (PTH)
3. Multilayer PWB (MLB)
4. Flexible circuit
5. Molded circuit
6. Additive circuit
7. Integrated circuit (IC)

Types of PWBs include print and etch, one-sided or two-sided, plated-through holes to form connections between two-sided boards and to provide mounting holes for components such as resistors, capacitors, transistors, inductors, and so on with wire leads. Pads are also formed to provide surface mounting of components. Additional types of PWBs are multilay circuits where patterns are formed on PWBs which then are stacked using, for example, B-stage epoxy glass between the layers to bond the PWBs together. After curing, new holes are drilled and plated through to form interconnections between layers as well as for component mounting, flexible circuits, often using polyamide, polyimide, Mylar,[‡] or other flexible plastics..

Print and Etch. Print-and-etch PWBs are largely made from copper-clad phenolic, paper-reinforced plastics, and glass-reinforced copper-clad epoxy. The process involves application of a screened-on pattern that masks the circuit elements for the etching process which removes unwanted copper. After the etching process, the maskant is removed, leaving the circuit pattern. Photoresists are another method used to produce print-and-etch PWBs where a photoresist is applied over the entire copper surface, a mask is placed over the resist, and the entire surface is exposed to a strong light or uv light. The mask is removed, the photoresist is developed, and the unwanted copper areas are exposed for subsequent etching.

[‡]Trade name of DuPont.

Plated-through-hole processes are used to provide interconnections between two sides of copper-clad boards. Plated holes are also used to provide mounting for components such as resistors, capacitors, transistors, and inductors, which have wire leads. Pads can also be formed to provide for surface mounting of components.

Multilayer PWBs. Multilayer PWBs are made by stacking double-sided boards with B-stage epoxy glass fabric between the layers and heating the boards in a press to bond the layers together. After curing, new holes are drilled and through-plated to form interconnections between layers as well as for component mounting. The process is the same for both plated-through holes and multilayer PWBs, except that drilling resin smear must be removed or the inner layers etched back to assure good connections (see Figs. 9.4 and 9.5). The following typical process sequence is shown below:

1. The copper-clad laminates are punched or drilled, first registration holes, then all the holes to be used for functional purposes for two-sided or single-layer boards.
2. The holes are deburred, and the boards are cleaned chemically.
3. A maskant or photoresist is applied to define the circuit patterns.
4. The unwanted copper is etched, leaving the desired pattern.
5. The resist is removed.
6. For multilayer boards the patterned layers are stacked using epoxy prepreg (B stage) between circuits, then bonded under heat and pressure.
7. Holes are drilled through the multilayer package and deburred.
8. Chemical, or plasma, drilling resin smear removal or etch-back is done to clean the holes and to remove some of the plastic smeared on the copper layers exposed in the hole. *Etch-back* is further etching to expose a larger area of copper which some believe improves reliability. Others believe etch-back results in lessening the thermal shock resistance.
9. Plate through the holes. This process consists of cleaning and applying a catalyst to the inside of the holes. A *catalyst* is a material which causes electroless copper or nickel to start plating.
10. Electroless plate. Copper is the most-used process.
11. Electroplate copper in the holes; plating also takes place on the copper foil surface. (The outer layer does not yet have circuit patterns formed.) There are two ways to proceed:
    a. *Pattern plate,* where a thin layer of copper is deposited (0.0002 to 0.0003 in). Photoresist or screened resist is applied on the unwanted areas; the copper plating is then continued on the exposed circuit patterns to the final thickness, usually 0.001 in inside the holes.
    b. *Panel plate,* where the plating continues over all surfaces to the final thickness.
12. Etch resist is plated onto the pattern areas. Gold and tin lead (solder alloy) are usually selected.
13. The masking is removed.
14. Unwanted copper is removed by chemical etching.

15. Nickel, gold, or other metals are deposited on contact fingers. Tin lead may be reflowed; final cleaning takes place.
16. Postplating operations such as milling slots, grooves, and routing take place (see Fig. 9.6).

Flexible Circuits. Flexible circuits are widely used for connectors, bus wiring, and circuits that must be mounted in small areas requiring bending to assemble. Flexible circuits often use polyamide, Mylar, TFE, PCTFE, PVC, or other flexible plastics. The plating process is similar to that of other types of PWB substrate.

Molded Circuits. Molded circuits are made in a variety of forms and often include conductor patterns molded onto and/or into the molding. These conductor patterns can be subsequently plated by any desired platable metal. Circuits can be plated on molded substrates in several ways. For example: (1) Catalyst can be applied in the circuit pattern, then electroless copper or electroless nickel plated, followed by electroplating of copper and any other platable metal. (2) A procedure used to produce additive circuits may be used.

Additive Circuits. *Additive PWBs* are made by plating directly onto a nonconductive substrate in the final pattern configuration.

Integrated Circuits. *Integrated circuits* (ICs) are circuits deposited on a semiconductor or base. Circuit patterns can be formed in one of several ways. For example: (1) Etch the surface to be plated; apply a catalyst; electroless plate and electroplate. Then apply a resist such as a photoresist, screened resist, or dry film which protects the circuit pattern, and etch away the remaining plated metal. (2) Selectively catalyze the circuit pattern, electroless, and electroplate.

9.2.4 PWB Drilling

Epoxy glass copper-clad (both sides) for single-layer plated-through-hole boards and multilayer PWBs are drilled using high-speed drills and bits. The drills tend to leave epoxy smear over copper surfaces exposed in the holes. Although steps can be taken to minimize smear, most of which are design considerations, the plater must be prepared to deal with any amount of smear. Some considerations are: drill design, drilling speed, drill life, back-up material, and the size and depth of the holes (related to the number of layers and the thickness of each layer, both the dielectric and copper) (see Figs. 9.7, 9.8, and 9.9).

Smear removal and etch-back can be done chemically or mechanically or by plasma etching. The chemicals used present safety problems. Some examples are concentrated sulfuric acid, concentrated sulfuric acid plus hydrofluoric acid, chromic acid at a very strong concentration ($8^1/_2$ lb/gal) with buffers added, sulfuric acid dichromate mixtures, chloroform-alcohol, and alkaline systems of strong sodium hydroxide with potassium permanganate added. Strong chemical-oxidizing materials attack the epoxy while fluoride compounds remove exposed glass. Mechanical methods consist of vapor blasting, liquid honing, or wet honing. The plasma method provides a more uniform etch where two or more different polymers are part of the board.

MULTILAYER BOARD WITH PLATED DETAILS

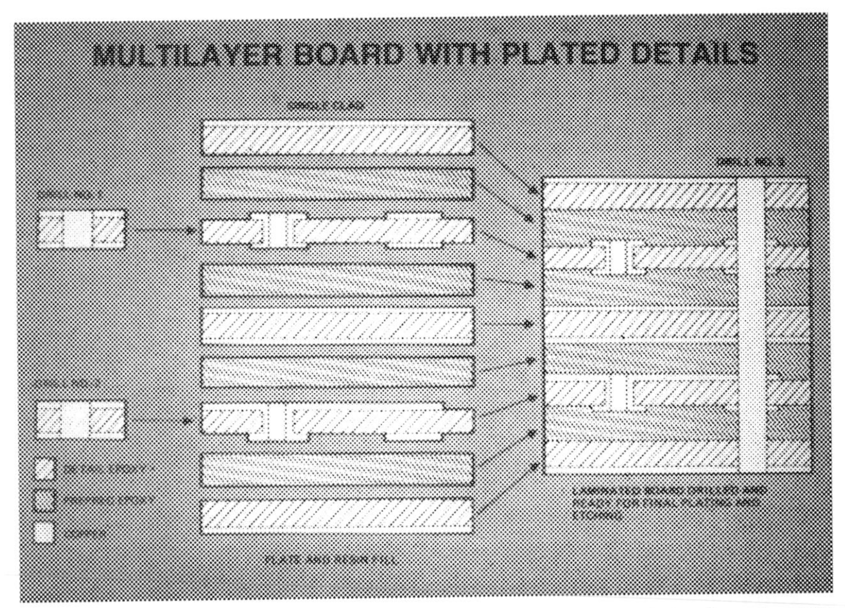

FIGURE 9.4 Multilayer board with plated details. (*Reprinted with permission from the* AESF Illustrated Lecture, Printed Circuit/Wiring Board Manufacture and Plating for Electronic Applications.)

Multilayer Board Fabrication
Copper Plate
Mask
Overplate
Strip Mask
Etch
Post Etch Finishing Operation
Final Machining
Final Inspection

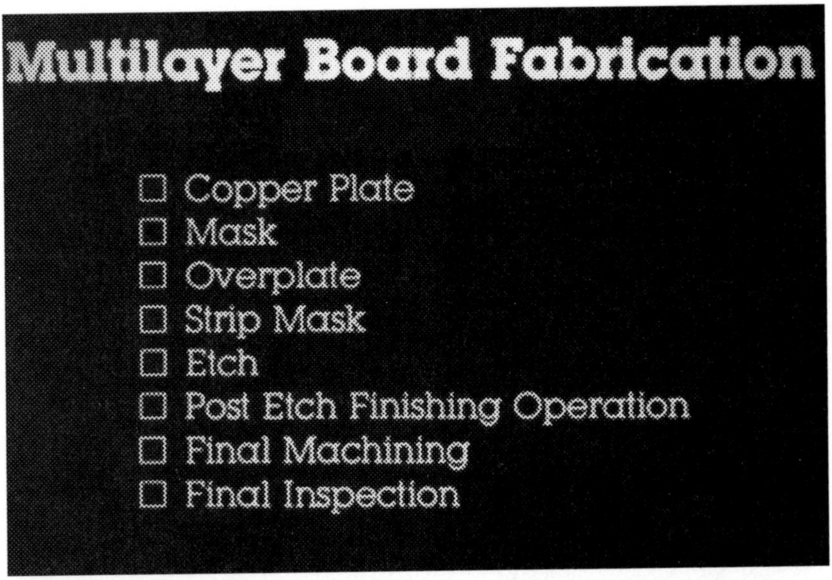

FIGURE 9.5 Multilayer board fabrication. (*Reprinted with permission from the* AESF Illustrated Lecture, Printed Circuit/Wiring Board Manufacture and Plating for Electronic Applications.)

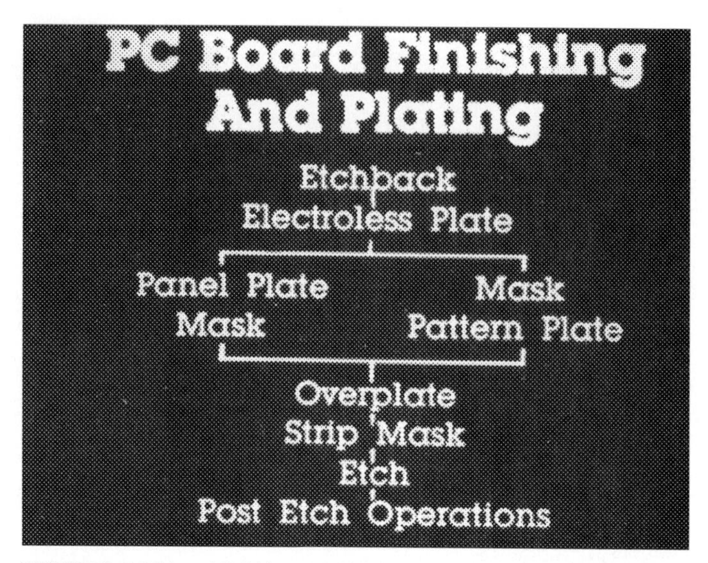

FIGURE 9.6 PC board finishing and plating. (*Reprinted with permission from the* AESF Illustrated Lecture, Printed Circuit/Wiring Board Manufacture and Plating for Electronic Applications.)

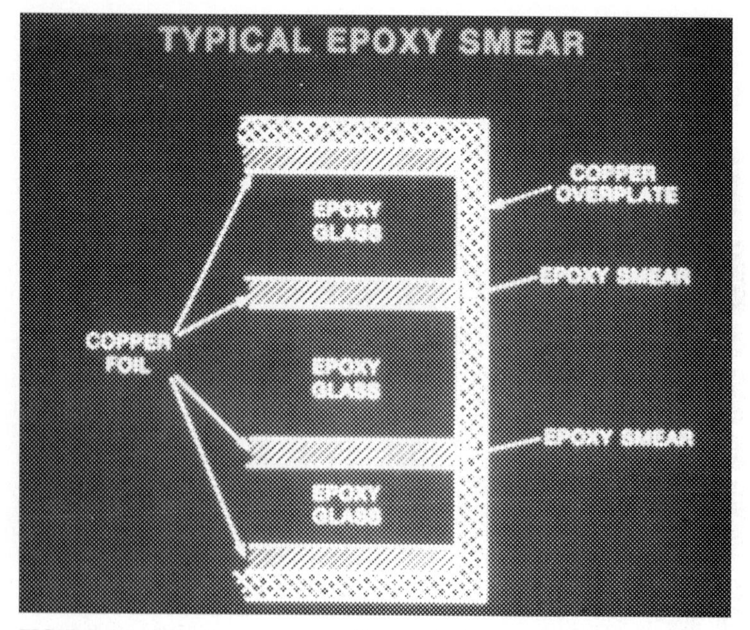

FIGURE 9.7 Typical epoxy smear. (*Reprinted with permission from the* AESF Illustrated Lecture, Printed Circuit/Wiring Board Manufacture and Plating for Electronic Applications.)

FIGURE 9.8 Epoxy smear and nail heading caused by poor drilling. (*Reprinted with permission from the* AESF Illustrated Lecture, Printed Circuit/Wiring Board Manufacture and Plating for Electronic Applications.)

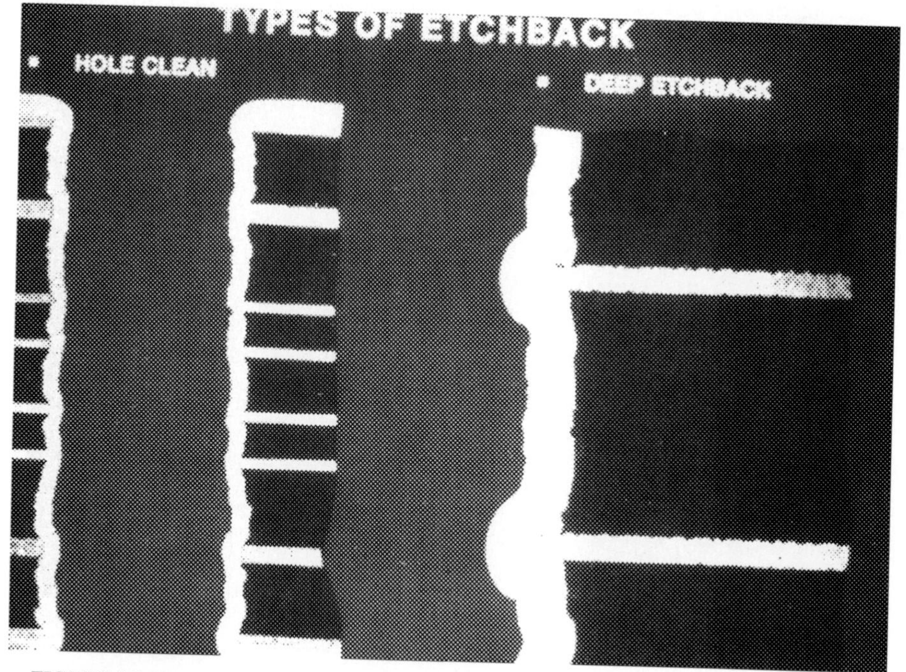

FIGURE 9.9 Types of etch-back. (*Reprinted with permission from the* AESF Illustrated Lecture, Printed Circuit/Wiring Board Manufacture and Plating for Electronic Applications.)

9.2.5 Electroless Plating-Through Holes

The process steps for electroless plating are:

- Cleaning
- Smear removal and/or etch-back
- Etching of copper (sometimes called *cleaning copper*)
- Activation or catalyzing
- Accelerator
- Electroless copper plating

Hot-alkaline cleaning is usually preferred since it wets and cleans well and neutralizes etch-back residue, if any. Acid cleaners are sometimes used. Thorough rinsing after cleaning is essential since some ingredients in the cleaners can poison subsequent catalyst steps.

The copper is then etched to assure complete cleanliness. The chemicals used are persulfate compounds, hydrogen peroxide–sulfuric acid, or cupric chloride. After 30 to 90 s of etching, the boards are rinsed and immersed in a 10% sulfuric acid solution to remove any copper oxide from the etching step.

After thorough rinsing, the polymer (nonconductor) surfaces are immersed in a catalyzing (activating) solution. Typical are tin-palladium combination catalysts or other nonprecious metal systems. These are generally colloidal in nature to prevent immersion (chemical replacement) deposits on the copper. Immersion deposits tend to produce poor adhesion when subsequently plated. The catalyst is adsorbed onto the polymer and is the material which starts the electroless deposition. The boards are rinsed after the catalyst step. An accelerator follows the catalyzing which dissolves the tin, leaving metallic palladium (or other noble or nonnoble metal) atoms on the polymer. For tin palladium systems the accelerator consists of either a dilute alkali solution or dilute acid; either can be used.

Following a rinse, electroless copper is deposited through the holes. The plating takes place in an alkaline solution in the presence of formaldehyde. Other reducing agents have also been used.

The electroless copper bath is usually made from copper sulfate (other copper salts may be used), a reducing agent (usually formaldehyde), sodium hydroxide to pH of 12 to 13, complexing agents such as ethylene diamine tetra acetate (EDTA) (or other similar complexors), stabilizers such as mercapto compounds or cyanides, and wetting agents.

The reaction is summarized as follows:

$$Cu^{2+} + 2HCHO + 4NaOH \xrightarrow{\text{catalyst}} Cu^{\circ} + 2HCO_2 + 2H_2O + 2Na^{+}$$

Other reducing agents have been used such as *dimethylamine borane* (DMAB) and glyoxylic acid.

Plating to a thickness of 20 to -40×10^{-6} in followed by electroplating from an acid copper bath to the final thickness, usually 0.001 in (panel plating) or electroplating to 0.0002 to 0.0003 in followed by masking areas to be etched later, then electroplating to the final thickness (pattern plating) are common techniques. Another is to use a "high-speed" electroless copper-plating solution to plate to 0.0001 in (100 millionths inch) followed by either panel or pattern plating. The copper-plating solution is usually an acid copper; however, pyrophosphate copper (pH 8 to 9) solutions have been used. Copper sulfate or fluoborate acid copper sulfate have also been used. The following chart lists typical operating data.

"Hi Throw" Copper Operating Data	
Copper sulfate $^1/_2H_2O$	8–12 oz/gal (60–90 g/liter)
Sulfuric acid	20–30 oz/gal (150–225 g/liter)
Chloride ion	30–60 ppm
Anode-to-cathode ratio	2:1
Temperature	Room
Current density	20–40 A/ft^2

The masking material called *resist* is available in several forms. Silk screen can define features of 12 to 20 mils line width and spacing. Photoresists can define features 2 to 3 mils (both negative and positive working). Photoresists exposed by ultraviolet light or x-ray can define features of 1 mil or less. Dry film resists are applied to the copper by pressure. Exposure is by active light (uv of 3000 to 4000Å). The polyester film is removed and the resist is developed, washing away the unexposed areas.

9.2.6 Plated Etch Resists

While the polymer resist is in place, exposing the circuit pattern and covering the unwanted copper, electroplated tin-lead, or nickel followed by gold, may be plated onto the circuit areas (0.0003 in to 1 mil). The polymer resist is then removed. The plated resist serves to protect the wanted copper areas, while the unwanted copper is dissolved by the etchant. *Note:* Polymer resists over the circuit areas are also used. Etchants are selected from a number of solutions, such as sulfuric acid–hydrogen peroxide, chromic acid–sulfuric acid, cupric chloride solutions, ferric chloride solutions, and alkaline etchants such as ammonium carbonate. The selection of the proper etchant is governed by the type of plated resist or organic resist, safety considerations, ease of copper recovery, and waste treatment.

9.2.7 Limitations

Design considerations take in limitations of the processes used to produce PWBs. For example, the plated-through-hole process presents limitations with respect to the hole diameter versus the depth or length of the hole, called the *aspect ratio.* Donaldson[1] asks us to consider a hole of 0.015 in (0.038 cm) diameter by 0.062 in (0.00254 cm) long as an aspect ratio of 4:1 (a low ratio). An acid copper solution contains 2.5 oz/gal (18.25 g/liter). When plating to the usual thickness specification [0.001 in (2.5 μm)] using the density of copper (8.96 g/cm^3), calculations show that the copper-plating solution must fill the hole (pass through) with fresh solution 634 times to deposit the required thickness. Horizontal agitation and solution impingement directly at the holes are required. Further, the current distribution is such that there will be less copper in the middle than at the outer portions (see the Theory Section for details). Designs have demanded aspect ratios of from 10:1 to as high as 30:1. These requirements make electroplated copper impossible to fulfill. Electroless copper and electroless nickel are used for via holes where interconnecting layers of circuitry is the only purpose. Components cannot be mounted in such holes; rather, components are surface mounted. This requires pads and depressions (i.e., blind holes, which are even more difficult to electroplate if the features are very small). Electroplating a conductor or pad results in a "dog bone" effect where edges and ends are plated thicker than the remaining areas as shown in Fig. 9.10. For best results, see Donaldson, "For Platers Only."[2]

Dimensional limitations exist for each type of resist. Screened resists can produce line width and spacing of 15 to 20 mils (6.0381 to 0.0508 cm). Dry film photoresist can produce 5 to 6 mil (0.0127 to 0.0152 cm) line width and spacing. Liquid photoresist is capable of 2 to 3 mils (0.00508 to 0.0076 cm) line width and spacing, and with special high-intensity radiation exposure, thinner lines and spacing are accomplished.

Line-width loss and undercutting due to the etching process must be considered when designing the circuit patterns (see Fig. 9.11).

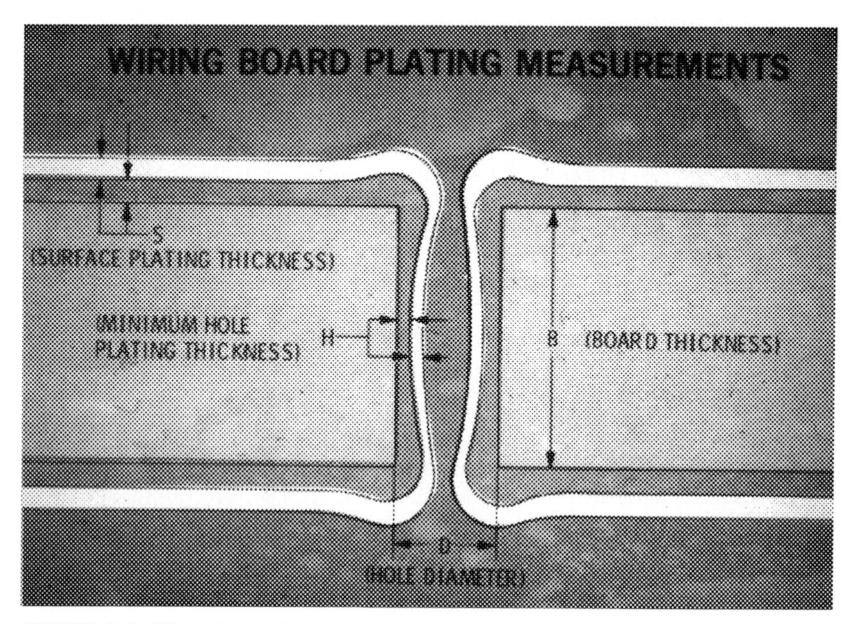

FIGURE 9.10 Wiring board plating measurements. (*Reprinted with permission from the* AESF Illustrated Lecture, Printed Circuit/Wiring Board Manufacture and Plating for Electronic Applications.)

9.3 NONELECTROLYTIC DEPOSITION OF METALS

Deposition on Metallic Substrates. The title of this section is somewhat misleading. Nonelectrolytic deposition is achieved without the purposeful imposition of an outside potential. Of course, in the total absence of any voltage, metal ions will not move in a preferred direction, and no deposit can form.

The source of such a potential in a system can arise in several ways and will be described rather loosely, since the actual theory involves electrochemical concepts that are rather abstruse.

Noncatalytic Deposition; Immersion Deposits Almost any metal can be satisfactorily deposited (coherently and reasonably adherently) from a properly formulated electrolyte on a more electropositive substrate without the use of an imposed electric

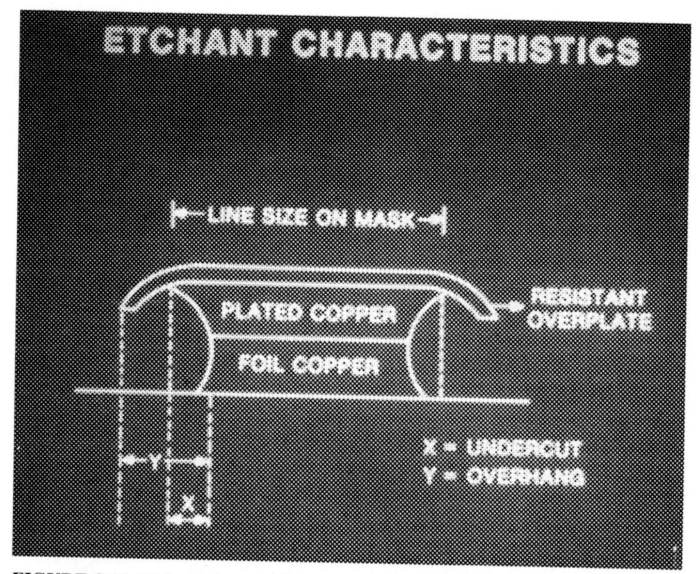

FIGURE 9.11 Etchant characteristics. (*Reprinted with permission from the* AESF Illustrated Lecture, Printed Circuit/Wiring Board Manufacture and Plating for Electronic Applications.)

potential. It must be emphasized that "electropositive," as used in this sense, is not a specific property of the metal itself. The polarity is determined by the relative potentials of the substrate and of the depositing metal in the particular solution in question. The familiar emf series of the metals (see Table 9.1), for instance, is only an arrangement of the metals in the order of their potentials against a standard specified concentration of simple solutions of their own salts at a definite temperature.

Plated plastics for functional and decorative purposes have become widely used. EMI and rf shielding produced by plating methods produce superior shielding compared to metal-filled polymer coatings and zinc-sprayed coatings.

Plating can provide strength, conductivity as in circuit patterns, moisture resistance, and surface hardness to plastic materials.

Plastic parts to be plated must be free of stresses if the best results are to be achieved. In some molded plastics such as ABS, stressed areas can crack when etched in preparation for plating. Stressed areas do not accept sensitization and catalyzing well enough, and skip plating can occur. To assure quality plating, quality plastics and molding techniques are essential. The design of molded parts must be such that wall thickness is sufficient and uniform; in addition, more gates than usual may be required, and careful attention to molding conditions such as time and temperature is critical if the parts are to be plated.

9.3.1 Adhesion of Metallized Coatings on Plastics

In no sense does the adhesion of a metallized coating on plastic approach that of an electrodeposited metal on a metal substrate. In the latter case the adhesion approaches

TABLE 9.1 Emf Series: Standard Electrode Potentials of Metals versus Solutions of Their Own Ions

(E_0 at unit molal activity and 25°C)

Metal/metal ion	E_0, potential, V
Base (or anodic) end of series:	
Magnesium/Mg^{2+}	−2.37
Beryllium/Be^{2+}	−1.85
Uranium/U^{3+}	−1.80
Aluminum/Al^{3+}	−1.66
Titanium/Ti^{2+}	−1.63
Zirconium/Zr^{4+}	−1.53
Manganese/Mn^{2+}	−1.18
Vanadium/V^{2+}	ca. −1.18
Columbium/Cb^{2+}	ca. −1.1
Zinc/Zn^{2+}	−0.763
Chromium/Cr^{3+}	−0.74
Iron/Fe^{2+}	−0.440
Cadmium/Cd^{2+}	−0.403
Indium/In^{3+}	−0.342
Cobalt/Co^{2+}	−0.277
Nickel/Ni^{2+}	−0.250
Tin/Sn^{2+}	−0.136
Lead/Pb^{2+}	−0.126
Iron/Fe^{3+}	−0.036
Hydrogen $H_2/2H^+$	0
Antimony/Sb^{3+}	0.2
Copper/Cu^{2+}	0.337
Silver/Ag^+	0.799
Palladium/Pd^{2+}	0.987
Plutonium/Pu^{2+}	ca. 1.2
Gold/Au^{3+}	1.50
Gold/Au^+	ca. 1.68

and usually exceeds the tensile strength of the weaker of the two metals involved. In the former case it is always possible to separate the metal coating from the plastic, but the measure of adhesion is only a measure of the force required to separate the two. Instead of the usual pounds per square inch measure, which describes the force necessary to effect separation between two areas supposed to be adherent, it is customary to measure adhesion to a plastic in terms of a peeling mechanism, and here the measure of adhesion is the pounds of pull per linear inch of the parting surfaces required to separate them. (Of course, this measure is in no way comparable to a true adhesion test performed in a tensile testing machine.)

Until fairly recently, a linear pull of 5 lb/in was considered acceptable for coatings plated on plastics. On certain formulations such as ABS, however, we are now looking for 10 to 15 lb/in as standard, while on others 25 to 35 is considered attainable. These adhesion values vary with the plastic, and such high adhesion values are achieved only with a few, such as polypropylene and some polyesters.

The degree of adhesion of plated plastics depends on the ability to produce microporosity in the surface of the plastic, such that, when plating takes place, it can fill the pores and form interlocking deposits. Each plastic requires a different etch procedure to produce micropores. An example of an etch solution is sulfuric acid 770 mL, potassium dichromate 15 g, water 230 mL. There are many variations on the preceding etch solution.

9.3.2 Plating Procedure

The plating procedure is usually one of the following two processes:

1. Clean using mild alkaline detergent cleaners.
2. Rinse.
3. Etch.
4. Rinse.
5. Reduce hexavalent chrome (mildly acidic sodium bisulfite is sometimes used).
6. Rinse.
7. *Sensitize:* Stannous chloride 10 g/liter, hydroquinone 5 g/liter, and hydrochloric acid 10 mL/liter is a typical formulation.
8. Rinse.
9. *Catalyze:* Palladium chloride 0.25 to 0.5 g/liter, and hydrochloric acid 10 mL/liter is a typical formulation.
10. Rinse.
11. 3% by volume hydrochloric acid.
12. Rinse.
13. Electroless nickel or electroless copper plate (or both for shielding 0.00005 to 0.0001 in nickel).
14. Rinse.
15. Electroplate the desired final coating(s).

Alternate process:

1. Clean using mild alkaline detergent cleaners.
2. Rinse.
3. Etch.
4. Rinse.
5. Reduce hexavalent chrome (mildly acidic sodium bisulfite is sometimes used).
6. Rinse.
7. 5 to 10% hydrochloric acid.
8. Mixed catalyst (proprietary mixtures of tin and palladium or mixtures of non-precious metal catalysts).
9. Rinse.
10. Activate (10% HCl or proprietary mixtures).
11. Rinse.
12. Electroless plate, and so on, as above.

9.4 TESTING OF PLATED DEPOSITS

Printed wiring boards can be tested automatically to determine thinning or cracking of plated-through holes by means of *radioscopy* (real-time x-ray). In radioscopy the CRT (video display) replaces photographic film. Microfocus can be used to closely examine any part of a plated-through-hole single- or multilayer PWB.[3]

9.4.1 Hardness Testing

"Hardness has been defined as resistance to indentation. Microhardness is probably an inaccurate term which actually refers to micro-indentation, a very small indent, not a very small 'hardness.' Hardness is generally measured by forcing an indenter into the surface of a test piece and measuring the depth of penetration or the imprint left by the indenter."[3]

Hardness testing of plated deposits should be done on cross sections of the test piece, unless the deposit is of sufficient thickness that there is no influence from the basis substrate. Practically, deposits used in electronics are thin. The Knoop indenter is specific for plated coatings and should be used in most cases. The Vickers indenter requires greater thickness of the deposit than the Knoop indenter (even when a cross section is used). The Vickers is said to be more tolerant to surface variations and is still sometimes used for plated deposits. "ASTM B578 presently describes only the Knoop indenter. Vickers is described along with Knoop in ASTM E384, but is not specific to coatings."[3] Hardness, solderability, heat resistance, ductility, braze, or weldability are characteristics sometimes specified.

9.4.3 Thickness Testing

Thickness testing is done in a variety of ways. Comparing weight before and after plating is sometimes used for small complex shaped components. Metallographic cross-sectioning and measuring the thickness using a microscope equipped with a filer eyepiece are also used. Most often, however, thickness testing instruments are used, such as x-ray, x-ray fluorescences, beta-backscattering, and anodic stripping. Adhesion testing is most often done by simple bend-and-tape tests. Grinding a selected area, cutting, and variations of these tests are sometimes used to determine adhesion.

A scanning electron microscope equipped with *electron dispersive x-ray* (EDAX) is a very useful tool used to examine and analyze surfaces for minor imperfections, to determine failure mode, and to identify materials, including trace elements.

9.5 ELECTROLESS NICKEL PLATING ON METALLIZED CERAMIC

An electroless nickel coating enhances the properties of the metallized patterns and elements of ceramic modules. Since the metallizing materials are not normally catalytic to electroless nickel, proper preplate treatment is important to avoid difficulties.

Metallized ceramic substrates overplated with electroless nickel are a logical response to the exacting design requirements of solid-state microelectronic components. Electroless nickel deposited on the metallized patterns and interconnects of the ceramic substrate plays an important part in producing sturdy, high-circuit-density electronic packages with increased reliability. For example, in hermetic sealing operations, a layer of electroless nickel-boron on the ceramic's metallized bonding areas provides the best hermetic seal possible. In high-circuit-density, *multilayer ceramic* (MLC-C) modules, which are characterized by narrow line width and spacing, an electroless nickel-boron deposit not only enhances the circuit's characteristics but avoids bridging between circuit patterns. Similarly, in hybrid circuitry, single and multiflipchip devices, memory arrays and logic circuits, the electroless nickel deposit not only assures continuous and dense conducting patterns and pads but increases their solderability and brazability and improves wire and diode bonding.

In this overview of the use of electroless nickel on metallized ceramic, popular metallizing materials and formulations are listed and their properties are generally described. The need for a subsequent electroless nickel coating is emphasized, and detailed pre-plate cleaning and treatment procedures are given for various metallizers to ensure proper adhesion of the electroless nickel.

9.5.1 Metallization

The electronics industry uses numerous types of conductive materials and formulations to form metallized circuit patterns and interconnects. (Detailed information on these metallizing materials, including mixing and firing procedures, can be found in Refs. 1 to 4.) The following are representative: molybdenum manganese silica; and tungsten manganese silica. Each of these materials is applied by making and brushing on a paste, using silk-screen for defining circuit patterns, and firing at high temperature. Although pure molybdenum and tungsten are popular choices, molybdenum manganese is the most widely used (see Table 9.2). There are no outstanding differences in the characteristics of these metallized coatings. All of these materials are capable of forming conductive, adherent coatings that exhibit high-temperature stability. Bond strengths are typically 70 to 105 MPa (10,000 to 15,000 psi).[5]
The refractory-metal-powder coating is nonmagnetic and thus has proven excellent for applications in the electronic industry. The process is also suited to the precision metallizing of many microminiature designs. Table 9.3 shows some practical tolerances applicable to these processes. Further electroplating tolerances must also be applied.

Because metallizing materials are not normally catalysts for electroless nickel plating, specific procedures must be used to make the metallized surface active and catalytic. When plating metals on ceramics, one must be aware of the differences in the thermal coefficient of expansion between the two materials. Large differences may result in debonding of the plating from the ceramic (see Table 9.4).

TABLE 9.2 Electroless Nickel Plating Cycles for Various Metallizing Materials (Rack and barrel plating)

Molybdenum-manganese-silica mixture (80% Mo, 10% Mn, 10% Si)	Molybdenum-manganese or molybdenum* (containing 1 to 3% Ni)	Tungsten-manganese-silica mixture (80% W, 10% Mn, 10% Si)
1. Soak clean in a strong alkaline cleaner for 5 min at 60°C (140°F) plus. Chelated cleaners similar to alkaline derusters are preferred. 2. Rinse for 2 min with deionized water at room temperature. 3. Immerse for 30 s in a 120 g/liter $NaHSO_4$ solution at room temperature. 4. Rinse for 2 min with deionized water at room temperature. 5. Catalyze in a palladium chloride solution (0.01 to 0.1% $PdCl_2$). To minimize the effect of an immersion deposit of palladium, use the lowest concentration of palladium that will provide an active catalytic surface. 6. Rinse for 2 min with deionized water at room temperature. 7. Dip in 1 percent hydrochloric acid for 2 min at room temperature. 8. Rinse with deionized water for 1 min at room temperature. 9. Plate with electroless nickel boron. (Nickel phosphorus can be used for some applications.)	1. Trichloroethylene vapor degrease, vapor blast, or alkaline clean. 2. Treat to remove traces of molybdenum manganese (or molybdenum) from the ceramic surface in areas between the circuit elements with a solution of 200 g/liter potassium ferricyanide and 100 g/liter potassium hydroxide, which also activates the surface for subsequent plating. Parts are immersed for 30 to 50 s at room temperature. A longer immersion time may cause loss of circuit dimensions as a result of excessive dissolution of metallizer materials. 3. Rinse for 1 min with deionized water. 4. Treat in a hot potassium hydroxide solution to remove traces of glass from the surface of the molybdenum manganese. Use a solution of 100 g/liter KOH at 100°C to boiling for 10 to 15 min. If the treatment time is too short, insufficient metal may be exposed for good plating adhesion; if treatment time is too long, a weakened plating bond may result. 5. Rinse thoroughly with deionized water. 6. Dip in 50% HCl for 8 to 15 s. 7. Rinse with deionized water. 8. Plate with electroless nickel boron.	1. Soak clean in a strong alkaline cleaner for 5 min at 60°C plus. Chelated cleaners similar to alkaline derusters are preferred. 2. Rinse for 2 min with deionized water at room temperature. 3. Immerse for 2 min in a 120 g/liter $NaHSO_4$ solution at room temperature. 4. Rinse for 2 min with deionized water at room temperature. 5. Immerse for 2 min in a solution of 20% by volume H_2SO_4, 100 g/liter $CuSO_4$, and 0.3% by volume HCl. 6. Rinse for 3 min with deionized water at room temperature. 7. Plate with electroless nickel boron in a bath especially designed to plate on copper.

*This process eliminates the use of palladium or other catalysts, thus reducing the risk of bridging between circuit elements. It should be noted that nickel-doped molybdenum-manganese accepts palladium better than molybdenum-manganese without nickel. Therefore, if palladium must be used, a more dilute catalyzing solution is recommended.

TABLE 9.3 Practical Metallizing Tolerances for Plating on Ceramics

Plating method	Pattern tolerance	Base metal thickness	Plate thickness	Solder coat thickness	Total metallized thickness
Spraying	±0.010	0.0004–0.0015	0.0002–0.001	0.0002–0.0015	0.0008–0.0035
Screening	±0.005 or 1%, whichever is larger	0.0004–0.0015	0.0002–0.001	0.0002–0.0015	0.0008–0.0035
Brushing	±0.020	0.0004–0.0015	0.0002–0.001	0.0002–0.0015	0.0008–0.0035
Banding by machine	±0.015	0.0004–0.0015	0.0002–0.001	0.0002–0.0015	0.0008–0.0035
Photoetching	±0.001	0.0004–0.0015	0.0002–0.001	0.0002–0.0015	0.0008–0.0035

Notes: 1. Tolerances apply to all low-temperature processes along with high-temperature refractory metal powder process.
2. Closer tolerances may be held on specific designs at increased cost.
3. Platings over this base metallizing affect these tolerances.

TABLE 9.4 Linear Coefficient of Thermal Expansion of Metals and Ceramics

Material	Linear expansion coefficient, in/in × 10^{-6}	Temperature range, °C
Ceramics	6.0–6.9	25–300
Ceramics	7.5–7.8	25–700
Ceramics	7.2–7.9	25–900
Chromium	6.2	20–200
Iridium	6.8	20–200
Rhodium	8.3	20–200
Titanium	8.4	20–200
Platinum	8.9	20–200
Beryllium	11.6	20–200
Palladium	11.8	20–200
Nickel	13.3	20–200
Cobalt	13.8	20–200
Gold	14.2	20–200
Copper	16.5–17.7	20–200
Silver	19.7	20–200
Aluminum	19.4–24.1	20–200
Magnesium	25.2–27.1	20–200
Cadmium	29.8	20–200
Zinc	39.7	20–200

9.5.2 Rack Plating Cycles

Cycle A: Ferricyanide Activation

1. Alkaline clean or vapor blast.
2. Treat to remove traces of molybdenum manganese (or molybdenum) from the ceramic surface in areas between the circuit elements. This is done with a solution of 200 g/liter potassium ferricyanide and 100 g/liter potassium hydroxide, which also activates the surface for subsequent plating. Parts are immersed for 30 to 50 s at room temperature. A longer immersion time may cause loss of circuit dimensions as a result of excessive dissolution of metallizer material.
3. Rinse for 1 min with deionized water; ultrasonic is preferred.
4. Immerse in ammonium bifluoride 16 oz/gal (120 g/liter) solution 5 to 8 min, or immerse in hot potassium hydroxide to remove traces of glass from the surface of the molybdenum manganese. Use a solution of 100 g/liter KOH at 100°C (212°F) to boiling for 10 to 15 min. If treatment time is too short, insufficient metal may be exposed for good plating adhesion; if treatment time is too long, a weakened plating bond may result.
5. Rinse thoroughly with deionized water.
6. Dip in 10% HCl for 6 to 10 s to remove metal oxides.
7. Rinse with deionized water.
8. Catalyze the surface for 30 s in a solution of 0.01 to 0.1% palladium chloride and 1 mL/liter concentrated HCl at room temperature. To minimize the effect of an immersion deposit, use the lowest concentration of palladium that will provide an active catalytic surface. The actual optimum concentration varies with the condition of the metallization and must be determined experimentally.
9. Rinse with deionized water.
10. Plate with electroless nickel boron to a thickness of at least 2.5 μm. This minimum thickness is required to assure that enough nickel is left on the surface for tinning and chip joining after brazing. Some nickel diffuses into the molybdenum during the brazing cycles (850°C; 1560°F). *Note:* Procedures for electroplating on molybdenum manganese or molybdenum are described in Refs. 6 and 7.

Cycle B: Hydrogen Fire Activation

1. Hydrogen fire in an atmosphere containing 25% hydrogen and 76% nitrogen (ammonia dissociation) for 18 min at 800°C (1470°F). This step is used to facilitate bonding of the metallized layer to the ceramic.
2. Catalyze with a solution of 0.01 to 0.1% palladium chloride and 1 mL/L concentrated HCl at room temperature. To minimize the effect of an immersion deposit, use the lowest concentration of palladium that will provide an active catalytic surface.
3. Rinse with deionized water.
4. Plate with electroless nickel boron.

9.5.3 Barrel Plating

1. Clean as in steps 1 through 5 of cycle A for rack plating. This should be done in a plastic basket.

2. Dip in 50% HCl at room temperature for 10 to 20 s.
3. Rinse with deionized water.
4. Transfer by pouring into plating barrel submerged in deionized water.
5. Add galvanic contact media which has been cleaned, dipped in 50% HCl, and rinsed.
6. Close barrel; dip in 20% HCl for 5 to 10 s.
7. Rinse with deionized water.
8. Plate with electroless nickel boron.

This procedure eliminates the use of palladium or other catalysts, thus reducing the risk of bridging between circuit elements. An efficient tumbling barrel is required. A polypropylene, octagonal-shaped barrel designed to provide maximum open area, fast draining, and steady delivery of fresh solution to all surfaces is recommended. Barrel rotation should be 6 to 8 rev/min. Galvanic contact can be made by using iron wire pieces 0.8 mm ($^1/_{32}$ in) long by 0.8 mm ($^1/_{32}$ in) in diameter, or steel balls (shot) 0.8 to 6.3 mm ($^1/_{32}$ to $^1/_4$ in) in diameter. The contact media can be reused.

Note: Tungsten is not used as widely as molybdenum manganese or molybdenum, and process details were not found in the literature. Generally, however, the same types of electroless nickel-plating solutions can be used, although the preparation steps may vary slightly.

9.6 DESIGN FOR PLATING

Since current distribution is not uniform over the work piece during electroplating, the thickness distribution which follows is not uniform (see Figs. 9.12 and 9.13). Design considerations should include allowances for differential thickness and should minimize variations. Corners and edges should be given a radius. Interior angles may be devoid of plating. Blind holes are difficult to plate. Gas entrapment causes voids in the deposit. Joints should be sealed by solder, braze, or welds so that solutions cannot be entrapped. Solution bleed-out from porosity, seams, or other sources will cause corrosion, unsightly spots, and possible functional failure of electronic devices. Rough surfaces make it difficult to provide corrosion protection to the basis metal.

Many metals are relatively easy to prepare and plate; others present difficulties which require very specialized preparation procedures and plating solutions. A very rough appraisal of the difficulty of plating upon various metals is indicated in the following listing.

9.6.1 Alloys

Alloying elements often make plating more difficult. Examples are beryllium copper or leaded brass. Since specialized preparation and plating procedures are necessary for different metals, the plater may be faced with an impossible task when two or more different metals are combined to make a single component which requires plating, or it may result in a very expensive plating operation. Always consult a plating expert in such cases. Hydrogen embrittlement of high-strength steels, spring steels, nickel-base alloys, and titanium can occur during preparatory steps and plating opera-

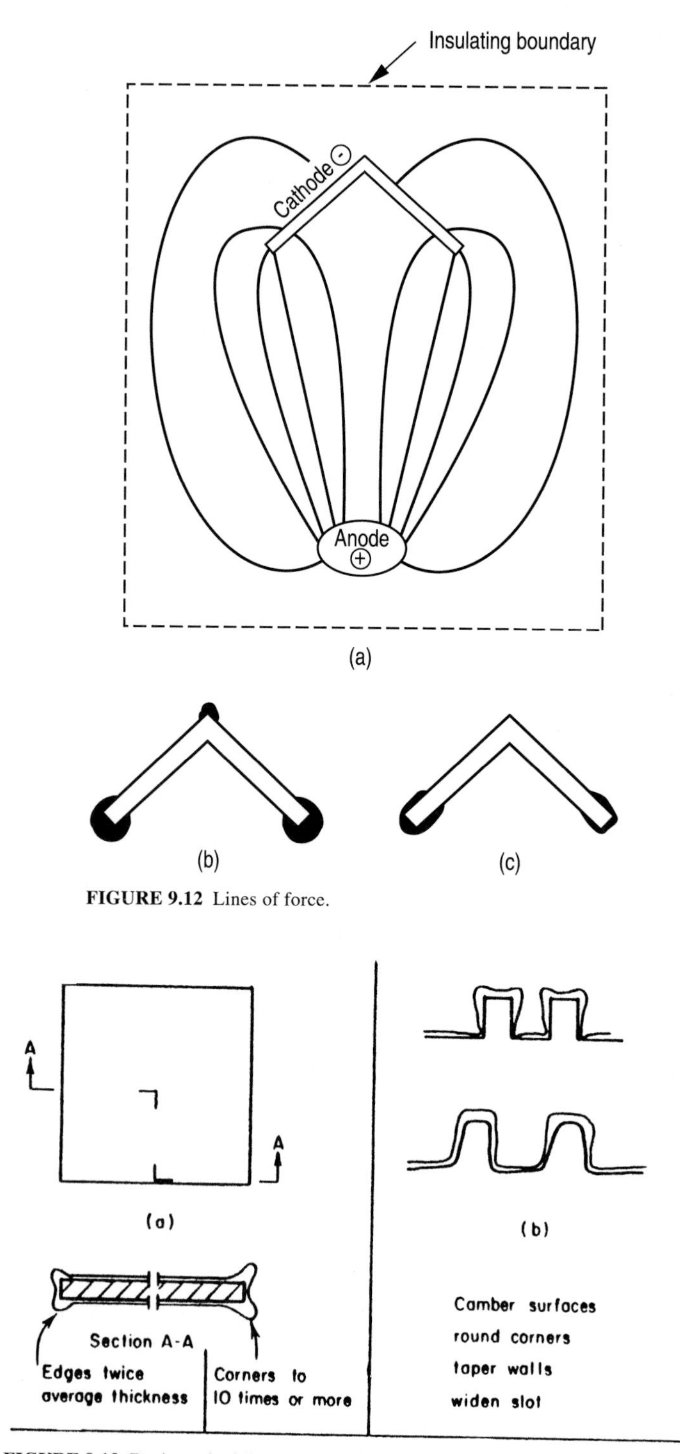

FIGURE 9.12 Lines of force.

FIGURE 9.13 Design principles.

tions. Embrittlement relief involves heating the part long enough (1 to 24 h) at a temperature sufficiently high (usually 275 to 400°F) to remove the hydrogen. To minimize the absorption of hydrogen, cathodic cleaning should be avoided and acid treatments should be short. Citric acid and sulfamic acid are mild, yet they effectively remove oxide so that they are often substituted for mineral acids such as hydrochloric acid and sulfuric acid. Hydrogen entrapment in the plated basis metal can also lead to blistering of plated deposits due to extreme pressure generated during elevated temperatures.

9.7 ELECTRODEPOSITION OF METALS

9.7.1 Common Metals Deposited from Aqueous Solutions

Only 13 of the common metals are readily electrodeposited from aqueous solution. As will be seen from Table 9.5, some can be deposited only from a single bath type and others from a choice of two or three types. The operating conditions shown are typical. Many other bath types are available, some of which serve special purposes. Speaking generally, the best baths are proprietary and are offered and serviced by a few specialized supply houses.

9.7.2 Other Metals Deposited from Aqueous Solutions

A number of other metals can be deposited (some with difficulty) from aqueous solutions and are of varying interest and commercial importance:

1. Platinum, palladium, iridium, and ruthenium are of some commercial interest.
2. Molybdenum and tungsten have not been deposited in the pure state, but alloys with other metals (chiefly iron, cobalt, and nickel) can be deposited.
3. Antimony and manganese are deposited chiefly for electrowinning purposes.
4. Mercury, gallium, thallium, arsenic, bismuth, selenium, tellurium, polonium, technetium, and rhenium can be deposited but are presently of no interest.
5. The deposition of osmium has received no attention, owing to scarcity, cost, and lack of demand.

Aluminum and germanium can be deposited from nonaqueous organic solutions, but not from aqueous solutions. The technical difficulties are many, including the problem of maintaining a water-free bath.

The following metals have not been deposited from either aqueous or nonaqueous baths (but most can be deposited from fused salt baths): beryllium, magnesium, calcium, strontium, barium, radium, scandium, yttrium, the lanthanide and actinide series, titanium, zirconium, hafnium, vanadium, niobium, tantalum, and the alkali metals.

Plating from fused salt baths is of marginal use only. For many applications, the high temperatures involved are too limiting, as for plating on plastics or metals of low melting point.

TABLE 9.5 Major Bath Types for Plating Common Metals from Aqueous Solutions

		Usual range of			
Metal	Chief salt(s)	pH	CD, asf*	Operating temperature, °F	Cathode efficiency, %
Cadmium	Fluoborate				
	Cyanide				
	Sulfate	<1	15–50	<80	95+
Chromium	Chromic acid				
	Sulfate (trivalent)	<1	50–300	100–130	14 ± 2
Cobalt	Sulfate				
	Chloride				
	Fluoborate				
	Sulfamate	3.5–5	50–300	120	98
Copper	Cyanide				
	Pyrophosphate				
	High-metal cyanide				
	Fluoborate				
	Sulfate				
Gold	Cyanide				
	"Acid" cyanide				
	Sulfite	8.5–9.5	1–15	95–130	98–100
Indium	Cyanide				
	Fluoborate				
	Sulfate				
Iron	Sulfate				
	Chloride				
	Fluoborate				
	Sulfamate	3.5–4.5	50–300	120	98
Lead	Fluoborate				
	Methane sulfonic acid	<1	5–300	70–100	100
Nickel	Sulfate chloride				
	Chloride				
	Fluoborate				
	Sulfamate				
Rhodium	Sulfate				
	Phosphate				
Silver	Cyanide				
	Cyanide nitrate				
Tin	Stannate				
	Fluoborate				
	Methane sulfonic acid	<1	5–75	70–100	100
Zinc	Cyanide				
	Sulfate				
	Sodium hydroxide	>13	10–50	Room–100	50–70
	Chloride	4.8–5.5	5–30	Room–110	60–9 0

9.8 POSTPLATING TREATMENTS

9.8.1 Bright Dipping

Zinc, cadmium, copper, and silver are sometimes "bright dipped" to enhance the appearance and leave a surface which fingerprints less readily.

9.8.2 Conversion Coatings

Chromate conversion coatings are widely used on zinc, cadmium, copper, and silver to provide added corrosion protection.

9.8.3 Water-Dip Lacquers

Water-dispersed polymers are used to provide a clear, thin protective coating. They are sometimes used over conversion coatings. Some commercial products allow ease of soldering by melting back in areas where heat is applied.

9.9 CHARACTERISTICS OF ELECTRODEPOSITED COATINGS IN GENERAL

The general characteristics desirable in plated deposits are soundness, good adhesion free from undesirable impurities, an acceptable level of intrinsic stress, and corrosion resistance.

9.9.1 Soundness

Soundness of the deposit refers to lack of such defects as pits, blisters, cracks, inclusions, roughness, burnt plate, and *skip plate* (isolated areas of no or very little plating scattered over a surface). These defects are caused by poor mechanical surface condition of the basis metal, poor preparatory steps, or a dirty or contaminated plating bath or one that is being operated at too high a current density.

When the basis metal is pitted, it is rare that the deposit fills the pits. Usually either the pit is bridged over with plate or else no metal deposits at all if a pit is filled with nonconductive debris. In the former case, gases trapped under the bridged plate may expand (e.g., in the hot rinse) and cause blisters. There are also many other causes of pits and blisters, including the mechanical inclusion of solid particles from the bath.

Cracking of other metal deposits may arise from flexure of a brittle coating, such as brittle nickel, or from a combination of stress and poor adhesion, in which case the deposit often exfoliates and curls up from the basis metal, or pops off. Generally roughness is caused by solid particles not removed by the cleaning cycle, or free solid matter floating in the plating bath itself, which particles are then plated over. Too high a plating current density causes burnt plate, while skip plating is caused by local areas that are either not clean or are passive.

9.9.2 Adhesion

The adhesion of an electrodeposit to the basis metal, *when* conditions are proper, is usually determined by the tensile strength of the weaker metal which will tear when an attempt is made to separate the two. For certain combinations of plate and basis metal, an intermediate alloy layer forms at the interface and sometimes, when this is weaker than either metal, rupture occurs at the interface even when the adhesion, per se, is perfect. But this degree of adhesion is almost always satisfactory, in contrast to the situation where preparation of the basis metal is poor. When metallographic examination of the cross section reveals some continuity of crystal structure (even if there may be an area of distortion due to differing crystal structure or of lattice parameters), adhesion will be perfect.

9.9.3 Crystal Structure

The structure of electrodeposits is often different from the usual structure of the massive metal. Also, changes in the deposition variables may cause a change from the normal crystal habit of the deposit. Nickel, for example, usually deposits in the face-centered cubic arrangement, but under certain conditions it can be plated in a hexagonal close-packed structure.

It is important that the designer avoid specifying (and that the plater avoid using) any procedures involving temperatures which would degrade the effects of any prior heat treatment. Operations which involve high temperatures, such as polishing and buffing, or special heat treatments for the purpose of hydrogen embrittlement relief and the like, are suspect, and a competent metallurgist should be consulted.

9.9.4 Internal Intrinsic Stress

All deposited metals (electroless as well as electrolytic) are laid down in a condition of more or less internal stress, which varies with the bath composition as the controlling variable and is further influenced by bath temperature, part geometry, and (for electrodeposits) by current density. These stresses are most often tensile, tending to crack the deposit, but are sometimes compressive, whereby the deposit tends to expand. If the adhesion is poor, a deposit with tensile stress may exfoliate and curl up, while a compressively stressed deposit will blister (this is only one of several causes of blistering, however).

9.9.5 Mechanical Properties

The mechanical properties of electrodeposits depend chiefly on bath composition and to a lesser extent on the conditions of operation within the operable range. Table 9.6 shows the range of mechanical properties for various metals obtained from different bath types.

Low-temperature mechanical properties of electrodeposits have particular importance for aerospace designers, but, unfortunately, few data have been published in this area.

9.9.6 Physical Properties

In contrast to mechanical properties, physical properties of electrodeposits are characteristic of the metal itself and are independent of bath and deposition variables. Table 9.7 lists a number of the physical properties of the important electrodepositable metals.

TABLE 9.6 Range of Mechanical Properties of Some Electrodeposits

Metal	Bath	Ultimate tensile strength, psi $\times 10^{-3}$	Elongation in 2 in, %
Cadmium	Cyanide		
Chromium†	Conventional	10–30	0–0.1
Copper	Sulfate	20–68	15–40
	Sulfate, with additions	69–90	1–20
	Fluoborate	17–50	6–25
	High-speed cyanide		30–50
	Above, with PR	100	6–9
Gold	Cyanide (conventional)		
	Proprietary		
Iron	Chloride	50–110	3–15
	Fluoborate, annealed	35–47	25–40
Nickel‡	Watts, no additions	50–88	20–25
	Watts, organic additions	170–230	0–5
	Cobalt-type baths	200	4
	All-chloride	95–140	2–20
	Fluoborate	55–120	5–30
	Sulfamate, no Cl	60–130	6–30
	Sulfamate with Cl	94–155	3–5
Palladium	Various		
Platinum	Amminonitrite		
Rhodium	Acid sulfate		
Silver	Cyanide	35–50	10–22
Tin	Stannate		
Zinc	Cyanide		

*Brinell hardness.

†For usual thin deposits used as topcoat.

‡See also Fig. 8.

TABLE 9.7 Physical Properties of Electrodeposits

Electrodeposit	Melting point, °C	Electrical resistivity, μΩ/cm	Thermal conductivity, % AG	Coefficient of linear expansion, 10^6 per °C	Modulus of elasticity, $\times 10^{-6}$
Cadmium	321	6.83 (0°)	22.2	29.8	8
Copper	1083	1.692 (20°)	91.8	16.6	17
Gold	1063	2.44 (20°)	70.0	14.2	10.8
Indium	157	8.8 (22°)	6	24.8	1.57
Nickel	1452	7.8 (20°)	14.2	13.5	30
Nickel (electroless)*	890	60	14	13	30
Platinum	1769	10.58 (20°)	16.6	8.9	22
Rhodium	1960	4.7 (0°)	21.0	8.5	41.2
Silver	961	1.59 (20°)	100	19.68	11
Tin	232	11.5 (20°)	15.5	20	5.9

*Nickel-phosphorous alloy—typical values.

9.10 STRIPPING OF ELECTRODEPOSITS

Stripping of defective deposits is probably the last thing the designer should have to worry about, except to realize that chemically inert metals such as rhodium and platinum are difficult, if not impossible, to strip from the more active substrates such as aluminum, zinc, or steel, unless there is an intermediate deposit more reactive than rhodium or platinum, such as nickel or (better) copper in the finishing system. Then the plater can arrange to attack and dissolve the more soluble undercoating so that the noble metal falls off mechanically.

9.11 MECHANISM OF CORROSION

9.11.1 Metallic and Chemical Finishes on Metals and Nonconductors

Except in cases where a metal is readily soluble in the corroding medium, most corrosion is due to galvanic action. Broadly, galvanic corrosion occurs when, for any reason, an electric potential exists between two areas of a part or of an assembly, while at the same time these areas are covered with or immersed in an electrolyte. This electrolyte may be only moisture condensed from the ambient atmosphere which, although it is intrinsically not a very conductive electrolyte, may quickly become one by dissolving carbon dioxide, chlorides, sulfates, nitrates, and the like from the air. Worse, if because of poor rinsing, residues of soldering fluxes or of plating solutions are left on the part, such condensation will *at once* form highly conductive electrolytes, which may be corrosive in themselves, and are also capable of carrying very appreciable currents at low voltages. Without the imposition of any external potential, a battery cell is formed, and drastic corrosion can occur. If, in addition, there is an externally applied direct voltage (which is often the case in an electronic assembly), the corrosion can be catastrophic.

Complete Inhibition of Corrosion. Only in a hermetically sealed space (e.g., with a metal-to-metal or metal-to-glass seal), which is completely free of both moisture and/or materials that may give off moisture or corrodents, can freedom from corrosion be assured. Other "seals" that may be used, such as rubber or elastomeric gaskets, leather, tapes, and shrinkable tubing, as well as liquid or paste, which solidifies in place to form seals, do not constitute true hermetic seals. All will "breathe" to a greater or lesser extent with changes in atmospheric pressure, and so draw moisture into the enclosure.

In every instance, the effect of moisture is damaging. Since a molecule of water has a diameter of only 3.4 Å, it can penetrate a very small opening, and so will diffuse through seals of rubber, paint, and many other materials.

9.11.2 Galvanic Corrosion

This arises from the use of dissimilar metals in contact and is probably the greatest single source of corrosion difficulties in sophisticated electronic gear, in which it is almost a mandatory feature of the design that many different metals must be used.

Numerous tables have been prepared for assisting the designer to avoid the most damaging combinations of metals. These tables, however, are not universally applicable, because each is based on one selected set of conditions. The familiar electromotive force (emf) series (Table 9.1) is probably the least useful, since it is based on the *single* electrode potential of one metal against a rather concentrated solution of its own ions, nor does it consider the other metal, which in practical cases forms the second electrode of the battery. The *galvanic series,* on the other hand, can be used to estimate the voltage generated between two immersed metals, but *it is necessarily limited to one electrolyte,* such as seawater, for example. Thus, we find one galvanic series for seawater (Table 9.8), another for the atmosphere applicable to guided missiles (Table 9.9), and so on. Table 9.10 is another listing taken from MIL-STD-454, and there are many more. All have been compiled for a specific set of circumstances, and the best design procedure is to select or find a table most closely describing the environmental conditions to be met. Unfortunately, no clearer guidelines can now be drawn.

Undesirable dissimilar metal combinations on different pieces can be made tolerable by plating each part. It is *extremely dangerous,* however, to extrapolate this freedom to the design of a *single* component which is constructed of several widely dissimilar metals joined together for plating. Such an agglomeration may quite often be

TABLE 9.8 Galvanic Series in Seawater

Noble (*cathodic*)	Naval brass
Platinum	Manganese bronze
Gold	Muntz metal
Graphite	
	Tin
Silver	Lead
Passive 316 stainless (18-8-3)	Active 316 stainless (18-8-3)
Passive 304 stainless (18-8)	Active 316 stainless (18-8)
Titanium	Active 410 Cr stainless
Passive 410 Cr stainless (13%)	
67 Ni 33 Cu	Cast iron
	Wrought iron
Passive 76 Ni 16 Cr 7 Fe	Mild steel
Passive nickel	
Silver solder	Aluminum 2024
M-B bronze	
G-bronze	Cadmium
70–30 cupronickel	
	Alclad
Silicon bronze	Aluminum 6053
Copper	
Red brass	Galvanized steel
Aluminum brass	Zinc
Admiralty brass	Magnesium alloys
Yellow brass	Magnesium
Active 76 Ni 16 Cr 7 Fe	*Base or active* (*anodic*)
Active nickel	

Note: Within a group, galvanic effects will usually not be appreciable. With other combinations, intensity of galvanic effects will vary with the distance apart in the series and with the relative areas of the materials (small anodic area worst).

TABLE 9.9 Galvanic Couples as Listed in MIL-STD-186B for Guided Missiles

Group	Metallurgical category	emf, V	Permissible couples
1	Gold, gold-platinum, platinum	0.15	
2	Rhodium, graphite	0.05	
3	Silver	0	
4	Nickel, Monel,* high nickel copper alloys, titanium	−0.15	
5	Copper, low brass or bronze, silver solder, German silver, high copper nickel alloys, nickel chromium, austenitic (type 300) stainless steels	−0.20	
6	Commercial yellow brass and bronze	−0.25	
7	High brass and bronze; naval brass, Muntz metal	−0.03	
8	18% chromium steels	−0.35	
9	Chromium, tin, 12% chromium steels	−0.45	
10	Tin-plate, tin-lead solders, terneplate	−0.50	
11	Lead, high-lead alloys	−0.55	
12	Aluminum, 2000 series wrought	−0.60	
13	Iron, low-alloy steels, Armco iron	−0.70	
14	Aluminum 3000, 6000, and 7000 series; aluminum silicon castings	−0.75	
15	Aluminum castings (other than silicon alloys); cadmium	−0.80	
16	Hot-dip zinc and galvanized steel	−1.05	
17	Zinc	−1.10	
18	Magnesium	−1.60	

O = cathodic ● = anodic

Note: Groups joined with arrows are permissible. Other combinations (usually over 0.1-V potential difference) can result in harmful corrosion. Magnesium, for instance, *must* be isolated from any other metal, preferably by a combination of plating and painting.

*Trademark of Huntington Alloy Products Division, The International Nickel Co., Inc., Huntington, W. Va.

impossible to plate satisfactorily, because the *required* pretreatment for one metal may cause catastrophic destruction of another, or at the very least may be responsible for the plate blistering on one of the other metals. When only two metals are involved, a compromise may often be worked out. Proposed untested combinations should first be cleared with a fully experienced plating engineer. The farther apart the metals are in a galvanic series, the greater is the likelihood of serious trouble.

TABLE 9.10 Groups of Compatible Metals in MIL-STD-454

Group I / Group III	Group II / Group IV
Group I	*Group II*
Magnesium and alloys	Aluminum and alloys
Aluminum 5052, 5056, 5356, 6061, 6063	Zinc
Tin	Tin
Group III	Cadmium
Zinc	Stainless steel
Cadmium	Tin lead
Steel	Solder
Tin	*Group IV*
Lead	Copper and alloys
Stainless steel	Nickel and alloys
Nickel and alloys	Chromium
Tin lead	Stainless steel
Solder	Silver
	Gold

Note: All metals in one group are considered *similar* and unlikely to corrode from galvanic action, but are *dissimilar* to metals in another group.

This listing is similar to that in MIL-E-5400.

9.11.3 Crevice Corrosion

Under spot-welded joints, in crevices at threaded fasteners, or under washers there will be a lack of available oxygen. In the presence of moisture, an electric cell is formed, and there will be metal attack. The attack can occur either where the two metals meet, or wholly underneath the joint. The latter is the most dangerous, since the progress of the corrosion is not visible until it is so far advanced that the joint may be in danger of failure.

Sealing the joint to exclude moisture will prevent this attack. (Sealing is necessary even when nonmetallic washers are used, since these keep out oxygen just as a metal would.) An epoxy, polyurethane, or silicone rubber sealant is preferred. For "stainless" steels in marine atmospheres, effective seals are 50:50 lead-tin solder or 10 percent copper powder in petrolatum.

9.11.4 Surface Contamination

Corrosion contributable to surface contamination may be wholly unrelated to galvanic corrosion (e.g., caused by a fingerprint), but when the opportunity for galvanic corrosion also is present, the effect of the contamination is enhanced. Even mild corrosion of contact surfaces can be ruinous to electronic components, causing noise in potentiometers, poor solderability, and other defects.

To avoid this type of attack, the designer should:

1. Select inherently resistant metals.
2. Use protective coatings where practical.
3. Improve resistance by passivating.

9.11.5 Localized Differences in Metal Surface (or Surfaces)

Commercial metals and alloys are not homogeneous. In aluminum alloys, the grain boundaries are anoded to the grain centers and are thus subject to corrosion. To avoid this:

1. Use the 5000 or 6000 series rather than the 2000 or 7000 series.
2. In heat treating, use a rapid quench.
3. Use clad aluminum rather than bare.
4. Anodize where practical.

Another type of local cell is formed when a small surface of one metal (e.g., a rivet head) is surrounded by a large area of another metal. If the smaller area is anodic, the total corrosion current results in a high current density and rapid corrosion at the anode (e.g., an aluminum rivet in a steel sheet). The reverse situation is not unduly troublesome, unless a mild diffused corrosion on the large area cannot be tolerated.

9.11.6 Galvanic Corrosion as Applied to Aerospace Components

Special problems are also introduced when aircraft or aerospace conditions are considered.

9.11.7 Cycling Temperature Changes and Pressure Conditions

These changes can be a major problem, since with a drop in temperature, relative humidity rises, and water condenses rapidly. A return to high temperature accelerates the corrosion as long as the water remains. There can also be "breathing" in closed but not hermetically sealed compartments.

Cycling pressure changes (as in aircraft alternately aloft and aground) cause massive breathing effects, with similar results.

Plastics. Plastics, normally thought of as inert and noncorrosive (especially when incompletely cured), give off corrosive vapors by outgassing at high altitudes or in space, which vapors attack metals in the vicinity, or metals plated on them (see Table 9.11). Cadmium is especially susceptible to this form of attack.

Exaggerated Galvanic Corrosion. Exaggerated galvanic corrosion is likely to occur in an electronic assembly, where an externally applied direct voltage, sometimes of considerable magnitude, may be imposed between parts subject to wetting with an electrolyte. Currents may pass through the electrolyte and influence the corrosion type and rate (as well as having other undesirable effects; for example, causing silver migration and a subsequent short circuit). These currents may act to reinforce a galvanically generated current and so accelerate the corrosion, or they may oppose it, actually reversing the current direction, to cause corrosion of the normally more noble member of the couple. Intermittent passage of such a current could conceivably cause corrosion of both the base and noble metals comprising the couple.

When other factors, such as vibration may be involved, entirely similar considerations apply.

TABLE 9.11 Organics as Source of Vapor Corrosive to Metals

Material	Severely corrosive	Somewhat corrosive	Not corrosive
Adhesive	Urea-formaldehyde	Phenol-formaldehyde	Epoxy
Gasket	Neoprene*-asbestos	Nitrile-asbestos	
	Resin-cork	Glue-cellulose	
Insulation (wire)	Vinyl	Teflon*	Polyurethane
	Polyvinyl chloride	Nylon	Polycarbonate
	Vinylidene fluoride	Polyimide	
Sealer	Polysulfide	Epoxy	Silicone
Sleeving	Vinyl	Silicone	
	Polyvinyl chloride		
Tubing	Neoprene,* shrinkable		
Plastics	Melamine	Polyester	Silicone
	ABS	Diallyl phthalate	Epoxy
	Phenolic		Polyurethane
Varnish	Vinyl	Alkyd	

Note: Extent of attack is worse when plastic is incompletely cured; at high temperatures of operation, it may vary with the plastic used.

*Trademark of E. I. du Pont de Nemours & Co., Wilmington, Del.

9.11.8 Other Causes of Failure

At least two other types of equipment failure may be caused by metal-finishing operations: (1) hydrogen embrittlement and (2) metal whiskers. When problems of this nature arise, they are best dealt with by metallurgical experts, but the designer should at least be aware of the conditions that usually give rise to each.

Hydrogen Embrittlement. Some metals and alloys containing absorbed hydrogen become quite brittle, particularly when subjected to alternating stresses. This is primarily a problem associated with spring steels, high-strength iron-base and nickel-base alloys, and titanium; particularly when electrocleaned cathodically; or pickled in acids; or plated with chromium or cyanide copper, zinc, or cadmium. Since hydrogen may be evolved in the corroding process, embrittlement may also occur along with the corrosion.

The relief of hydrogen embrittlement involves a baking stress relief before and *immediately* after plating, usually for 3 or preferably 24 h at 180°C (375°F), but this operation must be conducted with due regard for the metallurgical history of the workpiece. Embrittlement is most likely to occur in heat-treated parts, and every precaution must be taken to avoid baking times and temperatures that may deteriorate prior heat treatment.

Metal Whiskers. Metal whiskers may grow in storage (even in hermetically sealed units). Electroplated tin, zinc, cadmium, copper, and iron exhibit this phenomenon which, though fairly rare and not well understood, can cause serious short-circuiting of electronic circuits, especially when the physical size of the circuits is reduced as in microcircuitry elements. Whisker growth is not prevented but is retarded under conditions of low moisture and low temperature. If tin plating is used (the phenomenon seems to be most common with tin), it should be hot-dipped when practical, or reflowed by fusing after electroplating. Alloying tin with lead or bismuth (1 to 5 percent) can prevent whisker growth.

9.11.9 Designing for Corrosion-Resistant Finishes

Table 9.12 has been prepared to assist the designer in preventing galvanic corrosion. When doubts or uncertainties arise, some tests described below may assist in deciding whether certain combinations in the design are feasible or not.

Salt-Spray Test. This test does not correlate with service life, but it can be a valuable tool for locating major weaknesses in the selection of materials. It should be noted that salt deposits may bridge insulating barriers and cause failure in any subsequent electrical testing, especially under humid conditions. Unless the part is specifically designed to function under conditions of salt plus humidity, it should be well rinsed and dried after salt spray before any electrical testing is done.

Testing beyond 50 hr rarely causes more corrosion spots; usually each existing spot merely becomes larger.

Tests with Imposed Current and Other Factors. Although the ultimate answer to a design problem must be an environmental performance test of the complete assembled unit under simulated service conditions, it may sometimes be possible to impress the operating voltage on a subassembly still in the experimental design stage, while this unit is in a controlled test environment, such as cycling temperatures and humidity. When this can be done in advance of the final performance test (when both time and cost will be matters of more pressing concern), considerable savings can be achieved.

It is to be anticipated that such a test will be more significant than a simple environmental exposure without an imposed current, and the sooner the combined effect of galvanic and imposed currents can be established, the less need there will be for hasty, last-minute design changes and their associated cost and inefficiency.

9.11.10 Plating Thickness

Assuming that the designer has selected the optimum combination of coatings for minimizing galvanic effects, it remains for him or her to specify an adequate *thickness* for the plated coating. Two general problems must be considered:

Coating Anodic to the Substrate (e.g., zinc on aluminum, or cadmium on steel). Corrosion of the basis metal is prevented by the sacrificial protection of the coating. Even if the coating is porous initially or (as it will) corrodes away to form pits or pores, protection is afforded unless the coating is too severely eaten away. Exposure to service conditions can therefore be expected to result at first in worsening the appearance of the part. The duration of protection of the substrate is a direct function of coating thickness. In electronic parts, however, *no* corrosion products of either substrate or coating can be tolerated, as a rule, and this combination is thus generally to be avoided.

TABLE 9.12 Designing for a Corrosion-Resistant Finish

Problem	Solution	Example
Dissimilar metals	1. Select metals from appropriate table of permissible couples (Tables 9.8 to 9.10).	Use nickel or rhodium, not brass or bronze, next to silver.
	2. Plate with compatible metal to reduce potential difference.	Tinplate aluminum and bronze used together.
	3. Keep affected area of less-noble metal as large as possible.	Stainless steel hardware in sheet aluminum *may* be satisfactory because of large area of aluminum (but *not* the reverse).
Contact	4. Apply corrosion inhibitors such as zinc chromate paste.	Assemble dissimilar hardware with zinc chromate paste.
	5. Interpose inert barrier or gaskets to prevent contact (extend 1/4 in beyond joint).	Vinyl tape, rubber gasket (and sometimes a plated washer).
	6. Paint both metals (or cathode at least) with *alkali-resistant* organic coating.	MIL-P-52192 or MIL-P-15930.
Electrolyte	7. Avoid designs where moisture can be trapped.	Use sealant bead on crimped, spot welded, and threaded joints.
	8. Use dessicant.	Useful only in hermetically sealed compartment.
	9. Seal joint with organic insulation.	MIL-S-7124.
	10. Seal metal faces against contact with electrolyte.	Primer, paint, or sealant.
General	11. Where possible, avoid use of magnesium.	Protection of magnesium requires very special attention.
	12. Do not use zinc-plate on aluminum; use cadmium.	
	13. Avoid using cadmium in high vacuum.	

Coating Cathodic to (i.e., nobler than) the Substrate (e.g., lead tin on aluminum; silver on nickel, or on bronze; brass or nickel on steel). The protection offered by the coating in this case depends solely on its integrity. Once perforated, the coating then causes accelerated attack on the basis metal. Since it can only protect the basis metal as long as it can exclude the corrosive environment from contact with the substrate, a minimum deposit thickness is required to avoid the formation of pinholes.

Because the porosity of deposits varies widely and because the corrosivity of environments cannot be strictly defined, it is not possible to make a general statement as to the minimum deposit thickness required in a given case. However, less than 0.0010 in of any commercially electrodeposited metal is almost certain to show some porosity, and hence to be of very doubtful value in an aggressive atmosphere (see Table 9.13).

Effects of Thickness. Whenever, for structural, electrical, or other reasons, dissimilar metals must be used in parts having electric contact with one another whereby an unacceptable corrosion voltage could be set up, the effects of the voltage can be minimized or negated by plating the two parts with a sufficient thickness of the same metal. The metal to be plated will be chosen with a view to its functional properties of conductivity, magnetism, and so on, and will, as has been stated, usually be cathodic to the substrate. Whether the effect of the corrosion potential is wholly or only partly suppressed will depend on the porosity of the deposit, which in turn is a direct function of the plating thickness. A thickness of 0.5 mil of most commercial electrodeposits is only relatively pore-free, and 1.0 mil is far superior. In severely aggressive environments and especially when a superimposed exterior potential exists, 3 to 5 mils may be required.

When considerations of weight and closeness of dimensional tolerance are paramount, plate thicknesses of marginal corrosion resistance value may have to be considered, and the probable consequences of plating failure must be weighed against the weight or tolerance penalty.

Variations in Thickness. In design, due allowance must be made for the variations in plating thickness that must be expected. Unless special precautions are taken (which are usually not practiced in production tanks), the edge-to-center plating thickness ratio of a flat plate will be of the order of 2:1. For parts with only reasonably complex geometry (e.g., a topless cubical box) a ratio of 10 or more to 1 will be likely without special fixturing. By using suitable fixtures, which must be designed and tested *in advance* of production (just as any other machine tool or fixture must be), this ratio can be made to approach more closely the flat plate ratio of 2:1.

Except when electroless nickel is being used, another factor to consider when specifying a thicker plate is that the variation in plating thickness over a part is increased at least in proportion to the increase in thickness, and thus dimensional tolerances may be exceeded, even though the designer has made due allowance for the *average* thicknesses called out.

Specifying Thickness

General Considerations. Whenever a single metal can be deposited directly on a substrate and is able to perform its function over a useful service life without additional protection, there is no need for a multilayer plate. For various reasons, however, multilayer plates are frequently used and are called out in finish specifications as required. For estimating the probable relative corrosion resistance of such plates, neglect individual metal thicknesses of 0.0002 in or less, and consider that corrosion resistance will be afforded by the remaining major constituent(s), of which the thickest layer is most significant.

Service Conditions. The expected service conditions under which a component must operate elude precise definition, but obviously the opportunity for massive "breathing" exists, and for at least part of the time the atmosphere thus drawn in will be highly corrosive to electronic gear.

Consequence of Malfunction Due to Corrosion. The designer is in the position to decide whether corrosion will cause a malfunction that will be *critical* (immediate and

TABLE 9.13 Guide for Plating Thickness Selection

Plated metal	Thickness, mils	Probable protection rating*	Notes
Cadmium	0.1–0.3	0	
	0.5	1	
	1.0	2	
	2.0	3	
	0.2	0	Cd on aluminum
	0.3–0.5	1	
	0.2	0	
	0.3	1	Chromated Cd
	0.5	2	
Gold	Corrosion resistance depends on undercoat rather than on gold (up to 0.1 mil).		
Nickel	0.1–0.3	0	
	0.5–0.1	1	Electroplated
	1.5	1+	
	2.0	2	
	0.5	1	
	1.0–1.5	2	Electroless
	2.0	3	
	1.0	2	Electrodes on Al
	2.0	3	
Silver	0.1–0.2	0	
	0.3–0.5	1	
	0.7–1.0	2	
	2.0–5.0	2–3	
	0.1–0.2	0	
	0.3–0.5	1	On aluminum with 0.4-mil Cu strike
	0.7–1.0	2	
	2.0–3.0	2–3	

**Ratings:* 0 Minimal protection.
1 Marginal protection.
2 Good protection.
3 Excellent protection.

total loss of function, as by a dead short circuit), *major* (tolerable loss of function), or *minor,* or none at all.

For purposes of ensuring reliability, it is better for the designer to err on the side of too thick rather than too thin a plate, and the cost differential is minimal compared to the incalculable cost of malfunction of a critical component in military or space equipment. Naturally, as the criticality of the component diminishes, the cost of a heavier plating assumes a greater importance.

Selecting Thickness. Table 9.13 may be used as a rough guide in selecting thicknesses of deposits, but the final choice may have to be based on other factors, and the final responsibility is the designer's.

It should be emphasized that this table is necessarily only the roughest of guides. It is based on the usual porosity level of deposits and is related to the anticipated corrosion, as this is considered by the designer to affect the reliability of performance:

Rating	Probable protective value	Likelihood and effects of corrosion
0	Minimal	Corrosion may very likely occur, but it has no functional relationship. Any deterioration will relate to aesthetic values only.
1	Marginal	If corrosion occurs, no appreciable loss of functional reliability is expected.
2	Good	Some corrosion may occur in time. When it does, there may be a tolerable partial loss of function.
3	Excellent	Corrosion will require an extended period of time under severe conditions. If it occurs, corrosion may cause complete component failure. Periodic inspection can be expected to detect onset of corrosion before massive failure of the component.

9.12 PRECIOUS METAL PLATING

The design of electronic devices requiring precious metal plating is particularly critical. The high cost of these metals makes it important to conserve the amount of metal deposited. The current distribution, and thus metal distribution, is an important consideration. The relative difficulty of handling parts for plating must be considered. More gold is deposited on edges and high-current-density regions than on other surfaces (see Theory section). For example, some small parts are stamped, resulting in individual parts. The only technique for plating small parts is barrel plating. Overplating occurs at ends and outer surfaces. If the contact area is an inner surface, gold is wasted. It is more economical to use parts produced from flat stock where parts are connected together in strip form. Precious metals can be selectively plated on areas where needed. Shields can be utilized to aid metal distribution, and the parts can be plated on a strip line, reel to reel if the strip will tolerate coiling. There will be some gold recovery on the connecting areas if not masked.

Barrel plating is a viable alternative to racking of parts when parts cannot be made into strips.

9.12.1 Gold Plating

Gold plating is used for low-load contacts. Pure gold deposits are soft, malleable, and wear easily, so alloying metals are used to harden the gold. A small amount of cobalt or nickel is used to harden the deposits. Gold in pure form has the ability to form a low-temperature eutectic with silicon and is used on critical silicon chip carrier devices. Silver-filled epoxy bonding to silver plating has replaced gold for the majority of semiconductor applications. "Gold is applied to the surfaces of electronic components in one of four ways: (1) by electroplating; (2) by thermal decomposition of a screen-printed paste (thick-film technology); (3) by vacuum deposition or thermal decomposition of a metallo-organic compound (thin-film technology); or (4) by mechanical rolling to form a strip stock from which components can be stamped or punched (inlay materials)."[12]

Electroplating accounts for the greatest consumption of gold among these techniques since it is generally more versatile than the others and offers more precise control of the precious metal used.

Plating solutions usually contain potassium gold cyanide, $KAu(CN)_2$, conductive salts such as sodium citrate and phosphates, and hardening agents such as cobalt or nickel; pH is about 3.8 to 4.5. Pure gold deposits are usually pH 5.5 to 8. Alkaline-free–cyanide gold solutions are not generally used for electronic applications. Grain-refining additives are often used, including lead, arsenic, and thallium in trace amounts (up to 20 mg/liter). Thallium is prohibited for certain wire-bonding applications because of the risk of fatigue failure of the bond on aging.

Pure gold electrodeposits have a density of 19.3 g/cm^2. Alloy hard gold deposit density ranges from 17.3 to 17.7 g/cm^2, and has a hardness of 130 to 200 Knoop at a 25-g load. Hard alloy gold deposits should not be heated above 485°F (250°C). The "polymer" content decomposes at high temperatures.

It is sometimes necessary to solder to gold-plated components. Gold dissolves in molten solder. Gold-contaminated solder results in a weak, dull soldered joint. Gold thickness must be a minimum 20 μin (0.5 μm), and the solder monitored for gold content. Dewetting is a common failure when soldering to gold-plated parts. Causes include a dirty surface, passive nickel beneath the gold, or impurities in the gold deposit from the plating solution—usually organic contaminants. Carbon treating the plating solution will often help.

9.12.2 Silver Plating

Because silver has very low electrical resistance (1.59 μΩ • cm), it is used for many electronic devices. Wave guides use silver plating to minimize signal loss. Silver is used on other devices such as lead frames, contacts, and electrical connectors that are designed to operate at higher and less critical electrical loadings compared to gold-plated devices. Bonding silicon chips using silver-filled epoxy has led to an increase in silver-plating use for lead frames. Silver is a very ductile metal, allowing forming and bending operations without damage to the deposit.

Silver has limitations which must be considered. Silver tarnishes easily and reacts with sulfide to form films which increase contact resistance. The silver sulfide films are difficult to remove, being only soluble in oxidizing acids such as nitric acid and hot sulfuric acid. Silver is sometimes overplated with gold or rhodium for contact devices. Silver migration occurs under positive direct current potential where moisture is present, such as on a dielectric—a printed wiring board, for example. Silver can migrate across the dielectric, resulting in a short circuit or a low-resistance leakage path.

Plating solutions are primarily alkaline cyanide formulations using potassium silver cyanide, $KAg(CN)_2$, free potassium cyanide up to 16 oz/gal (120 g/liter), and potassium carbonate 2 oz/gal (15 g/liter), to provide solution conductivity. Brighteners and grain refiners are often added. The process uses pure silver anodes which dissolve to replenish the silver metal content of the solution. Other electrolytes have been used such as sulfamates, iodide complex, and fluoborates. These solutions do not have the excellent throwing power of cyanide solutions. These electrolytes are also affected by light and produce coarse deposits. Research is currently working to produce cyanide-free plating solutions which are commercially suitable.

Silver-plating solutions will form immersion or replacement deposits on many metals, including copper, iron, and their alloys. A preplating solution called a *silver strike* must be used to provide adequate adhesion. Immersion deposits tend to have poor adhesion. The strike solution consists of a very low silver concentration of 0.2 oz/gal (1.5 g/liter) to 0.4 oz/gal (3 g/liter) of silver cyanide and 10 to 12 oz/gal (75 to 90 g/liter) potassium cyanide. For steel and stainless steel, a prestrike containing a small amount of copper cyanide can be used for best adhesion.

Postplating treatments are often used to protect silver from tarnishing or fingerprinting during assembly operations. Chromate conversion coatings provide this protection without affecting solderability or contact resistance to any significant degree. Lacquers that can be soldered (water-dispersible and solvent types) and epoxy and acrylic polymers applied electrophoretically are good protectors but cannot be used where electrical contact or soldering is required.

9.12.3 Rhodium Plating

Plated rhodium is very hard and wear resistant. Its stable contact resistance, tarnish resistance, high melting point, and resistance to arcing provide good reasons to use rhodium for electronic contact applications. Rhodium is the most expensive of precious metals. It is, therefore, plated in thin deposits over nickel, nickel boron, silver, or ruthenium.

Plating solutions are usually sulfuric acid based for electronic applications and phosphate based for jewelry due to its whiter deposit color. Plating solutions are inefficient, producing hydrogen. Postbaking to remove hydrogen is used when embrittling metals are plated with rhodium.

9.12.4 Palladium Plating

The precious metal palladium is lower priced than gold and can be substituted for gold in some applications. Like nickel boron deposits, palladium and palladium-nickel alloys have a lower specific gravity than gold, are harder than gold, and in some cases reduce thickness of gold when used as an underlayer: on printed wiring board fingers, edge-card connectors, switches, lead frames for plastic packaged ICs, solderable contacts, end terminations for multilayered ceramic capacitors, etch resists for PWBs, glass-to-metal sealed contacts, and battery contacts.[12] Palladium silver alloys are used as well as nickel alloys. In addition to electroplating, palladium is applied as a paste and sintered on ceramic substrates. Vacuum deposition is used for thin-film ceramic circuits.

The plating solutions used to deposit palladium are either amine based or ammonia based. Organic addition agents are used in the ammonia-based solutions to produce brighteners, lower-stress and crack-free deposits. Palladium nickel alloys are plated from ammonia-based electrolytes. Nickel content can vary from 10 to 40 percent by weight of the deposit; 20 percent is typical. Amine-based solutions usually do not use additives and produce semibright deposits. Sulfonic acid-based electrolytes have been reported.

Postplating processes typically are 1 to 5 millionths of an inch (0.025 to 0.125 μm) of gold, usually hard gold. Lubricants are sometimes applied to improve wear resistance.

9.13 NONPRECIOUS METAL PLATING

9.13.1 Copper Plating

The primary use of copper plating for electronic industry applications is plated-through holes in printed wiring boards. Specifications require 0.0001 in (25 μm) minimum copper inside the holes. Both sulfuric acid and pyrophosphate electrolytes are used for this purpose and meet the requirements of adequate throwing power and ductility for most PWBs. The acid sulfate copper is the most widely used process. See the section "Printed Wiring Boards" for details.

Pyrophosphate copper solutions are mildly alkaline and not as corrosive as sulfuric acid copper electrolytes. A typical formulation for pyrophosphate copper plating is as follows:

Copper, as metal	3.5 to 4.5 oz/gal (26 to 33.5 g/liter)
Pyrophosphate	22 to 34 oz/gal (164 to 253 g/liter)
Ratio $N_{B2}P_20_7$	Cu by weight 7.0:1 to 8.0:1
Ammonia	0.1 to 0.4 oz/gal (0.75 to 3.0 g/liter)
Nitrate	0.6 to 1.3 oz/gal (4.5 to 9.7 g/liter)
Oxylate	2 to 4 oz/gal (15 to 30 g/liter)
Proprietary additives	As required

Air agitation from oil-free positive displacement blowers is required for both types of solutions. Anodes for pyrophosphate are oxygen free and pure, while anodes for acid copper require pure copper containing 0.02 to 0.08 percent phosphorus. Continuous filtration is required for both, with bath turnover rates of 10 times per hour or more recommended.

9.13.2 Electroless Nickel Plating

Because there are many types of electroless nickel solutions producing different deposit characteristics, a wide variety of electronic components can be plated with electroless nickel to enhance their functional characteristics. Deposits are uniform in thickness, hard (500 to 750 Knoop at a 100-g load as deposited), and corrosion resistant. There are solutions which produce deposits that are solderable, that can be brazed or welded, that are receptive to wire bonding and die bonding, and have low electrical resistance (6 to 13 μΩ • cm).

Nickel phosphorus deposits from different solution compositions produce alloy deposits of low phosphorus (1 to 4 percent), medium phosphorus (5 to 9.5 percent), and high phosphorus (10 to 13 percent). Each level of phosphorus content results in deposit characteristics different from the other. Likewise, nickel boron deposits which range from 0.2 to 3 percent boron provide a variety of deposits useful for electronic applications. See Table 9.14 and *Electroless Nickel Plating.*[13]

TABLE 9.14 Autocatalytic Nickel Systems Most Suitable for Obtaining Specific Deposit Characteristics

Characteristics desired	Autocatalytic system most suitable
Wear resistance	1. Nickel phosphorus, acid solution.
Corrosion resistance	1. Nickel phosphorus, acid solution formulated for maximum corrosion resistance on the basis metal selected. 2. A polyalloy of nickel tin phosphorus, nickel tin boron, nickel tungsten phosphorus, nickel tungsten boron, nickel tungsten tin phosphorus, nickel tungsten tin boron, or nickel copper phosphorus.*
Hardness	1. Nickel phosphorus, acid solution—heat treated. 2. If heat treatment is not practical, use nickel boron of 3% or more boron.
Lubricity	1. Nickel phosphorus, acid solution—highest phosphorus content.
Chemical resistance	1. Nickel phosphorus, acid solution. 2. Polyalloy system.
Solderability	1. Nickel boron—low boron content, less than 1%. 2. Polyalloys—exhibit excellent shelf-life solderability.
Diode-bonding	1. Polyalloys. 2. Nickel boron, less than 1% boron.
Nonmagnetic	1. Polyalloys.
Magnetic (for memory)	1. Nickel cobalt phosphorus. 2. Nickel cobalt boron. 3. Cobalt phosphorus. 4. Nickel cobalt iron phosphorus.
Electrical conductivity	1. Nickel boron with boron less than 0.3% (resistivity is about 5.8–6.0 $\mu\Omega$ cm/cm^2).
Electrical resistance	1. Some polyalloys. 2. Nickel phosphorus—high phosphorus content.
Rhodium replacement	1. Nickel boron, 1–3% boron.
Gold replacement	1. Nickel boron—low boron for soldering, 0.1–0.3% B; higher boron for contacts, 0.5–1% B. 2. Polyalloys of phosphorus or boron if below 0.5% P or B.

*Selection depends on severity of requirements and on economics. Boron-reduced systems are about five times more costly in terms of dollars per mil per square foot. However, thinner deposits may be used.

Examples. Uniform thickness of deposit makes nickel phosphorus an ideal process to protect aluminum heat sinks. It would be impossible to electroplate a fin-type heat sink due to the extreme recessed areas—no current could reach these areas. Hardness and corrosion protection of electroless nickel deposits on aluminum allow greater use of aluminum for many hardware applications.

Silicon transistor chips are difficult to solder and often use electroless nickel deposits to provide adherent solderable contact areas used for common grounds. Silicon wafers are plated directly using alkaline electroless nickel solutions of low phosphorus or boron content. Nickel-boron deposits are used for their excellent wire bonding of aluminum wire using ultrasonic bonding techniques. Gold wire is bonded using thermalsonic methods.

Electroless nickel deposits provide good diffusion barriers between copper and basis metal and gold-plated deposits. Medium to high phosphorus content and medium to high boron-containing deposits provide even better diffusion barriers. In some cases electroless nickel-boron deposits can replace gold entirely.[13]

Electroless nickel deposits are used to protect molybdenum manganese, molybdenum, and tungsten metallized ceramics and to improve electrical conductivity. The nickel deposit also provides the basis metal for precious metal deposits such as electroless gold and electrodeposited noble metals.

Electroless processes can be used for isolated areas where it would be difficult or impossible to make electrical contact for electroplating.

9.13.3 Nickel Plating

Nickel plating is used in electronics for its corrosion protection, hardness, good electrical conductance, and as a diffusion barrier between basis metals and precious metal deposits. The edge-board connectors (plug-in fingers) are almost always nickel plated (or electroless nickel plated) prior to gold or rhodium or other precious metal plating. Nickel deposits have a hardness of from about 140 to 250 Knoop hardness number at a 100-g load. Hardeners can be added to increase the hardness to as high as 550 Knoop at a 100-g load. However, deposits with organic hardening agents cannot be used above 500°F for prolonged times. The organic compounds tend to char or decompose at elevated temperatures, leaving a weak nickel structure. The electrical resistivity of a pure nickel deposit is 6.84 μΩ • cm. The melting point is 2647.4°F (1453°C).

There are a number of different plating solutions which can be used. The most popular for electronic applications is sulfamate nickel. Its high purity, low intrinsic stress, and ease of operation make it the solution of choice. The plating solution contains 10 oz/gal (75 g/liter) of nickel as metal added as a liquid nickel sulfamate, usually containing 24 oz/gal (178.8 g/liter) of nickel. Boric acid is added from 5.5 to 6.5 oz/gal (41 to 48.5 g/liter) to buffer the solution to a pH of 3.5 to 4.5. Wetting agents are used to prevent pitting, and a small amount of chloride [0.3 oz/gal (2.2 g/liter) or more up to 3 oz/gal (22 g/liter)] is added for barrel plating. Sulfur depolarized anodes must be used.

Watts nickel solutions are also used. They consist of nickel sulfate at about 40 oz/gal (298 g/liter) as $NiSO_4 \cdot 7H_20$, nickel chloride at from 6 to 10 oz/gal (45 to 75 g/liter), and boric acid at 5 to 6.5 oz/gal (37.5 to 48.5 g/liter). Wetting agents are added to prevent pitting.

Fluoborate nickel is sometimes used for very high speed plating. *Caution:* Fluorides are dangerous toxic compounds, sometimes used for special purposes but rarely for electronic applications.

9.13.4 Tin Plating

Tin deposits are widely used for electronic applications. Tin is nontoxic, ductile, and corrosion resistant. The excellent ductility allows tin-plated basis metals to be bent or formed into various shapes without damage to the tin coating. Tin's ability to protect copper, nickel, and tin nickel from oxidation preserves the solderability of the basis metal. Tin has excellent lubricity, making it suitable for sliding contacts. Tin is used also to eliminate incompatible metal couplings which could lead to accelerated corrosion of dissimilar metals. Tin is an effective etch resist for copper etchants used in making printed wiring boards, except for ferric chloride.

Pure tin is subject to spontaneous metallic whisker growth when an electric current flows through it or on long-time standing, unless the tin has been reflowed or unless additives, such as a small amount of lead, bismuth, or certain organic brighteners, are in the plating solution and are codeposited with the tin. Reflowing is accomplished by momentarily heating the tin deposit to slightly above its melting point of 450°F (232°C) and quenching quickly in acid-free kerosene over water. The heating can be done in a suitable oil bath or by infrared heating followed by rapid cooling. The resistivity of tin-plated deposits is approximately 11.5 $\mu\Omega \cdot$ cm.

Tin plating is done from six basic types of solutions: acid sulfate, acid sulfonate, acid fluoborate, Ferrostan, Halogen, and alkaline stannate. Tin-plating solutions have excellent throwing power. The alkaline potassium stannate solution has the best throwing power of all the tin solutions. The deposits from alkaline stannate solutions are pure tin; thus they must be reflowed for use in electronic applications. Aluminum heat sinks are often copper and tin plated in these solutions and then reflowed.

Plating solutions based on stannous sulfate can be bright or matte deposits, depending on the addition agents used. The bright deposits have improved corrosion resistance, reduced porosity, resistance to fingerprints, and improved solderability over the matte deposits. Soldering or reflowing may leave very slight carbon deposits due to the decomposition of a small amount of codeposited organic material. Matte deposits are more prone to whisker formation and tin pest. *Tin pest* is a recrystallization at very low temperatures which results in an unsatisfactory, brittle, weak structure. Deposits from alkaline stannate solutions are also subject to tin pest. The acid sulfate process operates at nearly 100 percent anode and cathode efficiency and has excellent throwing power.

Tin fluoborate solutions plate pure deposits at much higher current densities (up to 100 A/dm^2) than other processes. These, too, have about 100 percent anode and cathode efficiencies. Fluorides are dangerous toxic materials and should be handled with extreme care.

Tin-plating solutions based on methane sulfonic acid (15 to 20 percent by volume) require simple waste treatment procedures. They contain no fluorides or boron. The solutions are less corrosive than fluoborate processes. High-speed plating is possible, but the cost is much higher than any of the other processes.

9.13.5 Tin Lead Plating

An alloy of 63 percent tin and 37 percent lead forms the eutectic (lowest melting point) at 358°F (181°C) and can be electroplated from two types of solutions, acid fluoborate and acid sulfonate. Alloys ranging from 2 to 60 percent lead are also used. The eutectic alloy is used on printed wiring boards to enhance solderability, to act as an etch resist, and for corrosion protection. Ten to 40 percent tin alloys are plated onto wire to prevent oxidation of the basis metal and to enhance solderability.

Acid fluoborate solutions are the most common. They must be carefully maintained to control the alloy, by the ratio of metals in the bath and addition agents concentration. Free fluoboric acid maintains conductivity of the plating solution, while boric acid prevents formation of free fluoride which causes precipitation of lead fluoride.

The acid sulfonate solution utilizes alkane (usually methyl), sulfonic acid, lead and tin alkane sulfonate, and addition agents for grain refinement or brightness. Lead is a very toxic substance.

Reflowing of tin lead deposits is common. Temperatures above 400°F (204°C) melt the deposit, forming a true solder alloy. Hot vapor, hot oil, or infrared processes are used to reflow tin lead alloys.

Lead compounds and solutions should be handled with care. Avoid breathing vapors from melted lead.

9.14 OTHER PLATING AND SURFACE-FINISHING PROCESSES

9.14.1 Brush Plating

Brush plating, now sometimes called *electrochemical metallizing* (EM) technology, has changed considerably from its early beginnings. Its original junction was primarily touch-up and repair, but now production methods are common. Brush plating is performed with hand-held or portable tools rather than in tank plating. It consists of an anode covered with a swab of cotton or other cloth material soaked in a concentrated plating solution—or the solution can be metered or pumped into the swab or stylus. The part to be plated is made cathodic by connecting the negative pole of the power supply to the part; the anode (positive) pole is connected to the swab. When the plating tool (swab) is placed on the part, plating occurs. Continuous motion of the tool assures a sound deposit.

The equipment consists of a power supply which includes microprocessing controls for more accurate thickness control, a selection of styluses (swabs) and replaceable anodes, and pumps, filters, and coolers, when required. Plating proceeds up to 50 times faster, thus more concentrated plating solutions and special power supplies are required. Styluses are often pure graphite or a platinum iridium alloy.

Preparation of basis metals for plating consists of precleaning, electrocleaning, activation, and bonding (see Table 9.15).

Solutions for brush plating are available for most metals: for electronic applications, copper (acid or alkaline), cobalt, lead, tin, and alloys of tin and lead, bismuth, antimony, nickel (acid or alkaline), and zinc (acid, neutral, alkaline); platable precious metals are gallium, gold (acid, neutral, alkaline), indium, palladium, platinum, rhenium, rhodium, and silver. Alloys can be plated, and both sulfuric acid and chromic acid anodizing of aluminum is done.

9.14.2 Chromate Conversion Coatings for Aluminum

Conversion coatings are applied by first cleaning the aluminum, deoxidizing, and immersing the aluminum in a chromate solution for 1 to 3 min (appropriate rinsing must be done between each step and after chromating). The films are thin but very corrosion resistant, and they provide excellent paint adhesion. MIL-C-5541D requires 168 h of 5 percent salt-spray resistance, in accordance with ASTM B117.

9.14.3 Phosphate Coatings

Phosphate coatings provide paint adhesion for ferrous alloys and improve salt-spray resistance of painted steel. Iron phosphate is a very thin, inexpensive process step in preparation for painting. Zinc phosphate coatings offer more corrosion resistance for painted and oiled surfaces. Manganese phosphate is used primarily to improve lubrication. Phosphate coatings are sometimes used to promote adhesion of paint to aluminum (and to hold lubricant for extruding aluminum), zinc, and cadmium. However, chromate conversion coatings generally provide better adhesion and more corrosion protection to these metals.

TABLE 9.15 Preparation of Basis Metals Prior to Brush Plating

Basis metal	Electrocleaning	Activating	Bonding
Copper and its alloys, mild steel	Electroclean 8–12 V; rinse	Not usually needed	A nickel preplate is recommended under gold. For silver, use of gold or palladium flash.
Aluminum	Electroclean 8–12 V; rinse	Acid activator, 8–12 V; reverse current; rinse	Nickel (0.0003 in), 6–14 V; rinse
Aluminum alloys containing silicon	Electroclean 8–12 V; rinse	Acid activator, 8–14 V; reverse current; rinse; electroclean, 8–12 V; reverse current; rinse	Nickel (0.0003 in), 6–14 V; rinse
Nickel chrome alloys, stainless steel	Electroclean 10–15 V; rinse	Acid activator, 10–12 V; direct current; no rinse	Special nickel (0.00005 in), 8–14 V; rinse
High-carbon steels	Electroclean 10–12 V; rinse	Acid activator, 10–15 V; reverse current; rinse; neutral activator, 10–20 V; reverse current; rinse	Nickel (0.0003 in), 8–14 V; rinse
High-alloy steels containing carbon	Electroclean 10–15 V; rinse	Acid activator, 10–15 V; reverse current; rinse; neutral activator, 10–20 V; reverse current; rinse; acid activator, 10–12 V; direct current; no rinse	Special nickel, 8–14 V; rinse
Ultra-high-strength steels	Al_2O, dry abrasive blast	Optional: Quick, mild reverse electroclean, 10–12 V; rinse	Immediately brush plate with a low hydrogen embrittlement (LHE) cadmium
Lead tin alloys, white metals	Abrade area lightly, electroclean, 6–8 V; rinse		Brush plate using alkaline solutions. If a hard metal coating is desired, pre-plate with neutral nickel

Source: *AESF Shop Guide,* 9th ed., p. 42.

9.14.4 Plating and Surface Treatment of Magnesium

Magnesium is a very light metal; molecular weight is 24.32, atomic number 12. It is also very active and must be protected. Dow Chemical Company has published methods of plating, chromate coating, and anodizing processes. In addition, there are proprietary chromate coatings called out in Military Specification MIL-M-3171 and MIL-M-54202.

9.14.5 Anodized Aluminum

Anodizing is the process of electrolytically producing an adherent, hard, abrasion-resistant oxide coating using anodic (positive) current. The coating is porous prior to sealing, which allows dye colors to be applied. The final step, whether dyed or not, is to seal the coating in hot deionized (DI) water or nickel acetate solutions or other proprietary sealing materials. The oxide coatings produced by anodizing are electrically resistant coatings with high breakdown potential (up to 500 V dc depending on the type of solution and process used to produce the coating).

Sulfuric acid is the most widely used. A 15 percent wt/vol solution used at 15 to 20 V (positive), and a current density of 12 to 16 A/ft^2 is used. Coating thickness of 600 mg/ft^2 (0.6 mg/cm^2) or more is usual; 1 mil [0.001 in (25 μm)] is formed at 12 A/ft^2 at 70°F. These deposits are most suitable to accept dye colors.

Chromic acid anodizing solutions are specified for high-corrosion-resistant applications. A 5 to 10 percent by weight solution of chromic acid (CrO_3) operated at 120°F (49°C) and a potential of 40 to 60 V, anodic, for 40 to 60 min, is used. When the coating is to be dyed, a solution of 50 to 100 g/liter chromic acid, at a potential of about 40 V and a temperature of 95 to 105°F (35 to 40°C). Anodize time is 30 to 40 min.

Hard anodizing is done from modified sulfuric acid solutions operated at lower temperature and higher voltage, for example, 30°F (1.1°C) and 24 V. These solutions are used to produce thicker coatings—2500 mg/ft^2 (2.7 mg/cm^2).

Other anodizing solutions, less often used, are oxalic acid, boric acid, and phosphoric acid. Thin, tight-grained deposits are usually produced from phosphoric acid and phosphoric acid solution, while oxalic acid is used to produce up to 2-mil coatings of golden bronze color on most alloys.

9.15 PLATING SOLUTION CONSTITUENTS AND THEIR FUNCTIONS

9.15.1 Major Constituents

The plating bath must contain a soluble metal cation and an anion which usually serves to dissolve the anode. The bath may also contain buffer salts which are designed to maintain it in the proper pH range; salts to improve the conductivity; wetting agents to prevent pitting; and additives for brightness, leveling, or other purposes.

Cations. The soluble metal salt is usually present in a fairly high concentration because this permits the use of higher current densities and hence faster plating. In the precious metal baths such as gold and rhodium, low concentrations are usually used because of the high metal cost; plating speeds are therefore slower in these baths.

Anions. The anion is chosen on the basis of a number of properties: solubility of the particular metal salt; the degree to which the metal is complexed (held in a complex radical, such as a cyanide, in contrast to a simple, readily disassociated molecule, such as a chloride); the corrosivity of the anion for the anode metal; and its chemical stability. The most commonly used anions are listed in Table 9.5. For various reasons, nitrates, nitrites, chlorates, perchlorates, iodides, bromides, borates, carbonates, and organic acid radicals are not used as the metal salts, although some of these may be used for other purposes.

9.15.2 Minor Constituents

Conductive salts are sometimes added for the sole purpose of improving bath conductivity, which is important for two reasons: A highly conductive bath materially reduces the operating cost, and the factors that control the secondary current distribution will have more effect, as may be seen from Eqs. (9.1) and (9.2).

With reference to the cathode, the electrochemical properties which determine the secondary current distribution are the cathode polarization and the conductivity of the solution. The *primary* current distribution between two points 1 and 2 on a cathode is given by the equation

$$\frac{i_1}{i_2} = \frac{d_2}{d_1} \tag{9.1}$$

where i_1 and i_2 are the current densities at points 1 and 2, respectively, at distances d_1 and d_2 from the anode *measured along the lines of force traversed by the current.* The secondary current distribution is generally accepted to be given by the equation

$$\frac{i_1'}{i_2'} = \frac{d_2 + k(dE/dI)}{d_1 + k(dE/dI)} \tag{9.2}$$

where k is the solution conductivity and dE/dI is the slope of the cathode polarization curve between the current densities i_1 and i_2. The second terms in the numerator and denominator are identical and have the dimensions of a length. The effect of polarization is therefore equivalent to adding an equal solution length to the path lengths d_1 and d_2. Also, therefore, the secondary current distribution is always more or less of an improvement over the primary distribution.

Wetting Agents. Surfactants are added to certain baths to avoid pitting. Pitting of a deposit is disastrous from every point of view. It occurs most often in the acid baths and can be traced to several causes, the most common of which are: suspended insoluble matter in the bath, organic contamination of the solution, and (sometimes associated with this) gas pitting. Gas pitting occurs most often on nonvertical surfaces in the baths with less than 100 percent cathode efficiency. The evolved hydrogen may cling to the surface (sometimes even a vertical one), and the bubble sits there preventing metal deposition, while a small round depression is formed in the plate, often with a streak above it caused by gas release as the bubble grows larger. Surface tension holds the bubble on the metal; consequently, by reducing the surface tension of the bath, we can prevent it from clinging and causing an observable pit. The surface tension of a plating bath is usually close to that of water—about 72 dyn/cm; by reducing this to 30 to 40 dyn/cm, pitting can often be prevented. For low levels of organic contamination, pitting from this source can also be relieved by lowering surface tension.

Buffers. Buffers are compounds used in a number of (usually acidic) baths to help maintain pH at the desired level. These are simply compounds which (over a specific narrow pH range for each buffer) resist normal pH changes that would otherwise occur due to local depletion of metal content and other chemical imbalances due to electrolysis.

Additives. Specially selected compounds are added in minor amounts (sometimes in the range of parts per million) to achieve special objectives, of which the most common are brightness and leveling. These compounds are discovered purely empirically, often by evaluating hundreds until one is found which works. The best of them produce bright plates, and some will produce a plate that is smoother than the surface on which it is deposited, whereas in the normal course of events, the plate is always rougher than the substrate.

9.15.3 Bath Operation and Control

Solution Control. All solutions in a plating line must be analytically controlled on a scheduled basis. The most critical solution is the plating or processing bath, which should be analyzed daily and maintained within specified limits.

Impurities. In a plating bath that is otherwise within control limits, impurities can cause every defect known to the art. Modern practice, therefore, is to use specially designed tools (Table 9.16) for their detection and control.

Analytical instruments used for plating solution control include atomic absorption (some elements require a graphite furnace for accurate determinations), liquid chromatographs, Knoop hardness tester, polarograph, and automatic titrators.

It is important that all constituents of the plating solution be maintained within specified limits designed for best performance. Automatic feed systems help greatly but do not replace analytical testing and control.

TABLE 9.16 Control Tests Used to Monitor Bath Contamination

Test	Method of operation	Notes
Hull cell	Sample is plated under controlled conditions of cell geometry to give a defined range of current density variations over the panel. Various refinements and variations of cell type are available.	Defects are shown at a lower impurity level than that which would cause rejected work. Experience is required to evaluate test panels.
Bent cathode	Sample is plated on a specially formed cathode (U-, J-, or pigtail). More limited than Hull cell, but widely used for certain baths.	Same.

Note: Both tests can detect the presence of impurities that would defy any analytical method.

SOURCE: Reprinted with permission from the November 1979 issue of *Plating and Surface Finishing*, the official journal of the American Electroplaters and Surface Finishers Society, Orlando, FL.

Operating temperature should be controlled closely [±2°F (±1°C)] for best results. Agitation is important. For some parts and solutions, cathode (the part being plated) movement provides the desired agitation. For others, particularly for high-speed plating, rapid solution motion is desirable. Jets providing impingement of solution over the parts are sometimes used. The solution is pumped through a filter into a sparger and recirculated continuously.

Filtration is important to provide defect-free plated deposits. To maintain particle-free (as much as possible) solutions, provide continuous recirculation through a filter of selected micrometer size to suit the application, with a flow rate sufficient to equal at least 10 times the solution volume. Fifteen or 20 times turnover rate may be required for certain applications, e.g., memory disks and semiconductors. For these applications, nominal 0.2- or 0.2-μm absolute filters are recommended. The micrometer size indicates that particles smaller than the rated size will pass through; larger particles are stopped. As the filter is used, clogging takes place, the flow rate declines, pressure in the filter chamber increases, smaller particles are retained. If allowed to continue, there is danger of breakthrough, allowing particles of all sizes to pass. Therefore, careful attention to filters is required.

9.16 PREPARATION OF METAL SUBSTRATE FOR PLATING

9.16.1 Cleaning

Cleaning the surface is critical to successful plating. The surface to be plated must be chemically clean, that is, totally free from any type of foreign material including oxides and any other molecule reacted with the substrate metal. This condition cannot be achieved by solvent degreasing alone. Surfactant (detergent) materials in neutral, mildly alkaline, strongly alkaline, or acid medium is required, depending on the metal to be cleaned, and is the most common first step. An acid dip is used for most metals to remove oxides. Acid selection depends on the metals in the substrate. Dilute mineral acids such as sulfuric acid or hydrochloric acid may be used for many steel alloys, except nickel stainless steel. Some metals such as lead and silver form insoluble salts with mineral acids; therefore, acids such as sulfamic acid, methane sulfonic acid, citric acid partly neutralized with ammonia, or fluoboric acid must be used.

9.16.2 Nickel and Cobalt Alloys

Nickel, stainless steels, and cobalt alloys require a nickel strike. Options are: sulfamate nickel with the pH adjusted to 1.5 with sulfamic acid and a small amount of hydrochloric acid, or a Woods nickel strike with 32 oz/gal (240 g/liter) nickel chloride and 10 to 12 percent by volume hydrochloric acid. These solutions are operated at high current density of 50 to 100 A/ft^2 for 1 to 3 min cathodic to deposit a thin, active layer of nickel. Rinsing should follow quickly. Subsequent plating should start immediately after the rinse to prevent oxidation of the deposited nickel.

9.16.3 Aluminum

Aluminum is prepared differently from other metals. Mild alkaline or acid cleaning is used to start the process. A "deoxidizing" step follows after rinsing, often consisting of 50 to 60 percent by volume of 42°Be′ nitric acid. Numerous proprietary deoxidizers are available. After thorough rinsing, a zinc immersion step takes place. A "zincate solution" which is strongly alkaline, or a mildly acetic zinc solution may be used to produce a thin zinc layer. For wrought alloys, the first zinc layer is removed in nitric acid or proprietary acid mixtures and reapplied, leaving a very thin, uniform layer of zinc which prevents oxidation of the aluminum long enough to rinse thoroughly and transfer to the plating solution.

9.16.4 Rinsing

Rinsing is a critically important process. Chemicals from one tank must be prevented from entering any subsequent treatment or plating solution. Thorough rinsing in clean water, free from organic matter, is required. This requires sterile water free from algae or mold. Deionized water is best to use. Sterilizing the DI water system should be done based on culture tests made on the water. Organic-material-free water is maintained by continuous recirculation of all the water in the system, passing through ultraviolet light and clean activated carbon.

Counterflow, two- or three-stage rinsing provides the most efficient method and conserves water. "Drag-out" rinses are suggested for all precious metal plating and can be effectively used for other plating processes. A *drag-out rinse* is a still rinse in which the plating solution accumulates. The rinse can often be returned to the plating solution, or metal recovery procedures can be used to prevent waste and minimize environmental impact.

9.17 BASIC THEORY OF ELECTROPLATING

9.17.1 Faraday's Laws

Michael Faraday (1791–1867) published his laws of electrochemistry in 1833 to 1834. In an electrolyte:

1. The amount of chemical change produced by a current at an electrode-electrolyte interface is proportional to the quantity of electricity used.

$$W \propto I \cdot t$$

2. The amounts of chemical change produced by the same quantity of electricity in different substances are proportional to their equivalent weights. (An *equivalent weight* is the gram formula weight divided by the valence.)

$$W \propto \frac{A}{n}$$

where W = chemical change as a weight
I = current, A
t = time, s
A = atomic weight
n = valence, gain or loss of electron(s)

The Faraday constant F is

$$E = I \cdot t \cdot \frac{A}{n} \cdot \frac{1}{F}$$

In Faraday's law

F is the quantity of electricity that will cause a chemical change of 1 equivalent weight.

The value of F is typically 96,486.75 coulombs per gram equivalent. A coulomb is 1 A for 1 s.

Faraday's law was derived experimentally using copper and silver electrolytes at efficiencies near 100 percent.

Another approach is to use the physical properties of electrons:

$$F = Ne$$

where N = Avogadro's number, 6.022174×10^{23} atoms per mole
e = charge on the electron, 1.602192×10^{-20} emu

F, then, is 9.648675×10^4 C/mole.

The U.S. Bureau of Standards and the National Physical Laboratory in London have reported updates from time to time.

9.17.2 Metal Distribution

Metal electrodes are much more conductive than the plating solution; therefore, the anode can be considered to have equal potential on its surface. The cathode (the work piece being plated), however, exhibits unequal potential. Current flows via the shortest conductive path and at right angles to the work piece surface, as illustrated in Fig. 9.12. Sharp edges receive more current than other areas. There are two factors of current (thus metal distribution): primary current distribution, which has the major influence on metal distribution and is controlled by the electrical field, and the secondary current distribution, which is controlled by cathode polarization. *Cathode polarization* is the sum of concentration polarization (metal ion depletion near the cathode in the Hemboltz double layer), activation polarization [the voltage required to move metal ions through the field (varies with the metal)], gas polarization (evolution of hydrogen at the cathode), and chemical additives (and some impurities) which can cause general or localized polarization. Some substances are attracted to high-current-density regions, on a micro scale, causing polarization which forces current to flow to adjacent areas. The "leveling" ability of some plating solutions depends on this phenomenon.

The rate of metal distribution at any point on a cathode is determined by the current density existing at that point, the secondary current, and by the cathode efficiency of the solution at that current density.

Throwing power describes the ability of certain plating solutions to deposit more metal into areas of low current density that other solutions would under the same circumstances (Figure 9.12). Throwing power is quantified in a test apparatus known as the *Haring Cell.*

Covering power is a term to describe a solution's ability to plate any small thickness in low-current-density areas.

Other factors influencing the deposition of metals are deposition potential (activation voltage) and hydrogen potential (over voltage). Hydrogen often codeposits with the desired metal, lowering the efficiency of metal deposition. Cathode efficiency is also altered by impurities in the solution, the condition of the basis material, and varies with current density. The details of theory are contained in Refs. 15 and 16.

Anode: The positive electrode often serves as a source of metal replenishment to the solution. Anode efficiency may not equal the cathode efficiency, creating an imbalance.

Insoluble anodes are often used for noble metal plating, e.g., gold and rhodium plating. Metal salts must be added continuously or at frequent intervals to maintain proper solution composition.

Auxiliary anodes are often necessary where recessed (low-current-density) areas would not otherwise plate to the specified thickness.

Bipolar auxiliary anodes can also be used to direct current to otherwise low-current-density regions. A *bipolar anode* is a conductive metal present in the solution, not connected directly, whereby the part nearest the anode becomes cathodic and the part nearest the cathode becomes anodic.

9.17.3 Current Density Limitations

There is a practical range of current density for each plating solution formulation. Anode and cathode current density limitations usually differ. Adherence to the anode-to-cathode area ratio specified is essential for good results during extended periods of solution operation. The limiting current density is a function of the solution composition and operating conditions of temperature and agitation.

Excessive anode current density results in lowered anode efficiency and evolution of oxygen. In the case of soluble anodes, an oxide film can raise the resistance of the surface, resulting in lower-current-density–higher-voltage operation, which can inhibit dissolution of the replenishing metal. This condition is known as a *polarized anode* (meaning excessive anode polarization).

Excessive cathode polarization causes dark, rough corners and edges (high-current-density regions), or *burnt deposits,* spongy and nonadherent.

In addition to lowering the current, there are other expedients for correcting the situation:

1. Increase agitation, particularly around the cathode.
2. Raise the solution temperature.
3. Change the position of the cathode.
4. Increase the metal content of the solution.
5. Use shields of "thieves."

A *thief* is a metal piece (wire or other) located near the high-current-density areas and connected to the cathode to rob current from the troubled areas. Since metal is deposited on the thief, it is not often used for precious metal plating.

Shields are nonconductive materials placed strategically to force the lines of current to take a longer path, thus equalizing the cathode current density.

The ripple percentage in the *direct current* (dc) rectifier should not exceed 5 percent. Otherwise, the current distribution problems referred to above are exaggerated. Pulse plating, however, using a carefully designed duty cycle, can enhance more even current distribution for some plating solutions. It is often used for gold plating.

9.18 SPECIFICATIONS

9.18.1 Alloy Plating

ASTM B200*	Standard specification for electrodeposited *lead tin alloys* on steel and ferrous alloys
ASTM B605	Standard specification for electrodeposited coatings of *tin nickel alloy*
ASTM B635	Coatings of *cadmium tin mechanically deposited*
AMS 2416 F	*Nickel cadmium* plating, diffused
AMS 2417 D	*Nickel zinc alloy* plating
AMS 2419 A	*Cadmium titanium* plating
AMS 2433	*Electroless nickel thallium boron* plating
PWA 259 C	*Electroless nickel boron* plating
ISO 2179	Electroplated coatings of *tin nickel* alloy
MIL-C-85455	*Chromium molybdenum* plating (electrodeposited)
MIL-STD-1500A (USAF)	*Cadmium titanium* plating (low embrittlement)

9.18.2 Aluminum Treatments

AMS 2473	Chemical treatment for aluminum-base alloys: General-purpose coating
AMS 2474	Low electrical resistance
ASTM D1730	Preparation of aluminum and aluminum alloy surfaces for painting
MIL-C-5541	Chemical films and chemical film material for aluminum and aluminum alloys
MIL-C-81706 and aluminum alloys	Chemical conversion materials for coating aluminum
MIL-W-6858	Welding resistance: aluminum, magnesium, etc.; spot and seam

9.18.3 Anodizing Specifications

ASTM B 580	Standard specification for anodic oxide coatings on aluminum
AMS 2468D	Hard coating treatment of aluminum
AMS 2469D	Hard coating, process, and performance requirements
ISO 7599	Anodizing of aluminum and its alloys

**Note:* ASTM: American Society for Testing Materials.
AMS: Aerospace Material Specification.
ISO: International Standards Organization.
MIL: Military Specifications.

ISO 8076	Aerospace process chromic acid process 40 V
ISO 8077	Aerospace process chromic acid process 20 V
ISO 8078	Aerospace process sulfuric acid process, undyed
ISO 8079	Aerospace process sulfuric acid process, dyed
MIL-A-8625E	Anodic coatings for aluminum and aluminum alloys
NAS 1192	Performance specification for hard anodic coatings on aluminum and aluminum alloys

9.18.4 Brush Plating

MIL-STD-865C (USAF)	Selective electrodeposition (brush plating)
MIL-STD-2197 (SH)	Brush electroplating

9.18.5 Copper Plating

ASTM B734	Electrodeposition of copper for engineering uses
AMS 2418D	Copper plating
MIL-C-14550 (with Amendment 3)	Copper plating (electrodeposition)

9.18.6 Electroless Processes

ASTM B733	Autocatalytic nickel phosphorus
AMS 2399	Electroless nickel boron
AMS 2404 C	Electroless nickel plating
AMS 2405 B	Electroless nickel plating, low phosphorus
ISO 4527	Autocatalytic nickel phosphorus coatings
MIL-C-26074C	Coatings: Electroless nickel, requirements for

9.18.7 Gold Plating

ASTM B 488	Electrodeposited coatings of gold for engineering uses
AMS 2422C	Gold plating for electronic and electrical applications
AMS 2425D	Gold plating for thermal control
ISO 4523	Metallic coatings electroplated gold and gold alloys for engineering purposes
MIL-G-45204C	Gold plating, electrodeposited

9.18.8 Lead Plating

AMS 2414C	Lead plating
AMS 2415E	Lead and indium plating
MIL-L-13808B	Lead plating, electrodeposited

9.18.9 Magnesium Treatments

AMS 2475	Protective treatments, magnesium-base alloys
MIL-M-3171	Magnesium alloy, process for pretreatment and prevention of corrosion on magnesium and magnesium alloys
MIL-W-6858	Welding, resistance: aluminum, magnesium, etc.; spot and seam

9.18.10 Nickel Plating

ASTM B689	Electroplated engineering nickel coatings
AMS 2403H	Nickel plating, general purpose
AMS 2423B	Nickel plating, hard deposits
AMS 2424C	Nickel plating, low-stress deposits
ISO 1458	Metallic coatings: Electrodeposited coatings of nickel
ISO 4526	Metallic coatings: Electroplated coatings of nickel for engineering purposes
QQ-N-290A	Nickel plating, electrodeposited
MIL-STD-868A (USAF)	Nickel plating, low embrittlement, electrodeposition

9.18.11 Palladium Plating

ASTM B679	Electrodeposited coatings of palladium for engineering use
MIL-P-45209	Palladium plating, electrodeposited

9.18.12 Printed Wiring Boards and Circuits

IPC A 600	Acceptability standards for printed wiring systems
IPC ML-950	Performance specification for multilayer printed wiring boards
MIL-F-55561	Foil, copper, cladding for printed wiring boards
MIL-P-55110	Printed wiring boards
MIL-STD-275	Printed wiring for electronic equipment
MIL-STD-429	Terms and definitions of printed wiring boards
NSA 68-8	Specification for multilayer printed wiring boards

9.18.13 Rhodium Plating

ASTM B634	Electroplated coatings of rhodium for engineering use
MIL-R-46085B	Rhodium plating, electrodeposited

9.18.14 Silver Plating

ASTM B700	Electrodeposited coatings of silver for engineering use
AMS 2410G	Silver plating: nickel strike, high bake
AMS 2411D	Silver plating for high-temperature applications
AMS 2412F	Silver plating, copper strike, low bake
AMS 2413C	Silver and rhodium plating
ISO 4521	Metallic coatings: electrodeposited silver and silver alloy coatings for engineering purposes
QQ-S-365D	Silver plating, electrodeposited: general requirements for

9.18.15 Tin Plating

ASTM A624, A624M, A626, A626M	Tin mill products, electrolytic tin plate, single reduced, and tin mill products, electrolytic tin plate, double reduced
ASTM B545	Electrodeposited coatings of tin
AMS 2408E	Tin plating
AMS 2409E	Tin plating, immersion
ISO 2093	Electroplated coatings of tin
MIL-T-10727B	Tin plating: Electrodeposited or hot dipped, for ferrous or nonferrous metals
MIL-T-81955(0S)	Tin plating: Immersion for copper and copper alloys
QQ-T-425B	Tin plate (electrolytic)

9.18.16 Tin Lead Plating

ASTM-B579	Electrodeposited coatings of tin lead (solder plate)
ISO 7587	Electroplated coatings of tin lead alloys

9.18.17 Zinc Plating

ASTM B633	Electrodeposited coatings of zinc on iron and steel
ASTM B695	Coatings of zinc mechanically deposited on iron and steel
AMS 2402F	Zinc plating
AMS 2420B	Plating aluminum for solderability: zinc immersion process
ISO 2081	Metallic coatings: Electrodeposited coatings of zinc on iron and steel

ISO 4520	Chromate conversion coatings on electroplated zinc and cadmium coatings
MIL-A-81801	Anodic coatings for zinc and zinc alloys
MIL-T-12879	Treatments—chemical, prepaint, and corrosion inhibitive—for zinc surfaces
QQ-A-325	Zinc coating, electrodeposited, requirements for

REFERENCES

1. John Donaldson, "For Platers Only," *Metal Finishing,* Vol. 90, No. 9, September 1992, p. 43.
2. John Donaldson, "For Platers Only," *Metal Finishing,* Vol. 90, No. 8, August 1992, p. 24.
3. James H. Brim, *AESF International Technical Conference,* Vol. II, 1992, pp. 1043–1046.
4. John Horner, *ASTM B578/AESF International Technical Conference,* Vol. 1, June 1992, pp. 263–276.
5. E. F. Duffek, *Plating,* Vol. 57, 1970, pp. 37–42.
6. H. D. Kaiser, F. J. Pakulski, and A. F. Schmeckenbecher, *Solid State Technology,* Vol. 15, No. 5, 1972, p. 39.
7. H. Rochow, *Ber. Dtsch, Keram, Ges.,* Vol. 44, 1967, p. 224.
8. V. Jirkovsky, J. Mikulickova, and K. Balik, *TESLA Electronics,* Vol. 4, 1976, p. 107.
9. V. Jirkovsky, J. Mikulickova, and K. Balik, *TESLA Electronics,* Vol. 3, 1977, p. 75.
10. R. R. Freeman and J. Z. Briggs, "Electroplating on Molybdenum Metal," *Climax Molybdenum Co. Technical Note,* September 1958.
11. "Procedures for Electroplating Coatings on Refractory Metals," *Battelle Memorial Institute, DMIC Memo 35,* October 1959.
12. Joseph A. Abys, *AESF Shop Guide, 9th Edition,* 1991, p. 106.
13. Donald W. Baudrand, "Use of Electroless Nickel to Reduce Gold Requirements," *Plating & Surface Finishing,* December 1981.
14. Donald W. Baudrand, *Plating & Surface Finishing,* Vol. 11, November 1979.
15. H. E. Haring, and W. Blum, *Trans. Electrochem Society,* Vol. 44, pp. 313–345.
16. A. K. Graham, E. A. Anderson, H. L. Pinkerton, and C. E. Reinhard, *Plating,* Vol. 36, 1949, pp. 702–709.

CHAPTER 10

LITHOGRAPHY AND PHOTOFABRICATION

Jurgen Diekmann

10.1 INTRODUCTION

Lithography is the process of producing an image on a flat, specially prepared stone or plate so that it will absorb and print with special hydrophobic inks. This process was discovered nearly 200 years ago. When the lithographic image is produced through photography, it is called *photolithography*. Photolithography represents a significant advance in resolution over letterpress and stencil printing. It is now the dominant process in the printing and publishing industry. The terms *lithography* and *photolithography* are also widely used to describe fabrication steps in the processes of making many electronic components, even though they do not really absorb and print lithographic ink images. Here, they are defined as the process of drawing or printing an image, usually a resist image, on a planar substrate. These process steps can also be grouped under the broader term *photofabrication*.

Photofabrication employs the concepts of electromagnetic radiation from suitable light sources as the writing tool along with photosensitive media and chemical or physical processes to produce the desired, three-dimensional circuit patterns in the electronic components. It encompasses a wide range of disciplines, materials, and processes. The complete fabrication processes for printed wiring boards (PWBs) and semiconductor integrated circuits (ICs) are described in Chaps. 2 and 4. This chapter focuses on the underlying principles of photofabrication used to make these components with the intent to show the commonalities as well as the essential differences.

The phenomenal growth of the electronics industry has largely been driven by the miniaturization of component feature sizes, which has reduced cost per function continuously. As a result, a continuous stream of new applications has provided this growth as illustrated in Fig. 10.1.

Photofabrication, with its high-resolution capability, has been the enabling process in this miniaturization. It has allowed the accurate placement of smaller and smaller feature sizes in both active components as well as their associated interconnect circuitry. Today, critical dimensions are expressed in micrometers; but in the future, they will be expressed in nanometers.[1] The subject of nanostructures has recently been reviewed.[2]

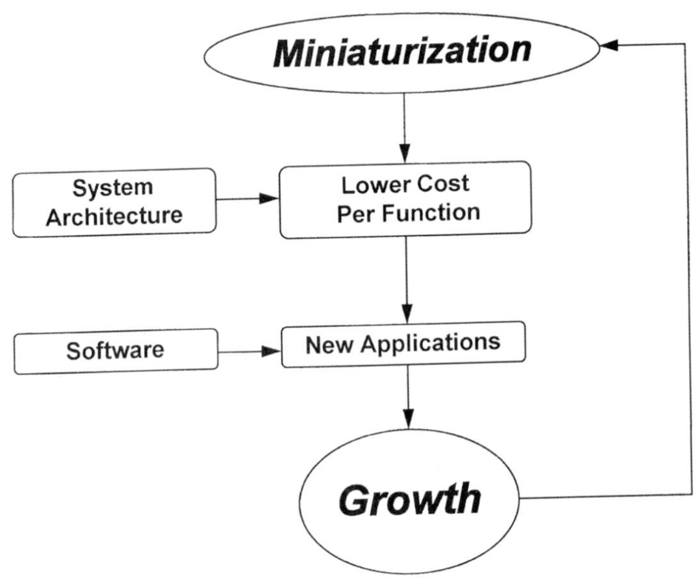

FIGURE 10.1 Miniaturization in electronics drives growth through new applications.

An additional driver for photofabrication is short production cycles, frequently coupled with the need for fast turnaround time. The tooling cost for alternate patterning processes with lower-resolution capability such as stamping or stencil printing requires adequate lead time and sufficient volume to be economical.

10.2 GENERAL CONSIDERATIONS IN PHOTOFABRICATION

10.2.1 Light Sources

The electromagnetic spectrum useful for photofabrication is shown in Fig. 10.2. Nearly all photofabrication processes are currently based on the ultraviolet (uv) part of the spectrum. There are a limited number of useful sources for radiation within this range. Important considerations in selecting a proper light system are its spectral output, its intensity over the output range, and the collimation and declination of its rays.

Spectral Output and Intensity. The most popular light sources for photofabrication are lamps filled with xenon gas. Their spectral output can be modified with the use of metal additives to the xenon fill. A common additive is mercury. They are favored because they combine spectral output useful for most media with reasonable intensity in a cost-effective manner. In use, they are activated by microwaves or a discharge from high-voltage electrodes. This activation ionizes and transfers energy to the gaseous molecules and induces them to give off this transferred energy in the form of radiation. The output of the xenon-mercury lamp immediately after ignition is essen-

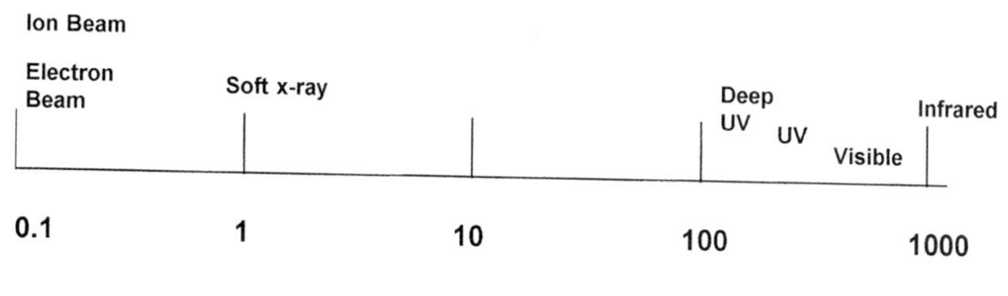

FIGURE 10.2 The electromagnetic spectrum.

tially the same as the pure xenon lamp. As the mercury vaporizes over a period of several minutes, the spectral output becomes that of mercury vapor.

The power of these sources is usually expressed in kilowatts on an input basis, and milliwatts per square centimeter for the intensity of their output in the useful part of the spectrum at the exposure plane. Xenon and xenon-mercury lamps are relatively efficient in energy conversion and therefore useful for large-area exposure. The output spectra for typical xenon and mercury-xenon arc lamps are shown in Figs. 10.3 and 10.4. The area under the curve is a measure of the relative light intensity for that part of the spectrum. The lamp envelopes are made from high-quality quartz, which can transmit radiation down to 180 nm. To avoid ozone formation in the deep uv, the lamps normally use specially doped quartz which starts to cut off transmission at 260 nm.

The above light sources have been the dominant workhorses in *optical* photofabrication, which employs area-wise, parallel exposure through prepatterned master images as opposed to serial exposure to create the desired patterns on the planar substrates. Their resolving power is diffraction limited, however, near feature sizes of 0.2 to 0.3 μm. The diffraction or spreading of light at the edges of patterns leads to undesired loss of edge definition from interference effects. Such effects are most pronounced when the wavelength of the radiation is comparable to the feature size, which is the case for these light sources at the above limit. As a result, smaller feature sizes require alternate forms of radiation with shorter wavelengths such as excimer lasers, x-rays, and electron and ion beams.

In addition to the light source, an exposure system relies on optical elements such as mirrors and lenses to afford the desired optical path for the light. The choice in materials of sufficient transparency for such elements is adequate for the visible portion of the spectrum but becomes quite limited at shorter wavelengths. An overview of commonly used materials is shown in Table 10.1.

Gaseous argon lasers also provide spectral output useful for photofabrication as shown in Fig. 10.5. But their low conversion efficiency of input energy to useful uv output energy makes them inefficient light sources for area-wise, optical photofabrication. Their principal use, with the aid of high-speed modulators, is as a *direct write device* for serially imaging onto photosensitive media using digitally stored data. Here, the available power is focused into a narrow diameter beam, which confers high intensity to the beam spot area. The term *nonoptical photofabrication* is often used for such a direct, spot-based write process, even though the optical path contains many optical elements. An overview of these light sources is shown in Table 10.2.

Spectral Distribution

Wavelength (Nanometers)	Percent Output
200–280	.09
280–330	1.33
330–380	2.98
380–430	2.13
430–480	3.60
480–530	3.92
530–580	3.76
580–630	3.60
630–680	3.26
680–730	3.60
730–780	3.76
780–830	12.83
830–880	2.37
880–930	13.65
930–980	10.22
980–1030	4.66
1030–1080	3.84
1080–1130	1.71
1130–1180	1.22
1180–1230	4.65
1230–1280	1.39
1280–1330	2.69
1330–2500	8.74
	100.0

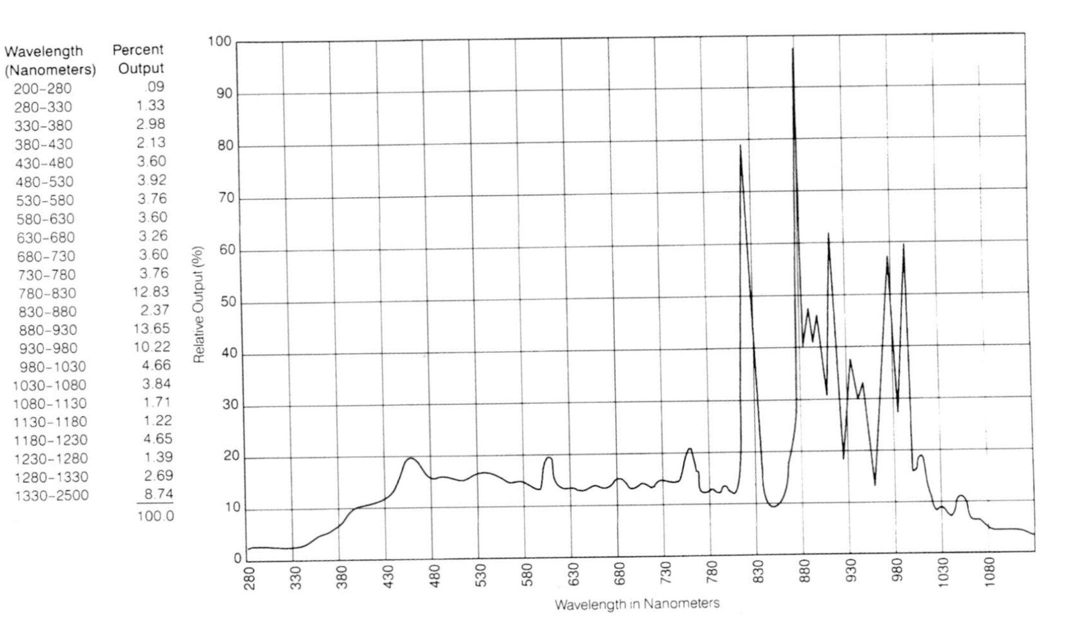

FIGURE 10.3 Spectral output of xenon lamp. (*Courtesy of Optical Radiation Corporation.*)

Spectral Distribution

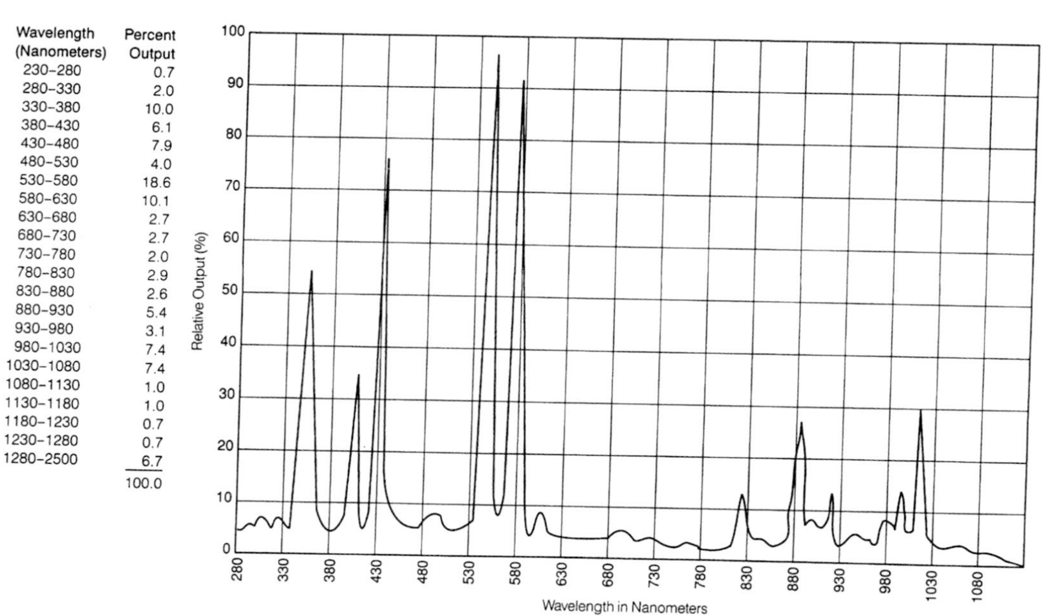

Wavelength (Nanometers)	Percent Output
230–280	0.7
280–330	2.0
330–380	10.0
380–430	6.1
430–480	7.9
480–530	4.0
530–580	18.6
580–630	10.1
630–680	2.7
680–730	2.7
730–780	2.0
780–830	2.9
830–880	2.6
880–930	5.4
930–980	3.1
980–1030	7.4
1030–1080	7.4
1080–1130	1.0
1130–1180	1.0
1180–1230	0.7
1230–1280	0.7
1280–2500	6.7
	100.0

FIGURE 10.4 Spectral output of mercury-xenon lamp. (*Courtesy of Optical Radiation Corporation.*)

TABLE 10.1 Optical Materials of Construction

Region	λ, nm	Optics
Visible	400–750	Soda lime glass
Near uv	350–400	White glass
Mid uv	300–350	Pyrex/quartz
Deep uv	200–300	Quartz
Far uv	150–200	Calcium fluoride

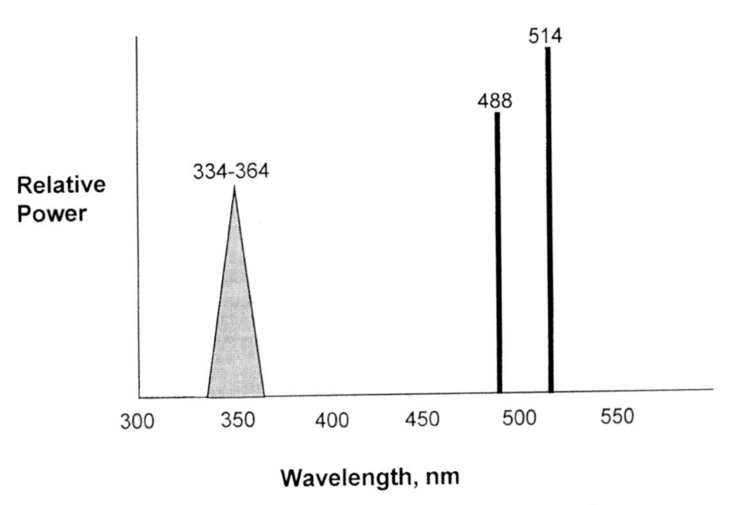

FIGURE 10.5 Spectral output of argon laser. (*Courtesy of Coherent, Inc.*)

TABLE 10.2 Comparison of Light Sources

Source	λ, nm	Intensity
Mercury arc	280–600	High
Xenon flash	280–800	High
Continuous Ar laser	350–560	Low
Pulsed excimer lasers	193–353	Very high

Collimation and Declination. Imaging media possess a finite thickness. For high resolution, it is therefore important to use parallel light rays and have them strike the image plane in a perpendicular manner to provide straight side walls, as illustrated in Fig. 10.6. Lamp sources inherently provide nonparallel light unless they are very far away from the image plane, such as the sun is from the earth. *Collimation* is the process of collecting the rays from a light source in a curved mirror in order to produce parallel—or collimated—light rays. Such a collimated optical system including a mercury-xenon short-arc lamp mounted in an elliptical collector is shown in Fig. 10.7. The reflected light is directed from the collector to a dichroic mirror which transmits the unwanted longer wavelengths and reflects the desired ultraviolet spectrum. An optical integrator composed of a multilens optical array then uniformly distributes the energy at the exposure plane. The light is collimated through a spherical mirror, which directs the light toward the substrate at right angles over the entire exposure plane. As a result, there is minimal feature growth and image displacement.

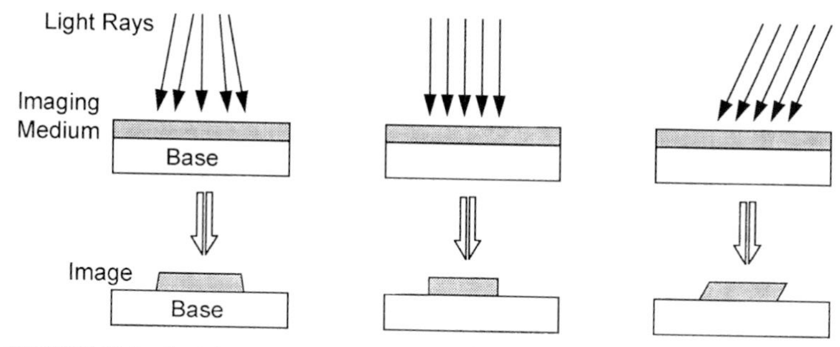

FIGURE 10.6 Straight side walls require perpendicular rays.

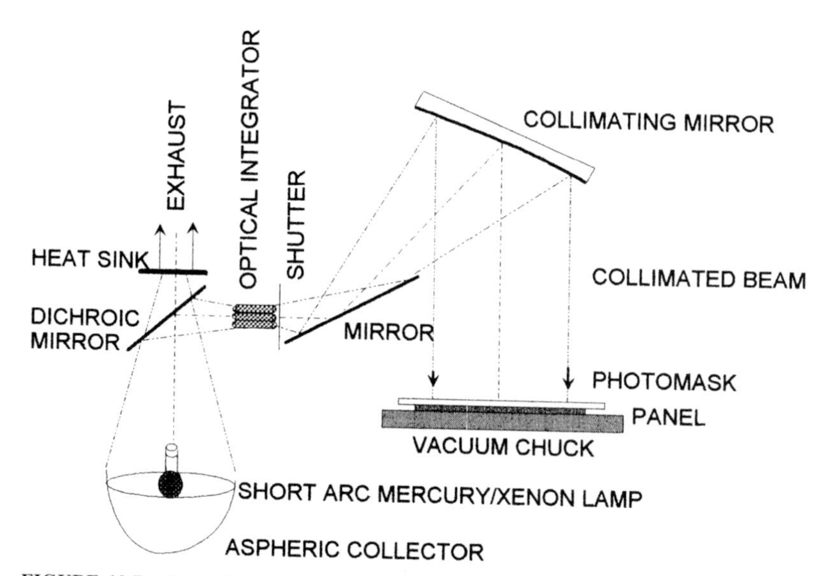

FIGURE 10.7 Optical path for collecting, integrating, and collimating light. (*Courtesy of Polyscan, Inc.*)

Collimation is measured in units of degrees half angle, as shown in Fig. 10.8. The smaller the half angle, the more the light is parallel. The coherent light from lasers, where the light beams are all in phase, is inherently collimated. But even perfectly collimated light with a 0-degree half angle still needs to strike the image plane in a perpendicular manner to achieve high resolution, as illustrated in Fig. 10.6. The deviation from the perpendicular is called *declination* and is measured as the declination angle, as shown in Fig. 10.8. The lower the declination angle, the less is the deviation from the perpendicular.

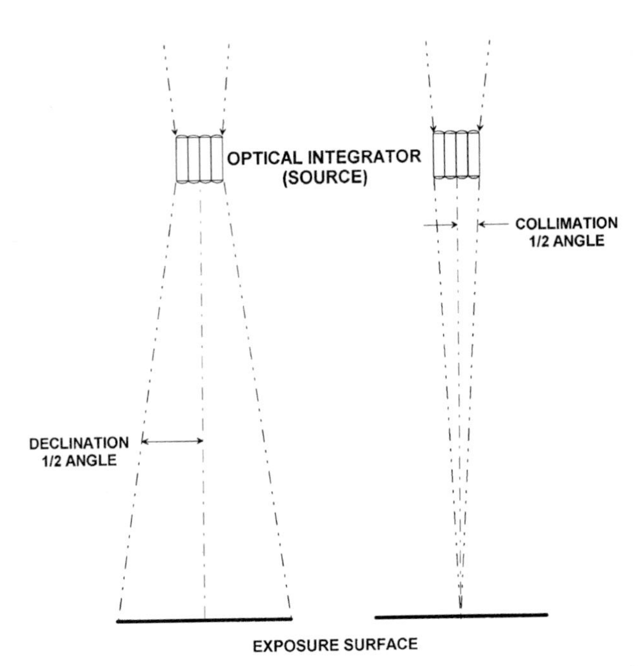

FIGURE 10.8 Declination and collimation half angles. (*Courtesy of Polyscan, Inc.*)

10.2.2 Photoresponse of Medium

The basic photoresponse processes in any medium are

- Absorption of the light rays, which presumes a reasonable match between the spectral output of the light source and the spectral absorption of the medium
- A physical and/or chemical change in the exposed areas
- A subsequent development step, for nearly all media, to clearly differentiate exposed from unexposed areas

Since the availability of suitable light sources is quite limited, the photosensitive medium usually bears the burden of providing for a matching spectral absorption. Key attributes of a given medium are sensitivity, contrast, and resolution.

Sensitivity. The exposure energy E interacting with the photosensitive medium is the product of the power or intensity I per unit area of the light source at the image plane and the exposure time t. It is customary to measure intensity in milliwatts per square centimeter and time in seconds, which then expresses the exposure energy in millijoules per square centimeter:

$$E\ (\text{mJ/cm}^2) = I\ (\text{mW/cm}^2) \times \text{t (s)}$$

Sensitivity expresses the relationship between exposure energy and degree of useful change in the exposed area of the medium after development. Such a relationship for a typical photosensitive medium is shown in Fig. 10.9, where the variable of exposure energy is plotted on the x-axis on a log scale. The y-axis shows the desired, normalized change in the medium. The S-shaped curve of Fig. 10.9 is commonly called the *characteristic curve,* or *H&D curve* after its inventors Hurter and Driffield. The *straight-line* portion of the curve is essentially linear; equal increments of change in exposure energy produce equal increments of change in the medium. The upper part of the straight line is usually considered the region of correct exposure. The characteristic response curve of a medium can vary to some extent with the method of exposure. Some media are unable to respond proportionately to very high intensities from sources such as pulsed xenon lamps or lasers; as a result, they appear to require higher exposure energy at very high intensity levels. Such a response is known as *reciprocity failure.* Alternatively, exposure at a very low intensity level can lead to loss of resolution because of undesirable side reactions during the long exposure time.

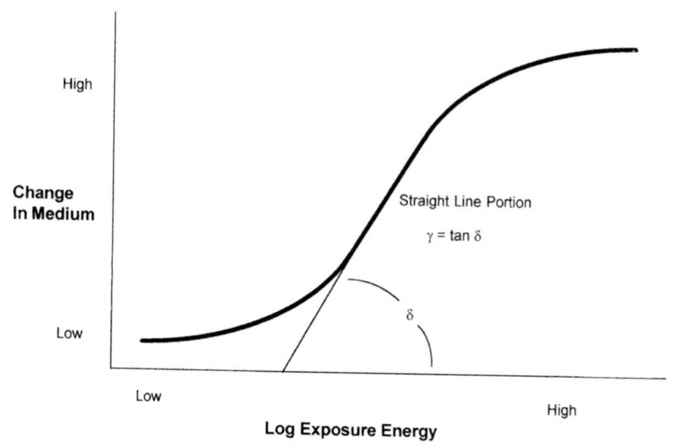

FIGURE 10.9 Characteristic curve for photosensitive medium.

The exposure efficiency of photofabrication can vary over many orders of magnitudes. If one photon of light results in the useful change of one molecule in the photosensitive medium, the quantum yield is 1, or 100 percent. Side reactions typically reduce the quantum yield to less than 1. Many photosensitive media, however, are capable of physical and/or chemical amplification with quantum yields much larger than 1. They are discussed in Sec. 10.3.

Sensitivity and light source intensity are two of the key factors in limiting the throughput of a practical exposure system.

Contrast. *Contrast* defines the potential resolution of a medium. The contrast response of a medium is apparent from its characteristic curve. The slope of the straight-line portion of the characteristic curve of the medium in Fig. 10.9 is called *gamma.* A high slope means high contrast. For photofabrication, high contrast is desirable as it has a direct impact on the edge sharpness and wall profile of fine features, important for high resolution. A common formula for defining contrast C is:

$$C = \frac{1}{\log (E_{min}/E_{max})}$$

The minimum energy E_{min} is the threshold energy to produce a change, and the maximum energy E_{max} is the energy needed to produce the desired maximum change in the medium. The maximum and minimum light energies, for instance, can correspond to the light transmitted through the clear and opaque areas of a patterned image master onto the medium to produce a useful resolution pattern. Alternatively, they can define attainable resist thickness on a resist-coated substrate.

But high contrast also means smaller exposure latitude. Exposure latitude is usually expressed as the amount of over- or underexposure that can be tolerated without a sacrifice in image quality. The contrast, or gamma, can be affected by the formulation of the medium as well as its subsequent exposure and development conditions. Gamma for developed images can range from less than 1 to 6 or greater, depending on the medium and processing conditions.

Resolution. Resolution is affected by the previously discussed factors such as light collimation, declination, and contrast of the medium. High contrast can compensate for several undesirable effects. These are unexposed space between two exposed lines gradually receiving exposure from diffraction in the patterned master, backscatter through reflections from the substrate, refraction in the medium, and insufficient *depth of focus* (DF). An additional important factor is medium thickness. In general, the thickness of the medium limits the attainable feature size to a 1:1 aspect ratio. In practice, the thickness is usually less than the feature size.

In area-wise optical photofabrication with a prepatterned master, the limit for resolution is set by diffraction. The ultimate limit is set by the wavelength of the light source. For lens-based optical systems, the resolution W of a lens at its barely detectable Rayleigh limit is dependent on its numerical aperture NA, the wavelength, and a medium and process dependent factor k_1:

$$W = \frac{k_1 \lambda}{\text{NA}}$$

The numerical aperture in turn is a function of lens diameter D and focal length f.

$$\text{NA} = \frac{D}{2f}$$

The value of k_1 for media ranges from 0.3 to 1.[3] A practical limit for resolution is 3 Rayleigh limits. Shorter wavelengths and higher numerical apertures give better resolution, but unfortunately higher apertures also mean smaller exposure fields and decreased depth of focus DF. DF is also dependent on a process factor k_2:

$$\mathrm{DF} = \frac{\pm k_2 \lambda}{(\mathrm{NA})^2}$$

The general tradeoff is between resolution and DF. For a given resolution, the DF will increase with decreasing wavelength of the light source. At a resolution of 0.5 μm, the DF is less than 1.5 μm for a NA of 0.5, a wavelength of 365 nm, and typical values of $k_1 \approx 0.6$ and $k_1 \approx 1$. This tradeoff between resolution and DF is illustrated in Fig. 10.10 for a wavelength of 248 nm. The DF decreases with higher resolution. The flatness and surface topography of the substrate with its photosensitive medium become very important considerations to overcome the DF limits.

FIGURE 10.10 The depth of focus for this positive deep uv resist changes from 1.2 to 0.6 μm, as the feature size critical dimensions change from 0.40 and 0.35 to 0.275 μm. (*Courtesy of OCG Microelectronic Materials, Inc.*)

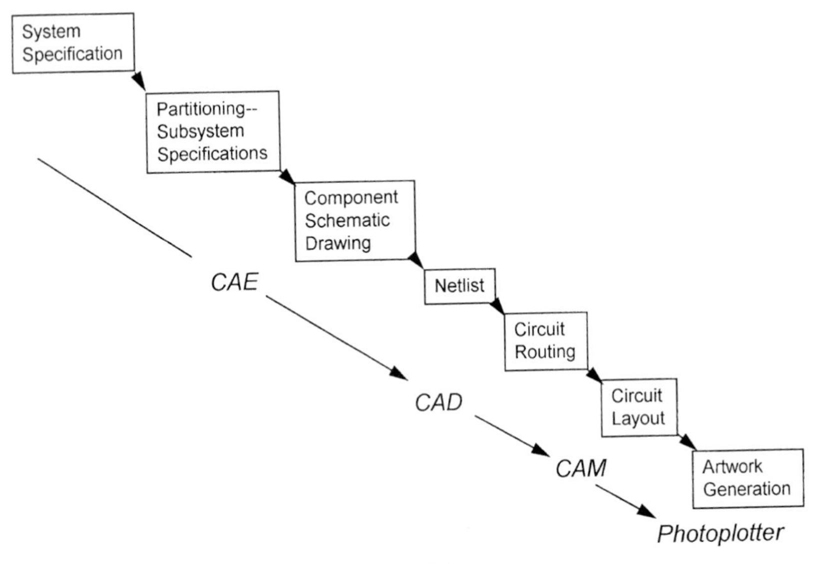

FIGURE 10.11 Flowchart for generation of circuit layout.

10.2.3 Imaging

Information Transfer. A flowchart for a typical photofabrication process is shown in Fig. 10.11. It begins with a definition of the system and its subsystems with the aid of *computer-aided engineering* (CAE). The electronic circuits are then generated as electrical schematics of the circuit design on a *computer-aided design* (CAD) system. The schematics are then converted to actual circuit layouts layer by layer, adhering to established design rules. The data are then transferred to a *computer-aided manufacturing* (CAM) system and converted to a form useful for actual fabrication. The computerized circuit layer patterns are now ready for output onto a photosensitive medium.

The information content of a circuit layer pattern can be expressed on a picture element, or *pixel,* basis. The number of pixels per unit length is usually 3 to 10 times the finest feature size. For IC designs the information content is near 1 billion pixels, depending on feature, pixel, and chip sizes. Interestingly, printed wiring board panel layers have similar information content as their larger size formats offset their feature sizes, which are typically 100 times larger than IC chip features. A comparison of the essential features and information content is shown in Table 10.3.

TABLE 10.3 Comparison of Information Content for ICs and PWBs

	Integrated circuit	Printed wiring board
Feature size, μm	0.5–2	40–300
Circuit size, mm	4–12	150–750
Information content, pixels	10^8–10^9	10^8–10^9

The large information content of the circuit patterns has favored the generation of these circuit patterns on an intermediate *photomask,* or *phototool,* master, so that the subsequent repeat information transfer of billions of pixels can be done in parallel on an area-wise basis as opposed to the serial transfer for the direct write, nonoptical mode. Recent advances in multibeam laser scanning and modulation (a form of parallel imaging), faster electronic image transfer rates, greater laser power in the uv range as shown in Fig. 10.12, and more sensitive photoresists, however, are now making the direct exposure of resist-coated PWB panels without intermediate phototools more cost-effective.[4]

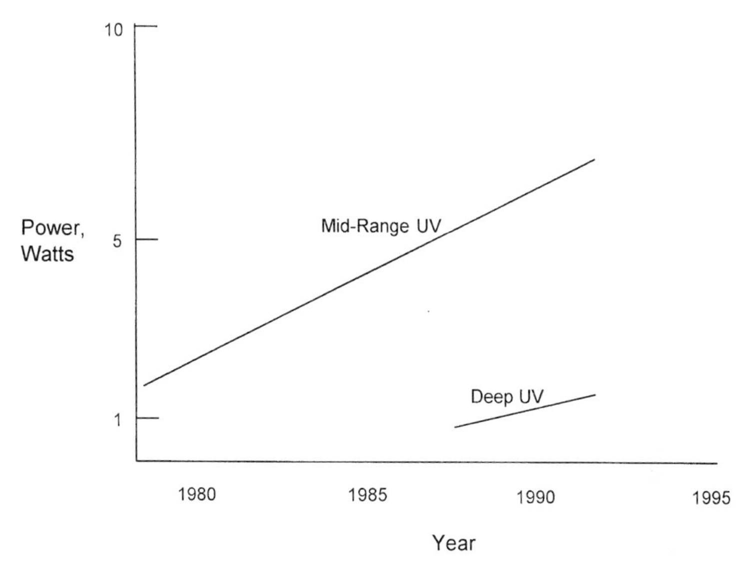

FIGURE 10.12 Evolution of argon ion uv laser power. (*Courtesy of Coherent, Inc.*)

The term *photomask* is largely used for the intermediate in IC fabrication, whereas the term *phototool* is typically used for the intermediate in PWB fabrication. Tradeoffs among feature and substrate sizes with their associated tolerances and dimensional control, optical field, throughput, and yield have led to three methods to accomplish this subsequent optical image transfer from the intermediate to the final product. They are contact, off-contact, and projection printing.

Contact Printing. For contact printing, the photomask-tool and the photoresist-coated substrate are brought into direct contact with each other, often with a vacuum assist. The latter is also called *hard contact.* Light then passes through the phototool's transparent areas to expose the photoresist. The hard-contact method affords the highest resolution in optical transfer printing. It suffers no loss in edge definition as illustrated in Fig. 10.13. But in practice, small defects introduced in this contact process can severely impact yield. Contact printing was the original method for patterning ICs. While contact printing is still the dominant method for PWB manufacture, it is no longer able to pattern most IC chips with their much finer feature sizes without excessive yield losses from the contact damage, contamination of the mask, and general photomask-tool wear.

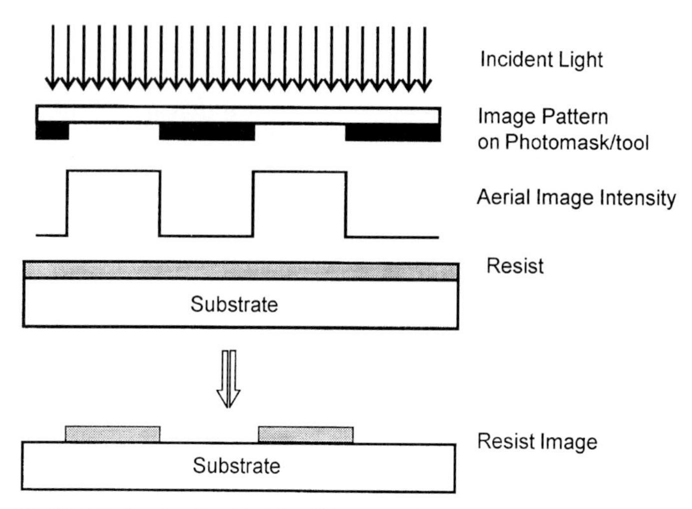

FIGURE 10.13 Profile of incident light energy in contact printing.

Contrast-enhancement layers (CEL) can be considered a special form of contact printing. Here, a thin, special photosensitive layer is deposited on top of the main photosensitive medium. Upon exposure, the CEL forms an intermediate *in situ* image on the medium, which then reinforces the exposure of this medium.[5]

Off-Contact Printing. In off-contact, or proximity, printing, the phototool is positioned at a narrow gap from the surface of the sensitized substrate. This method largely avoids the contamination problem of contact printing, but it also degrades resolution capability significantly because of insufficient depth of focus. With conventional light sources, this method is useful for nondemanding ICs but wholly unacceptable for VLSI chips. It has found limited acceptance for volume production of PWBs.

Off-contact printing is expected to again become important for IC fabrication with the advent of x-ray lithography, where the very short wavelength of the source is two orders of magnitude smaller than the feature size of the image. Such an exposure system provides a much greater depth of focus at the desired resolution and thus overcomes the loss of resolution in off-contact printing encountered by conventional light sources with longer wavelengths.

Projection Printing. In projection printing, the photomask pattern is projected onto the sensitized substrate through a system of optical elements such as lenses and mirrors. The advantages are freedom from the contamination problems of contact printing and better resolution over off-contact printing, even though the use of lenses in the optical path reduces the contrast of the aerial image. This loss of contrast is illustrated in Fig. 10.14. A high-contrast medium is needed to restore the original edge definition.

Projection printing is now the dominant method for making IC chips. The two basic approaches are *refractive* image projection through a lens and *reflective* image projection in a scanning mode. Early projection printers used refractive lens projection to image the entire wafer at once in a full-field exposure. At that time, wafer diameter was 50 mm and feature sizes were near 4 to 5 μm. But with increasing wafer size and decreasing feature size, image distortion from the refractive lens system became

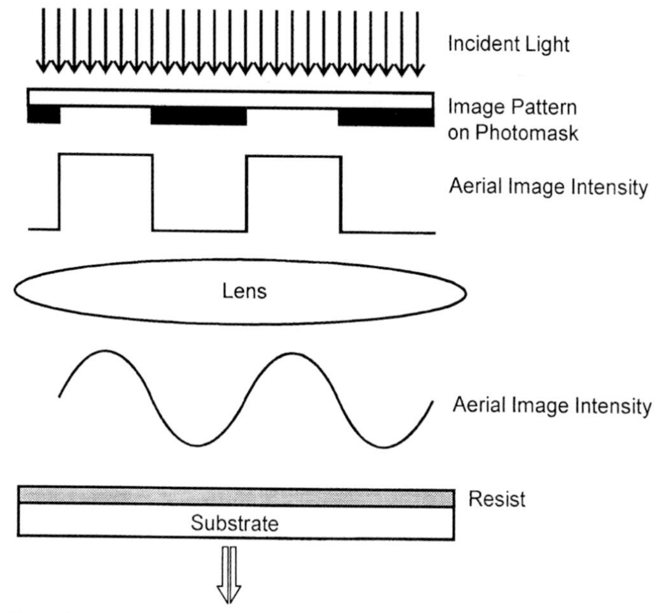

FIGURE 10.14 Profile of incident light energy in projection printing.

excessive in full-field exposure. Furthermore, a refractive system is also limited in use to one, or at most two, closely spaced wavelengths of the spectrum of the light source, which detracts from its available total energy.

Reflective, mirror-based ring field scanning exposure began to displace full-field refractive exposure in the 1970s. The principle is illustrated in Fig. 10.15 for a two-mirror system. The centers of curvature for the primary and secondary mirrors have a common axis. Each object point is reflected to a diametrically opposed image position. In the center of the common axis, there is a ring-shaped field of approximately 1 mm in width with almost no optical distortion. This field defines the permissible slit width for a scanning exposure. Reflective projection printing allows a wide range of exposing wavelengths, with little loss of energy from scattering and absorption. It also has less heat buildup because infrared energy is filtered out. Projection scanning from a 1:1 photomask is characterized by good throughput and very low image distortion but limited in practically attainable resolution. This is partly due to the need to require a very precise 1:1 photomask, wholly free of even minor defects.

The need to maintain good dimensional tolerances in the face of shrinking feature sizes as well as increasing wafer sizes led to the emergence of lens-based, refractive step-and-repeat projection printing of each individual die on the wafer. The smaller area of the individual die now allows good tolerance control of its features. Such lens-based steppers have two important advantages. They can use photomasks with the individual die patterns magnified up to 10 times, which greatly reduces the impact of flaws and distortion errors in the photomask. The second advantage is the ability to individually focus each die pattern to compensate for flatness deviations in the wafers. The disadvantage of steppers is slow throughput, typically one-third to one-half of a reflective scanner.

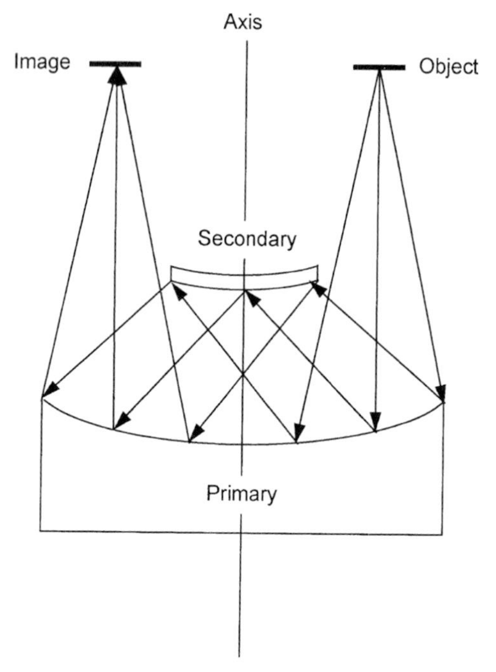

FIGURE 10.15 Principle of ring field reflection scanning.

The high cost of optical systems for the large PWB formats as well as limitations on handling energy intensity, however, have precluded projection printing from penetrating PWB fabrication to date.

Alignment. The accurate alignment of the aerial image with the registration marks on the substrate to ensure proper image placement is a key requirement for an exposure system. A prerequisite for accurate layer-to-layer alignment, of course, is precision placement of the features for each layer pattern.

For VLSI chips[6] as well as PWBs, the ability to precisely register the images of successive layers to each other is now more important than fabricating the features to their absolute sizes. As a rule of thumb, the required alignment accuracy of the printed image is 20 to 25 percent of the minimum feature size of the device, or 10 to 100 ppm for every image feature over the entire substrate area. For the VLSI chips of today, this means an accuracy of 0.15 to 0.25 μm, whereas for the larger PWBs the required accuracy typically is 50 to 100 μm. This required accuracy represents the alignment budget *for the entire photofabrication process,* which is also influenced by substrate size changes from relief of built-in stresses as well as variations in temperature and humidity. The alignment of the aerial image therefore needs to be better than the total budget.

Alignment can be done by optical or mechanical means. Optical alignment is essential to achieve the high degrees of precision required for IC fabrication. Mechanical pin registration is still widely used in PWB fabrication, but optical alignment is now becoming important for the exposure of high-density PWBs.

Precision stage metrology systems are used to perform the alignment. Measurement and feedback transducers for positioning typically use inductive, metal or glass scales. Laser interferometers are the most accurate devices, with a routine accuracy of 0.1 μm.

10.3 PHOTOSENSITIVE MATERIALS

Photosensitive media can be divided into the two basic types: positive acting and negative acting. After exposure and development, a positive-acting medium will retain the desired image pattern in the areas where it was not exposed to light, whereas a negative-acting medium will retain the desired image in the exposed areas.

Current photosensitive materials can also be divided by use into three groups. The first one consists of media for directly making the intermediate photomasks and phototools. The desired principal property here is high opacity in the patterned areas to provide high contrast during the subsequent image exposure step of the sensitized substrate. *Opacity* is defined as the inverse of light transmission. It is usually expressed as light absorbance or optical density, which is the logarithm of opacity, and measured with the aid of densitometers:

$$\text{Optical density} = \log \text{opacity} = \frac{\log 1}{\text{transmission}}$$

Table 10.4 shows the relationship between optical density and the percentage of light absorbed or blocked.

TABLE 10.4 Optical Density and Light Absorption

Optical density	% light absorbed
3.0	99.9
2.0	99
1.0	90
0.5	70
0.1	20

The second group are the media—commonly called *photoresists*—which are used to coat and thus photosensitize a substrate surface. After imaging on the substrate, they are subjected to subsequent chemical processes to modify the adjacent, nonimaged areas of the substrate. They simply serve as an aid to fabrication and are removed after use. The principal property here is the ability to *resist* the subsequent physical and/or chemical modifications of the areas adjacent to the image through metal evaporation, etching, or plating processes.

The third group are media, which remain on the substrate after imaging to provide desired functionality to the part, such as a dielectric layer, conductor, soldermask, or color filter for displays.

10.3.1 Silver Halide

Silver halide is the most widely used general-purpose imaging technology. It dominates medical and consumer applications. In electronics, its principal use is to make the intermediate photomasks and phototools. The desired properties are high contrast through high opacity to light in the image areas, coupled with high light transmission in the adjacent areas.

A photographic emulsion, consisting of silver halide crystals or grains dispersed in a gelatin binder, is coated onto a suitable support such as a glass plate or polyester film. The nature of the halide influences film sensitivity, or speed. Slow-speed emulsions are generally made from chloride or chloride plus some bromide. Fast emulsions are generally made from bromide with a few percent iodide. Photographic sensitivity is frequently increased through additives such as sensitizers, dyes, and antifoggants.

After light exposure of the emulsion, a few free silver atoms are formed in the exposed crystals. They constitute the latent image. These latent image crystals can now be differentiated from unexposed crystals by chemical reducing agents called *developers.* These developers preferentially attack the latent image crystals and reduce the entire crystal to free silver. This developing process embodies a 100-million-fold light amplification and accounts for the low light energy requirements when exposing silver halide emulsions. The sensitivity of silver films typically is in the range of mJ/cm^2. If the developer is allowed to act on the emulsion long enough, all crystals will eventually be reduced. The reduction of unexposed crystals results in undesirable fog in the unexposed areas.

After development, the remaining silver halide in the emulsion is dissolved with a complexing agent such as thiosulfate and then washed out of the film. An emulsion thickness of 6 μm is usually enough to produce an optical density greater than 2.

Solid silver halide absorbs blue light only. Dyes are added to the emulsion to expand the absorption spectrum. Light of longer wavelengths is then absorbed by these dyes, and the energy is transferred to the silver halide crystal to form the latent image. The spectral response can thus be extended to green and red light. Image exposing light not fully absorbed within the emulsion coating can spread by scattering and reflections from the silver halide crystals as well as reflections from the support. The latter is called *flare* or *halation.* The effect is minimized by putting a light absorber as a nonhalation layer on or inside the support.

Silver halide emulsions respond to light in a negative-acting manner. They retain the image in the exposed areas when normally developed. But through the use of special "reversal" developers, silver halide can respond in a positive-acting manner as well.

10.3.2 Diazo

Photosensitive diazo emulsions belong to the class of molecular dye imaging media. Images are formed by a molecular dye process rather than granular silver crystals. The diazo emulsions consist of a suitable binder, transparent diazo salts, sensitizers, and color couplers in the form of dye precursors. The diazo molecules, when exposed to light, are destroyed in direct relation to the amount of light striking them. Subsequent processing in a controlled atmosphere of ammonia then permits the color coupler to react with the unexposed and nondestroyed diazo salts to form an azo dye image. Diazo is a positive-working process because the image is formed in the unexposed areas. Since there is no molecular amplification in the image-forming steps, diazo requires very high exposure energy near 1 J/cm^2.

The principal use for diazo is in PWB fabrication as a working phototool copy from a silver halide original for the subsequent exposure of uv sensitive photoresists. Diazo is not only less expensive than silver halide but its image is also translucent in the visible spectrum. This allows better visual registration when exposing the photoresists on the PWB panels. The optical density of the azo dye image in the uv region, however, is very high and can readily reach values near 4.

10.3.3 Photoresists

Photoresists are the key materials for photofabrication. They permit the photopatterning of the substrate surfaces, onto which they are applied in intimate contact. The application can occur as coating of a *liquid solution*, or as lamination of a *dry film* precoated on a polyester film carrier. The exposed and developed photoresist image relies on a polymeric binder of reasonable molecular weight as its principal constituent to provide the desired *resist* properties for the subsequent modifications in the adjacent unimaged areas. For IC fabrication, this modification occurs in the form of metal evaporation and silicon oxidation. For PWB fabrication, this modification occurs in the selective removal of metal through etching or selective addition through plating.

Photoresists can be formulated to be positive or negative acting. For positive-acting resists, exposure renders the resist soluble to developers in the exposed areas. The exposed resist is then washed off with the developer. Negative-acting photoresists form the less soluble, polymeric compositions in the exposed areas. Exposure for a negative resist results in the polymerization and cross-linking of an initially low-molecular-weight composition. The unexposed, low-molecular-weight resist is then washed off with the developer. Negative-working resists are less costly than positive-working resists. They dominate in PWB fabrication. Positive-working resists provide excellent resolution and are used extensively in IC fabrication.

The sensitivity of some photoresists can depend strongly on the temperature of the medium. It is desirable to filter out the infrared portion of the light in exposure systems to avoid warming the resist.

Positive-Working Resists. Positive photoresists have been formulated in many forms to meet the tradeoffs of specific applications in IC fabrication, such as spectral sensitivity, contrast and resolution, latitude, adhesion to different surfaces, topography, and ability to withstand the environment necessary for modification. The deposited thickness is in the range of 0.5 to 1.5 μm. The photochemical reactions for producing the chemical transformations in the exposed areas come from relatively small amounts of a *photoactive component* (PAC) in the formulation. The transformation can be from nonpolar to polar, or polymeric to oligomeric-monomeric molecules. After full exposure, the resists typically develop at least 10 times faster in the exposed than the unexposed regions.

The most widely used positive resist system consists of a *novolac* (N) resin binder and a PAC sensitizer. The PAC is based on the *diazonaphthaquinone* (DQ) family. A common abbreviation for this system is DQN. It is shown in Fig. 10.16. The two principal ingredients, along with special additives, are dissolved in organic solvents. The resulting solution can now be coated onto the desired substrate surfaces and then baked to remove residual solvent.

Upon exposure to near uv light, the sensitizer is decomposed and converted from a nonpolar DQ molecule to a polar, base-soluble carboxylic acid.[7] It is readily washed out with the resin in an aqueous alkaline development. This process is illustrated in

Photoactive Diazoquinone (Base Insoluble)

Novolac Resin (Base Insoluble)

FIGURE 10.16 Principal constituents of DQN resist.

Fig. 10.17. The unexposed portion with its undecomposed PAC sensitizer remains as the resist image, since it dissolves an order of magnitude slower in the developer. The sensitivity of the DQN family is in the range of 250 to 500 mJ/cm^2, since there is no chemical amplification. It requires approximately 10 mol of photons (1 J = 1 mol of photons or 1 Einstein, 10^{18} photons) per mole of DQ.

The DQN resists are the workhorse positive resists for microphotofabrication because of their wide spectral sensitivity, excellent etch resistance, aqueous development, and fair thermal stability. But the high opacity of the novolac resin in the deep uv interferes with the light absorption by the DQ sensitizer. Two alternate systems have evolved for the deep uv. The first uses *polymethyl methacrylate* (PMMA) resin as the polymeric binder and relies on its photoinduced chain scission,[8] as shown in Fig. 10.18. The sensitivity is also upward of >250 mJ/cm^2 but can be enhanced through the addition of sensitizers.

The second uses polymeric binders with pendant or main chain carbonate and aldehyde groups. The sensitizers are composed of *onium* groups, which form Lewis acids upon exposure.[9] The Lewis acids then catalyze the chain scission of the polymeric binder as shown in Fig. 10.19 for a *sulfonium* ion-based sensitizer. This catalytic process provides for chemical amplification. One mole of photons generates 1 mol of acid, which can then cleave 10 or more polymer chains. The sensitivity of this system is very high at 5 mJ/cm^2.

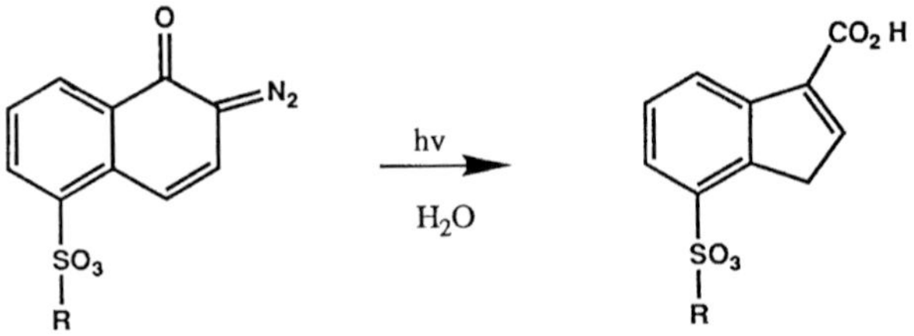

Base Insoluble

Base Soluble

FIGURE 10.17 Solubilization of DQ upon exposure and aqueous development.

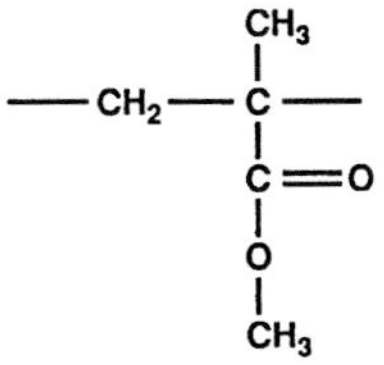

(-MMA-MMA-MMA-MMA-)$_n$ $\xrightarrow{hv}$ (-MMA-MMA-)+(-MMA-MMA-)+CO_2, CH_4, H_2

FIGURE 10.18 Chain scission of PMMA.

$Ar_3S^+X^-$ $\xrightarrow{hv}$ → $Ar_2SH^+X^-$

Sensitizer — Acid

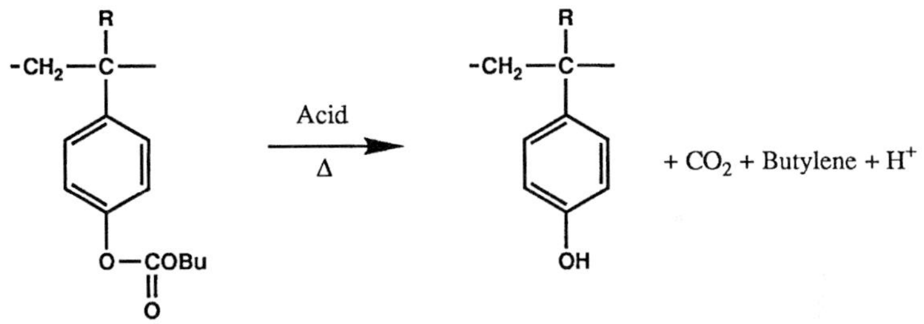

FIGURE 10.19 Photoinduced reaction sequence for onium-sensitized resist.

Negative-Working Photoresists. Negative-working photoresists also come in many forms to meet the tradeoffs in specific applications. The prevalent negative-working photoresists are based on random free radical polymerization and cross-linking of main chains or pendant side chains of monomer and/or oligomer binders, as shown in Fig. 10.20. Efficient chain propagation without premature termination can result in high chemical amplification. In the liquid solution state, direct monomer polymerization can actually be propagated efficiently to result in a 10,000-fold chemical amplification. A theoretical sensitivity of 1 mJ/cm^2 has been calculated for such a system.[10] The tacky nature of such a resist requires off-contact printing, however, which has severely limited its commercial use. The more practical formulations for amplification rely on monomers dispersed in the applied solid film coating. The high viscosity of the medium now limits chemical amplification to approximately a tenfold gain.

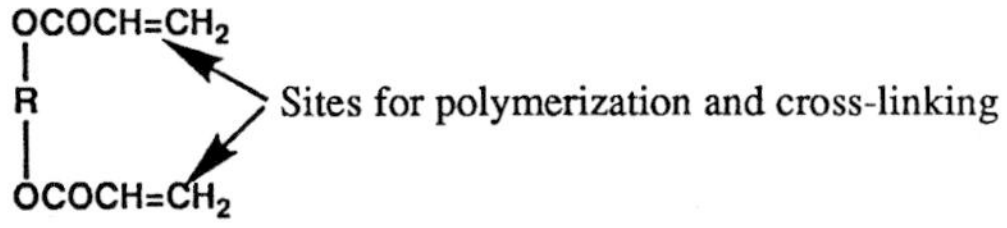

FIGURE 10.20 Multifunctional acrylate.

Negative-working, cross-linkable resists were used to fabricate all the early ICs with their feature sizes of 3 to 10 μm. They are still used in special IC applications such as in deep uv exposure or as high-temperature resists. Their principal ingredients consist of a photoinitiator and a polymerizable, low-molecular-weight polymer binder. This binder is unsaturated and usually responds to the action of free radicals generated by exposure of the photoinitiator.

The volume use for negative resists is now in PWB fabrication, where they are used in thickness ranges from 15 to 100 μm. The binder system usually consists of a higher-molecular-weight, saturated binder along with a reactive monomer such as a multifunctional acrylate[11] with two or more reactive sites. The acrylate monomer also polymerizes by a free radical mechanism and forms cross-linked, three- dimensional networks with the binder. Additional photoresist ingredients typically are adhesion promoters, thermal stabilizers, oxygen inhibitors, plasticizers, antiblocking agents, and dyes. Dyes can be used as spectral sensitizers as well as for identification and contrast of the exposed image. The use of chain transfer agents enables the initiated polymerization to proceed more efficiently and thus requires less light energy for exposure. Typical reactions for a negative-working resist are shown in Fig. 10.21. The polymeric binder contains carboxylic acid groups to permit aqueous alkaline development. An overall model for the photopolymerization process is shown in Fig. 10.22.

In the first step, uv radiation acts on the sensitizer-initiator system to form active radicals. These radicals will now preferentially react with any oxygen present to form inert products. This step is controlled by the oxygen diffusion rate and is responsible for an induction period and initial delay of polymerization during exposure. After local oxygen depletion, some of the radicals will then react with monomer to form monomer radicals. These monomer radicals then propagate polymerization by reacting with additional monomer to form an extensive network of cross-linked polymer chains. This is the desired result from exposure. But propagation can also be inhibited by oxygen. Polymerization can only proceed to completion when all the originally dissolved oxygen has been consumed. Final termination again occurs through reaction with ambient oxygen.

Alternate systems can be formulated, in which the initiation and propagation occurs by an ionic mechanism. Such systems are not inhibited by the presence of oxygen. But they have not yet found commercial acceptance at this time.

Resist Application. Resist adhesion to the substrate, thickness, and uniformity are the most important considerations in selecting a resist application process. A clean substrate surface is necessary for good adhesion. Inadequate or improper cleaning can lead to many defects in the subsequent processing steps. Accepted cleaning methods to remove organic soils are based on solvents or alkaline soak cleaners. Particulate soils can also be removed by them with the aid of agitation. Other methods for removing particulate soils are electrocleaners, which scrub the substrate surface through the generation of gas, and abrasive scrubbing cleaners such as pumice. Metal oxides on the surface can be removed through a chemical acid cleaning and/or mild acid etching step.

The application of the resist to the cleaned substrate surface can occur in *liquid* or in precoated *dry film* form. Commercial methods for applying liquid resists are spin coating, spray coating, roller coating, electrophoretic dip coating, screen coating, and curtain coating.

In spin coating, the resist is deposited in bulk on the substrate and then spun at speeds up to 10,000 rev/min. This method allows very uniform deposition of thin coatings and is the preferred method for IC wafer fabrication. In spray coating, a spray gun applies the resist as small droplets, which then coalesce to a continuous coating. In a special embodiment, the droplets can be electrostatically charged in the spray

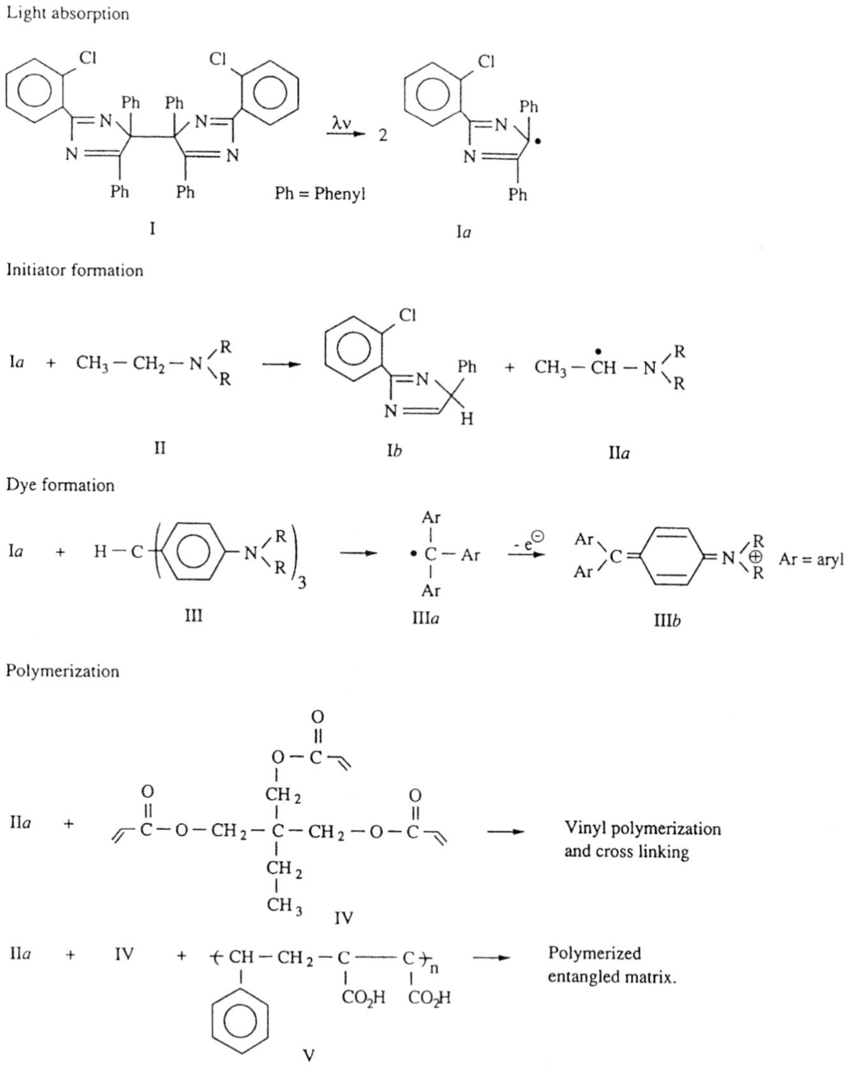

Light absorption

I $\xrightarrow{\lambda\nu}$ 2 I*a*

Ph = Phenyl

Initiator formation

I*a* + $CH_3-CH_2-NR_2$ (II) $\longrightarrow$ I*b* + $CH_3-\dot{C}H-NR_2$ (II*a*)

Dye formation

I*a* + $H-C(C_6H_4-NR_2)_3$ (III) $\longrightarrow$ $\bullet C(Ar)_3$ (III*a*) $\xrightarrow{-e^{\ominus}}$ $Ar_2C=C_6H_4=N^{\oplus}R_2$ (III*b*) Ar = aryl

Polymerization

II*a* + IV $\longrightarrow$ Vinyl polymerization and cross linking

II*a* + IV + $-(CH-CH_2-C(CO_2H)-C(CO_2H))_n-$ (V) $\longrightarrow$ Polymerized entangled matrix.

FIGURE 10.21 Photoinduced reaction sequence for a typical negative-acting resist.

gun. The spray method allows good surface coverage over raised conductors but affords only fair uniformity. It is now used commercially to deposit soldermasks over the circuit patterns of PWBs.

The principle of roller coating is illustrated in Fig. 10.23. Resist flows continuously between the doctor and the coating rolls and is transferred to the substrate in the nip formed by the coating roll and the supporting drive roller. Viscosity, nip roll pressure, and coating speed are the basic control parameters for coating thickness. The surface

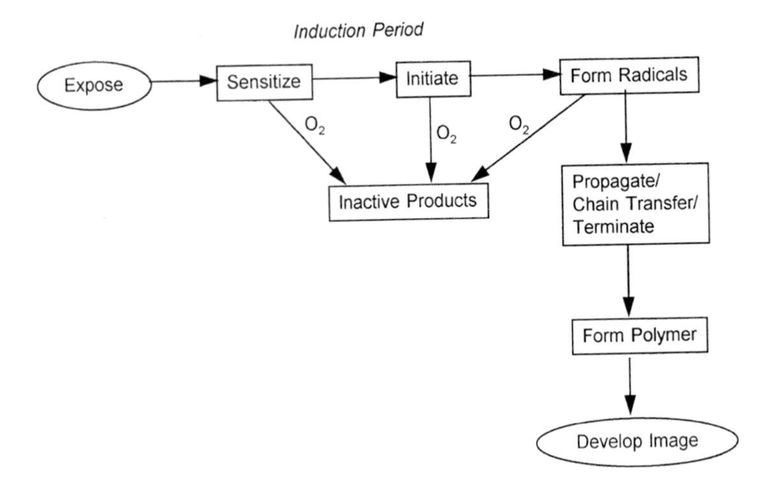

FIGURE 10.22 Model for photopolymerization.

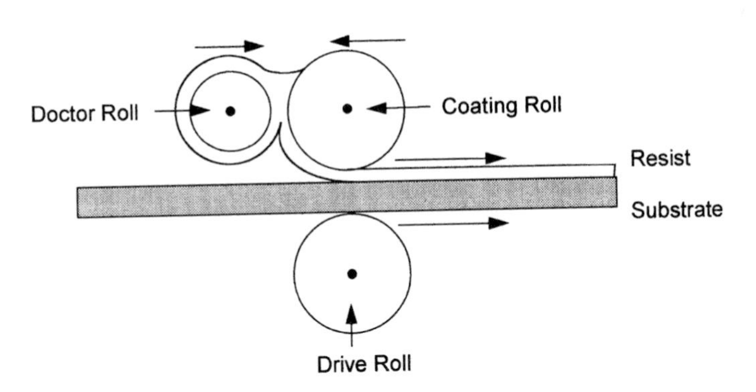

FIGURE 10.23 Roller coating process.

of the coating roller can be varied to accommodate a range of resist viscosities. Smooth rolls are used with high- viscosity resists, whereas threaded or serrated rollers are preferred with low-viscosity resists. A doctor blade is sometimes used in place of the doctor roll. Roller coating, in contrast to spin and spray coating, is very efficient in resist usage. Very little resist is wasted.

Electrophoretic coating has been used for some time in the automotive industry to prime car bodies. It has recently been adopted for applying resist to copper laminate for PWB fabrication. The liquid resist is in the form of an emulsion, similar to paint. The resist can be anodic or cathodic. The process shown in Fig. 10.24 deposits electrophoretically charged resist micelles in an electroplating tank onto the copper surface, which becomes one of the electrodes in this plating process. The micelles coalesce after deposition. This process provides very good coating uniformity because local deposition becomes very slow after a certain coating thickness has been attained with the nonconducting resist. The thickness is primarily controlled by the coales-

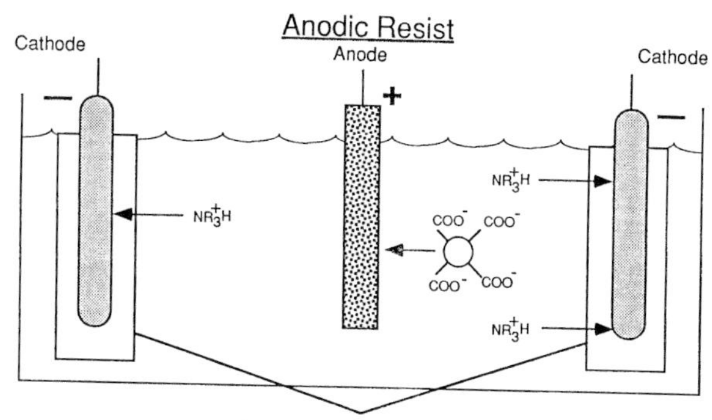

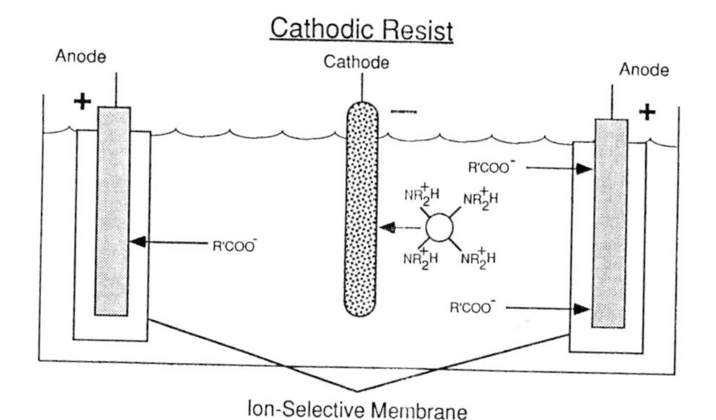

FIGURE 10.24 Mechanism for anodically electrodepositing resist. (*Courtesy of DuPont Electronics.*)

cence rate of the resist formulation, and secondarily by the applied current. This process is now used in PWB fabrication, where its ability to coat both sides simultaneously and deposit resist in the via and through holes give it a unique advantage.

In screen coating, the photodefinable resist is applied through a blank screen with the aid of a squeegee. The deposited thickness depends on the size of the mesh openings in the screen. This process has become popular for depositing soldermask resists over the raised circuitry of PWBs.

In curtain coating, the liquid resist is applied through an extrusion die. This process is shown in Fig. 10.25. Thickness is controlled by resist viscosity, die lip opening, and substrate speed. Curtain coating currently is the most popular commercial method for applying liquid, photodefinable soldermask onto PWBs, and its application is now being extended to deposit photoresist onto copper laminate for subsequent patterning of circuitry. It is also being developed to deposit polyimide dielectrics onto the square shapes of multichip modules to avoid the waste of material inherent in spin coating.[12]

The thickness of liquid resists deposited in a single pass for spin coating ranges from 0.01 to 10 μm and ±1 percent uniformity. The thickness for the other liquid application processes ranges from 1 to 20 μm and up to ±10 percent uniformity. This resist thickness range, when applied by a mechanical coating process without electro-

FIGURE 10.25 Soldermask is applied in a curtain coater. (*Courtesy of Ciba-Geigy.*)

static assist, generally has been insufficient in the past to allow the fabrication of PWB circuitry with high yields. The principal causes for the yield losses have been coating irregularities due to poor laminate topography and lack of cleanliness.

Dry film resist was formulated to overcome these irregularities in PWB fabrication. Here, the resist is precoated by the resist supplier under clean room conditions onto a polyester film carrier and then protected with a polyolefin cover sheet. The resist thickness ranges from 15 to 100 μm, with very good uniformity in the range of 1 to 3 percent. In use, the protective polyolefin sheet is removed, and the resist is then applied to the copper laminate with the aid of a laminator, as shown in Fig. 10.26. The polyester sheet protects the applied resist during subsequent exposure and serves as an oxygen barrier to avoid undesirable side reactions in the unexposed resist.

Resist Development. Exposure of a resist to a suitable light source induces a chemical reaction in the exposed areas. This chemical reaction can be the scission and rearrangement of a molecule such as DNQ for a positive resist, or the polymerization and cross-linking of lower-molecular-weight molecules for a negative resist. The readout of this change occurs in the development step, which distinguishes exposed from unexposed areas. The developing process is usually carried out with a liquid developer and relies on the differential solubility of exposed versus unexposed resist in this developer. In positive resist, the exposed portion becomes significantly more soluble

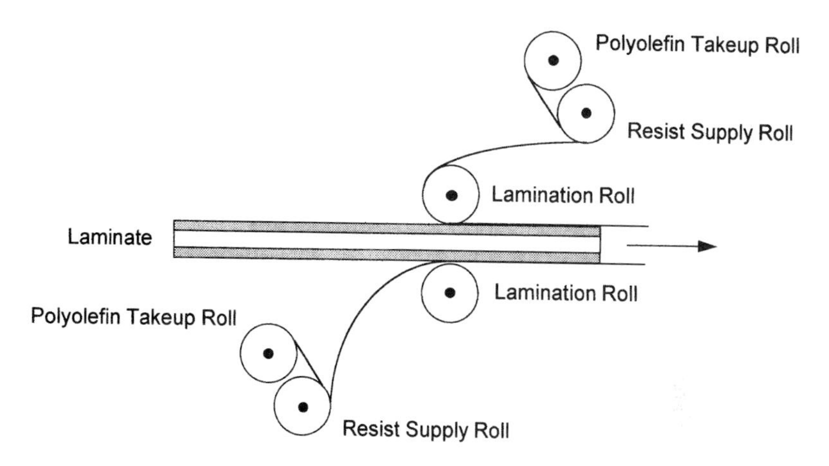

FIGURE 10.26 Dry film resist laminating process.

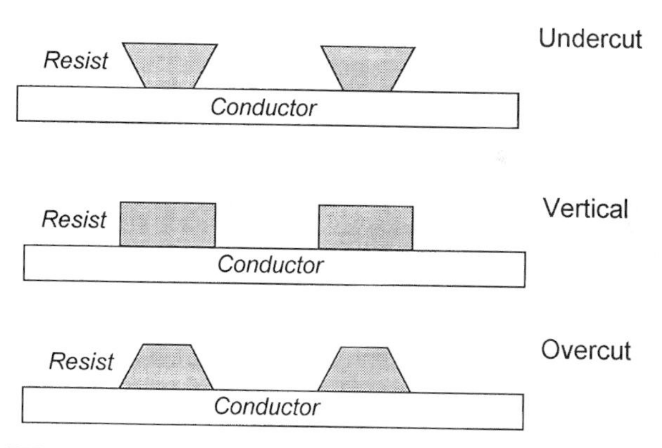

FIGURE 10.27 Resist side wall profiles.

in the developer, whereas the exposed portion of a negative resist becomes significantly less soluble. The object is to remove the solubilized resist portion consistently and reproducibly at a faster rate than the other portion.

The shape of the developed resist image depends strongly on the interaction of the exposure dose and the developer. Three different resist profiles are possible for the resist sidewalls after exposure and development. They are vertical, undercut, and overcut, as shown in Fig. 10.27.

A vertical sidewall can be attained through use of a properly formulated resist with a uniform, bulk response to the correct energy level from an exposure source with suitable collimation and declination. An undercut for a positive resist can result from overexposure, which leads to line narrowing at the substrate surface from reflected exposure rays. Conversely, a negative resist can have an undercut profile because of underexposure. Some negative resists have actually been formulated to absorb light at the surface only, which intrinsically results in such an undercut. An overcut for a positive resist is typically associated with underexposure, whereas overcut in a negative resist can be due to overexposure.

10.3.4 Photodefinable Dielectrics

Photodefinable dielectric materials, which form the dielectric layers in the finished product, are becoming more prominent. They serve many functions such as interlayer dielectrics, passivation layers, stress buffers, planarizing layers, and protective masks for subsequent soldering operations in assembly. These materials can be pattern defined through laser ablation or application of photoresists as described in Sec. 10.3.3, or they can be rendered intrinsically photosensitive through suitable formulation. Such photosensitive materials can be formulated to be positive or negative working. Current commercial use, however, is restricted to negative-working compositions.

Polyimides for Microcircuits. Polyimides are the preferred organic polymers dielectrics in current high-density interconnect applications because they combine low dielectric constants with very good thermomechanical properties. The basic monomers for this class of polymers are a dianhydride and a diamine. The composition of a PMDA/ODA polyimide system, which in film form has been commercialized as KAPTON* film, is shown in Fig. 10.28.

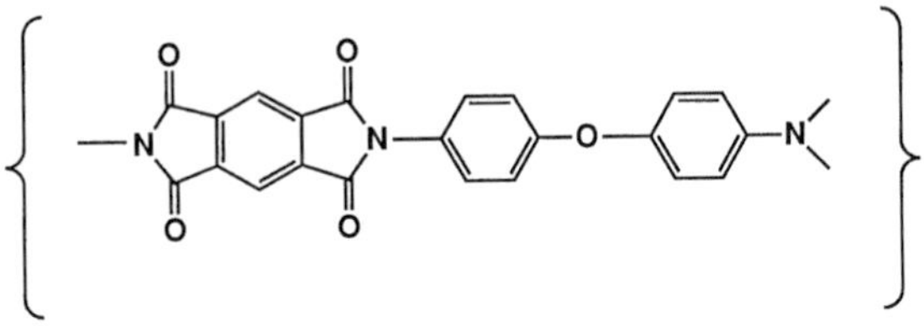

PMDA/ODA

FIGURE 10.28 PMDA/ODA polyimide.

For high reliability of the solder joints between the IC chips or packages and the interconnecting substrate, it is important to appropriately match their thermal expansion properties as measured by the respective *coefficients of thermal expansion* (CTE). A very good correlation exists between the rigidity of the monomers, the subsequent rigidity or modulus of the polyimide polymer, and a low CTE. The polyimide composition shown in Fig. 10.29 is based on BPDA and PPD as monomers and possesses a CTE near silicon.

The additional attribute of intrinsic photodefinability for easier use and lower cost often involves a tradeoff between good mechanical, thermomechanical (CTE), and photodefinable properties. Two major types of photosensitive polyimide systems in

*Tradename DuPont.

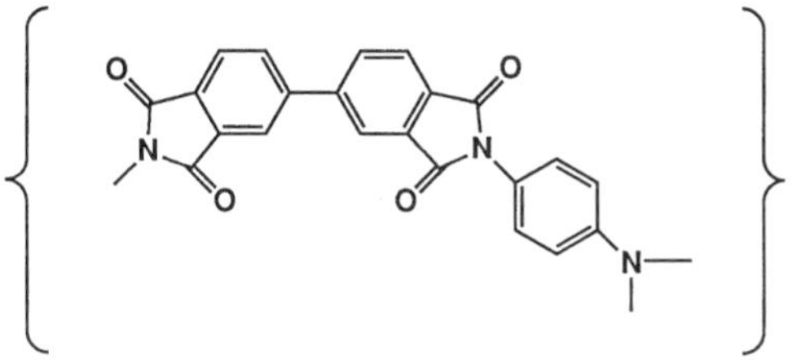

BPDA/PPD

FIGURE 10.29 PPDA/BPDA polyimide.

commercial use today employ photo-cross-linkable ester or salt-containing functions. The ester types[13,14] show low elongation but good photodefinition, whereas the ionic salt types[15] have good elongation but high CTEs and marginal photodefinition.

A new photodefinable polyimide system[16] based on the rigid monomers of BPDA and PPD has managed to combine the desirable attributes of low CTE and high elongation with very good photodefinition for high resolution. The high resolution benefits from the anisotropic development after exposure leading to straight sidewalls as shown in Fig. 10.30, whereas resist patterned polyimides will have sloped sidewalls after wet, isotropic etching.

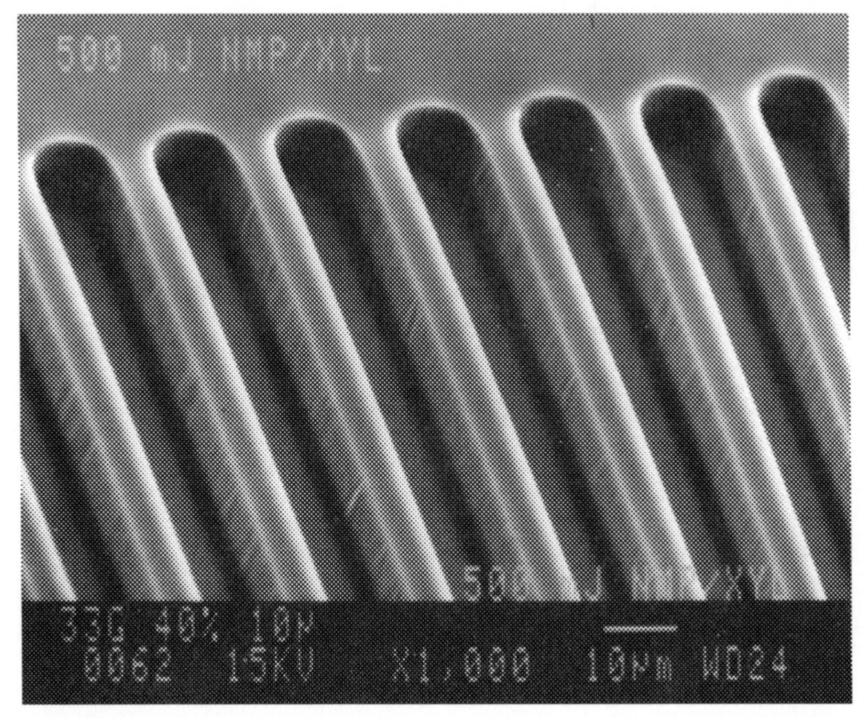

FIGURE 10.30 Photodefined PPDA/BPDA polyimide test pattern. (*Courtesy of DuPont Electronics.*)

Photodefinable Thick-Film Dielectrics. The dominant method for patterning via holes in thick-film dielectrics has been accomplished through stencil screen-printing. In practice, this limited via diameter to 200 μm.[17] A new development in negative-acting photodefinable thick-film dielectrics now allows via diameters as small as 75 μm.[18] An example of this via pattern is shown in Fig. 10.31. This dielectric, in conjunction with photodefinable thick-film conductors,[19] has allowed the photofabrication of all the layers in a thick-film multichip module.[20]

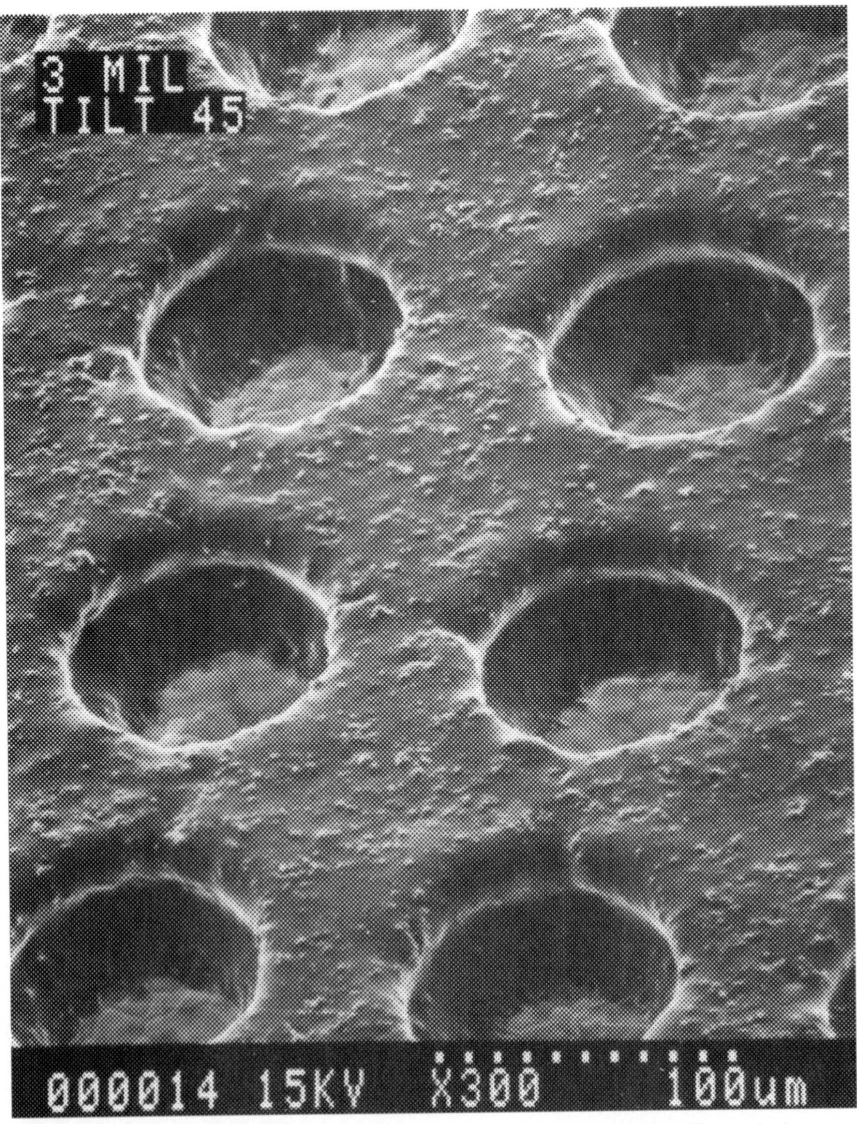

FIGURE 10.31 Photodefined thick-film 75-μm via pattern. (*Courtesy of DuPont Electronics.*)

Photoresists as Dielectrics and Soldermasks for Printed Wiring Boards. The fabrication process for PWBs relies on photoresists as fabrication aids for defining the more demanding conductor patterns on the circuit layers. This resist is usually removed after it has served its purpose. The defined conductor patterns on the outerlayers of PWBs are then usually covered with a protective, permanent soldermask to insulate them, provide mechanical protection, and avoid solder bridging in the subsequent soldering step in component assembly. Photodefinable soldermasks have become necessary for accurate mask image placement to accommodate the shrinking pad sizes for components. While the principal function of the mask is to protect the circuitry during component attachment, its dielectric constant also begins to influence the electrical characteristics of signal lines[21] such as noise, propagation delay, and characteristic impedance. This influence becomes more pronounced with the increasing clock speed of PWB assemblies.

The chemistry of these masks derives from the negative-working photoresists described in Sec. 10.3.3. The solvent or semiaqueous developable compositions are usually rich in epoxy binders, whereas the aqueous developable compositions typically are rich in acrylate based binders. Thermal cross-linking agents are added to the masks to allow a thermal cure in addition to a uv cure, so that the masks can suitably withstand the subsequent soldering and cleaning steps. This curing step for the photodefinable soldermasks has usually come at the expense of mask flexibility, which has precluded their use in flexible circuit applications, such as shown in Fig. 10.32. A recent advance in photodefinable, flexible films now combines accurate image definition and straight sidewalls as shown in Fig. 10.33 with flexibility in use after cure.

FIGURE 10.32 Flexible circuits used in camera. (*Courtesy of DuPont Electronics.*)

FIGURE 10.33 Photodefined flexible film solder dam pattern. (*Courtesy of DuPont Electronics.*)

IBM-Yasu in Japan has recently pioneered the use of a soldermask composition as an innerlayer dielectric for making a high-density PWB.[22] Here, the mask is deposited onto a conductor layer and then exposed and developed to generate the via holes, which then allow connections to a subsequent conductor layer formed in a sequential mode. As hole sizes become smaller in PWBs, the formation of these via holes through photofabrication with photodefinable dielectrics instead of mechanical drilling will undoubtedly receive more future attention.

10.3.5 Ablatable Materials

Ablatable systems represent the simplest photofabrication systems since they do not require the intermediate use of a photosensitive resist. They rely on the direct absorption by a given material of the very intense energy from a powerful radiation source, which results in the bond breaking and subsequent evaporation of the reaction products. The energy requirements typically are in the range of 100 to 10,000 J/cm^2. High-powered lasers are the preferred source of radiation. The wavelength of the laser light determines the nature of the reaction.

Carbon dioxide lasers, which emit in the infrared at 10.6 μm, heat materials locally at the point of light absorption. The hot spots then break down into volatile products, aided by air oxidation. The latter, however, can lead to undesired charring of the patterned edges in a material.

Excimer lasers produce very intense output in the deep uv region. Absorption of this light by a material will lead to volatilization through direct bond breakage, which results in good edge definition. This process is used commercially to generate small vias in multichip modules, formed with either thin-film polyimide dielectrics or polyimide film.

10.4 IMPORTANT PHOTOFABRICATION APPLICATIONS

10.4.1 Introduction

Two applications currently dominate the use of photofabrication in electronics. They are semiconductor integrated circuits and printed wiring boards. A third application, flat panel displays, is now widely believed to become similar in size and importance to the first two in the future.

The ability to constantly lower the cost per function on the IC chip through advances in photofabrication has served as a powerful driver in the quest to make such advances. The feature size reduction of PWBs, on the other hand, has had a much smaller influence in the past on reducing the cost per function at the system level. It has also been limited by the physical requirements and capabilities of the component assembly process. As a result, the sophistication of IC photofabrication is now several orders of magnitude higher than PWB fabrication. But as fine pitch and ultrafine pitch surface mount now become more prevalent to lower the system cost per function, PWB photofabrication will be driven to higher levels of sophistication including the use of substrates with greatly increased dimensional stability. Flat panel displays will be intermediate in sophistication between the two.

10.4.2 Semiconductors

The functional fabrication steps for IC chips are outlined in Fig. 10.34. This figure illustrates clearly the important role of optical photofabrication in defining the patterns for the gates, contacts, interconnections, and bonding pads. The starting point for optical photofabrication is the photomask.

Photomasks. The preferred substrate for photomasks is quartz glass. Quartz substrates transmit uv light and minimize thermally induced substrate expansion. Light scattering is minimized through antireflection coatings and tight flatness specifications.

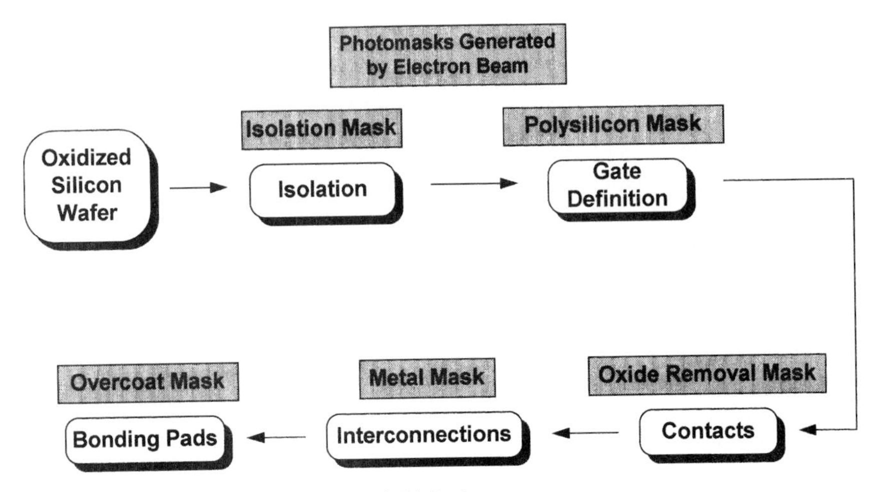

FIGURE 10.34 Outline of integrated circuit fabrication.

In the 1960s, the photomask pattern was formed through the use of silver halide emulsion coatings on the mask substrate to produce 1x masks for contact printing of 50-mm wafers. As features became smaller and wafer sizes grew in the 1970s from 75 to 100 mm and now in the range of 150 to 200 mm for the 1990s, silver emulsions began to be replaced by hard chrome coatings. Chrome offers superior durability and opacity in a thin coating. A liquid photoresist is then used to define the image pattern on the chrome-coated substrate, followed by subsequent etching.

Today, electron-beam exposure is the preferred method to generate the pattern on the photoresist for the 5*x* (most common) and 10*x* master of the desired circuit pattern. Such master patterns are also known as *reticles.* A commercial embodiment of an electron-beam unit is shown in Fig. 10.35. Automated inspection provides feedback on line width and defects. Repair can be carried out through use of ion beams[23] and lasers.[24] The 5*x* and 10*x* reticles are then ready for direct use in stepper projection printers. To produce the 1*x* photomasks used in scanners, the 10*x* master reticle is repeatedly imaged onto the photomask with optical photorepeaters on a precision stage. Accuracy in *x-y*[25] and rotational[26] alignment is extremely important in this step.

After fabrication of the mask it is usually covered with a pellicle to keep it from becoming contaminated. This pellicle needs to be transmissive since it stays on the mask during exposure. The pellicle also helps to defocus dust contamination on its surface.

Phase Shift Photomasks. Higher numerical apertures and shorter wavelengths give higher resolution, but at the expense of depth of focus as discussed in Sec. 10.2. In practice, this tradeoff sets a limit to higher resolution in projection printing through simple reduction in wavelength. Phase-shifting photomask technology shows great future promise to extend the resolution of current projection printing systems without loss of depth of focus.[27] This technology was pioneered by M. D. Levenson.[28]

Phase shifting is accomplished by depositing an additional patterned layer of transmissive material on the mask. Alternatively, the phase shift pattern can be added under the chrome or etched into the reticle substrate as shown in Fig. 10.36. During exposure, the light passing through the shift pattern follows an optical path different

FIGURE 10.35 E-beam machine. (*Courtesy of DuPont Electronics.*)

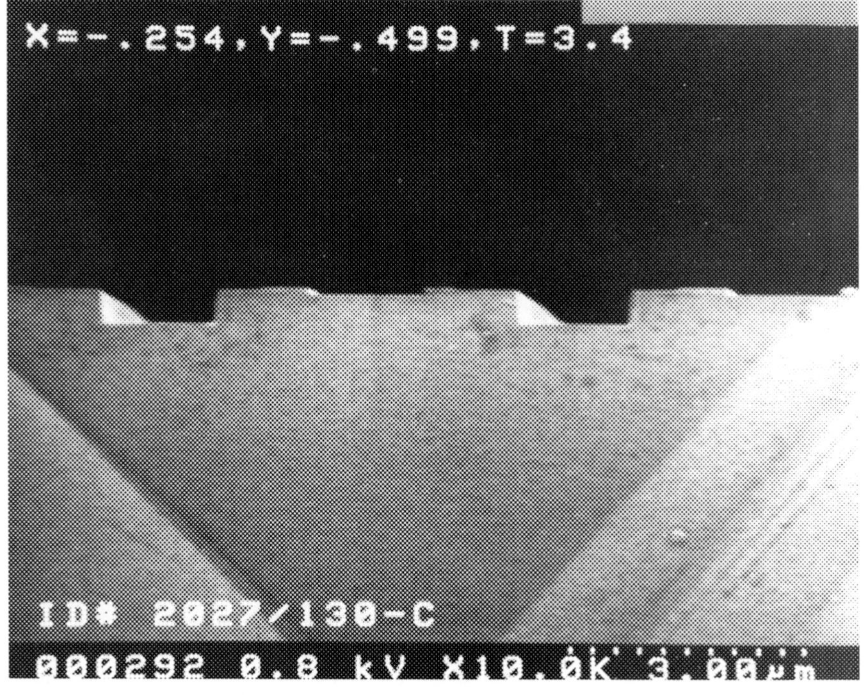

FIGURE 10.36 Cross section of photomask with phase shift pattern etched into quartz. (*Courtesy of DuPont Electronics.*)

from the light passing through the nonpatterned area. As a result, the light beams emerge 180° out of phase with each other. The interference effects of these two beams of partially coherent light now enhance edge contrast by cancelling, rather than reinforcing, the resolution, limiting edge diffraction. The results are higher resolution and greater depth of field. Several variations of this process are illustrated in Fig. 10.37.

The minimum feature size on a 5*x* phase shift reticle can be as small as 0.25 μm. The cost for phase shift reticles is estimated to be up to 10 times higher than conventional reticles. Defect prevention, detection, and repair are very important since phase shift defects are twice as printable as chrome defects for the same reason that these reticles double resolution.

Commercial Resist Exposure. Optical photofabrication currently is the dominant process for making IC chips. IC wafer fabrication started out as a process for imaging the entire wafer at once in full-field exposure through 1:1 contact printing with a photomask. As discussed in Sec. 10.2.3, contact printing soon gave way to proximity and then projection printing in order to print increasingly finer features and solve the problem of high defect levels without loss of resolution.

Today, the two established commercial printing technologies are scanning projection and refractive stepper projection, as discussed in Sec. 10.2.2. The refractive steppers cannot use polychromatic light. As a result, they have to optimize around one of the major output lines of the mercury light source. These lines are near 465, 405, and 365 nm and are known as the *g-line, h-line,* and *i-line,* respectively. They are capable of 0.50 μm resolution for patterning 16-Mbit DRAMs. Their resolution limit is judged to be 0.35 μm.[29] The relative merits of steppers and scanners are compared in Table 10.5.

Deep uv light from excimer lasers allows greater resolution without restrictions on pattern layout.[30] A key challenge for this technology is its need for quartz lens optics, which makes the system monochromatic and prevents its full utilization of the spectral output of the excimer source. Also, the low refractive index of quartz has currently limited the numerical aperture of the lens to 0.5, as contrasted with apertures of 0.6 and greater for current steppers. The ultimate numerical aperture capability is judged to be 0.7.[31] A refractive stepper is shown in Fig. 10.38.

The concepts of reflective scanning and refractive projection have also been recently combined into an exposure unit for deep uv exposure.

Advanced Exposure Options. The leading options for increasing resolution, in addition to the previously discussed phase shift masks and shorter wavelength light sources from deep uv light, are x-ray and electron beam.

X-ray lithography with its very short wavelength in the range of 1 to 20 nm can readily image feature sizes below 0.1 μm. A synchrotron serves as the area-wise energy source in conjunction with proximity printing. The key challenge here is fabricating the 1*x* photomask for proximity printing accurately without distortion and defects. The required pattern placement accuracy on the mask is 10 times greater than that for current optical reticles, but substrate distortion needs to be controlled as well. A 20*x* x-ray reflection projection reduction system, which promises to overcome the 1*x* mask fabrication problems, has been described recently.[32] Synchrotron radiation in the range of 10 to 15 nm was chosen as the optimum wavelength. But this projection system is faced with formidable problems in making the reflective mirrors. The surface roughness must be less than 0.1 to 0.2 nm, and some of the mirrors need to be aspherical. Thus, x-ray lithography needs to overcome a number of fabricating problems for the supporting elements to become technically viable as well as cost-effective.

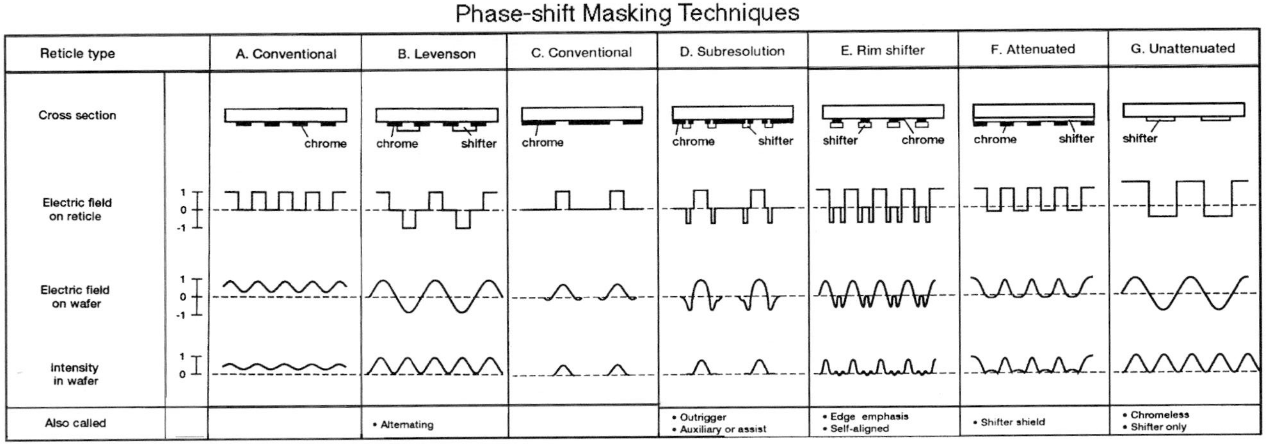

Phase-shift Masking Techniques

Reticle type	A. Conventional	B. Levenson	C. Conventional	D. Subresolution	E. Rim shifter	F. Attenuated	G. Unattenuated
Cross section	chrome	chrome, shifter	chrome	chrome, shifter	shifter, chrome	chrome, shifter	shifter
Electric field on reticle (1, 0, -1)							
Electric field on wafer (1, 0, -1)							
Intensity in wafer (1, 0)							
Also called		• Alternating		• Outrigger • Auxiliary or assist	• Edge emphasis • Self-aligned	• Shifter shield	• Chromeless • Shifter only

FIGURE 10.37 Several ways to add phase shifters to photomask patterns. (*Courtesy of Semiconductor International.*)

TABLE 10.5 Comparison of Steppers and Scanners

Performance	Scanner	Stepper
Reduction	1.1	1:1, 5:1, 10:1
λ (nm) for mercury arc	Polychromatic	435, 405, 365
Aberration	No	Yes
Field size	Full wafer diameter	5–20 mm square
Sensitivity to mask defects	Very high	Low to moderate
Throughput	Medium to high	Low to medium

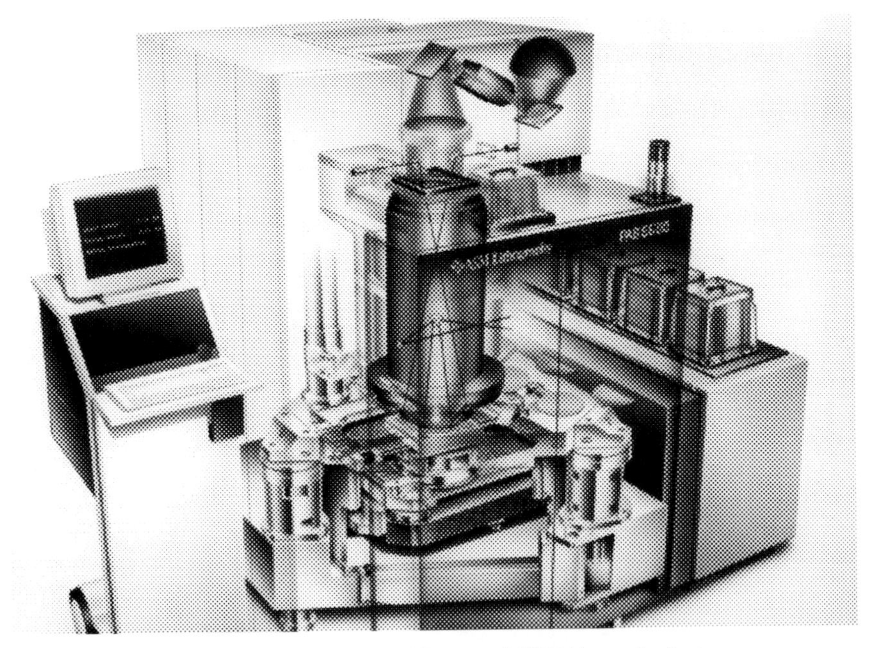

FIGURE 10.38 Refractive projection stepper. (*Courtesy of ASM Lithography, Inc.*)

Electron-beam nonoptical writing technology, used for fabricating photomasks, can also provide very high resolution in semiconductor fabrication.[33] But the high cost of the equipment and its slow throughput due to its serial exposure severely detract from its cost-effectiveness. It is not viewed as a serious contender for some time to come.

Today, it is difficult to forecast the winners in this technology race. The 1991 predictions of lithographers polled at the Fourth MicroProcess Conference in Kanazawa, Japan, are shown in Fig. 10.39 as applied to future DRAM memory chip production. The majority sees the road map as going from phase shift masks with currently used i-line light sources to phase shift with deep uv, and ultimately x-ray.

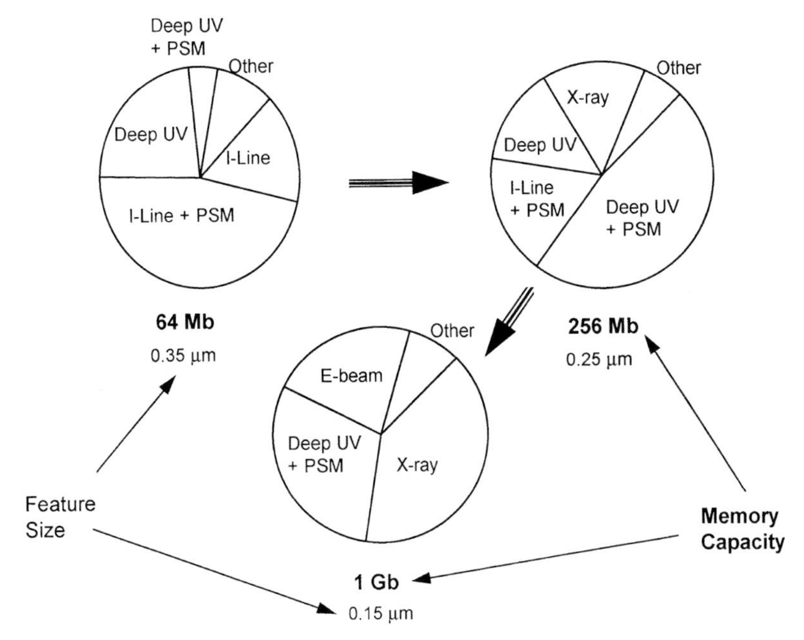

FIGURE 10.39 Which technology for...? Opinions polled at the Fourth Microprocess Conference, Kanazawa, Japan, July 1991. (*Courtesy of Semiconductor International.*)

10.4.3 Printed Wiring Boards

The functional fabrication steps for PWBs are shown in Fig. 10.40. There are two principal functions using photofabrication, namely, definition of the conductor patterns on the circuit layers and pattern definition of the soldermask over the external circuitry. These functions are described in Fig. 10.40 as *Image* and *Soldermask.* The process of defining the conductor patterns is frequently called *primary imaging,* and the soldermask definition is called *secondary imaging.*

Traditional PWB circuit imaging uses phototools for contact and off-contact printing onto photoresist-coated substrates. By contrast, laser direct imaging accepts CAM data to write with a raster-scanned laser beam directly onto the resist-coated substrates without the need for phototools. Laser direct imaging, although practiced for 20 years, has been inhibited by low resist sensitivities, limited laser reliability, lack of CAM data in standard formats, low throughput, and equipment designs incompatible with production environments. Recent advances in media sensitivities, improved laser reliability, optical design and manufacturing, and significant hardware and software improvements have now also made laser direct imaging viable.

Phototools. The most commonly used substrate for phototools is a polyester film base in a thickness range of 175 to 200 μm. This base combines good mechanical and optical properties with adequate dimensional stability at very reasonable cost. The phototools are used in a 1x format in both contact and off- contact printing. The thermal and humidity coefficients of expansion, however, require careful control of the temperature and humidity environment to maintain feature size tolerances, position-

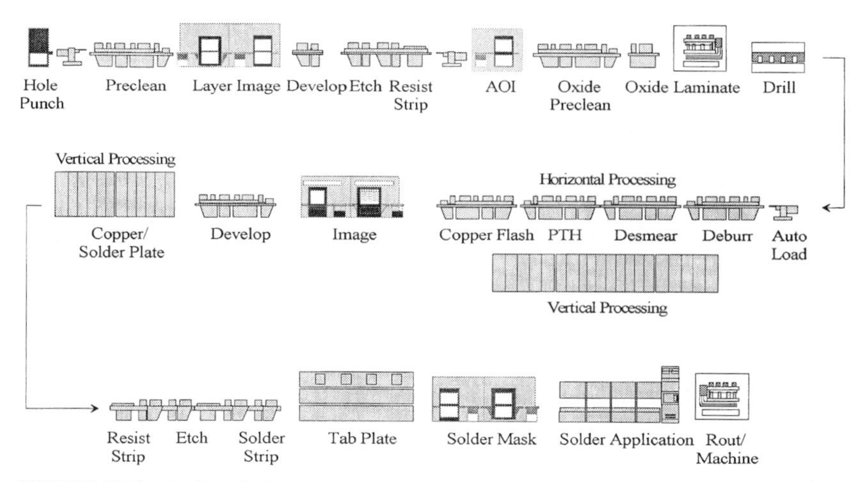

FIGURE 10.40 Outline of printed wiring board fabrication. (*Courtesy of MacDermid, Inc.*)

ing, and registration. A typical humidity coefficient of 14 ppm per percent *relative humidity* (RH), for example, means that a 500-mm-long phototool will expand by 70 μm for a 10 percent increase in RH. Humidity coefficients are slightly greater for rising RH than for falling RH, leading to a hysteresis curve for cyclical humidity levels. The published coefficients usually represent average values for rising and falling humidity over the range of 35 to 65 percent RH at room temperature.

In order to overcome dimensional stability problems from varying RH cycling, an improved polyester base has recently been commercialized.[34] It relies on a moisture-barrier coating to significantly slow the response to changes in RH. Glass, at a considerable increase in cost, provides the stable substrate necessary for fine-line work. It is quite insensitive to RH changes and less than half as sensitive as polyester film to thermal changes, as shown in Table 10.6.

TABLE 10.6 Comparison of Expansion Coefficients for Polyester and Glass

Photobase	Thermal, ppm/°C	Moisture, ppm/%RH
Polyester	14	11–14
Glass	5	<1

Silver halide emulsions are coated onto the above substrates to provide photosensitivity. The modern, laser-based photoplotters then create the master images in a raster format for the full-sized panel. They have largely displaced the slow process of vector plotting an individual circuit master and then forming the panel master image on a step-and-repeat camera. Flatbed plotters allow the use of both glass and polyester film substrates, whereas curvebed and drum plotters require the flexible film substrates. The minimum plotter pixel size currently is near 3 μm, with a minimum feature size of 25

FIGURE 10.41 Drum laser photoplotter with multiple beams. (*Courtesy of Orbotech, Inc.*)

μm. Prices range from $50,000 to $400,000 depending on plotter resolution capability and imaging speeds.[35] A commercial drum photoplotter is shown in Fig. 10.41.

Resist Application. Dry film resist has been the dominant technology for making the more advanced PWBs of the past 2 decades because it provides much greater latitude and therefore requires significantly less process control than liquid resists. Dry film resist is costlier than liquid resists, however. As a result, there have been intensive industry efforts underway to promote the use of liquid resists. The feature of greatest interest to the PWB fabricator is the ease of applying resist to a copper laminate for primary imaging, or raised circuitry for secondary imaging.

Resist Application for Primary Imaging. In primary imaging, the resist needs to be able to reliably cover the many irregularities of the copper surface. This need has favored the use of dry film. The precoated dry film resist is applied to the copper laminate in a hot-roll laminator. The hot roll warms the resist and makes it more conformable. This simple process provides good resist coverage of the thick dry film over the copper imperfections of the laminate. It has served as the workhorse of the industry. Additionally, the use of a thin liquid interface to further soften the resist during lamination can provide enhanced resist conformation in this lamination process. This approach has found extensive commercial acceptance over the past few years in the form of *wet lamination*. Here, a very thin layer of water is deposited on the copper laminate prior to lamination. The thin, aqueous layer now plasticizes the aqueous developable resist in the lamination nip for significantly better coverage of copper irregularities. The results from this process are illustrated in Fig. 10.42, which shows

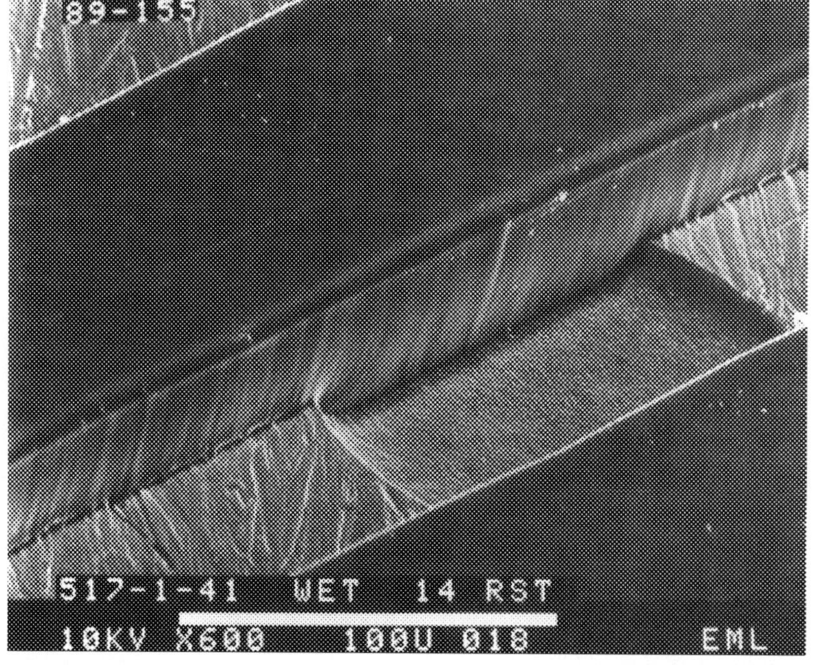

FIGURE 10.42 Wet laminated resist image conforms to preetched 15-μm defect. (*Courtesy of DuPont Electronics.*)

the resist conformation to a preetched defect. Conventional dry lamination will tent across the defect, as shown in Fig. 10.43, and lead to a subsequent etchout defect.

The leading commercial applications for applying liquid resist for primary imaging employ roller coating and *electrophoretic deposition* (ED). The roller coating process is illustrated in Fig. 10.23. In one commercial embodiment, the liquid resist is applied in 100 percent solids form to yield a coating thickness in the range of 15 to 25 μm. The tacky nature of the resist necessitates off-contact printing, however, which has limited attainable resolution to 125 μm in practice. Another embodiment applies a thinner, solvent-based resist which allows better resolution.

Liquid ED resist has found significant acceptance in Japan, and now has also begun to be more widely used in North America. The principle of the deposition process has been shown in Fig. 10.24. The sequential steps in its application are illustrated in Fig. 10.44, and a commercial line is shown in Fig. 10.45. The deposited thickness is in the range of 10 to 15 μm. ED resist is readily capable of good resolution, as shown in Fig. 10.46 for a 20 μm line/space pattern. The electrostatic nature of the deposition process ensures excellent conformation to the copper as shown in Fig. 10.47 for a developed resist image. The electrostatic deposition process also allows it to penetrate and protect metallized vias and through holes as shown in Fig. 10.48 for a negative-acting and Fig. 10.49 for a positive-acting ED resist.

Resist Application for Soldermask or Secondary Imaging. In primary imaging, the resist is only a temporary aid to fabricating the conductor pattern. It is stripped after it has served its purpose. Soldermask resist, however, stays on the PWB to pro-

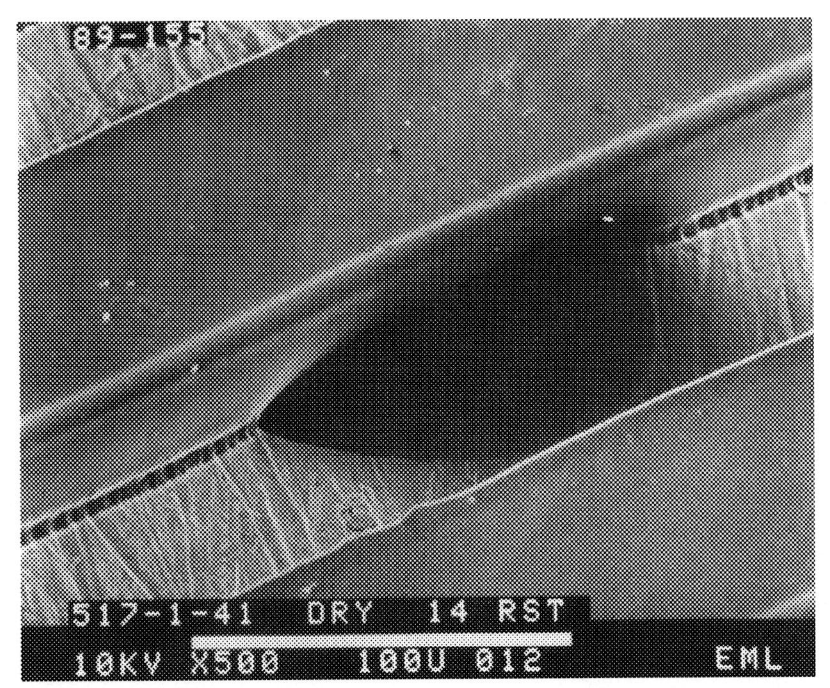

FIGURE 10.43 Dry laminated resist image does not conform to defect. (*Courtesy of DuPont Electronics.*)

vide an insulation coating over a wide area, to delineate component mounting areas and provide a base for mounting adhesives, and to protect against contamination and electrical shorts. The formulation and application of soldermask resist over the non-planar surfaces of PWBs with their raised conductor lines as well as through-hole and via connections is a significant challenge, however. Coatings equal to or thinner than the pad conductor heights are desirable for reliable component solderability, but thick coatings are often necessary to reliably cover high conductor lines and tent vias.

There are currently three commercial photodefinable soldermask technologies, namely, liquid resist, dry film, and a liquid–dry film combination. The liquid soldermask provides a thin coating of 20 to 25 μm which closely follows the PWB topography for spaces between lines greater than 125 μm, as shown in Fig. 10.50. The thin coating is necessary to permit reliable solder joint formation for fine pitch components with pad pitches of <500 μm. The high viscosity of the resist, however, can lead to mask bridging across narrow spaces between the conductor lines instead of uniform, void-free encapsulation when applied by mechanical means such as screen-printing, roller coating, or curtain coating. Spray coating, particularly with an electrostatic assist, is an alternate application method capable of encapsulating adjacent lines and/or spaces of less than 125 μm.

A thin soldermask resist cannot provide reliable encapsulation of circuitry with conductor height above 90 μm because of excessive thinning at the conductor edges. Such conductor heights are frequently encountered from a nonuniform plating process. Dry film soldermask is offered in thicknesses from 25 to 100 μm to provide reliable encapsulation independent of circuit height. It is also able to protect via holes by tent-

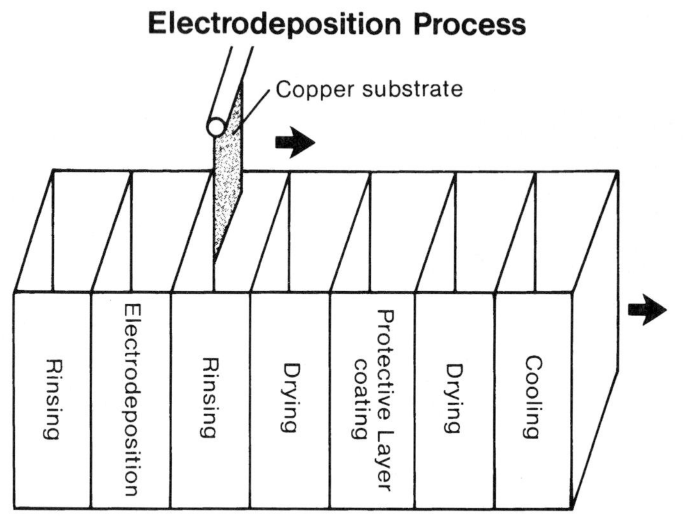

FIGURE 10.44 Process steps in applying ED resist. (*Courtesy of DuPont Electronics.*)

FIGURE 10.45 Commercial ED resist line. (*Courtesy of DuPont Electronics.*)

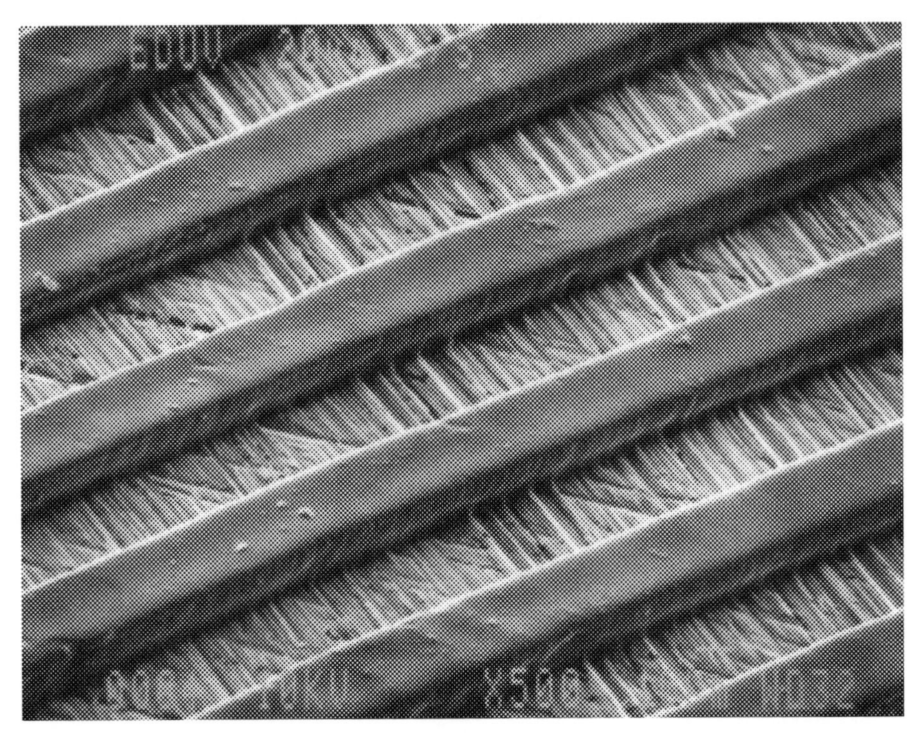

FIGURE 10.46 ED resist image at 20-μm lines and spaces. (*Courtesy of DuPont Electronics.*)

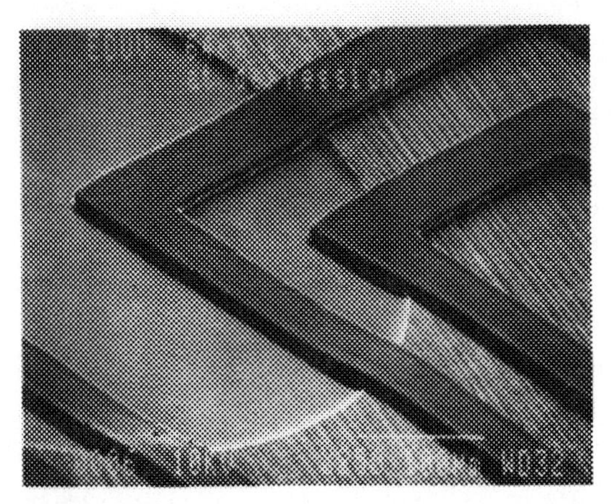

FIGURE 10.47 ED resist image conformation to preetched 15-μm defect. (*Courtesy of DuPont Electronics.*)

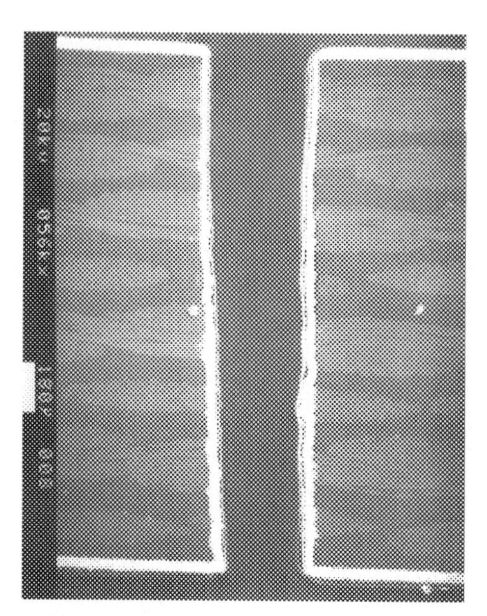

FIGURE 10.48 Negative ED resist in hole protects hole metallization through etching. (*Courtesy of DuPont Electronics.*)

FIGURE 10.49 Via coverage with positive acting ED resist. (*Courtesy of Ciba-Geigy.*)

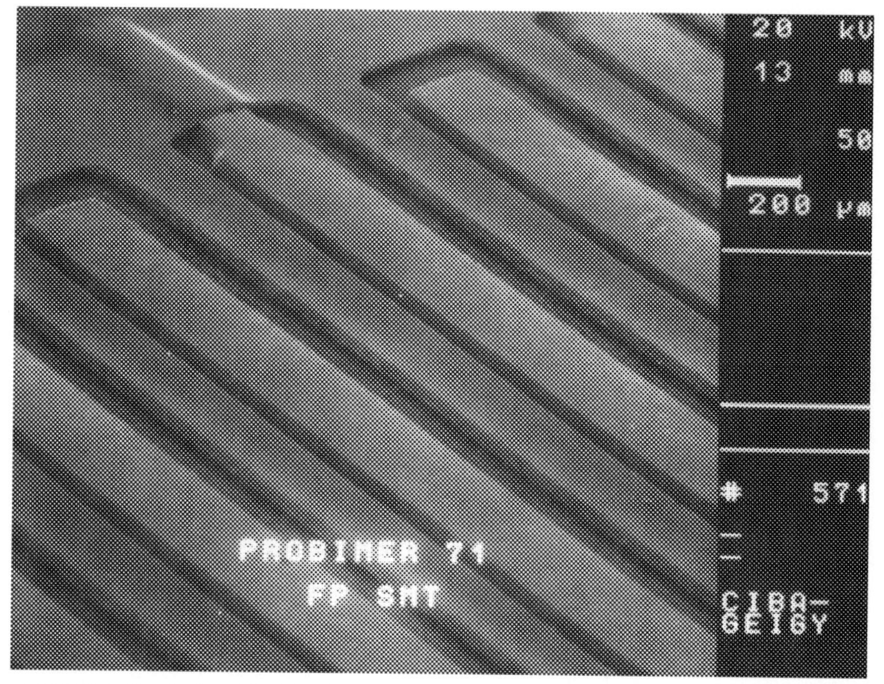

FIGURE 10.50 Profile of liquid soldermask over circuitry. (*Courtesy of Ciba-Geigy.*)

ing across them, which helps to prevent solder wicking and shorting during soldering in assembly. The lamination of the dry film to the PWB must be done with a vacuum assist to ensure good encapsulation of the circuitry, particularly with narrow spaces in the range of 75 to 100 μm. Its capability for providing solid encapsulation of a fine-line pattern along with smooth surface topography is shown in Fig. 10.51. But its thickness interferes with the reliable attachment of fine-pitch components.

The combination of liquid with dry film technology aims to combine the strong points of both.[36] It applies a solvent-free, low-viscosity liquid to the PWB followed by a thin capping layer of dry film. The results of this process are illustrated in Fig. 10.52. The low-viscosity liquid thoroughly fills in the surface topography while the dry film acts as a sealant and provides uniform coverage with relatively little edge thinning. The liquid–dry film combination can be applied as a thin coating. Dry film for this process is available in a range of 25 to 50 μm, with the liquid adding 4 μm to the finished coating thickness at its highest point. A commercial laminator is shown in Fig. 10.53. The combination of thinness with good encapsulation also allows this technology to create narrow dams between the closely spaced pads of fine-pitch surface mount technology as shown in Fig. 10.54 and to encapsulate closely spaced conductors between pads as shown in Fig. 10.55.

Resist Exposure. The challenge in any 1x phototool exposure step is to combine uniform, high illumination intensity from the light source and good alignment between substrate and phototool for high throughput with good resolution at reasonable cost. A commercial, noncollimated exposure system available with light sources up to 5 kW and a manual feed is shown in Fig. 10.56.

FIGURE 10.51 Cross section of dry film soldermask filling in 100-μm space. (*Courtesy of DuPont Electronics.*)

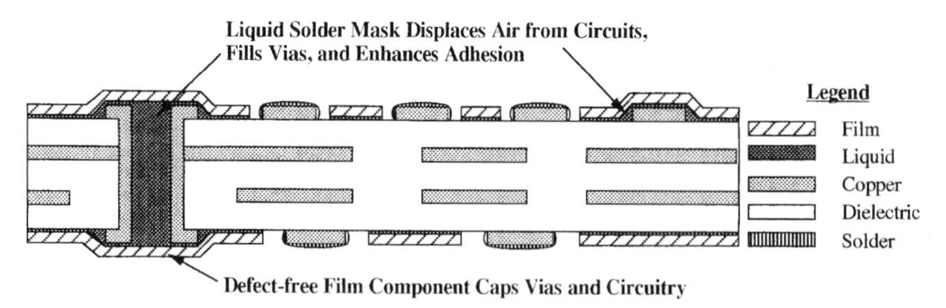

FIGURE 10.52 Schematic of liquid–dry film soldermask over circuitry. (*Courtesy of DuPont Electronics.*)

FIGURE 10.53 Liquid–dry film laminator. (*Courtesy of DuPont Electronics.*)

High illumination intensity allows short exposure times beneficial for both resolution and throughput. Lamps with 5 to 8 kW of energy are now also in common use. Dichroic mirrors minimize the transmittance of undesirable heat radiation to the image plane to avoid thermal expansion of the phototool as well as undesirable changes in the photoresponse of the resist. Water cooling is usually provided for lamp sources above 5 kW. Uniformity across the entire exposure area is achieved in an optical integrator with multilens fused silica elements. Good resolution is aided by collimation. A precision collimating mirror directs the uv radiation essentially perpendicular to the image plane. Since the light rays are nearly perpendicular, fine lines can be imaged even with soft contact or limited off-contact modes. The use of collimated light, however, also makes exposure more sensitive to contamination. Undesired dirt particles can now be imaged as well as the fine features. A clean room environment as well as good attention to cleanliness are becoming very important.

A diagram of the optical path in such a commercial exposure unit is shown in Fig. 10.57. The light from a mercury-xenon high-pressure short-arc lamp, concentrically mounted in an aspherical collector, passes from the collector at >80 percent efficiency as a 7.5-cm diameter spot to a dichroic mirror and then a multilens integrator. Each lens element of the integrator superimposes individual images at the collimating mirror and then the exposure plane. In this manner, the nonuniform intensity of the arc image is redistributed to achieve a uniform intensity distribution at the exposure plane. A beam splitter allows two-sided simultaneous exposure from a single light source. An exposure controller is used for the shutter to provide either a preset exposure energy level, a preset exposure time, or a measure of lamp intensity to monitor lamp aging.

Alignment of the phototools to the substrate is commonly done through pins for the less demanding PWBs. The hole locations can be varied and typically use up to four

FIGURE 10.54 Cross section of 62-μm solder dam for 250-μm pad pitch. (*Courtesy of DuPont Electronics.*)

FIGURE 10.55 Cross section of encapsulated conductors with 125-μm space. (*Courtesy of DuPont Electronics.*)

FIGURE 10.56 PWB exposure system with manual load and registration. (*Courtesy of DuPont Electronics.*)

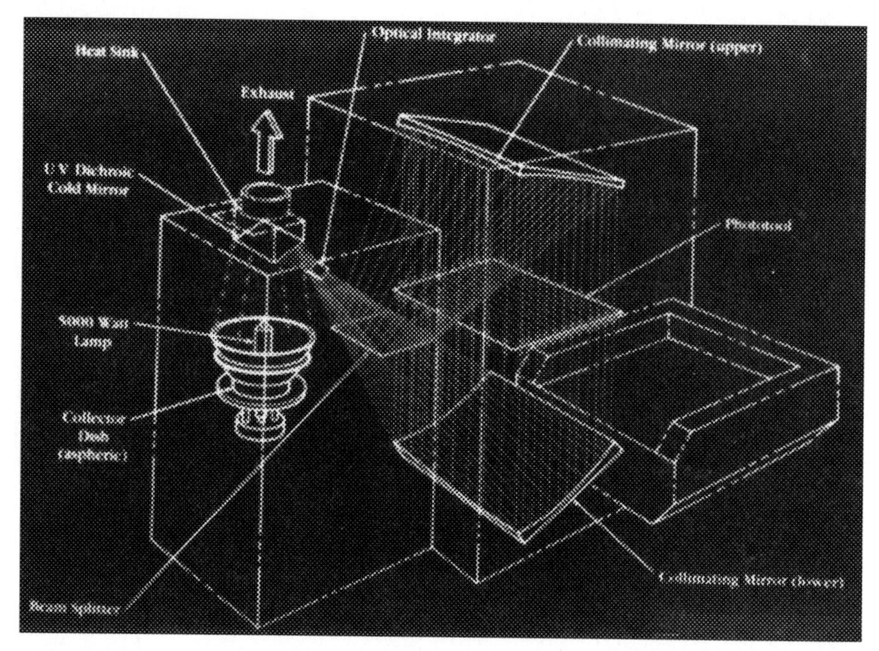

FIGURE 10.57 Optical system for two-sided, collimated exposure. (*Courtesy of Optical Radiation Corporation.*)

pins, which may be round or slotted. For equipment with automatic feed, the substrate panel is usually trimmed on two edges within 1.5 mm of specified hole locations for an initial, rough alignment. The alignment accuracy strongly depends on the pin and hole quality. In practice, it has been found to be near ±50 μm across 610 × 610 mm panels, or 80 ppm. For more precise alignment, machine vision systems are employed in conjunction with a computer and a precision closed-loop controlled stage to achieve an alignment accuracy of ±20 μm for panels up to 750 × 750 mm, or 30 ppm. They typically rely on aligning small dot fiducials in the artwork to larger holes in the panel after the vision system has computed the centroid and done the best fit computation for the appropriate x, y, and θ locations.

A leading exposure unit combining high resolution, accuracy, and throughput is shown in Fig. 10.58. It employs a 16-kW exposure system, collimated light, a *charge coupled device* (CCD) based vision system for double-sided registration, and automated panel load and unload. Such units are capable of imaging 100 to 200 panels per hour depending on panel size and resist sensitivity.

Laser direct imaging has also become commercially viable for generating fine-line circuit traces on printed circuit boards and multichip modules in a production environment. Utilizing polygon-scanned lasers for high-speed and high-resolution image generation, intense uv laser spots selectively expose image patterns on photoresist using digitally stored data rather than phototools. As the laser spots are scanned in the x-axis, the substrate translates at a constant velocity in the y-axis. A digitally controlled acousticoptic device acts as a shutter to turn the adjacent laser spots off and on to produce the desired circuit patterns. Substrates are thereby imaged in a noncontact, non-damaging mode without the requirement of expensive reticles or phototools. A commercial laser imager is shown in Fig. 10.59. It is capable of resolving 60-mm lines and

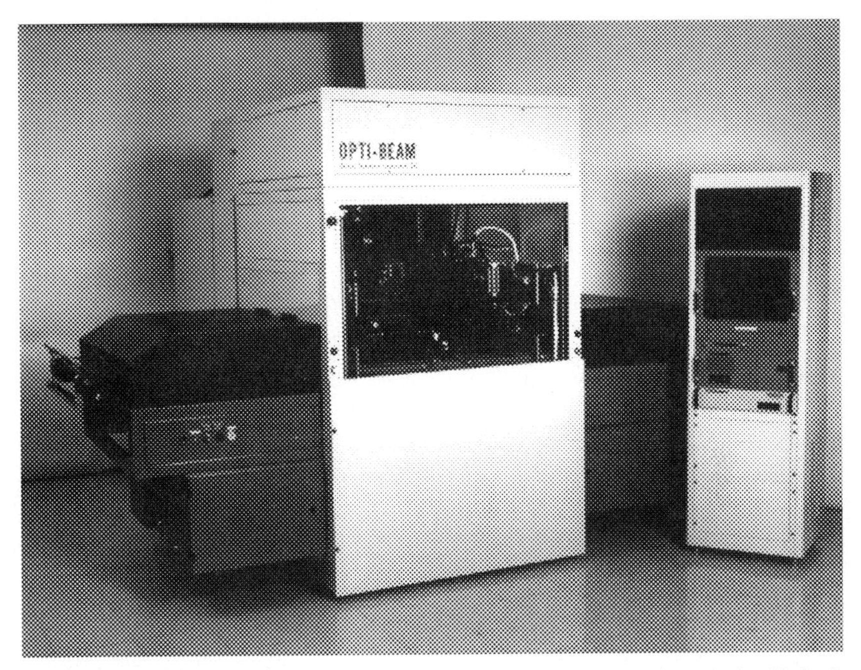

FIGURE 10.58 PWB exposure system with optical registration and automatic load/unload. (*Courtesy of Optical Radiation Corporation.*)

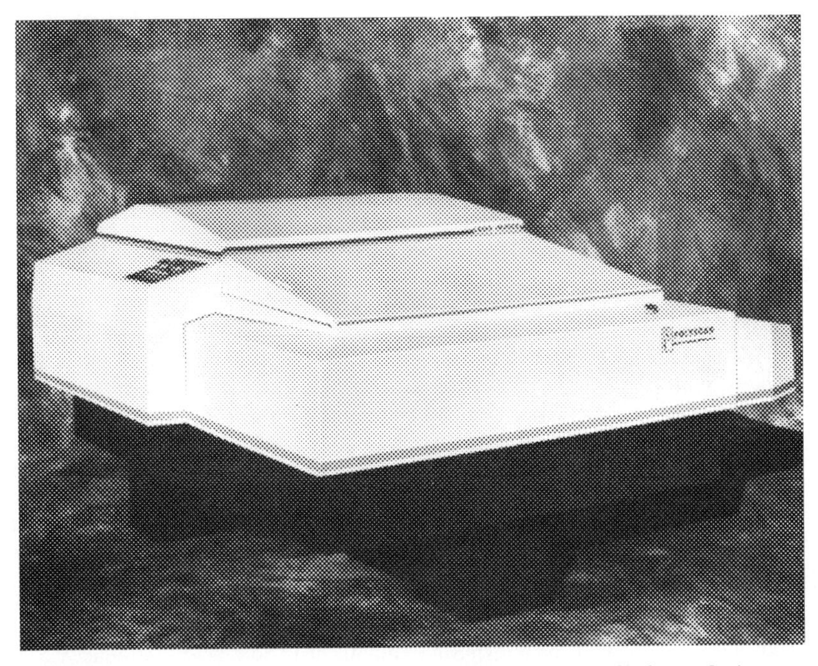

FIGURE 10.59 Laser imager for direct exposure of PWBs. (*Courtesy of Polyscan, Inc.*)

spaces when using a 12 μm spot size, and has an image location accuracy of ±12 μm. Its efficient use of laser power allows exposure times for a 600 × 600 mm format in the range of 15 to 30 s.

10.4.4 Flat Panel Displays

Cathode ray tubes have long been the dominant display technology. They offer high resolution, color, and large area format at reasonable cost. Their drawbacks have been size, weight, high power consumption, and concern about radiation. Alternate display technologies, led by liquid crystal displays, have been commercially available for the last 2 decades. But until recently, liquid crystals have largely been restricted to the low-resolution and small, monochrome formats used in watch, calculator, and radio displays. Portable and laptop computers are now driving the development of high-resolution, flat panel displays in a size range of 10 to 14 in. The visible picture element content for a color VGA display is 640 rows × 480 columns × 3 colors, or nearly 1 million visible pixels. High-definition television is a driver for larger future formats with higher visible pixel content.

The leading high-resolution technologies for flat panel displays are electroluminescence, gas plasma, passive liquid crystal, and active liquid crystal. The first three rely on the formation of rows and columns of aligned electrodes to activate the pixels through a multiplexed matrix. The pixel size typically is 300 × 300 mm for monochromatic displays. For color, the pixel is subdivided into three 100 × 300 mm red, blue, and green cells. Fabrication of the electrode patterns for these three display technologies has been relatively straightforward, but all three have specific limitations in reproducing either full-color or larger formats.[37]

Liquid crystal technology appears to offer the best options for full color, but the response characteristics of liquid crystals with regard to voltage and switching time make it very difficult to address them through a passive, multiplexed matrix for the larger, high-resolution formats.[38] Active matrix liquid crystals, where each pixel is activated by a diode or transistor fabricated as part of the pixel electrode structure on each respective electrode, are much easier to operate but much more difficult to fabricate. The most advanced displays use liquid crystals activated by individual *thin-film transistors* (TFT) behind each pixel, as shown in Fig. 10.60. The complexity of the TFT layout is comparable to that of dynamic memory chips. Such a display represents a tremendous challenge to photofabrication since it requires the accurate definition of transistor feature sizes of 3 to 5 μm over the large, 300- to 400-mm format sizes of PWBs. The history, status, and remaining challenges for this technology have been reviewed recently.[39]

A typical fabrication process for an active liquid crystal display is composed of three basic steps as shown in Fig. 10.61. The first step prepares the front glass substrate, which includes the deposition and patterning of the color filter elements. The three color filters are formed in repetitive steps, and a black border is then patterned around them for enhanced contrast. The rear glass substrate is prepared separately in a parallel step by forming the TFTs and the connecting conductor lines. Liquid photoresists are used to define the patterns on both the front and rear substrates. In the final step, the patterned front and rear substrates are assembled and the liquid crystals are injected between them.

The pattern resolution requirements are outside the current capabilities of exposure equipment for making PWBs. IC exposure equipment, however, lacks the size capability above 150 to 200 mm. New exposure equipment is now being introduced to fill

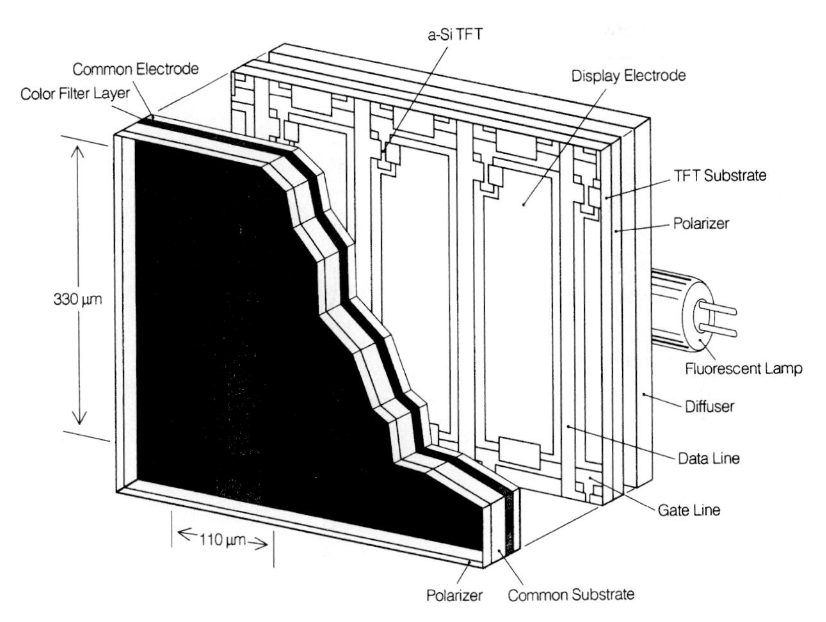

FIGURE 10.60 Illustration of active matrix liquid crystal display. (*Courtesy of IBM.*)

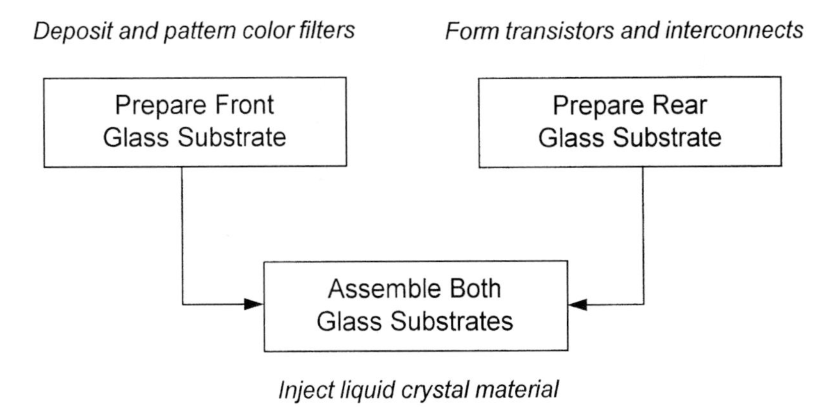

FIGURE 10.61 Process overview for active matrix liquid crystal display fabrication.

this gap. An example is shown in Fig. 10.62. The equipment is capable of 5 μm resolution in an off-contact mode and ±2 μm alignment accuracy over a 350 × 350 mm area with the aid of a video alignment system.

The value of the flat panel display industry is forecast to grow to tens of billion dollars. Active liquid crystal display technology is widely forecast to have a bright future once the current low fabrication yields for this new display technology are

FIGURE 10.62 Exposure system for flat panel displays with cassette load/unload. (*Courtesy of Optical Radiation Corporation.*)

overcome. The defects are mostly due to particle contamination.[40] Inspection and repair technology and equipment are now being developed to support the quest for improved yields.

10.5 OUTLOOK

It is clear that photofabrication will continue to be the key fabrication technology to enable the continuing reduction in electronic cost per function through miniaturization. Continuing miniaturization at all component levels will also mean continuing demands for higher standards in the current fabrication processes of these components. But as shown in Fig. 10.1, miniaturization and growth are interdependent. Without growth, there would only be limited funds available to drive the development of technologies for further miniaturization. Future growth thus depends not only on miniaturization but even more importantly on the development of new applications. Current advances in IC chip technologies, parallel system architectures, neural networks, and artificial intelligence lead this author to believe the best is yet to come.

ACKNOWLEDGMENTS

The author wishes to acknowledge the valuable help of Dan Amey and Hans-Karl Mueller of DuPont Electronics, Don van Arnam of Polyscan, and Gene Weiner of Gene H. Weiner Associates for their many suggestions to this manuscript.

REFERENCES

1. C. R. K. Marrian, E. A. Dobisz, and O. J. Glembocki, *R&D Magazine* (February 1992), pp. 23–126.
2. M. Hatzakis, *IBM Journal of Research and Development,* **32**(4), (1988), pp. 441–453.
3. M. Bolsen, G. Buhr, H. Merrem, and K. Van Werden, *Solid State Technology* (February 1986), p. 83.
4. J. Murray, *Printed Circuit Fabrication* (May 1992), p. 25.
5. P. West and B. Griffing, *SPIE,* **394** (March 1983), p. 33.
6. A. Reisman, *Proc. IEEE,* **71** (1983), p. 550.
7. J. Pacansky and J. R. Lyerla, *IBM Journal of Research and Development,* **23**(1) (1979), p. 42.
8. A. Gupta, R. Liang, F. Tsay, and J. Moacanin, *Macromolecules,* **13** (1980), p. 1696.
9. J. Frechet et al., *Journal of Imaging Technology,* **30** (1986), p. 59.
10. R. Jacobson, *Journal of Photographic Science,* **31** (1983), p. 1.
11. W. DeForest, *Photoresist Materials and Processes,* McGraw-Hill, New York, 1975, p. 168.
12. T. Snodgrass and G. Blackwell, *First International Conference on Multichip Modules,* Denver, CO, April 1992, p. 428.
13. A. E. Nader et al., "Synthesis and Characterization of a Low Stress Photosensitive Polyimide," *Fourth International Conference on Polyimides,* Ellenville, NY, November 1991.
14. Y. Matsuoka et al., "Ester Type Photosensitive Polyimide Precursor With Low Thermal Expansion Coefficient," *Fourth International Conference on Polyimides,* Ellenville, NY, November 1991.
15. T. Ozana et al., "Thin Film Hybrid IC," *ISHM '91 Proceedings,* pp. 6–10.
16. A. E. Nader et al., *First International Conference on Multichip Modules,* Denver, CO, April 1992, p. 410.
17. J. J. Felten, C. B. Wang, and J. Collins, *ISHM '91 Proceedings,* pp. 545–550.
18. D. I. Amey, *ISHM '92 Proceedings,* pp. 225–234.
19. N. Hagen, J. Henderson, W. Nebe, and J. Osborne, *The International Journal for Hybrid Microelectronics,* **12**(4) (1989), pp. 175–179.
20. L. Bernier, *Electronic Engineering Times* (June 15, 1992), pp. 4, 92.
21. Y. Belopolsky, E. Abramson, and A. T. Murphy, *Proceedings of the 1991 International Microelectronics Conference,* Tokyo, Japan.
22. Y. Tsukada, S. Tsuchida, and Y. Mashimoto, *Proceedings of the 42nd Electronic Components and Technology Conference* (May 1992), pp. 22–27.
23. U.S. Patent 4,503,329 (1985), Hitachi.
24. U.S. Patent 4,510,222 (1985), Hitachi.
25. H. Rottman, *IBM Journal of Research and Development,* ***24*** (1980), p. 461.
26. Y. Hashimoto, *Japan Semiconductor Technology News,* **2**(5) (1983), p. 18.

27. P. Burggraaf, *Semiconductor International* (February 1992), p. 42.
28. M. D. Levenson et al., *IEEE Trans. on Electr. Dev.,* **29**(12) (1982), p. 1828.
29. G. E. Flores and B. Kirkpatrick, *IEEE Spectrum* (October 1991), p. 24.
30. V. Pol et al., *Proc. SPIE,* **633** (1986), p. 6.
31. P. Burggraaf, *Semiconductor International* (March 1992), p. 54.
32. E. Bjorkholm et al., *Optics and Photonics News* (May 1991), pp. 27–30.
33. Y. Kawamoto et al., *Digest of Tech. Papers on 1990 VLSI Symposium,* Hawaii, 1990, p. 13.
34. K. Smith and N. Steinberg, *Circuits Manufacturing* (July 1988).
35. R. C. Henningsgard, *Printed Circuit Fabrication* (July 1992), pp. 18–27.
36. M. Weinhold, *Printed Circuit Fabrication* (February 1989), p. 134.
37. D. Pryce, *Electronic Design News* (October 11, 1990), pp. 79–88.
38. P. Brody, F-C Luo, and P. Malmberg, *Electronics* (July 12, 1984), pp. 113–117.
39. W. E. Howard, *IBM Journal of Research and Development,* **36**(1) (1992), pp. 3–10.
40. W. C. O'Mara, *Solid State Technology* (December 1991), pp. 65–70.

BIBLIOGRAPHY

Coombs, C. F., Jr., *Printed Circuits Handbook,* 3rd ed., McGraw-Hill, New York, 1988.

DeForest, W. F., *Photoresist: Materials and Processes,* McGraw-Hill, New York, 1975.

Elliott, D. J., *Microlithography: Process Technology for IC Fabrication,* McGraw-Hill, New York, 1986.

Moreau, W. M., *Semiconductor Lithography: Principles, Practices, and Materials,* Plenum Press, New York, 1988.

Reiser, A., *Photoreactive Polymers: The Science and Technology of Resists,* Wiley, New York, 1989.

CHAPTER 11

ADVANCED ELECTRONIC PACKAGING, MATERIALS, AND PROCESSES

Stephen G. Konsowski

11.1 INTRODUCTION

Some of the materials and processes discussed in the earlier chapters are presently undergoing changes which will render them more useful and less susceptible to obsolescence. In addition, new materials are appearing constantly, especially in the microelectronic packaging arena. While it is beyond the scope of this work to describe all these situations in depth, this chapter will cover those considered to be most important, such as digital and RF multichip modules, power-generating circuits, composites, interconnects, coatings, and advanced printed wiring boards. For the purposes of this chapter multichip modules are grouped into three major types: digital, RF, and power. The higher speeds being achieved in digital circuitry are blending some of the digital materials performance requirements with those of microwave circuits. Consequently there is a growing commonality in some materials selected for hybrid packaging of both types. Power generating and conditioning circuits continue to challenge materials by demanding more thermal conduction and sometimes higher operating temperature performance, as thermal densities increase. Composites fall into three major groups: polymer matrix, metal matrix, and ceramic. Composites play a role in electronic packaging by virtue of their tailorable physical properties which make them quite attractive for lightweight structural and thermal applications. Interconnects in electronic packaging are the means by which all the components of a circuit are electrically "tethered" together. Many electrical components or devices have high-performance capability which can be either enhanced or encumbered by the interconnection material and technique chosen to tie them together. Coatings or protective materials are applied to a circuit to isolate the circuit and its components from certain environmental influences, such as moisture and contamination, which could alter performance or diminish the expected life of the circuit. The developments taking place in coatings technology are changing the way circuits are packaged. Printed wiring materials are widely varied, and the materials selection depends on the application. There are, however, significant advances in high performance materials used to fabricate the board dielectrics which consist of a reinforcing fiber and a resin matrix. These are opening new applications to a technology which has existed for nearly 40 years. This chapter will describe technology developments in the categories listed with emphasis on performance characteristics and the relationship of these characteristics to the developments taking place in electronic circuitry applications.

11.2. MULTICHIP MODULE MATERIALS

Multichip modules (MCMs) have been around for some time. In fact, their use dates back to 1959 with the issuance of a patent to Jack Kilby of Texas Instruments for the integrated circuit concept. As soon as a circuit function had been incorporated on a chip of silicon, engineers began to place two or more of them in a hermetic box (package) with a few other components, creating a hybrid circuit. These early multichip packages or modules were precursors of more complex hybrid assemblies which followed. As integrated circuits became increasingly more integrated, they were called *large-scale integration* (LSI) and later, *very large scale integration* (VLSI). More and more functions became incorporated into chips, and engineers continued to place many of them into hybrid circuits. The hybrids grew in complexity by virtue of the devices they contained as well as the internal connecting circuitry and the number of leads or pins or simply, the *input-output* (I/O) count. This "race" between higher levels of chip integration and the combining of increased complexity chips into a higher-order circuit package or module will likely continue.

Today, the complex hybrid circuit is called a *multichip module.* Not only are the devices contained within it complex but the internal interconnect systems are sophisticated and varied. The dielectrics may be ceramic, silicon, organic, or even diamond, and the conductors may be aluminum, copper, gold, or other relatively high electrical conductivity metals.

There is a generally accepted classification of MCMs primarily by their dielectric materials. At this time it is as follows:

MCM-D: The substrate is silicon or ceramic plus a deposited dielectric for additional layers.

MCM-C: The substrate is ceramic only.

MCM-L: The substrate is a printed circuit board.

More recent developments of the MCM technology include RF in addition to the digital variety just described and power-conditioning circuits.

One reason for the existence of the new generation of digital hybrids and MCMs is electrical performance. A major use of the digital MCM is in personal computers and workstations, where clock rates are nearing 100 MHz and signal rise times are in picoseconds. At these performance levels, devices must be placed together as closely as possible, and the dielectric beneath signal conductors must have a dielectric constant as low as possible so as to minimize the times the signals require to get to their next location.[1]

The packaging of analog circuitry which deals with microwave and millimeter wave technology progressed in a similar fashion to the digital world. In this arena, techniques which preserve signal characteristics at high speeds have always been required. Thus concerns about the dielectric constant and loss tangent of a dielectric have resulted in very high quality materials development.

Power-conditioning circuits involve the conversion of ac to dc or the inversion of dc to ac. As might be expected, these circuits involve high current, at times high voltage, and almost always lost electrical power which must be dissipated. Weight and size reductions have occurred in this area because of modularity trends as well as the systematic elimination of bulky inductive components. This facilitates hybridization of circuit components. However, with this shrinking of size and increase of component density, operating temperatures of components become excessive. Consequently special materials and techniques are required to remove heat efficiently from these circuits.

11.2.1 Digital Hybrid Circuit Materials

MCM-D. The multichip module-D will be discussed first. The D represents the process of depositing alternating layers of conductors and dielectrics onto the base dielectrics in this type of module. A cross section is shown in Fig. 11.1. The selection of the base material—silicon (Si), alumina (Al_2O_3), aluminum nitride (AlN), silicon carbide (SiC), or mullite (a mixture of Al_2O_3 and SiO_2)—is somewhat a function of the performance required. Silicon wafers are typically polished for semiconductor applications with surface finishes of 3 μm, peak to peak or less. This in turn allows the subsequent dielectric and metal layer to be nearly as fine. The dielectrics currently used are polyimide, silicon dioxide, and benzocyclobutane (BCB). Polyimide and BCB are applied as a liquid to the substrate which is rotated at high speed to produce a thin-film coating, just as a photoresist is applied to a semiconductor wafer. The coated substrates are then baked in an oven to polymerize the coating. (Silicon dioxide is deposited from the vapor state with no polymerizing taking place.) The ensemble is

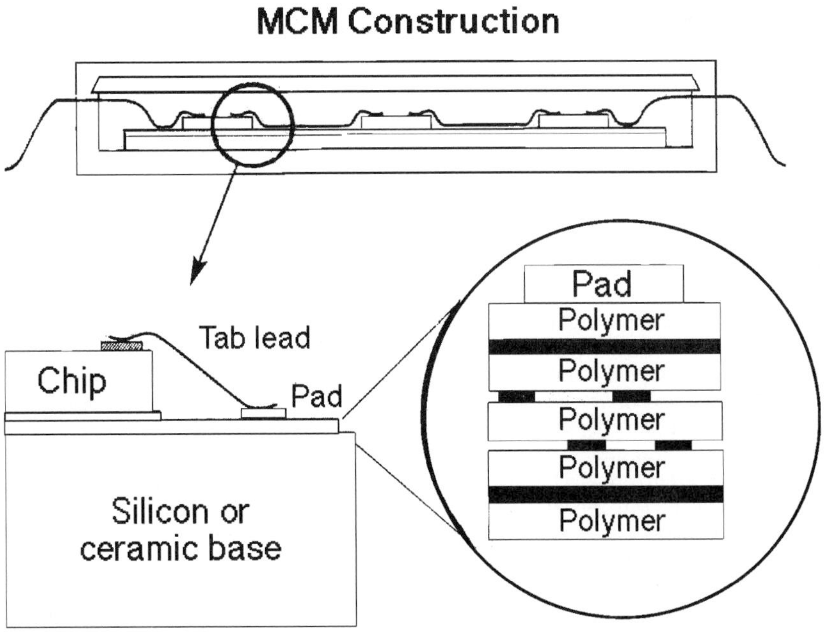

FIGURE 11.1 A typical MCM-D construction. (*Courtesy of Electronic Packaging and Production.*)

placed in vacuum chamber where a thin film of aluminum or copper is applied to the polymer surface. This metal layer is then coated with another layer of dielectric which is then defined as a sheet with holes etched in it to expose the layer of metal beneath. Another layer of metal is deposited atop this dielectric which makes contact to the first metal layer (a ground plane or power plane) at the sites which have been etched in the dielectric. A photoetch step divides this conductor layer into individual conductors. The next step involves coating this layer with a dielectric and opening up holes in the dielectric to expose the conductors beneath for a subsequent metal layer contact. The process continues, producing as many conductor and dielectric layers as required by the circuit design. Dielectric is applied, (cured) and etched; metal is applied and etched; dielectric is applied again, and so on. The initial coating of the substrate may be metal or dielectric according to the specific design.

The substrate as described earlier may be silicon or a ceramic. If silicon is used, it results in an assembly which requires containment in a package because silicon does not have sufficient strength to be a package material. Alumina and aluminum nitride, on the other hand, serve well as package materials.

There are two main constructions which are used in the MCM-D module. They are chip-first and chip-last. In the former case, the chips are mounted into cavities in the base MCM substrate or package floor and the interconnects are applied by depositing metals on top of a polyimide film which is placed on the chips and substrate. Holes are opened in the film for the metal to contact the chips. The metals are etched into a pattern, and polymers are applied atop the chips. Holes are opened in the polymers, on which vacuum-deposited metal layers are etched into conductors. The process is repeated as many times as necessary, but three to five conductor layers are typical. In the chip-last method the conductors and dielectrics are formed in the package first as described earlier, and chips are introduced after the interconnect layers have been verified to be operational. Neither chip-first nor chip-last has yet emerged as the preferred process, and both are being used in MCM foundries.

Regarding the add-on dielectrics, the polymers described do not have a high thermal conductivity K. Silicon dioxide, however, has a K several times that of most polymers. As such, it can contribute to a design which requires less reliance on thermal accommodations such as metal vias used for conducting heat away from the devices in the package. Typical physical properties of the polymer materials used in MCM-D technology are listed in Table 11.1.

From a processing standpoint, some MCM-D manufacturers prefer polymers as the dielectric for the built-up layers, while others depend on silicon dioxide because it is a vacuum-deposited thin film and can be applied using the same equipment which deposits the metal layers. The conformal coating which the silicon dioxide creates generally results in an irregular surface. Thin-film conductor lines are controlled best when they are etched from metal deposited on a regular flat surface. The irregularities caused by the conformal coat of silicon dioxide can be controlled so as not to interfere with the control of line widths. This has been demonstrated by n-Chip, a San Jose, California, MCM foundry. However, the chemical vapor deposition process used by n-Chip does not seem to be easily reproduced by others. Polymers, on the other hand, coat the metal lines and spaces beneath them but are not so conformal and can be made to have a flat surface. The difference is shown in Figs. 11.2 and 11.3.

From Table 11.1, it can be seen that the physical properties of the dielectrics differ sufficiently to allow choices based on performance. For example, if maximum thermal conductivity is a requirement for a chip-last design, one would prefer to choose silicon dioxide with a thermal conductivity nearly 10 times that of polyimide. Although material properties deficiencies can often be compensated by design ingenuity, the price to be paid always exists in the form of weight and/or size penalties, and very often cost

TABLE 11.1 Polymer Dielectric Material Properties

Polymer name	Polymer type	Dielectric constant	Dissipation factor	CTE, 10^{-6}/°C	T_g °C	Modulus, GPa	Tensile strength, MPa	Percent elongation	Moisture uptake, wt %
Amoco Ultradel 4212	Fluorinated polyimide	2.9	0.005	50	295	2.7			0.9
Amoco Ultradel 7501	Photosensitive polyimide	2.8	0.004	24	>400	3.4			3.4
Demota IP 200	Polyphenylquinoxaline	2.7	0.0005	55	>365	2.0	117	8–12	0.9
Dow BCB-13005	Benzocyclobutene	2.7	0.0008	65	>350	3.3	68	2.5	0.3
DuPont PI-2545	Standard polyimide	3.5	0.002	20	>400	1.4	105	40	2–3
DuPont PI-2545	Standard polyimide	3.3	0.002	40	>320	2.4	135	15	2–3
DuPont PI-2610D/2611D	Low-stress polyimide	2.9	0.002	3 (*x-y*)	>400	8.4	350	25	0.5
DuPont PI-2732/33	Photosensitive polyimide	2.9		25	>400	6.0	192	8	1.5
Hitachi PIQ-13	Standard polyimide	3.4	0.002	45	>350	3.3	130	20	2.3
Hitachi PIQ-L100	Low-stress polyimide	3.2	0.002	3	410	11	380	22	1.3
National Starch EL5010	Preimidized polyimide	3.2	<0.002	34	214	2.8	150	7	1.3
National Starch EL5512	Fluorinated preimidized polyimide	2.8	<0.002	35	225	2.8	150	6	0.8
OCG Probimide 400	Preimidized photosensitive polyimide	3.0	0.003	37	357	2.9	147	56	2.0
OCG Probimide 500	Low-stress polyimide	2.9	0.003	6–7	400	11.6	444	28	0.7
Toray UR-3800	Photosensitive polyimide	3.2	0.002	40	280	3.4	140	11	1.1

*Values given are for flexural modulus and flexural strength, respectively.

Source: Courtesy of Electronic Packaging and Production.

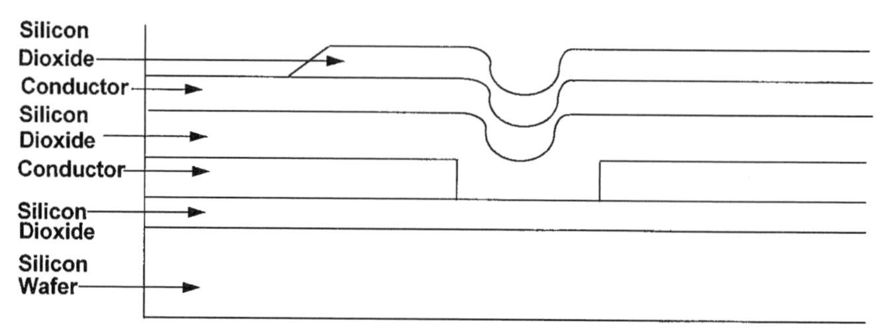

FIGURE 11.2 MCM silicon substrate with silicon dioxide dielectric inner layers.

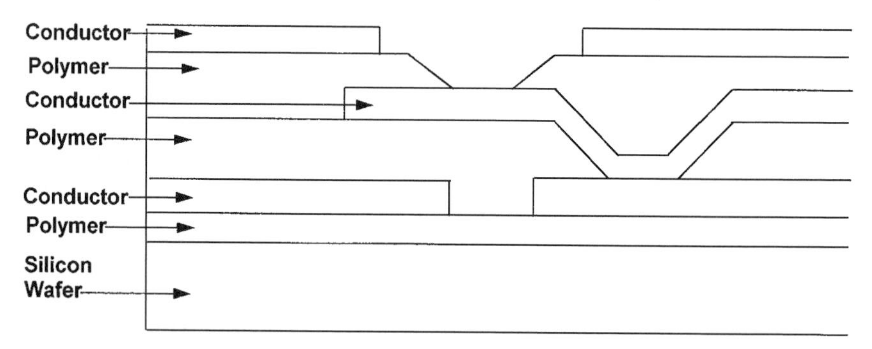

FIGURE 11.3 MCM silicon substrate with polymer dielectric inner layers.

as well. Continuing with the example, if the devices in this design generate more heat than can be conducted away by a polymer dielectric system alone, thermal vias consisting of metallization may be required. This is shown in Fig. 11.4. As in printed wiring board technology, vias penetrating all layers (through vias) consume valuable signal-routing real estate. More layers may be needed to complete the interconnect pattern with this system than would be required if a dielectric were chosen which could readily dissipate the heat these devices generate.

MCM-C. The MCM-C where the C stands for "ceramic" is the next type of multichip module material. The term *ceramic* is being applied to a variety of materials which include aluminum oxide, beryllium oxide, aluminum nitride, cordierite, silicon carbide, mullite, diamond, and glass ceramic. Thick-film ceramics where the conductors are made from glassy materials, which contain conductive and dielectric particles, and the dielectrics are mixtures of glasses also are considered ceramic in this scenario.

In the MCM-C type of multichip module, dielectrics consist of ceramic materials from the group just listed. MCM-C dielectrics can be divided into two categories: high-temperature firing (greater than 1500°C) and low-temperature firing (between 800 and 1000°C). While both processing temperatures are considerably above room temperature, the "low-temperature" material systems, that is, dielectric and metallization, can be processed in firing ovens and furnaces which process thick-film circuits.

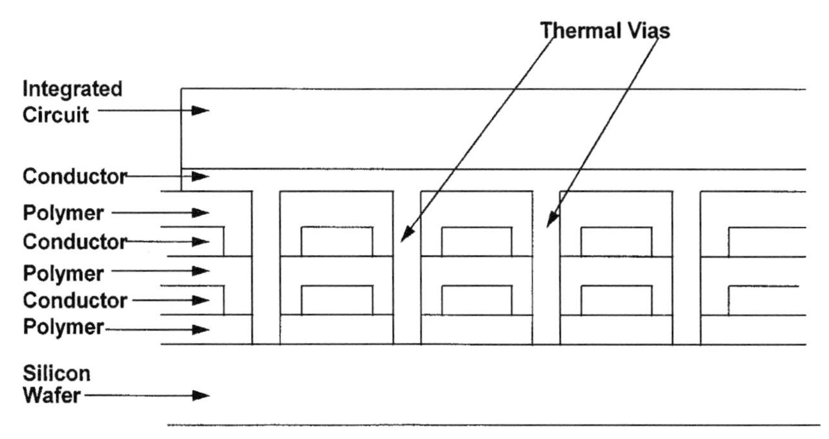

FIGURE 11.4 MCM substrate with polymer dielectric inner layers and thermal vias.

A more complete description of the firing processes and ceramic technologies can be found in Chaps. 3 and 8 of this handbook. However, for the purposes of this chapter, the following descriptions, though incomplete, should suffice.

The Cofired Ceramic Process. A cofired ceramic is one in which the dielectric and conductor are matched so that they "fuse" together at the same temperature to form the desired multilayer construction. In this approach, a ceramic powder is mixed with organic binders which hold the particles together and with solvents that allow fabrication of the ceramic into a flat tape. This is called a *slurry.* The slurry is placed on a ribbon of plastic (usually polyester) tape, and the plastic tape is drawn under a knifelike gate called a *doctor blade.* As the plastic tape is pulled under the blade, the slurry on top of the plastic tape is made to become a thin film about 6 to 8 thousandths of an inch thick. As the plastic tape is drawn past the doctor blade, the solvents in the slurry evaporate rapidly and leave the dried ceramic-binder combination attached to the plastic tape. This system is then rolled up into convenient lengths for further processing later.

When it is desired to process the ceramic, the tape is unreeled, and the ceramic layer is removed from the plastic support. The ceramic material is then called a *green,* or *unfired,* tape. It is flexible, yet able to support itself, and can be handled like a sheet of paper. This form factor is ideal for the fabrication of a multilayer conductor-dielectric plate which can be an MCM-C module interconnect system. The sheet of ceramic/binder is made into a multilayer conductor system as follows: Dielectric sheets are drilled or punched to open holes which will contain conductor vias. They are then metallized by printing a metal slurry pattern on both sides of the sheet. This pattern will form conductors when it gets fired. Another printing places a via metallization into the open holes for further processing.

If a multilayer construction is required, the layers of tape are placed one atop the other such that layer 1 is the topmost one and layer N is the bottommost. Each tape layer has two conductor layers on it, and each layer is aligned with the others during pressing through keying holes which fit over alignment pins in the press. The group is then pressed together so that the layers will become a monolithic slab after firing.

A major difference between the high-temperature and low-temperature systems is the metallization. In the high-temperature system, the metallization is usually tungsten. Tungsten is used because metals other than those which are refractory would either melt and/or oxidize so completely they would not retain the desired shape and dimensions nor would they any longer be conductors if indeed they were not burned off. The reason the conductors would oxidize or melt is that high-temperature ceramic must be fired at temperatures well beyond the metal's melting point.

The electrical resistivity of tungsten is quite high compared to metals used in thick films and with low-temperature firing ceramics. In applications requiring narrow or fine conductor widths measuring many millimeters in length, the total dc line resistance can measure tens of ohms. For circuits depending on low-resistance conductors, this limits high-temperature ceramics somewhat in their range of application. Thick-film and low-temperature ceramics allow the use of comparatively high conductivity metals which widen the range of applications. Consequently, MCM-C modules are being fabricated from low-temperature ceramic–metal systems. It should be noted that not all ceramics have the same strengths, and this can be a factor in the selection of a material. The high-temperature materials have greater flexural strength than low-temperature ceramics. Table 11.2 shows some mechanical and electrical properties of materials being used for MCM-C multichip modules as well as other ceramics of interest in electronic packaging. Strength is not the only property of importance in MCM-C material selection. Dielectric constant and loss tangent are properties which strongly influence electrical performance of high-speed digital and microwave circuits as well. In general, digital MCMs require a very low dielectric constant because the speed of a signal traveling along a conductor depends on the dielectric surrounding the conductor. The higher the dielectric constant, the slower the speed of the signal, and consequently, fewer computations may be performed in a unit of time if a signal must travel slowly. Microwave circuits, on the other hand, depend more on the dissipation factor of the dielectric than on the dielectric constant. Microwave signals become attenuated more per unit length of signal path when the dielectric has a high dissipation factor (usually a value above 0.005 is considered high). It is important to know

TABLE 11.2 Several Characteristics of MCM Packaging Materials

	Characteristic		
	Thermal conductivity	Coefficient of thermal expansion	Dielectric constant
		Units	
	W/mK at 25°C	ppm/°C; 40–400°C	at 1 MHz
Alumina, 92%	19–20	6.77	9.5
Alumina, 95%	34	7.0	9.9
Aluminum nitride	170	4.45	8.8
Beryllia	290	7.4	6.7
Silicon carbide	70–260	3.8	40.79
Mullite	5	4.2	6.8
Diamond	>1000–1500	1.0–4.0	5.7
Glass ceramic*	3–5	2.5–8	4.5–8

*Depending on substrate formulation.

Source: Courtesy of Electronic Packaging and Production.

the value of the dissipation factor at the frequency (or frequencies of interest because, as frequency increases, so does the dissipation factor in most cases). No dielectric material is a perfect insulator, but the more it behaves as a perfect insulator, not only in preventing electrical current from passing through it but also in being not easily polarizable, the less electrical energy it causes to be lost from signals and converted to heat. As the dissipation factor increases, the dielectric is said to be "lossy." The lower the loss, the higher is the quality of the dielectric. In power transmission, much energy can be lost in a poor dielectric. However, this loss is also important in low-signal-level microwave circuits such as receivers. If a received signal is weak, it will become even weaker as it travels through a lossy dielectric. Although amplification of the weak signal is possible, this consumes more power and adds noise and distortion to the signal. In high-performance circuitry it is usually advisable to select the lowest loss dielectrics consistent with other properties and cost.

It can be seen from the previous discussion on electrical conductivity of metallizations used in multilayer ceramics that a high-conductivity metal provides the greatest design flexibility. By the same token, a dielectric which has very low loss will afford the designer similar freedom. For these reasons many MCM-C designs have concentrated on dielectric materials which can be cofired with thick-film metallization systems. Among these are the LTCC modules which IBM has chosen for its 390/ES9000 computers.[2] These modules consist of as many as 64 buried conductor layers. There are available today materials for MCM-C designs from a variety of sources with a large range of properties. LTCC metallization systems offer high conductivity, and dielectrics are becoming low in loss characteristics. Compatible resistor systems can now be included in buried layers leading to lower-cost multilayer structures.

MCM-L. The MCM-L style of multichip module is used principally in packaging digital circuitry, with few exceptions. The L stands for "laminate." *Laminate* refers to the layering of strips of materials atop one another. In this case it is the layering of *multilayer printed circuit board* (MLPCB) materials. This multilayer technology is highly developed for MLPCBs but less so for MCM technology. The reason for the existence of MCM-L is that surface mounting of components with high lead count and leads spaced more closely has driven printed wiring board technology to finer conductors, more conductors per layer, and finally to a point of component density comparable to MCM-C technology. The number of PCB manufacturers currently producing surface mount boards with line widths of 0.003 in either from fully additive methods or from copper cladding through subtractive methods provides sufficient evidence that MCM-L technology is a viable competitor with MCM-C for many applications. In order to achieve 0.003-in etched-line widths it is necessary to begin with a cladding which is thinner than conventional PCB claddings, generally about one-quarter ounce (0.0035 in thick). Thicker claddings produce a pyramidal shape in cross section which can penetrate the dielectric alone during lamination and cause an electrical short circuit. This is shown in Fig. 11.5. The dielectric material most used for standard MLPCB technology is epoxy-glass. Since the dielectric constant ϵ_r strongly affects signal transmission speeds, it is very important for digital MCMs that the dielectric constant be as low as possible. Some typical MCM-L candidate dielectrics are shown in Table 11.3.[3] Most of the signal layers of an MCM-L are buried within the multilayer structure and should be considered as stripline, rather than microstrip, for timing analyses. Table 11.3 shows that signals traveling in the stripline configuration require more time to traverse a given distance than those traveling in microstrip. Therefore, the importance of the dielectric constant cannot be emphasized too much. Care should be exercised, however, in the selection of the dielectric material because characteristics such as thermal-

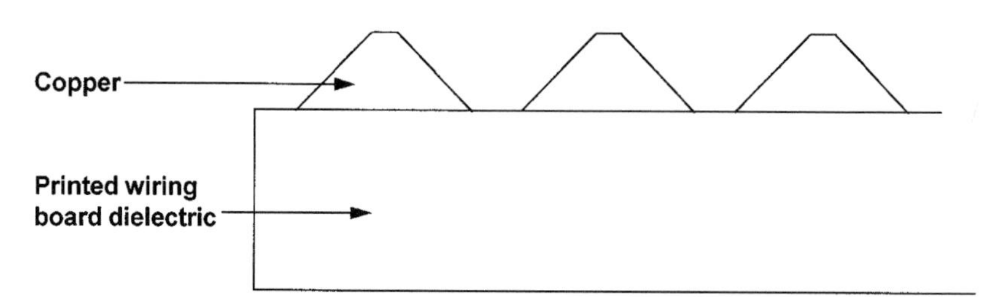

FIGURE 11.5 Cross section of etched copper conductors on PWB laminate.

TABLE 11.3 Dielectric Material Trends

Material type	Dielectric constant Er	Propagation delay Microstrip	Propagation delay Stripline	
Epoxy-glass	4.5	1.75	2.25	ns/ft
Polyimide/Kevlar	3.6	1.56	1.93	ns/ft
Polyimide/quartz	3.3	1.56	1.86	ns/ft
CE/aramid	3.0	1.53	1.78	ns/ft
PTFE/glass	2.2	1.40	1.51	ns/ft
Air	1.0	1.01	1.01	ns/ft

Note: The trend is toward a lower dielectric constant for faster signal propagation speeds.
Source: Courtesy of Electronic Packaging and Production.

cycling-induced microcracking of the resin and thermal coefficient of expansion can be major factors. The reason microcracking of the resin surface can be problematic is that fissures could open in the dielectric which would cause the fine conductors to become severed. This would in turn cause intermittent conduction of signals and power as well. The thermal coefficient of expansion is a major factor in MCM-L technology because of interconnections to the chips. Depending on how interconnections are made between chips and MCM-L substrates, differences in expansion coefficients between chips and the supporting dielectrics could cause failure of the chip-mounting material and/or electrical connection. A chip-mounting material which is tolerant of expansion differences will be more reliable if many temperature cycles can be expected in the application. Chip interconnects such as *flip chip* using solder bumps will experience considerable stress during thermal excursions and consequently could fail prematurely. The choice of not only the MCM-L substrate material but also the mounting material is very important from the standpoint of chemical compatibility with the devices to be packaged. Most of these devices are sensitive to potentially corrosive constituents in organic resins, particularly hardeners, as well as mobile ions which can influence device performance by virtue of their electrical charge.

A further consideration is the circuit line density requirement of MCM technology. Using through-hole approaches that require excessive routing space will force interconnects to require, in turn, an excessive number of layers. Staggered vias, on the other hand, can greatly reduce the number of layers, typically by 25 percent or more. In order to interconnect to devices, a line width and spacing consistent with pad-to-pad pitch on the devices is required. This could be 0.006 in or lower and would be an absolute requirement for flip-chip mounting technology since the mounting pads on the substrate must match those on the device. Currently this is not feasible with subtractive processes for producing conductors on PWB materials, considering plated metals required for soldering and wire bonding. Considerable PWB processing technology improvements will be required to make this approach viable for flip chip. Since flip chip is the densest device-packaging technology available today, MCM-L may not be able to compete head to head for the most dense packaging with MCM-D which easily accommodates flip-chip technology. On the other hand, because of the potentially lower cost to fabricate modules with MCM-L technology, MCM-D and even MCM-C may be the technologies of choice for higher end product, with MCM-L filling in lower end applications.

There is a technology just emerging which could, however, make the MCM-L approach viable for the densest known applications. This is known as conductive polymer technology. Although currently being developed for devices, this technology may be readily adaptable to PWBs for MCM-L applications. These polymers are made conductive by chemical reactions with reducing or oxidizing agents to create charge carriers.[4] The stability and processing characteristics of these materials as well as relatively low cost to produce them are causing them to be considered for a variety of usages, of which MCM-L would be a logical candidate.

Diamond Substrates. Technology to produce diamond coatings on substrates and free-standing diamond substrates exists now. However, the processes to assure an adherent metallization for conductors and the means to open holes in the diamond for producing metallized through holes is not yet captured. Consequently, multilayer technology with diamond material as the dielectric does not exist. Diamond is formed through the chemical vapor deposition process when thin films or coatings are desired. A gas is introduced into a vacuum chamber in which the substrate such as aluminum oxide has been placed. The gas must have carbon as part of its molecular makeup. A typical gas is methane. A plasma is created from this gas through the introduction of energy such as high voltage between several electrodes in the chamber. Microwaves are also used to assist in creating the plasma. The gas, being highly energized, is readily dissociated into its individual atoms. Through proper electrical biasing of the source and target materials in the chamber, the carbon condenses from the gaseous state onto the substrate. The process is made to continue until the desired thickness of coating is achieved. Nucleation of the condensate is essential to the formation of crystalline material. Although the deposit may be polycrystalline, it is of the allotropic form diamond, rather than graphite. In order for the coating to have high electrical resistance and be a good dielectric, only a small percentage of the carbon can exist as graphite (which is an electrical conductor).

The reason diamond holds so much attraction as a coating or as a substrate, for that matter, is that its thermal conductivity is far higher than that of the best thermally conductive dielectric. Other features or properties that are attractive are high insulation resistance and good flexural strength. The thermal conductivity which can be expected is between 600 and 800 W/mK. Higher values have been reported, but the designer would be wise to use a more conservative number. A value of thermal conductivity so high presents many application opportunities to the designer. Digital devices continue to become faster, performing more and more computations per second. Power supply devices (transistors) carry more current and become hotter, which limits their efficiency. More heat is being generated, and the circuit thermal density is increasing. The designer could spread out the heat-generating components, but this would make the assemblies larger, bulkier, and heavier. A method to carry away the generated heat quickly is what the designer needs, and this method comes in the form of high-thermal-conductivity diamond.

Several investigators have recently reported the achievement of diamond wafers as large as 15 cm in diameter and 2 mm thickness.[5] This makes it possible for substrates to be made entirely of diamond. Metallization tracks must be created on top of the diamond surface to form conductors, and herein lies the demand for a metallization process which can be depended on to produce an adherent metal surface. The low coefficient of expansion of diamond (between 2 and 3×10^{-6}/°C) presents a mismatch of at least two to one between itself and the metals used in thin-film circuits. Extensive thermal cycling of the metallized diamond could lead to metal lifting from the diamond surface with this mismatch if the adherent layer is not fused to the diamond. Sputtering the metal atop the diamond has been reported as a successful approach to achieving adherent metallizations.

11.3 PACKAGE MATERIALS

The materials which are used to package MCMs generally come from the classes of ceramic or metal enclosures that have been used to house hybrid circuits for some time. They are usually hermetic, that is, they meet or at least are intended to meet the provisions of MIL-STD-883, Method 1014.9. This means that they should not allow atmospheres to leak into or out of the cavity containing the circuit according to Table 11.4.

TABLE 11.4 Package Leak Rates As Specified in MIL-STD-883D, Method 1014.9

	Bomb condition			
Volume of package V, cm^3	psia ±2	Minimum exposure time hours, t_1	Maximum dwell hours, t_2	R1: Reject limit, atm cc/s He
<0.05	75	2	1	5×10^{-8}
≥0.05–<0.5	75	4	1	5×10^{-8}
≥0.5–<1.0	45	2	1	1×10^{-7}
≥1.0–<10.0	45	5	1	5×10^{-6}
≥10.0–<20.0	45	10	1	5×10^{-6}

Most MCMs would fall into the second or third category because of their size. The most favored metal for package enclosures has been Kovar* which is a combination of iron, nickel, and cobalt. Aluminum oxide (alumina) has served as the main ceramic package material with beryllium oxide (beryllia) being used for applications requiring thermal conductivity greater than that of alumina. Recent materials (aluminum nitride, silicon carbide, and diamond) have emerged as substitutes for beryllia because that material has been regarded as potentially toxic. According to Hudson,[6] the *Defense Electronic Supply Center* (DESC) has notified the electronics industry that it considers beryllium and its compounds sufficiently toxic that it has been elevated to a higher

*Tradename of Westinghouse Electric Corp.

position on the DESC's *hazardous materials list* (HAZMAT). This action has had two results. The first is that designers are somewhat reluctant to specify beryllium compounds in any new design. The second is that research into substitutes for beryllium compounds, and beryllium oxide in particular, has been intensified. This research has resulted in remarkable strides being made in aluminum nitride packages and substrates and in diamond dielectrics as coatings and substrates. Silicon carbide substrates have also undergone development and refinement.

11.3.1 Metal Packages

Kovar and other alloys, notably Alloy 51 and Dumet, were first developed and used to fulfill the need for metal leads of vacuum tubes to match the coefficient of expansion of the glass envelope. These leads penetrated the glass as feedthroughs, and they expanded and contracted with the glass as the operation of the tube caused them to experience both a rise and fall in temperature. The electrical conductivity of these metals was considered sufficiently high to serve most needs since they were generally used for short lengths, only to penetrate the hermetic envelope. Later it was found that miniature relays and similar electromechanical devices which encountered sparking as two conductors or electrodes at different electrical potentials were brought together had these electrodes eroded prematurely because of the presence of oxygen.

Glass-to-Metal Seals. The placement of the entire relay assembly into a hermetic envelope greatly enhanced relay life. The envelope at first was glass, but Kovar was later used to provide a rugged assembly which would withstand handling more readily. To provide electrical paths, Kovar leads were used to penetrate the envelope with matching expansion glass beads surrounding them which isolated them from the Kovar envelope or can. The leads were fused to the glass beads and then electroplated to provide protection from processing chemicals and the environments encountered in service, as well as to enhance solderability for wire attachment in subsequent assembly. This technology was adopted by the hybrid circuit industry and used for almost 30 years for hybrid circuits packages. Improvements in glass feedthroughs and lead design have continually taken place such that this packaging method has proven to be dependable when applied correctly. Certain features of glass-to-metal seals are prone to problems, however. The principal one is cracking of the glass meniscus on the lead. This is especially troublesome with rectangular leads because of the extremely small radius at the corners of the lead. Minor flexing of the lead can cause tiny flakes of the glass to be cracked away, leaving exposed a small amount of bare Kovar which was previously protected by the glass. This could lead to corrosion of the lead and eventual failure of the device. Another potential problem is the cracking of the glass bead which can be either radial or circumferential. Cracks can appear to be so deep as to penetrate all the way to the inside of the package. Radial cracks are specifically forbidden by MIL-STD-883. The ultimate tests of whether a crack is detrimental are the leak tests of this military standard. Correlation of visual criteria and leak testing is troublesome and often inconclusive. For these reasons the trend toward ceramic packages has become very strong, although Kovar enclosures continue to be used in both commercial as well as military microelectronics.

11.3.2 Alumina Packages

Alumina packages have been used to house individual die as well as hybrid circuits successfully for some time. They have been constructed in several different configurations. First of all, they have been used very much like the Kovar envelopes, with a base which has a cavity to contain a chip, or multiple chips, and a metal cover which fits over the walls of the cavity and is soldered to metallization atop the walls. This metallization is fired or sintered into the alumina surface to provide a hermetic seal. The metallization is a refractory metal or combinations of refractory metals. Typically, these are either tungsten or molybdenum manganese. The walls may be attached to the base through a sealing glass or directly as a cofired layer. Electrical paths are provided by metallization patterns which pass beneath the walls. In the case of the cofired wall, the metallization pattern is cofired as well, that is, sintered along with the alumina. Most of the packages available today are constructed by cofiring layers of ceramic tape laminated together rather than with sealing glass. Cofired technology is well refined and practiced broadly both domestically in the United States and abroad as well. Alumina ceramic package sizes range from single-chip carriers measuring 0.180 × 0.180 × 0.036 in thick, with 16 metallized and plated electrical contact pads (4 per edge) to multichip packages measuring over 2 in on a side with many hundreds of pads on all 4 edges brazed to tin-plated or gold-plated Kovar leads. Since multichip packages are very often used for MCMs, the chips they contain are frequently so output pin intensive and large that only a few will be able to fit into the package. System partitioning with the limitation of just a few chips per package often results in the need for several hundred outputs on the package. The output count therefore can exceed the number of available pads of a package which has leads brazed to the 4 edges. The solution to this problem is to place the output pads on the bottom of the package. In this way, the pad count increases quadratically rather than linearly with circumferential dimensions. For example, a typical package measuring 2 in on a side with leads on 0.040 in centers might have 46 leads per edge or 184 leads total. The theoretical limit to the number of output pads in this case is 2116.

Pin Grid Array and Area Array Packages. Two approaches are used to exploit this feature. One is called *pin grid array* (PGA), and the other is known as an *area array* or *pad grid array*. If a PGA configuration is desired, the center-to-center spacing is usually 0.100 in. The reason for this is that the pins which are brazed to the bottom of the package are nail headed and the nail head is, in the highest pin count package, approximately 0.032 in diameter. The pad to which such a pin is to be brazed must include an area around the head for the braze fillet, and this requires a pad dimension of 0.050 in, considering a few thousandths of an inch tolerance for pin placement. With a center-to-center spacing of 0.100 in, this example yields a total pin count of 324 for a package 2 in on a side. While this number is appreciable, it will not suffice for applications that require as much as 600 to 1000 output pads per module. Such applications exist in MCM usage and these require a solution. The use of a pad grid array greatly increases the pad count of an MCM without changing its dimensions. Referring to the example, if a center-to-center spacing of 0.050 in is selected, the pad array will be 36 × 36 = 1296, which would accommodate the requirement with almost 300 pads to spare. Pad array packages, however, present an interconnection problem. It is not feasible to solder connections reliably to each pad.

Interconnections to High Output Count Packages. Interconnecting to very large numbers of input-output pads of alumina MCM packages reliably at center-to-center spacings of 0.050 in in the pad grid or area array configurations can be a formidable task. Most of the pads on the MCM to be soldered to the supporting PWB are hidden from visual inspection. Most of the interconnect area of the MCM is not available to any visual scrutiny. X-ray inspection could provide some detail of the joints, but since the MCM packages contain several levels of metallizations which could interfere with the x-ray presentation of soldered pads, soldering is not the preferred interconnect technique.

Interconnection without Solder. A means of providing interconnections to these pad grid arrays lies in solderless technology. This technology has taken several forms which are listed here with a brief description of typical applications. They are the following:

- Epoxies filled with metal particles which are used with chips, leadless chip carriers, and hybrid circuits
- Elastomers with spring contacts or filled with metal particles and used with leadless chip carriers as well as hybrid circuits, key pads, and test sockets
- Button boards (carriers) with compliant springs used with area array packages

Further detail is provided later in this chapter.

11.4 MICROWAVE PACKAGE MATERIALS

This section will consider the special requirements for packaging microwave circuits. Microwave circuitry differs from dc and low-frequency circuits in that the signals are carried in both the conductors *and* in the dielectrics surrounding the conductors. For this reason the dielectric properties of packages which contain microwave circuits are quite important. A brief discussion of the development of materials for microwave circuits as well as the techniques known as *microstrip* and *stripline* and their foundation in coaxial (coax) signal transmission is presented here.

Early microwave printed wiring technology was derived from microwave power dividers for antennas and was then developed into flat coaxial configurations.[7] ITT introduced a configuration which had a single conductor and a single ground plane with a solid dielectric that was a variation of the air dielectric power divider scheme which used two ground planes. They called this configuration *microstrip.*[8] Later Sanders Associates developed a central conductor, two solid dielectric, two ground plane system they called *Tri-plate.*[9] This configuration has emerged into what is known today as *stripline,* although the original name "stripline" referred to an air dielectric arrangement introduced by Airborne Instrument Laboratory.[10] The dielectric materials which were used at that time were fiberglass followed by Teflon* and Rexolite.† The Rexolite and Teflon were unreinforced and as such were simply matrix materials. Reinforcement in the form of woven glass fabric provided mechanical strength but at the same time detracted somewhat from the quality of the dielectric. Less lossy reinforcement (quartz and randomly oriented short glass fibers) were introduced to restore the performance while still providing mechanical strength. This increased the design flexibility considerably for a wide variety of applications (microwave components such as couplers, filters, dividers, and combiners) and power distribution manifolding in phased array antennas. The component manufacturing community continued to seek higher performance materials and began to experiment with microstrip dielectrics such as quartz, aluminum oxide, sapphire, beryllium oxide, and some titanates as well. These materials not being organic printed circuit boards required a technique to deposit and adhere a conductor system to them. Thin-film

* *Teflon:* Tradename of DuPont.

† *Rexolite:* Tradename of C-Lec Plastics, Inc.

technology was the most convenient since it was being developed and refined to produce integrated circuits and hybrid components and was well characterized. Sapphire because of its original cost and the extensive machining required to render a usable substrate finish for thin-film metallizations did not survive as a major substrate material. Quartz had similar machining costs and found limited, though continued, usage. Aluminum oxide being easily produced in reasonable purity levels emerged as a major material of choice. Beryllium oxide because of its high thermal conductivity and availability was also used extensively.

As dielectric technology continued to be refined and a commercial market for microwave components developed in the communications field, considerable effort was spent on reducing the costs of both the materials and the processes to produce microwave circuit components. This required a different approach to metallizing the substrates. While an industry was under development for the production of analog and digital hybrid circuits, technologies emerged which began to be applied to microwave circuits. Thick-film metallizations which could be applied by stenciling techniques to alumina or beryllia substrate surfaces and subsequently baked and/or fired into the ceramic had potential usage in the microwave arena if the electrical conductivity of the metallization were sufficiently high for microwave applications. This issue was resolved with the introduction of high electrical conductivity thick-film pastes by Dupont, EMCA, and ESL in the late 1960s. At this point microwave circuitry moved into the realm of affordable high performance.

Solid-state microwave receivers and transmitters of low power (less than 500 W) had their hardware made up of many microwave components such as low-noise receivers, circulators, amplifiers, mixers, and phase shifters, connected to each other with miniature coaxial conductors, referred to as *coax*, or *hardlines*. This arrangement was bulky and costly to fabricate. A better integration of the components was required. Systems producers began to combine various subfunctions into the same package rather than interconnect individually packaged subfunctions with coax. This resulted in more compact electronics equipment that became lighter in weight and easier to install and maintain.

Subfunctions were generally constructed on ceramic substrates and placed together into large packages where they were interconnected together by ribbons of gold or by gold wire bonds. Solder and diffusion bonding were the methods of interconnecting the substrates together. Although this approach was economical, there were subfunction designs that needed to be larger than the standard ceramic sizes in vogue. Larger ceramic plates were available, but handling and assembly issues dictated 2 in on a side to be an optimum size. A solution to the size limitation of ceramic substrates was offered by the 3M Company with their Epsilam 10 which was a double-sided copper cladding on a proprietary dielectric composed of a high dielectric constant K filler dispersed within a lower K matrix with a combined effective K of approximately 10. The value of 10 was chosen to simulate the K of 99+ percent alumina, so the Epsilam 10 could be substituted directly for it. This clad material was available in sheets 10 by 10 in, making it possible to construct large integrated microwave assemblies because the copper cladding could be patterned into the desired conductor configurations through photoetching. The copper cladding on the bottom side served as the ground plane. This material did not have the handling and breakage characteristics associated with alumina and therefore offered more versatility. Irregular shapes and sizes could be achieved by the user through the use of printed circuit technology, namely, routing. Etching and plating was possible since the dielectric was compatible with standard processes. Several other manufacturers (Rogers and Keene) provided materials with similar characteristics later. A list of materials and properties of substrates is shown in Table 11.5.

TABLE 11.5 Properties of Substrates Used in Hybrid Microwave Integrated Circuits

Material	ϵ_r, approximate	$\tan\delta \times 10^{-4}$ at 10 GHz	Surface roughness, μm	Thermal conductivity K, W/(cm°C)	Remarks and applications
RT-duroid 5880	2.16–2.24	5–15	0.75–1.0* 4.25–8.75†	0.0026	Cu plating, flexible/stripline, microstrip
RT-duroid 6010	10.2–10.7	10–60	0.75–1.0* 4.25–8.75†	0.0041	Cu plating, flexible/stripline at L band, microstrip
Epsilam-10	10–13	20	—	0.0037	Cu plating, flexible/stripline at L band, microstrip
Alumina 99.5%	9.6–10.4	0.5–3	0.05–0.25	0.37	Cr-Au layer/microstrip, microstriplike lines, slot line, coplanar lines
Fused quartz 99.9%	3.75–3.8	1	0.006–0.025	0.01	Cr-Au layer, optical finish/microstriplike lines at millimeter wave frequencies
Beryllia (BeO)	6.6	1	0.05–1.25	2.5	High conductivity/compound substrate
Glass	5	20	0.025	0.01	Lossy/lumped element
Kapton	3–3.5	—	—	—	Flexible substrate/stripline
Cu-flon	2.1	4.5	—	—	Flexible substrate/stripline, microstrip
Rutile (TiO_2)	100	4	0.25–2.5	0.02	High dielectric constant/microstrip, slot line, coplanar lines
Ferrite/garnet	13–16	2	0.25	0.03	Porous/nonreciprocal devices in slot line, coplanar lines
Sapphire (single crystal)	$\epsilon_r = 9.4$ $\epsilon_r = 11.6$	0.4–0.7	0.005–0.025	0.4	Well-defined and repeatable electrical properties, anisotropic/microstrip
Pyrolytic boron nitride	$\epsilon_r = 3.4$ $\epsilon_r = 5.12$	—	—	—	Anisotropic/microstrip, suspended stripline

* Average peak-to-valley difference in height (rolled copper).

† Average peak-to-valley difference in height (electrodeposited copper).

Source: Courtesy of McGraw-Hill.

The previous discussion related primarily to microstrip circuitry. Since the ceramic substrates were only double-sided, that is, printed with a backplane on one side and a conductor pattern on the other, they conformed to the definition of microstrip. Continued microwave circuit materials development made it possible in the mid 1980s for designs to include several layers of conductors and dielectrics in the same structure, much as a printed wiring board is configured. This development concerns ceramiclike materials known as *low-temperature cofired ceramics* (LTCC) which are discussed in Sec. 11.2.1. The cofiring of conductors and dielectrics at less than 1000°C allows the use of high electrical conductivity metals for top surface and buried layers as well. This means that microstrip configurations can be constructed and used in conjunction with lower-frequency circuitry and digital circuits, all within the same substrate. Such flexibility offers the designer opportunities to combine, within the same substrate, circuit functions that should be located in close proximity to each other. It also allows more compactness and has the potential to improve circuit performance because of fewer interconnections. With a wide variety of dielectric properties, the LTCC materials are opening new vistas for their application. This includes components such as resistors and capacitors printed onto buried dielectric layers. Certain passive components can be included within the multilayer structure to increase overall component density because the materials that constitute resistors and low-value capacitors have demonstrated compatibility with the cofired LTCC systems.

11.4.1 Single-Chip Packaging Materials

The packaging of microwave circuits can vary from single-chip packages to highly integrated packages with dozens of devices contained and interconnected within one package. The most common single-chip microwave packages contain power transistors. These power transistors are mounted either onto the package base metallization or a metal which serves as the base of the package and is often used to provide mechanical mounting for the package. This metal is generally Kovar. The remainder of the package is alumina. A two-part frame of alumina is brazed to the Kovar base, and a cover of either Kovar or alumina is soldered to the upper part of the frame. Kovar leads emanate from the two-part frame—which is really two frames glassed together with the two transistor leads sandwiched between them. The third contact is through the base.

The variety of microwave components used today require packages which are sometimes larger than the single-chip package and contain a hybrid microwave circuit rather than a single chip. This variety is extensive and represents different configurations with axial and radial leads, leads on one side only, leads on two sides, leads on three sides, and leads on all four sides. An example of the wide variety of package options is shown in Fig. 11.6. Although the figure depicts many configurations, it by no means represents all the different packages which are being used. "Duplication" and "triplication" are words which could be used to describe the situation in hybrid microwave packaging. Standardization of packages and their materials is required to bring order and to provide a means of lowering costs.

11.5 POWER HYBRIDS

A power hybrid circuit is one which uses solid-state components and which controls a large amount of electrical power. This has implications in terms of power dissipation, high levels of current, and even high voltage.

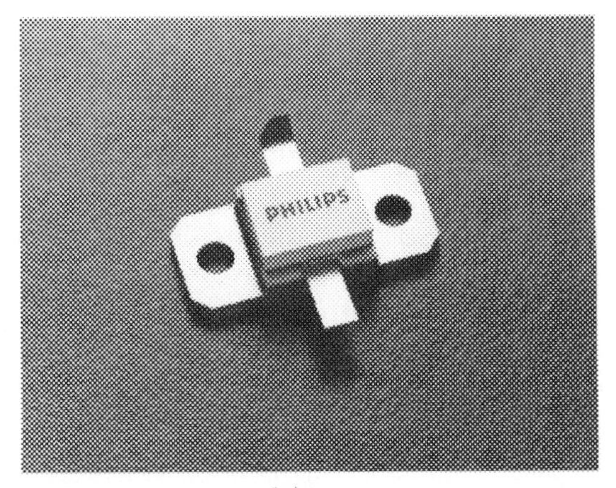

(a)

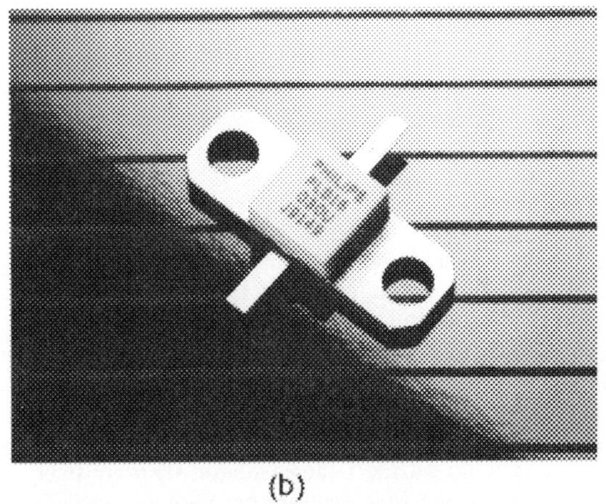

(b)

(c)

FIGURE 11.6 An example of the variety of microwave packages available. (*Courtesy of Phillips Components, Discrete Products Division.*)

Power dissipation is of prime importance to packaging engineers when they design power-handling circuits. Where power conditioning is involved, device inefficiencies and I^2R losses generate heat which must be removed so that the device temperatures do not rise above design limits. Although the requirement to control operating temperatures applies to all circuit designs, it is especially critical for power-handling circuits because the relatively large amount of electrical energy that is converted to heat in power conversion can cause rapid heating of the circuit if an efficient path is not provided for the heat to be removed.

The amount of electrical current carried in power-handling circuits varies from a few amps to dozens of amps. When these circuits are in hybrid form, particular attention must be paid to the capability of conductors to carry the imposed current without contributing significantly to the I^2R losses. This means that substrate conductors must have very little resistance (high conductivity) and, if the current is substantial, must be thick and perhaps wide as well. Feedthroughs or conductors that carry the current from the outside world through the hermetic walls must also have high current-carrying capability.

High voltage is not encountered as often as high current, but it is not unusual for power hybrids to carry several hundred volts. This also requires careful attention by the designer in the materials selection process since hybrids allow circuits to become smaller, bringing conductors closer together, both in-plane and between layers.

11.5.1 Power Hybrid Packages

Power hybrid packages are usually made of copper because of its high thermal conductivity. This kind of package can readily conduct away the heat generated within it by the controller circuitry. It does this efficiently by allowing the heat to spread quickly across its base so that the heat can then move through the interface between the package base and the heat sink beneath it. Recently aluminum nitride has emerged as a viable package material as well. Beryllium oxide served as a hybrid package material for some time before aluminum nitride made its debut. Copper packages are electrically conductive, and aluminum nitride is an insulator. This must be considered in the specific application because the first material beneath the package might be thermal vias consisting of *plated-through holes* (PTHs) in a *printed wiring board* (PWB). Some or all of these holes might serve as electrical paths for the circuits in the PWB. If a copper package were to be used in this case, an electrical insulator might need to be placed between the package and the plated vias. This would raise the thermal resistance of the heat path because most insulator materials are poor thermal conductors. In this case an aluminum nitride or beryllium oxide package might merit consideration. The thermal characteristics of several package materials are shown in Table 11.6. The power-generating components inside the package may be attached directly to the package base if the package is an insulator. If it is conductive, the components may be attached first to a substrate that insulates the components from the package since the component bottoms are often conductive and part of the circuit. For instance, power transistors have their collector contact point as the device bottom. If the transistor were electrically connected to the copper package and the package in turn were in contact with another part of the circuit such as ground, a short circuit would be created.

Copper was the early choice for this style of package because *oxygen-free, high-conductivity* (OFHC) copper was well characterized and had a long history of use in microwave power-generating devices such as klystrons and traveling wave tubes. The packages were formed by machining them from copper sheet. Impact extrusion was also a method employed to produce a near net shape at lower cost than machining the

TABLE 11.6 Typical Ceramic Packaging Materials and Characteristics

Property	Units	CC-92 771/777	CC-99.5	CC-AN10	CC-LT20
Material	—	92% alumina*	99.5 alumina*	AlN	Glass ceramic (LTCC)†
Color		White or black	White	Translucent gray	White
Density	g/ml (#/in^3)	3.62 (0.131)	3.89 (0.141)	326 (0.118)	2.70 (0.098)
Hardness	kg/mm^2	1207 (Knoop)	87 (Rockwell)	1200 (Knoop)	†
Flexural strength	MPa (psi × 10^3)	443 (64)	572 (83)	280 (41)	124 (18)
Young's modulus	GPa (psi × 10^6)	275 (40)	372 (40)	340 (49)	128 (19)
Shear modulus	GPa (psi × 10^6)	112 (16)		140 (20)	50 (7.3)
Surface finish	μm	<1.02	<0.15	<0.76	†
	(μin)	(<40)	(<6)	(<30)	†
	CLA (as fired)				
Thermal expansion:					
25–300°C	10^{-6}/°C (10^6/°F)	6.73 (3.73)	6.7 (3.7)	4.0 (2.2)	2.8 (1.6)
25–500°C	10^{-6}/°C (10^6/°F)	7.16 (3.98)	7.5 (4.2)	4.5 (2.5)	†
Thermal conductivity:					
25°C	W/m K	20.03	34.7	170–190	2.3
	(Btu-in/ft^2-h°F)	(141)	(241)	(1179–1318)	(16)
Dielectric strength	kV/mm (Vmil)	11.6 (295)	26.3 (670)	13.0 (330)	†
Volume resistivity	Ω-cm^2/cm	>10^{14}	>10^{14}	>10^{14}	>10^{14}
Dielectric constant:					
1 KHz			9.9	8.7	
100 KHz					5.4
1 MHz		9.11	9.9	8.5	5.4
10 MHz					5.4
Dissipation factor:					
1 KHz			0.0005	0.0002	
1 MHz		0.0003	0.0002	0.0001	0.001
10 MHz					0.001
Loss factor:					
1 MHz		0.003	0.002		0.005
10 MHz					0.005

*Developmental material.

†Material under development—additional values to be published.

Source: Courtesy of Coors Electronic Package Company.

part completely. With this method the outline shape of the package is created by stamping, and the remaining dimensions are achieved by machining, including boring holes for the feedthroughs. A finish plating is required since bare copper will corrode. This is usually nickel followed by gold. Feedthroughs are soldered into the holes with a high temperature solder such as gold tin. Copper is still used extensively because of its economies and because there is virtually no limitation on the size of the package.

Beryllium oxide (beryllia) has been used for high-power hybrid packages for several years in cases where an insulating package was required. Its thermal conductivity k is high with respect to aluminum oxide (alumina), and several manufacturers provided packages to specifications so that the user was not required to process any beryllia parts beyond soldering covers onto the packages. Packages as large as 4 in on a side can be fabricated from this material.

Aluminum nitride (AlN) is the recent material innovation for high thermal conductivity packaging. Packages can be made which measure over 4 in on a side and which have as many as 10 layers of circuitry. The metallization is cofired tungsten and molybdenum which yields high-resistivity conductors. It should be noted that aluminum nitride has been successfully bonded to LTCC materials to form packages that have high electrical conductivity conductors for applications that require multilayer circuitry as well as high package thermal conductivity. This allows the use of a substrate with a low k (LTCC has a k of approximately 2 to 5 W/m-K) in conjunction with a package that has a k of 150. That combination greatly extends the applicability of AlN to a variety of high power circuits.

High-Power Package Feedthroughs. In place of the familiar fine leads usually seen on a hybrid package, power hybrid packages have 0.040- to 0.090-in-diameter pins emanating from the package walls. Electrical connections are made to the circuit contained within the package by means of these pins which consist of a central conductor insulated from an outer metal shell by a bead of glass. The arrangement of conductor, glass, and shell is a delicate one by virtue of the glass which must maintain a hermetic seal between itself and the shell as well as between itself and the central pin. The glass whose coefficient of expansion is very close to that of the pin is attached to the pin by a chemical bond between the oxides on the pin and the glass. In certain pin designs where the shell is not matched in thermal expansion to the glass, the shell and pin are assembled to the glass by heating all of them together until the glass flows. The shell contracts more rapidly than the glass, forming a compression seal between itself and the glass. The shell is sometimes called a *ferrule*. The ferrule material is of course different from the central pin material since it is desired to have the shell constrict more rapidly than the glass as they both cool. If the pin were the same material as the ferrule, it would reduce its diameter more rapidly than the glass, causing the bond between them to rupture. This kind of seal is reputed to be more achievable than a matched seal in which all elements of the feedthrough have approximately the same coefficient of expansion. For the copper package, it is probably more advantageous to use a compression seal with the ferrule having a coefficient of expansion less than copper but greater than the glass. This design distributes the expansion differences to each element (material) of the seal so as to minimize the tensile and compressive loads on any one interface.

Attempting to avoid the glass-cracking issues associated with the design just described, some packages have recently been made with alumina used in place of the glass. This type of feedthrough generally consists of only the ceramic and the pin. The ceramic (alumina) is metallized and plated on its outer circumference and the walls of the central hole to allow it to be brazed to the pin and soldered or brazed to the copper package.

The central pin is by design required to carry high current, and therefore it should be made from a good electrical conductor. Copper would make a good choice from the conductivity standpoint, but its coefficient of expansion would preclude its selection for use with glass seals. This apparent dilemma is solved by the use of a metal cladding over a copper pin. This cladding is a match to the glass (usually Kovar or a similar alloy such as HAI-52 alloy* is used) yet bonded metallurgically to the copper pin. The cladding thickness is chosen to have sufficient strength to maintain a close match to the expansion of the glass while not reducing the electrical conductivity of the pin below design limits.

---

*Tradename of Harrison Alloys Inc.

On the inside of the package, electrical connections are usually made by wire bonds between the devices and the pins and/or substrate tracks and the pins. To accommodate wire bonding, the pins are frequently coined to have a flat spot on the top side of the pin inside the package. (Outside the package the pin cross section is round.) Wires from the substrates or devices are bonded to the pins on the flat spots. Both gold and aluminum wires are used, but the costs of using gold in the diameters required for the high currents carried in these packages dictate the use of aluminum wire.

11.5.2 Substrates Used in Power Hybrid Packages

There are a variety of substrates which can be used in power hybrid packages. They are alumina, beryllia, silicon carbide, and aluminum nitride. Diamond could be considered at this time since free-standing diamond substrates have been synthesized which would support power-dissipating circuits such as those used in power-conditioning circuits, but its present cost places it in a special category. Diamond substrates are discussed in Sec. 11.2.1 and Chap. 3 of this handbook.

Alumina substrates are the lowest-cost substrate materials, although alumina's thermal conductivity usually restricts its use in these applications to supporting control and conditioning circuits which do not dissipate large amounts of heat. In many cases alumina is used in conjunction with beryllia or silicon carbide or aluminum nitride substrates within the same package. Those substrates are more capable of dissipating the heat than components such as power transistors mounted upon them will generate. In essence, those substrates spread the heat rapidly within themselves so that more heat per unit of time is passed through the material below (usually solder) to the package base. The properties of these substrate materials germane to power hybrids are listed in Table 11.6. The power hybrid can have from one to as many as a dozen individual substrates which contain and interconnect up to 100 components. Power transistors are almost always mounted on and soldered to beryllia substrates as was mentioned earlier. In most cases all the substrates are mounted to the base by solder. For copper packages this is the preferred technique.

It is significant to note that the coefficients of thermal expansion of these substrates are widely varied, and compared to copper which is the package material in most cases, they present a mismatch that merits concern for the designer.

Soldering Substrates to Power Hybrid Packages. Components are attached to substrates first, and the populated substrates are then attached to the package. This sequence requires a solder system which uses one melting point material for component-to-substrate attachment and a lower-melting-point material for the attachment of the substrate to the package. In this way, reflow of the component-substrate joint will not occur during substrate attachment. Since bare silicon devices (diodes and transistors) are being packaged, it is desirable to use soldering processes that do not use flux. This minimizes the chances of possible contamination. Components which require solder attachment in power hybrids can have a variety of backing metallizations, such as chrome silver and chrome gold. The beryllia substrates to which the power devices are attached usually are metallized first with molybdenum manganese followed by a plating of nickel and then gold. Wire bonds of either aluminum or gold can be made to the substrate gold plating, and devices can be soldered to this plating as well. Gold tin solder is commonly used for silicon device attachment to substrates. The main requirements of a solder for this application are that it wet the various chip backing metallizations and form a joint with a minimum of voids, have a relatively high melting point, and not degrade during subsequent process steps and thermal aging experienced during service. A study of commonly used solders concluded that gold tin solder with a melting point of 280°C exhibited excellent wetting to the various backing materials and formed very few voids. The same study evaluated solders for substrate attachment to copper packages.[11]

Fluxless vacuum soldering was used to reflow the solder-attaching substrates to packages. Substrate metallizations included the gold, nickel, and molybdenum manganese cladding of the beryllia substrates, thick-film gold backing on alumina substrates, and gold nickel plating on copper packages. A very important requirement of these solder joints is that they withstand the stresses imposed during thermal cycling caused by the large mismatch in coefficients of expansion between the substrates and packages. Tin lead eutectic solder (SN 63 as controlled by Federal Specification QQ-S-571) and indium lead were evaluated as attachment solders. The indium lead was 70 percent indium and 30 percent lead. The indium lead combination was selected because of the reported superior fatigue resistance of indium-bearing solders. Testing involved extensive thermal cycling and high-temperature storage followed by centrifuge testing to determine the durability of the solder attachment system for the particular metallization combinations encountered in power hybrids. A centrifuge failure would consist of the substrate-to-package bond being destroyed with the substrate departing the package base. No centrifuge failures occurred.

At the end of testing, cross sections of the tin lead solder showed fatigue cracking in the case of gold over nickel electroplate. The thick-film backing metallization did not show any cracking in the solder. Indium lead solder did not exhibit fatigue cracking in any combination of substrate metallizations soldered to the copper package. For this reason the indium lead combination was selected as the preferred solder attachment material.

It bears mentioning that indium solders used to solder gold-plated materials must be protected from humidity which could induce corrosion of the joints. Sealing the parts thus soldered in hermetic packages is the best way to assure joint integrity.

11.6 COMPOSITES IN ELECTRONIC PACKAGING

Composites are well known for their use in structural applications such as reinforced concrete and control surfaces on aircraft and in early electrical applications such as glass-reinforced phenolic cases for housing small transformers. A printed wiring board is a form of composite and is an early example of composite usage in electronics. In a broad sense, a composite is a combination of materials which has properties that none of the constituent materials have by themselves. Materials are selected to be combined with each other to take advantage of one or more characteristic properties that the partner material possesses. The properties can be mechanical, such as tensile strength or thermal conductivity, and electrical, such as high resistivity or dielectric constant. A brittle material can be made less so by combining it with a softer one. A material's tensile strength can be improved greatly by combining it with a reinforcement which binds itself well to the weaker material, thereby imparting some of its strength to the combination.

Composites today are classed into three groups according to the matrix material or main body of the combination. These groups are *polymer matrix* (PMC), *metal matrix* (MMC), and *ceramic matrix* (CMC). Polymer matrix composites are by far more numerous in variety or types than either metal matrix or ceramic matrix materials, but in spite of this, there are significant contributions that have been made to the advancement of electronic packaging by metal matrix and ceramic matrix technologies. This section will describe some of the developments in all three types of composites with special emphasis on the materials and technologies that are available for the packaging of electronics.

11.6.1 Polymer Matrix Composites in Electronic Packaging

Polymers have found extensive use in all aspects of electronics packaging, and the employment of polymer matrix composites is a natural evolution to be expected as polymer technology continued to advance. Since printed wiring board technology was well entrenched as an interconnect medium for a large variety of electronic components, it was also used for supporting and interconnecting leadless ceramic chip carriers (LCCCs) shortly after their introduction. Ceramic substrates with leads attached for connection to sockets or, more commonly, printed wiring boards (PWBs) were used to a minor extent. The initial leadless chip carriers were small (0.250 × 0.250 in) and had metallization tracks underneath the carriers which were intended to be soldered to the supporting substrate. This substrate could be ceramic or a PWB. The ceramic substrate (which could carry a small number of LCCCs) was itself a high-temperature cofired multilayer circuit and could be used to provide some of the interconnection required for the LCCCs. As long as the LCCCs remained small, the popularity of the ceramic substrates was very low. However, as the LCCCs grew in size and pinouts with the complexity of the devices they carried, the practice of soldering them to PWBs began to experience difficulties. Larger LCCCs grew by 0.05 in on a side for every additional four input-output connections that the packaged chip contained. At 40 leads per carrier a critical junction was reached. The package had grown to 0.550 in per side, and the expansion difference between the chip carrier and the PWB beneath it which contained the solder pads began to cause cracking of the solder joints. This was not immediately noticed because it took many hundreds of thermal cycles between −55°C and +125°C to cause cracks that were observable to the unaided eye. After several hundred of these thermal cycles, expansion-difference-induced stress lines began to appear in the solder joints. After further thermal cycling, these stress lines developed into actual cracks in the solder.

Chips continued to grow in complexity and size and so did the LCCCs. The number of thermal cycles required to cause solder joint cracks went down, causing the problem to surface more rapidly. Consequently the awareness of this expansion difference problem grew quickly, particularly among military equipment suppliers and their customers. Surface mount technology (which is the name that evolved to describe the practice of using leadless devices mounted onto PWBs) began to be used more extensively, and this caused both users and furnishers of PWBs to seek materials for constructing the PWBs which could have lower coefficients of expansion than standard epoxy-glass and polyimide. Such board materials would provide a better match to the ceramic packages which contained integrated circuit chips. Another approach was to restrain the expansion of the PWB with a core material. These cores consisted of several combinations of metals whose coefficients of thermal expansion were intended to match that of (LCCCs) which have a coefficient of thermal expansion of between 6 and 10 parts per million per degree Celsius.

Composite PWB Materials. A large variety of PWB materials were formulated (and continue to be formulated) to match the needs of surface-mounted components. Both the matrix materials and the reinforcements underwent change and development. Some of them are listed in Table 11.7 with measured data for the "warp" and "fill" directions of the woven reinforcement corresponding to in-plane measurements (x and y). The glass transition temperature T_g is listed as well because a high value means

TABLE 11.7 Physical, Mechanical, and Thermal Data on Advanced PWB Materials

	Resin content		CTE × 10^{-6}/°C*		CTE × 10^{-6}/°C†	T_g,
Material type	Wt %	Vol %	Warp	Fill	z direction	°C
Aramid, mod. polyimide	58	61	6.5	7.2	66	167
Quartz, polyimide	37	46	8.5	9.8	34.6	188
Aramid, mod. epoxy	59	61	5.5	5.6	100.4	137
Aramid, mod. epoxy	62	63	7.4	7.2	114.3	106
Glass, polyimide	38	55	12.6	13.0	41	249
Quartz, FR-4 epoxy	50	64	11.0	13.6	62.6	125
Quartz, FR-4 epoxy with phenolic	51	65	10.2	12.4	39.6	172
Quartz, epoxy/polyimide (80/20)	50	54	12.2	14.0	60.3	109
Aramid, low-T_g epoxy	66	68	7.7	9.1	107.4	82
Aramid, high-T_g epoxy	58	60	5.8	6.2	118	157

*Data in these columns were derived using a horizontal quartz dilatometer.

†Data in this column were derived using a thermal mechanical analysis technique.

Source: Courtesy of McGraw-Hill.

that the thermal coefficient of expansion is relatively constant and predictable up to that temperature. A material with a high T_g has additional advantages such as resistance to chemical etchants used in the processing of PWBs and their assembly and can withstand repeated component assembly and disassembly with very little damage to the bond between the copper cladding and the laminate material.

Aramid reinforcements had held the coefficients of thermal expansion of polymers to their design values but at the expense of cracking of the surface matrix material (epoxy, polyimide). The cracking was limited to the surface or buttercoat of the PWB but nevertheless presented a serious drawback to the full acceptance of this promising material. Consequently, considerable research was expended on lowering the *coefficient of thermal expansion* (CTE) of the resins or matrix materials. Table 11.8 lists some materials which resulted from this research. The values of the CTE in the out-of-plane direction (z) are made to rise when the in-plane expansion is constrained because the effective bulk coefficient of expansion is a constant. This aggravated expansion can be detrimental to the copper plating on the walls of the plated-through holes because of severe stressing during thermal cycling in service. Reinforcements other than woven fabric may be used for the electrical properties they impart. Microspheres of glass or plastics and powdered fillers are sometimes used to lower the dielectric constant. The CTE of this kind of PWB is isotropic because the fillers are not tied together.

11.6.2 Solid-Core PWB Expansion Restraint

Solid cores have been laminated to PWBs to serve as expansion limiters. Printed wiring boards may be bonded to one or to both sides of the restraining material, thus the name "core." The out-of-plane expansion of the PWB is increased with their use just as it is when a woven fabric is the board reinforcement. The core is usually

TABLE 11.8 Materials Developed for CTE–Matching to Ceramic Packages

Material type	Resin content, wt %	CTE × 10^{-6}/°C			T_g, °C
		Warp	Fill	z direction	
Aramid (matte/fabric), FR-4 epoxy	60	5.5	6.0	66	125
Glass, epoxy with compliant layer	50	16.0	12.7	59	115
Aramid, BT epoxy	52	7.7	6.9	114.3	185
Quartz, polytetrafluoroethylene	67	7.5	9.4	88	19
Aramid (matte/fabric), mod. polyimide	61	10	10.4	113	177
Hybrid (aramid/quartz), polyimide	41	8.3	8.4	60	249

Source: Courtesy of McGraw-Hill.

designed to serve as a heat sink as well, but the cost and weight penalties their use extracts limits their use. Table 11.9 describes typical materials used in this manner. For a time the copper-Invar-copper was used extensively for PWB assemblies which contained large LCCCs of 48 or more pinouts. This combination exploited the high thermal conductivity of copper and the very low CTE of Invar to its advantage. The MMCs listed in Table 11.9 have an appealing specific thermal conductivity, but their costs reduce that appeal. Although the relative costs today are about one-quarter of the values shown in that 1986 table, they are nevertheless substantial. MMCs for the most part are not used simply as reinforcements and heat sinks but rather as complex shapes and special microwave packages. These are discussed in Sec. 11.6.4.

11.6.3 Graphite Epoxy Heat Sinks

The thermal conductivity of graphite has long attracted interest in its use for managing the problem of heat removal from components which generate considerable heat as they operate. Just how to utilize carbon efficiently in electronic assemblies remained a major impediment until work began on graphite epoxy composites. This combination has had a long history of applications in replacing or reinforcing metals in the aerospace industry. As such, it was developed and matured and well quantified. Its use in combination with other materials such as aluminum and titanium was well established. This considerable experience provided a reasonable foundation for graphite epoxy to serve as a principal mechanism for heat removal from electronic assemblies. Some research was funded by the U.S. government in support of the Standard Electronic Module activity at the Naval Weapons Support Center in Crane, Indiana. Additional work has been done by industry to supplement the Navy studies. Much of this work has been focused on using graphite epoxy composites as heat sinks, thermal paths, and module frames. In certain constructions the module frame serves as the heat removal mechanism because the frame is clamped to rails which are the principal mechanical support as well as the heat conduction medium. This is depicted in Fig. 11.7 in which

TABLE 11.9 Solid Core Heat Sink Mechanical Parameters

Material	CTE × 10^{-6}/°C	Thermal conductivity, W/°C in	Density, lb/in^3	Specific thermal conductivity, W in^2/lb°C	Relative cost	Young's modulus E, psi × 10^6
Aluminum	23.6	5.49	0.098	56.0	1	10
Copper	17.6	9.92	0.323	30.7	1.2	17
Alloy 42	5.3	0.396	0.28	1.41	3.3	21
Molybdenum	4.9	3.71	0.369	10.1	5.0	47
Invar	1.3	0.263	0.30	0.88	3.3	21
Copper-Invar 20-60-20	6.5	4.12*	0.31	13.3	2	18.5
Copper molybdenum 13-74-13	6.5	5.32†	0.36	14.8	4	31.2
Boron aluminum MMC‡ 20%	12.7	4.65§	0.095	48.9	100	27
Graphite aluminum MMC, 0–90° crossply	5.8	4.08	0.087	46.9	400	21
SiC aluminum MMC	16.2	3.51	0.103	34.1	200	15

*Conductivity in normal direction is 0.430 W/°C in.

†Conductivity in normal direction is 4.43 W/°C in.

‡MMC stands for *metal matrix composites,* a class of exotic metal matrices reinforced with ceramics such as boron, silicon carbide, or graphite.

§Conductivity in normal direction is 1.93 W/°C.

Source: Courtesy of McGraw-Hill.

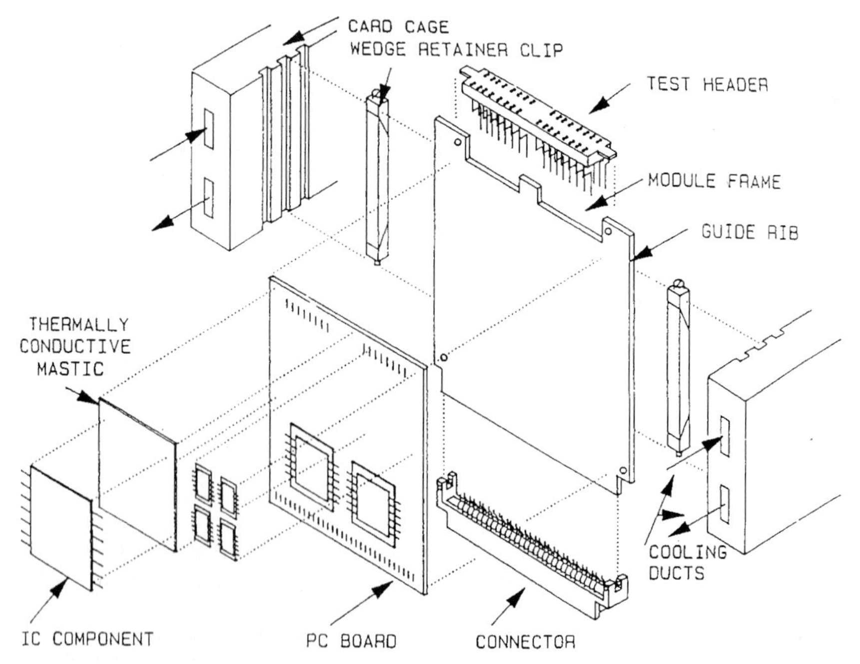

FIGURE 11.7 Components for guide rib conduction cooling. (*Courtesy of Society for the Advancement of Materials and Process Engineering.*)

the module frame supports the printed wiring assembly and conducts heat from itself to the chassis with the help of card cage wedge retainers.

Heat can be conducted along the carbon fibers very efficiently. The difficulty in using this composite lies in transferring heat to the fibers through the epoxy matrix (epoxy has a very poor thermal conductivity) and out of the fibers to the interface (chassis). The issue of heat transfer to the fibers perpendicular to their length is one of minimizing the amount of epoxy resin between the fibers and the heat source, such as a high-power package. If the module frame is made from a lay-up of graphite epoxy, the layers are typically placed one on another at 90 or 45 degrees so that the strength is more or less uniform in all directions in the plane of the frame. Placing the first several and last several layers in the direction of the heat exchangers assures that the heat will flow more toward the exchangers rather than at right angles. The relatively poor thermal conductivity of the composite at right angles (from fiber to fiber) helps to assure that the heat is conducted away from the source to the exchanger efficiently. Removal of this heat is accomplished in the standard electronic module by the wedge retainer clips which clamp the frame firmly to the exchanger rib. Heat is then transferred from the uppermost fibers to the rib. An improvement to this approach is described in U.S. Patents 4849858, 4867235, and 5002715. In those patents the inventors depict a method of bending the fibers near the heat exchangers in order to have the ends of the fibers make contact with the exchangers. Since most of the heat is conducted along the fibers rather than at right angles to them, removing the heat from the ends rather than at right angles to the ends results in a greater flux from the fibers to the exchangers.

11.6.4 Metal Matrix Composites

High strength, high thermal conductivity, low density, good toughness—all available at low risk and low cost. This sounds hard to believe, but for the most part, these features which designers seek in the materials they specify for electronic packaging are possible to realize through metal matrix composites. While no one material has all the characteristics listed, it is possible to pick and choose how much of each feature one can obtain in a given material. This happens when one or more materials and reinforcements are combined. For example, a metal with high thermal conductivity can be combined with a reinforcement which has a high elastic modulus and low density to achieve a lightweight, high-strength, high-thermal-conductivity composite. In practical terms, combinations like these exist. They are combinations of copper and graphite or aluminum and graphite, or aluminum and silicon carbide. These composites are being manufactured and marketed as electronic packaging materials.

To see why MMCs are so attractive as a design consideration, consider the properties of materials which match the CTEs of silicon devices. The CTE of aluminum is 22×10^{-6}/K, and silicon is 3.8×10^{-6}/K. If one must mount the silicon on aluminum and expect the bond between the two to survive thermal cycling, it is necessary to use a number of different materials between the silicon and aluminum which have CTEs somewhere between those of aluminum and silicon. This is not a desirable solution. Another approach would be to modify the CTE of the aluminum with a material which would constrain the aluminum so that the combined CTE would more closely match that of silicon. One could use carbon fibers or particles of silicon carbide (SiC) for this purpose. The choice of which to use could be governed somewhat by factors other than the resultant CTE, such as cost. While commercially available carbon fibers have a Young's modulus of 130×10^{-6} psi (this is about 12 times that of aluminum) and a thermal conductivity of 600 W/m-K (one and one-half times that of copper and three times that of aluminum), the cost of carbon fibers is many times higher than the cost of high-purity silicon carbide particles. (Carbon fibers are sold for over 100 U.S. dollars per pound while SiC particles are being sold for between 10 and 25 U.S. dollars per pound. This in itself accounts for the popularity of SiC as an MMC reinforcement over carbon fibers.) Other features to consider are: (1) Carbon fibers are generally used in long lengths and laid in the matrix they are placed and (2) the thermal conductivity of the fibers is much higher along the major axis than it is perpendicular to that axis. This causes the composite to have orthotropic physical properties. Particles of SiC are used to reinforce aluminum and to endow it with isotropic properties. By varying the volume percentage of the SiC in the matrix, it is possible to tailor the composite's CTE from roughly 10 to over 20×10^{-6}/°C.

Metal matrix composites have been used in a variety of applications for electronics. The principal examples are for substrate supports or carriers and hermetic packages. An emerging application is the standard electronic module heat sink. Comparison testing between MMCs and PMCs for this purpose is under way. The main reasons for using MMCs as substrate carriers are (1) tailored matching to the CTEs of substrates which carry devices such as gallium arsenide and silicon and (2) improved thermal conductivity over the composite matrix material. Although these features are addressed to some extent by other technologies (PMCs), metal matrix composites are also required to become hermetic housings, and this brings several additional requirements into the picture. These are low cost, isotropy of physical properties such as expansion characteristics and thermal conductivity, and isolation from the surrounding environment through hermetic sealability. Two material combinations have extensive usage, and the experience gained with their use has been valuable in influencing further development. These are aluminum silicon carbide and aluminum graphite. These composites have helped to spur further research resulting in carbon- or graphite-reinforced copper which is a newer MMC with promise of outstanding thermal conductivity and very low in-plane CTE. This CTE is a close match to silicon and makes this composite attractive as a candidate for packages containing high-power-dissipating silicon devices if the in-plane CTE is the same in all directions. Uniform in-plane CTE can be achieved by constructing the composite so that the reinforcing continuous fiber tows are placed at zero degrees and the next layer at 90 degrees, the next at zero degrees, and so on. If the layers are all placed at zero degrees, the in-plane CTE will be quite low in the fiber length direction but much higher (nearly that of copper) in the transverse direction. In order to take full advantage of the benefits and properties the reinforcements can bring, it is important to use them in an orientation that results in essentially uniform behavior unless anisotropy is desired for the application.

11.6.5 Reinforcements for Metal Matrix Composites

Several popular MMC reinforcements were described in Sec. 11.6.4. These are in common use, and the properties of the resultant composites are quantified to a large extent. It is important to realize that the values in the tables represent measurements of specific loading or volume percentages of reinforcements which can vary from 10 to as much as 60 percent. Composite properties vary as the amount of reinforcement in them is increased or decreased, whether that reinforcement be in the form of short filaments or fibers, continuous fibers, or particles.

More materials are used as reinforcements than listed earlier. Some of these additional reinforcements are boron, aluminum oxide (alumina), and refractory metals. Bonding between the matrix material and the reinforcement is the key to performance of the MMC. This performance depends on the bond integrity which in some materials combinations is enhanced by interface coatings to decrease reactivity between the matrix and reinforcement. Treatments of the reinforcement are sometimes used to decrease lubricity and prevent premature pull out of the fiber from the matrix.

An example of a reactive combination is the combination of graphite fibers and aluminum. This composite has enjoyed success in structural applications, but its use in electronics as a hybrid package material is nonexistent. Its low density (weight) and high thermal properties render it a candidate for heat sinks in electronic assemblies. To avoid the reaction between aluminum and graphite at liquid aluminum temperatures, vacuum deposition of the matrix onto the graphite fibers is used.

Alumina fibers are another material used to reinforce aluminum. To strengthen the bond between matrix and reinforcement in this case, small amounts of lithium are added to the melt.[12]

11.7 INTERCONNECTS

The term *interconnect* is used here to connote the electrical connection of devices such as semiconductors and passive components like capacitors, inductors, and resistors to an electrically conductive medium that in turn is connected to the next higher level of assembly. Semiconductor devices (chips) are varied in configuration and size and in number of inputs and outputs, and, as might be expected, there are a variety of techniques that are used to interconnect them. Semiconductor devices are generally either packaged in housings that contain one chip or are part of a multichip assembly which serves as the package. Another technique avoids the use of a package for the chip and places it directly on the interconnecting medium, for example, a printed wiring board. The other passive devices can exist in "chip" form or as encapsulated components. The reason these passive components can be procured as chips is that they can be combined with semiconductor devices to produce circuits which have components that are compatible in size and form factor.

The principal semiconductor devices that comprise most circuits are integrated circuits. These devices can be either analog or digital types. Analog types are those that are used in applications that are not specifically computing functions. These include receivers, control circuits, amplifiers, and many microwave and millimeter wave circuits. Digital devices, on the other hand, are principally memory devices, gates, counters, buffers, and arithmetic units. The main feature that all these devices have in common is that many individual transistors, diodes, and passive components such as resistors and capacitors make up the devices; that is, a large number of components (sometimes many thousands of transistors, for example) are contained within a single integrated circuit. In the case of very complex digital devices, hundreds of electrical contact or attachment points are required for each integrated circuit. If a computer function circuit contains many of these integrated circuits, several thousand interconnects will be needed to tie all the points to their respective component's pads via the connecting substrate or printed wiring board. This can be an expensive task not only due to the large number of interconnects but also because, with so many individual contacts, it is possible to make mistakes in wiring. In addition to mistakes, it is possible to make the bond incorrectly, that is, with the bonding parameters not controlled tightly. Repair of such device interconnects is very seldom accomplished successfully, and the entire device must be removed and replaced. For these reasons it is necessary to use highly repeatable electrical connection techniques. It is also desirable that these techniques be capable of interconnecting many of the integrated circuits' pads simultaneously.

11.7.1 Device Interconnection Techniques

Devices can be connected by a variety of techniques. The main ones used today for interconnecting unpackaged devices are wire bonding, flip chip, *tape automated bonding* (TAB), and Z-direction adhesives. A new technique recently introduced to interconnect individually packaged devices is solder-free interconnect.

Wire Bonding. *Wire bonding* refers to the very early method of connecting the three portions of transistors—emitter, base, and collector—to the package containing the devices. There are three common types of wire bonding: thermocompression, thermosonic, and ultrasonic. The first to be described is thermocompression bonding. In this technique a gold wire previously formed to a very fine diameter (0.001 in and less) and wound on a spool is fed through a tungsten carbide tip which is part of a wire bonding machine. The wire protrudes below the tip (called a *capillary*) and is formed into a ball by moving the capillary over a tiny hydrogen flame or causing a spark to jump to the wire and melt it into a ball. The ball is several wire diameters in size and is too large to be pulled up into the capillary hole which is only about twice the wire diameter. This feature is important because, as the capillary moves down vertically, it pushes down on the gold ball and causes the gold wire to unreel from the spool. The spool tension is adjusted to be low enough to keep the wire from paying out too swiftly and becoming tangled while not exceeding the elastic limit of the very soft gold wire. Early capillaries were energized by heating alone, and this type of bond is called a *thermocompression bond.*

A design improvement which added ultrasonic energy to the thermal energy of the tip has been in place for some time. This type of bonding is called *thermosonic bonding.* The improvement allows tip temperatures to be lowered and thereby improves process yield. The process of attaching the wire to the bonding pads of the devices does not differ for either thermocompression or thermosonic bonding except in the application of ultrasonic energy for the latter.

Connections are made to the devices by causing the ball of the gold wire to be pressed onto the pads of the devices. These pads are almost always aluminum. The heat—and, in the case of thermosonic bonding, heat and ultrasonic energy—causes the gold and aluminum to alloy, forming a metallurgical bond. Purity of the metallic constituents is a stringent requirement for a successful long-term bond. Typically the gold wire is 99.999 percent pure, and the aluminum pad is 99.99 percent pure. Impurities can cause the formation of intermetallics which will, with elevated temperature, cause the bond joints to become embrittled and ultimately fail.

As just described, the ball of the gold wire is pressed onto the aluminum pad of the device to form the electrical connection. However, this bond is only one-half of the connection. The other half is formed when the capillary is lifted from the bond on the device. At this point the capillary can be moved in any direction over the plane of the work area. This feature of ball bonding is especially important for producing wire bonds on substrates that contain large numbers of devices. The capillary is then moved to the desired bond pad on the supporting substrate or package and made to press the wire onto the pad. This time there is no ball under the capillary but rather the wire emanating from the hole in the capillary and bent under a portion of the capillary bottom surface (parallel to the substrate). After the capillary makes the bond, which is called a *stitch bond,* it is raised up from the substrate. As this transpires, the wire is clamped above the capillary and held tightly. This causes the wire to separate from the flattened wire portion of the bond. A portion of the wire separated from the bond protrudes below the capillary and is made to form a ball as indicated at the beginning of the cycle.

Ultrasonic bonding is the third form of wire bonding. This type uses aluminum wire instead of gold wire. Aluminum wire was introduced as an alternate to gold in the 1960s because many gold wire bonds were failing to remain attached to the aluminum pads on devices. The symptom of the failure was a condition known as *purple plague* wherein a small portion of the ball bond at the interface to the chip pad would become purple-colored. Not long after the purple material would appear, the bond would fail. The cause of the failure was quite elusive (it was actually a combination of causes which included contaminants in the wire introduced during wire production combined with high-temperature storage of the bonded devices), and, while investigators struggled to find the cause, the alternate, aluminum ultrasonic bonding, emerged. In ultrasonic bonding, the tip is a grooved foot with the wire, which in this case is aluminum, passing under the foot and guided along the groove. The wire again is one-thousandth of an inch in diameter or larger. Whereas the aluminum bonding pads of devices are 99.99 percent pure, the wire contains about 1 percent silicon to improve its lubricity so that it does not drag on and become hung up on the extrusion die when it is being drawn into wire. This does not affect the quality of the wire bond. The reason thermocompression bonding cannot be used for aluminum wire bonding is that aluminum is an extremely active metal that forms oxides on its surface which do not get broken up sufficiently during thermocompression bonding, and interfere with the formation of a diffusion bond between the wire and the pad. Ultrasonic energy, however, breaks some of these oxide bonds, but more importantly, the action of the tip scrubbing the wire back and forth opens up fresh unoxidized aluminum in both the wire and pad and permits an aluminum-aluminum diffusion bond to form readily between the wire and pad. No ball is formed at the free end of the aluminum wire. A disadvantage to this type of wedge bond to wedge bond lies in the requirement that the wire be dressed almost in a direct line away from the chip bond. Bonding hybrid circuits in which wires must go from the chip to the substrate in any direction is severely restricted when this type of bonding is used. Ultrasonic bonds can be made closer together than ball bonds (0.003 in as opposed to 0.004 in). This allows chip pads to be placed 33 percent closer, which on a device with hundreds of outputs saves a great deal of chip area that can be used instead for the active portion of the device.

Aluminum wire brings with it certain economies over gold wire. Although the cost is less than that of gold wire, it is a small difference but can be significant when large amounts are used as in devices with hundreds of outputs or with power devices that carry large amounts of current requiring significantly increased diameters and multiple bonds to the same pad (it is not uncommon to use 0.020-in-diameter aluminum wire). Multiple bonds may serve the purpose for carrying high current in low frequency and power circuits, but the excessive inductance associated with fine wires causes performance to become degraded in high-frequency circuits. In these instances relatively large, gold-plated copper ribbons are frequently used instead.

Both gold and aluminum wire are used extensively today in packaging, with aluminum use being higher in individually packaged chips, and gold use higher in hybrid-style and MCM packages.

Flip-Chip Bonding. In flip-chip bonding, the devices are not mounted to the package or substrate right side up, as they are in wire bonding. Instead, they are mounted face down with their pads serving as the electrical and mechanical connections. The principal pioneer, developer, and user of this packaging technology is the IBM Corporation. Flip-chip packaging is the densest packing technique that exists because it uses no real estate on the substrate beyond the boundaries of the chip to make the electrical connections. This fact is extremely significant for very high input-output devices used in hybrid multichip modules (MCMs). Extremely high speed circuits can benefit from the proximity of chip pads to substrate lands. The bonding pads on the chip are placed directly over the corresponding pads on the substrate.

Solder Ball Flip Chip. The mounting of the chip is accomplished by an ingenious use of solder "bumps," or balls, which are plated onto the chip pads and reflowed when the chip is placed onto the substrate. The flip-chip technology described here is the main interconnect method between chips and substrates in use at IBM. Flip chip was introduced in 1964 by IBM as a *solder ball flip-chip* (SBFC) interconnect, a main feature of its *solid logic technology* hybrid modules. This approach was very revolutionary and was watched by all packaging engineers but practiced essentially by only IBM until recently. Initially a copper ball was contained within the solder ball to act as a standoff and establish a minimum height or separation between the chip and thick-film land or pad on the substrate because the edges of the chips were unpassivated. This was improved upon later by printing glass dams onto the substrates and eliminating the copper balls.[13]

At the time of its introduction, SBFC held promise of delivering large improvements in reliability and cost reductions for chip interconnects by virtue of the simultaneity of joint creation for all the chips on a substrate. The promise was essentially fulfilled, and several other features have been established as well that make the technique attractive, especially for MCMs. Robustness of the joints in many environments and very efficient packaging densities have caused engineers to reconsider earlier rejections of the technique for their applications. Much work has been accomplished to mitigate concern over CTE differences between the chips and some of the substrates on which they are mounted. Stacked bumps which distribute the stresses involved in thermal cycling are reported to improve fatigue life substantially.[14,15]

A serious deterrent to the use of SBFC is the requirement that the chips be metallized with chrome and then chrome and copper (in a phased manner) followed by copper and tin. This metallization must be applied at the wafer stage of semiconductor fabrication. Since most hybrid circuits and MCMs are designed to use various manufacturer's chips, it is necessary that those manufacturers agree to process the devices with preparatory metallizations or with solder bumps.[16] This requirement makes it very difficult to have an SBFC technology under control because industry demand has not established a strong market for device manufacturers to address. To circumvent this problem, a Japanese company has developed a process that involves a solder wire which can be ball bonded to aluminum chip pads to avoid the requirement for wafer metallizations applied atop the aluminum.[17] At this time, however, the wire is not available for sale, but a wafer bumping service is. The company involved is Tanaka Denshi Kogyo, and its U.S. agent is Carmel Chemicals of San Jose, California. The success of this approach depends first of all on the ability of the wafer manufacturer and chip purchaser to negotiate an agreement that is mutually acceptable. Having achieved such an agreement, the chip user must still compare the additional costs of the services required for bumping with conventional bump production techniques to conventional wire bonding to determine the viability of this approach. On a production scale, this technique may be feasible in certain instances where the wafer producer is not able to furnish bumped devices. On a prototype scale, it appears more plausible if there can be an economically feasible approach to the issue when production ensues.

Z-axis Adhesive Flip Chip. Flip-chip innovations have begun to appear in the 1990s because of the difficulties involved with producing solder bumps on devices. Although the SBFC process has been proven to be a high-performance technology, the requirements for special metallizations to be applied at the wafer step in device production and special agreements with device producers have driven some to explore alternatives. One of these alternatives uses electrically conductive adhesives in place of the solder bumps.

Tape Automated Bonding. *Tape automated bonding* (TAB) is a technique for interconnecting semiconductor chips of all kinds to substrates and packages by the attachment of beamlike conductors to both chips and substrates. The beams are photolithographically formed from clad or plated metal (usually copper) atop a polyimide film. Parts of the beams are cantilevered over openings or windows in the polyimide film. The beams are bonded to the pads of chips in a simultaneous fashion (*inner lead bonding,* or ILB), and later the other ends of the beams are bonded to substrate lands or package leads (*outer lead bonding,* or OLB). This technique emerged as a means to produce packaged chips economically on a large scale because, as the cost to produce the chips themselves became very low, producing the connections between the chip pads and the package pads had then become a major part of the total cost. It was necessary to reduce the cost of that part of the assembly procedure to keep pace with the automation which took place with the introduction of larger wafers (from 3-in diameter to 5- and 6-in diameter). Even as chips became more complex and larger, the wafer sizes grew to accommodate more devices per wafer and, consequently, more devices per lot. Device yields increased as better diffusion, epitaxy, photolithography, and clean room techniques were introduced. Therefore, lower costs were incurred to produce the devices as chips. As long as labor-intensive manual wire bonding remained as the primary chip interconnection technique (automatic wire bonders of that time were slow and cumbersome to use, not to mention very expensive even for manufacturers of millions of chips per year), packaged chips would remain expensive and the market for them would not grow quickly. This problem was partially solved by the introduction of TAB. There were two types of TAB. The first was used with chips that had plated bumps applied to the pads after the final metallization step. The second type used chips with no special metals applied after the final metallization and, instead, had bumps applied to the interconnect beams. This second type is called *bumped tape automated bonding* (BTAB).

As devices became more complex and required many inputs and outputs, some manufacturers sought to reduce wafer real estate by using TAB because TAB generally allows chip pads to be smaller and placed closer together. This results in less chip area devoted strictly to interconnects and frees up some precious area for more active components to be placed on the wafer. There are electrical considerations contributing to improved performance such as better impedance control and lower inductance over wire bonds. Other features are more consistent physical placement of the beams with respect to each other and greater bond strength resulting in higher reliability. TAB also allows more testing of the chip to take place because test pads can be printed onto the polyimide film that are consistent with larger dimensions required by probe pins. The conductors on the film can be made to fan out to pads on any desired centers.

Bumped Chip TAB. The technology of bumped chip TAB depended on postprocessing of device wafers after the aluminum bonding pads were deposited. The specific processing involved deposition of several additional metal layers over the aluminum. These included a barrier metal such as titanium followed by nickel and gold. This is shown in Fig. 11.8. The bumps are required either on the chip pads or on the beams because it is necessary to have clearance between the remaining part of the beam and the chip surface. The most common bump material was gold plating which was chosen to permit thermocompression bonds to be made between the beams and the bumps. It was necessary to have highly uniform bump height because the bonding

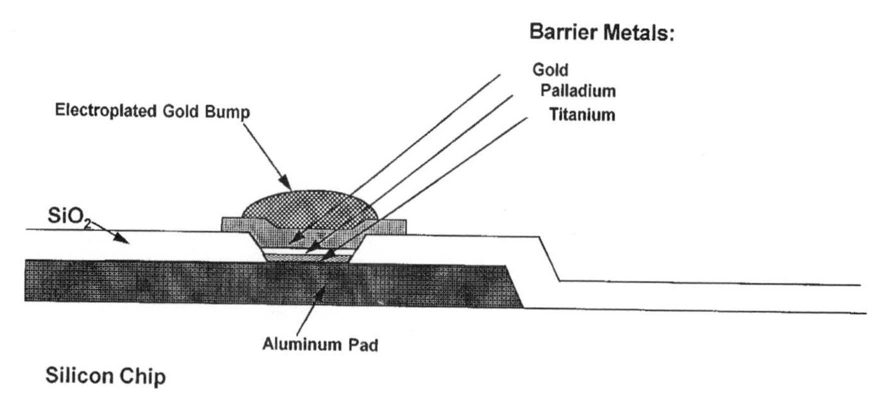

FIGURE 11.8 Cross section of plated bump on a chip.

tool consisted of a tungsten block made to press on top of all the beams on a chip simultaneously. Out-of-plane bumps, particularly short bumps, would not be bonded with the same force and would form weaker bonds or no bonds at all. Electroplating uniform-height bumps across an entire 5- or 6-in wafer represented a formidable technological challenge that was not met successfully by many chip manufacturers. In the first place, not many chip manufacturers were interested in providing bumped wafers for sale because the demand was very small. Of those who would provide this service, only a few invested in special equipment to meet the challenge and so the availability of bumped wafers was poor for hybrid manufacturers who did not have a captive or in-house chip manufacturing capability.

The second type of chip bump material used was solder. To avoid the TAB thermocompression bonding problems associated with lack of bump height uniformity, several companies instituted a solder technology for bumping wafers. The most prominent was Honeywell in both the United States and abroad which had a large manufacturing operation producing chips with solder bumps for their computers. Honeywell Bull in France produced large numbers of such devices and interconnected them with TAB solder reflow technology onto hybrid substrates.

Bumped Lead TAB (BTAB). The alternate method to bumped chips was bumped lead or bumped tape. This method evolved when it appeared too expensive and too limiting to use bumped chips. With a seriously limited bumped chip supplier field, this method emerged. It was devised to avoid the plating issue which plagued bumped chips. Instead of depending on bumped chips, the beams were configured to have bumps on themselves. Figure 11.9 is a scanning electron micrograph of a bumped lead bonded to the pad on a silicon chip. The bump can be seen clearly beneath the lead. Since the difficulty in producing uniform bumps resided in electroplating technology that was stretched beyond the state of the art, the approach was to avoid using electroplating to achieve uniform bump height. The thickness uniformity of copper foil on polyimide film was exactly what spawned this approach. Instead of plating up the bump, the bump was created by etching away the beam from the underside except in the location where the bump was desired. This created the bump *in situ* on the bottom of the beam, and a subsequent thin overplating of nickel and gold protected the copper. The beams or leads were formed by photolithography after windows were opened in the polyimide film beneath the copper foil. Another variation of this technique

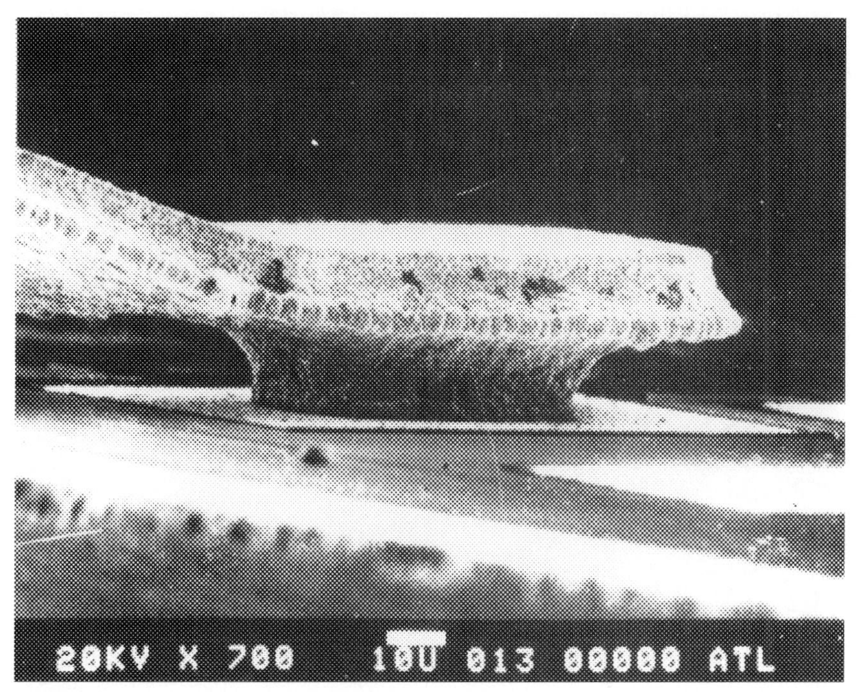

FIGURE 11.9 Scanning electron micrograph of a bumped lead.

avoided the use of polyimide film at this step and instead used only gold-plated copper foil. The bumped beams and sprocket holes were formed in the foil to create strips of beam sites. Sprocket holes were required for bonding to the chip with a TAB ILB machine. This provided a means of indexing to the next site as bonds were made to the chips. An example of such a bond is shown in Fig. 11.10.

The TAB Process of Bonding to Chips (Inner Lead Bonding). TAB bonding to chips involves first of all the preparation of a wafer containing chips. Whether a wafer is bumped or not, if the chips are to be bonded using TAB, the wafer is waxed to a silicone rubber or similar elastomer on its backside. It is then scribed in the avenues between the chips, and the chips are separated from each other although still held onto the elastomer by the wax. The wafer is then placed in a TAB ILB machine where the foil strip containing the beams is brought over it. After the ends of the beams are aligned to their appropriate chip bonding pads, the heated bonding tool presses the beams onto the chip and a thermocompression bond (or a soldered joint in the case of solder bumps) is made. As the bonding tool is retracted, the chip is released from the wax bond beneath it because the wax melts as the ILB is made, and the chip is suspended from the beams on the foil. The foil is indexed to the next site and brought over the chip to which it is to be bonded, and the process is repeated. As the chips are bonded, the foil containing them is reeled up. The reel is later taken to an outer lead bonder where the other ends of the beams are bonded to packages or substrates.

The TAB Process of Bonding to Packages or Substrates (Outer Lead Bonding). The process of *outer lead bonding* (OLB) begins with the placement of the reel of ILB

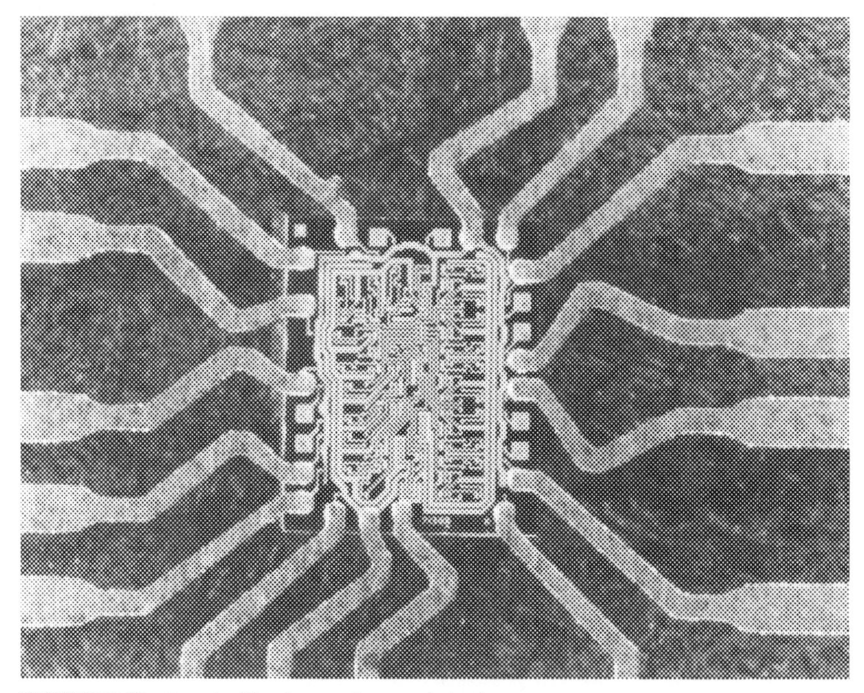

FIGURE 11.10 Inner lead bonds on an integrated circuit.

bonded chips onto the outer lead bonding machine. The reeled up foil is fed to an excising tool which cuts the beams away from the foil near the edge of the window in the foil. At this time, a vacuum pickup placed directly above the chip picks up the inner lead bonded chip with its beams which have been freed from the foil and moves to a location directly above the site on the substrate or package where the outer leads are to be bonded. After alignment is complete, the OLB tool comes down and bonds the beams to lands or pads on the substrate or package. This operation is shown in Fig. 11.11 where a chip's outer leads have just been bonded to a substrate. The chip is shown on the OLB equipment television monitor.

Maximizing Testability Features of TAB. One of the advantages of TAB for hybrid circuits which can significantly reduce the danger of bonding nonfunctional chips into the hybrid is the ability to test the chip after it has been separated from the wafer. This includes testing at temperature extremes and also burn-in. A universal TAB chip test carrier and a means of making nonstandard chip sizes standard was devised by F. A. Lindberg.[18] This approach uses copper foil etched from both sides to form the leads, bumps, and sprocket holes as well as the tape edges into a 35-mm format. The technique is not limited to 35-mm tape width, but, when it was devised, all chips were small enough to fit easily within that format. Inner lead locations and bumps are customized to match the pads of each chip. Outer leads continue to a length sufficient to allow bonding to a polyimide test tape and, after test, excision from that tape, bending, and bonding to a substrate. Although chips come in various sizes, a window is punched in the test tape commensurate with the chip size. In a 35-mm format, a 64-pad device is the maximum size that can be accommodated.

FIGURE 11.11 Outer lead bonder with a view of chip outer leads bonded to a substrate.

The test tape is configured to have pads in an arrangement that allows pads on all four edges of a chip to be bonded to leads which in turn will be bonded to lands on the tape. The tracks on the tape are fanned out and lead to pads on the periphery of the tape for accommodation into a test fixture. A process flow of the major steps used in this approach is depicted in Fig. 11.12.

The foil is gold plated after etching which protects the copper from oxidation and enhances thermocompression bonding to aluminum pads on the chip. Since the lands on the polyimide test tape are electroless tin plating over copper, the outer lead bonds to the test tape are accomplished by reflowing the tin and allowing it to alloy with the gold. After being tested, the leads are cut from the tape, and the chip with its leads are then picked up and brought to the hybrid substrate where the leads are bonded to matching lands.

Solder-Free Interconnects. Solder has served as the principal interconnection technique for electronic assemblies for over 70 years. At the time of electron tubes and other large components, it was an ideal method to effect a reliable electrical contact between two or more wires and socket pins or other component mechanical supports and connection terminals. The diameters of the wires and other terminals were large with respect to the amount of solder required to form the electrical connection. Even as components became miniaturized and had some increases in their pin count (from one dozen to about four dozen), their size and spacing remained large enough so that they could be located 0.100 in from the center of one lead to the center of the other. However, the trend toward increased device speed and computational power caused

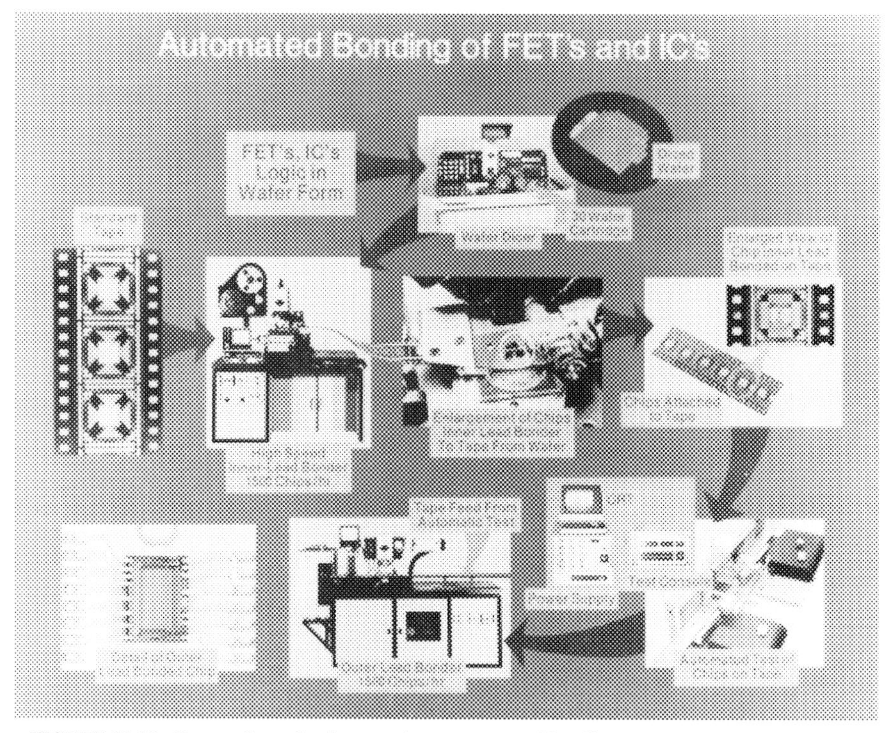

FIGURE 11.12 Process flow of major steps in tape automated bonding.

the lead counts for individual chips to increase significantly. To retain the same lead sizes and spacing would cause the packages containing the chips to grow excessively large and thereby require correspondingly large footprint areas on the printed wiring boards on which they would be placed. Therefore, it became necessary to reduce both the lead width and spacing. Leads were brought out of the package on all four edges or sides to minimize overall package size, and their size was reduced from 0.020 in width to 0.013 in. In addition, lead center-to-center spacing went to 0.050 in. This change in lead spacing was not sufficient to keep up with the increase in lead count so lead center spacing was driven downward to 0.025 in and then to 0.020 in. At this point it was discovered that soldering leads so small (and even smaller in special cases with corresponding spacing of only 0.006 in) presented assembly problems. Handling the devices required special carriers because the leads were very fragile and easily became entangled with each other. New techniques for soldering such fine leads had to be developed. These techniques included inert gas assisted lead tinning and solder reflow procedures such as vapor phase soldering, hot bar soldering, and infrared reflow.

As these technologies were developing, some effort was devoted to pin grid array and pad grid array packaging. Both approaches depended on bringing the electrical contacts out of the package through the bottom rather than the edges. In both cases, the contacts were placed on 0.050-in centers and formed an array of contacts on the bottom surface of the package. In the case of the pin grid array configuration, plated pins were brazed to the pads on the package bottom surface. The pad grid array configuration has only pads on the bottom of the package. Pin grid array packages were more acceptable to designers than pad arrays because the pins represented a known interconnect technology, that is, pin and socket. Pad arrays would be interconnected through soldering to their respective lands on printed wiring boards. This approach, although used in limited fashion, did not satisfy the requirement of military specifications that the joints be accessible for inspection. A further drawback was that the solder joints would be too short in height to allow flux and cleaners captured between the printed wiring board and the package to be flushed away. This condition would make it very difficult if not impossible for such populated assemblies to pass the low ionic contaminant residue requirement of MIL-P-28809 (currently known as MIL-C-28809). Consequently this type of package did not receive much support.

There are several solder-free interconnect technologies that do not necessarily deal with high-pin-count packages which should be mentioned. Aside from wire bonding, which is discussed in Sec. 11.7.1 of this chapter, these interconnects are elastomers, epoxies, and memory metals. These technologies will be discussed in order followed by a new method that addresses high-lead-count packages.

Elastomeric Interconnects. The most often used elastomeric interconnect, and the oldest, consists of a strip of silicone rubber slices that are alternatively insulative and conductive. The conduction is achieved by filling silicone with conductive metallic particles. Sheets of both conductive and nonconductive silicone are cast into uniform thicknesses (roughly 0.010 in) and cured. They are then sliced into squares typically measuring 0.100 in on a side and assembled and glued alternately together to form a stack of squares of insulator/conductor/insulator, and so on. They function as a connection medium because they are placed between a conductor such as a land on a printed wiring board and a contact for a key pad in a calculator. Their application is not limited to the example, however. A variation of this technique has the conductive portions made smaller to accommodate fine-pitch applications.[19] Alignment between the elastomer and pads to be connected is not critical because the elastomeric connections are usually at a much finer pitch than the contacts being made. A disadvantage to the use of elastomers is the requirement to use relatively high pressure to maintain sufficiently low resistance to constitute a reliable joint. Another problem that ensues is that all elastomers relax with time, that is, they lose their elasticity to some degree, and therefore the pressure initially applied to them is reduced. This in turn causes the resistance in the connection to increase as the pressure becomes lowered. Care must be taken to determine both the amount of relaxation expected in the elastomer with its corresponding increase in contact resistance and the effect of that increased contact resistance on the performance of the circuit under consideration.

Loss of contact force at cold temperatures was reported in recent studies done by Rajendra Pendse at Hewlett Packard. This work cautions against using elastomers alone if significantly low temperatures will be encountered while the equipment is operating.[20] A drop or loss of contact force of over 37 percent on the elastomer was measured from room temperature to minus 55°C. Thermal hysteresis is another phenomenon to consider with elastomeric connections. When a clamped elastomer is heated from room temperature and then cooled back to the starting temperature, the force on the interconnect increases with temperature increase. When cooling takes place, however, the loss of force is greater for each degree of temperature change than the increase was for the same delta temperature. Under moderate temperature changes, typical losses of force were in the neighborhood of 50 percent.

To obviate these design difficulties, Pendse suggests using a metal spring (the softer, the better) in series with the elastomer. In this way, when the elastomer relaxes under the fixed load, the spring pushes on it to minimize the loss of contact force. Comparative experiments demonstrated that force retention was as high as 92 percent with a soft spring (low spring constant), and 74 percent with a harder (1.6 higher spring constant) spring. With sufficiently low ratios of metal to elastomer spring constants, open circuits can be avoided even at temperatures as low as minus 55°C.

Epoxies As Interconnects. Metal-filled epoxies are seeing increased usage as a means of effecting interconnects between surface-mounted components and printed wiring boards. At first used to provide an electrical as well as mechanical function for use with chip capacitors, their success has led to use with leadless chip carriers and other leadless components as well. When only several components are being mounted onto printed wiring boards, the conductive epoxy is generally dispensed from a tube into which it has been placed after mixing of the constituents. Hypodermic syringes are used to perform this step because they can be depended upon to control the dispensing of small amounts of resin at one time. Although time-consuming and labor intensive, the approach is quite effective if only a few sites are required to have the resin applied. On the other hand, if many components are to be connected to the printed wiring board by this technique, a method is called for which can dispense resin to many or all sites simultaneously. Stenciling or silk-screening are such methods. They are used to apply the resin to multiple sites on printed wiring boards. Masks are created with openings in them to allow the epoxy to be squeegeed onto the boards. These masks are made from thin brass or stainless-steel shim stock, usually about 0.002 to 0.003 in thick. Holes are photolithographically etched into the metal to match the pads on the printed wiring boards. Masks are then placed in silk-screen frames and secured onto a silk-screen printer. Conductive epoxy is spread across the mask at one end, and a mechanically driven squeegee then prints the resin onto the board. Pot life of the resin is important if more than several boards are to have the resin applied at the same time. While it may be convenient to have the epoxy become cured rapidly after components have been assembled to the boards, the speed of the cure should be selected so that enough time is allowed to place all the components onto their printed wiring board pads before a significant amount of curing occurs. The most advantageous type of epoxy to use is one which will not cure rapidly at room temperature but rather at elevated temperatures. The types of epoxies used to mount and/or interconnect surface mount components are listed in Table 11.10.

Memory Metal Interconnects. Memory metals fall into two classes which, when either heated or cooled to a specific transition temperature, change their crystalline structure and hence their volume. This feature is put to use in certain connectors which, when taken through the temperature range of −55 to +125°C do not change crystalline state but, at a transition temperature safely below the range in the one case, or safely above in the other case, will exhibit a change in structure. In the first case, within the temperature range noted, a connector socket staked into a printed wiring board will retain an inner diameter slightly less than the component lead or pin intended to fit within it. In order to assemble the component, one cools the sockets to the temperature of liquid nitrogen and places the component pins within the sockets which, upon cooling, have opened their inner diameters sufficiently to accept the component pins. Upon heating back to room temperature, the component pins are tightly grasped by the connector sockets' inner diameters. That which causes the socket to open and then close tightly around the pin is a memory metal collar which surrounds a beryllium copper socket. The behavior is similar for connectors that depend upon

TABLE 11.10 Adhesives Used for Mounting and Interconnecting Devices

Type	Unidirectionally conductive preform	Epoxy paste
Contact resistivity, mΩ/mm^2	<5	<5
x-y dielectric strength, V/mil separation measured at 5 mils	60	60
Glass transition temperature, °C	−25	60
Lap-shear strength psi, N/mm^2	1000 (6.9)	2000 (13.8)
Linear thermal coefficient of expansion, ppm/°C	110	50
Maximum continuous operating temperature, °C	150	150
Thermal conductivity, Btu-in/ft^2-h-°F (W/m-K)	6 (0.87)	25 (3.6)
Typical use	Solder replacement, component attachment, tab outerlead bonding	Matched CTE component attachment, fine pitch lead attach

Source: Courtesy of A. I. Technology, Inc.

heating to be taken to a transition temperature except that the procedure is reversed. In this case, heating relaxes the socket's grasp on the pin.

One strong feature of memory metals is their ability to form gas-tight connections without the use of solder or welding. This is attributed to the high forces of the connection which are in the neighborhood of twice those of friction fits. Another feature is the lack of a requirement for special tools since the assembly is accomplished strictly with application of heat or the withdrawal of it. On the negative side, the cost of complicated sockets or other grasping devices make the approach attractive for very special applications but not for common designs which can be accomplished more readily by standard pin and socket technology.

Button Board Interconnects. Button board technology is a new approach to solderless interconnects based on fine-wire buttons contained within a thin plastic sheet populated with these buttons of fine wire wound around itself to form springlike cylinders which are reliable interconnects. The wire is randomly woven to form cylinders measuring 0.020 in diameter by 0.040 in long. This random weave produces a spring which does not relax after repeated loading or under thousands of temperature cycles. The wire consists of a molybdenum core with nickel and gold plating and measures about 40 mm in length before it is formed into the cylinder shape. The buttons are placed into holes in the plastic sheet and protrude above and below the sheet even when compressed. The buttons are placed at locations in the sheet matching the lands on the printed wiring board which are intended to be connected to the pads on leadless ceramic chip carriers. The sheet with the buttons in it is dropped into a hole in a structure bonded to the board, and the leadless chip carrier is then placed onto the sheet. A clamp is then applied to a plate above the leadless chip carrier, and the electrical connection is made between the leadless chip carrier and the printed wiring board. Figure 11.13 shows a button contact from the button board, and Fig. 11.14 is a photograph of the button board. Some of the inherent features of this approach are listed here:

- Compliance to out-of-plane conditions of the printed wiring pads is an inherent characteristic of the button board. Camber existing in the printed wiring board can be compensated by the button board because of the range of compression designed into the buttons.
- The connection between the button and the leadless chip carrier as well as the printed wiring pad is reliable because the compression pressure is 2 oz distributed over a very small area (between 500 and 800 psi).
- Continuous connections have been maintained during high vibration levels using this technique. Since the button is completely or nearly completely compressed as assembled, it is very unlikely to have its contact with either connecting member interrupted during vibration.

Depending on the design, the cost to implement the button board approach can be considerably lower than a comparable soldered assembly. On a one-to-one basis, the cost has been found to be one-third lower using button boards over solder. In addition, since solder fluxes are not used, there is no need to use ozone-depleting or environmentally friendly cleaning solvents either. The cost of such cleaners including disposal and cleaning labor steps are not necessary. Solder inspection, a costly and no-value-added operation, is also eliminated.

Modules (printed wiring board pairs) were subjected to environmental tests to determine survivability of the hardware. No provisions were made to protect the modules from the full effects of the environmental conditions such as would normally be

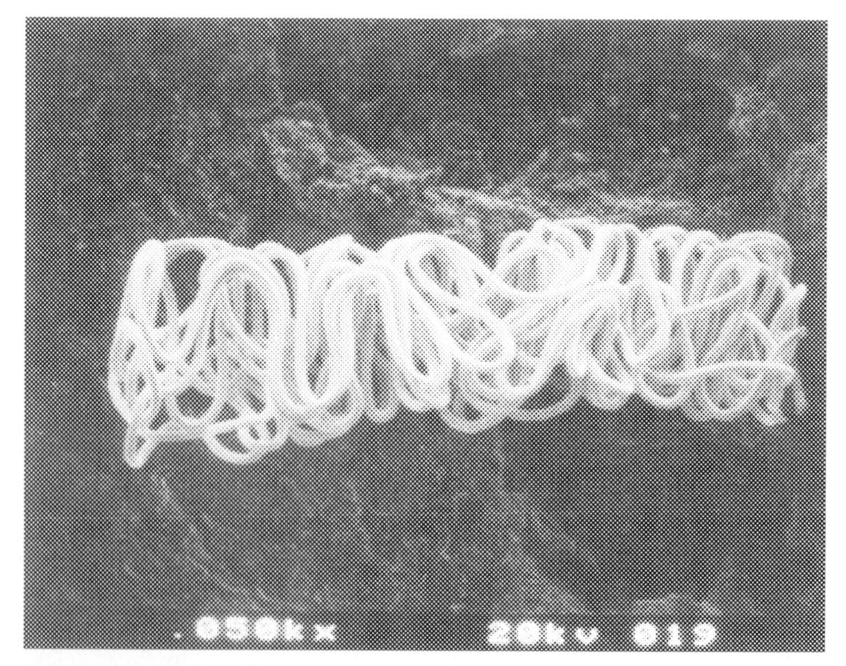

FIGURE 11.13 Fuzz button contact depicting the button wire.

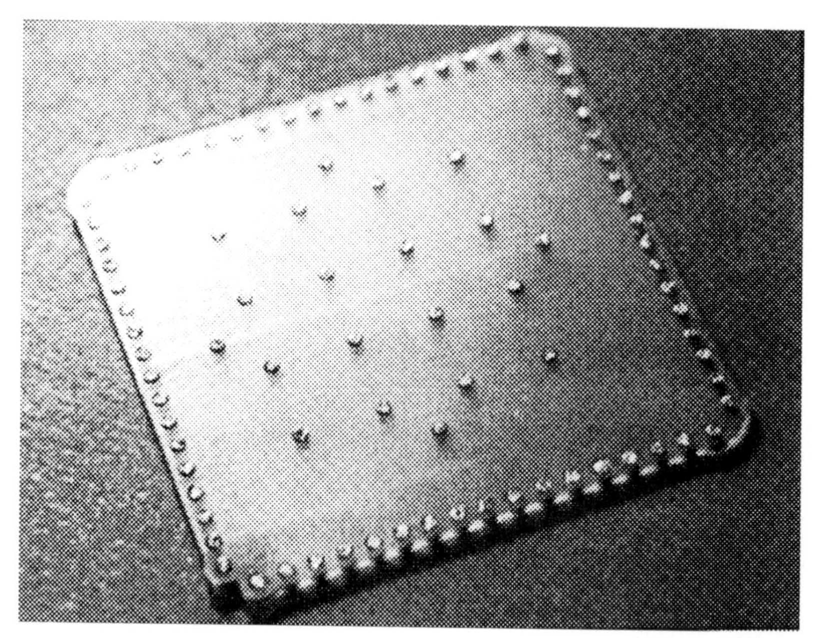

FIGURE 11.14 Compliant solderless button board.

afforded by the chassis housings if the environmental tests were performed on higher assemblies. The following tests were passed:

Thermal cycling	69 cycles from −55 to +125°C
Vibration	22 g (rms-input), 20 to 1500 Hz
Humidity	120 h at 31 to 60°C, 59 to 99 percent RH (MIL-STD-810D, Method 507.2-1)
Salt fog	48 h at 35°C; 5 percent salt concentration (MIL-STD-810D)

In all tests except salt fog, bias and constant current were applied, indicating that the gold-plated button contact is highly robust to the rigors of environmental testing and consequently to conditions expected to be encountered in actual field experience.

Chip on Board. *Chip on board* (COB) refers to the practice of placing the bare die either directly on the printed wiring board or on a substrate which is then placed directly on the printed wiring board. The main users of this technology are portable electronics such as pagers, laptops, and personal organizers. The main purpose of doing this is to avoid the baggage which comes with using packaged devices. This baggage is costly and a much larger device footprint. With COB, assemblies are much smaller than they would be if all devices were placed in packages before assembly to printed wiring boards. The actual board size reduction can be as much as 50 percent. Packages in the average in typical computer card applications account for slightly over 50 percent of total device costs, so the savings in both cost and area can be substantial. Package height also can be greater than the space available, and COB offers lower-profile assembly possibilities as well. In 1992, COB on both sides of the board emerged as a means of further increasing component density.

Substrate materials consist of FR-4 epoxy-glass, ceramic (alumina), silicon, and even films such as polyimide. Choices depend on performance and application. The die are attached most often by epoxy, and sometimes the conductive type is used if the circuit requires it. Die are often attached directly to the main printed wiring board as just mentioned, for simplicity. At times as the design requires, cavities are prepared in the board into which devices are placed, thus allowing the board height to be reduced by the depth of the cavities.

Electrical connections are made by wire bonding, with either gold or aluminum wire being the medium. Higher reliability is achieved, however, with gold.[21]

Gold has been found to resist thermal cycling better than aluminum. More cost is involved with the use of gold because of the need to gold plate the bonding pads on the substrate or board. Newcombe[22] states that the maximum number of wire bonds to be made on one assembly should not exceed 1000. This is presumably because the reliability of one bond is a certain value, and the reliability of an assembly containing 1000 bonds is one one-thousandth of that value.

Environmental protection is usually provided by an overcoat of epoxy on the die and the wire bonds which also gives some ruggedization to the board for handling during test. Since COB assemblies are highly unrepairable, ruggedization and process control are quite important.

REFERENCES

1. Howard Markstein, *Electronic Packaging and Production,* Vol. 31, No. 10, October 1991, p. 40.
2. Rao R. Tummala, and J. Knickerbocker, "Advanced Co-Fire Multichip Technology at IBM," *Proceedings of the International Electronics Packaging Conference,* San Diego, CA, September, 1991, p. 39.
3. John A. Biancini, "MCM-L; A Dynamic Evolution for the Multilayer PCB," *Electronic Packaging and Production,* May 1992, p. 37.
4. H. E. Saunders, and K. F. Schoch, Jr., *Proceedings of the 8th International SAMPE Conference,* Baltimore, MD, June 1992.
5. E. Frederick Borcheit, and Grant Lu, "Use of CVD Diamond Substrates in Electronic Applications," *Proceedings of the 1992 SAMPE Conference,* Baltimore, MD.
6. Timothy L. Hudson, "Selecting a Ceramic Substrate for Multichip Modules," *Electronic Packaging and Production,* June 1992, p. 68.
7. R. M. Barrett, "Microwave Printed Circuits—An Historical Survey," *IRE Transactions on Microwave Theory and Techniques,* MTT-3, March 1955, p. 1.
8. D. D. Grieg, and H. F. Englemann, "Microstrip—A New Transmission Technique for the Kilomegacycle Range," *Proceedings of the IRE,* December 1952, p. 1644.
9. N. R. Wild et al., *Handbook of Tri-Plate Microwave Components,* Sanders Associates, Nashua, NH, 1956, ASTIA No. AD110157.
10. H. S. Keen, "Scientific Report on the Study of Strip Transmission Line," *AIL Report 2830-2,* December 1955.
11. S. G. Konsowski, R. C. Pearson, and M. R. Lucas, "Materials and Process Considerations for Reliable High Wattage Hybrids," *Proceedings of the IEEE 1980 National Aerospace and Electronics Conference,* Dayton, OH.
12. John V. Folz, and Charles M. Blacmon, "Metal Matrix Composites," *ASME Engineered Materials Handbook,* Vol. 1.

13. Karl J. Puttlitz, Sr., *International Journal of Microcircuits and Electronic Packaging,* Vol. 15, No. 3, Third Quarter, 1992, p. 113.

14. R. Satoh, M. Oshima, H. Komura, I. Ishi, and K. Serizawa, "Development of a New Microsolder Bonding Method for Vlai," *Proceedings of the International Electronics Packaging Society Conference,* September 1987.

15. N. Matsui, S. Sasaki, and Tl Ohsaki, "VLSI Interconnection Technology Using Stacked Solder Bumps," *Proceedings of the 37th Electronic Components Conference.*

16. P. A. Totta, and R. P. Sopher, "SLT Device Metallurgy and Its Monolithic Extension," *IBM Journal of Research and Development,* Vol. 13, 1969, p. 226.

17. John Tuck, "Concurrent Product Development," *Circuits Assembly,* August 1992, p. 20.

18. F. A. Lindberg, "Hybrid Tape Bonding with Fast Turn-around Standard Cell Reusable Tape," *Proceedings of the ERADCOM Hybrid Microcircuit Conference,* 1980.

19. Robert A. Bourdelaise, "Solderless Alternatives to Surface Mount Component Attachment," *4th International SAMPE Electronics Conference,* June 1990.

20. Rajendra D. Pendse, "Low Relaxation Elastomeric Pressure Contact System for High Density Interconnect," *International Journal of Microcircuits and Electronic Packaging,* Vol. 15, No. 3, Third Quarter, 1992.

21. Elliott H. Newcombe, "COB Increases SMT Density," *Electronic Packaging and Production,* December 1992.

22. Ibid.

INDEX